Advances in Intelligent Systems and Computing

Volume 949

The series "Advances in Intelligent Systems and Computing" contains publications on theory, applications, and design methods of Intelligent Systems and Intelligent Computing. Virtually all disciplines such as engineering, natural sciences, computer and information science, ICT, economics, business, e-commerce, environment, healthcare, life science are covered. The list of topics spans all the areas of modern intelligent systems and computing such as: computational intelligence, soft computing including neural networks, fuzzy systems, evolutionary computing and the fusion of these paradigms, social intelligence, ambient intelligence, computational neuroscience, artificial life, virtual worlds and society, cognitive science and systems, Perception and Vision, DNA and immune based systems, self-organizing and adaptive systems, e-Learning and teaching, human-centered and human-centric computing, recommender systems, intelligent control, robotics and mechatronics including human-machine teaming, knowledge-based paradigms, learning paradigms, machine ethics, intelligent data analysis, knowledge management, intelligent agents, intelligent decision making and support, intelligent network security, trust management, interactive entertainment, Web intelligence and multimedia.

The publications within "Advances in Intelligent Systems and Computing" are primarily proceedings of important conferences, symposia and congresses. They cover significant recent developments in the field, both of a foundational and applicable character. An important characteristic feature of the series is the short publication time and world-wide distribution. This permits a rapid and broad dissemination of research results.

** Indexing: The books of this series are submitted to ISI Proceedings, EI-Compendex, DBLP, SCOPUS, Google Scholar and Springerlink **

More information about this series at http://www.springer.com/series/11156

R. Venkata Rao · Jan Taler
Editors

Advanced Engineering Optimization Through Intelligent Techniques

Select Proceedings of AEOTIT 2018

 Springer

Editors
R. Venkata Rao
Sardar Vallabhbhai National
Institute of Technology, Surat
Surat, Gujarat, India

Jan Taler
Cracow University of Technology
Kraków, Poland

ISSN 2194-5357   ISSN 2194-5365 (electronic)
Advances in Intelligent Systems and Computing
ISBN 978-981-13-8195-9   ISBN 978-981-13-8196-6 (eBook)
https://doi.org/10.1007/978-981-13-8196-6

This Springer imprint is published by the registered company Springer Nature Singapore Pte Ltd.
The registered company address is: 152 Beach Road, #21-01/04 Gateway East, Singapore 189721, Singapore

Conference Patron

Director, SVNIT, Surat

Conveners

Dr. D.Sc. R. Venkata Rao
Professor, Department of Mechanical Engineering, Sardar Vallabhbhai National Institute of Technology, Surat, India
Dr. D.Sc. Jan Taler
Professor, Faculty of Mechanical Engineering, Cracow University of Technology, Cracow, Poland

International Advisory Committee

Dr. Dan Simon, Cleveland State University, USA
Dr. A. Gunasekaran, California State University, USA
Dr. Daizhong Su, Nottingham Trent University, UK
Dr. Atulya Nagar, Liverpool Hope University, UK
Dr. D.Sc. Pawel Oclon, Cracow University of Technology, Poland
Dr. Leandro S. Coelho, Pontifícia Universidade Católica do Paraná, Brazil
Dr. Viviana C. Mariani, Pontifícia Universidade Católica do Paraná, Brazil
Dr. Joze Balic, University of Maribor, Slovenia
Dr. Franc Cus, University of Maribor, Slovenia
Dr. V. S. Kovalenko, National Technical University of Ukraine, Ukraine
Dr. S. H. Masood, Swinburne University of Technology, Australia
Dr. Syed J. Sadjadi, Iran University of Science and Technology, Iran

Dr. Husam I. Shaheen, Tishreen University, Syria
Dr. David K. H. Chua, National University of Singapore, Singapore
Dr. Manukid Parnichkun, Asian Institute of Technology, Thailand
Dr. H. T. Luong, Asian Institute of Technology, Thailand
Dr. Samuelson W. Hong, Oriental Institute of Technology, Taiwan
Dr. Liang Gao, Huazhong University of Science and Technology, China
Dr. Wenyin Gong, China University of Geosciences, China

National Advisory Committee

Dr. S. G. Deshmukh, ABV-IIITM, Gwalior
Dr. Souvik Bhattacharya, BITS, Pilani
Dr. R. P. Mohanty, SOA University, Bhubaneshwar
Dr. V. K. Jain, MANIT, Bhopal (formerly with IIT Kanpur)
Dr. P. K. Jain, Indian Institute of Technology, Roorkee
Dr. B. K. Panigrahi, Indian Institute of Technology, Delhi
Dr. P. V. Rao, Indian Institute of Technology, Delhi
Dr. J. Ramkumar, Indian Institute of Technology, Kanpur
Dr. Amit Agrawal, Indian Institute of Technology Bombay, Mumbai
Dr. S. K. Sharma, Indian Institute of Technology (BHU), Varanasi
Dr. A. K. Agrawal, Indian Institute of Technology (BHU),Varanasi
Dr. B. Bhattacharya, Jadavpur University, Kolkata
Dr. S. Chakraborty, Jadavpur University, Kolkata
Dr. S. K. Mohapatra, Thapar University, Patiala
Dr. Dixit Garg, National Institute of Technology, Kurukshetra
Dr. B. E. Narkhede, NITIE, Mumbai
Dr. Manjaree Pandit, MITS, Gwalior

Preface

Optimization may be defined as finding the solution to a problem where it is necessary to maximize or minimize a single or set of objective functions within a domain which contains the acceptable values of variables while some restrictions are to be satisfied. There might be a large number of sets of variables in the domain that maximize or minimize the objective function(s) while satisfying the described restrictions. They are called as the acceptable solutions, and the solution which is the best among them is called the optimum solution to the problem. An objective function expresses the main aim of the model which is to be either minimized or maximized. For example, in a manufacturing process, the aim may be to maximize the profit or minimize the cost. In designing a structure, the aim may be to maximize the strength or minimize the deflection or a combination of many objectives. The use of optimization techniques helps the engineers in improving the system's performance, utilization, reliability, and cost.

An international conference on "Advanced Engineering Optimization Through Intelligent Techniques (AEOTIT 2018)" was held during August 03–05, 2018, at Sardar Vallabhbhai National Institute of Technology, Surat, India. The objective of the conference was to bring together experts from academic institutions, industries, and research organizations and professional engineers for sharing of knowledge, expertise, and experience in the emerging trends related to advanced engineering optimization techniques and their applications. There had been an overwhelming response to the call for papers. More than 200 research papers were received from the researchers and academicians of the leading institutes and organizations. However, only 76 good-quality papers have been selected based on the recommendations of the reviewers for inclusion in the proceedings. These papers have covered various intelligent optimization techniques including meta-heuristics, neural networks, decision-making methods, and statistical tools.

We are extremely thankful to the authors of the papers, national and international advisory committee members, session chairmen, faculty and staff members of SVNIT, Surat, and CUT, Cracow, and student volunteers for their cooperation and help. We are grateful to the team members of Springer Nature for their support and help in producing these proceedings. We are confident that these proceedings would benefit the optimization research community.

Surat, India   R. Venkata Rao
Kraków, Poland   Jan Taler

Contents

About the Editors

Dr. R. Venkata Rao is a Professor at the Department of Mechanical Engineering, S. V. National Institute of Technology, Surat, India. He has more than 28 years of teaching and research experience. He completed his B.Tech. in 1988, M.Tech. in 1991, Ph.D. in 2002, and obtained his D.Sc. in 2017. Dr. Rao's research interests include: advanced optimization algorithms and their applications to design, thermal and manufacturing engineering, and fuzzy multiple attribute decision-making (MADM) methods and their industrial applications. He has more than 300 research papers to his credit, published in national and international journals and conference proceedings. He has received national and international awards for his research efforts. He is a reviewer for more than 80 national and international journals and serves on the editorial boards of several international journals. He has also been a Visiting Professor at Cracow University of Technology, Poland in January 2016, November 2016, and June 2018, at BITS Pilani Dubai campus in 2017, and at the Asian Institute of Technology, Bangkok in 2008, 2010, 2015 and 2018. He has authored six books entitled "Decision Making in the Manufacturing Environment Using Graph Theory and Fuzzy Multiple Attribute Decision Making Methods" (Volume 1 (2007) and Volume 2 (2013)), "Advanced Modeling and Optimization of Manufacturing Processes: International Research and Development" (2010), "Mechanical Design Optimization Using Advanced Optimization Techniques" (2012), "Teaching-Learning-Based Optimization Algorithm And Its Engineering Applications" (2016), and "Jaya: An Advanced Optimization Algorithm And Its Engineering Applications" (2018), all of which were published by Springer.

Dr. Jan Taler is a Professor and Director of the Institute of Thermal Power Engineering of Cracow University of Technology, Poland. He has more than 40 years of teaching and research experience, having completed his M.Sc. in 1974, Ph.D. in 1977, and D.Sc. in 1987. He has published about 300 articles in scientific journals. He has authored 10 books and over 20 chapters in scientific monographs and entries in the Encyclopedia of Thermal Stresses. He conducts research in the field of heat transfer engineering and thermal power engineering with special interest in the following areas – inverse heat conduction problems, measurement of heat flux and heat transfer coefficient, ash fouling and slagging in steam boilers, dynamics of large steam boilers, and optimization of thermal systems. He has participated in many research projects funded by industry and the Polish Committee for Scientific Research. Many of his innovative technical solutions have been implemented in power plants.

Combined Intelligent and Adaptive Optimization in End Milling of Multi-layered 16MnCr5/316L

Uros Zuperl◉ **and Franc Cus**◉

Abstract In this work, a new intelligent and adaptive optimization for end milling of four-layer functionally graded steel is presented in order to maximize the machining performance, minimize the production costs, and maximize the metal removal rate and ensuring the surface quality requirements. The proposed optimization consists of intelligent modeling of cutting quantities and particle swarm optimization (PSO) algorithm. Particle swarm optimization method is employed in real time to find optimum cutting conditions considering the real tool wear. Adaptive neural inference system is used to predict the tool flank wear timely on the basis of estimated cutting forces. Cutting forces and finally surface roughness were estimated during machining by using artificial neural networks (ANNs). The experimental results show that the proposed approach found an optimal solution of cutting conditions which improved the metal removal rate and improved the machining performance for 24% compared to conventional machining with off-line optimized parameters.

Keywords Adaptive optimization · Cutting conditions · Ball-end milling · Multi-layered metal material · Flank wear · PSO · ANFIS · Neural networks

1 Introduction

Machining of functionally graded steels fabricated by LENS technology [1] has been commonly applied in the industry of sheet metal forming tools. The machinability of these materials is problematic due to their composition that is not homogenous. This is the main reason for sudden and unexpected cutting tool wear which results in immense tool costs.

Cutting tool wear is directly connected to the set of chosen cutting conditions. Therefore, there is a considerable practical interest in searching of economically optimum cutting conditions. Many researchers were concerned with increasing the

U. Zuperl (✉) · F. Cus
Faculty of Mechanical Engineering, University of Maribor, Maribor, Slovenia
e-mail: uros.zuperl@um.si

© Springer Nature Singapore Pte Ltd. 2020
R. Venkata Rao and J. Taler (eds.), *Advanced Engineering Optimization Through Intelligent Techniques*, Advances in Intelligent Systems and Computing 949, https://doi.org/10.1007/978-981-13-8196-6_1

productivity of machining processes [2] and improving surface quality [3] of the final part through cutting parameters' optimization.

In order to prevent excessive tool wear and its negative effects, the cutting conditions are often selected conservatively and without taking into account the real cutting tool condition. Progressive cutting tool wear is not considered in optimization routine. Even if the cutting conditions are optimized by sophisticated, state-of-the-art optimization algorithms such as teaching learning-based optimization (TLBO), evolutionary algorithms (EA), and swarm intelligence-based optimization (PSO) algorithms, they cannot be optimized–adjusted in real time during the machining [4].

In most cases, such activities result in lowering the productivity and increasing manufacturing costs.

Therefore, many optimization systems with an adaptation of cutting parameters during machining have been developed and tested. They are designated as the adaptive optimization (AC) systems.

Zuperl and Cus [5] employed the ANN and three types of limitation equations for multi-objective optimization of machining parameters. Ko and Cho [6] tested an adaptive optimization technique to maximize the material removal rate in milling based on artificial flank wear model.

Liu and Wang [7] developed an adaptive control with optimization for the milling process based on the neural network for modeling and the neural network for control. Chiang et al. [8] developed an artificial neural network for online determination of optimal cutting parameters by maximizing material removal rate.

In this work, a new intelligent adaptive optimization is introduced for end milling with the aim to maximize the machining performance, minimize the production costs, and ensure the surface quality requirements.

The progressive cutting tool wear is incorporated into optimization algorithm. Optimization structure is composed of (1) an intelligent prediction models that predict cutting forces, flank wear, surface roughness and (2) an optimization module that maximize the milling performance by consideration of cutting constraints.

2   Combined Intelligent and Adaptive Optimization Routine

The aim of the combined intelligent adaptive optimization is to improve the ball-end milling performance by adjusting the feed rate (f) and cutting speed (v_c) in real time. The four-layer functionally graded steel is used in the cutting process.

The optimization takes into account gradually increasing tool wear, a machining cost and a machining rate (represented by metal removal rate—MRR). Therefore, a multi-objective optimization process with two conflicting objectives and with seven limitation non-equations has to be performed to solve the defined task.

In order to find optimal cutting parameters, the particle swarm optimization (PSO) algorithm was integrated with intelligent models that generalize the cutting forces,

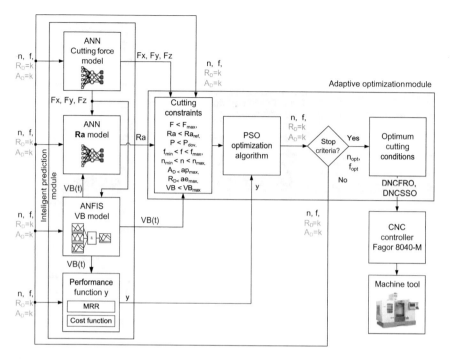

Fig. 1 Structure of intelligent adaptive optimization routine

tool wear, and surface quality based on experimental measurements. An optimization scheme is shown in Fig. 1.

The intelligent models are grouped together in the intelligent prediction module, and its outputs are fed into the adaptive PSO module.

The PSO starts the optimization process by generating random population which consists of cutting speeds, feed rates, and depth of cuts. Then, the trained ANN assigns cutting forces for entire population.

The cutting forces are inputted to an adaptive neural fuzzy inference system (ANFIS) which predicts the flank wear second by second for every particle. The output from the ANFIS model is the input to the surface roughness (Ra) model.

The ANN is trained to model the surface roughness for entire population. In the last phase, the performance function (y) is calculated which serves as objective function and represents the solution domain. The optimization algorithm moves the particles of the swarm in order to find the maximal value of the performance function. The position of the found performance function extreme represents the optimal cutting conditions for particular second of machining.

Seven constraints are listed in Table 1.

The particles that result in higher cutting force and surface roughness than defined threshold are eliminated from the population.

Table 1 List of used
constraints in adaptive
optimization routine

Name	Limitation equation
Spindle speed (n)	$n_{\min} \leq \frac{1000}{\pi \cdot D} v_c \leq n_{\max}$
Feed rate (f)	$f_{\min} \leq \frac{1000 \cdot z}{\pi \cdot D} v_c \cdot f_z \leq f_{\max}$
Cutting width (R_D)	$R_D \leq ae_{\max}$
Cutting depth (A_D)	$A_D \leq ap_{\max}$
Required cutting power (P)	$\frac{MRR \cdot Kc}{60} \leq P_{dov}$
Surface roughness (Ra)	$Ra \leq Ra_{\mathrm{ref}}$
Cutting force (F)	$F(f, n) \leq F_{\max}$

Where z represents the of cutting edges, D is cutting tool diameter,
Kc is specific cutting force constant, and ae$_{\max}$/ap$_{\max}$ is allowable
depth/width of cut for specific cutting tool

The performance function considered in this research is the quotient of the MRR
and machining costs. The performance index is calculated according to [9]:

$$y = \mathrm{MRR}(C_1 + (C_1 \cdot t_1 + C_1 \cdot \beta) \cdot (\mathrm{TWT})/W_0)^{-1} \tag{1}$$

The machining costs for milling operation consist of machining cost per effective
time of cutting (C_1) and cost of tool and regrind per change (C_2); t_1 is tool change
time. W_0 is maximum flank wear. TWR is tool wear rate. β parameter determines
the type of performance function [9].

Figure 2 presents block diagram of the intelligent adaptive optimization routine.
The adaptive optimization takes into consideration produced surface roughness and
actual value of tool flank wear. The adaptive optimization stops when the tool is
worn.

The ANFIS system is applied for predicting the values of flank wear VB(t) for
the whole tool life of the cutting tool during machining.

The ANFIS is chosen for developing the tool wear model due to its ability to
quickly determine the connections between machining parameters, cutting force,
and VB at a given time point.

The first developed ANFIS-based flank wear model is described in the work of
Zuperl et al. [10]. The inputs to the model are: $v_c, f, A_D, R_D, F_x, F_y, F_z$ and machining
time (t). The output is flank wear (VB(t)). The model predicts the flank wear with
maximal 4% error.

Firstly, a statistics is employed to simulate the Ra. In the statistical model, the
functional relations between the Ra, VB(t), F_x, F_y, F_z and cutting conditions are
provided. Statistical model is obtained at 75% confident level. Validation results
reveal inaccuracies in modeling of the Ra at small feed rates.

Therefore, a simple three-layered feedforward network based on the backpropagation learning algorithm is used instead to predict the Ra. The developed ANN
has one output neuron for Ra, eight input neurons for $v_c, f, A_D, R_D,$ VB(t), and three

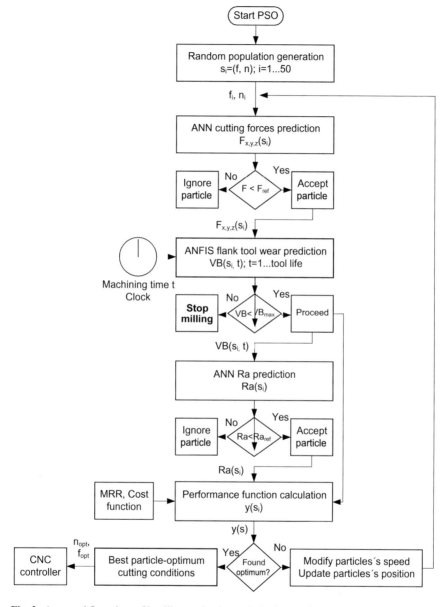

Fig. 2 A general flow chart of intelligent adaptive optimization routine

cutting force components F_x, F_y, F_z. The ANN architecture has seven neurons in other two layers. The output from the ANN is Ra. The validation results revealed that the maximum percentage prediction Ra error is less than 3.2%.

A simple three-layered feedforward network based on the backpropagation learning algorithm is used to predict F_x, F_y, and F_z. Therefore, three output neurons for three cutting force components are needed.

The inputs to the ANN are: v_c, f, A_D, R_D, D, the cutter rotation angle Θ, the height of the deposited stainless steel layer (d), the deposited layer hardness (HV). The developed ANN architecture has 12 hidden neurons in other two layers. An ArcTangent activation function was used to transform the signals between the neurons. The validation test results revealed that the ANN can forecast components of cutting forces with the prediction error 3.9% [11].

The optimum number of hidden neurons and hidden layers was determined by simulations. The aim of the simulations was to find the most appropriate architecture of the neural network, which will have the lowest prediction error at the lowest number of completed training iterations. In simulations, the number of hidden neurons and layers was systematically varied in order to find the best compromise between the training speed and the accuracy of the predicted values. For this purpose, 150 neural networks with different architecture were trained and tested. The results of the simulations were shown on the diagrams from which it was possible to determine the optimal number of hidden neurons and layers.

The optimum number of hidden neurons was determined in two steps. The goal of the first step is to identify a number of hidden neurons, in which the network prediction error was the lowest after the 500 performed training iterations. It has been found that the increase in the number of neurons leads to the decrease of prediction error. In the second step, the simulations determined the number of hidden neurons in which the network reached the prediction error of 5% with the smallest number of training iterations. It has been found that the increase in the number of neurons reduces the training speed. Furthermore, by increasing the number of hidden neurons, the number of necessary training iterations to achieve the 5% prediction error first decreases until it reaches a minimum (12 neurons) and then begins to increase again.

In the same way, in two steps, the effect of the number of hidden layers on the performance of a neural network was determined. The simulation results indicate: The optimal number of hidden neurons is 12. The neural network with 12 hidden neurons in two hidden layers reaches the smallest prediction error after 500 training iterations and needs the lowest number of training iterations to achieve the 5% prediction error.

3   Experimental Testing and Results

The proposed adaptive optimization routine has been validated with two machining tests. In these experimental tests, the machining with constant A_D and R_D has been carried out on a HELLER BEA02 milling machine.

The first test is conventional milling with constant cutting conditions. In this test, the cutting conditions are determined with the PSO and intelligent models before machining (off-line optimization), and then, these conditions are kept constant during the rest of machining.

The second test is milling with employed adaptive optimization routine which is after that compared with machining experiment using constant cutting conditions in order to analyze the obtained y and examine the efficiency of developed adaptive optimization method. The sum of y was employed as a comparison criterion.

The objective of the adaptive optimization is to maximize the y and to maintain the Ra at the defined Ra_{ref}. The PSO algorithm starts with a population of 50 candidate solutions (particles), which are continuously moving on function y and searching for the maximum value of y.

In order to achieve the faster convergence of the PSO algorithm, the acceleration constants c_1 and c_2 was initially set to 1.94 according to past experiences. The population size of 50 was found to be appropriate. The w_{min} was set to 0.4, the w_{max} to 0.9 and the number of iteration to 150 in order to calculate the inertia weight factor w. A uniform probability distribution was used to generate random numbers for updating the velocity. The V_{max} (maximum velocity) is determined by considering 10% of the variable dynamic range. After the initial PSO parameters were set, the PSO algorithm was run 10 times until convergence. The identical results are obtained in all runs; therefore, the PSO parameters and swarm population are appropriate. Furthermore, the perceived scattering of the particles in the swarm during all performed iterations indicated that the PSO parameters were appropriate and should be used for optimization of cutting conditions. The laser engineered functional graded steel material with dimension 400 mm × 60 mm was used for machining tests. The four 0.8 mm thick layers of the 316L powder were deposited on the 16MnCr5 substrate by the LENS 850-R 3D metal printer with 0.7 laser beam. The hardness of individual deposition was 290 HV.

The test piece was machined by two flute ball nose end mill with 4 mm nose radius. The material of the cutter was a sintered tungsten carbide K88UF coated with PVD-TiAlN coating. The cutting material hardness was 1770 HV.

The second experiment is started with the following initial parameters and constraints: $R_D = 4$ mm, $A_D = 1$ mm, $v_c = 100$ m/min, $f = 250$ mm/min, $F \leq 300$ N, $100 \leq f \leq 400$ mm/min, and $n \leq 4100$ min^{-1}. Maximum tool wear (VB_{max}) was set to 0.2 mm. Figure 3 shows the response of cutting conditions and objective function for milling with adaptive optimization. In this experiment, cutting speed and feed rate are optimized adaptively in order to maintain the value of performance function y at the maximal value. Figure 3a shows the response of cutting speed when adaptive optimization is carried out and the course of cutting speed during conventional machining. In the first experiment, the cutting speed is constant and set to 85 m/min. In second experiment, the cutting speed varies between 85 and 98 m/min. At the beginning of machining, when the TWR is low, the adaptive optimization increases the cutting speed up to 100 m/min in order to maximize the MRR and consequently the y. During machining, tool wear increases and to compensate the higher TWR, the system decreases the cutting speed. The lower cutting speeds decrease the values of

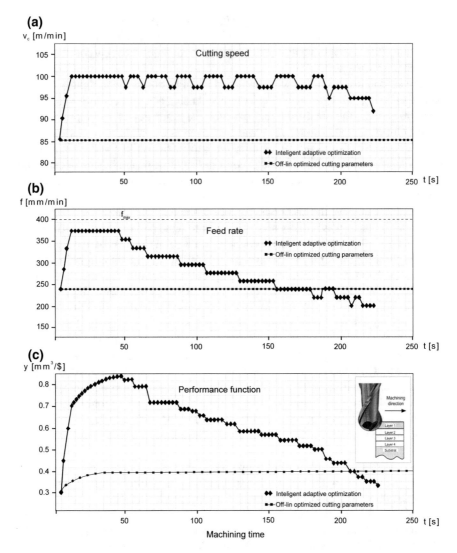

Fig. 3 Response of feed rate and cutting speed during milling with intelligent adaptive optimization and during conventional milling with off-line determined parameters

performance function y. Figure 3b shows the two responses of feed rate for adaptive optimization and for conventional machining. From Fig. 3b, it can be seen that the feed rate for the conventional milling is constant (235 mm/min) during the machining process. When milling with intelligent adaptive optimization, the feed rates have reached the values that were significantly above the feed rates during the first 155 s of conventional machining.

The test results have confirmed the significant influence of feed rate on the Ra. The increasing tool wear during machining also affects the Ra. Therefore, the optimization system decreases the feed rates to maintain the Ra at the desired level and to compensate the effect of the increased flank wear on the Ra. Figure 3c shows the two responses of performance function y for both the adaptive optimization and conventional machining with constant parameters.

In milling with fixed cutting conditions (Fig. 3c), the y exceeds the value of adaptive optimization only after 205 s of machining. Before that, during milling with intelligent adaptive optimization, the calculated value of y is about 120% higher. First, the adaptive optimization routine abruptly increases the y up to 0.84 mm^3/$. After the peak is reached, the value of the performance function is slowly decreasing due to the reduction of v_c and f.

In the first experiment, the flank wear rate is higher in the beginning of machining and then became linear. Consequently, the y increases at the beginning of machining and then remains almost constant during the rest of operation. The calculated sum of y for proposed adaptive optimization routine was 46.4, compared to 35.5 for milling with constant cutting conditions. That indicates that the total sum of y is for 24% higher during milling with employed intelligent adaptive optimization. The sum of y has increased despite the fact that the tool wear was much more intensive. During the entire machining, the Ra was maintained at the desired level. Milling with the proposed optimization routine increases the material removal rate and improves the machine tool capacity.

In this paper outlined, adaptive optimization routine was compared with three non-traditional adaptive control optimization systems (ACO) and conventional machining with constant cutting conditions. The comparison results show that the intelligent adaptive optimization based on a PSO method and intelligent prediction algorithms outperforms the optimization strategy which is not connected to the machining process. A cost reduction of 14% is established compared to conventional machining. A cost reduction of 25% is established compared to adaptive control optimization approaches. The proposed approach found an optimal solution of cutting conditions which improved the MRR by 26% and reduced the machining time up to 18%.

Chiang et al. [8] used in adaptive control optimization (ACO) system a neural network estimation module and an optimization module with evolutionary optimization algorithms to minimize the production cost. A production cost reduction of 28% is established compared to conventional machining with off-line optimized parameters.

Zuperl et al. [12] merged the off-line optimization algorithm with adaptive cutting force control in end milling. In his work, the cutting force surface is the objective function and the PSO is employed to adjust feed rate and thus to improve the MRR up to 27% in comparison with the conventional machining.

4   Conclusion

In this work, the new intelligent adaptive optimization for milling of multi-layered metal materials is developed. In machining of multi-layered metal materials, the classical off-line optimization methods are not efficient due to intensive and uncharacteristic cutting tool wear. Therefore, it is needed to online optimize the cutting parameters throughout the machining and to include the progressive tool wear in optimization routine. The proposed optimization integrates two ANN models to predict cutting forces and Ra, the ANFIS model to predict the progressive flank wear, the algorithm to asses the performance function and the PSO to find the appropriate cutting conditions during the machining operation.

A test case has been presented to demonstrate effectiveness of the proposed intelligent adaptive optimization. The experimental results show that the proposed approach found an optimal solution of cutting conditions which improved the MRR, reduced the machining time up to 17%, and thus improved the machining performance for 24% compared to conventional machining with off-line optimized parameters. The results of optimization are also compared and analyzed using methods of other researchers. The results indicate that the proposed optimization is effective compared to other techniques. The comparison results show that the adaptive optimization based on a PSO method and intelligent prediction algorithms outperforms off-line optimization strategy and a larger cost reduction of 14% compared to conventional machining is achieved. The study's main findings are also that the ANN and ANFIS methods are effective in precise modeling of milling quantities and can be successfully integrated in the adaptive optimization algorithm. The proposed intelligent adaptive optimization provides a novel way for maximizing machining performance in end milling of multi-layered metal materials.

References

1. Palčič, I., Balažic, M., Milfelner, M., Buchmeister, B.: Potential of laser engineered net shaping (LENS) technology. Mater. Manuf. Processes **24**(7–8), 750–753 (2009). https://doi.org/10.1080/10426910902809776
2. Hamdan, A., Sarhan, A.A., Hamdi, M.: An optimization method of the machining parameters in high-speed machining of stainless steel using coated carbide tool for best surface finish. Int. J. Adv. Manuf. Technol. **58**(1–4), 81–91 (2012). https://doi.org/10.1007/s00170-011-3392-5
3. Chandna, P., Kumar, D.: Optimization of end milling process parameters for minimization of surface roughness of AISI D2 steel. World Acad. Sci. Eng. Technol. Int. J. Mech. Aerosp. Ind. Mech. Manuf. Eng. **9**, 3 (2015)
4. Khan, W., Raut, N.: Optimization of end milling process parameters for minimization of surface roughness of AISI P20 steel. Asian J. Sci. Technol. **7**(4), 2777–2787 (2016)
5. Zuperl, U., Cus, F.: Optimization of cutting conditions during cutting by using neural networks. Robot. Comput. Integr. Manuf. **19**, 189–199 (2003). https://doi.org/10.1016/S0736-5845(02)00079-0
6. Ko, T.J., Cho, D.W.: Adaptive optimization of face milling operations using neural networks. J. Manuf. Sci. Eng. **120**(2), 443–451 (1998). https://doi.org/10.1115/1.2830145

 7. Liu, Y., Wang, C.: Neural network based adaptive control and optimisation in the milling process. Int. J. Adv. Manuf. Technol. **15**(11), 791–795 (1999). https://doi.org/10.1007/s001700050133
 8. Chiang, S.T., Liu, D.I., Lee, A.C., Chieng, W.H.: Adaptive-control optimization in end milling using neural networks. Int. J. Mach. Tools Manuf. **35**(4), 637–660 (1995)
 9. Koren, Y.: Adaptive control systems for machining. In: American Control Conference, pp. 1161–1167. IEEE (1988)
10. Zuperl, U., Cus, F., Kiker, E.: Adaptive network based inference system for estimation of flank wear in end-milling. J. Mater. Process. Technol. **209**(3), 1504–1511 (2009). https://doi.org/10.1016/j.jmatprotec.2008.04.002
11. Zuperl, U., Cus, F.: Tool cutting force modeling in ball-end milling using multilevel perceptron. J. Mater. Process. Technol. **153**, 268–275 (2004). https://doi.org/10.1016/j.jmatprotec.2004.04.309
12. Zuperl, U., Cus, F., Kiker, E., Milfelner, M.: A combined system for off-line optimization and adaptive adjustment of the cutting parameters during a ball-end milling process. Strojniski Vestnik **51**(9), 542 (2005)

Jaya: A New Meta-heuristic Algorithm for the Optimization of Braced Dome Structures

Tayfun Dede⬤, Maksym Grzywiński⬤ and R. Venkata Rao

Abstract A new algorithm called Jaya is presented for the design of the braced dome structures by taking into account the objective function as least weight with frequency constraints. The size optimization is considered for the 3D truss elements. The performance of Jaya algorithm is presented through benchmark 120-bar braced dome. This study indicated that the proposed technique is a powerful technique for the optimal design of domes with the constrained problem. The developed computer program for the analysis and optimization of the dome structure and the optimization algorithm for Jaya are coded in MATLAB.

Keywords Jaya algorithm · Size optimization · Frequency constraints · Dome structure

1 Introduction

It has been the goal of the researchers to do designs in a short time and with a fewer number of analyses. For this aim, many optimization algorithms are proposed until now. Some of them are bat algorithm (BA), teaching–learning-based optimization (TLBO), evolution strategies (ES), Jaya algorithm (JA), artificial bee colony (ABC), simulated annealing (SA), Grey wolf optimization algorithm (GWO), and genetic algorithm (GA). These algorithms were used in many engineering problems.

T. Dede (✉)
Department of Civil Engineering, Karadeniz Technical University, Trabzon, Türkiye
e-mail: tayfundede@gmail.com

M. Grzywiński
Faculty of Civil Engineering, Czestochowa University of Technology, Czestochowa, Poland
e-mail: mgrzywin@bud.pcz.pl

R. Venkata Rao
Department of Mechanical Engineering, Sardar Vallabhbhai National Institute of Technology, Surat 395007, Gujarat, India
e-mail: ravipudirao@gmail.com

© Springer Nature Singapore Pte Ltd. 2020
R. Venkata Rao and J. Taler (eds.), *Advanced Engineering Optimization Through Intelligent Techniques*, Advances in Intelligent Systems and Computing 949,
https://doi.org/10.1007/978-981-13-8196-6_2

13

Dome structures like 3D trusses are considered in this study as a benchmark problem to test the Jaya algorithms. In the literature, many papers are presented related to the optimization with frequency constraints for the truss structures.

Bellagamba and Yang [1] were the first who studied structural optimization with frequency constraints. Least weight design of structure by taking into account dynamic and static constraints was studied by Lin et al. [2].

The genetic algorithm (GA) for optimum structural design applications of trusses structure was investigated in [3–5]. Talaslioglu [6] has developed an optimization algorithm with many populations. Hybridized genetic algorithm (GA) with Niche techniques is used by Lingyum et al. [7]. Firefly algorithm (FA) and harmony search (HS) were firstly used by Miguel and Miguel [8]. The particle swarm optimization (PSO) algorithm is implemented by Gomes [9] for geometry and size optimization. Kaveh and Zolghadr [10, 11] presented a study by using the democratic particle swarm optimization (DPSO), harmony search, and a ray optimizer to improve the particle swarm optimization algorithm (PSRO). Kaveh and Javadi [12] developed a hybrid algorithm (HRPSO). They applied this developed algorithm for optimal shape and size design of trusses by taking into account the constraints such as natural frequency.

Colliding bodies optimization (CBO) was developed by Kaveh and Mahdavi [13]. Symbiotic organisms search (SOS) was proposed by Tejani et al. [14] Kaveh and Ghazaan [15] studied cascade sizing optimization utilizing a series of design variable configurations (DVCs). Baghlani and Makiabadi [16], Dede and Toğan [17] developed an algorithm called teaching–learning-based optimization (TLBO). The education process from a single classroom to a school with multiple parallel classes is simulated by Farshchin et al. [18] as a new algorithm named MC-TLBO.

The aim of this study is to present an implementation of the newly developed optimization algorithm named Jaya. In 2016, the first study related to this algorithm was presented by Rao [19] and its engineering applications are given in the study Rao [20]. Jaya is based on the assumption that the solution obtained for a given problem should be directed toward the best solution and should avoid the worst solution. Dede [21] had studied the optimum design of steel grillage structure by taking into account this algorithm. Sizing, layout, and large-scale optimization problems were previously studied by Degertekin et al. [22] for the weight minimization of truss structures by using this algorithm. Using different shapes of X-bracing structure is studied by Özdemir and Ayvaz [23].

2 The General Structure of Optimization Problem

The main target of solving braced dome optimization problem is to reduce the total volume or weight of a structure without violating constraints such as natural first five frequencies of the dome structure. The optimization problem can be described as given below:

$$\text{Design variables: } \{D\} = \left\{d_1, d_2, \ldots, d_{ng}\right\} \tag{1}$$

$$\text{Objective function: } F = \sum_{g=1}^{ng}\left(D_g * \rho_g * \sum_{m=1}^{nm}(L_m)\right) \tag{2}$$

$$\text{Constraints: } \omega_j \geq \omega_j^* \text{ for natural frequencies } j$$
$$\omega_k \leq \omega_k^* \text{ for natural frequencies } k \tag{3}$$

$$\text{Bounds: } L_b < \{D\} < U_p \tag{4}$$

where "D" is the design variable vectors which are cross-sectional areas of the bar elements of dome, "ng" is the number of grouping if the design variables are categorized, "F" is the objective function representing the total volume or weight of the whole truss structure, and "nm" is the number of member in the grouping of the structure. "ρ," "L," and "D" are the density of the material, the length of bar element, and cross-sectional area of the bar element, respectively. "ω_j" and "ω_k" are the natural frequencies of the structure that some of them must be greater or little than the specified frequency (ω_j^* or ω_k^*) depending on the engineering problem. "L_b" and "U_b" are the lower and upper bounds of the design variables, respectively.

To take into the constraints, the objective function must be changed as a penalty function (ϕ) including the constraint. The penalized objective function can be written in a basic form as;

$$\phi = F * (1 + P * C) \tag{5}$$

where "C" is the sum of the violations of the constraints and "P" is a constant value. For the frequency constraint "ω_j," a violation can be calculated as;

$$g_i = \frac{\omega_j}{\omega_j^*} - 1 \geq 0 \quad \text{if } g_i < 0 \text{ then } c_i = g_i \tag{6}$$

In Eq. (6), the natural frequency is calculated in term of the mass "m" and the rigidity of the structure "k" by using the following equation.

$$\omega = \frac{1}{2 * \pi}\sqrt{\frac{k}{m}} \tag{7}$$

3 Optimization with Jaya Algorithm

Jaya prefers to use randomly created initial population. The main principle of this algorithm is that the updated solutions (possible candidates for the objective of the problem) are created by getting close the best candidate and moving away the worst

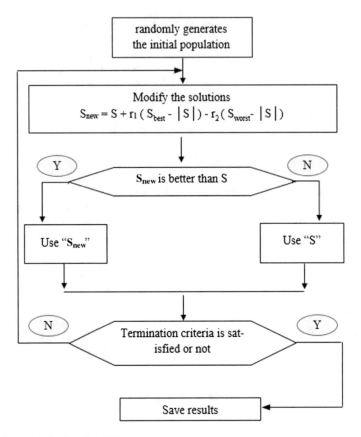

Fig. 1 Jaya optimization algorithm

candidate. The best solution is defined as the combination of the design variables which give the least weight of dome structure for this study. The superiority of this method is that only the common check parameters are required. To explain the general process of the Jaya algorithm, the flowchart is given in Fig. 1.

Where "S" is any solution, S_{best} is the best solution, S_{worst} is the worst solution, and r is the random number between from 0 to 1.

4   Numerical Example

In this study, frequency constraint is taking into account. The 120-bar dome structure is optimized for size parameters. The design variables are selected as continuous by taking into account the lower and upper. In the optimization process, 20 independent runs are carried. Figure 2 illustrates the configuration of the 120-bar dome structure.

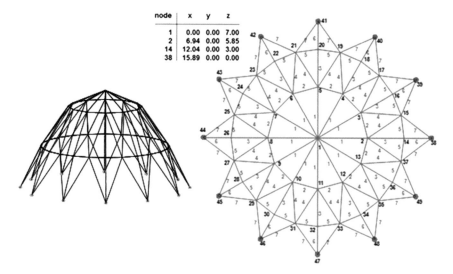

node	x	y	z
1	0.00	0.00	7.00
2	6.94	0.00	5.85
14	12.04	0.00	3.00
38	15.89	0.00	0.00

Fig. 2 Plan and 3D view with element grouping of the dome structure

Table 1 Structural constraints and material properties for the dome structure

Properties/Constraints	Symbol	Value/Notes
Elasticity modulus	E	2.1×10^{11} (N/m^2)
Density of material	ρ	7971.81 (kg/m^3)
Non-structural mass at nodes	m	3000 for 1 (kg) 500 for 2:13 100 for 14:37
Bounds of cross-sectional area	A	$0.0001 \leq A \leq 0.01293$ (m^2)
Frequency constraints	ω	$\omega_1 \geq 9, \omega_{2,3} \geq 11$ (Hz)

To obtain the geometry of this structure, initial nodal coordinates and grouping of the elements can be seen from this figure. This example was investigated by Kaveh and Zolghadr [11] using (PSRO), Tejani et al. [14] using symbiotic organisms search (SOS-ABF), Kaveh and Mahdavi [13] using colliding-bodies optimization (CBO), and Kaveh and Zolghadr [10] using democratic particle swarm optimization (DPSO).

The structural element of this example is classified into seven groups for the size optimization. The properties of the material, constraints for frequency, and the additional mass on the free nodes masses are given in Table 1. The size of the population is 30 and the iteration is 600 for this example.

Table 2 shows the optimal solutions obtained by using different algorithms. As seen in this comparison, the best solution is given by using the proposed algorithm.

The history of the optimal solution, mean solution, and the standard deviation are given in Fig. 3. To show the convergence in more detail, a large-scaled graphic

Table 2 Optimal results and comparison for the dome structure

Design variables		[11]	[14]		[13]	[10]	This study Jaya	
		PSRO	SOS-ABF1	SOS-ABF2	CBO	DPSO	Pn = 20	Pn = 30
Cross-sectional area (cm²)	A1	19.972	19.5449	19.5715	19.6917	19.607	19.300	19.309
	A2	39.701	40.9483	39.8327	41.1421	41.290	40.861	40.763
	A3	11.323	10.4482	10.5879	11.1550	11.136	10.697	10.791
	A4	21.808	21.0465	21.2194	21.3207	21.025	21.107	21.272
	A5	10.179	9.5043	10.0571	9.8330	10.060	9.989	9.943
	A6	12.739	11.9362	11.8322	12.8520	12.758	11.779	11.695
	A7	14.731	14.9424	14.7503	15.1602	15.414	14.743	14.579
Weight (kg)		8892.33	8712.11	8710.33	8889.13	8890.48	8712.67	8709.35
Mean (kg)		8921.30	8727.42	8725.30	8891.25	8895.99	8730.17	8713.21
Std (kg)		18.54	16.55	10.64	1.79	4.26	12.78	2.97
nfe or Pn/Gn		20/200	4000	4000	6000	30/200	20/200	30/600
Run		20	–	–	20	30	20	20
Frequency ω (Hz)		9.000	9.0011	9.0012	9.0000	9.0000	9.0016	9.0000
		11.000	11.0003	11.0023	11.0000	11.0000	11.0013	11.0002
		11.005	11.0003	11.0023	11.0000	11.0052	11.0013	11.0002
		11.012	11.0015	11.0056	11.0096	11.0134	11.0044	11.0008
		11.045	11.0674	11.0720	11.0494	11.0428	11.0716	11.0674

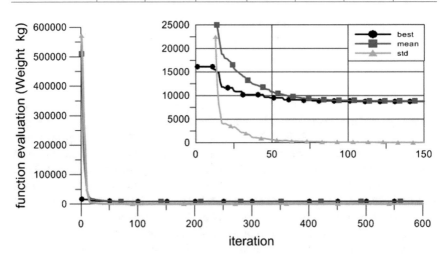

Fig. 3 Convergence history of the best solutions by using Jaya algorithm

Table 3 Diversity of the run for the dome structure

Run	Best	Run	Best
1	8709.3539	11	8714.4637
2	8715.9286	12	8715.7458
3	8711.9890	13	8712.5166
4	8709.5981	14	8715.1362
5	8710.2483	15	8713.9430
6	8710.9132	16	8719.6570
7	8711.6594	17	8719.5643
8	8713.1122	18	8713.9493
9	8711.6286	19	8715.0774
10	8710.0645	20	8709.7543
Pn/Gn	30/600	Best	8709.3539
	Mean CPU time (s)		1237.9688

is added to the same figure. Table 3 presents the variety of the dome structure with different runs.

5 Conclusion

Sizing optimization with the frequency constraints of 3D dome structure is investigated in this study. To optimize the dome structure, a new and efficient algorithm called Jaya is coded in the MATLAB. The results obtained from the optimization process of the example taken from the literature as a benchmark problem are compared with the other solutions obtained from different studies. The results of this study indicated that the Jaya algorithm gives the best solution among the other algorithms. As a result, it can be stated that the Jaya optimization algorithm can be used as an effective algorithm to find the best solution for the 3D dome structures.

References

1. Bellagamba, L., Yang, T.: Minimum-mass truss structures with constraints on fundamental natural frequency. AIAA J. **19**(11), 1452–1458 (1981). https://doi.org/10.2514/3.7875
2. Lin, J.H., Chen, W.Y., Yu, Y.S.: Structural optimization on geometrical and element sizing with static and dynamic constraints. Comput. Struct. **15**, 507–515 (1982)
3. Bekiroglu, S., Dede, T., Ayvaz, Y.: Implementation of different encoding types on structural optimization based on adaptive genetic algorithm. Finite Elem. Anal. Des. **45**, 826–835 (2009). https://doi.org/10.1016/j.finel.2009.06.019
4. Grzywiński, M.: Optimization of single-layer braced domes. Trans. VSB Tech. Univ. Ostrava **17**(1), paper #6 (2017). https://doi.org/10.1515/tvsb-2017-0006
5. Salam, S.A., El-shihy, A., Eraky, A., Salah, M.: Optimum design of trussed dome structures. Int. J. Eng. Innov. Technol. **4**, 124–130 (2015)

6. Talaslioglu, T.: Design optimization of dome structures by enhanced genetic algorithm with multiple populations. Sci. Res. Essays **7**, 3877–3896 (2012)

7. Lingyum, W., Mei, Z., Guangming, W., Guang, M.: Truss optimization on shape and sizing with frequency constraints based on genetic algorithm. Comput. Mech. **35**, 361–368 (2005). https://doi.org/10.1007/s00466-004-0623-8

8. Miguel, L.F.F., Miguel, L.F.F.: Shape and size optimization of truss structures considering dynamic constraints through modern metaheuristic algorithms. Expert Syst. Appl. **39**, 9458–9467 (2012). https://doi.org/10.1016/j.eswa.2012.02.113

9. Gomes, H.M.: Truss optimization with dynamic constraints using a particle swam algorithm. Expert Syst. Appl. **38**, 957–968 (2011). https://doi.org/10.1016/j.eswa.2010.07.086

10. Kaveh, A., Zolghadr, A.: Democratic PSO for truss layout and size optimization with frequency constraints. Comput. Struct. **130**(3), 10–21 (2014). https://doi.org/10.1016/j.compstruc.2013.09.002

11. Kaveh, A., Zolghadr, A.: A new PSRO algorithm for frequency constraint truss shape and size optimization. Struct. Eng. Mech. **52**(3), 445–468 (2014). https://doi.org/10.12989/sem.2014.52.3.445

12. Kaveh, A., Javadi, S.M.: Shape and size optimization of trusses with multiple frequency constraints using harmony search and ray optimizer for enhancing the particle swarm optimization algorithm. Acta Mech. **225**, 1595–1605 (2014). https://doi.org/10.1007/s00707-013-1006-z

13. Kaveh, A., Mahdavi, V.R.: Colling-bodies optimization for truss with multiple frequency constraints. J. Comput. Civ. Eng. **29**(5), 04014078-10 (2015). https://doi.org/10.1061/(asce)cp.1943-5487.0000402

14. Tejani, G.G., Savsani, V.J., Patel, V.K.: Adaptive symbiotic organisms search (SOS) algorithm for structural design optimization. J. Comput. Des. Eng. **3**, 226–249 (2016). https://doi.org/10.1016/j.jcde.2016.02.003

15. Kaveh, A., Ghazaan, M.I.: Optimal design of dome truss structures with dynamic frequency constraints. Struct. Multidisc. Optim. **53**, 605–621 (2016). https://doi.org/10.1007/s00158-015-1357-2

16. Baghlani, A., Makiabadi, M.H.: Teaching-learning-based optimization algorithm for shape and size optimization of truss structures with dynamic frequency constraints. Iran J. Sci. Technol. **37**, 409–421 (2013). https://doi.org/10.22099/IJSTC.2013.1796

17. Dede, T., Toğan, V.: A teaching learning based optimization for truss structures with frequency constraints. Struct. Eng. Mech. **53**, 833–845 (2015). https://doi.org/10.12989/sem.2015.53.4.833

18. Farshchin, M., Camp, C.V., Maniat, M.: Multi-class teaching-learning-based optimization for truss design with frequency constraints. Eng. Struct. **106**, 356–369 (2016). https://doi.org/10.1016/j.engstruct.2015.10.039

19. Rao, R.V.: Jaya: a simple and new optimization algorithm for solving constrained and unconstrained optimization problems. Int. J. Ind. Eng. Comput. **7**, 19–34 (2016). https://doi.org/10.5267/j.ijiec.2015.8.004

20. Rao, R.V.: Jaya: An Advanced Optimization Algorithm and Its Engineering Applications. Springer, Berlin (2019)

21. Dede, T.: Jaya algorithm to solve single objective size optimization problem for steel grillage structures. Steel Compos. Struct. **26**, 163–170 (2018). https://doi.org/10.12989/scs.2018.26.2.163

22. Degertekin, S.O., Lamberti, L., Ugur, I.B.: Sizing, layout and topology design optimization of truss structures using the Jaya algorithm. Appl. Soft Comput. **70**, 903–928 (2018). https://doi.org/10.1016/j.asoc.2017.10.001

23. Özdemir, Y.I., Ayvaz, Y.: Earthquake behavior of stiffened RC frame structures with/without subsoil. Struct. Eng. Mech. **28**(5), 571–585 (2008)

Damage Detection of Truss Employing Swarm-Based Optimization Techniques: A Comparison

Swarup K. Barman, Dipak K. Maiti and Damodar Maity

Abstract Swarm-based optimization techniques are very popular and well known in the field of damage detection of structures. Present paper evaluates the performance of three different variants of particle swarm optimization (PSO) and continuous ant colony optimization (ACOr) to detect damages in plane and space truss structure based on frequency and mode shapes-based objective function. The algorithms considered for the comparison are: unified particle swarm optimization (UPSO), ageing leader challenger particle swarm optimization (ALC-PSO), enhanced PSO with intelligent particle number (IPN-PSO) and continuous ant colony optimization (ACOr). A 25 member plane truss and a 25 member space truss are considered for the comparison among the algorithms. The numerical study reveals the superiority of UPSO over other algorithms in terms of minimum computational effort and success rate.

Keywords PSO · ALC-PSO · UPSO · ACOr · Frequency · Modeshapes

1 Introduction

Frequent occurrence of various kinds of damages in structures has made the damage detection of structure a very active area of research in the field of structural health monitoring. Presence of various kinds of damages brings into undesirable displacements, stresses into the structures. Even small damages grow into large damages in time if neglected during the service period of the structures. This can lead to major failure of the entire structure in the later period. So, damage should be detected at

S. K. Barman (✉) · D. K. Maiti
Department of Aerospace Engineering, IIT Kharagpur, Kharagpur, India
e-mail: Swarup.jadavpur@gmail.com

D. K. Maiti
e-mail: dkmaiti@aero.iitkgp.ac.in

D. Maity
Department of Civil Engineering, IIT Kharagpur, Kharagpur, India
e-mail: dmaity@civil.iitkgp.ac.in

© Springer Nature Singapore Pte Ltd. 2020
R. Venkata Rao and J. Taler (eds.), *Advanced Engineering Optimization
Through Intelligent Techniques*, Advances in Intelligent Systems and Computing 949,
https://doi.org/10.1007/978-981-13-8196-6_3

early stage to avoid such kind of mishaps. Among various other non-destructive methods vibration-based methods of damage detection have become very popular in the recent years. Presence of damages alters the stiffness and mass of the structures and thus reduces the dynamic response parameters (frequency response function (FRF), natural frequencies, mode shapes, damping) in the structures. Thus, damage can be detected from these changes by solving finite model updating-based inverse problem utilizing optimization algorithms. Natural frequencies [1–4], FRF [5], combined natural frequency and mode shapes [6–10] have been used as damage indicator of beam truss and plate-like structures. Optimization techniques also are a major part of solving vibration-based inverse problem. Throughout the years, researchers have used different optimization algorithms for damage detection of structures. Particle swarm optimization (PSO) [5, 11], genetic algorithm (GA) [12], neural network (NN) [13], artificial bee colony (ABC) [9], ant colony optimization (ACO) [4], modified charge system search (CSS) [6], fuzzy cognitive map (FCM) [14] and unified particle swarm optimization (UPSO) [8] are such optimization techniques, which have been used for damage detection of structures effectively.

In the present study, three modifications of PSO and ACOr have been considered for a comparative study on the performance of each of them in detecting damages in truss structures. They are: UPSO, ALC-PSO, IPN-PSO, ACOr. Combined natural frequencies and mode shapes-based objective function have been used as damage indicators.

2   Mathematical Background

2.1   Finite Element Formulation

Finite element formulation of truss structures have been formulated using two-dimensional two-noded truss element with two (δ_x, δ_y) and three $(\delta_x, \delta_y, \delta_z)$ degrees of freedom (Fig. 1), respectively, for plane and space truss. Axial rigidity EA and cross-sectional area A are considered same throughout the length of the element. Element stiffness matrix and element mass matrix in element coordinate system, respectively, are given by [15].

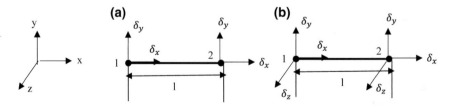

Fig. 1 **a** Plane truss element **b** Space truss element

$$[k]_e = \frac{EA}{l}\begin{bmatrix} 1 & -1 \\ -1 & 1 \end{bmatrix} \quad [m]_e = \frac{\rho Al}{6}\begin{bmatrix} 2 & 1 \\ 1 & 2 \end{bmatrix}$$

where l is the length of the element and ρ is the mass density of the material used. These matrices can be transformed into global coordinate system using the following equation, respectively.

$$[K]_e = [T]^T[k]_e[T] \quad [M]_e = [T]^T[m]_e[T]$$

where T is the transformation matrix. For plane truss,

$$[T] = \begin{bmatrix} c & s & 0 & 0 \\ -s & c & 0 & 0 \\ 0 & 0 & c & s \\ 0 & 0 & -s & c \end{bmatrix}$$

where $c = \cos\alpha$, $s = \sin\alpha$. α is the inclination of the element axis with global X axis.

For space truss,

$$[T] = \begin{bmatrix} m_1 & m_2 & m_3 & 0 & 0 & 0 \\ n_1 & n_2 & n_3 & 0 & 0 & 0 \\ p_1 & p_2 & p_3 & 0 & 0 & 0 \\ 0 & 0 & 0 & m_1 & m_2 & m_3 \\ 0 & 0 & 0 & n_1 & n_2 & n_3 \\ 0 & 0 & 0 & p_1 & p_2 & p_3 \end{bmatrix}$$

$\{m_1, n_1, p_1\}$ are the direction cosines of global X axis with respect to local x, y, z axis, respectively. Similarly, $\{m_2, n_2, p_2\}$ and $\{m_3, n_3, p_3\}$ are the direction cosines of global Y and Z axis, respectively, with respect to local x, y, z axis.

Element stiffness and mass matrices were obtained for each elements and assembled together to calculate global stiffness and mass matrices. Consistent mass matrix was used here to calculate local and global mass matrices, as it is easy to construct in a same manner as the element stiffness matrix using shape functions. Finally, the obtained analytical model of the structures was used to compute the modal parameters by solving the eigenvalue equation:

$$([K] - \omega_i^2[M])\{\varphi_i\} = 0 \tag{1}$$

where $[K]$, $[M]$, ω_i and φ_i are the global stiffness matrix, global mass matrix, ith natural frequency and mode shape corresponding to ith natural frequency, respectively.

2.2 Damage Model

Damaged element was modelled as percentage reduction of the axial stiffness of the particular element. Only one damage parameter per structural elements is considered as all the materials are considered isotropic. So, the global damaged stiffness matrix ($[K_d]$) of the structure can be written as

$$[K_d] = \sum_{i=1}^{ne}(1 - \alpha_i)[Ke_i], \quad \alpha \in [0, 1] \tag{2}$$

where $[Ke_i]$ is undamaged element stiffness matrix for ith element. ne denotes the number of elements in the structure. α_i is the damage parameter for ith element and it can vary from 0 to 1. The value of zero means no damage and one means complete damage.

2.3 Frequency and Modeshapes-Based Objective Function

The objective function considered here is given below, Nanda et al. [8]:

$$F(\% \text{ damage in each element}) = \sqrt{\frac{1}{n}\sum_{i=1}^{i=n}\left(\left(\frac{f_i^m}{f_i^a}\right) - 1\right)^2 + \sum_{i=1}^{i=n}(1 - \text{MAC}_{ii})} \tag{3}$$

where

$$\text{MAC}_{ii} = \frac{\left|\varphi_{mi}^T\varphi_{ai}\right|^2}{\left(\varphi_{mi}^T\varphi_{mi}\right)\left(\varphi_{ai}^T\varphi_{ai}\right)} \tag{4}$$

where n stands for number of modes, f and φ denote the frequency and corresponding mode shape, respectively. Suffix m and a stand for actual response of damaged structures and updated response through FEM, respectively. MAC denotes the modal assurance criteria (MAC).

3   Optimization Algorithms

3.1   Particle Swarm Optimization (PSO)

Particle swarm optimization (PSO) was developed by Kennedy and Eberhart [16] as a stochastic optimization algorithm based on concepts and rules that govern socially organized populations in nature, such as bird flocks, fish schools and animal herds to search for food or to avoid predators. Initially, each particle assumes random positions and velocities within the search space. During the search process, position and velocity of each particle get updated in each iteration: based on three factors:

(i) *pbest*—Best position visited by individual particle ($p(t)$)
(ii) *gbest*—Overall best position ever visited by all the particle ($g(t)$) [Global-PSO]
(iii) *lbest*—Best position visited by neighbours of each particle $l(t)$ [Local-PSO]

Velocity update ($V(t + 1)$) and position update ($X(t + 1)$) in each iteration of basic PSO can be calculated from the following equations (Eqs. 5–7).

$$X(t + 1) = X(t) + V(t + 1) \tag{5}$$

For Global-PSO:

$$V(t + 1) = wV(t) + c_1 \, \text{rand}(0, 1)(p(t) - X(t)) + c_2 \, \text{rand}(0, 1)(g(t) - X(t)) \tag{6}$$

For Local-PSO:

$$V(t + 1) = wV(t) + c_1 \, \text{rand}(0, 1)(p(t) - X(t)) + c_2 \, \text{rand}(0, 1)(l(t) - X(t)) \tag{7}$$

where w is inertia weight, c_1 is cognitive parameter and c_2 is social parameter. A flowchart of PSO has been presented in Fig. 2. In the present paper, three different modifications of PSO have been considered for comparison.

3.1.1   Unified Particle Swarm Optimization (UPSO)

UPSO is proposed by Parsopoulos and Vrahatis [17]. UPSO combines the features of both Global-PSO and Local-PSO through a unification factor. The unification factor ensures the high exploration capability in the earlier stage and high exploitation capability in the later stage. Position update ($X(t + 1)$) in each iteration can be calculated according to Eq. 5, whereas, Velocity update ($V(t + 1)$) can be obtained as per Eqs. 8–10.

$$V(t + 1) = \mu G(t + 1) + (1 - \mu)L(t + 1) \tag{8}$$

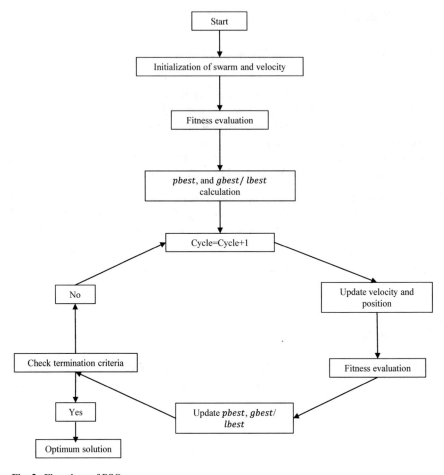

Fig. 2 Flowchart of PSO

$$G(t + 1) = \chi(V(t) + c_1 \text{rand}(0, 1)(p(t) - X(t)) + c_2 \text{rand}(0, 1)(g(t) - X(t)))$$
(9)

$$L(t + 1) = \chi(V(t) + c_1 \text{rand}(0, 1)(p(t) - X(t)) + c_2 \text{rand}(0, 1)(l(t) - X(t)))$$
(10)

where χ is constriction factor and μ is unification factor. The main steps of UPSO algorithm are given below:

1. Initialize swarm
2. **repeat**
3. Fitness evaluation
4. *pbest, gbest, lbest* calculation

5. Velocity and position update
6. **until** the termination criteria met

3.1.2 Ageing Leader Challenger Particle Swarm Optimization (ALC-PSO)

ALC-PSO is proposed by Chen et al. [18]. In nature, when a leader of a colony gets aged, it is replaced by some other challenger capable of leading the colony. ALC-PSO is inspired by this concept. It is an upgradation of the basic Global-PSO. In basic Global-PSO, the entire swarm follow the leadership of the *gbest* particle, irrespective of its leadership capability for advancement in the search space in order to find the optimum position. Thus, as a result, if the *gbest* particle falls in local minima, it can lead the entire swarm to local minima. So, ALC-PSO concept of *gbest* particle is replaced by concept of *leader* to overcome the aforementioned drawback of the Global-PSO. So, velocity update can be calculated according to the Eq. 11.

$$V(t+1) = wV(t) + c_1 \text{ rand}(0, 1)(p(t) - X(t)) + c_2 \text{ rand}(0, 1)(\text{leader}(t) - X(t)) \tag{11}$$

Moreover, instead of following same *leader* throughout the entire process, leadership capability of the *leader* is evaluated in every iteration. Initially, a fixed lifespan is assigned to the *leader*. After each iteration, lifespan of the *leader* is adjusted (increased, kept same, decreased) based on its leading capability [18]. When lifespan of the *leader* is exhausted, a *challenger* is decided to challenge the present *leader* for the position of next *leader*. The *challenger* is chosen according to following equation:

$$\text{challenger}_j = \begin{cases} \text{random}(X_{\min}, X_{\max}), & \text{rand}_j < \text{pro} \\ \text{leader}_j, & \text{otherwise} \end{cases} \quad j = 1, 2, \ldots, N \tag{12}$$

where pro $= 1/$problem dimension. Then, the *challenger* is given an opportunity for a few iteration to prove its superiority over the previous *leader* to lead the swarm. If it becomes successful in leading the swarm, the *challenger* is chosen as the new *leader*. Otherwise, the previous *leader* is given another opportunity to lead the swarm for a single iteration. So, the main steps of ALC-PSO algorithm are given below:

1. Initialize swarm
2. Initialize *age* and *lifespan* of leader
3. **repeat**
4. Fitness evaluation
5. *pbest, leader* calculation
6. Velocity and position update
7. $age = age + 1$, adjust *lifespan*
8. If $age \geq life - span$,

(a) generate *challenger*
(b) Capable of leading: *leader = challenger*
      Otherwise: *leader = leader (age = lifespan-1)*

9. **until** the termination criteria met

3.1.3 Particle Swarm Optimization with Intelligent Particle Number (IPN-PSO)

IPN-PSO is proposed by Lee et al. [19] as an upgradation of Global-PSO. In each iteration, status of the optimization is determined through comparing the *gbest* value of the current iteration with the previous iteration (Eq. 13).

$$D = \begin{cases} |g(t) - g(t-1)| \leq \varepsilon, \text{ in exploration} \\ |g(t) - g(t-1)| > \varepsilon, \text{ in exploitation} \end{cases} \tag{13}$$

ε is the tolerance value. If it is in exploration stage, swarm size is kept unchanged. On the other hand, if it is in exploitation stage, the second best particle is selected and removed from the swarm in a very intelligent manner. The second best particle is decided based on Euclidean distance (E_n). Euclidean distance for nth particle in the swarm can be calculated according to Eq. 14.

$$E_n = \sqrt{\sum_{i=1}^{\dim} (X_i^{\text{best}} - X_i^n)} \tag{14}$$

Velocity and position update are same as Global-PSO. The main steps of IPN-PSO algorithm are given below:

1. Initialize swarm
2. **repeat**
3. Fitness evaluation
4. *pbest, gbest* calculation
5. Velocity and position update
6. Adjust swarm size
7. **until** the termination criteria met

3.2 Ant Colony Optimization (ACO)

Ant colony optimization (ACOr) for continuous problems was developed by Socha and Dorigo [20] as a stochastic optimization algorithm based on the behaviour of the ants inside an ant colony. Initially, each ant assumes random positions within the search space to create a solution archive. During the search process, position of each particle gets updated in each iteration based on the transition probability. Transition probability of ith ant p_i can be calculated according to Eq. 15.

$$p_i = \frac{w_i}{\sum_{i=1}^{NA} w_i} \tag{15}$$

$$w_i = \frac{1}{q * NA\sqrt{2\pi}} e^{-\frac{(i-1)^2}{2q^2 * (NA)^2}} \tag{16}$$

q is the selection pressure and NA is the number of ants in the archive. New solutions are created based on the mean and standard deviation of the existing ant-based solutions with the help of gaussian PDF kernel. Thus, in case of ith ant X_i, mean (μ_i) and standard deviation (σ_i) can be calculated according to Eqs. 17–18.

$$\mu_i = X_i \tag{17}$$

$$\sigma_i = \xi \sum_{t=1}^{NA} \frac{|X_i - X_i|}{NA - 1} \tag{18}$$

ξ is the deviation-distance ratio. Then, newly generated solutions are added to the solution archive and then the same number of worst solutions is deleted from the archive. Through this operation, pheromone update is achieved in case of continuous ACO (ACOr). Schematic representation of ACO is presented in Fig. 3. Basic steps of ACOr are as follows:

7. Initialize solution
8. Fitness evaluation
9. **repeat**
10. Ant-based solution construction
11. Pheromone update
12. Daemon action
13. **until** the termination criteria met

4 Numerical Results

A comparative study among four optimization algorithms named UPSO, ALC-PSO, IPN-PSO and ACOr has been conducted in terms of their damage detection capability. A plane and a space truss have been considered for this purpose. For each damage case, ten numerical experiments were conducted. Parameters of algorithms considered for both the structural problems are mentioned in Table 1. Tolerance limit of objective function has been considered as 1×10^{-5}. If the objective function value reaches the tolerance limit within the specified number of iteration for a particu-

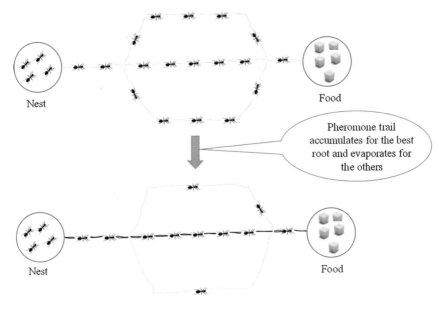

Fig. 3 Schematic representation of ACO

lar numerical experiment, the numerical experiment is recognized as successful to predict the damage.

4.1 25 Member Plane Truss

Finite element program has been developed for the plane truss structure (Fig. 4) [21, 22] in MATLAB environment to find out the natural frequencies and mode shapes. A comparison has been drawn among aforementioned four algorithms in terms of their performance in detecting damage of the aforementioned truss structure. Only the first six natural frequencies and mode shapes have been considered for this purpose. Specific parameters used for each optimization are mentioned in Table 1. Maximum iteration for each algorithm is considered as 2500. Damage in a particular element has been modelled as percentage reduction in axial stiffness. Two damage cases have been considered as mentioned in Table 2.

Damage prediction by all four algorithms has been plotted in Fig. 5a, b for damage cases E1 and E2, respectively. Thus, all three algorithms have predicted the damage successfully for both the damage cases. Mean convergence curve of ten experiments are plotted in Fig. 6a, b for damage cases E1 and E2, respectively.

Standard deviation of objective function value has been plotted in Fig. 7a, b, respectively, for damage case E1 and E2. Summary of results has been presented in Table 3. UPSO is found to be the best among four algorithms in terms of convergence

Table 1 Parameter setting for optimization algorithms

Optimization algorithm	Parameters
UPSO	Swarm size $= 375$ Cognitive parameter $(c_1) = 2.05$ Social parameter $(c_2) = 2.05$ Constriction coefficient $(\chi) = 0.729$ Neighbourhood radius (NR) $= 1$
ALC-PSO	Swarm size $= 375$ Cognitive parameter $(c_1) = 2.05$ Social parameter $(c_2) = 2.05$ Inertia weight $(w) =$ decreasing linearly in each iteration from 0.9 to 0.4 Lifespan of leader $= 60$ Trial period for challenger $= 2$
IPN-PSO	Swarm size $= 375$ Cognitive parameter $(c_1) = 2.05$ Social parameter $(c_2) = 2.05$ Inertia weight $(w) = 0.8$
ACOr	Archive size $= 375$ Number of ants used in an iteration $= 375$ Selection pressure $(q) = 0.5$ Deviation-distance ratio $(\xi) = 1$

Fig. 4 Details of the 25 member plane truss [22]

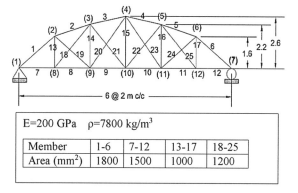

E=200 GPa   ρ=7800 kg/m³

Member	1-6	7-12	13-17	18-25
Area (mm²)	1800	1500	1000	1200

Table 2 Details of damage cases

Damage case	Description
E1	10% in member 5 +15% in member 8 +25% in member 17
E2	10% in member 5 +15% in member 8 +25% in member 17 +20% in member 15 +10% in member 25

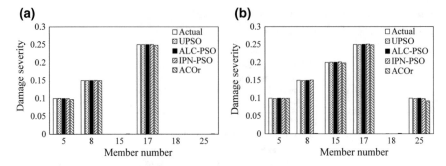

Fig. 5 Damage prediction by the algorithms: **a** damage case E1 **b** damage case E2

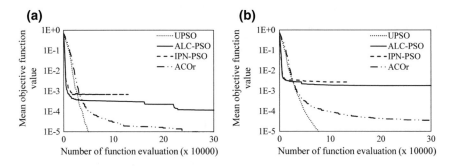

Fig. 6 Convergence curve: **a** damage case E1 **b** damage case E2

rate and success rate. UPSO is found to have least mean function evaluation, and ACOr is found to have highest mean function evaluation. So, computational effort required for UPSO is least among four algorithms. In terms of standard deviation of objective function IPN-PSO and ALC-PSO have shown better performance than other two (Fig. 7). However, their success rate is lower in comparison with other two algorithms. UPSO has shown 100% (10/10) success rate in both damage cases. ALC-PSO, IPN-PSO and ACOr have given, respectively, 60% (6/10), 90% (9/10) and 90% (9/10) success rate for damage case E1. In case of damage case E2, success rate of ALC-PSO, IPN-PSO and ACOr are 30% (3/10), 50% (50/100) and 70% (7/10), respectively. So, in terms of success rate and computational effort, UPSO has outperformed the other algorithms.

4.2 25 Member Space Truss

A space truss structure (Fig. 8) is chosen as second example to compare the performance of four aforementioned algorithms in the field of damage detection. Material properties of the truss members are as follows:

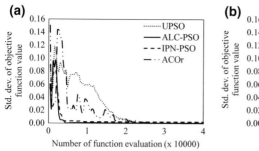

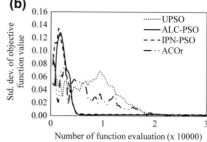

Fig. 7 Standard deviation of objective function: **a** damage case E1 **b** damage case E2

Table 3 Summary of damage detection for plane truss

Damage case	Optimization algorithm	Number of function evaluation		Objective function value		Success rate
		Mean	Standard deviation	Mean	Standard deviation	
E1	UPSO	50,025.0	7181.1	9.47E−07	3.15E−06	10/10
	ALC-PSO	145,687.5	17,686.0	1.17E−04	2.93E−04	6/10
	IPN-PSO	68,694.3	16,130.6	6.85E−04	2.28E−03	9/10
	ACOr	401,625.0	139,947	2.41E−06	4.82E−06	9/10
E2	UPSO	92,400.0	40,535.5	1E−06	3.33E−06	10/10
	ALC-PSO	501,751.5	319,791.2	1.31E−03	8.81E−04	3/10
	IPN-PSO	119,006	6388.0	2.79E−03	2.31E−03	5/10
	ACOr	601,250	150,996	1.99E−05	4.22E−06	7/10

Table 4 Details of damage cases

Damage case	Description
E3	25% in member 5 +20% in member 10 +10% in member 15
E4	25% in member 5 +20% in member 10 +10% in member 15 +15% in member 20 +10% in member 25

Young's modulus $(E) = 6.98 \times 10^{10}$ N/m^2, Area of each member $= 0.0025$ m^2
Density $= 2770$ kg/m^3

Finite element program has been developed in MATLAB environment based on the presented formulation in the previous sections to calculate the natural frequencies and mode shapes. Only the first six natural frequencies and mode shapes have been considered for this purpose. Damage in a particular element has been modelled as percentage reduction in axial stiffness. Table 4 is presented with two damage cases.

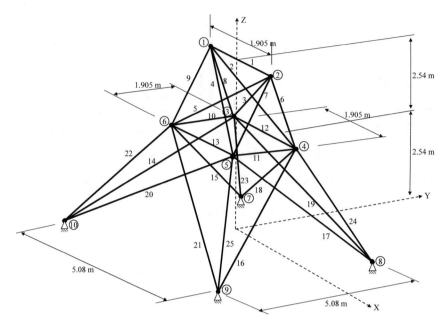

Fig. 8 Details of the 25 member space truss

Damage prediction by the algorithms has been plotted in Fig. 9a, b for damage case E3 and E4, respectively. So, all of the algorithm have predicted the damage successfully for both the damage cases. Mean convergence curve of ten experiments is plotted in Fig. 10a, b for damage case E3 and E4 ,respectively. Standard deviation of objective function value has been plotted in Fig. 11a, b, respectively, for damage case E3 and E4. Summary of results has been presented in Table 5. UPSO is found to be best among four algorithms in terms of convergence rate and success rate. UPSO is found to have least mean function evaluation for damage case E4, while IPN-PSO has least mean function evaluation for damage case E3. In terms of standard deviation of objective function, IPN-PSO and ALC-PSO have shown better performance than other two (Fig. 7). However, their success rate is lower in comparison with other two algorithms. UPSO has shown 100% (10/10) success rate in both damage cases. ALC-PSO, IPN-PSO and ACOr have given, respectively, 50% (5/10), 90% (9/10) and 90% (9/10) success rate for damage case E3. In case of damage case, E4 success rate of ALC-PSO, IPN-PSO and ACOr are 30% (3/10), 50% (50/100) and 90% (9/10), respectively. So, in terms of success rate and computational effort, UPSO has outperformed the other algorithms.

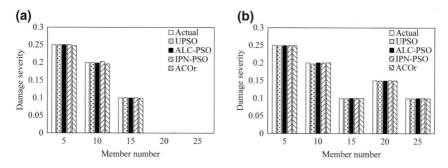

Fig. 9 Damage prediction by the algorithms: **a** damage case E3 **b** damage case E4

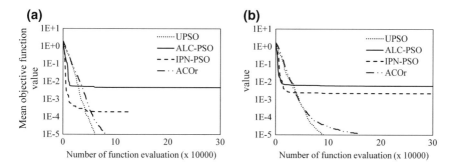

Fig. 10 Convergence curve: **a** damage case E3 **b** damage case E4

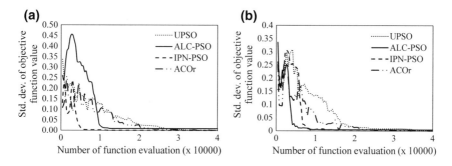

Fig. 11 Standard deviation of objective function: **a** damage case E3 **b** damage case E4

5 Conclusions

In the present paper, a finite element formulation is developed in MATLAB environment to calculate natural frequencies and mode shapes of a 25 member plane truss and a 25 member space truss in undamaged and damaged condition.

An inverse problem is developed based on frequency and mode shapes to detect damages in the truss structures. Four algorithms, ALC-PSO, UPSO, IPN-PSO and

Table 5 Summary of damage detection for space truss

Damage case	Optimization algorithm	Number of function evaluation		Objective function value		Success rate
		Mean	Standard deviation	Mean	Standard deviation	
E1	UPSO	58,275.0	10802.3	1.99E−06	4.45E−06	10/10
	ALC-PSO	199,500.6	239,172.3	7.74E−04	2.09E−03	5/10
	IPN-PSO	56,524.0	29,099.0	1.92E−04	4.29E−04	9/10
	ACOr	84,900.0	3254.08	1.91E−06	4.26E−06	9/10
E2	UPSO	99,750.0	17,839.1	1.97E−06	4.41E−06	10/10
	ALC-PSO	210,300.3	159,132.2	5.49E−03	4.38E−03	3/10
	IPN-PSO	248,603.0	145,578.4	2.24E−03	4.41E−03	5/10
	ACOr	182,175	27,581.5	1.97E−06	4.41E−06	9/10

ACOr are used to solve the inverse problem to locate and quantify the damages. IPN-PSO has been found very interesting due to its process of reducing the swarm size in an intelligent manner with the progress of the optimization. However, UPSO is found to more successful and robust in comparison with other three algorithms to detect and quantify the damages in the truss.

Acknowledgements This research work is financially supported by ISRO (Indian Space Research Organisation) IIT Kharagpur cell. The authors are grateful to ISRO cell for their financial support to carry out the research work at Department of Aerospace Engineering, IIT, Kharagpur.

References

1. Adams, R.D., Cawley, P., Pie, C.J., Stone, B.J.: A vibration technique for non-destructively assessing the integrity of structures. J. Mech. Eng. Sci. **20**, 93–100 (1978). https://doi.org/10.1243/JMES_JOUR_1978_020_016_02
2. Cawley, P., Adams, R.D.: A vibration technique for non-destructive testing of fibre composite structures. J. Compos. Mater. **13**, 161–175 (1979). https://doi.org/10.1177/002199837901300207
3. Messina, A., Williams, E.J., Contursi, T.: Structural damage detection by a sensitivity and statistical-based method. J. Sound Vib. **216**, 791–808 (1998). https://doi.org/10.1006/jsvi.1998.1728
4. Majumdar, A., Maiti, D.K., Maity, D.: Damage assessment of truss structures from changes in natural frequencies using ant colony optimization. Appl. Math. Comput. **218**, 9759–9772 (2012). https://doi.org/10.1016/j.amc.2012.03.031
5. Mohan, S.C., Maiti, D.K., Maity, D.: Structural damage assessment using FRF employing particle swarm optimization. Appl. Math. Comput. **219**, 10387–10400 (2013). https://doi.org/10.1016/j.amc.2013.04.016
6. Kaveh, A., Zolghadr, A.: An improved CSS for damage detection of truss structures using changes in natural frequencies and mode shapes. Adv. Eng. Softw. **80**, 93–100 (2015). https://doi.org/10.1016/j.advengsoft.2014.09.010

7. Kang, F., Li, J., Xu, Q.: Damage detection based on improved particle swarm optimization using vibration data. Appl. Soft Comput. **12**, 2329–2335 (2012). https://doi.org/10.1016/j.asoc.2012.03.050
8. Nanda, B., Maity, D., Maiti, D.K.: Modal parameter based inverse approach for structural joint damage assessment using unified particle swarm optimization. Appl. Math. Comput. **242**, 407–422 (2014). https://doi.org/10.1016/j.amc.2014.05.115
9. Ding, Z.H., Huang, M., Lu, Z.R.: Structural damage detection using artificial bee colony algorithm with hybrid search strategy. Swarm Evol. Comput. **28**, 1–13 (2016). https://doi.org/10.1016/j.swevo.2015.10.010
10. Nhamage, I.A., Lopez, R.H., Miguel, L.F.F.: An improved hybrid optimization algorithm for vibration based-damage detection. Adv. Eng. Softw. **93**, 47–64 (2016). https://doi.org/10.1016/j.advengsoft.2015.12.003
11. Seyedpoor, S.M.: A two stage method for structural damage detection using a modal strain energy based index and particle swarm optimization. Int. J. Non-Linear Mech. **47**, 1–8 (2012). https://doi.org/10.1016/j.ijnonlinmec.2011.07.011
12. Maity, D., Tripathy, R.R.: Damage assessment of structures from changes in natural frequencies using genetic algorithm. Struct. Eng. Mech. **19**, 21–42 (2005). https://doi.org/10.12989/sem.2005.19.1.021
13. Zapico, J.L., González, M.P., Friswell, M.I., Taylor, C.A., Crewe, A.J.: Finite element model updating of a small scale bridge. J. Sound Vib. **268**, 993–1012 (2003). https://doi.org/10.1016/S0022-460X(03)00409-7
14. Beena, P., Ganguli, R.: Structural damage detection using fuzzy cognitive maps and Hebbian learning. Appl. Soft Comput. **11**, 1014–1020 (2011). https://doi.org/10.1016/j.asoc.2010.01.023
15. Chandrupatla, T.R., Belegundu, A.D.: Introduction to Finite Elements in Engineering. Prentice Hall, Upper Saddle River, New Jersey, USA (2002)
16. Kennedy, J., Eberhart, R.: Particle swarm optimization. In: Proceedings of IEEE International Conference on Neural Networks, Piscataway, NJ, pp. 1942–1948 (1995)
17. Parsopoulos, K.E., Vrahatis, M.N.: Unified particle swarm optimization for solving constrained engineering optimization problems. In: Proceedings of the First International Conference on Advances in Natural Computation, vol. Part III, pp. 582–591. Springer, Berlin, Heidelberg (2005)
18. Chen, W.N., Zhang, J., Lin, Y., Chen, N., Zhan, Z.H., Chung, H.S.H., Li, Y., Shi, Y.H.: Particle swarm optimization with an aging leader and challengers. IEEE Trans. Evol. Comput. **17**, 241–258 (2013). https://doi.org/10.1109/TEVC.2011.2173577
19. Lee, J.H., Song, J.-Y., Kim, D.-W., Kim, J.-W., Kim, Y.-J., Jung, S.-Y.: Particle swarm optimization algorithm with intelligent particle number control for optimal design of electric machines. IEEE Trans. Ind. Electron. **65**, 1791–1798 (2017). https://doi.org/10.1109/TIE.2017.2760838
20. Socha, K., Dorigo, M.: Ant colony optimization for continuous domains. Eur. J. Oper. Res. **185**, 1155–1173 (2008). https://doi.org/10.1016/j.ejor.2006.06.046
21. Esfandiari, A., Bakhtiari-Nejad, F., Rahai, A.: Theoretical and experimental structural damage diagnosis method using natural frequencies through an improved sensitivity equation. Int. J. Mech. Sci. **70**, 79–89 (2013). https://doi.org/10.1016/j.ijmecsci.2013.02.006
22. Barman, S.K., Maiti, D.K., Maity, D.: A new hybrid unified particle swarm optimization technique for damage assessment from changes of vibration responses. In: Proceedings of ICTACEM 2017 International Conference on Theoretical, Applied, Computational and Experimental Mechanics, pp 1–12. IIT Kharagpur, Kharagpur, India (2017)

Multi-objective Optimization of Wire-Electric Discharge Machining Process Using Multi-objective Artificial Bee Colony Algorithm

P. J. Pawar and M. Y. Khalkar

Abstract The selection of optimum process parameters in wire-electric discharge machining process such as peak current, wire tension, water pressure and duty factor plays a significant role for optimizing the process performance measure. In this paper, multi-objective optimization of wire-electric discharge machining (wire-EDM) is done using recently developed evolutionary optimization algorithm. The advanced algorithm developed for the purpose of optimization is known to be multi-objective artificial bee colony (MOABC) algorithm. The objectives considered in this work are material removal rate and wear ratio and subjected to the constraint of surface roughness.

Keywords Wire-electric discharge machining · Multi-objective artificial bee colony · Material removal rate · Surface roughness · Wear ratio

1 Introduction

The wire-electric discharge machining (wire-EDM) is an important widely accepted advanced machining processes for machining intricate-shaped components with complex profiles. Manufacturers and users of W-EDM always want higher productivity without compromising desired machining accuracy, surface finish, etc. However, the performance of wire-EDM is affected by process parameters like peak current, pulse-on and pulse-off times, water pressure, wire tension, etc. The performance of W-EDM is highly susceptible to the selection of these parameters in complex way. The W-EDM process includes many variables with complex and stochastic nature. Hence to achieve optimal performance of this process, for a skilled machinist with

P. J. Pawar (✉) · M. Y. Khalkar
K. K. Wagh Institute of Engineering Education and Research, Nashik, Maharashtra, India
e-mail: pjpawar@kkwagh.edu.in

M. Y. Khalkar
e-mail: mykhalkar@kkwagh.edu.in

Savitribai Phule Pune University, Pune, Maharashtra, India

© Springer Nature Singapore Pte Ltd. 2020
R. Venkata Rao and J. Taler (eds.), *Advanced Engineering Optimization Through Intelligent Techniques*, Advances in Intelligent Systems and Computing 949,
https://doi.org/10.1007/978-981-13-8196-6_4

39

a high state-of-the-art W-EDM machine, is highly difficult. Various investigators have proposed optimization techniques, both traditional like Taguchi method [1–3], response surface method [4], grey relational analysis [5], non-traditional like genetic algorithm [6], simulated annealing [7, 8] and artificial bee colony [9] for optimizing process parameters of wire-EDM process.

The traditional methods of optimization, over a broad spectrum, do not farewell over wide problem domains. Moreover, traditional optimization techniques lack robustness and may obtain a local optimal solution as they do not perform well over multi-modal problems. This drawback of traditional algorithms to be trapped into local optima necessitates the use of evolutionary optimization techniques for optimization. The evolutionary methods use instead of the gradient-based derivatives, and the fitness information makes them much more robust, efficient and more effective. These methods are so enabled to obtain an almost global optimum solution. Therefore, efforts are being made to use recently developed optimization algorithms, being robust, more powerful, and able to provide near best solution. In this paper, therefore an attempt is made to use one of the recent evolutionary optimization algorithms, i.e. multi-objective artificial bee colony algorithm (MOABC), to optimize process parameters of wire-EDM process.

It is also revealed from the literature that very few efforts are made to achieve the optimum parameter setting in W-EDM process considering multiple responses simultaneously. The parameter setting for single response will obstruct the other responses significantly. Thus, to achieve separate parameter setting for each individual response is of very limited practical use. Hence in this work, the attempt has been made to present multi-objective optimization aspects of W-EDM process by using the recently developed one of the non-traditional optimization algorithm known as MOABC. The next section describes briefly the mathematical model for optimization of wire-EDM process.

2 Modelling of Wire-EDM Process for Optimization

Optimization model for W-EDM process is based on the mathematical model which has been taken from Hewidy et al. [4]. The response surface methodology is used by Hewidy for development of the model. The objectives considered in this work are maximization of volumetric material removal rate (VMRR) and of wear ratio (Wr) as given by Eqs. (1) and (2), respectively. The process variables are peak current (I_p), wire tension (T), duty factor (D) and water pressure (P). The values of the variables I_p, D, T and P are coded and considered as y_1, y_2, y_3 and y_4, respectively, within the bounds of coded levels -2 to $+2$.

$$\text{VMRR} = 6.96 - 0.466y_1 + 0.149y_4 + 0.362y_1y_2 - 0.386y_1y_3 - 0.253y_3y_4$$
$$- 0.316y_1^2 - 0.27y_2^2 - 0.216y_4^2 \text{ mm}^3/\text{min} \qquad (1)$$

$$Wr = 2.73 + 0.508y_1 + 0.249y_4 \qquad (2)$$

where $VMRR_{max}$ = Maximum value of volumetric material removal rate. $VMRR_{max}$ is obtained considering single-objective optimization for VMRR as an objective and was solved with given constraint.

Wr_{max} = Maximum value of wear ratio obtained considering single-objective optimization problem of wear ratio as an objective for the given constraint. $p1$ and $p2$ are the weightages assigned to the two objectives VMRR and Wr, respectively.

The third objective, surface roughness, is considered in the form of constraint as usually it is the customer requirement. The constraint on surface roughness is shown by Eq. (3)

$$R_{max} - SR \geq 0 \qquad (3)$$

where R_{max} is the maximum value of surface roughness. Equation (4) shows the surface roughness (SR).

$$
\begin{aligned}
SR = {} & 2.06 + 0.684x_1 - 0.0967x_2 - 0.399x_3 - 0.0992x_4 \\
& + 0.0975x_1x_2 - 0.355x_1x_3 + 0.249x_xx_4 + 0.284x_2x_3 \\
& - 0.095x_2x_4 - 0.295x_3x_4 + 0.334x_1^2 - 0.129x_2^2 + 0.0354x_3^2 + 0.233x_4^2 \quad (4)
\end{aligned}
$$

The bounds of these process variables are:

$$3 \leq I_p \leq 7 \, (A)$$

$$0.375 \leq D \leq 0.75$$

$$7 \leq T \leq 9 \, (N)$$

$$0.3 \leq P \leq 0.7 \, (MPa)$$

The MOABC algorithm based on advanced optimization is used to optimize wire-EDM parameters. The following section briefly details the optimization aspects of wire-EDM process using MOABC algorithm.

3 Parameter Optimization of Wire-EDM Process Using Multi-objective Artificial Bee Colony (MOABC) Algorithm

Hewidy et al. [4] suggested the parameters setting for maximum VMRR as $I_p = 6 \, A$, $D = 0.5$, $P = 0.5 \, MPa$ and $T = 7 \, N$ so that VMRR $= 6.97 \, mm^3/min$, SR $= 4.727 \, \mu m$ and Wr $= 3.238$, and that for maximum Wr as $I_p = 7 \, A$, $D = 0.75$, $T = 7 \, N$ and $P =$

0.7 MPa so that VMRR = 7.16 mm³/min, Wr = 4.244 and SR = 8.197 μm. From these results, it is observed that while the individual responses are optimized, the other responses are hampered seriously. For maximization of the responses VMRR and Wr, the surface roughness values obtained are 4.727 and 8.197 μm, respectively, which may not be practically acceptable. Hence, these responses must be optimized simultaneously. The multi-objective optimization is therefore carried out using the ABC algorithm for $R_{max} = 0.4$ μm which is the best possible value of surface roughness for the present optimization problem [4].

The discussed below are the steps (S1 to S12) of artificial bee colony algorithm with non-dominating sorting base [10]:

S1: Algorithm-specific parameters' selection

The algorithm-specific parameters which are number of onlooker bees (N), population size that is number of food sources $N2$ (=number of employed bees), number of scoutbees $N3$ (6–30% of the colony-size) are to be calculated. These are as follows.

- $N1 = 50$
- $N2 = 20$
- $N3 = 1$

S2: The nectar amount that is quality calculated for every food source

The amount of nectar is the actual fitness value of the solution.
Nectar amount of each source of food source is calculated, obtaining objective function values using Eqs. 1, 2 and 3.

S3: Non-dominated sorting

Non-dominated sorting of the solutions is done using the criteria mentioned in Eq. (5). Then the solutions are ranked based on their non-dominated status.

$$\text{Obj.1}[i] < \text{Obj.1}[j], \quad \text{Obj.2}[i] < \text{Obj.2}[j], \quad i \neq j \tag{5}$$

S4: Obtain for every solution the normalized Euclidean distance

Equation (6) gives normalized Euclidean distance of every solution from others

$$d_{ij} = \sqrt{\sum \frac{\left(z_s^i - z_s^j\right)}{z_s^{max} - z_s^{min}}} \tag{6}$$

where z_s is sth decision variable value.
i and j are solution numbers.
z_s^{max} is upper limit and z_s^{min} is lower limit of the sth decision variable.

S5: Obtain niche count of the solutions

Niche counts for each solution are obtained by Eq. 7.

$$nc_i = \sum Fs(d_{ij}) \tag{7}$$

$Fs(d_{ij})$: first front solution sharing function values. This is as given by equation:

$$Fs(d_{ij}) = \left\{ 1 - \left(\frac{D_{ij}}{\sigma_{share}} \right)^2 \right\} \text{ if } D_{ij} < \sigma_{share}$$
$$= 0 \qquad\qquad \text{otherwise} \tag{8}$$

σ_{share}: the max. distance which is allowed between any two values of solutions. The value of σ_{share} is to be appropriately chosen.

S6: Calculate shared-fitness value of all solutions

Assuming dummy-fitness value as 50 for rank 1 solution, and shared-fitness is computed for solutions by Eq. (9).

$$\text{Sharefitness } (F_i) = \text{Dumyfitness } (f)/nc_i \tag{9}$$

For second rank solutions, the dummy-fitness value is calculated to ensure that it should be minimum shared-fitness value and less than rank 1 solution. İn the same way, the dummy-fitness values of ranks 2, 3 and 4 are established.

S7: Using shared-fitness values, calculate the probabilities which are calculated in step S6

The following shared-fitness Eq. 10 is used to assign probability (P_i) so that onlooker bee is assigned to employed bee.

$$P_i = \frac{\sum_{k=i}^{R} (1/f_k)^{-1}}{f_i} \tag{10}$$

where R is number of sources of food.

S8: Determine the number of onlooker bees

Following Eq. 11 gives the number of onlooker bees to be sent to an employed bee (M).

$$M = P_i \times \text{Ob} \tag{11}$$

where 'Ob' is the number of onlooker bees.

S9: Find the new position of every onlooker bee

To obtain position of onlooker bees, every employed bee is updated for N times using Eq. (12)

$$\theta_i(g + 1) = \theta_i(c) \pm k_i(c) \tag{12}$$

'g' is the number of generation, $k_i(c)$ is a randomly selected step, in order to find a source of food having more nectar surround 'θ_i'. If better position is obtained, each food source is updated by onlooker.

The comparison is then made between onlooker bees for every source of food and the employed bee assigned to that source of food. The equal weights are considered for all the objective functions, and combined function is obtained for this comparison.

The combined objective function is evaluated as:

$$\min Z = \frac{SR}{SR^*} - \frac{Wr}{Wr^*} - \frac{VMRR}{VMRR^*} \tag{13}$$

where SR^*, Wr^* and $VMRR^*$ are threshold values of surface roughness, wear ratio and volumetric material removal rate, respectively. This process keeps the superiority as best amongst new solutions. İt is then compared with previous solution, and the superior solution gets selected.

S10: Evaluation of the best solution

For every source of food, best onlooker bee position is identified. Subsequently, honeybee swarm global best is obtained in every generation, and at previous generation, it replaces the global best, if it has better value of fitness function.

S11: Updation of the scout bee

The equal number of worst employed bees, and scoutbees within the population, is compared against the scout solutions. If the scout solution found better compared to employed solution, then scout solution is updated by employed solution. Else employed solution passes without any change to the next generation.

S12: The set of Pareto-optimal solution

Pareto-optimal solution set is as shown in Table 1.

To facilitate the user to select the optimal combination of process parameters based on his specific choice, the set of non-dominated solutions shown in Table 1. Convergence graph is as shown in Fig. 1.

The results of optimization obtained using MO-ABC algorithm are found superior to those obtained using NS-GA, and the average improvement in the combined objective function is 37%. Also MO-ABC algorithm requires only 20 iterations to converge.

Table 1 Outcomes of the optimization process

S. No.	I	D	T	P	SR	Wr	VMRR
1	3.0820	0.7307	8.7715	0.5762	0.643	1.945	5.371
2	4.3220	0.4614	8.6945	0.4530	1.126	2.269	7.479
3	6.3530	0.7492	7.5815	0.5577	3.001	3.561	6.256
4	4.7450	0.5952	8.7090	0.6078	1.166	2.869	6.650
5	4.3260	0.6944	8.7965	0.6296	0.786	2.710	5.962
6	4.7903	0.5038	9.0000	0.6243	0.514	2.933	6.366
7	4.5103	0.6400	8.6307	0.5751	1.149	2.668	6.760
8	4.8917	0.4420	8.8733	0.6158	0.429	2.963	6.054
9	5.3760	0.4047	8.9333	0.5810	0.332	3.123	5.079
10	4.5180	0.7500	8.6915	0.6479	0.754	2.853	5.161
11	5.0083	0.4763	8.9857	0.5467	0.654	2.851	6.508
12	4.6710	0.5921	8.7438	0.5894	1.106	2.786	6.821
13	4.1157	0.7500	8.5430	0.6268	0.574	2.596	5.251
14	4.4207	0.7500	7.6442	0.5419	0.846	2.540	5.554
15	4.0413	0.7400	7.7703	0.5637	0.784	2.402	5.384
16	5.2420	0.5228	9.0000	0.5647	1.030	3.014	6.240
17	3.7497	0.6948	7.8175	0.5088	1.144	2.117	5.691
18	3.0000	0.7038	7.0000	0.5688	1.486	1.885	3.689

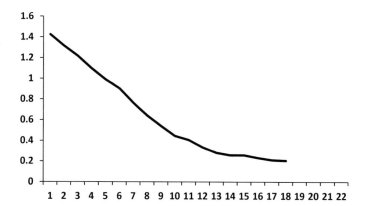

Fig. 1 Convergence of MO-ABC algorithm

4 Conclusion

The multi-objective optimization aspects of wire-electric discharge machining process parameters are considered in the present work. The objectives considered in this work are maximization of material removal rate and wear ratio. The maximization of objectives is subjected to constraint of surface roughness. The recently developed evolutionary algorithm known as multi-objective artificial bee colony (MOABC) algorithm is applied to obtain optimum set of the process parameters. The complex decision could be taken by analysing the Pareto's optimal solutions, depending upon the specific conditions of wire-EDM process. The convergence rate of MOABC algorithm is observed to be much higher as it required only 20–22 iterations to converge to the optimal solution. The convergence of the process is as shown in Fig. 1. The convergence rate proves its robustness, which enables it to effective use in the optimization of multi-objective and multi-modal problems. Moreover, the flexibility permits modifications easily to suit for process parameters optimization of other advanced and non-traditional manufacturing processes such as laser beam machining (LBM), abrasive waterjet machining, ultrasonic machining, electrochemical machining (ECM) and plasma arc machining.

References

1. Scott, D., Boyina, S., Rajurkar, K.P.: Analysis and optimization of parameter combination in wire electrical discharge machining. Int. J. Prod. Res. **29**, 2189–2207 (1991)
2. Liao, Y.S., Huang, J.T., Su, H.C.: A study on the machining parameters optimization of wire electrical discharge machining. J. Mater. Process. Technol. **71**, 487–493 (1997)
3. Anand, K.N.: Development of process technology in wire-cut operation for improving machining quality. Total Qual. Manag. **7**, 11–28 (1996)
4. Hewidy, M.S., El-Taweel, T.A., El-Safty, M.F.: Modelling the machining parameters of wire electrical discharge machining of Inconel 601 using RSM. J. Mater. Process. Technol. **169**, 328–336 (2005)
5. Huang, J.T., Liao, Y.S.: Optimization of machining parameters of wire-EDM based on grey relational and statistical analyses. Int. J. Prod. Res. **41**, 1707–1720 (2003)
6. Kuriakose, S., Shunmugam, M.S.: Multi-objective optimization of wire-electro discharge machining process by Non-Dominated Sorting Genetic Algorithm. J. Mater. Process. Technol. **170**, 133–141 (2005)
7. Tosun, N., Cogun, C., Tosun, G.: A study on kerf and material removal rate in wire electrical discharge machining based on Taguchi method. J. Mater. Process. Technol. **152**, 316–322 (2004)
8. Tarng, Y.S., Ma, S.C., Chung, L.K.: Determination of optimal cutting parameters in wire electrical discharge machining. Int. J. Mach. Tools Manuf. **35**, 1693–1701 (1995)
9. Rao, R.V., Pawar, P.J.: Modeling and optimization of process parameters of wire electric discharge machining. J. Eng. Manuf. **223**, 1431–1440 (2009)
10. Pawar, P.J., Vidhate, U.S., Khalkar, M.Y.: Improving the quality characteristics of abrasive water jet machining of marble material using multi-objective artificial bee colony algorithm. J. Comput. Des. Eng. **5**, 319–328 (2018)

Optimization of Process Parameters in Pulsed Electrochemical Honing Process Using Evolutionary Algorithms

Sunny Diyaley and Shankar Chakraborty

Abstract This paper aims in the optimization of process parameters for straight bevel gear finishing by pulsed electrochemical honing (PECH) process using four evolutionary algorithms. The controllable parameters selected for optimal setting in PECH are the applied voltage, pulse-on time and pulse-off time, whereas finishing time, interelectrode gap and the rotary speed of workpiece gear are set as constant parameters. Theoretical model of material removal rate and surface roughness in PECH process developed by the past researchers are considered for a comparative analysis of the optimization problem by using four different algorithms, i.e. firefly algorithm, particle swarm optimization algorithm, differential evolution algorithm and teaching-learning-based algorithm, for arriving at the most global optimal settings of PECH process parameters. Teaching-learning-based algorithm attains the best optimal setting value within the range of different input process parameters for both single- and multi-objective optimization problems.

Keywords Optimization · Pulsed electrochemical honing · Material removal rate · Surface roughness · Evolutionary algorithms

1 Introduction

Pulsed electrochemical honing (PECH) process is a combination of pulsed electrochemical finishing (PECF) process and mechanical honing process and has the advantage to overcome the limitations caused by traditional gear finishing operations. Traditional gear finishing operations were expensive and have limitations such

---

S. Diyaley (✉)
Department of Mechanical Engineering, Sikkim Manipal Institute of Technology, Majitar, Sikkim, India
e-mail: sdiyaley@gmail.com

S. Chakraborty
Department of Production Engineering, Jadavpur University, Kolkata, West Bengal, India
e-mail: s_chakraborty00@yahoo.co.in

© Springer Nature Singapore Pte Ltd. 2020
R. Venkata Rao and J. Taler (eds.), *Advanced Engineering Optimization Through Intelligent Techniques*, Advances in Intelligent Systems and Computing 949,
https://doi.org/10.1007/978-981-13-8196-6_5

47

as grinding burn, noise and vibration and incorrect gear teeth profile manufacturing. There are even certain limitations of PECF and mechanical honing process. During the electrolytic dissolution in PECF, there is a major problem of the evolution of oxygen at anode due to the passivation of anodic workpiece surface which results in no electrolytic dissolution of the workpiece further. Mechanical honing process suffers some drawbacks, like incapability of finishing hard materials and mechanical damages to the workpiece material while finishing. Low tool life of honing tool is also a serious problem in the mechanical honing process. Thus, the proper combination of the PECF process and the mechanical honing process is essential to have superior quality and durable gears. In order to study the working process, and the ability to achieve better results in various machining and finishing process, theoretical models are formulated over a broad range of process parameters. Development of these theoretical models is an extensive work and requires a detailed knowledge of the process. These theoretical models are standard models for a process and have wide applicability. Misra et al. [1] analysed the effects of various parameters for quality finishing in helical gears by the electrochemical honing process. Pathak and Jain [2] also developed mathematical models, and prediction was done with experimental values for MRR and surface roughness in pulsed electrochemical honing process. Shaikh et al. [3] developed mathematical models for material removal rate (MRR) and surface roughness for the electrochemical honing process and predicted its values with close values with its experimental results. Thus, from the various literature reviews, it is concluded that no work has so far been reported in the area of application of evolutionary algorithms for the optimal parametric setting in the electrochemical honing process using theoretical models of MRR and surface roughness. This leads to the motivation of the present study. This paper highlights the optimal process parameter selection in PECH process by using four popular evolutionary algorithms, namely firefly algorithm (FA), particle swarm optimization (PSO), differential algorithm (DA) and teaching-learning-based algorithm (TLBO). It is observed that TLBO algorithm is capable of providing the most global optimal solution in comparison to the other algorithms. It has a unique feature that it requires only two algorithmic parameters, like population size and number of iterations. The optimal results obtained with other population-based algorithms, i.e. FA, PSO, and DA, are compared with the results obtained by TLBO algorithm. These algorithms are selected for comparison with the TLBO algorithms as they are mostly adopted population-based algorithms by the researchers nowadays for complex optimization problems. These algorithms are also capable of providing the most optimal solutions. But these algorithms requires many algorithmic parameters which are to be tuned properly, otherwise may lead to a local optimal solution. The paper also compares the ability of the four algorithms for their convergence towards global optimum value. Both single- and multi-objective optimization is performed in which the TLBO algorithm arrives at the best optimum global solution than the other algorithms and proves itself better than the rest algorithm with respect to accuracy and consistency of the derived solutions. Pareto optimal fronts are also provided to facilitate the operator/process engineer to select the combination of output responses based on his/her order of importance of the objective. These Pareto fronts also eliminate the task of assigning weights to each of the responses, which is usually infeasible.

2 Methodology

2.1 Total Volumetric MRR Theoretical Model in PECH Process While Straight Bevel Gear Finishing

In PECH process, for finishing straight bevel gears the total volumetric MRR is obtained by the summation of the volumetric MRR resulting from PECF and mechanical honing.

The proposed model is shown below as proposed by Pathak et al. [2]:

$$V_{\text{PECH}} = \left[\frac{C_{\text{PECF}} \eta E_{\text{w}} k_{\text{e}} A_{\text{s}} (1 - \lambda)}{F \rho_{\text{w}} Y} \right] (V - \Delta V) \frac{T_{\text{on}}}{T_{\text{on}} + T_{\text{off}}}$$

$$+ \left[\frac{2 C_{\text{h}} k F_{\text{n}} L_{\text{iw}} T}{H} \right] N_{\text{s}} (\text{mm}^3/\text{s}) \tag{1}$$

2.2 Depth of Surface Roughness Theoretical Model in PECH Process While Straight Bevel Gear Finishing

The mathematical model for minimization of the surface roughness depth in PECH process finishing of a bevel gear tooth at its flank surface as proposed by Pathak et al. [2] is expressed as:

$$R_{\text{ZPECH}} = R_{\text{Zi}} - 10^{-3} (1 - 2k) f \frac{[C_{\text{PECF}} \eta E_{\text{w}} k_{\text{e}} A_{\text{s}} (1 - \lambda)}{F \rho_{\text{w}} Y} (V - \Delta V) \frac{T_{\text{on}}}{T_{\text{on}} + T_{\text{off}}} t$$

$$+ \frac{2 C_{\text{h}} k F_{\text{n}} L_{\text{iw}} T}{A_{\text{f}} H} N_{\text{s}} t \bigg] (\mu\text{m}) \tag{2}$$

$$R_{\text{zi}} = H_{\text{v}} - H_{\text{p}} \tag{3}$$

where H_{v} and H_{p} are the valley depth and peak height before finishing of the flank surface in bevel gear tooth, respectively.

The variable bounds used for three decision variables, i.e. applied voltage (V), pulse-on time (T_{on}) and pulse-off time (T_{off}) by Pathak et al. [2], are given below: $8 \leq V \leq 14$ (V), $2 \leq T_{\text{on}} \leq 5$ (ms) and $4.5 \leq T_{\text{off}} \leq 9$ (ms).

2.3   Evolutionary Algorithms

Evolutionary algorithms are based on the idea of biological evolutions and have proved itself as a powerful search process. In evolutionary computation, random generation of a population of the candidate solution is carried out. Each candidate solution is assigned a fitness value, and their selection as individuals is governed by the fitness value. The selected individuals are now the parents based on its fitness evaluation. Then the process of parents' selection is based on its fitness value, and they produce offsprings. New offsprings produced are replaced if they are found to be weaker as compared to other offsprings produced until a specified number of iteration. The iteration is terminated if an optimal or near optimal solution is discovered.

2.3.1   Firefly Algorithm

FA is based on the flashing behaviour of real fireflies developed by Yang [4] in 2008. It uses the following idealized rules: every firefly is unisex, and they get attracted towards each other without considering their sex. The brightness of the fireflies decreases as the distance between each firefly advances. Attractiveness value is proportional to the brightness of the fireflies. For any two flashing fireflies, the less bright one will travel towards the brighter one and there will be a random movement of the fireflies if no brighter firefly is available. The firefly's brightness is evaluated by the landscape of the objective function. The various parameters used in FA algorithm are maximum number of iteration = 500, number of fireflies = 500, initial randomness = 0.90, randomness factor = 0.91, absorption coefficient = 1 and randomness reduction = 0.75.

2.3.2   Particle Swarm Optimization

PSO is a direct search population-based method proposed by Kenndy and Eberhart in 1995 [5]. It is based on the swarm intelligence in which swarm reaches the food source comfortably. Each swarm is called a "particle" which is made to move randomly in a space which is multi-directional space. Every particle moved has its own position and velocity. The particles communicate with each other, and they update their velocity and position. This upgrading process will lead to the better position of the particles. The best solution (fitness) obtained with the help of each particle which keeps a track of its coordinate in the search space is called the personal best (*pBest*). The overall best value in a population where the new position is formed based on any particle in the population so far is termed global best (*gBest*). The inertia factor is introduced to preserve the balance between the searching abilities of the particles. The various parameters used in PSO algorithm are maximum number of iteration =

500, population size $= 500$, inertia weight factor $= 0.65$, acceleration coefficients: c_1 and c_2 as 1.65 and 1.75, respectively.

2.3.3 Differential Evolution

DA is a population-based search algorithm developed by Storn and Price in [6]. It is similar to a genetic algorithm (GA) that applies crossover, mutation and selection operators. It is self-adaptive in nature which makes it different from GA as the selection and mutation process works differently in these two algorithms. The various steps in DA involve the initialization of independent variables and defining the variable bounds for each parameter. Mutation process deals with the creation of a donor vector to change the population member vector. It also expands the search space. Next is the recombination operator which incorporates good solutions from the previous generation. A trial vector is formed by implementing a crossover strategy for each pair of target vector and equivalent donor vector. The various elements of the donor vector enter the trial vector with crossover probability. Selection operator deals with maintaining the constant population in the coming generations by determining the existence of the child and the parent. If better fitness value of the child is obtained, then they will become parents in the next generation, else the parents are retained. Thus, solution obtained based on the fitness value does not get worse; they can either get better or remain constant. Target vector and the trial vector are compared, and the one with the minimum function value is used in the next generation. Mutation, recombination and selection continue until the criteria for stopping are satisfied. For DE algorithm, the controllable parameters are set as the following: maximum number of iterations $= 500$, population size $= 500$, crossover probability $= 0.9$ and weight factor $= 0.5$.

2.3.4 Teaching-Learning-Based Algorithm

TLBO algorithm is a teaching-learning-based optimization algorithm proposed by Rao et al. [7]. It imitates the teaching and learning ability of the teachers and learners in a classroom. It emphasizes the two modes of teaching, i.e. the teacher phase and the learner phase. The teacher phase includes learning through a teacher and in learner phase, there is an interaction of learners with the rest learners. The grades and results of the learners are decided by the quality of teachers and with the interaction of learners among themselves. The group of learners in TLBO are considered as the population. The various design variables are the different subjects offered to the learners, and the learner's result is comparable to the fitness value of the optimization problem. Over the entire population, the teacher signifies the best solution. In the teacher phase, the mean of the entire class is improved which greatly depends upon of the learners in the class. In the learner phase, the learners upgrade their knowledge by interaction among the learners. The various controllable parameters used in TLBO algorithm are as maximum number of iterations $= 500$ and population size $= 500$.

2.4 Optimization Results Using Evolutionary Algorithms

Comparative analysis for single-objective optimization of MRR and surface roughness was performed using four evolutionary algorithms using various parameters in each case. The values of the constant and parameters are shown in Tables 1 and 2. Single-response optimization results for MRR and surface roughness derived while employing FA, PSO, DE and TLBO algorithms in MATLAB (R 2013 a) are exhibited in Tables 3 and 4, respectively. With the comparative analysis of four algorithms, it is seen that TLBO algorithm performs better in comparison to the other algorithms which proves its potential to derive at the best optimal solution. The consistency of the solution which is measured by the standard deviation (SD) value for the four algorithms also proved the better performance of TLBO algorithm. The variation of output parameters with various input parameters is shown in Fig. 1. The superiority of TLBO algorithm can be validated from the convergence diagrams shown in Fig. 2. In multi-objective optimization, the two responses are simultaneously optimized instead of optimizing the responses individually. The objective function developed for such case is given by

$$\text{Min}\ (Z) = \frac{w_1(R_{\text{ZPECH}})}{R_{\text{ZPECH}_{\text{min}}}} - \frac{w_2(V_{\text{PECH}})}{V_{\text{PECH}_{\text{max}}}} \tag{4}$$

Table 1 Values of constants and parameters [2]

Notation	Description (units)	Values
k_e	Electrolyte electrical conductivity ($\Omega^{-1}\text{mm}^{-1}$)	0.025
η	Current efficiency (%)	0.4
λ	Percentage of pulse-on time to attain the set voltage (%)	0.85
ΔV	Total voltage loss in the gap between the electrodes (V)	3.5
Y	Gap between the electrodes (mm)	1
C_{PECF}	Contribution factor of electrolytic dissolution	0.8
C_h	Contribution factor of mechanical honing	0.1
t	Finishing time (min)	6
N_s	Rotational speed of workpiece gear (rpm)	40
k	Factor indicating the proportion of material removed from the valleys on workpiece gear in one cycle of PECF or honing	0.1
K	Coefficient of wear	1.2×10^{-4}
F	Faraday's constant (C)	96,500
f	Height conversion factor for converting rectangle height to a height of a triangle with same base length and area	2

Table 2 Values of various parameters of workpiece bevel gear tooth [2]

Notation	Description (units)	Values
F_n	The total normal force acting on gear tooth in its line of action (N)	90.9
D_w	Working depth (mm)	9.74
F_w	Face width (mm)	16
r_b	Base circle radius (mm)	35.7
W	Base width (mm)	8.25
α	Pressure angle (°)	22.5
E_w	Electrochemical equivalent	17.16
H	Brinell's hardness number (BHN) (N/mm^2)	513
ρ_w	Density (g/mm^3)	0.00777
W_t	Top land width (mm)	2.8
T	Number of teeth	16

Table 3 Results of the single-objective optimization for MRR

Method	Mean	SD	Voltage (V)	Pulse-on time (ms)	Pulse-off time (ms)	MRR (mm^3/s)
FA	3.6435	0.00055	13.6370	4.9575	4.5752	3.6438
PSO	3.6404	0.00035	13.4564	4.7595	4.9201	3.6407
DE	3.6497	0.00041	13.8648	4.9991	4.5000	3.6499
TLBO	3.6510	0.00030	13.9123	5	4.5000	3.6512

Table 4 Results of the single-objective optimization for surface roughness

Method	Mean	SD	Voltage (V)	Pulse-on time (ms)	Pulse-off time (ms)	Surface roughness (μm)
FA	7.2787	0.0077	10.6	3.54	8.96	7.2785
PSO	7.2782	0.0074	9.11	2.27	8.26	7.2781
DE	7.2785	0.0088	11.21	3.65	5.67	7.2784
TLBO	7.2779	0.0073	8.0003	2	9	7.2778

where $R_{ZPECH_{min}}$ and $V_{PECH_{max}}$ are the minimum value of surface roughness and the maximum value of MRR obtained from single-objective optimization in PECH process. Here, equal weights are considered taking w_1 and w_2 as 0.5. The results for the multi-objective optimization are shown in Table 5.

TLBO algorithm is capable of achieving the most global optimal solution with respect to consistency and accuracy of the derived solutions for both single- and multi-objective optimization. Figure 3 represents the Pareto optimal fronts of the two responses, i.e. MRR and surface roughness. It can be seen from the figure that the solutions are non-dominated and none of the solutions are better than any other. It

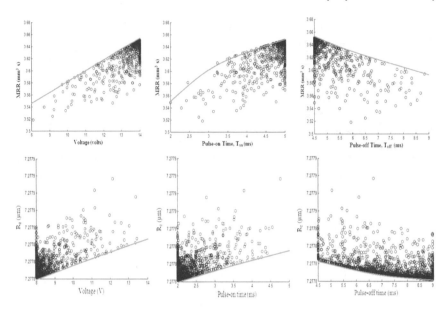

Fig. 1 Variation of MRR and surface roughness with different PECH process parameters

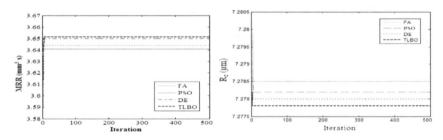

Fig. 2 Convergence of FA, PSO, DE and TLBO for output responses

totally depends upon the process engineer to select the solution on his/her choice to enhance the machining performance. The corresponding values of objective functions and input parameters of these no dominated solutions are exhibited in Table 6.

3 Conclusion

It has been observed that the desired value of MRR and surface roughness can only be achieved by a correct setting and selection of the various process parameters in PECH process. In this paper, the four different optimization algorithms were tested for single- and multi-response optimization in PECH process, as a comparative analysis. Both single- and multi-objective optimization results prove that TLBO

Table 5 Results of the multi-objective optimization

Optimization method	Response	Mean	SD	Optimal value	Z	Parameter		
						Voltage (V)	Pulse-on time (ms)	Pulse-off time (ms)
FA	MRR	0.004049	0.000006	3.6512	0.004048	13.945	5	4.5
	Surface roughness			7.2799				
PSO	MRR	0.002808	0.000007	3.6465	0.002789	13.911	4.925	4.671
	Surface roughness			7.2642				
DE	MRR	0.003641	0.000005	3.6474	0.003638	13.64	5	4.5
	Surface roughness			7.2839				
TLBO	MRR	0.001646	0.000001	3.6515	0.001643	13.8757	5	4.5
	Surface roughness			7.2579				

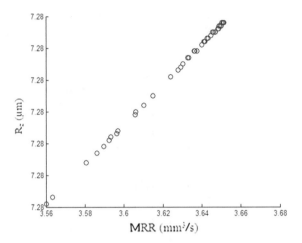

Fig. 3 Pareto optimal front of the output responses

Table 6 A few set of optimal values

MRR (mm^3/s)	Surface roughness (μm)	Voltage (V)	Pulse-on time (ms)	Pulse-off time (ms)
3.580678	7.279998	12.59192	4.192554	7.103972
3.580678	7.279998	12.59192	4.192554	7.103972
3.589767	7.279998	12.68829	4.239361	6.428345
3.596778	7.279998	11.37223	4.54899	4.715953
3.630316	7.279998	13.18972	4.679549	4.597541
3.642909	7.279998	13.84982	4.89132	4.71184
3.648502	7.279998	13.8485	5	4.507739
3.648957	7.279998	13.86639	5	4.5

algorithm arrives at the best solution with respect to the consistency of the solution, objective function value and convergence speed. In PECH process, the simultaneous optimization yields applied voltage as 13.8757 V, pulse-on time as 5 ms and pulse-off time as 4.5 ms. With this, the MRR obtained is 3.6515 mm^3/s and surface roughness as 7.2579 μm, which completes the optimization aspects in the theoretical modelling proposed by the past researchers. Also, the obtained values of MRR and surface roughness are better than the predicted values reported by Pathak and Jain [2].

References

1. Misra, J.P., Jain, N.K., Jain, P.K.: Investigations on precision finishing of helical gears by electrochemical honing process. In: Proceedings of the Institution of Mechanical Engineers, Part B: Journal of Engineering Manufacture, vol. 224(12), pp. 1817–1830 (2010)
2. Pathak, S., Jain, N.K.: Modeling and experimental validation of volumetric material removal rate and surface roughness depth of straight bevel gears in pulsed-ECH process. Int. J. Mech. Sci. **124**, 132–144 (2017)
3. Shaikh, J.H., Jain, N.K.: Modeling of material removal rate and surface roughness in finishing of bevel gears by electrochemical honing process. J. Mater. Process. Technol. **214**(2), 200–209 (2014)
4. Yang, X.S.: Firefly algorithm, Nature-Inspired Metaheuristic Algorithms. **20**, 79–90 (2008)
5. Kennedy, J., Eberhart, R.: Particle swarm optimization. In: Proceedings of IEEE International Conference on Neural Networks, pp. 1942–1948. Perth, Australia (1995)
6. Storn, R., Price, K.: Differential evolution—a simple and efficient heuristic for global optimization over continuous spaces. J. Global Optim. **11**, 341–359 (1997)
7. Rao, R.V., Savsani, V.J., Vakharia, D.P.: Teaching learning-based optimization: a novel method for constrained mechanical design optimization problems. Comput. Aided Des. **43**(3), 303–315 (2011)

Modeling and Simulation of Huge AC Power Network for Optimization of Corona Power Loss Through TLBO Algorithm

Manan Pathak and Ishita Bhatt

Abstract Corona loss happens most in transmission half in a power network. In this paper, a new methodology for estimating corona power loss along with the intention of optimum choice of transmission lines in local or state dispatching centers with high response rate is suggested. During this methodology, convergence characteristics of the TLBO have been assessed. Several trials have been conducted with differential values so as to justify the robustness of the planned methodology. Considering the equality of the solution and convergence speed obtained, this technique looks to be able to calculate corona for estimating the corona loss. Designed formula which supported this suggested model in India's 220, 400, and 765 kV power network with 1296 exploited samples for coaching TLBO network is implemented. Results of the algorithm indicate the generalized chance and better response rate of the suggested model.

Keywords Corona · Transmission line network · Dispatching center · TLBO algorithm

1 Introduction

Precise quantitative data regarding corona loss is inadequate even though Indian electrical business has been dealing with high voltages since 30 years. The data mentioned in various statistical reports has been monthly or yearly average which has been calculated from distinction between generated and utilized power at different voltage levels [1]. Monitory loss of ample amount occurs due to corona as suggested by various researches which indicates that the corona loss amplifies to 30% of total consumption peak from 20% during peak hours [1]. Large improvement in corona

M. Pathak (✉) · I. Bhatt
Aditya Silver Oak Institute of Technology, Ahmedabad, India
e-mail: mananpathak.gn@socet.edu.in

I. Bhatt
e-mail: ishitabhatt.ee@socet.edu.in

© Springer Nature Singapore Pte Ltd. 2020
R. Venkata Rao and J. Taler (eds.), *Advanced Engineering Optimization Through Intelligent Techniques*, Advances in Intelligent Systems and Computing 949,
https://doi.org/10.1007/978-981-13-8196-6_6

losses can be achieved by applying various rigorous methods which exploit, protect, and design the useful algorithm as indicated by recent studies [1]. If we consider difference between generated and consumed energy as corona losses in the India, following statics are obtained:

1. Generating station causes 5.6% of total production as corona.
2. Transmission network causes 3.8% of total production as corona.
3. Subtransmission network causes 11.1% of total production as corona.

This paper represents comparative study for a generating station to deliver optimized power to the transmission network and further from transmission network to the distribution network.

Reduction of corona loss fields

Substantial losses caused by corona have to be decreased, which can only be achieved by identifying various areas for improving the efficiency of the sections of the power system and managing them individually to obtain a reduced corona system that will achieve the earlier set target of the power delivery [2]. Reduction of corona loss can be bifurcated as below sections:

- Internal consumption of power stations,
- Copper loss in transmission and distribution power lines,
- Corona loss in over-head transmission lines,
- Corona loss in transformers and generator,
- Losing energy due to human errors.

Optimization of the transmission lines can be found out to reduce the corona loss in it by applying teaching-learning-based optimization (TLBO) algorithm. The algorithm would suggest optimized dimensions for conductors used in transmission network such that corona losses are reduced to the possible minimum value.

Reasons for implementing TLBO

Natural phenomena of teaching-learning are implemented by representing it in the form of algorithm by TLBO. Only common controlled parameters are required to be tuned leaving the user hassles of tuning algorithm controlled parameters. Advantages such as simpler algorithm, less complexity, and straightforward approach to the problem make TLBO noteworthy than other algorithms. Hence, the TLBO is utilized for solving the corona loss drawback [3].

2 Corona Development

Corona development causes radio interferences in the neighboring sensitive devices such as communication lines due to development of ion deposits on the surfaces conductors and generation of noise takes place during occurrence of the corona [4].

Electric discharges are also created due to the same, and it happens in both AC and DC lines. Calculation of the corona losses must be carried out for various situations like normal load and peak load to account for total losses caused by the corona and to effectively design the corona optimization method.

2.1 Problem Formulation

Calculating corona loss:

$$Pc = \frac{244}{\delta}(f + 25) \frac{E_n - E_0}{} \text{ kw/km/phase} \tag{1}$$

where

Pc corona power loss
F frequency of supply in Hz
Δ air density factor
E_n r.m.s phase voltage in kV
E_o disruptive critical voltage per phase in kV
R radius of the conductor in (cm)
D spacing between conductors in meters.

It is also to be noticed that for a single-phase line, $E_n = 1/2$ line voltage and for a three-phase line voltage

$$E_0 = G_0 m_0 r \delta^{\frac{2}{3}} \ln \frac{\text{Deq}}{r} \text{V/phase} \tag{2}$$

G_o maximum value of disruptive critical voltage gradient in V/m
M_o roughness factor of conductor.

Assumptions for finding optimal value of corona power loss

E_n r.m.s phase voltage in kV [taken as 220, 400, and 765 kV for three different cases],
F Frequency of supply in Hz [taken frequency in between 45 and 55 Hz for different cases],
m_o Roughness factor of conductor [taken as $m_o = 1$ for polished conductor $m_o = 0.92$–0.98 for dirty conductor $m_o = 0.8$–0.87 for standard conductor],
r Radius of the conductor in meters [fixed depending on voltage level],
D Spacing between conductors in meters [fixed depending on voltage level].

3   Teaching-Learning-Based Optimization

3.1   TLBO Algorithmic Rule

Influence of teacher on learners is the driving force in the development of the TLBO algorithm which is a population based algorithm. Output of the learner is affected by the teacher which is the concept behind the TLBO algorithm [5]. TLBO stands out among other optimization parameters because it is free from the burden of the standardization of the parameters [6]. We have used TLBO to compute corona loss in an exceedingly transmission network with variable voltage ranges from 220, 400, and 765 kV. As mentioned, higher and lower bounds for all the parameters were identified to solve the given drawback of calculative corona loss and getting the optimized parameters. The steps concerned within the search procedure of the TLBO algorithmic rule for the projected corona loss drawback are summarized as follows.

3.2   Initialization of Corona Loss Drawback

To be able to turn out the economical corona calculation, we have a tendency to propose the corona power loss calculation steps as displayed below. The algorithmic rule is predicated on a TLBO methodology for computing corona power loss that permits reducing iteration calculation of the corona power loss. The algorithmic rule includes variety of mechanisms making certain its correctness and stability [6]. The planned algorithmic rule operation starts by inputting the model configuration with many input data together with conductor and alternative connected components and follows the following steps:

Step 1: Outline the optimization drawback as reduction drawback.
Step 2: Population size (Ps), range of design variables (Nd) that represents range of generating units, most and minimum generation limits (limits of design variables), and stopping criteria (maximum number of iterations) are defined in this step.
Step 3: Teacher phase: Measure the distinction between existing mean result and best mean result by utilizing Tf.
Step 4: Learner phase: Update the learner's generation value with the help of teacher's generation.
Step 5: Update the learner's generating value by utilizing the generating value of some other learner.
Step 6: Termination criteria: Repeat the procedure from Steps 2 to 5 until the most number of iterations is met.

3.3 Corona Loss Problem Handling by TLBO

Corona loss for three completely different voltage levels has been calculated initially, for 230, 400, and 765 kV for each voltage level; the spacing between the conductors is predefined according to the standards. The variables taken to resolve drawback through TLBO are m_0, frequency starting from 47 to 52 Hz and diameters of the conductors. Supported this variable with appropriate higher and lower limits, 150 iterations were performed to get optimized values of tumultuous voltages and corona loss. One calculation is additionally conducted statically, supported formula to possess minimum value of loss, so it is compared to the TLBO results [7] (Table 1).

Table 1 Optimized value of corona power loss by TLBO

Corona loss at 400 volts, spacing 11.25 m							
m_0	Diameters	47 Hz	48 Hz	49 Hz	50 Hz	51 Hz	52 Hz
0.92	0.569	0.345902	0.350706	0.35551	0.360314	0.365118	0.369922
	0.805	0.391015	0.396446	0.401877	0.407308	0.412739	0.418169
	0.795	0.389409	0.394818	0.400226	0.405635	0.411043	0.416452
	0.889	0.403632	0.409238	0.414844	0.42045	0.426056	0.431662
	1.036	0.422395	0.428261	0.434128	0.439994	0.445861	0.451727
	1.265	0.44484	0.451018	0.457196	0.463375	0.469553	0.475731
	1.473	0.459589	0.465972	0.472356	0.478739	0.485122	0.491505
	1.664	0.469365	0.475884	0.482403	0.488922	0.495441	0.50196
1	0.569	0.341663	0.346409	0.351154	0.355899	0.360645	0.36539
	0.805	0.384362	0.3897	0.395039	0.400377	0.405715	0.411054
	0.795	0.382862	0.388179	0.393497	0.398814	0.404132	0.40945
	0.889	0.396078	0.401579	0.40708	0.412581	0.418082	0.423583
	1.036	0.413228	0.418967	0.424706	0.430445	0.436185	0.441924
	1.265	0.433093	0.439108	0.445123	0.451138	0.457154	0.463169
	1.473	0.445456	0.451643	0.45783	0.464017	0.470204	0.476391
	1.664	0.453025	0.459317	0.465609	0.471901	0.478193	0.484485
0.85	0.569	0.349631	0.354487	0.359343	0.364199	0.369055	0.373911
	0.805	0.396884	0.402396	0.407909	0.413421	0.418933	0.424445
	0.795	0.395184	0.400673	0.406161	0.41165	0.417139	0.422627
	0.889	0.4103	0.415998	0.421697	0.427396	0.433094	0.438793
	1.036	0.430498	0.436477	0.442456	0.448435	0.454415	0.460394
	1.265	0.455247	0.46157	0.467893	0.474216	0.480539	0.486861
	1.473	0.472137	0.478694	0.485252	0.491809	0.498367	0.504924
	1.664	0.4839	0.490621	0.497342	0.504063	0.510784	0.517505

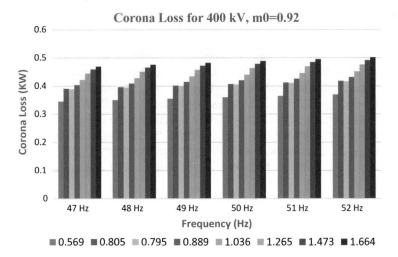

Fig. 1 Corona loss for 400 kV, $m_0 = 0.92$

Table 2 Optimized value of corona power loss considering all values

Diameter (m)	m_0	Frequency (Hz)	Conductor spacing (cm)	Voltage (kV)	Corona loss (kw/km/phase)
1.750	0.940	45	13.87	230	45.678

4 Simulation Result

See Fig. 1 and Table 2.

5 Conclusion

Various mathematical and heuristic approaches have been applied for the calculation of the corona loss issue in the optimum power flow over the past half a century. A new approach to the nonlinear functions has been used to find out realistic relation of the corona losses with the various dimensions of conductors used on the power system. Three unique cases have been studied and solved using proposed algorithm, and obtained results agree with the desired result predictions. The noteworthy points of the research carried out are as follows:

1. A three objective-based corona loss has been considered.
2. An attempt has been made to utilize TLBO algorithm for the corona power loss dilemma.
3. Proposed algorithm works satisfactorily for large power system and composite corona power loss function.

References

1. Epri.: Power transmission line Reference Manual
2. Gh-Heidari.: Experimental/ mathatical model for loss factor. IEEE—NAPS, NEVADA, USA (1992, October)
3. Rao, R.V., Savsani, V.J., Vakharia, D.P.: Teaching-learning-based optimization: a novel method for constrained mechanical design optimization problems. Comput. Aided Des. **43**, 303–315 (2011)
4. Maruvada, P.S.: Corona performance of high-voltage transmission lines, pp. 179–233. Research Studies Press, Baldock, UK (2000)
5. Rao, R.: Review of applications of TLBO algorithm and a tutorial for beginners to solve the unconstrained and constrained optimization problems. Decis. Sci. Lett. **5**(1), 1–30 (2016)
6. Rao, R.V., More, K.C.: Optimal design of the heat pipe using TLBO (teaching–learning-based optimization) algorithm. Energy **80**, 535–544 (2015)
7. MATLAB 2016

Optimization of Water Distribution Networks Using Cuckoo Search Algorithm

Maduukuri Naveen Naidu, Pankaj Sriman Boindala, A. Vasan and Murari R. R. Varma

Abstract Optimization of water distribution network is one of the popular problems faced by hydrologists and water distribution designers. The nonlinear complexity involved in this problem makes it challenging for optimization. To solve this complexity, several researchers have studied various approaches involving numerous optimization techniques. This paper focuses on using one such approach involving a simulation and optimization models. The optimization technique used is Cuckoo Search (CS), and the simulation model is EPANET. This proposed model is tested for two water distribution network benchmark problems, namely two-loop network and Hanoi network. The obtained results when compared with previous studies show significant improvement.

Keywords Optimization · Cuckoo Search algorithm · Water distribution networks

1 Introduction

Water is one of essential requirements for life's sustainability. With the growing population as well as pollution, satisfying the demand needs of providence makes the design of water distribution network a challenging task. To satisfy the economic as well as resilience constrains makes this an even more challenging. Practical difficulties like limited discrete diameters available in the market and the nonlinearity involved between the head loss and discharge make this a complex nonlinear problem.

M. Naveen Naidu (✉) · P. S. Boindala · A. Vasan · M. R. R. Varma
Department of Civil Engineering, BITS Pilani Hyderabad Campus, Pilani, India
e-mail: vimpnaveen1234@gmail.com

P. S. Boindala
e-mail: srimanpankaj@gmail.com

A. Vasan
e-mail: vasan@hyderabad.bits-pilani.ac.in

M. R. R. Varma
e-mail: murari@hyderabad.bits-pilani.ac.in

© Springer Nature Singapore Pte Ltd. 2020
R. Venkata Rao and J. Taler (eds.), *Advanced Engineering Optimization Through Intelligent Techniques*, Advances in Intelligent Systems and Computing 949, https://doi.org/10.1007/978-981-13-8196-6_7

These entangled issues intrigued researchers to develop various solutions. With the development of evolutionary and meta-heuristic optimization techniques, researchers focused on implementing these to solve this problem in recent past [1–5].

2 Problem Formulation

Pipe cost is the function of diameter and length

$$\text{cost} = \sum_{i}^{\text{nop}} l(i) * a(d(i)) \tag{1}$$

where $l(i)$ = length of ith pipe, $a(d(i))$ is the cost of diameter with unit length, and nop = number of pipes.

2.1 Constraints

$$\sum q_{\text{in}} = \sum q_{\text{out}} \tag{2}$$

Equation 2 shows that inflow is equal to outflow

$$\sum_{k \in l} \Delta h = 0, \forall l \in \text{nl} \tag{3}$$

Equation 3 shows that summation of head losses around a loop must be equal to zero

$$(H_{\text{avl}})_i \geq (H_{\text{min}})_i, \forall i \in \text{nn} \tag{4}$$

Equation 4 shows that available heads should be greater than or equal to minimum head

$$D_k \in \{D\} \tag{5}$$

Equation 5 shows the diameter should be discrete and available commercially.

Where q_{in} = inflow, q_{out} = outflow, H_{avl} = available head, H_{min} = minimum head, D = commercially available diameters, Δh = head loss in a pipe, nn = number of demand nodes, nl = number of loops, D = commercially available diameters, D_k = output diameters.

3 Cuckoo Search Optimization

Cuckoo Search (CS) is one of the latest meta-heuristic algorithms proposed by [6]. The main advantage/key element in this algorithm is its search walk. It uses a levy-flight-based search walk making it more robust in searching.

The algorithm pseudo-code is given below:

- Formulation of objective function: $f(v)$, $v = (v1, v2, v3....) T$ where f is the function and v is the variable set (nest in this Cuckoo Search).
- Decide the population number (nn) (i.e., number of nests): in general 10–20 times the variables involved.
- Assign limits or boundaries to the search space: upper and lower bounds to each variable.
- Initialize population: Generate nn nests with values within the boundaries.
- Finding best nest: Evaluate all the nests, and find the best nest.
- Generate or update the population: Using this best nest, generate new set of nests using the levy-flight random walk.
- Evaluate these new nests, and update the best nest.
- Continue this process until the termination condition.

4 CSNET

An optimization simulation model called Cuckoo Search NET (CSNET) was developed in which optimization is done by Cuckoo Search algorithm and simulation is carried in EPANET using EPANET-MATLAB toolkit. This simulation tool kit is used to find out the heads obtained at each junction in the network by satisfying the continuity and energy constraints. As the search space is limited to discrete values, the diameters obtained from the optimization model are rounded off the nearest discrete value. These new diameters are used for simulation in EPANET. The minimum head constraint is added to the cost using exterior penalty function. The optimization is continued until the best set of diameters produce sufficient heads with minimum cost.

5 Design Examples

5.1 Two-Loop Network

The information related to different types of diameter available in the market, elevation, minimum pressure, and demand for two-loop network are taken from [7]. Totally eight pipes with Hazen–Williams coefficient of 130 are used in networks.

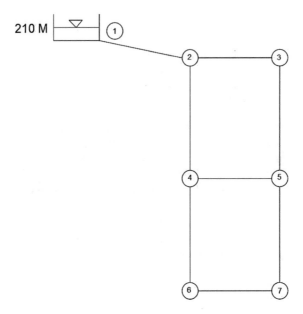

Fig. 1 Layout of two-loop network [7]

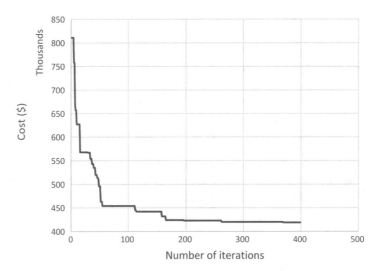

Fig. 2 Convergence graph for optimization of two-loop network using Cuckoo Search

One reservoir with head 210 m and six demand nodes with minimum required pressure 30 m (Figs. 1, 2; Tables 1, 2).

Table 1 Optimal solutions for two-loop network for minimization of cost of diameters in inches

Pipes	1	2	3	4	5	6	7	8	Cost ($)
Savic and Walters (1997) [1]	18	10	16	4	16	10	10	1	419,000
Cunha and Sousa (1999) [2]	18	10	16	4	16	10	10	1	419,000
Eusuff and Lansey (2003) [3]	18	10	16	4	16	10	10	1	419,000
Present work	18	10	16	4	16	10	10	1	419,000

Table 2 Evaluation number for different types of techniques used

Sl. No	Techniques used	Authors	Evaluation function numbers
1.	Genetic algorithm	Savic and Walters (1997) [1]	65,000
2.	Simulated annealing algorithm	Cunha and Sousa (1999) [2]	25,000
3.	Shuffled leap frog algorithm	Eusuff and Linsey (2003) [3]	11,155
4.	Shuffled complex algorithm	Liong and Atiquzzaman (2004) [8]	25,402
5.	Cuckoo Search algorithm	Naveen and Pankaj (2018)	15,300

5.2 Hanoi Water Distribution

The information related to different types of diameter available in the market, demand, and lengths for Hanoi network are taken from [7]. Totally 36 pipes with Hazen–Williams coefficient of 130 are used in networks. One reservoir with head 100 m and 32 demand nodes with minimum required pressure 30 m (Figs. 3, 4; Table 3).

6 Conclusions

The efficiency of Cuckoo Search algorithm was tested by solving two benchmark problems of water distribution networks, namely two-loop network and Hanoi water distribution network. The number of Function Evalutions obtained by cuckoo search algorithm when compared with Genetic Algorithm, Simulated Annealing algorithm showed significant improvement (i.e. found better solutions in few number of iterations). From the results, it is concluded that CSNET is efficient alternative evaluation model to optimize water distribution networks. CSNET can support the engineers to design of water distribution networks.

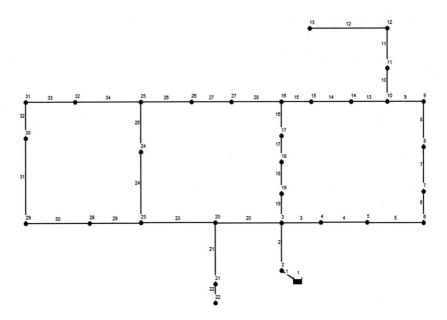

Fig. 3 Layout of Hanoi water distribution network [7]

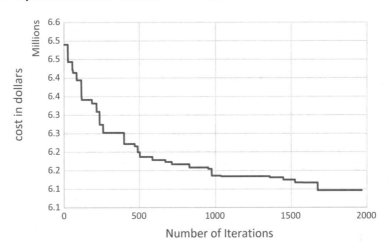

Fig. 4 Convergence graph for optimization of Hanoi water distribution network

Table 3 Optimal solutions for Hanoi water network for minimization of cost of diameters (inches)

Link	Savic and Walters (1997) [1]	Cunha and Sousa (1999) [2]	Liong and Atiquzzaman (2004) [8]	Current work Naveen and Pankaj (2018)
1	1016	1016	1016	1016
2	1016	1016	1016	1016
3	1016	1016	1016	1016
4	1016	1016	1016	1016
5	1016	1016	1016	1016
6	1016	1016	1016	1016
7	1016	1016	1016	1016
8	1016	1016	762	1016
9	1016	1016	762	1016
10	762	762	762	1016
11	609.6	609.6	762	762
12	609.6	609.6	609.6	609.6
13	508	508	406.4	609.6
14	406.4	406.4	304.8	508
15	304.8	304.8	304.8	304.8
16	304.8	304.8	609.6	304.8
17	406.4	406.4	762	304.8
18	508	508	762	406.4
19	508	508	762	609.6
20	1016	1016	1016	508
21	508	508	508	1016
22	304.8	304.8	304.8	508
23	1016	1016	762	304.8
24	762	762	762	1016
25	762	762	609.6	762
26	508	508	304.8	762
27	304.8	304.8	508	508
28	304.8	304.8	609.6	406.4
29	406.4	406.4	406.4	304.8
30	406.4	304.8	406.4	406.4
31	304.8	304.8	304.8	304.8
32	304.8	406.4	406.4	304.8
33	406.4	406.4	508	304.8
34	508	609.6	609.6	508
Cost ($)	6,073,000	6,056,000	6,220,000	6,096,565

Acknowledgements The authors acknowledge the financial support provided by Council of Scientific and Industrial Research (CSIR) through project No. 22/0723/17/EMR-II, dated May 16, 2017.

References

1. Savic, D.A., Walters, G.A.: Genetic algorithms for least cost design of water distribution networks. J. Water Resour. Plan. Manag. **123**, 67–77 (1997)
2. Cunha, M.C., Sousa, J.: Water distribution network design optimization simulated annealing approach. J. Water Resour. Plan. Manag. **125**, 215–224 (1999)
3. Eussuf, M.M., Lansey, K.E.: Optimization of water distribution networks design using shuffled frog leaping algorithm. J. Water Resour. Plan. Manag. **129**, 210–225 (2003)
4. Fujiwara, O., Khang, D.B.: A two phase decomposition method for optimal design of looped water distribution networks. J. Water Resour. Res. **27**, 985–986 (1990)
5. Vasan, A., Slobodan, P.: Simonovic: optimization of water distribution networks using differential evolution. J. Water Resour. Plan. Manag. **136**(2), 279–287 (2010)
6. Yang, X.S.: Nature-inspired metaheuristic algorithms. Luniver Press (2010)
7. Center for water systems university of Exeter. http://emps.exeter.ac.uk/engineering/research/cws/resources/benchmarks/design-resiliance-pareto-fronts/medium-problems/
8. Liong, S.Y., Atiquzzaman, M.: Optimal design of water distribution network using shuffled complex evolution. J. Inst. Eng. **44**(1), 93–107 (2004)

GA-Based Hybrid Approach to Solve Fuzzy Multi-objective Optimization Model of Multi-application-Based COTS Selection Problem

Anita Ravi Tailor and Jayesh M. Dhodiya

Abstract Due to the quick growth of the modular software development, the commercial off-the-shelf (COTS) selection model of optimization technique becomes more popular in a component-based software system (CBSS). In order to realize the benefits of the COTS product, it is necessary to select the right products for various software systems. This paper proposed a genetic algorithm (GA)-based hybrid approach with fuzzy exponential membership function for best fit of COTS components. In this proposed approach, decision-maker (DM) is required to specify the different aspiration levels as per his/her preference to obtain an efficient allocation plan with different shape parameters in the exponential membership function. A real-world scenario of developing two financial applications for two small-scale industries is provided to represent the importance of the proposed algorithm with data set from a realistic situation.

Keywords Intra-modular coupling density · Multi-objective optimization · Cohesion · Coupling · Genetic algorithm

1 Introduction

Nowadays, many critical problems are being faced by many industries in real-world decision-making to develop an effective complex software system. The software system is working to improve the effectiveness of structured programming, design patterns, modelling languages, etc. With such a growing requirement of complex software systems, the utility of COTS products has grown progressively. COTS-based

A. R. Tailor · J. M. Dhodiya (✉)
S.V. National Institute of Technology, Surat, India
e-mail: jdhodiya2002@yahoo.com

A. R. Tailor
e-mail: anitatailor_185@yahoo.com

© Springer Nature Singapore Pte Ltd. 2020
R. Venkata Rao and J. Taler (eds.), *Advanced Engineering Optimization Through Intelligent Techniques*, Advances in Intelligent Systems and Computing 949, https://doi.org/10.1007/978-981-13-8196-6_8

system development gives assurances of lower resource cost with quicker delivery. COTS selection optimization models are way of managing cost, effort and development time which requires less time to be planned and executed by the developers.

In software development system, the component-based approach has created more attention in today's world as it is concerned with the system development by incorporating components called as component-based software system (CBSS). With this idea of component based software system based approach, software systems easily developed by proper COTS components and can apply in the application field with minimum resource. Over the years, there have been several developments available in solution process of COTS selection problem in CBSS development [1–5].

In a practical situation, the COTS components are provided by different dealers, so their functions are different and they are diverse from each other. To develop the modular software system, minimizing the coupling and maximizing the cohesion criteria of software modules are generally used in which 'coupling is concerning gauge of interactions among software modules' and 'cohesion is concerning the gauge of interactions among software components inside a software module'. Software modules with low coupling and high cohesion are characteristics of high-quality software system as high cohesive module shows a high reusability and loose coupled system allow straightforward maintenance [1]. The relations of cohesion and coupling of modules can be précised by using intra-modular coupling density (ICD). Moreover, in developing modular software systems, multiple applications and reusability of components in different applications must be considered.

The main objective considered in this paper is how to select the proper and suitable component and arrange them into different applications that maximize the functional performance and minimize the total cost based on high cohesion and low coupling of modular software system using genetic algorithm (GA)-based hybrid approach. If problem is of big size, then non-dominated sorting genetic algorithm (NSGA) plays a significant role to find the solution of such a problem. In this paper, the problem is of 2 applications with 6 modules and 12 components only, so with GA, it also provides better solutions. Hence in this paper, we have utilized GA only but if problem is of big size then NSGA is more convenient for solving such a problem.

To find the solution of multi-objective optimization model (MOOM) for COTS selection problem, the multi-application of MOOM for COTS selection problem is transformed into a single-objective nonlinear optimization problem using fuzzy exponential membership function with different shape parameters and therefore such a problem becomes NP-hard problem. GA is a proper technique to solve such 'NP-hard' problems. Thus, in this paper, we proposed genetic algorithm-based hybrid approach to find the solution of fuzzy MOOM for COTS selection problem for modular software system using fuzzy exponential membership function. GA is a famous random search and global optimization method taking into account thought of evolution and natural selection of the fitness in the biological system. It is also an appropriate method for solving discrete, nonlinear and non-convex large-scale optimization problems as it searches the optimal solution by simulating the natural evolution procedure and work by mimicking the evaluating principle of natural genetics. It has verified numerous major advantages such as convergence to global optimum and strong robustness [6–8].

2 Multi-objective Optimization Model

In this paper, we have adopted multiple applications of MOOM for COTS selection problem for modular software system to optimize the functional requirement and total fuzzy development cost (procurement and adaption cost) under some realistic constraints [1].

Objective functions:

The function performance $F(x)$ and total cost $C(x)$ objective functions for software system can be written as follows:

$$\text{Max } F(x) = \sum_{j=1}^{M} \sum_{k=1}^{L} f_{jk} x_{jk}, \text{ Min } C(x) = \left(\sum_{k=1}^{L} \widetilde{c_k^p} y_k + \sum_{j=1}^{M} \sum_{k=1}^{L} \widetilde{c_{jk}^a} x_{jk} \right)$$

Realistic constraints of the model:

$$(ICD)_i = \frac{\sum_{j=1}^{M} s_{ij} \left(\sum_{k=1}^{L-1} \sum_{k'=k+1}^{L} r_{kk'} x_{jk} x_{jk'} \right)}{\sum_{k=1}^{L-1} \sum_{k'=k+1}^{L} r_{kk'} \left(\sum_{j=1}^{M} s_{ij} x_{jk} \right) \left(\sum_{j=1}^{M} s_{ij} x_{jk'} \right)} \tag{1}$$

$$\sum_{j=1}^{M} s_{ij} x_{jk} \leq 1; \ i = 1, 2, \ldots, N, \ i = 1, 2, \ldots, L \tag{2}$$

$$\sum_{j=1}^{M} x_{jk} \leq y_k . N; \ k = 1, 2, \ldots L \tag{3}$$

$$x_{jk} \leq b_{jk}; \ j = 1, 2, \ldots, M, \ k = 1, 2, \ldots, L \tag{4}$$

$$\sum_{k \in s_t} \sum_{j=1}^{M} s_{ij} x_{jk} = 1; \ i = 1, 2, \ldots, N, \ t = 1, 2, \ldots, T \tag{5}$$

$$\sum_{k=1}^{L} x_{jk} \geq 1; \quad j = 1, 2, \ldots, M \tag{6}$$

$$x_{jk} \in \{0, 1\}; \quad j = 1, 2, \ldots, M, \quad k = 1, 2, \ldots, L \tag{7}$$

where 'N', 'M', 'L' and 'T' defined the number of application, number of modules, number of alternative and number of sets of alternative COTS components, respectively; sc_k, s_t and H_i defined the kth COTS component, the set of alternative COTS components for tth functional requirement of software system and a threshold value of ICD of ith application, respectively; $k = 1, 2, \ldots L$, $t = 1, 2, \ldots T$, $i = 1, 2, \ldots, N$; f_{jk}, c_k^p, c_{jk}^a and $r_{kk'}$ defined functional rating of kth COTS component for the jth module, the procurement cost of the kth COTS component, the adaption cost if kth COTS component is adapted into jth module and the number of interaction between kth and k'th COTS components, respectively; $k, k' = 1, 2, \ldots, L$, $j = 1, 2, \ldots, M$ and $f_{jk} \in [0, 1]$. Here, s_{ij}, y_k, b_{jk} and x_{jk} are some binary parameters [1].

Multi-objective Decision Problem:

The multi-objective optimization model for COTS selection problem of modular software system is defined as follows:

Model-1: $\text{Max } F(x) = \sum_{j=1}^{M} \sum_{k=1}^{L} f_{jk} x_{jk}, \; \text{Min } C(x) = \left(\sum_{k=1}^{L} c_k^p y_k + \sum_{j=1}^{M} \sum_{k=1}^{L} c_{jk}^a x_{jk} \right)$

Subject to the constraints: (1)–(7).

3 Formulation of Fuzzy Multi-objective COTS Optimization Models Using Possibility Distribution

3.1 Formulation of Fuzzy Multi-objective COTS Optimization Model-1

To convert the model-1 into a crisp multi-objective optimization model, the triangular possibility distribution strategy is used. Thus, the cost objective function can be written as follows [7, 8]:

$$\min \widetilde{Z}_2 = \min(Z_2^o, Z_2^m, Z_2^p) = \left(\sum_{k=1}^{L} \widetilde{c_k^p} y_k + \sum_{j=1}^{M} \sum_{k=1}^{L} \widetilde{c_{jk}^a} x_{jk} \right) = (\min Z_{21}, \min Z_{22}, \min Z_{23})$$

$$= \min \left(\begin{matrix} \left(\sum_{k=1}^{L} \left(\widetilde{c_k^p}\right)^o y_k + \sum_{j=1}^{M} \sum_{k=1}^{L} \left(\widetilde{c_{jk}^a}\right)^o x_{jk} \right), \left(\sum_{k=1}^{L} \left(\widetilde{c_k^p}\right)^m y_k + \sum_{j=1}^{M} \sum_{k=1}^{L} \left(\widetilde{c_{jk}^a}\right)^m x_{jk} \right), \\ \left(\sum_{k=1}^{L} \left(\widetilde{c_k^p}\right)^p y_k + \sum_{j=1}^{M} \sum_{k=1}^{L} \left(\widetilde{c_{jk}^a}\right)^p x_{jk} \right) \end{matrix} \right) \tag{8}$$

Optimistic, most likely, and the pessimistic scenarios of both objectives are represented by Eq. 8. Each C can be stated as $(C)_\alpha = \left((C)_\alpha^o, (C)_\alpha^m, (C)_\alpha^p\right)$, where $(C)_\alpha^o = C^o + \alpha(C^m - C^o)$, $(C)_\alpha^m = C^m$ and $(C)_\alpha^p = C^p - \alpha(C^p - C^m)$ by using the α-level set concepts $(0 \le \alpha \le 1)$. Hence, Eq. 8 can be written as:

$(\min Z_{21}, \min Z_{22}, \min Z_{22})$

$$
= \left(
\begin{array}{l}
\left(\sum_{k=1}^{L} \left(\widetilde{c_k^p}\right)_\alpha^o y_k + \sum_{j=1}^{M}\sum_{k=1}^{L} \left(\widetilde{c_{jk}^a}\right)_\alpha^o x_{jk}\right), \left(\sum_{k=1}^{L} \left(\widetilde{c_k^p}\right)_\alpha^m y_k + \sum_{j=1}^{M}\sum_{k=1}^{L} \left(\widetilde{c_{jk}^a}\right)_\alpha^m x_{jk}\right), \\
\left(\sum_{k=1}^{L} \left(\widetilde{c_k^p}\right)_\alpha^p y_k + \sum_{j=1}^{M}\sum_{k=1}^{L} \left(\widetilde{c_{jk}^a}\right)_\alpha^p x_{jk}\right)
\end{array}
\right)
\tag{9}
$$

3.2 Crisp Multi-objective Optimization Model

To reflect the three different scenarios with $\alpha-$ set concept, the fuzzy MOOM of COTS product selection problem is converted into a crisp MOOM of COTS product selection problems defined as follows:

Model-1.1:

$(\max Z_1, \min Z_{21}, \min Z_{22}, \min Z_{23})$

$$
= \left(
\begin{array}{l}
\sum_{j=1}^{M}\sum_{k=1}^{L} f_{jk} x_{jk}, \left(\sum_{k=1}^{L} \left(\widetilde{c_k^p}\right)_\alpha^o y_k + \sum_{j=1}^{M}\sum_{k=1}^{L} \left(\widetilde{c_{jk}^a}\right)_\alpha^o x_{jk}\right), \left(\sum_{k=1}^{L} \left(\widetilde{c_k^p}\right)_\alpha^m y_k + \sum_{j=1}^{M}\sum_{k=1}^{L} \left(\widetilde{c_{jk}^a}\right)_\alpha^m x_{jk}\right), \\
\left(\sum_{k=1}^{L} \left(\widetilde{c_k^p}\right)_\alpha^p y_k + \sum_{j=1}^{M}\sum_{k=1}^{L} \left(\widetilde{c_{jk}^a}\right)_\alpha^p x_{jk}\right)
\end{array}
\right)
$$

Subject to the constraints: (1)–(7).

4 Solution Approach for Solving MOOM of COTS Selection Problem

This section presented genetic algorithm-based hybrid approach for fuzzy MOOM of COTS selection problem for modular software system to determine a best efficient solution with the use of exponential membership function to differentiate the indefinite aspiration levels of DM.

Algorithm for finding the solution of fuzzy MOOM of COTS product selection problem using GA-based hybrid approach:

Input: *Parameters*: $(Z_1, Z_2, \ldots, Z_m, n)$
Output: *To find the solution of fuzzy* MOOM of COTS selection problem for modular software system.
Solve fuzzy MOOM of COTS selection problem for modular software system
$(Z_k \downarrow, X \uparrow)$
  begin
       read: *example*
         while example = *fuzzy* MOOM of COTS selection problem
do
            for k=1 to m do
            enter matrix Z_k

            end
  -| *Find triangular possibilities distribution for cost objective function.*
  -| *Define the crisp multi-objective COTS selection problem according to* α – *level.*
  -| *Convert the maximization problem into minimization problem.*
  -| *determine the positive ideal solution (PIS) and negetive ideal solution (NIS) for each objective.*
            for k=1 to m do
         $Z_k^{\text{PIS}} = \min(Z_k)$
            Under the given constraints,
            end
            for k=1 to m do
         $Z_k^{\text{NIS}} = \max(Z_k)$
            Under the given constraints
            end
         -| *Define exponential membership function for each objective.*
         for k=1 to m do

$$\mu_{Z_k}^E(x) = \begin{cases} 1; & \text{if } Z_k \leq Z_k^L \\ \dfrac{e^{-S\psi_k(x)} - e^{-S}}{1 - e^{-S}}, & \text{if } Z_k^L < Z_k < Z_k^U \\ 0; & \text{if } Z_k \geq Z_k^U \end{cases}$$

         end
         -| *find single objective optimization model under given constraints.*
         for k=1 to m do
         $\max W = prod(Z_1, Z_{21}, Z_{22}, Z_{23})$
         Subject to the constraints: (1) to (7)
         $\mu_{Z_k}(x) - \overline{\mu_{Z_k}}(x) \geq 0; k = 1, 2$
         end
         |- *find the solution SOP using GA.*
       end

If decision-maker accepts the obtained solution, then consider it as the ideal compromise solution and stop solution process; else, change value of shape parameters and confidence level and repeat the above algorithm till a satisfactory solution is achieved for model-1.

5   Problem and Result Analysis

To validate the appropriateness of the proposed approach for MOOM of COTS selection problem for modular software system, the problem has been adopted from the article of the Mukesh and Pankaj [1] with only one change. Here, we have considered cost as a fuzzy number and other data are shown in Table 1.

In Table 1, '–' denotes that to execute the jth modules, the kth COST component cannot be reused. The degrees of functional contribution of COTS components towards the software modules are expressed by the functional rating which ranges 0–1 where 0 is pointed as zero degree of contribution and 1 shows high degree of contribution.

To evaluate the fuzzy MOOM of COTS selection problem, the model is coded. The following aspects of the parameter for solving the problem are as follows: $N = 2$, $M = 6$, $L = 12$, $t = 6$, $H_1 = 0.3$ and $H_2 = 0.3$ as each COTS component is used two times in applications. Table 2 gives the PIS and NIS for each objective function of model-1 for a fixed value of ICD at $\alpha = 0.1, 0.5$ and 0.9. Here, PIS and NIS are used to display a more reliable and simple way which ensures that the preferred solution is closer to the PIS and more distant from the NIS. As a result, a compromise solution can be found, so the closeness coefficient value of each alternative for the PIS and NIS can also be considered, while maintaining the objectivity with respect to the criteria of ups and downs of points.

Here for the uniformity, the maximization objective function is converted into minimization objective function and then the solution is found. The optimal allocation plans for fuzzy MOOM of COTS selection problem are obtained by solving the model-2 with the various shape parameters and aspiration levels which are indicated by the DM for a fixed value of ICD. According to different shape parameters, the efficient solutions of functional requirement and fuzzy cost objective and its optimal allocation plans of model-2 are reported in Table 3 for different values of confidence level α. Each combination of the shape parameters and its corresponding aspiration level is discussed in Table 3 for finding the optimal COTS selection plans. For optimal COTS selection plans, we have used the crossover and mutation rate as 0.2 and 0.3, respectively.

Table 3 gives the optimal allocation plans for fuzzy MOOM of COTS selection problem at $\alpha = 0.1$, $\alpha = 0.5$ and $\alpha = 0.9$ with various shape parameters and its corresponding aspiration level. In this study, optimal COTS selection plans were obtained by taking different shape parameters and aspiration levels at different confidence levels in the exponential membership function by using GA-based hybrid approach. Table 3 indicates efficient solutions of functional requirement and fuzzy

Table 1 Fuzzy cost and functional performance parameters for COTS components

Functional Require-ment	sc	c_k^p	Application-1						Application-2					
			M_1		M_2		M_3		M_4		M_5		M_6	
			c_{1k}^a	f_{1k}	c_{2k}^a	f_{2k}	c_{3k}^a	f_{3k}	c_{4k}^a	f_{4k}	c_{5k}^a	f_{5k}	c_{6k}^a	f_{6k}
R_1	sc_1	(44, 49, 53)	–	0	(17, 20, 23)	0.32	(12, 18, 24)	0	–	0	(11, 17, 23)	0.35	(7, 13, 19)	0.45
	sc_2	(62, 66, 69)	–	0	(15, 19, 24)	0.22	(6, 12, 18)	0.63	–	0	(13, 18, 23)	0.24	(3, 9, 15)	0.65
	sc_3	(50, 54, 58)	–	0	(14, 18, 22)	0.15	(11, 13, 15)	0.72	–	0	(10, 14, 19)	0.13	(6, 11, 16)	0.49
	sc_4	(52, 55, 58)	–	0	(12, 17, 22)	0.23	(8, 11, 13)	0.57	–	0	(10, 15, 19)	0.21	(5, 13, 20)	0.54
R_2	sc_5	(60, 62, 65)	(18, 22, 26)	0.35	(9, 14, 19)	0.94	–	0	(16, 20, 24)	0.15	(6, 11, 15)	0.85	–	0
R_3	sc_6	(57, 61, 65)	–	0	(5, 13, 20)	0.68	(15, 21, 27)	0.45	–	0	(8, 13, 18)	0.7	(17, 20, 23)	0.4
R_4	sc_7	(70, 74, 79)	–	0	–	0	(10, 17, 24)	0.94	–	0	–	0	(14, 18, 22)	0.9
	sc_8	(72, 74, 77)	–	0	–	0	(8, 19, 29)	0.86	–	0	–	0	(9, 14, 19)	0.65
	sc_9	(75, 79, 83)	–	0	–	0	(11, 16, 21)	1	–	0	–	0	(5, 13, 21)	0.97
R_5	sc_{10}	(40, 47, 54)	(9, 12, 15)	0.98	–	0	–	0	(9, 12, 15)	0.9	–	0	–	0
	sc_{11}	(35, 40, 45)	(10, 14, 18)	0.89	–	0	–	0	(13, 16, 19)	0.83	–	0	–	0
R_6	sc_{12}	(45, 49, 54)	(12, 16, 20)	0.75	–	0	–	0	(10, 15, 20)	0.6	–	0	–	0

Table 2 PIS and NIS solutions for functional performance and fuzzy cost objective function at $\alpha = 0.1$, 0.5 and 0.9

α−value	Solutions	Objectives			
		Functional performance	Cost		
			Z_{21}	Z_{22}	Z_{23}
$\alpha = 0.1$	PIS	2.35	430.9	509	583.7
	NIS	3.29	443.6	521	598.4
$\alpha = 0.5$	PIS	2.35	466.5	509	550.5
	NIS	3.29	478	521	564
$\alpha = 0.9$	PIS	2.35	500.7	509	517.3
	NIS	3.29	512.4	521	529.6

cost objective as [9.62, (434.6, 512, 587.6)], [9.58, (436.3, 511, 585.7)], [8.96, (434.3, 509, 583.7)] and [9.58, (469.5, 511, 552.5)] at $\alpha = 0.1$; [9.42, (469, 510, 551)], [9.58, (469.5, 511, 552.5)], [8.96, (467.5, 509, 550.5)] and [9.58, (469.5, 511, 552.5)] at $\alpha = 0.5$; and [9.42, (501.8, 510, 518.2)], [9.58, (502.7, 511, 519.3)], [8.96, (500.7, 509, 517.3)] and [9.58, (502.7, 511, 519.3)] at $\alpha = 0.9$ for (−5, −5), (−1, −3), (−5, 1) and (5, −2) shape parameters and its corresponding (0.8, 0.9), (0.75, 0.85), (0.7, 0.55) and (0.6, 0.7) aspiration levels, respectively.

Here, confidence level is used to reflect the various situations of DM's assurance on fuzzy judgment. Table 3 also indicates that the changes in value of shape parameters, directly the influence of each objective and all obtained solutions, are reliable with the preference of DM. Table 3 shows that the proposed solution approach gives flexibility and the large collection of information in the sense of changing the shape parameters. It also provides the analysis of the different scenarios to DM for allocation strategy [6–8]. If DM is not satisfied with obtaining an assignment plan, other plans can be produced by varying the values of shape parameters and the ICD values. Moreover, the DM can choose any one of the achieved solutions in view of the different criteria according to their own preferences or customer's budgets.

In this study, optimal COTS selection plans were obtained by taking different shape parameters and aspiration levels in the exponential membership function by using the GA-based hybrid approach. Table 3 indicates that for (−5, −5, −5) shape parameter and (0.7, 0.75, 0.85) aspiration level, COTS products for different modules are $m_1 \rightarrow sc_{13}$, $m_2 \rightarrow sc_{21}$, $m_3 \rightarrow sc_{32}$ and $m_4 \rightarrow sc_{43}$.

6 Conclusion

In this paper, GA-based developed hybrid approach with some specific parameter provided the solution of fuzzy multi-objective COTS selection problem using exponential membership function easily and effectively with sensitivity analysis. Several

Table 3 Summary results of different scenarios at $\alpha = 0.1$, 0.5 and 0.9

α	Case	Shape parameter and aspiration level	Degree of satisfaction level	Membership function	Objective values	Optimal allocations					
						M_1	M_2	M_3	M_4	M_5	M_6
0.1	1	$(-5, -5)$ $(0.8, 0.9)$	0.9777	0.9900, $(0.9777, 0.9831, 0.9812)$	9.62 $(434.6, 512, 587.6)$	sc_{10} sc_{12}	sc_5 sc_6	sc_4 sc_9	sc_{10} sc_{12}	sc_5 sc_6	sc_4 sc_9
	2	$(-1, -3)$ $(0.75, 0.85)$	0.8648	0.9550, $(0.8648, 0.9660, 0.9736)$	9.58 $(436.3, 511, 585.7)$	sc_{10} sc_{12}	sc_5 sc_6	sc_3 sc_9	sc_{10} sc_{12}	sc_5 sc_6	sc_3 sc_9
	3	$(-5, 1)$ $(0.7, 0.55)$	0.6285	0.7405, $(0.6285, 1, 1)$	8.96 $(434.3, 509, 583.7)$	sc_{11} sc_{12}	sc_5 sc_6	sc_3 sc_9	sc_{11} sc_{12}	sc_5 sc_6	sc_3 sc_9
	4	$(5, -2)$ $(0.6, 0.7)$	0.6869	0.6869, $(0.7902, 0.9381, 0.9510)$	9.58 $(436.3, 511, 585.7)$	sc_{10} sc_{12}	sc_5 sc_6	sc_3 sc_9	sc_{10} sc_{12}	sc_5 sc_6	sc_3 sc_9
0.5	1	$(-5, -5)$ $(0.8, 0.9)$	0.9837	0.9837, $(0.9867, 0.9965, 0.9986)$	9.42 $(469, 510, 551)$	sc_{11} sc_{12}	sc_5 sc_6	sc_3 sc_9	sc_{11} sc_{12}	sc_5 sc_6	sc_3 sc_9
	2	$(-1, -3)$ $(0.75, 0.85)$	0.9378	0.9550, $(0.9378, 0.9660, 0.9707)$	9.58 $(469.5, 511, 552.5)$	sc_{10} sc_{12}	sc_5 sc_6	sc_3 sc_9	sc_{10} sc_{12}	sc_5 sc_6	sc_3 sc_9
	3	$(-5, 1)$ $(0.7, 0.55)$	0.7405	0.7405, $(0.8682, 1, 1)$	8.96 $(467.5, 509, 550.5)$	sc_{11} sc_{12}	sc_5 sc_6	sc_3 sc_9	sc_{11} sc_{12}	sc_5 sc_6	sc_3 sc_9

(continued)

Table 3 (continued)

α	Case	Shape parameter and aspiration level	Degree of satisfaction level	Membership function	Objective values	Optimal allocations					
						M_1	M_2	M_3	M_4	M_5	M_6
	4	(5, −2) (0.6, 0.7)	0.6869	0.6869 (0.8928, 0.9381, 0.9460)	9.58 (469.5, 511, 552.5)	sc_{10} sc_{12}	sc_5 sc_6	sc_3 sc_9	sc_{10} sc_{12}	sc_5 sc_6	sc_3 sc_9
0.9	1	(−5, −5) (0.8, 0.9)	0.9873	0.9837 (0.9959, 0.9965, 0.9970)	9.42 (501.8, 510, 518.2)	sc_{11} sc_{12}	sc_5 sc_6	sc_3 sc_9	sc_{11} sc_{12}	sc_5 sc_6	sc_3 sc_9
	2	(−1, −3) (0.75, 0.85)	0.9550	0.9550 (0.9649, 0.9660, 0.9671)	9.58 (502.7, 511, 519.3)	sc_{10} sc_{12}	sc_5 sc_6	sc_3 sc_9	sc_{10} sc_{12}	sc_5 sc_6	sc_3 sc_9
	3	(−5, 1) (0.7, 0.55)	0.7405	0.7405 (1, 1, 1)	8.96 (500.7, 509, 517.3)	sc_{11} sc_{12}	sc_5 sc_6	sc_3 sc_8	sc_{11} sc_{12}	sc_5 sc_6	sc_3 sc_8
	4	(5, −2) (0.6, 0.7)	0.6896	0.6896 (0.9362, 0.9381, 0.9398)	9.58 (502.7, 511, 519.3)	sc_{10} sc_{12}	sc_5 sc_6	sc_3 sc_9	sc_{10} sc_{12}	sc_5 sc_6	sc_3 sc_9

choices of the shape parameter with the various combinations of desired aspiration level in the exponential membership functions described different fuzzy utilities of the DM and also described the behaviour in best-fit COTS selection in terms of low coupling and high cohesion which play a significant role to take a decision for DM.

References

1. Gupta, P., Verma, S., Mehlawat, M.K.: Optimization model of COTS selection based on cohesion and coupling for modular software systems under multiple applications environment. In: International Conference on Computational Science and Its Applications, 7335, pp. 87–102. Springer, Berlin, Heidelberg (2012)
2. Jha, P.C. et al.: Optimal component selection of COTS based software system under consensus recovery block scheme incorporating execution time. Int. J. Reliab. Qual. Saf. Eng. **17.03**, 209–222
3. Kwong, C.K. et al.: Optimization of software components selection for component-based software system development. Comput. Ind. Eng. **58.4**, 618–624 (2010)
4. Mehlawat, M. K., Gupta, P.: Multiobjective credibilistic model for COTS products selection of modular software systems under uncertainty. Appl. Intell. **42.2**, 353–368 (2015)
5. Mohamed, A., Ruhe, G., Eberlein, A.: COTS selection: past, present, and future. In: 14th Annual IEEE International Conference and Workshops on the Engineering of Computer-Based Systems (ECBS'07), IEEE, pp. 103–114 (2007)
6. Jayesh, D.M., Tailor, A.R.: Genetic algorithm based hybrid approach to solve uncertain multi-objective COTS selection problem for modular software system. J. Int. Fuzzy Syst. **34.4**, 2103–2120 (2018)
7. Dhodiya, J.M., Tailor, A.R.: Genetic algorithm based hybrid approach to solve fuzzy multi-objective assignment problem using exponential membership function. SpringerPlus **5.1** 2028, 1–29 (2016)
8. Tailor, A. R., Dhodiya, J.M.: Genetic algorithm based hybrid approach to solve optimistic, most-likely and pessimistic scenarios of fuzzy multi-objective assignment problem using exponential membership function. Br. J. Math. Comput. Sci. **17.2**, 1–19 (2016)

Optimization of Parameters for Steel Recycling Process by Using Particle Swarm Optimization Algorithm

S. Allurkar Baswaraj and M. Sreenivasa Rao

Abstract In steel recycling process, a lot of scrap is generated and it is used widely for producing steel. Nearly 40% of the world steel is produced by scrap steel recycling process. The main issue in recycling process is producing a quality steel out of scrap with minimum energy consumption. The parameters which influence the recycling process are furnace temperature, sponge steel addition percentage, scrap steel composition, TDS number of water used, and quenching temperature of steel. In this paper, an optimization model for maximizing tensile strength and hardness number value as objective function and energy consumption rate as a constraint is obtained by using response surface methodology. The detailed optimization model is solved by using state-of-the-art optimization technique called particle swarm optimization algorithm. Optimization values obtained are evaluated using experimentally and compared with other optimization techniques like Grey Taguchi. The optimum results obtained by particle swarm optimization (PSO) method outperform other techniques.

Keywords Input parameters · Steel recycling · Tensile strength · Swarm

1 Introduction

It is observed that there is a large variation in tensile strength and hardness number values for steel rod manufactured by different steel recycling industries based on test results of specimen. Based on the literature and consultation with experts from secondary steel manufacturing industries, five significant affecting input process parameters were identified with their levels. They are furnace temperature, amount of sponge steel addition, quality of scrap steel, TDS number of water used for cool-

S. Allurkar Baswaraj (✉) · M. Sreenivasa Rao
JNTU College of Engineering, Hyderabad, Telangana State, India
e-mail: allurkar@gmail.com

M. Sreenivasa Rao
e-mail: raoms@yahoo.com

© Springer Nature Singapore Pte Ltd. 2020
R. Venkata Rao and J. Taler (eds.), *Advanced Engineering Optimization Through Intelligent Techniques*, Advances in Intelligent Systems and Computing 949,
https://doi.org/10.1007/978-981-13-8196-6_9

Table 1 Input variables at three levels

Input parameter	Level 1	Level 2	Level 3
Ft (°C)	1650	1675	1700
SS (%)	10	12.5	15
SCS (%)	75	80	85
TDS (No.)	30	35	40
Qw (°C)	500	525	550

ing, and quenching temperature of steel as shown in Table 1. For analysis steel rod specimen of diameter 16 mm and 2 ft length was used on universal testing machine and for hardness number value steel rod specimen of 1 cm length with its flat face polished with emery paper of different grades was used on Brinell hardness testing machine. The process parameters involved are interdependent, and the entire process of secondary steel making is complex [1]. Only main influencing parameters are considered for analysis starting scrap sorting to storing of final steel bars at the store yard. Investigation of steel recycling process is observed, weaknesses are identified, and the same is incorporated in the process. The prime aim is to enhance tensile strength along with hardness number value with less consumption of energy. L46 orthogonal array is chosen, experiments are conducted, and the results for performance measure are noted for each test run. For the experimental results, particle swarm optimization algorithm is applied.

Kennedy and Eberhart developed the above algorithm before three decades. It is an advanced technique applied for solving optimization problems in manufacturing industries with less computations [3]. The main feature of this algorithm is to get the best solution for multiresponse objective function and reaches optimal solution with less computations as compared to other methods in this domain. It conducts a search using a population, and the results are improved with every iteration. Each particle in the swarm is indicated by its position and velocity. This helps in determining the best position of a particle [2].

2 Model Formulation

The optimization model for the given problem can be formulated as shown below.

$$\text{Find } X \begin{bmatrix} \text{Ft} \\ \text{SS} \\ \text{SCS} \\ \text{TDS} \\ \text{QW} \end{bmatrix} \tag{1}$$

$$\begin{aligned}
\textbf{Max.T.S} = {} & -12105 + 13.37\,\text{Ft} + 82.0\,\text{SS} - 3.2\,\text{SCS} + 8.7\,\text{TDS} + 1.72\,\text{Qw} \\
& - 0.00377\,\text{Ft} * \text{Ft} - 0.017\,\text{SS} * \text{SS} - 0.0042\,\text{SCS} * \text{SCS} \\
& - 0.0142\,\text{TDS} * \text{TDS} - 0.00137\,\text{Qw} * \text{Qw} - 0.0400\,\text{Ft} * \text{SS} \\
& + 0.0060\,\text{Ft} * \text{SCS} - 0.0020\,\text{Ft} * \text{TDS} + 0.00000\,\text{Ft} * \text{Qw} \\
& - 0.280\,\text{SS} * \text{SCS} + 0.100\,\text{SS} * \text{TDS} + 0.0160\,\text{SS} * \text{Qw} \\
& - 0.0200\,\text{SCS} * \text{TDS} - 0.0040\,\text{SCS} * \text{Qw} - 0.0080\,\text{TDS} * \text{Qw} \geq 580. \quad (2)
\end{aligned}$$

Such that

$$\begin{aligned}
\text{Hs} = {} & -1364 + 1.516\,\text{Ft} + 5.42\,\text{SS} - 0.07\,\text{SCS} + 0.66\,\text{TDS} + 0.271\,\text{Qw} \\
& - 0.000397\,\text{Ft} * \text{Ft} - 0.0370\,\text{SS} * \text{SS} - 0.00258\,\text{SCS} * \text{SCS} - 0.00692\,\text{TDS} * \text{TDS} \\
& - 0.000237\,\text{Qw} * \text{Qw} - 0.00280\,\text{Ft} * \text{SS} + 0.00000\,\text{Ft} * \text{SCS} + 0.00080\,\text{Ft} * \text{TDS} \\
& - 0.000160\,\text{Ft} * \text{Qw} - 0.0120\,\text{SS} * \text{SCS} - 0.0020\,\text{SS} * \text{TDS} + 0.00280\,\text{SS} * \text{Qw} \\
& - 0.02000\,\text{SCS} * \text{TDS} + 0.00260\,\text{SCS} * \text{Qw} + 0.00020\,\text{TDS} * \text{Qw}. \quad (3)
\end{aligned}$$

$$\begin{aligned}
\text{En.con.} = {} & 4397 - 4.91\,\text{Ft} + 9.5\,\text{SS} + 2.34\,\text{SCS} - 0.44\,\text{TDS} - 1.08\,\text{Qw} \\
& + 0.001510\,\text{Ft} * \text{Ft} + 0.0923\,\text{SS} * \text{SS} + 0.0214\,\text{SCS} * \text{SCS} + 0.0118\,\text{TDS} * \text{TDS} \\
& + 0.000977\,\text{Qw} * \text{Qw} - 0.00920\,\text{Ft} * \text{SS} - 0.00080\,\text{Ft} * \text{SCS} + 0.00240\,\text{Ft} * \text{TDS} \\
& + 0.000400\,\text{Ft} * \text{Qw} + 0.0100\,\text{SS} * \text{SCS} + 0.0200\,\text{SS} * \text{TDS} + 0.00280\,\text{SS} * \text{Qw} \\
& - 0.0320\,\text{SCS} * \text{TDS} - 0.00660\,\text{SCS} * \text{Qw} - 0.00380\,\text{TDS} * \text{Qw}. \quad (4)
\end{aligned}$$

Hardness number value must be high, energy consumption rate should be low, and corresponding regressions Eqs. (3) and (4) were used.

3 Algorithm for Particle Swarm Optimization with Sample Calculation

The main objective was to maximize tensile strength and hardness number with minimum energy utilization. The algorithm flow is as shown below:

Input:
Number of particles in population (population size) = 5,
Number of iterations = 1.
Step 1. Generate initial population.

$$\text{Initial population} = \begin{bmatrix} 1675 & 15 & 80 & 40 & 525 \\ 1675 & 12.5 & 75 & 30 & 500 \\ 1700 & 15 & 80 & 30 & 525 \\ 1675 & 12.5 & 80 & 35 & 550 \\ 1675 & 15 & 85 & 35 & 500 \end{bmatrix}$$

Step 2. Find fitness values (tensile strength) of all particles.

$$
\text{Tensile Strength} = \begin{bmatrix} 621.5125 \\ 612.7325 \\ 632.6088 \\ 607.6875 \\ 618.9788 \end{bmatrix}
$$

Step 3. Generate initial velocity matrix randomly which has size same as population.

$$
\text{Initial Vel.} = \begin{bmatrix} 3.2261 & 0.8329 & 3.4097 & -3.0644 & -1.2736 \\ -3.7062 & 0.3070 & 1.8979 & -2.5502 & -0.5900 \\ -3.2147 & -1.5742 & 2.2410 & 0.2957 & 2.1526 \\ 1.1115 & 3.1452 & -3.5145 & -2.5938 & -0.6693 \\ 1.9183 & 3.1436 & -3.7932 & -2.8995 & -0.6073 \end{bmatrix}
$$

Step 4. Assign the particle best and global best. Assign each particle to its Pbest. Assign particle with maximum throughput to Gbest.

$$
\text{Particle best} = \begin{bmatrix} 1675 & 15 & 80 & 40 & 525 \\ 1675 & 12.5 & 75 & 30 & 500 \\ 1700 & 15 & 80 & 30 & 525 \\ 1675 & 12.5 & 80 & 35 & 550 \\ 1675 & 15 & 85 & 35 & 500 \end{bmatrix},
$$

$$
\text{Fitness} = \begin{bmatrix} 621.5125 \\ 612.7325 \\ 632.6088 \\ 607.6875 \\ 618.9788 \end{bmatrix},
$$

$$
\text{Global best} = \begin{bmatrix} 1700 & 15 & 80 & 30 & 525 \end{bmatrix},
$$

Fitness of Global best = 632.6088

Step 5. Calculate new velocity for each particle i by using Eq. (3)

$$
\begin{aligned}
V_{i\,\text{new}} = {} & \omega * V_i + C_p * \text{rand}(0, 1) * (\text{Pbest} - X_i) \\
& + C_g * \text{rand}(0, 1) * (\text{Gbest} - X_i)
\end{aligned} \tag{3}
$$

$$
\begin{bmatrix} 3.1910 & 0.1666 & 0.6819 & -4.0000 & -0.2547 \\ 2.0636 & 3.4877 & 3.6969 & -0.5100 & 4.0000 \\ -0.6429 & -0.3148 & 0.4482 & 0.0591 & 0.4305 \\ 4.0000 & 4.0000 & -0.7029 & -4.0000 & -4.0000 \\ 4.0000 & 0.6287 & -4.0000 & -4.0000 & 4.0000 \end{bmatrix}
$$

Step 6. Prepare new population as follows.

$$
\text{New vel.} =
\begin{aligned}
&\left.\begin{array}{l}
\text{if } V_{i\,\text{new}}(\text{Furnace\_temp}) > 0 \quad \text{then Furnace\_temp} \leftarrow \text{Furnace\_temp} + 25 \\
\text{if } V_{i\,\text{new}}(\text{Furnace\_temp}) > 0 \quad \text{then Furnace\_temp} \leftarrow \text{Furnace\_temp} - 25
\end{array}\right\} \forall\,\text{particles} \\
&\left.\begin{array}{l}
\text{if } V_{i\,\text{new}}(\text{Sponge\_steel}) < 0 \quad \text{then Sponge\_steel} \leftarrow \text{Sponge\_steel} + 2.5 \\
\text{if } V_{i\,\text{new}}(\text{Sponge\_steel}) > 0 \quad \text{then Sponge\_steel} \leftarrow \text{Sponge\_steel} - 2.5
\end{array}\right\} \forall\,\text{particles} \\
&\left.\begin{array}{l}
\text{if } V_{i\,\text{new}}(\text{Scrap\_steel}) < 0 \quad \text{then Scrap\_steel} \leftarrow \text{Scrap\_steel} + 5 \\
\text{if } V_{i\,\text{new}}(\text{Scrap\_steel}) > 0 \quad \text{then Scrap\_steel} \leftarrow \text{Scrap\_steel} - 5
\end{array}\right\} \forall\,\text{particles} \\
&\left.\begin{array}{l}
\text{if } V_{i\,\text{new}}(\text{TDS\_of\_water}) < 0 \quad \text{then TDS\_of\_water} \leftarrow \text{TDS\_of\_water} + 5 \\
\text{if } V_{i\,\text{new}}(\text{TDS\_of\_water}) > 0 \quad \text{then TDS\_of\_water} \leftarrow \text{TDS\_of\_water} - 5
\end{array}\right\} \forall\,\text{particles} \\
&\left.\begin{array}{l}
\text{if } V_{i\,\text{new}}(\text{Quench\_temp\_steel}) < 0 \quad \text{then Quench\_temp\_steel} \leftarrow \text{Quench\_temp\_steel} + 25 \\
\text{if } V_{i\,\text{new}}(\text{Quench\_temp\_steel}) > 0 \quad \text{then Quench\_temp\_steel} \leftarrow \text{Quench\_temp\_steel} - 25
\end{array}\right\} \forall\,\text{particles}
\end{aligned}
$$

$$
\text{New population} =
\begin{bmatrix}
1675 & 15 & 80 & 40 & 525 \\
1675 & 15 & 80 & 30 & 525 \\
1675 & 12.5 & 85 & 35 & 550 \\
1700 & 15 & 75 & 30 & 525 \\
1700 & 15 & 80 & 30 & 525
\end{bmatrix},
$$

Make adjustments to fulfil the necessary conditions. If new tensile strength value is better than earlier, replace earlier tensile strength with new one. Evaluate new population.

$$
\text{Fitness} =
\begin{bmatrix}
621.5125 \\
632.6088 \\
606.4725 \\
635.3638 \\
632.6088
\end{bmatrix},
$$

Step 7. Find new particle best and global best.

Step 8. Upgrade particle best and global best. If new particle is better than earlier, replace particle best with new particle. Find the best particle from new population. If it is better than Gbest, replace with the best particle of new population.

$$
\text{Updated Fitness} =
\begin{bmatrix}
621.5125 \\
632.6088 \\
632.6088 \\
635.3638 \\
632.6088
\end{bmatrix}
$$

Updated global best $= \begin{bmatrix} 1700 & 15 & 75 & 30 & 525 \end{bmatrix}$,

Maximum tensile strength $= 635.36$ N/mm^2, Hardness number value $= 104.38$, Energy consumption/ton $= 366$ kWh.

Step 9. Assign new velocity to old velocity.

Step 10. If the termination condition is satisfied, go to step eleven; otherwise, go to step five.

Step 11. It is the maximum tensile strength.

4   Results

It was observed that tensile strength increases in first few iterations and then reaches
highest value and remains steady for further computations as seen in above figure.
The algorithm converges, and at that point, optimal input parameters for furnace
temperature were 1700 °C, sponge steel addition of 15%, scrap steel composition of
75%, TDS number of water used at 35, and quenching temperature of steel was at
500 °C.

The optimal input variables obtained by using PSO algorithm are
$\begin{bmatrix} 1700 \ 15 \ 75 \ 35 \ 500 \end{bmatrix}$ (Fig. 1).

4.1   Experimental Verification

The optimal input parameters $\begin{bmatrix} 1700 \ 15 \ 75 \ 35 \ 500 \end{bmatrix}$ obtained by applying particle
swarm optimization algorithm are considered for experimentation. The optimized
input parameters obtained were set, and the experiment was conducted for a steel
rod of 16 mm diameter. A sample from the above lot is picked and tested for tensile
strength, hardness number, and energy consumption rate. The results of experiments
are as shown below, and the comparison of results is shown in Table 2.

Tensile strength = 659.382 N/mm^2

Hardness number value = 104.96

Energy consumption/ton of steel = 365.19 kWh

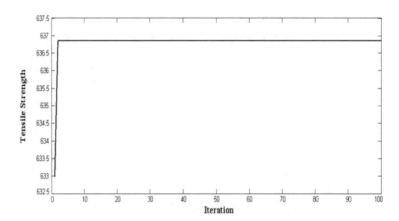

Fig. 1 Tensile strength variation with iterations

Table 2 Comparison of Grey Taguchi results with PSO results

Optimal input parameters	Output results obtained by Grey Taguchi method	Experimental results obtained by PSO output values
Grey Taguchi	$Ts = 645$ N/mm^2	$Ts = 652.382$ N/mm^2
$\begin{bmatrix} 1700 & 12.5 & 80 & 35 & 500 \end{bmatrix}$	$Hs = 107$ BN	$Hs = 105$ BN
	En.con. $= 367.5$ kWh	En.con. $= 365.372$ kWh
PSO $\begin{bmatrix} 1700 & 15 & 75 & 35 & 500 \end{bmatrix}$		

5 Conclusion

Tensile strength and hardness number of steel rods utilized in civil construction works play a key role for assuring safety of the structure. Input variables are related to a complex way in steel recycling process. By using the regression equation, the optimization model was framed and it was solved by applying particle swarm optimization (PSO) algorithm. The best combination of input variables was found out with corresponding performance measures. The optimum input parameter levels obtained by the above method were analysed by conducting the experiment in a secondary steel recycling industry. The output responses in tensile strength and energy consumption rate show significant improvement, whereas the hardness number value remains nearly the same.

References

1. Khoeai, A.R.: Design optimization of steel recycling processes using Taguchi method. J. Mater. Process. Technol. **127**(1), 97–106 (2007)
2. Patel, V.K., Rao, R.V.:Design optimization of shell and tube heat exchanger by PSO method. Appl. Thermal Eng. **30**(11–12), 1417–1425 (2010)
3. Anil, G., Singh,H., Aggarwal, A.: Taguchi method for multi output optimization high speed CNC turning of AISI steel. Expert Syst. Appl. **38**(6), 6823–6827 (2011)
4. Haupt, M., Vadenbo, C.: Influence of input-scrap quality on the environmental impact for secondary steel. J. Ind. Ecol **21**(2), 391–401 (2017)
5. Feng, M., Yi, X.: Grouping particle swarm optimization algorithm for job shop problem with constraints.In: 2008 IEEE Pacific-Asia Workshop on Computational Intelligence and Application, pp. 332–335
6. Liu, Z.: Investigation of particle swarm optimization for job shop scheduling problem. In: Third International Conference on Natural Computation (2007)
7. Kennedi, J., Eberhart, R.: Particle swarm optimization. In: Proceedings of IEEE International Conference on Neural Networks, pp. 1941–1949 (1995)
8. Zhu, H., Ye, W.: A particle Swarm optimization for integrated process planning and scheduling. In: IEEE Proceedings, pp. 1070–1074 (2009)
9. Xu, X.-H., Zeng, L.-L.: Hybrid particle Swarm optimization for flexible job shop scheduling problem and its implementation. In: Proceedings of 2010 IEEE International Conference on Information and Automation, pp. 1155–1159

Modeling 3D WSN to Maximize Coverage Using Harmony Search Scheme

Deepika Sharma and Vrinda Gupta

Abstract This paper develops a 3D WSN (three-dimensional wireless sensor network) to solve point coverage problem at minimum cost. Coverage in WSN is crucial as it measures the quality of monitoring. This requires finding of optimal number of sensors (cost of the network) at optimal locations to cover the given 3D space. Here, Harmony search scheme has been employed for maximizing coverage and reducing the network cost. The Network coverage ratio, the minimum distance between two sensors and the number of sensors are the parameters that have been incorporated in the objective function. Simulation results have proved the capability of proposed scheme by providing maximum network coverage by placing lesser number of sensor nodes at optimal locations. In addition, incorporation of minimum distance parameter in objective function has reduced overlapping of sensor nodes and reduced the cost of the network by 25%. Furthermore, scalability of the proposed scheme has also been investigated. Also, significant performance enhancement has been observed in terms of network lifetime when the performance of proposed scheme is compared with random-based deployment scheme using Low-Energy Adaptive Clustering Hierarchy (LEACH) protocol.

Keywords 3D wireless sensor network · Harmony search algorithm · Coverage ratio · Network cost and node placement

1 Introduction

Wireless sensor networks (WSNs) comprise of large number of small, low-powered, multifunctional sensor nodes which communicate with each other through a wireless interface over a small distance [1], and they perform various surveillance and

D. Sharma (✉) · V. Gupta
Department of Electronics and Communication Engineering, National Institute of Technology, Kurukshetra, Kurukshetra, India
e-mail: deepikasharmavatss@gmail.com

V. Gupta
e-mail: vrindag16@gmail.com

© Springer Nature Singapore Pte Ltd. 2020
R. Venkata Rao and J. Taler (eds.), *Advanced Engineering Optimization Through Intelligent Techniques*, Advances in Intelligent Systems and Computing 949, https://doi.org/10.1007/978-981-13-8196-6_10

monitoring tasks [2]. For studying real-life applications, a 3D (three-dimensional) model of WSN is useful. Some of the applications of 3D WSNs are during the natural calamities; 3D WSN coverage is required [3], for example, in space exploration [4], to monitor volcano eruption [5], undersea monitoring [6], for detecting targets in the battle field [7], building aerial defense system, etc. These 3D applications face the problem of high density of sensors in region of interest (ROI) resulting in increase in the cost of the network in terms of manufacturing, deploying and maintenance cost. In addition, low network coverage appears which results in more un-sensed space in ROI [8]. Therefore, an attempt has been made in this paper to minimize these problems.

Cao et al. [9] consider directional WSN (DWSN) in which node deployment optimization issue in 3D terrain is solved using a modified differential evolution (DE) algorithm. To reduce computation time, message passing interface (MPI) parallelism has been implemented. A cost function has been defined [10], which maximizes the network coverage with a minimal number of sensors placed in the 2D WSN using HS (Harmony search)-based scheme. Simulation results have shown that HS algorithm performs better than genetic algorithm and random deployment technique. To form a connected network along with maximized coverage and reduced number of sensors, an objective function is defined with the highest value being calculated using HS algorithm [11]. Temel et al. [12] have used wavelet transformation to find the coverage cavities before the deployment of sensor nodes so that coverage is enhanced along with a minimal number of the sensors and called it as Cat Swarm Optimization with Wavelet transformation (CSO-WT).

Harmony search (HS) algorithm is a metaheuristic approach and has been applied in a number of optimization problems [13, 14]. The basic idea of HS algorithm is that when an instrumentalist (decision variable) plays musical notes, they generate every possible combination (new solution) which are stored in the memory, for obtaining most pleasing musical note (optimal result). In this, the optimum solution of the objective function is determined through the iterations just as the pleasing note is found by the music creator through practice [13, 14].

The proposed scheme is the 3D extension of the work done by Alia and Al-Ajouri [10]. It aims at solving point coverage problem in 3D WSN, which is defined as every point of interest in the given space to be under the coverage range of at least one sensor node [15]. Network coverage has been used as the main performance metric for measuring the space covered by the deployed network [2]. Deterministic deployment technique has been employed in the proposed scheme [16]. Here, harmony search (HS) scheme is utilized to solve the above-mentioned coverage problem in 3D WSN.

This paper provides the following findings:

- An optimal number of sensor nodes and their optimal locations in the deployed network.
- Better network coverage is obtained.
- Reduced network cost due to the inclusion of performance metrics viz. minimum distance and number of nodes in the objective function.
- Scalability of the proposed scheme is tested.

- Network lifetime is improved in terms of first dead node, first 10% dead nodes and 50% dead nodes.

The remainder of the paper is organized as follows: Problem formulation and proposed approach are explained in detail in Sect. 2. Simulation results are shown in Sect. 3. Section 4 gives the conclusion of the paper and discusses the future work.

2 Problem Formulation and Proposed HS-Based Scheme

The proposed scheme finds an optimal number of sensors and places them in 3D volume thus giving enhanced network coverage by employing HS algorithm. This paper considers ROI in 3D space i.e., length (L), width (W) and height (H) and an obstacle-free space. The ROI is divided into $l \times w \times h$ blocks of equal size as is done in multi-objective territorial predator scent marking algorithm (MOTPSMA) [17] and Monte Carlo GA (MGA) [18]. Center of each block is represented by x, y, z-coordinates and is named as demand point where the entire weight of the cell is concentrated and is shown in Fig. 1. A binary sensing model is used as a sensor coverage model [16], in which a point is said to be under coverage range if the distance between the point and sensor node $(\mathrm{dis}(s, n))$ is less than the sensing range (R_{sens}). The coverage function is given by the following function as shown in Eq. (1)

$$F_{\mathrm{cov}}(\mathrm{dis}(s, n)) = \begin{cases} 1, & \text{if } \mathrm{dis}(s, n) \leq R_{\mathrm{sens}} \\ 0, & \text{otherwise} \end{cases} \tag{1}$$

The performance of metaheuristic algorithms is based on two important components i.e., diversification (exploration) and intensification (exploitation). HS scheme

Fig. 1 Schematic of the region of interest in 3D space

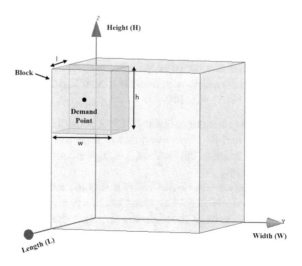

has the advantage that there is a proper balance between intensification and diversification in it; diversification is provided by pitch adjustment rate (*PAR*) and bandwidth (*bw*), whereas intensification is provided by the harmony memory consideration rate (*HMCR*) [14]. HS-based scheme used is discussed in the following subsections [10, 13, 14] and the steps of execution are shown in Fig. 2.

2.1 Define an Objective Function ($O_{HSA}(n)$) and Initialize HS Parameters

An objective function ($O_{HSA}(n)$) is defined initially, the value of which is to be maximized to solve the optimization problem. An objective function is given by the ratio of the product of the network coverage ratio (Cov_{ratio}), minimum distance (D_{min}) and constant k to the number of sensors (S) in the given network [9].

$$O_{HSA}(n) = \frac{Cov_{ratio} \times D_{min} \times k}{S} \tag{2}$$

Here, coverage ratio (Cov_{ratio}) is defined as the ratio of the number of demand points covered by the deployed sensor nodes (P_{cov}) to the total number of the demand points (P_{total}) [10]. The significance of incorporating coverage ratio in the objective function is that it provides the amount of coverage provided by the sensor nodes placed in the network for different solution vector. It is calculated using Eq. (3).

$$Cov_{ratio} = \frac{P_{cov}}{P_{total}} \tag{3}$$

Moreover, the minimum distance (D_{min}) is the distance between two sensors present in the given network. To avoid wastage of energy due to overlapping between the sensors coverage range, this factor is included. Here, k is a random number and $k \in (0, 1)$, and it is used to avoid unwanted convergence of $O_{HSA}(n)$ to zero [9]. In addition to this, number of sensors (S) is considered in the objective function so that the optimal number of sensors for providing maximum coverage in the given 3D WSN is determined with minimum cost. Various HS parameters used are defined as follows [10, 11, 13, 14]:

- Harmony memory (*HM*): *HM* is a performance matrix which keeps the solution vector and its respective value of objective function. Solution vector is a vector which contains the location (i.e., x, y, z-coordinates) of the sensor nodes placed in the network.
- Harmony memory size (*HMS*): It is the total number of solution vectors which can be stored in *HM*. Optimal solution is given by solution vector whose objective function value is the highest after improvisation.
- Harmony memory vector length (*HMVL*): It is the number of the sensors in each solution vector.

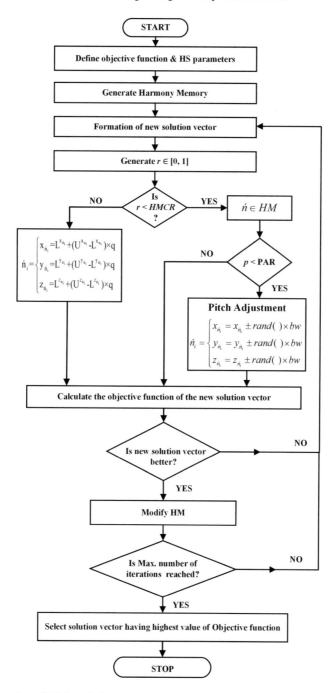

Fig. 2 Flow chart of HS-based scheme

- Number of improvement (*NI*): It is the maximum number of improvisation steps, in which new solution vector is generated for improvisation in *HM*.
- Harmony memory consider rate (*HMCR*): It is the probability of choosing the element from the *HM* for new solution vector, here $HMCR \in (0, 1]$.
- Pitch adjustment rate (*PAR*): This is a probability employed to find the need of location adjustment for the elements stored in new solution vector, here $PAR \in (0, 1]$ and is performed using a constant known as bandwidth (*bw*).
- Bandwidth (*bw*): It is a parameter used to perform pitch adjustment.

2.2 Generate Harmony Memory

In this step, HM is generated which keeps the solution vector and its respective value of objective function and is shown in Eq. (4). Solution vectors generated should be in between the upper limit (U^{n_i}) and lower limit (L^{n_i}).

$$
HM = \begin{bmatrix} n_1^1 & \cdots & n_N^1 & \bigg| & O_{HSA}(n^1) \\ \vdots & \ddots & \vdots & \bigg| & \vdots \\ n_1^{HMS} & \cdots & n_N^{HMS} & \bigg| & O_{HSA}(n^{HMS}) \end{bmatrix} \tag{4}
$$

In Eq. (4), *i*th sensor node is represented by n^i and $O_{HSA}(n^i)$ is an objective function of *i*th solution vector. If sensors are deployed at the boundary of ROI, sensing energy is wasted due to sensing of the unwanted region. Hence, to avoid this wastage, lower and upper limits are defined, which are calculated using the following equations:

$$
U^{x_n} = L - (R_{sens} - R_{uc}) \tag{5}
$$

$$
L^{x_n} = R_{sens} - R_{uc} \tag{6}
$$

$$
U^{y_n} = W - (R_{sens} - R_{uc}) \tag{7}
$$

$$
L^{y_n} = R_{sens} - R_{uc} \tag{8}
$$

$$
U^{z_n} = H - (R_{sens} - R_{uc}) \tag{9}
$$

$$
L^{z_n} = R_{sens} - R_{uc} \tag{10}
$$

Here, R_{uc} is the uncertainty range of the deployed sensor node, and locations of sensors in *ROI* are calculated using the following equations:

$$n_i = \begin{cases} x_{n_i} = L^{x_{n_i}} + (U^{x_{n_i}} - L^{x_{n_i}}) \times \text{rand}() \\ y_{n_i} = L^{y_{n_i}} + (U^{y_{n_i}} - L^{y_{n_i}}) \times \text{rand}() \\ z_{n_i} = L^{z_{n_i}} + (U^{z_{n_i}} - L^{z_{n_i}}) \times \text{rand}() \end{cases} \tag{11}$$

A maximum number of sensor nodes ($MaxNo$) and a minimum number of sensor nodes ($MinNo$) are defined by applicant to obtain the optimal number of sensors, and harmony memory vector length ($HMVL$) gives number of the sensors in each solution vector. It is calculated using Eq. (12), and each solution vector has different vector length [10].

$$HMVL = MinNo + (MaxNo - MinNo) \times rand() \tag{12}$$

In the above equation, $rand()$ is a variable whose value lies in the range 0–1.

2.3 Improvisation

After formation of HM, a new solution vector ($\acute{n}$) is computed for improvement in HM and is shown in Eq. (13).

$$\acute{n} = \left(x_{\acute{n}_1}, y_{\acute{n}_1}, z_{\acute{n}_1} x_{\acute{n}_2}, y_{\acute{n}_2}, z_{\acute{n}_2} \ldots, x_{\acute{n}_{\text{MaxN}}}, y_{\acute{n}_{\text{MaxN}}}, z_{\acute{n}_{\text{MaxN}}} \right) \tag{13}$$

A random $r \in [0, 1]$ is generated such that if r is less than $HMCR$, then the location of an element of new solution vector belongs to the HM, else location of an element of new solution vector will be selected from limits defined. The above-mentioned process is explained by the following equation:

$$\acute{n}_i = \begin{cases} \acute{n} \in \left\{ x_{n_i^1}, y_{n_i^1}, z_{n_i^1} x_{n_i^2}, y_{n_i^2}, z_{n_i^2} \ldots, x_{n_i^{\text{HMS}}}, y_{n_i^{\text{HMS}}}, z_{n_i^{\text{HMS}}} \right\} & \text{HMCR} \\ x_{\acute{n}} \in [L^{x_n}, U^{x_n}]; \; y_{\acute{n}} \in [L^{y_n}, U^{y_n}]; \; z_{\acute{n}} \in [L^{z_n}, U^{z_n}] & (1 - \text{HMCR}) \end{cases} \tag{14}$$

If the random number r is greater than $HMCR$, then elements of the new solution vector are generated using Eq. (15).

$$\acute{n}_i = \begin{cases} x_{\acute{n}_i} = L^{x_{\acute{n}_i}} + (U^{x_{\acute{n}_i}} - L^{x_{\acute{n}_i}}) \times \text{rand}() \\ y_{\acute{n}_i} = L^{y_{\acute{n}_i}} + (U^{y_{\acute{n}_i}} - L^{y_{\acute{n}_i}}) \times \text{rand}() \\ z_{\acute{n}_i} = L^{z_{\acute{n}_i}} + (U^{z_{\acute{n}_i}} - L^{z_{\acute{n}_i}}) \times \text{rand}() \end{cases} \tag{15}$$

The element for new solution vector selected from HM is examined for pitch adjustment. A random number, $p \in [0, 1]$ generated. If p is within the probability range of PAR, then pitch adjustment (i.e., adjustment in location) of the element of new solution vector is done. The pitch adjustment process is done using bandwidth (bw) which displaces the sensor node from the given position so that more space is explored using the following equation:

$$\acute{n}_i = \begin{cases} x_{\acute{n}_i} = x_{\acute{n}_i} \pm \text{rand}() \times \text{bw} \\ y_{\acute{n}_i} = y_{\acute{n}_i} \pm \text{rand}() \times \text{bw} \\ z_{\acute{n}_i} = z_{\acute{n}_i} \pm \text{rand}() \times \text{bw} \end{cases} \tag{16}$$

2.4 Modification in Harmony Memory

In this step, new solution vector objective function $(O_{\text{HSA}}(\acute{n}))$ is calculated and compared with the lowest objective function stored in *HM*. If the value of $O_{\text{HSA}}(\acute{n})$ is greater than the selected objective function from *HM*, it is replaced with the new solution vector, else the next step is executed.

2.5 Check for Maximum Number of Iterations

In this step, the number of improvements required is checked. If the number of iterations (*NI*) reaches to its highest value, then the process of improvement is terminated and the solution vector having the highest value of objective function stored in the *HM* is checked, giving the optimal solution to the problem. Otherwise, go to step C.
    Pseudo code for above-mentioned approach is shown in Fig. 3.

3 Simulation and Results

The proposed scheme is simulated using MATLAB 2013a and its results are discussed in the following parts.

3.1 Network Settings and Results

A sensor node coverage area is modeled as a sphere. Sensor nodes are static and homogenous in nature (i.e., all sensor nodes have equal sensing range, $R_{\text{sens}} = 10\ m$ and uncertainty range, $R_{\text{uc}} = 5\ m$). The 3D space is an obstacle-free space. The dimensions of the 3D space and other values of the HS parameters used are shown in Table 1 [10]. A total 50 successful runs of proposed scheme are conducted.
    Initially, locations for sensor nodes are generated randomly and are stored in *HM*, and after this improvisation is performed as shown in Fig. 2. Simulation results obtained are shown in Table 2 considering two cases. In Case I, result of highest coverage ratio is given and in Case II, result of highest coverage with the least numeral of sensors is given which are taken from 50 observations. It has been observed that

```
1.  Begin
2.  Define objective function O_HSA(n)
3.  Define HMCR, PAR, HMS, NI // Harmony Search parameters
4.  Define L, W, H, Cell size, R_sens, R_un // dimensions of the
    given space
5.  Generate Harmony Memory randomly
6.  Calculate objective for each solution vector stored in
    Harmony Memory
7.  While iter ≤ Max. number of iterations do
        {improvise a new solution}
8.  Calculate length for new solution vector
    N = MinNo + (MaxNo - MinNo) × rand()
9.      if (N ≤ N_max) then
10.         if (N ≥ N_min) then
11.             while (i ≤ N) do // Here, i ∈ [1, N]
12.                 if (r < HMCR) then // Checking for element
    of new solution vector belongs to HM
                        [row, column] = size(HM)
                        R = randi(1, row), C = randi(1, column)
13. x_n̂_i = X(R,C), y_n̂_i = Y(R,C), z_n̂_i = Z(R,C)
14. if (p < PAR) then // checking for pitch adjustment
```

$$\acute{n}_{in} = \begin{cases} x_{\hat{n}_i} = x_{\hat{n}_i} \pm rand() \times bw \\ y_{\hat{n}_i} = y_{,} \pm rand() \times bw \\ z_{\hat{n}_i} = z_{\hat{n}_i} \pm rand() \times bw \end{cases}$$

```
15. end if
16.                     else
```

$$\acute{n}_{in} = \begin{cases} x_{\hat{n}_i} = x_{\hat{n}_i} \pm rand() \times bw \\ y_{\hat{n}_i} = y_{,} \pm rand() \times bw \\ z_{\hat{n}_i} = z_{\hat{n}_i} \pm rand() \times bw \end{cases}$$

```
17.                     end if
18.                 end while
19.             end if
20.         end if
21. find the worst objective function solution vector stored
    in HM
22. replace the worst solution vector with new solution vector
    if better than worst
23. end while
24. Best= find the solution vector having maximum value of the
    solution vector
```

Fig. 3 Pseudo code of HS-based scheme

Table 1 Dimensions and parameters used

S. No.	Parameter	Value	S. No.	Parameter	Value
1	Monitored area size ($L \times W \times H$)	$50 \times 50 \times 50$ m^3	9	U^{x_n}	$45\ m$
2	Cell size ($l \times w \times h$)	$10 \times 10 \times 10$ m^3	10	L^{x_n}	$5\ m$
3	Demand points	125	11	U^{y_n}	$45\ m$
4	HMS	30	12	L^{y_n}	$5\ m$
5	NI	100	13	U^{z_n}	$45\ m$
6	HMCR	0.9	14	L^{z_n}	$5\ m$
7	PAR	0.3	15	MaxNo	150
8	bw	0.2	16	MinNo	50

Table 2 Simulation results of the proposed scheme

Performance metrics	Case I	Case II	Average value	SD
Coverage ratio (Cov_{ratio}) (%)	92	83.2	80.70	3.25
Number of sensors (S)	70	51	58	7.5
Minimum distance (D_{min}) (m)	3.49	3.83	3.134	0.6293

average coverage ratio value obtained is 80.70% with standard deviation (SD) of 3.25% and an average number of sensors placed is 58 with a standard deviation of 7.5. Figures 4 and 5 show the sensor node placement for Case I and the sensor node placement for Case II in the given 3D space.

3.2 Study of Minimum Distance

This part discusses the importance of minimum distance parameter incorporated in the objective function. In this study, two cases have been considered: Case I with the consideration of minimum distance parameter and Case II without considering minimum distance parameter. Here, NI is set to 10,000. The Case I results in approximately 100% coverage ratio with 120 sensors, and Case II results in approximately 100% coverage ratio with 150 sensors. It can thereby be concluded that the number of sensor nodes has decreased in Case I and hence cost of network gets reduced by 25%. In addition to this, overlapping of coverage range of sensor nodes has been minimized due to insertion of minimum distance factor.

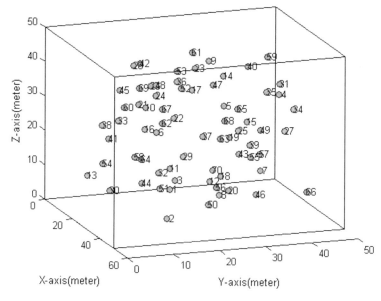

Fig. 4 MATLAB output image of sensors placement in *ROI* for Case I

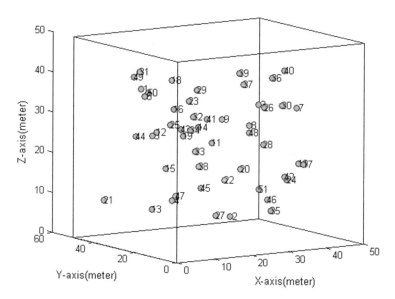

Fig. 5 MATLAB output image of sensors placement in *ROI* for Case II

Table 3 Simulation results to check scalability of the proposed scheme

Performance metric	Vol. I	Vol. II	Vol. III	Vol. IV
Volume (m^3)	$50 \times 50 \times 50$	$100 \times 100 \times 100$	$150 \times 150 \times 150$	$200 \times 200 \times 200$
MaxNo.	200	1000	2000	4000
MinNo.	50	400	1400	3350
No. of demand points	125	1000	3375	8000
Average coverage ratio (%)	80.59	80.55	80.56	80.87
Average sensor ratio (%)	30.5	45.70	73.70	86.57
SD of average coverage ratio	4.82	3.17	2.15	1.07

Fig. 6 Measure of scalability of the proposed scheme

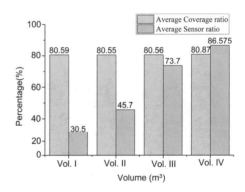

3.3 Scalability of the Model

To study the scalability of the proposed approach, four different volumes have been considered: Vol. I, Vol. II, Vol. III and Vol. IV. Here, average sensor ratio represents the ratio of an average number of sensors divided by maximum number of sensors. The value obtained of various performance metrics are shown in Table 3.

It is observed from Table 3 and Fig. 6 that with the increment in the volume, the number of sensors is increasing but providing approximately same average network coverage. In addition to this, standard deviation is decreasing with the increment in volume. Hence, the proposed scheme can be said to be scalable.

Table 4 Measure of death time of nodes

Round of nodes death	Random-based deployment	HS-based deployment
First node	493	516
First 10% nodes	703	783
Half nodes	1265	1286

3.4 Study of Network Performance Using LEACH Protocol

LEACH is a hierarchical protocol which aggregates the data so that the energy consumption is reduced and the data packet transmissions to the base station minimized to improve the deployed network lifetime. It is a routing protocol that uses cluster-based routing [19, 20]. Importance of LEACH protocol is that the whole data is aggregated by the cluster heads resulting in reduced traffic in whole network and reduction in congestion at base station, single hope routing leads to network lifetime improvement, no sensor location information is required in the formation of clusters and is distributed in nature [21].

The motive of using LEACH protocol is to study the impact of the node placement on the network lifetime as the distant nodes when communicate with each will consume more energy in data transmission; if nodes are closer to each other, they will waste their energy by overlapping sensing area by them. Hence, an attempt has been made to find out the effect of node placement on the network lifetime.

In this part, performance of deployed network using HS-based scheme is compared with random-based deployment using LEACH protocol [20, 22]. Here, for comparison, number of nodes are set to 150 because random-based deployment achieves 100% network coverage by deploying 150 nodes as per the results obtained in part 3.2, and network lifetime and remaining total energy in each round are the parameters considered for comparison.

From Table 4 and Fig. 7, it is observed that the network lifetime is improved for HS-based deployment in terms of the first dead node, first 10% dead nodes and half dead nodes because of balanced energy consumption due to optimal placement of sensor nodes. Also, total energy consumption is less in case of HS-based deployment. Hence, it is observed that optimal placement results in the improvement of the network lifetime (Fig. 8).

3.5 Formation of Connected Network

In this part, an objective function has been defined so that a connected network is formulated by incorporating network connectivity parameter in objective function as per the previous work done [10]. Following is the objective function defined for

Fig. 7 Number of alive
nodes in each round

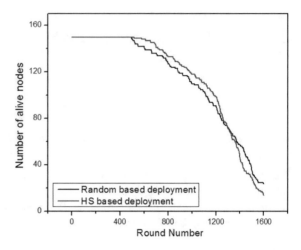

Fig. 8 Comparison of Total
remaining energy in HS and
Random based algorithm

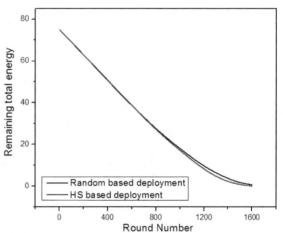

the formation of connected network in which network coverage is maximized using
minimum number of sensor nodes:

$$O_{\mathrm{HSA}}(n) = \frac{\mathrm{Cov}_{\mathrm{ratio}} \times \mathrm{Con}_{\mathrm{ratio}} \times D_{\mathrm{min}} \times k}{S} \qquad (17)$$

Initially the communication range (R_{comm}) is set to twice of the sensing range
[23] for simulation, i.e., $R_{\mathrm{sens}} = 10\ m$ and $R_{\mathrm{comm}} = 20\ m$. Here, connectivity ratio
is given by the ratio of the number of the connected nodes to the total number nodes
deployed in the given network [24]. Following observations are obtained for the
above-mentioned settings.

In Table 5, two cases are defined in which Case III shows the highest value of
the coverage ratio obtained over 50 successful runs and Case IV shows the highest

Table 5 Simulation results $R_{sens} = 10 \, m$ and $R_{comm} = 25 \, m$

Performance metrics	Case III	Case IV	Average value	SD
Coverage ratio (Cov_{ratio}) (%)	86.4	82.4	79.25	3.72
Connectivity ratio (Con_{ratio}) (%)	100	100	100	0
Number of sensors (S)	65	50	56	4

coverage ratio obtained by deploying the lowest number of sensor nodes. From the above table, it is observed that connected network is formed which gives average coverage ratio 79.25% and average connectivity ratio 100% while deploying 56 number of sensors on average. Here, 100% signifies that each node is connected with at least one sensor node.

4 Conclusions and Future Scope

The designed 3D WSN has been able to ensure coverage with minimum number of sensors using multi-objective node placement problem. The average coverage ratio obtained is 80.70% with standard deviation of 3.25%. Furthermore, with the insertion of minimum distance factor in the objective function, the network cost has been reduced by 25%. In addition, scalability of developed 3D WSN was also established. Also, network performance is checked using LEACH protocol which shows that network lifetime is improved because of balanced energy consumption due to optimal placement of sensor nodes. It is observed that with the modification in objective function, the proposed scheme can be used for the enhancement in the connectivity of the deployed network.

In the future work, the performance of the proposed work can be evaluated in terms of the effect of increasing communication range and sensing range on the coverage ratio, connectivity ratio and number of sensors deployed in the region of interest. In addition, 3D WSN performance can be evaluated by considering 3D space with obstacles, and the effect of obstacles on coverage can be studied. Also by considering different environment like underwater, the proposed scheme can be modified.

References

1. Akyildiz, I.F., Su, W., Sankarasubramaniam, Y., Cayirci, E.: A survey on sensor networks. IEEE Commun. Mag. **40**(8), 102–114 (2002). https://doi.org/10.1109/MCOM.2002.1024422
2. Fei, Z., Li, B., Yang, S., Xing, C., Chen, H., Hanzo, L.: A survey of multi-objective optimization in wireless sensor networks: metrics, algorithms, and open problems. IEEE Commun. Surv. Tutor. **19**(1), 550–586 (2017). https://doi.org/10.1109/COMST.2016.2610578

3. Commuri, S., Watfa, M.K.: Coverage strategies in wireless sensor networks. Int. J. Distrib. Sens. Netw. **2**(4), 333–353 (2006). https://doi.org/10.1080/15501320600719151
4. Iovanovici, A., Topirceanu, A., Udrescu, M., Vladutiu, M.: Design space exploration for optimizing wireless sensor networks using social network analysis. In: 18th International Conference on System Theory, Control and Computing (ICSTCC), pp. 815–820. IEEE Press (2014)
5. Chien, S., Tran, D., Doubleday, J., Davies, A., Kedar, S., Webb, F., Shirazi, B.: A multi-agent space, in-situ Volcano Sensor Web. In: International Symposium on Space Artificial Intelligence, Robotics, and Automation for Space, Sapporo, Japan (2010)
6. Akyildiz, I.F., Pompili, D., Melodia, T.: Underwater acoustic sensor networks: research challenges. Ad Hoc Netw. **3**(3), 257–279 (2005). https://doi.org/10.1016/j.adhoc.2005.01.004
7. Lian, X.Y., Zhang, J., Chen, C., Deng, F.: Three-dimensional deployment optimization of sensor network based on an improved Particle Swarm Optimization algorithm. In: 10th World Congress on Intelligent Control and Automation (WCICA), pp. 4395–4400. IEEE Press (2012)
8. Zhang, C., Bai, X., Teng, J., Xuan, D., Jia, W.: Constructing low-connectivity and full-coverage three-dimensional sensor networks. IEEE J. Sel. Areas Commun. **28**(7), 984–993 (2010). https://doi.org/10.1109/JSAC.2010.100903
9. Cao, B., Kang, X., Zhao, J., Yang, P., Lv, Z., Liu, X.: Differential evolution-based 3D directional wireless sensor network deployment optimization. IEEE Internet Things J. **5**(5), 3594–3605 (2018). https://doi.org/10.1109/JIOT.2018.2801623
10. Alia, O.M., Al-Ajouri, A.: Maximizing wireless sensor network coverage with minimum cost using harmony search algorithm. IEEE Sens. J. **17**(3), 882–896 (2017). https://doi.org/10.1109/JSEN.2016.2633409
11. Sharma, D., Gupta, V.: Improving coverage and connectivity using harmony search algorithm in wireless sensor network. In: Emerging Trends in Computing and Communication Technologies (ICETCCT), pp. 1–7, IEEE (2017). https://doi.org/10.1109/icetcct.2017.8280297
12. Temel, S., Unaldi, N., Kaynak, O.: On deployment of wireless sensors on 3-D terrains to maximize sensing coverage by utilizing cat swarm optimization with wavelet transform. IEEE Trans. Syst. Man Cybern.: Syst. **44**(1), 111–120 (2014). https://doi.org/10.1109/TSMCC.2013.2258336
13. Zong, W.G., Joong, H.K., Loganathan, G.V.: A new heuristic optimization algorithm: harmony search. Simulation **76**(2), 60–68 (2001). https://doi.org/10.1177/003754970107600201
14. Yang, X.S.: Harmony search as a metaheuristic algorithm. In: Music-Inspired Harmony Search Algorithm, pp. 1–14. Springer, Berlin (2009)
15. Sangwan, A., Singh, R.P.: Survey on coverage problems in wireless sensor networks. Wireless Pers. Commun. **80**(4), 1475–1500 (2015). https://doi.org/10.1007/s11277-014-2094-3
16. Deif, D.S., Gadallah, Y.: Classification of wireless sensor networks deployment techniques. IEEE Commun. Surv. Tutor. **16**(2), 834–855 (2014). https://doi.org/10.1109/SURV.2013.091213.00018
17. Abidin, H.Z., Din, N.M., Yassin, I.M., Omar, H.A., Radzi, N.A.M., Sadon, S.K.: Sensor node placement in wireless sensor network using multi-objective territorial predator scent marking algorithm. Arabian J. Sci. Eng. **39**(8), 6317–6325 (2014). https://doi.org/10.1007/s13369-014-1292-3
18. Yoon, Y., Kim, Y.H.: An efficient genetic algorithm for maximum coverage deployment in wireless sensor networks. IEEE Trans. Syst. Man Cybern. B **43**(5), 1473–1483 (2013). https://doi.org/10.1109/tcyb.2013.2250955
19. Heinzelman, W. R., Chandrakasan, A., Balakrishnan, H.: Energy efficient communication protocol for wireless sensor networks. In: Proceeding of the 33rd Hawaii International Conference System Sciences, Hawaii (2000). https://doi.org/10.1109/hicss.2000.926982
20. Heinzelman, W.R., Chandrakasan, A., Balakrishnan, H.: An application-specific protocol architecture for wireless microsensor networks. IEEE Trans. Wireless Commun. **1**(4), 660–670 (2002). https://doi.org/10.1109/TWC.2002.804190
21. Gill, R.K., Chawla, P., Sachdeva, M.: Study of LEACH routing protocol for wireless sensor networks. In: International Conference on Communication, Computing & Systems (ICCCS-2014) (2014)

22. Kohli, S., Bhattacharya, P., Jha, M.K.: Implementation of homogeneous LEACH Protocol in three-dimensional wireless sensor networks. Int. J. Sens. Wirel. Commun. Control **6**(1), 4–11 (2016). https://doi.org/10.2174/2210327905666150903214939
23. Zhang, H.C., Hou, J.: Maintaining sensing coverage and connectivity in large sensor networks. Ad Hoc Sens. Wirel. Netw. **1**, 89–124 (2005). https://doi.org/10.1201/9780203323687
24. Guangjie, H., Chenyu, Z., Lei, S., Joel, J.P., Rodrigues, C.: Impact of deployment strategies on localization performance in underwater acoustic sensor networks. IEEE Trans. Ind. Electron. **62**(3), 1725–1733 (2015). https://doi.org/10.1109/tie.2014.2362731

An Efficient Jaya Algorithm for Multi-objective Permutation Flow Shop Scheduling Problem

Aseem Kumar Mishra, Divya Shrivastava, Bhasker Bundela and Surabhi Sircar

Abstract The Jaya algorithm is a novel, simple, and efficient meta-heuristic optimization technique and has received a successful application in the various fields of engineering and sciences. In the present paper, we apply the Jaya algorithm to permutation flow shop scheduling problem (PFSP) with the multi-objective of minimization of maximum completion time (makespan) and tardiness cost under due date constraints. PFSP is a well-known NP-hard and discrete combinatorial optimization problem. Firstly, to retrieve a job sequence, a random preference is allocated to each job in a permutation schedule. Secondly, a job preference vector is transformed into a job permutation vector by means of largest order value (LOV) rule. To deal with the multi-objective criteria, we apply a multi-attribute model (MAM) based on Apriori approach. The correctness of the Jaya algorithm is verified by comparing the results with the total enumeration method and simulated annealing (SA) algorithm. Computational results reveal that the proposed optimization technique is well efficient in solving multi-objective discrete combinatorial optimization problems such as the flow shop scheduling problem in the present study.

Keywords Jaya algorithm · Permutation flow shop · Multi-objective · Multi-attribute model

1 Introduction

Permutation flow shop scheduling problem (PFSP) has been proved to be the most popular non-deterministic polynomial-time (NP)-hard with an extensive engineer-

A. K. Mishra (✉) · D. Shrivastava · B. Bundela · S. Sircar
Department of Mechanical Engineering, Shiv Nadar University, Dadri, India
e-mail: am712@snu.edu.in

D. Shrivastava
e-mail: divya.shrivastava@snu.edu.in

B. Bundela
e-mail: bb907@snu.edu.in

© Springer Nature Singapore Pte Ltd. 2020
R. Venkata Rao and J. Taler (eds.), *Advanced Engineering Optimization Through Intelligent Techniques*, Advances in Intelligent Systems and Computing 949,
https://doi.org/10.1007/978-981-13-8196-6_11

ing relevance [1]. PFSP is based on the identification of an optimum sequence of given n jobs to be processed at m machines in order to achieve the desired objective function in a most efficient way under the given constraints and limited resources. At a time, one job can be processed on one machine, and also, one machine can handle at most one job. In the last few decades, many heuristics and meta-heuristics (SA, PSO, GA, etc.) are developed by researchers for PFSP in order to optimize the desired single-objective function [2]. Since most of the practical scheduling problems include many objectives, multi-objective flow shop scheduling problems (MOFSPs) have been studied by the researchers in the last few decades. Some of the important objectives, including makespan, total flow time, tardiness cost, mean flow time, total tardiness, mean tardiness, the total number of tardy jobs, have been considered, and several approaches and optimization techniques are developed to deal with MOFSPs [3]. For instance, Yagmahan and Yenisey [4] applied ant colony algorithm for flow shop scheduling with multi-objective of total flow time and makespan. The results reveal the effectiveness of the proposed algorithm over benchmarks. Minella et al. [5] developed an effective restarted Pareto greedy algorithm for considering three objectives: makespan, tardiness, and flow time. They used a graphical method to compare the performance of random Pareto fronts depending upon empirical attainment functions. MOFSP has been extensively studied in the literature, and it is beyond the scope of this paper to cover all contributions. Therefore, we refer to some related surveys by Yenisey and Yagmahan [3] and Sun et al. [6] which provide a comprehensive review of all pioneer works in this area.

A new meta-heuristic named Jaya algorithm recently proposed by Venkata Rao [7] has received a wide variety of applications in the areas of engineering and sciences. The specialty of Jaya as against other classical meta-heuristics (GA, PSO, SA) is that it does not contain any algorithm-specific parameter. As a consequence, optimization of algorithm parameters is not required which would be rather tricky and time-consuming. It has been successfully applied to both unconstrained and constrained optimization problems as well as single and multi-objective functions. Rao and Saroj [8] applied the Jaya algorithm for cost minimization of heat exchanger considering maintenance as a constraint. Computational results proved the effectiveness of Jaya when compared with other heuristics. A similar problem was addressed by Rao et al. [9] for the optimization of design parameters for a heat sink with a micro-channel by the Jaya algorithm. The decision variables were pumping power and thermal resistance. The efficiency of the Jaya algorithm is examined with case studies, and objectives were optimized simultaneously. In the manufacturing area, Rao et al. [10] proposed Jaya for multi-objective optimization of four machining processes including a laser cutting process, ion beam micro-milling process, an electrochemical machining process, and wire electrodischarge machining process. For these many machining processes, various confining objectives such as cutting velocity, surface quality, material removal rate, and dimensional accuracy were considered in multi-objective optimization. The computational results of various algorithms such as PSO, GA, and NSGA available in the literature are compared. Jaya showed better performance as compared to these meta-heuristics.

The application of Jaya in some other multi-objective optimizations can be found in the literature [11, 12]. However, its applications are still limited to discrete combinatorial optimization problems such as flow shops and job shops. Buddala and Mahapatra [13] proposed an improved Jaya and teaching–learning-based optimization algorithm to solve flexible flow shop scheduling problem with the single objective of makespan minimization. The search technique inspired by the mutation strategy of genetic algorithm is embedded, and computational experiments were conducted based on public benchmarks. The performance of both TLBO and Jaya was superior when compared with other efficient meta-heuristics. Gao et al. [14] applied Jaya to solve flexible job shop scheduling problem and concluded that it is superior from available heuristics. From the discussion above, it is clear that the scheduling problems are NP-hard in nature, and thus, it is essential for new any efficient meta-heuristic to perform good for these problems as well. Although the Jaya algorithm has been applied for solving scheduling-related problems, most of them are based on single objectives. Thus, the application of Jaya for multi-objective optimization particularly in permutation flow shop scheduling problems is missing in the literature and has needs to be addressed.

In the present paper, we apply the Jaya algorithm for multi-objective flow shop scheduling problem considering total tardiness cost and makespan as objective functions. To the best of our knowledge, the Jaya algorithm has not been yet applied for solving MOFSP. A multi-attribute model (MAM) is applied to evaluate the present multi-objective optimization problem. The robustness of the Jaya algorithm is evaluated by comparing with the results obtained by the total enumeration method and simulated annealing algorithm (SA). The detailed approach for the solution to the problem is discussed in proceeding sections.

2 Statement of Problem

The permutation flow shop scheduling problem involves the scheduling n jobs on m machines. Each job $i \epsilon \{1, 2, \ldots, n\}$ has to process on machine $j \epsilon \{1, 2, 3, \ldots, m\}$ in the same order; i.e., permutation sequence follows. The objective of the problem is to obtain the optimum production sequence which simultaneously minimizes the makespan and tardiness cost. The first objective is referred to the maximum completion time (makespan) of all the jobs in the production schedule. Therefore, the aim is to retrieve an optimum production sequence with minimum makespan. On the other hand, the tardiness cost is incurred if the job i is completed beyond its given due date. Hence, the objective is to identify the production schedule with minimum total tardiness cost of all the jobs. As common in multi-objective scheduling, some of the objectives are conflicting in nature; that is, the betterment in one objective may worsen another. Therefore, initially, the problem is solved independently considering the single-objective functions. Then, a multi-attribute model (MAM) depending on multiple attribute theory [15] is applied to define the multi-objective function in order to determine the optimum production schedule.

2.1 Single-objective Optimization

In the present paper, we select the two most relevant objective functions for the permutation flow shop scheduling problem (time and cost minimization). The mathematical models of the individual objective functions can be expressed as follows:

Makespan minimization (X_k)

$$CT_{1,j,k} = P_{1,j} + P_{1,j-1} \tag{1}$$

$$CT_{i,1,k} = P_{i,1} + P_{i-1,1} \tag{2}$$

$$\text{CT}_{i,j,k} = P_{i,j} + \max\left\{\text{CT}_{i,j-1,k}, \text{CT}_{i-1,j,k}\right\} \tag{3}$$

$$X_k = \text{CT}_{n,m,k} \tag{4}$$

$$X^* = \min\{X_1, X_2, X_3, \ldots, X_k\} \tag{5}$$

Tardiness cost minimization (Y_k)

$$\text{DD}_i = \sum_{j=1}^{m} P_{i,j} \times n \times \left(1 - t_f\right) \tag{6}$$

$$Y_k = \sum_{i=1}^{n} \text{TC}_i \times \max\left\{\left(\text{CT}_{i,m,k} - \text{DD}_i\right), 0\right\} \tag{7}$$

$$Y^* = \min\{Y_1, Y_2, Y_3, \ldots, Y_k\} \tag{8}$$

$$P_{i,j}, \text{CT}_{i,j,k}, X_k, X^*, \text{DD}_i, t_f, \text{TC}_i, Y_k, Y^* \geq 0 \tag{9}$$

where $P_{i,j}$ is the processing time of ith job on jth machine, $\text{CT}_{i,j,k}$ is the completion time of ith job on jth machine in schedule k, X_k is the completion time of the job scheduled at the last position in the production sequence, i.e., makespan, X^* is the production schedule with the minimum makespan, DD_i is the due date of job i, t_f is the tardiness factor, TC_i is the tardiness cost of job i, Y_k is the total tardiness cost of schedule k, Y^* is the minimum tardiness cost among k production schedules. Inequality (9) resembles the non-negativity constraint for both known and decision variables.

2.2 Multi-objective Optimization

The multi-objective flow shop scheduling problem (MOFSP) has received great attention in the design of multi-objective optimization methods and solution techniques. Combining the above two single-objective models together, the multi-objective model based on MAM attribute theory can be defined as follows:

$$Z_k = w_1 \times \frac{X_k}{X^*} + w_2 \times \frac{Y_k}{Y^*} \tag{10}$$

$$Z^* = \min\{Z_1, Z_2, Z_3, \ldots, Z_k\} \tag{11}$$

where w_1 and w_2 are the weights given to the single-objective functions, Z_k is the value of schedule k while considering both objective functions simultaneously, Z^* signifies the value of optimum schedule with the multi-objective approach. Note that in the present study, equal weights of 0.5 are assigned.

3 Jaya Algorithm

3.1 Brief Description of Jaya Algorithm

The Jaya algorithm is defined on the principle that the particular solution obtained in the population should shift toward the best solution and moves away from the worst solution. The peculiarity of the Jaya algorithm as against other meta-heuristics is that it is independent of algorithm-specific parameters. The basic procedure of the Jaya algorithm is depicted in Fig. 1.

3.2 Implementation of Jaya Algorithm to MOFSP

In the present paper, firstly the single-objective functions are optimized individually to retrieve X^* and Y^*, followed by multi-objective optimization (Z^*). Moreover, Jaya is coded in MATLAB 8.6.1 and results are analyzed on a 3.15 GHz i5-4670 Processor. The procedure followed in the application of the Jaya algorithm for MOFSP is summarized in the following steps as follows:

Step 1: Initialize decision variables, i.e., number of jobs (n), population size, i.e., number of schedules (NP), and termination criterion, i.e., generation number (GEN).

Step 2: Create a job preference vector, $V_{k,l} = \{v_{1,k,l}, v_{2,k,l}, \ldots, v_{i,k,l}, \ldots, v_{n,k,l}\}$, where $V_{k,l}$ is an n-dimensional vector which represents the sequence of jobs in the kth schedule at lth iteration and $v_{i,k,l}$ is the preference value

Fig. 1 General approach of
the Jaya algorithm

Begin (Jaya algorithm procedure)
　Initialize:
　Decision variables, x_i
　Population size, np
　Generation number, gen
　Evaluate objective function, $f(x_i)$
　While (iteration $\leq$ gen)
　　Generate random numbers (r_1, r_2), U [0, 1]
　　Identify *best* and *worst* solution in the population
　　Update the decision variables of other solutions as:
　　$x'_{i,} = x_{i,} + r_1 * (x_{best} - |x_i|) - r_2 * (x_{worst} - |x_i|)$
　　If $(f'(x_i) < f(x_i))$ // for minimization problem
　　　$f(x_i) = f'(x_i)$
　　End
　　iteration = iteration+1
　End
　Optimum solution = $f(x_i)$
End

---

assigned to an ith job in the kth schedule at lth iteration. The preference value is randomly generated with a uniform random number distribution as per Eq. (12).

$$v_{i,k,l} = 1 + rand\,(0,1) * (n-1) \tag{12}$$

Step 3:　Convert the job preference vector $(V_{k,l})$ into job permutation vector $(\Pi_{k,l})$ by largest order value (LOV) rule; i.e., jobs are sequenced in non-increasing order of their preference values $(v_{i,k,l})$. Therefore, $\Pi_{k,l} = \{\pi_{j,1,k,l}, \pi_{j,2,k,l}, \ldots, \pi_{j,i,k,l}, \ldots, \pi_{j,n,k,l}\}$, where $\pi_{j,i,k,l}$ is jth job placed at ith position in kth schedule at lth iteration.

Step 4:　Calculate the values of X_k, Y_k, and Z_k for each permutation schedule $(\Pi_{k,l})$. Compare the values of each schedule with the corresponding values obtained in the previous iteration and update the better solution.

Step 5:　Identify the schedule with minimum objective function and maximum objective function and their corresponding preference vectors $(V_{k,l\min} \,\&\, V_{k,l\max})$, respectively.

Step 6:　Update the preference values of all vectors based on the preference values of $V_{k,l\min} \,\&\, V_{k,l\max}$ as per Eq. (13)

$$v'_{i,k,l} = v_{i,k,l} + r_{1,i,l} * \left(v_{i,k,l\min} - |v_{i,k,l}|\right) - r_{2,i,l} * \left(v_{i,k,l\max} - |v_{i,k,l}|\right) \tag{13}$$

where

$v_{i,k,l}$	the preference value of ith job in kth schedule during lth iteration
$v_{i,k,l}$ min	the preference value of the ith job in the schedule having minimum objective function
$v_{i,k,l}$ max	the preference value of the ith job in the sequence with maximum value of objective function
$v'_{i,k,l}$	updated value of $v_{i,k,l}$
$r_{1,i,l}, r_{2,i,l}$	random numbers generated with uniform distribution function U[0,1]

Step 7: Convert the updated preference vector into job permutation vector and identify the objective function values of the new schedules obtained (as per steps 3 and 4).

Step 8: Compare the values of the new schedule with the corresponding previous schedule and update the better solution (replace or retain the schedule with minimum value).

Step 9: Repeat all the above steps until the optimum solution is reached and the termination criteria are satisfied.

Step 10: Address the optimum solution (optimum schedule with minimum makespan and tardiness cost subject to multi-objective function, i.e., X^*, Y^*, and Z^*).

$$i\epsilon\{1, 2, 3, \ldots, n\}, k\epsilon\{1, 2, 3, \ldots NP\}, l\epsilon\{1, 2, 3, \ldots, GEN\}$$

3.3 Illustrative Example

For illustration purpose, we consider a simple three-job and three-machine problem. The input parameters of the problem are given in Table 1a. For a three-job problem, there could be 3! possible sequences. Table 1b presents the solutions for all six job schedules considering both single and multi-objective functions. In Table 1b, it can be seen that while optimizing the single-objective function, sequence 2-1-3 yields minimum makespan of 232.42 h, whereas 2-3-1 has minimum tardiness cost of 3682.4 ₹. However, 2-3-1 is the optimum schedule for multi-objective function as well. This is due to the fact that the corresponding value of makespan in multi-objective is only 2% more than the single objective. On the other hand, the corresponding total tardiness cost is 54% less when the only makespan is acknowledged as the objective function. Therefore, multi-objective optimization can provide a better-compromised solution considering both time and cost as objective functions.

4 Results and Comparisons

To evaluate the efficiency of the Jaya algorithm, nine problem sets from 8 jobs and 5 machines to 10 jobs and 20 machines are taken and the optimized results are compared with that obtained by total enumeration method as given in Table 3.

Table 1 **a** Input parameters for evaluation of objective function and **b** results for various instances of the single and multi-objective function

(a)

Jobs	Processing time on machines (h)			Due date (h)	Tardiness cost (₹/h)
	M1	M2	M3		
1	99	31	71	181	30
2	15	24	86	113	40
3	20	80	7	96	42

(b)

Sequence	Single-objective		Multi-objective
	Makespan (X_k) (h)	Total tardiness cost (Y_k) (₹)	Objective function value (Z_k)
3-2-1	280.59	7329.5	1.60
3-1-2	306.51	9384.2	1.93
2-3-1	*235.56*	*3682.4*	*1.01*
2-1-3	*232.42*	*5282.9*	1.49
1-2-3	293.98	15867.3	2.79
1-3-2	320.40	13986.6	2.59

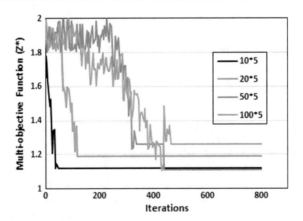

Fig. 2 Convergence of multi-objective function with the Jaya algorithm for different problem sets

Total enumeration method provides the exact solution by calculating all possible permutation sequences. The population size is set as 100, and for each instance, 10 independent runs were conducted. Figure 2 shows the convergence of objective function with iterations using the Jaya algorithm for problem sets of different sizes.

The processing times are randomly generated with uniform distribution in the range U[1,100] and the tardiness cost as U[10,50], and the tardiness factor (t_f) is set as 0.8. Table 2 represents the maximum, minimum, and average values of the two single-objective functions. Note that in Table 2, X_k^* and Y_k^* represent the

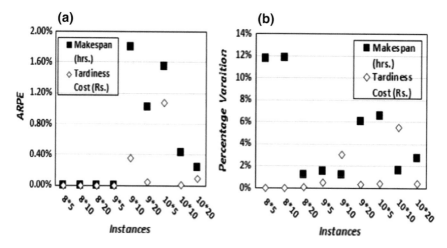

Fig. 3 **a** ARPE versus instances and **b** variation of objectives in single and multi-objective functions

corresponding values of makespan and total tardiness cost in the multi-objective optimization. From Table 2, it is seen that the minimum values obtained by Jaya are equal to that retrieved by the total enumeration method for all instances.

Figure 3a depicts the average relative percentage error (ARPE) from the best solutions for different instances. The ARPE can be calculated as follows:

$$\text{ARPE} = \frac{\left(S_{\text{best}} - S_{\text{avg}}\right)}{S_{\text{best}}} \times 100 \qquad (14)$$

Figure 3a shows a maximum of 2% variation in the solutions obtained by the Jaya algorithm for various instances. Figure 3b depicts the percentage increase in the value of makespan and tardiness cost for multi-objective function when compared with the individual objectives. It shows a maximum of 12% increase in makespan and 6% in tardiness cost. This concludes that MAM approach for multi-objective optimization provides the near optimum solution with a best possible combination of individual objectives. Above all, it can be articulated that the Jaya algorithm has considerable potential in solving multi-objective permutation flow shop scheduling problems.

It is a well-known fact that meta-heuristics may not guarantee optimal solutions. Secondly, since *PFSP* is NP-hard problem, enumeration method could not obtain the optimal solution for large size problems (beyond 10 jobs). Therefore, to identify the effectiveness of Jaya for large size problems, a popular algorithm named simulated annealing (SA) is applied and the computational results are compared. SA is an iterative search technique developed by Kirkpatrick et al. [16]. It is a generalized probabilistic method which mimics the simulation of thermal annealing of critically heated solids. The general approach of SA is shown in Fig. 4. It basically consists of four basic parameters, viz. initial temperature, loop factor, temperature reduction

Table 2 Comparison of results for a multi-objective function for various instances

Instances (n * m)	Total enumeration		Jaya					
	X_K^*	Y_K^*	X_K^*			Y_K^*		
			Min	Max	Avg	Min	Max	Avg
8 * 5	704.35	8595.10	**704.35**	**704.35**	**704.35**	**8595.10**	**8595.10**	**8595.10**
8 * 10	1072.89	67584.32	**1072.89**	**1072.89**	**1072.89**	**67584.32**	**67584.32**	**67584.32**
8 * 20	1571.3	193010.4	**1571.3**	**1571.3**	**1571.3**	**193010.4**	**193010.4**	**193010.4**
9 * 5	800.97	12916.11	**800.97**	**800.97**	**800.97**	**12916.11**	**12916.11**	**12916.11**
9 * 10	1061.12	79543.92	**1061.12**	**1061.12**	**1061.12**	**79543.92**	**79543.92**	**79543.92**
9 * 20	1778.18	261145.6	**1778.18**	1814.32	1796.25	**261145.6**	261353.81	261249.70
10 * 5	921.95	7313.92	**921.95**	950.47	936.21	**7313.92**	7436.78	7358.26
10 * 10	1030.35	1659.82	**1030.35**	1074.02	1034.72	**165982**	166028.22	166005.11
10 * 20	1721.77	330522.9	**1721.77**	1731.88	1725.81	**330522.9**	331164.77	330843.84

Bold represents the best solutions obtained by the Jaya algorithm

Fig. 4 General mechanism of SA algorithm

```
Begin (Simulated annealing mechanism)
    Intial temperature, Tᵢ
    Iterations
    temperature reduction factor α (0<α<1)
    Initial Solution, f(xᵢ)
    While (stopping criteria)
        While (iterations)
            Generate few solution= f'(xᵢ)
            ΔF= f'(xᵢ) - f(xᵢ)
            If (ΔF≤0) // for minimization problem
                f(xᵢ) = f'(xᵢ)
            else
                Calculate E= exp(- ΔF/ Tᵢ)
                Generate a random number, U[0,1]
                if (U<E)
                    f(xᵢ) = f'(xᵢ)
                end
            end while loop (inner)
        Tᵢ = Tᵢ * α
        end while loop (outer)
End
```

factor, and final temperature. In the present study, the initial and final temperatures are set as 100,000 and 1, respectively. The loop factor is 20, and the reduction factor is 0.95. Three large problem sets, viz. 20, 50, and 100 jobs, are selected for the evaluation of the effectiveness of the Jaya algorithm. The machine size is set as 5 for each problem set. Figure 2 depicts the convergence of Jaya w.r.t. iterations for multi-objective function (Z^*) for large size problems as well (20, 50, and 100 jobs). For each set, ten independent runs are conducted. The comparison of results obtained by SA and Jaya for large size problems (20, 50, and 100 jobs) is given in Table 3. The mean values of tardiness and makespan (obtained in multi-objective function) are represented. The results show that for all problem sizes, Jaya yields better results as compared from SA. However, the results obtained by SA are fairly comparable with Jaya. The percentage variation among the results of the two algorithms is also represented in Table 3 which shows a very less variation (less than 5%) in the optimal solution obtained by algorithms. Thus, it can be concluded that Jaya is well efficient in solving MOFSPs.

5 Conclusions

In the present study, we apply a recently proposed, simple, and efficient meta-heuristic optimization technique named Jaya algorithm to a multi-objective permutation flow shop scheduling problem. The objective is to minimize the makespan and total tardi-

Table 3 Comparison of results for a multi-objective function for large size problems

Instances (n * m)	SA		Jaya		Percentage variation	
	$X_K{}^*$	$Y_K{}^*$	$X_K{}^*$	$Y_K{}^*$	$X_K{}^*$ (%)	$Y_K{}^*$ (%)
20 * 5	1822.4	14198.7	**1783.6**	**13779.9**	2.13	3.03
50 * 5	5569.3	75667.2	**5473.8**	**75023.6**	1.74	0.86
100 * 5	12328.6	182364.6	**11836.2**	**180393.4**	4.15	1.19

Bold demonstrates the best solution obtained by algorithm

ness cost. Simultaneous minimization of time and cost is the most important aspect of real-life industrial scheduling problems. Computational results reveal that the Jaya algorithm is well efficient in dealing with discrete combinatorial problems. An exhaustive study along with statistical analysis is performed to explore the potential of Jaya for solving other multi-objective optimization problems.

As a scope for future research, Jaya can find its applications in other scheduling problems such as job shops, parallel flow shops, flexible flow shops, considering multi-objectives. Also, it can be modified or hybridized with other state-of-the-art algorithms in order to achieve more efficient results.

References

1. Rinnooy Kan, A.H.G.: Machine scheduling problems : classification, complexity and computations. Nijhoff (1976)
2. Arora, D., Agarwal, G.: Meta-heuristic approaches for flowshop scheduling problems: a review. Int. J. Adv. Oper. Manag. **8**, 1 (2016). https://doi.org/10.1504/IJAOM.2016.076203
3. Yenisey, M.M., Yagmahan, B.: Multi-objective permutation flow shop scheduling problem: literature review, classification and current trends. Omega (United Kingdom) **45**, 119–135 (2014). https://doi.org/10.1016/j.omega.2013.07.004
4. Yagmahan, B., Yenisey, M.M.: A multi-objective ant colony system algorithm for flow shop scheduling problem. Expert Syst. Appl. **37**, 1361–1368 (2010). https://doi.org/10.1016/j.eswa.2009.06.105
5. Minella, G., Ruiz, R., Ciavotta, M.: Restarted Iterated Pareto Greedy algorithm for multi-objective flowshop scheduling problems. Comput. Oper. Res. **38**, 1521–1533 (2011). https://doi.org/10.1016/j.cor.2011.01.010
6. Sun, Y., Zhang, C., Gao, L., Wang, X.: Multi-objective optimization algorithms for flow shop scheduling problem: a review and prospects. Int. J. Adv. Manuf. Technol. **55**, 723–739 (2011). https://doi.org/10.1007/s00170-010-3094-4
7. Venkata, Rao R.: Jaya: A simple and new optimization algorithm for solving constrained and unconstrained optimization problems. Int. J. Ind. Eng. Comput. **7**, 19–34 (2016). https://doi.org/10.5267/j.ijiec.2015.8.004
8. Rao, R.V., Saroj, A.: Economic optimization of shell-and-tube heat exchanger using Jaya algorithm with maintenance consideration. Appl. Therm. Eng. **116**, 473–487 (2017). https://doi.org/10.1016/j.applthermaleng.2017.01.071
9. Rao, R.V., More, K.C., Taler, J., Ocłoń, P.: Dimensional optimization of a micro-channel heat sink using Jaya algorithm. Appl. Therm. Eng. **103**, 572–582 (2016). https://doi.org/10.1016/J.APPLTHERMALENG.2016.04.135

10. Rao, R.V., Rai, D.P., Balic, J.: A multi-objective algorithm for optimization of modern machining processes. Eng. Appl. Artif. Intell. **61**, 103–125 (2017). https://doi.org/10.1016/j.engappai. 2017.03.001
11. Radhika, S., Ch, S.R., Krishna, N., Karteeka Pavan, K.: Multi-objective optimization of master production scheduling problems using Jaya algorithm 1729–32 (2016)
12. Rao, R.V., Saroj, A.: Multi-objective design optimization of heat exchangers using elitist-Jaya algorithm. Energy Syst (2016). https://doi.org/10.1007/s12667-016-0221-9
13. Buddala, R., Mahapatra, S.S.: Improved teaching–learning-based and Jaya optimization algorithms for solving flexible flow shop scheduling problems. J. Ind. Eng. Int. 1–16 (2017). https://doi.org/10.1007/s40092-017-0244-4
14. Gao, K., Sadollah, A., Zhang, Y., Su, R.: Discrete Jaya algorithm for flexible job shop scheduling problem with new job insertion. In: 14th International Conference on Control Automation Robot Vis 13–5 (2016)
15. Xia, T., Xi, L., Zhou, X., Lee, J.: Dynamic maintenance decision-making for series-parallel manufacturing system based on MAM-MTW methodology. Eur. J. Oper. Res. **221**, 231–240 (2012). https://doi.org/10.1016/j.ejor.2012.03.027
16. Kirkpatrick, S., Gelatt, C.D., Vecchi, M.P.: Optimization by simulated annealing. Science (New York,) **220**(4598), 671–680 (1983)

Multi-objective Optimization of EDM Process Parameters Using Jaya Algorithm Combined with Grey Relational Analysis

Ashish Khachane and Vijaykumar Jatti

Abstract Electrical discharge machining is a controlled machining process used to machine parts of complex geometry with high precision and low tolerance having many applications in various fields. Therefore, for effective utilization of available resources and to improve overall process performance, it calls for an optimization of EDM process parameters. In this study, tool wear rate (TWR) and material removal rate (MRR) are optimized for the three independent parameters such as magnetic field, pulse on time and gap current by using Jaya algorithm combined with grey relational analysis. From the grey relational grade, using regression analysis optimization model is developed and optimized using Jaya algorithm. The purpose of this work is to provide a platform to the researchers and manufacturing industries for determining the optimal setting of machining parameters involving multiple target responses. In general, this approach can be used for other manufacturing processes and even for other application fields.

Keywords Jaya algorithm · Multi-objective optimization · Grey relational analysis · Electrical discharge machining · Tool wear rate · Material removal rate

1 Introduction

Selection of machining process parameters is very important and crucial when it comes to high productivity, high precision and fine surface finish. This invites an optimization problem of machining parameters to increase the productivity and tool life with simultaneously minimizing machining time and cutting force by improving the process performance. A non-conventional process, electrical discharge machining

A. Khachane (✉) · V. Jatti
Department of Mechanical Engineering, D.Y. Patil College of Engineering, Akurdi, Pune, Maharashtra, India
e-mail: ashishkhachane3@gmail.com

V. Jatti
e-mail: vksjatti@gmail.com

© Springer Nature Singapore Pte Ltd. 2020
R. Venkata Rao and J. Taler (eds.), *Advanced Engineering Optimization Through Intelligent Techniques*, Advances in Intelligent Systems and Computing 949, https://doi.org/10.1007/978-981-13-8196-6_12

(EDM), is widely used for the machining of parts for automobile and aerospace applications. Range of process parameters is involved while machining of parts by electrical discharge machining process. Because of the multiple process variables, it becomes very difficult to get the optimum setting of process parameters by traditional optimization techniques. Therefore, non-traditional optimization techniques play a vital role in such cases. Jaya algorithm is a newly developed optimization technique which can be used to solve the various optimization problems in the area of traditional and non-traditional machining processes.

There have been lots of research work reported over the past few years in the area of optimization of machining process parameters to improve the machining part quality, tool life, material removal rate and to reduce the machining cost, surface roughness so as to enhance the productivity and overall process performance.

Radhika et al. [1] performed multi-objective optimization using grey relational analysis to determine the best setting of process parameters of AlSi10Mg aluminium hybrid composite in electrical discharge machining process. They have considered three objective functions in multi-objective optimization problem: minimum tool wear rate, minimum surface roughness and maximum material removal rate. In their study, the peak current was found to be the most influential parameter. Raghuraman et al. [2] optimized the EDM process parameters using Taguchi method with grey relational analysis for mild steel using copper electrode. Ohdar et al. [3] studied the optimization model to investigate the effects of the peak current, pulse on/off time and flashing pressure in electrical discharge machining on tool wear rate and material removal rate in machining of mild steel with copper as an electrode. In their study, they have found that pulse on time is the most significant factor for material removal rate (MRR) while the peak current found to be the most significant factor for tool wear rate (TWR). Pawade and Joshi [4] performed multi-objective optimization of cutting forces and surface roughness in high-speed turning of Inconel 718 using Taguchi grey relational analysis. They have demonstrated the step-by-step calculation of grey relational analysis for the two-objective optimization. Somashekhar et al. [5] performed multi-objective optimization on process parameters of micro-electrical discharge machining using Taguchi and grey relational analysis method. They have considered maximum material removal rate, minimum overcut and surface roughness as the three target responses. Vijayanand and Ilangkumaran [6] used hybrid grey-based fuzzy logic with Taguchi method to find out the optimum setting of process parameters of a micro-electrical discharge machining process. They performed the optimization on control complex nonlinear system. Kolahan et al. [7] optimized turning process parameters using grey relational analysis and simulated annealing algorithm. They developed an optimization model by combining grey relational analysis and regression modelling to convert grade values of multi-response system as the function of process parameters and optimized the developed model by simulated annealing method.

Rao [8] has proposed Jaya algorithm, a simple and an effective search optimization technique for solving constrained and unconstrained optimization problems. Jaya algorithm is basically a search optimization technique which does not require any algorithm-specific control parameter. It just requires random population sets to find the optimum solution of the given problem. While testing on benchmark problems, Jaya algorithm is found to be promising and hence can be used for the optimization in various application fields. Shabgard et al. [9] made use of fuzzy logic with full factorial design to select optimal machining parameters in electrical discharge machining. They concluded that MRR and surface roughness increased by increasing pulse duration and discharge current. Teimouri et al. [10] in their research used rotary tool accompanied by magnetic field. They considered different levels of energy. Continuous ACO algorithm is applied to optimize electrical process parameters, to increase MRR and to reduce SR. Chinke [11] performed an experiment on Electronica Machine Tools Limited make of C400 × 250 model of die sink-type electrical discharge machine. The measurement results obtained in the experimentations are used in this study for the calculation of the weighted grey relational grade. Beryllium copper alloy was employed as workpiece material. In the experiment, process parameters were set according to response surface methodology.

Box–Behnken design was used for the experimentation design. In the design, the response surface is fitted by choosing the levels of input variables, namely magnetic field, gap current, pulse on time at three levels. The reason for the selection of three input parameters is as follows: 1. during EDM process, debris is generated in the machining zone from the unwanted material removed from the workpiece. If they are not expelling out from the cutting zone, leading to arcing instead of sparking, this reduces the machining efficiency. External magnetic strength facilitates to remove the debris from the cutting zone. This also prevents the clogging of particles in the cutting zone. This increases the electrical discharge machining process stability. 2. Gap current is directly proportional to spark energy, and increase in gap energy is increasing the spark energy which tends to increase the surface temperature of workpiece. This causes the material to melt, and this molten metal is then flushed away by dielectric fluid. 3. Pulse on time is the time duration in which current flows per cycle. And the amount of energy applied during pulse on time is directly proportional to material removed. The values of material removal rate and tool wear rate calculated from the experimental data for all different sets of EDM process parameters are given in Table 1.

2 Grey Relational Analysis

Grey relational analysis basically converts a multi-objective optimization problem into a mathematically equivalent single-objective optimization problem. It is basically a very simple multi-objective decision-making technique that has found applications in many engineering fields. The following steps are to be followed to calculate the values of the weighted grey relational grade for each experimental setting.

Step [I] Normalization of S/N Ratio:

In the first step, absolute values of the sequence are converted into a comparable sequence by normalization of the original sequence. This is necessary because objec-

Table 1 Design layout with observed values

Exp. no.	Magnetic field (T)	Gap current (A)	Pulse on time (μs)	MRR (mm³/min)	TWR (mm³/min)
1	0	8	26	2.322984141	0.071456447
2	0.496	8	26	2.221590062	0.08000488
3	0	16	26	6.002886003	0.332907212
4	0.496	16	26	6.547848015	0.315534057
5	0	12	13	1.936314784	0.158468387
6	0.496	12	13	2.042160738	0.160269164
7	0	12	38	4.662896912	0.149856031
8	0.496	12	38	5.003223727	0.166595269
9	0.248	8	13	0.966942149	0.068639414
10	0.248	16	13	1.882378543	0.195799646
11	0.248	8	38	2.897500905	0.044564471
12	0.248	16	38	7.400932401	0.215109276
13	0.248	12	26	4.862787164	0.164980565
14	0.248	12	26	4.643257977	0.201223915
15	0.248	12	26	4.789644013	0.206337829

tive functions in multi-objective optimization problems operate between different ranges and are having different units. Because of this, the original objective function values cannot be compared directly. A linear normalization is therefore required in the range of zero to unity to enable the comparative data sets. Normalization [7] is done by using the following expressions:

(a) If the target response falls under the category of "higher the better", i.e. of the maximization type, then the normalization is done as:

$$x_i^*(k) = \frac{x_i^0(k) - \min x_i^0(k)}{\max x_i^0(k) - \min x_i^0(k)} \tag{1}$$

(b) If the target response falls under the category of "lower the better", i.e. of the minimization type, then the normalization is done as:

$$x_i^*(k) = \frac{\max x_i^0(k) - x_i^0(k)}{\max x_i^0(k) - \min x_i^0(k)} \tag{2}$$

Step [II] Finding the Deviation Sequence:

The deviation sequence, $\Delta^{0i}(k)$, is the difference between the reference sequence $x_0^*(k)$ and the comparability sequence $x_i^*(k)$ after normalization. It can be calculated by the following expression given as:

$$\Delta^{0i}(k) = |x_0 * (k) - x_i * (k)| \tag{3}$$

Step [III] Calculation of Grey Relational Coefficient (GRC):

The grey relational coefficient for all the sequences is calculated by using the following expression:

$$\gamma(x_0 * (k), x_i * (k)) = \frac{\Delta\min + \zeta * \Delta\max}{\Delta^{0i}(k) + \zeta * \Delta\max} \tag{4}$$

$\Delta\min$ is the smallest value of $\Delta_{0i}(k)$ and is given by $\Delta\min = \min_i \min_k (|x_0^*(k) - x_i^*(k)|)$. $\Delta\max$ is the largest value of $\Delta_{0i}(k)$ and is given by $\Delta\max = \max_i \max_k (|x_0^*(k) - x_i^*(k)|)$. The value of ζ is defined in the range of zero to unity but generally taken as 0.5 [4].

Step [IV] Determination of the Weighted Grey Relational Grade (WGRG):

The weighted grey relational grade is the average sum of the weighted grey relational coefficients which is given by

$$\gamma(x_0, x_i) = \frac{1}{m} \sum_{i=1}^{m} W_i * \gamma_i(x_0(k), x_i(k)) \tag{5}$$

where $\sum_{i=1}^{m} W_i = 1$.

The weighting factor is assigned as per the degree of importance of respective objective function in intended application compared to all other objective functions. The weighted grey relational grade values for all the experiments with different sets of weighting factor are calculated and given in Table 2. If the objective functions are having the same relative importance, then both of the objective functions would be assigned the same weighting factor as "0.5".

Step [V] Development of Optimization Model by Regression Analysis:

At this stage, values of the weighted grey relational grade obtained by grey relational analysis are used to develop an optimization model as a function of original design variables with their respective search range. Based on the data given in Table 2, the optimization function is determined by regression analysis using Minitab statistical software. The value of R^2 for the model is 91.79% which sounds great. The maximization of the function is equivalent to optimization of MRR and TWR together. The developed first-order optimization model is as follows:

$$\begin{aligned} \text{WGRG} = {} & 0.251268 + 0.00750929 * \text{Magnetic Strength} - 0.00422563 * \text{Current} \\ & + 0.00300459 * \text{Pulse on time} \end{aligned} \tag{6}$$

Step [VI] Formulation of Pptimization Problem:

The optimization of a newly developed model is equivalent to the multi-objective optimization of original objective functions in the same search range of all design

Table 2 Weighted grey relational grade (WGRG)

Expt. no.	Magnetic field (T)	Gap current (A)	Pulse on time (μs)	WGRG
1	0	8	26	0.307655312
2	0.496	8	26	0.296456483
3	0	16	26	0.25760019
4	0.496	16	26	0.284420996
5	0	12	13	0.232298361
6	0.496	12	13	0.23247399
7	0	12	38	0.279535434
8	0.496	12	38	0.278636254
9	0.248	8	13	0.297559978
10	0.248	16	13	0.214077177
11	0.248	8	38	0.354171547
12	0.248	16	38	0.364524911
13	0.248	12	26	0.275967212
14	0.248	12	26	0.254420755
15	0.248	12	26	0.255799021

variables. Therefore, the original problem now transforms to the single-objective optimization problem as follows:

Maximize, WGRG = 0.251268 + 0.00750929 * G − 0.00422563 * I + 0.00300459 * T_{on}

Subjected to, $0 \leq G \leq 0.496, 8 \leq I \leq 16, 13 \leq T_{on} \leq 38$.

3   Optimization by Using Jaya Algorithm

Optimization of the above function is done by using Jaya algorithm. Jaya is an algorithm-specific-parameterless algorithm [8] is very promising and simple optimization technique to solve various optimization problems. The weighted grey relational grade is considered as the fitness functions of Jaya algorithm which is to be optimized for the process parameters such as magnetic field strength (G), gap current (I) and pulse on time (T_{on}) as the design variables. The algorithm always tries to converge towards the best solution and avoids the worst solution [8]. If $X_{j,k,i}$ is the value of the jth variable for the kth candidate for the ith iteration, then the corresponding population set gets the new values as follows [8]:

$$X'_{j,k,i} = X_{j,k,i} + R_{1,j,i}\left(X_{j,\text{best},i} - \left|X_{j,k,i}\right|\right) - R_{2,j,i}\left(X_{j,\text{worst},i} - \left|X_{j,k,i}\right|\right) \quad (7)$$

Table 3 Search range for the design variables

Serial no.	Design variables	Lower bound	Upper bound
1	Magnetic field (T)	0	0.496
2	Gap current (A)	8	16
3	Pulse on time (μs)	13	38

where $X_{j,\text{best},i}$ and $X_{j,\text{worst},i}$ are the values of the variable j for the best and worst candidates, respectively.

In every iteration, all population sets are either updated or kept same as the previous iteration based on the minimization and maximization problem. If the problem is of minimization type, the solution which has lower fitness function value is carried forward for the next iteration. Similarly, if the problem is of maximization type, the solution which has higher fitness function value is carried forward for the next iteration. Sometimes for any population set, the value of one or more design variables could fall beyond the search range, in that situation the value of respective design variable for that population set is assigned equal to the bound value nearer to it and the population sets are updated accordingly discussed previously.

In the presented work, three design variables: magnetic field (G), gap current (I) and pulse on time (T_{on}) are optimized within their respective search range. The lower bound and upper bound considered for the respective design variables are given in Table 3.

Optimization Results

The optimized results obtained through Jaya algorithm are as follows:

Magnetic strength (G) = **0.496** T,
Gap current (I) = **8.0**A,
Pulse on time (T_{on}) = **38.0** μs.

The corresponding optimum value of WGRG = **0.335362** mm³/min. The optimum conditions as per Table 2 belong to experiment 12. However by Jaya algorithm, the optimum settings come out to be different. This is because in Jaya algorithm all possible parameter settings of design variables have been considered while in Table 2 only 15 different conditions are considered as per surface response methodology.

The convergence graph of Jaya algorithm to the optimum solution is represented in Fig. 1. It can be observed from the above plot that Jaya algorithm just needed six iterations to reach the optimum solution. However, a number of iterations may be different every time as random numbers are involved in the algorithm flow.

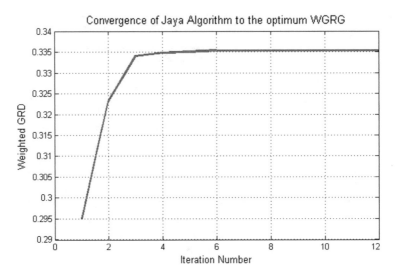

Fig. 1 Convergence graph of Jaya algorithm to the optimum WGRG

4   Conclusion

This work demonstrates the Jaya algorithm optimization technique combined with grey relational analysis for optimizing the EDM process parameters involving multiple target performance characteristics. The optimum setting of design parameters found to be 0.496 T, 8 A, 38 μs for magnetic strength (G), gap current (I), pulse on time (Ton), respectively. It can be deduced from the results that Jaya algorithm converged to the optimum solution in just six iterations with ten initial population sets and hence it required very less function evaluations to reach the optimized solution. Considering all the discussed points, it is clear to conclude that the approach discussed in this paper for multi-objective optimization problems can be extensively applied to the other manufacturing processes. However, higher-order regression model instead of the single-order regression model could be used to increase the reliability of optimization results by this approach.

References

1. Radhika, N., Chandran, G., Shivaram, P., Karthik, K.T.: Multi-objective optimization of EDM parameters using grey relation analysis. J. Eng. Sci. Technol. **10**, 1–11 (2015)
2. Raghuraman, S., Thiruppathi, K., Panneerselvam, T., Santosh, S.: Optimization of edm parameters using taguchi method and grey relational analysis for mild steel is 2026. Int. J. Innov. Res. Sci. Eng. Technol. **2**(7), 3095–3104 (2013)
3. Ohdar, N.K., Jena, B.K., Sethi, S.K.: Optimization of EDM process parameters using taguchi method with copper electrode. Int. Res. J. Eng. Technol. (IRJET) **4**, 2428–2431 (2017)

4. Pawade, R.S., Joshi, S.S.: Multi-objective optimization of surface roughness and cutting forces in high-speed turning of Inconel 718 using taguchi grey relational analysis (TGRA). Int. J. Adv. Manuf. Technol. **56**, 47–62 (2011). https://doi.org/10.1007/s00170-011-3183-z

5. Somashekhar, K. P., Mathew, J., Ramachandran, N.: Multi-objective optimization of micro wire electric discharge machining parameters using grey relational analysis with taguchi method. J. Mech. Eng. Sci. **225**(Part C), 1742–1753 (2011). https://doi.org/10.1177/0954406211400553

6. Vijayanand, M.S., Ilangkumaran, M.: Optimization of micro-EDM parameters using grey-based fuzzy logic coupled with the taguchi method. Mater. Technol. **6**, 989–995 (2017). https://doi.org/10.17222/mit.2017.048

7. Kolahan, F., Golmezerji, R., Moghaddam, M.A.: Multi-objective optimization of turning process using grey relational analysis and simulated annealing algorithm. Appl. Mech. Mater. **110–116**, 2926–2932 (2012). https://doi.org/10.4028/www.scientific.net/AMM.110-116.2926

8. Rao, R.V.: A simple and new optimization algorithm for solving constrained and unconstrained optimization problems. Int. J. Ind. Eng. Comput. **7**, 19–34 (2016). https://doi.org/10.5267/j.ijiec.2015.8.004

9. Shabgard, M.R., Badamchizadeh, M.A., Ranjbary, G., Amini, K.: Fuzzy approach to select machining parameters in electrical discharge machining and ultrasonic assisted EDM processes. J. Manuf. Syst. **32**, 32–39 (2013)

10. Teimouri, R., Baseri, H.: Optimization of magnetic field assisted EDM using the continuous ACO algorithm. Appl. Soft Comput. **14**, 381–389 (2014)

11. Chinke, S.: Experimental investigation and modelling of productivity aspects in EDM of BeCu Alloys. M.Tech. Thesis, Symbiosis International University, Pune, pp. 1–55 (2015)

Application of Fuzzy Integrated JAYA Algorithm for the Optimization of Surface Roughness of DMLS Made Specimen: Comparison with GA

Hiren Gajera, Veera Darji and Komal Dave

Abstract In today's competitive era, the quality of the die or mold plays a significant role in tooling industries. The surface roughness of the die or mold drastically affects the final product and requires many post-processing processes which consume considerable time and money. These problems need to be minimized by controlling process parameters of direct metal laser sintering (DMLS). In order to establish the relationship between the surface roughness of DMLS made parts and process parameters, i.e., laser power, hatch spacing, scan speed, and layer thickness, Box–Behnken design (BBD) of response surface methodology (RSM) was used. This study demonstrates the application of optimization by a combination of nonlinear regression modeling in terms of a fuzzy inference system with the JAYA algorithm to select an optimal parameter set. The result obtained from the JAYA has been compared with the genetic algorithm (GA). A very good agreement has been found between JAYA and GA.

Keywords DMLS · RSM · JAYA · GA · CL50WS · Surface roughness

1 Introduction

Nowadays, additive manufacturing (AM) is widely adopting by industries for their unique capabilities to produce direct parts from a computer-aided design (CAD) file. Direct metal laser sintering (DMLS) is emerging technology among various AM methods like fused deposition modeling, electron beam welding, laminated object

H. Gajera (✉) · V. Darji
C U Shah University, Surendranagar, India
e-mail: gajera.hiren684@gmail.com

V. Darji
e-mail: veera.jani@rediffmail.com

K. Dave
L D College of Engineering, Ahmedabad, India
e-mail: dave_komal@yahoo.co.in

© Springer Nature Singapore Pte Ltd. 2020
R. Venkata Rao and J. Taler (eds.), *Advanced Engineering Optimization Through Intelligent Techniques*, Advances in Intelligent Systems and Computing 949, https://doi.org/10.1007/978-981-13-8196-6_13

modeling, selective laser sintering. DMLS is used to produce metal parts which are used as a functional unit, end-user application, or tooling purpose [1]. Concept laser 50 (grade) work steel (CL50WS) has attained a lot of attention in the area of the tooling industries, aerospace, automotive, medical, and other industries. This is possible because of its superior properties like high strength, high fracture toughness, good weldability at elevated temperatures [2]. The surface quality of the die or mold is the main concern in tooling industries. The surface roughness of die or mold plays a vital role in the performance and durability of die or mold.

Surface roughness is the reasonable parameter for the corrosion and pitting of die and mold. Hence, surface roughness must be reduced by several post-processing methods in order to prevent such effects. Also, the dimensional accuracy of the die or mold is becoming poor due to surface roughness. In addition, time and cost of post-processing process (like super-finishing process) will be increased due to surface roughness. Hence, all these issues encourage to carry out research toward to optimize the surface roughness of die or mold. In that direction, Benedetti et al. [3] have observed that the surface roughness of the built specimen is 7 μm which can be reduced by post-treatment. Also, cavities and pores existed in the unpolished sample increases surface roughness. Corrosion attacks in terms of pitting which develop a crack. Hence, corrosion fatigue failure of AlSi10Mg alloy due to these cracks [4]. Mengucci et al. [5] have revealed that laser power, scan speed, and orientation are the significant parameters for the variation in surface roughness of parts. Mengucci et al. [6] have revealed that the surface roughness is independent of the orientation of part. Snehashis et al. [7] and Casalino et al. [8] have reported that minimum surface roughness (9.39 μm to 10 μm) are obtained at laser power of 150 W and scan speed of 900 mm/s, and also they have revealed that the surface roughness is increased while increasing the value of laser power and decreasing the value of scanning speed. Bandar et al. [9] have concluded that low-cost shot-peening treatment is an effective method to reduce the surface roughness of an untreated sample. Evren et al. [10] have suggested that the large inclination angle reduces surface roughness.

From the literature review, it has been noticed that combined effect of independent parameters like laser power, scan speed, hatch spacing, and layer thickness of the DMLS processed CL50WS material for the surface roughness was not properly explored till date. Also, the systematic optimization method did not explore for the surface roughness of DMLS parts. Hence, it is very essential to address study with an aforesaid parameter for the surface roughness. This attempt is made to determine the parameters set for the surface roughness of DMLS made specimen. In this study, individual specific weight has been assigned to the responses. These weights might be changed as per the responses. Generally, these responses do not have the same priority or weight. The intensity of this weight relies on the application area of part and demand of the parts. It is expected that the surface roughness of DMLS made specimen should be less in order to reduce the post-processing cost as well as to prevent the dimensional accuracy. The designer has not able to set priority weight of response in a proper manner. To avoid these limitations, fuzzy inference system (FIS) has been proposed.

In this study, all specimens were fabricated by the DMLS machine using the Box–Behnken design (BBD) by influencing aforesaid process parameters to examine the surface roughness of two surfaces of the DMLS specimen. In this attempt, the fuzzy inference system (FIS) was applied to convert both the surface roughness value into a single output parameter called as multi-performance characteristic index (MPCI). After that, it developed a nonlinear regression equation which depicts the relationship between the aforesaid process parameter and calculated MPCI. The nonlinear equation has been considered as a fitness function in the JAYA algorithm and genetic algorithm (GA).

2 Experimental Details

Generally, the high carbon steel possesses high carbon content which is responsible for corrosion and cracking causing the unexpected failure of parts. But, CL50WS material is having very less content of carbon and higher amount of nickel content. Hence, CL50WS material is not suffering from corrosion or cracking problem [11]. The base material in the present research work is CL50WS in the form of powder material which is manufactured by German-based Concept Laser Inc. The chemical composition of material was estimated according to the ASTM E-1086-2014 by utilizing spectral analysis at the Divine Metallurgical lab, Ahmedabad. The chemical composition of the CL50WS material is shown in Table 1. The setup consists of an ytterbium (Yb) fiber laser system (peak power 200 W) and an inert gas chamber. The specimens were manufactured on concept laser (Model: M1 Cusing) machine (as shown in Fig. 1).

Design of experiment (DOE) gives a systematic approach to carry out research experiments to determine the best combination of process parameters for achieving a desirable result for research characteristic. BBD is a well-known test outline system used to optimize the process parameter [12, 13]. The BBD system has turned out to be a to a great degree important tool, allowing the exact optimum value of experimental parameters to be determined and additionally the likelihood to assess the communication between factors with a decreased number of trail or experiments [28]. In this study, the three levels of four parameters to be specific layer thickness, laser power, scan speed, and hatches' distance were chosen. Three-level design of Box and Behnken (1960) was utilized. The Box–Behnken design method requires only three levels to run an experiment. It is a special 3-level design because it does not contain any points at the vertices of the experimental region. In BBD, any points

Table 1 Chemical content of CL50WS materials

Material	Carbon	Sulfur	Phosphorous	Manganese	Nickle	Chromium	Molybdenum	Copper	Iron
CL 50 WS	0.03	0.010	0.010	0.15	17–19	0.25	4.50–5.20	8.50–10.0	Balanced

Fig. 1 Direct metal laser sintering machine tool (Courtesy: IGTR, Vatva, Ahmedabad)

Table 2 Range of input parameters

Sr. no.	Process parameter	Unit	Level 1	Level 2	Level 3
			Low level (−1)	Center level (0)	High level (+1)
1	Laser power ($x1$)	W	110	120	130
2	Scanning speed ($x2$)	mm/s	550	600	650
3	Layer thickness ($x3$)	mm	0.03	0.04	0.05
4	Hatch distance ($x4$)	mm	0.01	0.015	0.02

do not lie on the vertices of the experimental design. This is the main characteristic of the BBD method. When points lying on the corner of the cube which describing the level combination of experiment, it becomes very expensive or impossible to carry the test due to the limitation of the physical process. These "missing corners" may be useful when the experimenter should avoid combining factor extremes. This property prevents a potential loss of data in those cases [14]. The range of input parameters used in this study is shown in Table 2. In this study, 27 specimens of CL50WS material were fabricated with various combinations of process parameters as shown in Table 3. Three-level design of Box and Behnken (1960) method was used.

Table 3 Box–Behnken design for experimental work in the coded and actual form

Sr. no.	X1	X2	X3	X4	Laser power (W)	Scan speed (mm/s)	Layer thickness (mm)	Hatch distance (mm)
	Coded variable				Actual variable			
1	−1	−1	0	0	110	550	0.04	0.015
2	1	−1	0	0	130	550	0.04	0.015
3	−1	1	0	0	110	650	0.04	0.015
4	1	1	0	0	130	650	0.04	0.015
5	0	0	−1	−1	120	600	0.03	0.010
6	0	0	1	−1	120	600	0.05	0.010
7	0	0	−1	1	120	600	0.03	0.020
8	0	0	1	1	120	600	0.05	0.020
9	−1	0	0	−1	110	600	0.04	0.010
10	1	0	0	−1	130	600	0.04	0.010
11	−1	0	0	1	110	600	0.04	0.020
12	1	0	0	1	130	600	0.04	0.020
13	0	−1	−1	0	120	550	0.03	0.015
14	0	1	−1	0	120	650	0.03	0.015
15	0	−1	1	0	120	550	0.05	0.015
16	0	1	1	0	120	650	0.05	0.015
17	−1	0	−1	0	110	600	0.03	0.015
18	1	0	−1	0	130	600	0.03	0.015
19	−1	0	1	0	110	600	0.05	0.015
20	1	0	1	0	130	600	0.05	0.015
21	0	−1	0	−1	120	550	0.04	0.010
22	0	1	0	−1	120	650	0.04	0.010
23	0	−1	0	1	120	550	0.04	0.020
24	0	1	0	1	120	650	0.04	0.020
25	0	0	0	0	120	600	0.04	0.015
26	0	0	0	0	120	600	0.04	0.015
27	0	0	0	0	120	600	0.04	0.015

In this study, $x1$; $x2$; $x3$, and $x4$ were taken as a coded variable. Here, $x1$ is coded variable that describes the laser power, $x2$ is the coded variable that describes the scan speed, $x3$ is the coded variable that describes the layer thickness, and $x4$ is a coded variable that describes the hatch space. In RSM, the actual variable has been converted into the coded variable that dimensionless with mean zero and the same spread or standard deviation. The relationship between the coded variable and the natural variable is given by the following formula [15].

$$= \frac{A - \frac{(H+L)}{2}}{\frac{(H-L)}{2}} \tag{1}$$

where A = actual variable or natural variable (such as laser power, scan speed, layer thickness, and hatch space)

H = value of the high level of respective factor

L = value of the lower level of respective factor.

The surface roughness tests were carried out using surface roughness tester (Mitutoyo Model: SJ 201). The surface roughness of each specimen was measured at the three different points on the top surface and three different points on side lateral surface, and the average value of each of them was treated as a final surface roughness value of the specimen.

3 Optimization Methodology

3.1 Fuzzy Inference System

A fuzzy inference system (FIS) consists of four parts, namely knowledge base, fuzzifier, inference engine, and defuzzifier. In the fuzzifier, input value has been inserted into the FIS, and this input value is known as crisp. Generally, this crisp value contains the accurate information of the given parameter. The function of fuzzifier is converting the accurate information into the inaccurate information, namely low, medium, and high. After fuzzifier, knowledge base provides the fuzzy sets and operators which govern by the fuzzy logic. This fuzzy logic comprises if–then rule statements which used to establish the conditional set for the given information [16].

For single fuzzy, if–then rule assumes the following sentence:

If $x1$ is A, then $x2$ is B.

where A and B are linguistic values defined by fuzzy sets on the ranges (universes of discourse) $x1$ and $x2$, respectively.

The aforesaid operation or set of input has been run by the inference system. All the rules have been handled during the inference engine. In defuzzifier, a number of input fuzzy set have been inserted in the inference engine and the output is a specific value. In this study, the center of gravity method has been used for the defuzzification process among various defuzzification methods in order to calculate the output of FIS.

3.2 JAYA Algorithm Optimization

Rao [17] have proposed the algorithm-specific parameterless algorithm named the JAYA. It is a very recently developed algorithm which prime aim is found in the global

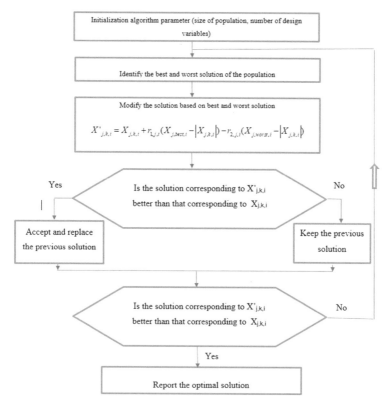

Fig. 2 Flowchart of JAYA algorithm [2]

optimal solution and avoiding the worst solution. The tendency of the JAYA algorithm is always trying to reach to best (success) and avoid the worst condition (failure). This tendency depicting the Sanskrit word JAYA means victory. The flowchart of JAYA algorithm is shown in Fig. 2 [18].

4   Results and Discussion

Table 4 depicts the individual normalization value of both responses and multi-performance characteristic index (MPCI). After that, FIS has been used to receive the individual normalization values of response as input and convert into the single output known as MPCI.

There are three fuzzy sets for each parameter like low, medium, and high which has been assigned as shown in Figs. 3 and 4; five fuzzy sets (very small, small, medium, large, and very large) had assigned for the output parameter (MPCI) as shown in Fig. 5. Here, the input parameters were fuzzified into the proper linguistic

Table 4 Plan of the experiment, the experimental value of the responses, individual normalization value, and MPCI

Sr. no.	Laser power (W)	Scan speed (mm/s)	Layer thickness (mm)	Hatch distance (mm)	Top average surface roughness (μm)	Side average surface roughness (μm)	N-Top RA	N-Side RA	MPCI
1	110	550	0.04	0.015	13.540	6.547	0.483133	0.906019	0.677
2	130	550	0.04	0.015	8.454	6.654	0.85131	0.891612	0.725
3	110	650	0.04	0.015	20.214	12.250	0	0.138145	0.179
4	130	650	0.04	0.015	10.994	6.548	0.667439	0.905884	0.704
5	120	600	0.03	0.010	6.400	5.898	1	0.993402	0.913
6	120	600	0.05	0.010	12.250	9.555	0.576517	0.50101	0.548
7	120	600	0.03	0.020	10.963	6.736	0.669683	0.880571	0.693
8	120	600	0.05	0.020	18.741	12.324	0.106631	0.128181	0.269
9	110	600	0.04	0.010	12.210	5.849	0.579412	1	0.754
10	130	600	0.04	0.010	11.987	7.256	0.595555	0.810556	0.655
11	110	600	0.04	0.020	15.998	8.123	0.305198	0.69382	0.5
12	130	600	0.04	0.020	14.857	7.453	0.387795	0.784031	0.565
13	120	550	0.03	0.015	8.994	6.124	0.812219	0.962973	0.767
14	120	650	0.03	0.015	10.122	7.444	0.730563	0.785243	0.665
15	120	550	0.05	0.015	18.654	10.245	0.112929	0.408106	0.311
16	120	650	0.05	0.015	15.898	11.842	0.312437	0.193079	0.333

(continued)

Table 4 (continued)

Sr. no.	Laser power (W)	Scan speed (mm/s)	Layer thickness (mm)	Hatch distance (mm)	Top average surface roughness (µm)	Side average surface roughness (µm)	N-Top RA	N-Side RA	MPCI
17	110	600	0.03	0.015	9.303	8.711	0.789851	0.614649	0.649
18	130	600	0.03	0.015	8.419	7.245	0.853844	0.812037	0.699
19	110	600	0.05	0.015	15.473	13.276	0.343203	0	0.237
20	130	600	0.05	0.015	13.241	8.032	0.504778	0.706072	0.607
21	120	550	0.04	0.010	11.751	6.335	0.612639	0.934563	0.714
22	120	650	0.04	0.010	12.365	7.212	0.568192	0.81648	0.655
23	120	550	0.04	0.020	17.892	9.086	0.16809	0.564158	0.543
24	120	650	0.04	0.020	12.368	9.894	0.567975	0.455366	0.513
25	120	600	0.04	0.015	14.095	8.745	0.442956	0.610071	0.527
26	120	600	0.04	0.015	14.758	8.245	0.394962	0.677393	0.531
27	120	600	0.04	0.015	14.645	8.889	0.403142	0.590683	0.497

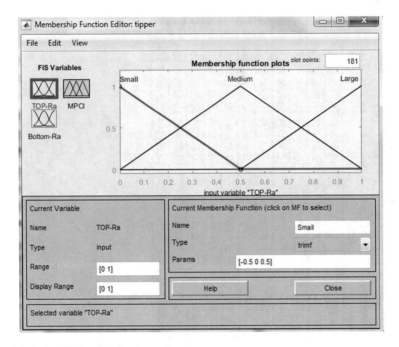

Fig. 3 Membership function for the top Ra

term and set 9 logic rules (as shown in Fig. 6) in the FIS. MPCI has been calculated using logic rules which are described in Fig. 7.

To obtain the fuzzy logic rules, a fuzzy set $\overline{A}$ is represented by triangular fuzzy number which is defined by the triplet (a, b, c). Membership function $\mu \overline{A}(x)$ is defined as:

$$\forall x, a, b, c \in R$$

$$\mu \overline{A}(x) = 0 \text{ if } x < a \text{ else} \left(\frac{x-a}{x-b}\right), \text{ if } a \leq x \leq b \text{ else} \tag{3}$$

$$\left(\frac{c-x}{c-b}\right), \text{ if } b \leq x \leq c \text{ else } 0, \text{ if } x > c \tag{4}$$

For a rule, Ri: If $x1$ is A_{ti} and x_2 is A_{ti} ... x_s is A_{si}, then yi is Ci, $i = 1,2,..., M$.

Here, M is the total number of the fuzzy rule, xj ($j = 1,2....,s$) is the input variable, yi is the output variable, and A_{ij} and Ci are the fuzzy sets modeled by membership functions μAij and $\mu c_i (y_i)$, respectively. The aggregated output for the M rules is:

$$\max\left[\left\{\min\left\{\mu_{A_{1i}}(x_1), \mu_{A_{2i}}(x_2), \ldots, \mu_{A_{si}}(x_s)\right\}\right]\right] \tag{5}$$

$$i = 1, 2, \ldots, M$$

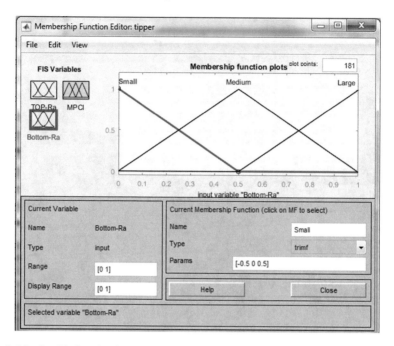

Fig. 4 Membership function for the bottom Ra

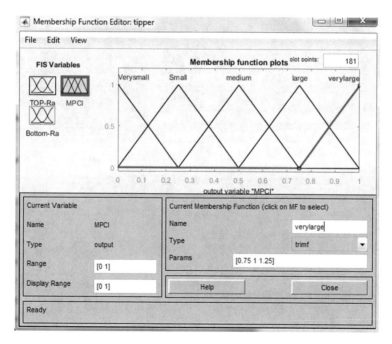

Fig. 5 Membership function of MPCI

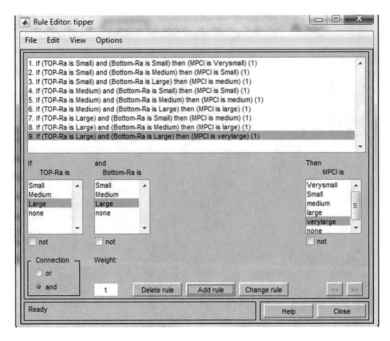

Fig. 6 Fuzzy rules' matrix

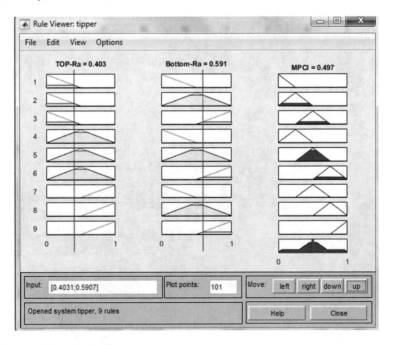

Fig. 7 Fuzzy logic rules' editor

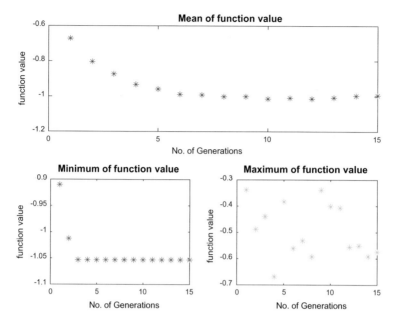

Fig. 8 Convergence plot by JAYA algorithm

In this study, a defuzzification method has been using the center of gravity method to calculate the output value. The following formula to find the centroid of the combined outputs is

$$\widehat{y_i} = \frac{\int y_i \mu_{ci}(y_i)\mathrm{d}y}{\int \mu_{ci}(y_i)\mathrm{d}y} \qquad (6)$$

In this work, the non-fuzzy value $\widehat{y_i}$ is called an output called as MPCI. Based on the above discussion, the larger the MPCI, the better is the response.

After completing the defuzzification, MPCI value has been considered as a single response value. The fitness function of MPCI as below has been obtained from the nonlinear regression analysis.

$$Z = 0.004 \times X1^{(1.366)} \times X2^{(-1.120)} \times X3^{(-1.052)} \times X4^{(-0.500)} \qquad (7)$$

The convergence plots for the JAYA algorithm and genetic algorithm are shown in Fig. 8 and 9, respectively. As shown in Table 5, optimal results were obtained through the JAYA and GA, are appeared the similar means a good agreement has been observed. The optimal value has been achieved at 1.00944 with as optimal process parameter set: laser power: 128.632 W, scan speed: 554.917 mm/s, layer thickness: 0.0307 mm, and hatch spacing: 0.010 mm.

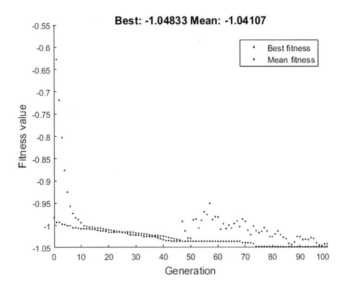

Fig. 9 Convergence plot for optimizing MPCI value by GA

Table 5 Results of optimizing MPCI value and comparison of JAYA and GA

Optimization results	Algorithm	
	JAYA	GA
Optimal parameter set	Laser power: 128.632 W	Laser power: 129 W
	Scan speed: 554.917 mm/Sec	Scan speed: 551 mm/Sec
	Layer thickness: 0.0307 mm	Layer thickness: 0.03 mm
	Hatch spacing: 0.010 mm	Hatch spacing: 0.010 mm
Fitness function value	1.00944	1.04833

5 Conclusion

From the past research literature, it can be said that of the optimization problem of surface roughness of DMLS made specimen with the same material and the same range of control parameters almost nil to address. The current attempt aims to optimize the responses, namely top and side surface roughness of DMLS made specimen. The Box–Behnken method was selected to determine the independent factors set to carry out the experiment and to obtain a multi-objective response by reducing the experimental turn as compared to the central composite design. Herein, JAYA and GA have been used to conduct optimization in which MPCI has been considered as a fitness function. It has been observed that JAYA gives results rapidly as it has only one phase. Here, the application of JAYA has been compared with the GA algorithm. In the case of GA, some specific parameter has been tuned properly to get an enhancement of the performance of the algorithm. The good agreement between the algorithms has been observed. Contrarily, JAYA was required only tuning of a common controlling parameter of the algorithm to implement successfully and get good performance of the algorithm. From the study, it can be said that optimal conditions for the surface roughness are laser power: 128.632 W, scan speed: 554.917 mm/s, layer thickness: 0.0307 mm, and hatch spacing: 0.010 mm.

References

1. Lee, J.H., Jang, J.H., Joo, B.D., Yim, H.S., Moon, Y.H.: Application of direct laser metal tooling for AISI H13 tool steel. Trans. Nonferrous Met. Soc. China (English Ed), **19**(SUPPL. 1), 2–5 (2009)
2. Yasa, E., Deckers, J., Kruth, J.-P., Rombouts, M., Luyten, J.: Charpy impact testing of metallic selective laser melting parts. Virtual Phys. Prototyp. (2010)
3. Benedetti, M., et al.: The effect of post-sintering treatments on the fatigue and biological behavior of Ti-6Al-4V ELI parts made by selective laser melting. J. Mech. Behav. Biomed. Mater. **71**(January), 295–306 (2017)
4. Leon, A., Aghion, E.: Effect of surface roughness on corrosion fatigue performance of AlSi10Mg alloy produced by selective laser melting (SLM). Mater. Charact. **131**(2016), 188–194 (2017)
5. Girardin, E., et al.: Biomedical Co–Cr–Mo components produced by direct metal laser sintering. Mater. Today Proc. **3**(3), 889–897 (2016)
6. Mengucci, P., et al.: Effects of build orientation and element partitioning on microstructure and mechanical properties of biomedical Ti-6Al-4V alloy produced by laser sintering. J. Mech. Behav. Biomed. Mater. **71**(February), 1–9 (2017)
7. Pal, S., Tiyyagura, H.R., Drstvenšek, I., Kumar, C.S.: The effect of post-processing and machining process parameters on properties of stainless steel PH1 product produced by direct metal laser sintering. Procedia Eng. **149**(June), 359–365 (2016)
8. Casalino, G., Campanelli, S.L., Contuzzi, N., Ludovico, A.D.: Experimental investigation and statistical optimisation of the selective laser melting process of a maraging steel. Opt. Laser Technol. **65**, 151–158 (2015)
9. AlMangour, B., Yang, J.M.: Improving the surface quality and mechanical properties by shot-peening of 17–4 stainless steel fabricated by additive manufacturing. Mater. Des (2016)
10. Yasa, E., Poyraz, O., Solakoglu, E.U., Akbulut, G., Oren, S.: A study on the stair stepping effect in direct metal laser sintering of a nickel-based superalloy. Procedia CIRP **45**, 175–178 (2016)
11. Yasa, E., Deckers, J., Kruth, J.-P., Rombouts, M., Luyten, J.: Charpy impact testing of metallic selective laser melting parts. Virtual Phys. Prototyp. **5**(2), 89–98 (2010)
12. Khajeh, M.: Application of Box-Behnken design in the optimization of a magnetic nanoparticle procedure for zinc determination in analytical samples by inductively coupled plasma optical emission spectrometry. J. Hazard. Mater (2009)
13. Kirboga, S., Öner, M.: Application of experimental design for the precipitation of calcium carbonate in the presence of biopolymer. Powder Technol. (2013)
14. Tekindal, M.A., Bayrak, H., Ozkaya, B., Genc, Y.: Box-behnken experimental design in factorial experiments: the importance of bread for nutrition and health. Turkish J. F. Crop (2012)
15. Abhishek, K., Datta, S., Mahapatra, S.S.: Multi-objective optimization in drilling of CFRP (polyester) composites: application of a fuzzy embedded harmony search (HS) algorithm. Meas. J. Int. Meas. Confed. **77**, 222–239 (2016)

16. Singh, A., Datta, S., Mahapatra, S.S., Singha, T., Majumdar, G.: Optimization of bead geometry of submerged arc weld using fuzzy based desirability function approach. J. Intell. Manuf. **24**(1), 35–44 (2013)
17. Venkata Rao, R.: Jaya: A simple and new optimization algorithm for solving constrained and unconstrained optimization problems. Int. J. Ind. Eng. Comput. **7**(1) 19–34 (2016)
18. Rao, R.V., More, K.C., Taler, J., Ocłoń, P.: Dimensional optimization of a micro-channel heat sink using Jaya algorithm. Appl. Therm. Eng. **103**, 572–582 (2016)

Comparative Analysis of Fruit Categorization Using Different Classifiers

Chirag C. Patel and Vimal K. Chaudhari

Abstract The aim of the paper is to measure the performance of fruit images on different mentioned classifiers based on color, zone, area, centroid, size, equvidi-ameter, perimeter and roundness features. Features are extracted automatically from the provided images and passed to classifiers such as multi-class SVM, KNN, Naive Bayes, random forest and neural network to train the models, classify test images and measure the performance of the classifiers on different data mining tools. The experimental results show that multi-class SVM obtained 87.5 and 91.67% accuracy which is the highest accuracy rather than other mentioned classifiers. KNN has 58.3, 62.5 and 45.83%, Naive Bayes has 62.5 and 58.3%, random forest has 70.8, 75 and 83.33%, and neural network has 66.7 and 75% accuracy based on different data mining tools. Confusion matrix and ROC analysis also show that multi-class SVM has an efficient performance on the proposed features than other mentioned classifiers.

Keywords Fruit classification · Image classification · Image recognition · SVM · KNN · Random forest · Naive Bayes · Neural network

1 Introduction

Automatically, fruit categorization using image processing has been proposed by various researchers. There are various methods for image classification. Traditional method for image classification is based on text-based image retrieving. It requires annotating images with keywords manually, so there is a possibility that the same images may contain different keyword annotations and different images may contain the same keyword annotations. This keyword ambiguity will reflect on the accuracy of image classification [1]. To overcome this limitation, content-based image

C. C. Patel (✉)
UCCC & SPBCBA & SDHG College of BCA & I.T, Udhna, Surat, India
e-mail: chirag3153@gmail.com

V. K. Chaudhari
Department of Computer Science, Veer Narmad South Gujarat University, Surat, India

© Springer Nature Singapore Pte Ltd. 2020
R. Venkata Rao and J. Taler (eds.), *Advanced Engineering Optimization Through Intelligent Techniques*, Advances in Intelligent Systems and Computing 949, https://doi.org/10.1007/978-981-13-8196-6_14

153

retrieval (CBIR) approach has been developed which works on extracting automatically visual features of images like color, appearance and outline from images [2, 3]. CBIR is required to identify the visual content of images, extract automatically it from provided images and classify images based on visual contents. Researchers need to find the features from provided images and automatically extract features from images [4]. Feature extraction process will create a dataset of variables with proper accuracy. Features are extracted from the images and transformed into class or categories and provide it to the machine learning algorithm. Classification process will process the unknown object in query image and compare to every sample of the objects that are previously provided at time of classifier training to train classifier algorithm researcher can use supervised training is the task of inferring a function from labeled training data or researcher can use unsupervised training is the task to put efforts to find masked structure in unlabeled data method to train classifiers. There have been approaches proposed previously based on color and shape of fruit images, but fruits may have similar color and shape which will lead to ambiguity to classify proper images and also it reflects on accuracy of the classifiers; therefore, the proposed approach is based on color, zone, area, centroid, size, equvidiameter, perimeter and roundness features of the fruit images. This paper provides an analysis of the fruit categorization on different machine learning classifiers.

2 Related Work

Fruit categorization is one of the complex tasks of the image recognition because fruits have many different types of properties. Many attempts have been made in the area of fruit recognition owing to its varied application like Veggie Vision produce recognize introduced in 1996, in that produce item placed on scale then image is taken and color, shape and density features are extracted from images afterward it compared with trained data depending upon certainty of the classification results are produced. Authors have found 95% accuracy of the system in the first four choices [5]. Arivazhagan and others have approached fruit recognition based on the fusion of color and texture features by using minimum distance classifier based on statistical and characterized features which are acquired from the wavelet transformed sub-bands. For experiment, they have used more than 2000 fruit images of 15 different categories and get 86.004% accuracy [6]. Shiv Ram Dubey and Anand Singh Jalal have suggested the state-of-the-art color and texture-based features for fruit recognition by using multi-class support vector classifier and K-mean clustering-based image segmentation to subtract the background and found 93.84% accuracy [7]. Yudong Zhang and Lenan Wu have suggested outline, pattern and histogram-based features, used multi-class kernel support vector machine algorithm, applied fivefold stratified cross-validation on training data and got 88.2% accuracy [8, 9]. Michael Vogl and others have the proposed color, shape, texture and intensity-based features; once features are extracted, they have generated a unique code for each image which is used as a search key; through search key distance table, the nearest match algorithm is applied for searching in the database. They have used 1108 images for 36 different classes and formed more than 98% of accuracy [10]. Hossam M. Zawbaa and others proposed shape, color and scale-invariant feature transform (SIFT)-based features using random forest classification. They have worked on apple, strawberry and orange fruits and formed 94.74% apple, 85.71% strawberry and 50% orange accuracy [11]. Woo Chaw Seng and Seyed Hadi Mirisaee have the proposed color, shape and size-based features using K-nearest neighborhood classifier and found 90% accuracy [12].

3 Methodology

The proposed content-based image retrieval method for fruit categorization is based on color, zone, area, centroid, size, roundness, perimeter and equvidiameter features of the fruit images. In this study, experiments are performed on various fruits: apple, banana, orange, pear, watermelon and mango fruit images; 138 images are used to training models, and 24 images are used for testing. Images were collected from the Internet freely available dataset.

3.1 Image Preprocessing

It is a process which is applied to all the images before extracting features from the images to accurately extract features from images; in the proposed approach, the following preprocessing is applied.

- **Image resized**: All images are resized to 200×200 pixel to create the same scale of the image.
- **RGB to gray scale**: Image is resized, and then it converts to gray scale to identify edges of the image.
- **Edges**: Find edges of the images using Canny edge detection algorithm.
- **Threshold**: Image threshold has been calculated by finding edges of the images.
- **Dilation**: By using image edge and threshold, apply dilation operation to remove backgrounds of images.
- **Binary**: After applying dilation, convert image to binary image.

3.2 Feature Extraction

Once preprocessing is applied on images, then features have been extracted from the images which transform into an array of the features to train classifier models and to test classifier performance.

- **Color**:
  To calculate the color feature, Mean of RGB fruits image is used, to identify color more accurately color images is cropped as per binary image which is used to find the region of the fruit in the image. To measure image color, crop five small parts from the different locations of cropped color image and calculate the mean of the color of small parts. Small parts of the color images have been taken from top, middle, bottom, left and right location of the cropped color image.
- **Zone**:
  Binary image contains fruit region that will be divided into four quadrants of pixels, and then calculate percentage of the each quadrants.
- **Centroid**:
  It is average mean pixel of rows and columns, and it is average mean row major axis length and average mean of column minor axis length.
- **Size**:
  Size of the fruit is height and width of the fruit area which have been calculated after converting image into binary image.
- **Equvidiameter**:
  Equvidiameter is to find the diameter of the fruit image region with the same area as region which will be found in binary image. To calculate equvidiameter, sqrt(4 * Area/pi) formulation is used.
- **Perimeter**:
  A perimeter is the distance of the region of the fruit. It calculates the area between each neighbor pair of pixels surrounding the border of the region. It is fetched from the fruit region and provided to the classifier.
- **Roundness**:
  Roundness contains the outline's gross features rather than it is border and region, or exterior roughness of created object. It will be calculated using (4 * Obj_area * pi)/Per.^ 2 formulations.

3.3   Classifier

An array of the features will be provided to classifiers: SVM, KNN, Naive Bayes, random forest and neural network to train models and to test images to measure the performance of classifiers in data mining tools.

3.3.1   Support Vector Machine

It is a machine learning algorithm for data categorization and regression analysis. It is designed for binary class data classifications, but it can work for the multi-class data. There are three approaches used for multi-class SVM; that is, one-against-all approach compares one class with all remaining classes; one-against-one approach builds $k(k-1)/2$ to create classifier model, it is the same as a model created by one-against-one, and for testing it uses source binary straight acyclic graph [13].

3.3.2 *K*-Nearest Neighborhood

It will use testing data to compare with training data. It will take small chunks of data that are closed with the test data; to compute closeness of data, Euclidean distance equation is used. *K*-nearest neighborhood classification is based on *K*, and it is a parameter for the training design models to compare with test data [14].

3.3.3 Naive Bayes

It encodes probability set for n variables as acyclic graph and set of conditional probability distributions (CPDs). Each node is linked with variable, and associated CPD will give the probability state of its parents. When unobserved data is used, then EM algorithm is used which computes alternative likelihood. The goal of inference in Naive Bayes network is to compute marginal distribution for query variables [15].

3.3.4 Random Forest

It uses tree-based architecture for classification, and each tree relies on random vector samples of all trees of forest. Each tree's intensity is relied on correlation between trees. If training datasets increased then numbers of trees are also increased after creating large trees they select the accepted classes [16].

3.3.5 Neural Network

It is made of various neurons which are correlated with each other with network architecture. Neural Network is a connection associated with weight. Neural network is a self-adaptive method which will adjust the weight during learning phase. It is a nonlinear model. It is able to estimate posterior probabilities which create rules and perform statistical analysis [17].

4   Results

For the evolution of classifiers, 138 images of six fruits are used to train classifier model and 24 images are used to test classifiers.

Figures 1 and 2 show sample images of the training and testing data to extract features and analyzed data in different classifiers. Following is the detailed accuracy of the classifiers on different parameters of the model performance.

Table 1 describes detailed accuracy of mentioned classifiers on different data mining tools. Its fields are area under ROC (AUC), classification accuracy (CA), $F1$ (balanced F-score or F-measure score). Precision is the proportion of instances calculated by the positive instances categorized by classifier divided by overall instances in that category, recall is instance classified as a given class divided by the actual total in that class, TP rate is true positive rate, FP rate is false positives rate, and MCC stands for Matthews correlation coefficient. Detailed accuracy of the classifiers shows that SVM has 87.5% and 91.67, KNN has 58.3, 62.5 and 45.83%, Naive Bayes has 62.5

Fig. 1   Sample images to train classifier model

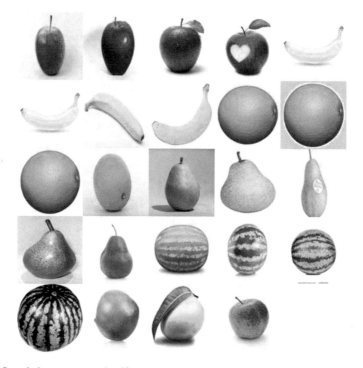

Fig. 2 Sample images to test classifier

Table 1 Detail accuracy of classifiers

Scores

Method	AUC	CA	F1	Precision	Recall
SVM	1.000	0.875	1.000	1.000	1.000
kNN	0.456	0.583	0.286	0.333	0.250
Naive Bayes	1.000	0.625	1.000	1.000	1.000
Random Forest	0.975	0.708	0.889	0.800	1.000
Neural Network	1.000	0.667	1.000	1.000	1.000

Method	CA	F-Measure	Precision	Recall
SVM	0.9167	0.9220	0.9333	0.9111
KNN	0.4583	0.5053	0.4083	0.6626
Naïve Bayes	0.5833	0.5584	0.5333	0.5861
Random Forest	0.8333	0.8289	0.8416	0.8166
Neural Network	0.7500	0.6969	0.6833	0.7111

	TP Rate	FP Rate	Precision	Recall	F-Measure	MCC	ROC Area	PRC Area
SVM	0.875	0.025	0.900	0.875	0.865	0.855	0.946	0.805
KNN	0.625	0.078	0.631	0.625	0.618	0.545	0.912	0.809
Naïve Bayes	0.583	0.078	0.561	0.583	0.547	0.498	0.964	0.895
Random Forest	0.750	0.036	0.806	0.750	0.751	0.724	0.976	0.876
Neural Network	0.750	0.039	0.765	0.750	0.745	0.714	0.946	0.846

Table 2 Confusion Matrix of classifiers

Confusion matrix for SVM (showing number of instances)

		Predicted						
		Apple	Banana	Mango	Orange	Pear	Watermelon	Σ
Actual	Apple	5	0	0	0	0	0	5
	Banana	0	4	0	0	0	0	4
	Mango	0	0	2	0	0	0	2
	Orange	0	0	1	3	0	0	4
	Pear	0	0	1	0	4	0	5
	Watermelon	0	0	0	1	0	3	4
	Σ	5	4	4	4	4	3	24

Confusion matrix for kNN (showing number of instances)

		Predicted						
		Apple	Banana	Mango	Orange	Pear	Watermelon	Σ
Actual	Apple	3	1	1	0	0	0	5
	Banana	0	1	0	2	1	0	4
	Mango	1	0	1	0	0	0	2
	Orange	0	0	1	3	0	0	4
	Pear	0	0	1	1	3	0	5
	Watermelon	0	1	0	0	0	3	4
	Σ	4	3	4	6	4	3	24

Confusion matrix for Naive Bayes (showing number of instances)

		Predicted						
		Apple	Banana	Mango	Orange	Pear	Watermelon	Σ
Actual	Apple	3	0	1	1	0	0	5
	Banana	0	4	0	0	0	0	4
	Mango	1	0	1	0	0	0	2
	Orange	0	0	3	1	0	0	4
	Pear	1	0	0	0	3	1	5
	Watermelon	0	0	1	0	0	3	4
	Σ	5	4	6	2	3	4	24

Confusion matrix for Random Forest (showing number of instances)

		Predicted						
		Apple	Banana	Mango	Orange	Pear	Watermelon	Σ
Actual	Apple	4	0	0	0	1	0	5
	Banana	0	4	0	0	0	0	4
	Mango	0	0	1	1	0	0	2
	Orange	0	0	3	1	0	0	4
	Pear	0	1	0	0	3	1	5
	Watermelon	0	0	0	0	0	4	4
	Σ	4	5	4	2	4	5	24

Confusion matrix for Neural Network (showing number of instances)

		Predicted						
		Apple	Banana	Mango	Orange	Pear	Watermelon	Σ
Actual	Apple	3	0	1	0	1	0	5
	Banana	0	4	0	0	0	0	4
	Mango	0	0	0	2	0	0	2
	Orange	0	0	0	4	0	0	4
	Pear	1	0	0	0	4	0	5
	Watermelon	1	0	1	0	1	1	4
	Σ	5	4	2	4	8	1	24

➢ Confusion Matrix for SVM

	Apple	Banana	Orange	Pear	Watermelon	Mango
Apple	5	0	0	0	0	0
Banana	0	4	0	0	0	0
Orange	0	0	4	0	0	0
Pear	0	1	0	3	1	0
Watermelon	0	0	0	0	4	0
Mango	0	0	1	0	0	1

➢ Confusion Matrix for KNN

	Apple	Banana	Orange	Pear	Watermelon	Mango
Apple	2	0	1	1	0	1
Banana	0	3	0	1	0	0
Orange	1	0	3	0	0	0
Pear	0	0	0	3	2	0
Watermelon	0	1	0	0	3	0
Mango	0	0	1	0	0	1

➢ Confusion Matrix for Naïve Bayes

	Apple	Banana	Orange	Pear	Watermelon	Mango
Apple	3	0	0	1	0	1
Banana	0	4	0	0	0	0
Orange	0	0	0	1	0	3
Pear	1	0	0	4	0	0
Watermelon	1	1	0	0	2	0
Mango	0	0	0	1	0	1

➢ Confusion Matrix for Random Forest

	Apple	Banana	Orange	Pear	Watermelon	Mango
Apple	4	0	1	0	0	0
Banana	0	4	0	0	0	0
Orange	0	0	1	0	0	3
Pear	0	0	1	3	1	0
Watermelon	0	0	0	0	4	0
Mango	0	0	0	0	0	2

➢ Confusion Matrix for Neural Network

	Apple	Banana	Orange	Pear	Watermelon	Mango
Apple	5	0	0	0	0	0
Banana	0	4	0	0	0	0
Orange	0	0	1	0	0	3
Pear	1	0	0	4	0	0
Watermelon	0	0	0	0	4	0
Mango	0	1	1	0	0	0

➢ Confusion Matrix for SVM

	Apple	Banana	Orange	Pear	Watermelon	Mango
Apple	5	0	0	0	0	0
Banana	0	4	0	0	0	0
Orange	0	0	4	0	0	0
Pear	0	0	1	3	0	1
Watermelon	0	0	0	0	4	0
Mango	0	0	0	0	0	2

➢ Confusion Matrix for KNN

	Apple	Banana	Orange	Pear	Watermelon	Mango
Apple	1	0	0	1	0	3
Banana	2	1	0	0	0	1
Orange	0	0	3	0	0	1
Pear	0	0	0	5	0	0
Watermelon	2	0	0	0	1	1
Mango	2	0	0	0	0	0

➢ Confusion Matrix for Naïve Bayes

	Apple	Banana	Orange	Pear	Watermelon	Mango
Apple	2	0	3	0	0	0
Banana	0	4	0	0	0	0
Orange	0	0	1	0	0	3
Pear	0	0	0	4	0	1
Watermelon	1	1	0	0	3	0
Mango	0	1	0	1	0	0

➢ Confusion Matrix for Random Forest

	Apple	Banana	Orange	Pear	Watermelon	Mango
Apple	4	0	1	0	0	0
Banana	0	4	0	0	0	0
Orange	0	0	1	0	0	3
Pear	0	0	0	5	0	0
Watermelon	0	0	0	0	4	0
Mango	0	0	0	0	0	2

(continued)

Table 2 (continued)

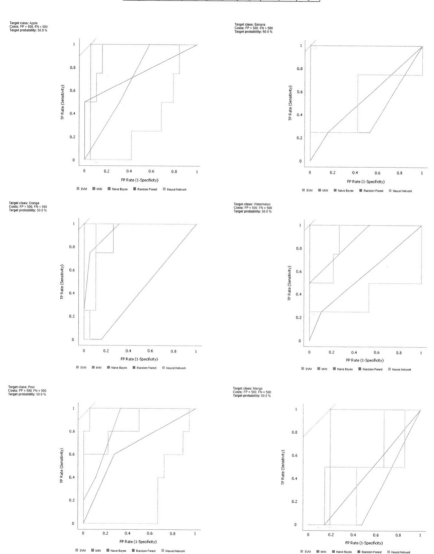

➤ **Confusion Matrix for Neural Network**

	Apple	Banana	Orange	Pear	Watermelon	Mango
Apple	4	0	0	0	0	1
Banana	0	4	0	0	0	0
Orange	0	0	2	0	0	2
Pear	0	1	0	4	0	0
Watermelon	0	0	0	0	4	0
Mango	0	0	1	1	0	0

Fig. 3 ROC analysis of fruit classes

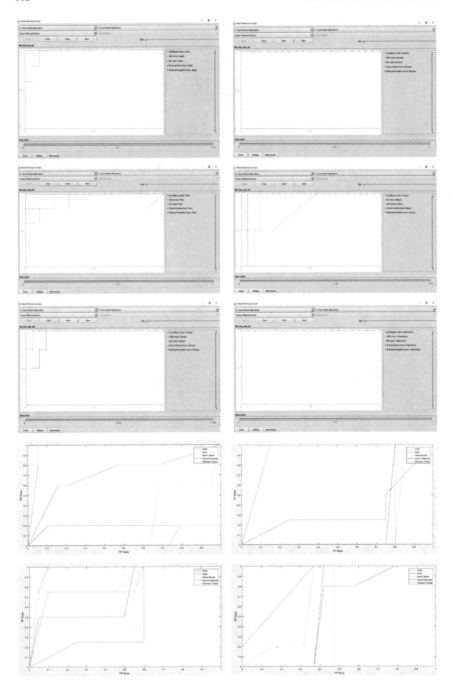

Fig. 3 (continued)

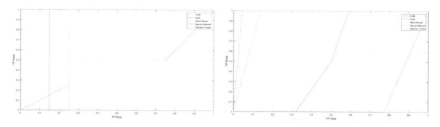

Fig. 3 (continued)

and 58.3%, random forest has 70.8, 75 and 83.33%, and neural network has 66.7 and 75% accuracy based on different data mining tools.

Table 2 describes confusion matrix of different categories of mentioned classifiers on different data mining tools. Confusion matrix will show occurrences in target category, and individual row shows occurrences in output class. Every element of rows and columns shows the occurrences of jth category that categorized as ith class. Confusion matrix defines that multi-class SVM is efficiently classifying fruits than other mentioned classifiers.

Figure 3 describes ROC analysis of fruit categories on different data mining tools. It depends on positive occurrences in difference of negative occurrences as per different thresholds. ROC analysis describes that multi-class SVM is more efficient than other classifiers taken for comparison.

5 Conclusion

In this research, analysis of the fruit categorization on different classifiers is performed based on color, zone, area, centroid, size, equvidiameter, perimeter and roundness features of fruit images and found that multi-class SVM has more efficient accuracy that is 87.5 and 91.67% on different data mining tools which are higher than other mentioned classifiers. Confusion matrix and ROC analysis charts are also showing that SVM provides more efficient performance based on the proposed features rather than other mentioned classifiers. From this experiment, it can conclude that the proposed parameters of fruit images will provide better performance on multi-class SVM than other mentioned classifiers.

References

1. Tang, X., Liu, K., Cui, J., Wen, F., Wang, X.: Intentsearch: capturing user intention for one-click internet image search. IEEE Trans. Pattern Analysis Mach. Intell. **34**(7) (2012)
2. Kwak, J.W., Cho, N.I.: Relevance feedback in content-based image retrieval system by selective region growing in the feature space. Signal Process Image Commun. **18**, 787–799 (2003)
3. Gode, C.S., Ganar, A.N.: Image retrieval by using colour, texture and shape features. Int. J. Adv. Res. Electr. Electron. Instrumentation Eng. **3**(4) (2014)
4. Smeulders, A.W., Worring, M., Santini, S., Gupta, A., Jain, R.: Content-based image retrieval at the end of the early years. IEEE Trans. Pattern Anal. Mach. Intell. **22**(12) (2000)
5. Bolle, R.M., Connell, J.H., Haas, N., Mohan, R., Taubin, G.: Veggievision: a produce recognition system: applications of Computer Vision, WACV96'. In: Proceedings 3rd IEEE Workshop on IEEE, 1996
6. Arivazhagan, S., Shebiah, R.N., Nidhyanandhan, S.S., Ganesan, L.: Fruit recognition using color and texture features. J. Emerg. Trends Comput. Info. Sci. **1**(2) (2010)
7. Dubey, S.R., Jalal, A.S.: Fruit and vegetable recognition by fusing colour and texture features of the image using machine learning. Int. J. Appl. Pattern Recognition, **2**(2) (2015)
8. Zhang, Y., Wu, L.: Classification of fruits using computer vision and a multiclass support vector machine. *Sensors* **12**(9) (2012)
9. Kshirsagar, M., Arora, P.: Classification techniques for computer vision based fruit quality inspection: a review. Int. J. Recent Adv. Eng. Technol. **2**(3) (2014)
10. Vogl, M., Kim, J.Y., Kim, S.D.: A fruit recognition method via image conversion optimized through evolution strategy. In: 2014 IEEE 17th International Conference on Computational Science and Engineering, 2014
11. Zawbaa, H.M., Hazman, M., Abbass, M., Hassanien, A.E.: Automatic fruit classification using random forest algorithm. In: 2014 14th International Conference on Hybrid Intelligent Systems, IEEE, 2014
12. Seng, W.C., Mirisaee, S.H.: A new method for fruits recognition system. In 2009 International Conference on Electrical Engineering and Informatics, 2009
13. Hsu, C.W., Lin, C.J.: A comparison of methods for multiclass support vector machines. IEEE Trans Neural Netw **13**(2) (2002)
14. Shukla, D., Desai, A.: Recognition of fruits using hybrid features and machine learning. In: 2016 International Conference on Computing, Analytics and Security Trends (CAST),IEEE, 2016
15. Lowd, D., Domingos, P.: Naive Bayes models for probability estimation. In Proceedings of the 22nd international conference on Machine learning, ACM, 2005
16. Breiman, L.: Random Forests. Kluwar Academic Publishers (2001)
17. Zhang, G.P.: Neural networks for classification: a survey. IEEE Trans. Syst. Man Cybern. Part C (Appl. Rev.) **30**(4) (2000)

Design Optimization of Plate-Fin Heat Exchanger by Using Modified Jaya Algorithm

Kiran Chunilal More and R. Venkata Rao

Abstract In the present paper, bi-objective optimization of plate-fin heat exchanger (PFHE) is considered. An advanced optimization algorithms called TLBO, Jaya algorithm (JA), and its modified version known as self-adaptive Jaya algorithms are applied for the design and optimization of PFHE. The minimization of cost and volume of PFHE is selected as an objective function and optimized both objective functions simultaneously. The outcomes achieved by TLBO, JA, and modified JA for considered optimization problems are found superior as compared to particle swarm optimization (PSO) and genetic algorithm (GA). In case of function evaluations, computational time and population size demanded by the modified JA are lesser than various algorithms. In case of combined objective function, the results found by TLBO, JA, and modified JA are 11.84% better in volume and 12.41% better in the total cost as associated with the outcomes using GA.

Keywords Plate-fin heat exchanger · Jaya algorithm · Modified Jaya algorithm · Bi-objective

1 Introduction

A heat exchanger (HE) is a heat transfer component. It used to retrieve thermal energy among binary or more fluids having dissimilar temperatures. Different types of HE are used for various industrial applications, and the most important types are STHE and PFHE. It is used in different industries like petroleum, refrigeration and

K. C. More (✉)
Department of Mechanical Engineering, DYPatil Institute of Engineering and Technology, 410506 Pune, India
e-mail: kiran.imagine67@gmail.com

R. Venkata Rao
Department of Mechanical Engineering, Sardar Vallabhbhai National Institute of Technology, 395007 Surat, India
e-mail: ravipudirao@gmail.com

© Springer Nature Singapore Pte Ltd. 2020
R. Venkata Rao and J. Taler (eds.), *Advanced Engineering Optimization Through Intelligent Techniques*, Advances in Intelligent Systems and Computing 949, https://doi.org/10.1007/978-981-13-8196-6_15

air conditioning, transportation, cryogenic, power, alternate fuels, heat recovery, and other industries. Parting sheets and corrugated fins are brazed together to form a compact heat exchanger which is called PFHE.

For the optimization of PFHE, different researchers [1–3] had used different algorithms like simulated annealing (SA), branch and bound algorithms (BBA), and GA for different objective functions. For bi-objective optimization of PFHE, Yu et al. [4] used PSO and Xie et al. [5] used GA. For thermodynamic optimization of a cross-flow PFHE, Rao and Patel [6] applied PSO algorithm. Sanaye and Najafi et al. [7] applied bi-objective genetic algorithm for minimization heat transfer rate and annual cost of PFHE. For bi-objective optimization of PFHE and STHE, Patel and Rao [8] used modified TLBO algorithm.

Ayala et al. [9], Turgut [10], and Wen et al. [11] used free search approach combined with differential evaluation (MOFSDE) method, hybrid PSO, and hybrid GA, respectively, for optimization of PFHE.

From this literature survey, it is noted that advanced optimization algorithms want common controlling parameters, viz number of population and number of iterations. Other than these common controlling parameters, the various algorithms need their individual algorithm-specific parameters. The unpleasant tweaking of these parameters raises the computational attempt or trapped to the local optimum result. The TLBO, JA, and modified JA does not require algorithm-specific parameters. Hence, these algorithms are known as an algorithm-specific parameterless algorithm [12]. In this paper, PFHE problem is attempted by using TLBO, JA, and modified JA to identify whether more enhancements are achievable in the results.

The key attribute of modified JA is that the automatic determination of number of population. This makes the designer free from the burden of selecting the number of populations. The modified JA is designated because of simple, parameter-free, and its capability to acquire optimal solutions. The construction of the paper is: introduction and the literature review of PFHE are presented in Sect. 1, second section expresses the optimization algorithms, the case study of PFHE is reported in Sect. 3, results and discussions are in Sect. 4, and the concluding remarks are expressed in Sect. 5 of paper.

To optimize total volume and cost of PFHE as well as the efficiency of modified JA are the main objectives of this work. The next section presents the proposed optimization algorithms.

2   Optimization Algorithms

2.1   Jaya Algorithm

The objective function is $f(x)$ which is maximized or minimized. At iteration i, design variable is 'l' and 'n' is number of populations (i.e., search space). The best population gains the finest value of $f(x)$, and the nastiest population obtains the nastiest value of $f(x)$ in the total population size. If the lth design variable value is $X_{l,n,i}$ for the nth population in the ith iteration, then the updated values are as Eq. (1) [13].

$$X'_{l,n,i} = X_{l,n,i} + r_{1l,i}\left(X_{l,best,i} - \left|X_{l,n,i}\right|\right) - r_{2,l,i}\left(X_{l,worst,i} - \left|X_{l,n,i}\right|\right) \quad (1)$$

All accepted values of objective function are reserved after the termination of iteration, and consider the input values to the next iteration [14].

2.2 Modified Jaya Algorithm

A modified edition of JA known as self-adaptive JA is projected in current work. The basic JA needs common controlling parameters. The selection of an exacting no. of populations for suitable examples is a very complicated job. Hence, very less research has been carried out on self-adaptive characteristics; this method does not need the selection of number of populations due to its self-adaptive characteristics. The key attribute of modified JA is that the no. of populations determined automatically. Let the primary population is used randomly $(10 * l)$, where design variables is 'l' and after that the fresh population is created as follows [15]:

$$P_{new} = round(P_{old} + R * P_{old}) \quad (2)$$

where R is the arbitrary number among -0.5 to 0.5 (Fig. 1).

3 Case Study

This case study is used from the earlier work of Xie et al. [5] and Rao and Patel [6]. Figure 2 shows the illustration of PFHE. The thickness of plate is 0.4 mm, and aluminum is used as a material for both fins and plates. The hot side maximum allowable pressure drop is 0.3 kPa, and cold side is 2 kPa. So, the aim is to determine the flow length values of hot stream (L_a), cold stream (L_b), and no flow (L_c) of PFHE which gives minimized annual cost and volume.

$$V = L_a \times L_b \times L_c \quad (3)$$

$$TAC = C_{in} + C_{op} \quad (4)$$

$$C_{in} = C_A \times A^n \quad (5)$$

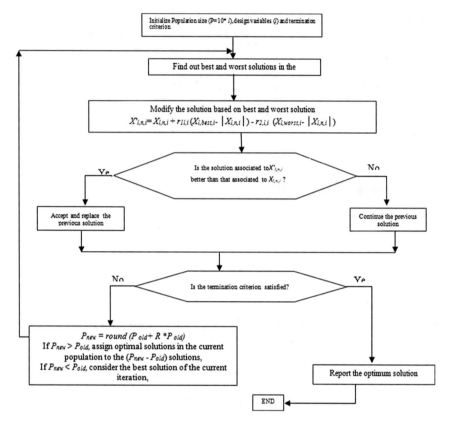

Fig. 1 Flowchart of self-adaptive Jaya algorithm [16]

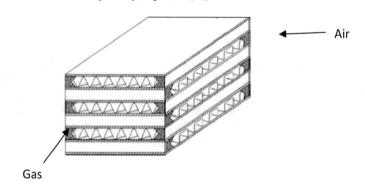

Fig. 2 Schematic diagram of PFHE [5]

$$C_{op} = \left\{ k_{el} \tau \frac{\Delta P V_t}{\eta} \right\}_h + \left\{ k_{el} \tau \frac{\Delta P V_t}{\eta} \right\}_c \qquad (6)$$

By using weight function, the two objectives get combined collectively in the form one objective function as shown in Eq. (7). The collective objective function (Z) is as follows:

$$\text{Minimize, } Z(x) = w_1 \frac{V}{V_{min}} + w_2 \frac{TAC}{TAC_{min}} \qquad (7)$$

The bounds of design variables are:
$L_a.-2$ to 0.4 m; L_b −0.5 to 0.7 m and L_c −0.75 to 1 m.
Constraints:
$\Delta P_h \leq 0.3$kPa and $\Delta P_c \leq 2$kPa

4 Results and Discussions

In the paper, the bi-objective optimization of PFHE is considered by using the TLBO, JA, and modified JA. The obtained results by using TLBO, JA, and modified JA for individual objective functions and also the assessment with the GA and PSO approaches are shown in Table 1.

To reduce the volume of PFHE, the best values gained by TLBO, JA, and modified JA show 9.67 and 3.66% improvement when compared with GA and PSO approach, respectively. Similarly, in terms of cost consideration, the above algorithms give 2.67 and 11.81% better results in contrast with GA and PSO approach, respectively. Figure 3 illustrations the convergence of TLBO, JA, and modified JA for minimum

Table 1 Results for individual objectives of PFHE

Parameter	Volume			Cost		
	GA [5]	PSO [6]	TLBO, JA, and self-adaptive JA	GA [5]	PSO [6]	TLBO, JA, and self-adaptive JA
L_a (m)	0.2390	0.2284	0.2356	0.2350	0.2190	0.2156
L_b (m)	0.5000	0.5131	0.5003	0.5000	0.5000	0.5000
L_c (m)	0.9840	0.9083	0.8696	1.000	1.0000	0.9601
ΔP_h (kPa)	0.2888	0.2999	0.1266	0.2979	0.2640	0.1002
ΔP_c (kPa)	1.9918	2.000	1.9975	1.8995	2.0000	1.9977
V (m^3)	0.1131	0.1064	**0.102510**	0.1175	0.1095	0.1055
Total annual cost ($/year)	3078.68	3072.2	2722.61	3047.67	3017.9	**2661.29**

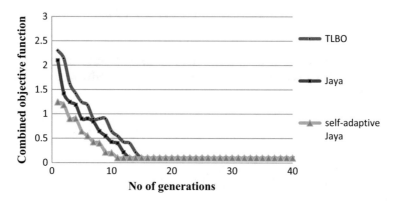

Fig. 3 Convergence of TLBO, JA, and modified JA for minimum volume

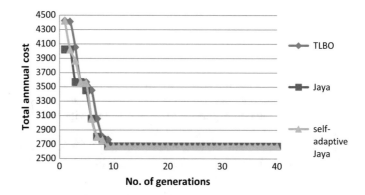

Fig. 4 Convergence graph of TLBO, JA, and modified JA for minimum cost of PFHE

volume of PFHE. The convergence takings in 16th, 13th, and 11th iteration for TLBO, JA, and modified JA, respectively.

The convergence graph of total annual cost for PFHE is shown in Fig. 4. The convergence takings in 14th, 10th and 9th iteration for TLBO, JA and Modified JA respectively.

The optimal results are presented in Table 2 which is achieved by TLBO, JA, and modified JA for the collective objective function with GA. The best possible solution reached by TLBO, JA, and modified JA shows 11.84% reduction in volume and 12.41% reduction in the cost when compared by using GA. The comparison with PSO is not feasible because details are not presented.

Figure 5 shows the convergence graph of collective objective function of PFHE. The convergence takes place in 16th, 15th, and 9th iteration for TLBO, JA, and modified JA, respectively. Modified JA needs fewer function evaluations in order to gain the best possible solution, and it decreases the computational attempt. It has been observed that different objective functions covered in the current work converge

Table 2 Validation results

Parameter	GA ([5]	TLBO, JA, and modified JA
L_a (m)	0.238	0.2286
L_b (m)	0.5	0.5008
L_c (m)	0.979	0.8971
ΔP_h (kPa)	0.2885	0.1163
ΔP_c (kPa)	1.9951	1.9999
V (m^3)	0.1165	**0.1027**
TAC ($)	3051.25	**2672.41**

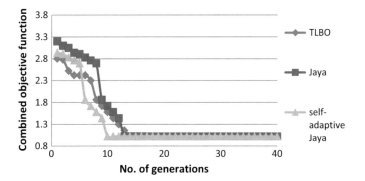

Fig. 5 Convergence graph of TLBO, JA, and modified JA for collective objective function

up to 20 iterations. This is the appreciable improvement when compared with 3000 iterations required by GA and 40 iterations required by PSO approach.

5 Conclusions

In this paper, the bi-objective optimization of PFHE is carried by using the TLBO, JA, and modified JA. In the case of PFHE, minimize cost and volume are considered as two objectives which are to be optimized as sole objective as well as dual objective functions simultaneously. The obtained results of TLBO, JA, and modified JAs are same, possibly because the obtained solution is a global solution. However, in TLBO and JA we want to carry out experiments by altering the number of populations to get the best possible solution, but in modified JA it is not essential to vary the number of populations manually; hence, modified JA requires lesser number of experiments and computational efforts. The ability of modified JA is verified, and the performance results are related with different algorithms. The modified JA may be easily suited for optimization of various thermal devices and systems.

References

1. Reneaume, J.M., Niclout, N.: MINLP optimization of plate-fin heat exchangers. Chem. Biochem. Eng. Q. **17**, 65–76 (2003)
2. Jorge, A.W., Gut, M., Pinto, J.M.: Optimal configuration design for plate heat exchangers. Int. J. Heat Mass Trans. **47**, 4833–4848 (2004)
3. Peng, H., Ling, X.: Optimal design approach for the plate-fin heat exchangers using neural networks cooperated with genetic algorithms. Appl. Therm. Eng. **28**, 642–650 (2008)
4. Yu, X.C., Cui, Z.Q.,Yu, Y.: Fuzzy optimal design of the plate-fin heat exchangers by particle swarm optimization. In: Proceedings of the Fifth International Conference on Fuzzy Systems and Knowledge Discovery, pp. 574–578, Jinan, China (2008)
5. Xie, G.N., Sunden, B., Wang, Q.W.: Optimization of compact heat exchangers by a genetic algorithm. Appl. Therm. Eng. **28**, 895–906 (2008)
6. Rao, R.V., Patel, V.K.: Thermodynamic optimization of cross flow PFHE using a particle swarm optimization algorithm. Int. J. Therm. Sci. **49**, 1712–1721 (2010)
7. Najafi, H., Najafi, B., Hoseinpoori, P.: Energy and cost optimization of a plate and fin heat exchanger using genetic Algorithm. Appl. Therm. Eng. **31**, 1839–1847 (2011)
8. Patel, V.K., Rao, R.V.: Design optimization of shell-and-tub heat exchanger using particle swarm optimization technique. Appl. Therm. Eng. **30**(11–12), 1417–1425 (2010)
9. Ayala, H.V.H., Keller, P., Morais, M.D.F., Mariani, V.C., Coelho, L.D.S., Rao, R.V.: Design of heat exchangers using a novel biobjective free search differential evaluation paradigm. Appl. Therm. Eng. **94**, 170–177 (2016)
10. Turgut, O.E., Çoban, M.T.: Thermal design of spiral heat exchangers and heat pipes through global best algorithm. Heat. Mass. Trans. 1–18 (2016)
11. Wen J, Yang H., Tong X, Li K., Wang S., LiY. Configuration parameters design and optimization for plate-fin heat exchangers with serrated fin by bi-objective genetic algorithm. Energ. Convers. Manage. 117, 482–489 (2017)
12. Rao, R.V., More, K.C., Taler, J., Ocłoń, P.: Bi-objective optimization of a thermo-acoustic engine using TLBO algorithm. Sci. Tech. Buil. Envir. **23**(8), 1244–1252 (2016)
13. Rao, R.V.: Jaya: a simple and new optimization algorithm for solving constrained and unconstrained optimization problems. Int. J. Indus. Engg. Comput. **7**(1), 19–34 (2016)
14. Rao, R.V., More, K.C., Taler, J., Ocłoń, P.: Dimensional optimization of a micro-channel heat sink using Jaya algorithm. Appl. Therm. Engg. **103**, 572–582 (2016)
15. Rao, R.V., More, K.C.: Optimal design and analysis of mechanical draft cooling tower using improved Jaya algorithm. Int. J. Refriger. **82**, 312–324 (2017)
16. Rao, R.V., More, K.C., Coelho L.S., Mariani V.C.: Optimal design of the Stirling heat engine through improved Jaya algorithm. J. Renew. Sustain. Energ. **9**, 033703 (2017)

Prediction of Factor of Safety For Slope Stability Using Advanced Artificial Intelligence Techniques

Abhiram Chebrolu, Suvendu Kumar Sasmal, Rabi Narayan Behera and Sarat Kumar Das

Abstract One of the major arising challenges in geotechnical engineering is to stabilize slopes for the sake of nature, economy, and valuable lives. In recent times, there has been a tremendous amount of developments in the field of computational geomechanics leading to the development of the slope stability analysis. The study explains the application of advanced artificial intelligence methods for finding the factor of safety of the slope. Multi-gene genetic programming (MGGP) and multivariate adaptive regression splines (MARS) are the two techniques used in predicting the factor of safety (FOS) for stability analysis of slopes. The present results are compared with Sah et al. [4] and the comparison seems to be reasonably good. The study finds that MGGP is more accurate than MARS in predicting the FOS.

Keywords Slope stability · Factor of safety · Artificial intelligence · Multi-gene genetic programming (MGGP) · Multivariate adaptive regression splines (MARS)

1 Introduction

Failure of slopes is something classified as a complex natural phenomenon that can cause damages ranging from waterlogging of nearby lands to potential environmental destruction. A lifetime of a slope is what heavily influences the nearby economy

A. Chebrolu · S. K. Sasmal · R. N. Behera (✉)
Department of Civil Engineering, National Institute of Technology Rourkela, Rourkela, India
e-mail: rnbehera82@gmail.com

A. Chebrolu
e-mail: ch.nagabhi@gmail.com

S. K. Sasmal
e-mail: suvendukumarsasmal@gmail.com

S. K. Das
Department of Civil Engineering, Indian Institute of Technology (Indian School of Mines), Dhanbad, India
e-mail: saratdas@rediffmail.com

© Springer Nature Singapore Pte Ltd. 2020
R. Venkata Rao and J. Taler (eds.), *Advanced Engineering Optimization Through Intelligent Techniques*, Advances in Intelligent Systems and Computing 949, https://doi.org/10.1007/978-981-13-8196-6_16

including all types of surrounding constructions and crop cycle. A slope is stable as long as the factor of safety (FOS) against failure is not less than one. However, the stability is tremendously influenced by the angle and height of slope, strength of the constituent soil, and the water table. The FOS obtained by Sah et al. [4] using maximum likelihood method had a high value of correlation value with the correlation coefficient (R). Yang et al. [7] proposed an equation to predict the value of FOS using 40 training and six testing data. Recently, Gandomi and Alavi [2] proposed a multistage model called multi-gene genetic programming (MGGP) which provides more accurate predictions than the conventional neural network techniques. In the present study, the FOS for slope stability is to be determined by MGGP and MARS (multivariate adaptive regression splines), the two advanced artificial intelligence techniques. The MGGP and MARS are used to find out the relation between various parameters that affect FOS in slope stability analysis. Also, the present results are compared with Sah et al. [4] and the comparison seems to be good.

2 Methodology

2.1 Genetic Programming

Genetic programming (GP) is the method in which a set of data is trained to provide the required outputs. The method works on the same principle of human brain. This method is based on the working principle of human genetic system and produces very good results. The technique was established by Koza [3]. In traditional regression methods, one has to give information about the structure of the model whereas the GP automatically provides the best mathematical fit for both structure as well as the parameters. It gives a solution as a tree structure or equation utilizing the given dataset.

2.2 Multi-gene Genetic Programming (MGGP)

MGGP, often referred as symbolic regression, is an advanced neural networking technique to predict the best suitable mathematical model to represent a given dataset. It deals with weighted linear combination of number of trees. Each tree called as gene represents lower order nonlinear transformations of input variables, the combination of which is referred to as multi-gene. With the increase in the maximum number of genes (G_{max}) and the maximum depth of tree (d_{max}), the efficiency of the model generally increases. However, the increase in the above two parameters is attributed to the increase in the complexity of the model. Relatively compact model can be constructed using appropriate (optimum) value of G_{max} and d_{max} (Searsons et al. [5]).

2.3 Multivariate Adaptive Regression Splines (MARS)

When a large number of input variables are concerned, the any type of calculation becomes complex. MARS is an advanced regression technique developed by Friedman [1] that works in accordance with the principle of 'divide and conquer.' MARS splits the input variables into several groups called as regions, providing each group its own equation using regression, hence easing the handling of large inputs. The present study emphasizes on predicting the FOS of slope by using statistical package 'earth [6]' available in 'R' using the concept of MARS.

2.4 Database Used

The database available in Sah et al. [4] including case studies of 23 dry and 23 wet slopes out of which 29 slopes are unstable and 17 slopes are stable, is analyzed in the present work. The numerical values and statistical parameters for inputs variable of the slope along with output are presented in Table 1. The database used for training and testing is shown in Tables 2 and 3, respectively.

3 Results and Discussions

3.1 MGGP Modeling

The MGGP modeling is done using GPTIPS as per Searson et al. [5]. Various factors like size of population, number of generations, reproduction, crossover, and mutation probability, G_{max} and d_{max} the performance of a neural model. In the present work, the best FOS model is obtained with population size of 1000 individuals at 100 generations with reproduction, crossover, and mutation probability of 0.05, 0.85, and 0.1, respectively, having tournament selection (tournament size of 2). The optimal result was obtained using G_{max} and d_{max} as 3 and 4, respectively. The proposed equation is mentioned below (Eq. 1). The comparison between the MGGP method and limit equilibrium method is presented in Fig. 1.

Table 1 Statistical values of the parameters

	γ (kN/m^3)	c (kPa)	ϕ°	β°	$H(m)$	r_u	FOS
Minimum	12.00	0.00	0.00	16.00	3.66	0.00	0.63
Maximum	28.44	150.05	45.00	53.00	214.00	0.50	2.05
Mean	19.72	20.48	27.51	32.93	43.91	0.17	1.24
Standard deviation	3.85	31.37	10.86	9.98	48.15	0.19	0.38

Table 2 Database used for training

SI. No.	γ (kN/m^3)	c (kPa)	ϕ^o	β^o	H (m)	r_u	FOS
1	18.68	26.34	15.00	35.00	8.23	0.00	1.11
2	18.84	14.36	25.00	20.00	30.50	0.00	1.88
3	18.84	57.46	20.00	20.00	30.50	0.00	2.05
4	28.44	29.42	35.00	35.00	100.00	0.00	0.83
5	28.44	39.23	38.00	35.00	100.00	0.00	1.99
6	20.60	16.28	26.50	30.00	40.00	0.00	1.25
7	14.80	0.00	17.00	20.00	50.00	0.00	1.13
8	14.00	11.97	26.00	30.00	88.00	0.00	1.20
9	25.00	120.00	45.00	53.00	120.00	0.00	1.30
10	18.50	25.00	0.00	30.00	6.00	0.00	1.09
11	18.50	12.00	0.00	30.00	6.00	0.00	0.78
12	22.40	10.00	35.00	30.00	10.00	0.00	2.00
13	21.40	10.00	30.34	30.00	20.00	0.00	1.70
14	22.00	0.00	36.00	45.00	50.00	0.00	0.89
15	12.00	0.00	30.00	35.00	4.00	0.00	1.46
16	12.00	0.00	30.00	45.00	8.00	0.00	0.80
17	12.00	0.00	30.00	35.00	4.00	0.00	1.44
18	12.00	0.00	30.00	45.00	8.00	0.00	0.86
19	23.47	0.00	32.00	37.00	214.00	0.00	1.08
20	16.00	70.00	20.00	40.00	115.00	0.00	1.11
21	20.41	24.90	13.00	22.00	10.67	0.35	1.40
22	21.82	8.62	32.00	28.00	12.80	0.49	1.03
23	20.41	33.52	11.00	16.00	45.72	0.20	1.28
24	18.84	15.32	30.00	25.00	10.67	0.38	1.63
25	21.43	0.00	20.00	20.00	61.00	0.50	1.03
26	19.06	11.71	28.00	35.00	21.00	0.11	1.09
27	18.84	14.36	25.00	20.00	30.50	0.45	1.11
28	21.51	6.94	30.00	31.00	76.81	0.38	1.01
29	14.00	11.97	26.00	30.00	88.00	0.45	0.63
30	18.00	24.00	30.15	45.00	20.00	0.12	1.12
31	23.00	0.00	20.00	20.00	100.00	0.30	1.20
32	22.40	100.00	45.00	45.00	15.00	0.25	1.80

Table 3 Database used for testing

SI. No.	γ (kN/m^3)	c (kPa)	ϕ°	β°	H (m)	r_u	FOS
1	22.40	10.00	35.00	45.00	10.00	0.40	0.90
2	20.00	20.00	36.00	45.00	50.00	0.25	0.96
3	20.00	20.00	36.00	45.00	50.00	0.50	1.78
4	20.00	0.00	36.00	45.00	50.00	0.25	0.79
5	20.00	0.00	36.00	45.00	50.00	0.50	0.67
6	22.00	0.00	40.00	33.00	8.00	0.35	1.45
7	20.00	0.00	24.50	20.00	8.00	0.35	1.37
8	18.00	5.00	30.00	20.00	8.00	0.30	2.05
9	16.50	11.49	0.00	30.00	3.66	0.00	1.00
10	26.00	150.05	45.00	50.00	200.00	0.00	1.02
11	22.00	20.00	36.00	45.00	50.00	0.00	1.02
12	19.63	11.97	20.00	22.00	12.19	0.41	1.35
13	18.84	0.00	20.00	20.00	7.62	0.45	1.05
14	24.00	0.00	40.00	33.00	8.00	0.30	1.58

Fig. 1 Plot showing performance of MGGP model

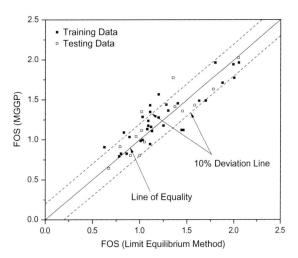

$$FOS = 0.4121 \tan h(C - \Phi + \beta \times r_u) - 0.004245(C - 2\beta + H) \tan h(c - r_u)$$
$$+ \frac{1.554\Phi\left(1 - r_u^2\right)}{\beta} + 0.2018 \qquad (1)$$

The above plots show us that the R value for training is 0.91 and R value of test data is 0.88, which show that the data is fairly correlated. The RMSE value of the

Table 4 Confusion matrix for training and testing data

Data type	Conditions	Predicted YES	Predicted NO	Accuracy
Training	Actual YES (26 cases)	23	3	$[(23 + 4)/(23 + 3 + 2 + 4)] \times 100$
	Actual NO (6 cases)	2	4	$= 84.38\%$
Testing	Actual YES (9 cases)	8	1	$[(8 + 4)/(8 + 1 + 1 + 4)] \times 100$
	Actual NO (5 cases)	1	4	$= 85.71\%$

model was found to be 0.16 for training data and a RMSE value of 0.19 was found for testing data. From the confusion matrices as shown in Table 4, it can be inferred that the accuracy of prediction of the slope stability in case of training data is 84.38% and in case of testing data is 85.71%.

3.2 MARS Modeling

The performance of MARS model is highly dependent on the number of basis functions used. Although a higher number of basis functions provide better performance, it provides complexity to the model at the same time. Sixteen basis functions and cross-validations using fivefold are found to be mathematically suitable for the development of the best MARS model. FOS is predicted for each set of basis functions and the correlation coefficient (R) was determined. The MARS model is written using the obtained coefficients and basis functions as:

$$\begin{aligned}
FOS = {}& 0.1239 + 0.0172h(21.4 - \gamma) - 0.1150h(\gamma - 21.4) \\
& - 0.0186h(24 - c) - 0.0015h(c - 24) - 0.0330h(26 - \Phi) \\
& - 0.0049h(\Phi - 26) + 0.0650h(45 - \beta) - 0.737h(\beta - 45) \\
& + 0.0185h(50 - H) + 0.0472h(H - 50) + 0.0027\gamma h(\Phi - 26) \\
& - 0.0017\Phi h(H - 50) + 0.0013Hh(\Phi - 26) - 0.0426r_u h(45 - \beta) \\
& - 0.0203r_u h(50 - H)
\end{aligned} \qquad (2)$$

The comparison between the MARS method and limit equilibrium method is presented in Fig. 2.

Upon prediction of factor of safety for slope stability using the above equation, it was observed that the R value for training is 0.91 and R value of test data is 0.86 through which it can be said that the data is fairly correlated. The RMSE value of the model was found to be 0.16 for training data and a RMSE value of 0.20 was found

Table 5 Confusion matrix for training and testing data

Data type	Conditions	Predicted YES	Predicted NO	Accuracy
Training	Actual YES (26 cases)	21	5	$[(21 + 5)/(21 + 5 + 1 + 5)] \times 100$
	Actual NO (6 cases)	1	5	$=81.25\%$
Testing	Actual YES (9 cases)	8	1	$[(8 + 4)/(8 + 1 + 1 + 4)] \times 100$
	Actual NO (5 cases)	1	4	$=85.71\%$

Fig. 2 Plot showing performance of MARS model

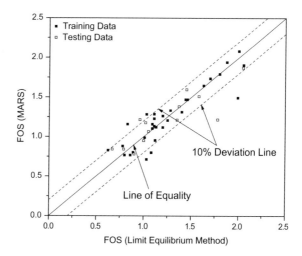

for testing data. From the confusion matrices shown in Table 5, it can be concluded that the accuracy of prediction of the slope stability in case of training data is 81.25% and in case of testing data is 85.71%.

3.3 Comparison with Sah et al. [4]

Figure 3 displays a comparison between prediction of factor of safety of both the models with Sah et al. [4]. It can be said that the prediction of FOS by MGGP model was more accurate than MARS model. From Fig. 3, it is evident that the predictions by using MGGP and MARS models are reasonably accurate compared to the maximum likelihood method used by Sah et al. [4].

Fig. 3 Plot showing
comparison of MGGP and
MARS results with Sah et al.
[4]

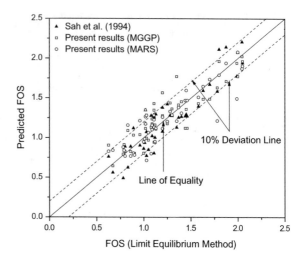

4 Conclusions

In the present study, forty-six numbers of dataset are used for the prediction of
FOS of the slope using advanced intelligence techniques such as multi-gene genetic
programming (MGGP) and multi-adaptive regression spline (MARS). Based on the
analysis of the developed models using two advanced intelligence techniques, the
following conclusions may be drawn:

- Two numbers of equations (i.e., Eqs. (1) and (2)) are developed using MGGP and
  MARS, which can predict the FOS of the slope with reasonable accuracy.
- The MGGP model predicts the FOS of the slope more efficiently than the MARS
  model.
- The comparison of present results with Sah et al. [4] seems to be reasonably good.

References

1. Friedman, J.H.: Multivariate adaptive regression splines. Ann. Stat. pp. 1–67 (1991)
2. Gandomi, A.H., Alavi, A.H.: Multi-stage genetic programming: a new strategy to nonlinear
   system modeling. Inf. Sci. **181**(23), 5227–5239 (2011)
3. Koza, J.R.: Genetic programming: on the programming of computers by natural selection. MIT
   Press, Cambridge, Mass (1992)
4. Sah, N.K., Sheorey, P.R., Upadhyama, L.W.: Maximum likelihood estimation of slope stability.
   Int. J. Rock Mech. Min. Sci. Geomech. Abstr. **31**, 47–53 (1994)
5. Searson, D.P., Leahy, D.E., Willis, M.J.: GPTIPS: an open source genetic programming toolbox from multi-gene symbolic regression. In: International Multi Conference of Engineers and
   Computer Scientists, vol. 1, March 17–19, Hong Kong (2010)
6. Milborrow, S.: Package 'earth'. R Software package (2019)

7. Yang, C.X., Tham, L.G., Feng, X.T., Wang, Y.J., Lee, P.K.: Two stepped Evolutionary algorithm and its application to stability analysis. J. Comput. Civil. Engg. **18**(2), 145–153 (2004)

The Use of Grey Relational Analysis to Determine Optimum Parameters for Plasma Arc Cutting of SS-316L

K. S. Patel and A. B. Pandey

Abstract Use of plasma arc cutting requires careful selection of parameters to control kerf and material properties. Use of multi-objective optimization technique like grey relational analysis (GRA) to optimize parameters in plasma arc cutting of SS-316L plate using five performance characteristics is discussed. The multi-objective optimization problem arises due to mutually conflicting nature of the responses in cutting. Experiments were conducted using L9 orthogonal array (OA), and generated data was used to apply GRA giving a current of 40 A, a pressure of 6 bar, a stand-off distance of 2 mm and a speed of 0.3044 m/min as the best parameters for plasma arc cutting of SS-316L 6 mm thick plate.

Keywords Plasma arc cutting · Grey relation analysis · Optimization

1 Introduction

Plasma arc cutting (PAC) became popular for cutting aluminium, high alloy steels and all metals that could not be gas cut due to benefits like fast cutting and small kerf widths compared to gas cutting. SS-316L material is widely used in industries. It has good application properties like corrosion resistant against chlorides, bromides, iodides, fatty acids, etc. that why it is used in coastal area plants. In plasma cutting, there are many process parameters like current, voltage, pressure, plasma gas type, stand-off distance and cutting speed. It is necessary to select the optimal process parameters to cut material with higher MRR, but there are also other response parameters like kerf width, straightness, bevel angle and heat affected zone which need to be reduced. It is a challenging issue to find the optimal combination for PAC.

K. S. Patel (✉) · A. B. Pandey
Department of Mechanical Engineering, The Maharaja Sayajirao University of Baroda, Vadodara, Gujarat, India
e-mail: Karanp34@gmail.com

A. B. Pandey
e-mail: akashpandey@gmail.com

© Springer Nature Singapore Pte Ltd. 2020
R. Venkata Rao and J. Taler (eds.), *Advanced Engineering Optimization Through Intelligent Techniques*, Advances in Intelligent Systems and Computing 949, https://doi.org/10.1007/978-981-13-8196-6_17

The Taguchi method [1–5] is used to overcome the drawbacks of detailed experiments with full factorial designs by performing the least number of experimental measurements. Many papers have reported effective methods that solve engineering optimization problems and obtain a robust output quality in applications [6–9]. Since optimizing multiple output qualities of a process requires the calculation of overall S/N ratios and cannot optimize the multiple output qualities simultaneously by using the Taguchi's method. Therefore, a grey relational analysis (GRA) is recommended and used to integrate and optimize the multiple output qualities of a process [10–15]. Application of GRA has been reported in many publications in field of machining [16, 17].

2 Experimental Procedure

2.1 Experimental Setup

Experiments are performed on plasma CUT40 with air plasma. Current, pressure, stand-off distance and travelling speed of the torch are controlled. Here current is regulated with the help of regulator provided on the machine set up. Pressure is controlled with the help of pressure regulator. Stand-off distance is set with the help of gauge. Cutting speed is set with the help of regulator provided on the machine. This regulator is calibrated by measuring travelling distance and travelling time. Here accuracy of input parameters is very high as all the inputs are given with the help of regulators and measuring gauge. Response parameters selected are MRR, kerf width at the top and bottom, straightness and bevel angle. MRR is calculated by weight difference before and after cutting and actual cutting time. Other parameters are calculated with the help of digital photography of cutting section of plate as shown in Fig. 1. Figure 1 shows the cut quality parameters. Kt = top kerf width, St = straightness, Kb = bottom kerf width and Θ = bevel angle. Figure 2 shows the experimental setup. A = ground Clamp, B = Worksupport, C = Plasma torch assembly and D = Work Piece.

Fig. 1 Plasma cut quality parameters

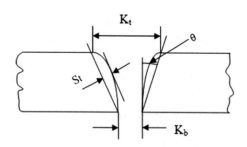

Fig. 2 Experimental setup

2.2 Design of Experiment

Three levels of all factors are selected for analysis as listed in Table 1 after conducting screening experiment. During screening experiment, worst and best cutting condition parameters are found, and from those results, three levels of input parameters are selected for final experimentation. The Taguchi L9 OA is used to plan experiments which give least trials for given cost and detection of parameter and their interaction effects [14] as listed in Table 2.

Table 1 Design factors and their level

Parameters		Current (A)	Sod (mm)	Pressure (bar)	Speed (m/sec)
Level	1	35	2	5	0.1964
	2	37.5	2.5	5.5	0.2439
	3	40	3	6	0.3044

Table 2 Experimental settings of an orthogonal array of L9

Exp. no.	Current (A)	Sod (mm)	Pressure (bar)	Speed (m/min)
1	35	2	5	0.1964
2	35	2.5	5.5	0.2439
3	35	3	6	0.3044
4	37.5	2	5.5	0.3044
5	37.5	2.5	6	0.1964
6	37.5	3	5	0.2439
7	40	2	6	0.2439
8	40	2.5	5	0.3044
9	40	3	5.5	0.1964

Table 3 Experimental results

Exp. no.	MRR (mm^3/min)	Top kerf width (mm)	Bottom kerf width (mm)	Straightness (mm)	Bevel angle (°)
1	2472	2.9637	1.6137	0.3972	12.68682
2	2695.431	2.9329	1.5837	0.3592	12.67954
3	3365.506	2.6193	1.2885	0.2432	12.5121
4	3441.456	2.7246	1.5825	0.3052	10.44272
5	2556.522	2.7765	1.6037	0.2706	10.8538
6	3043.421	3.0598	1.6299	0.2559	13.32779
7	3201.777	2.6722	1.6409	0.2023	9.757836
8	3675	2.8046	1.524	0.2421	12.05417
9	3303.261	3.4678	2.311	0.2301	10.91826

3   Results and Discussion

The experimental results for response parameters for each experiments are given in Table 3. All these response parameters are optimized in this study by formulating a multi-response optimization problem.

3.1   Grey Analysis

The multiple responses in Table 3 need to be combined to evaluate a single quality representation for desired cut conditions, and using Grey relational analysis, a single grey relational grade can be generated with appropriate weights being given to the responses for desired condition.

3.1.1   Normalization

The first step in grey relational analysis is normalization. The experimental results are scaled in the range of 0–1. For normalization of MRR data, higher-the-better (HB) criterion (Eq. 1) and for kerf width, straightness, bevel angle parameters, lower-the-better (LB) criterion (Eq. 2) are used as MRR is to be maximized and kerf width, straightness, bevel angle are to be minimized.

$$X_{ij} = \frac{Y_{ij} - \min Y_j}{\max Y_j - \min Y_j} \tag{1}$$

$$X_{ij} = \frac{\min Y_j - Y_{ij}}{\max Y_j - \min Y_j} \tag{2}$$

Table 4 Normalization of data

No.	Normalization of experiment data				
	MRR	KT	KB	ST	Angle
1	0	0.5941	0.682	0	0.1795
2	0.1857	0.6304	0.7113	0.195	0.1816
3	0.7427	1	1	0.7901	0.2285
4	0.8059	0.8759	0.7125	0.472	0.8081
5	0.0703	0.8147	0.6917	0.6496	0.693
6	0.475	0.4808	0.6661	0.725	0
7	0.6066	0.9377	0.6554	1	1
8	1	0.7816	0.7697	0.7958	0.3568
9	0.691	0	0	0.8574	0.6749

where X_{ij} is the value after normalization, min Y_j is the least value of Y_{ij} for the kth response and max Y_{ij} is the highest value of Y_{ij} for the kth response. An ideal sequence is x_{0j} ($j = 1, 2, 3, …, 9$) for the response. The normalized data is listed in Table 4.

3.1.2 Grey Relational Coefficient

Grey relational coefficients are evaluated to find the distance of given combination from the best combination. The grey relational coefficient γ_{ij} given by:

$$\gamma_{ij} = \frac{(\Delta\min + (\zeta \times \Delta\max))}{(\Delta ij + (\zeta \times \Delta\max))} \tag{3}$$

where $\Delta_{ij} = \|X_i - X_{ij}\|$ shows absolute gab between X_i and X_{ij}, $\Delta_{\min}$ and $\Delta_{\max}$ largest and smallest absolute distance. ζ is a differentiating coefficient such that, $0 \leq \zeta \leq 1$, to reduce the effect of $\Delta_{\max}$ when very large. The recommended level of the distinguishing coefficient ζ is 0.5 [11]. The grey relation coefficient of each performance characteristic is shown in Table 5.

3.1.3 Grey Relational Grade

The grey relational grade is treated as single quality parameter to decide the combination in place of the multiple responses. The grey relational grade γ_i is calculated as:

$$\gamma_i = \sum_{j=1}^{n} w_j \times \gamma_{ij} \tag{4}$$

Table 5 Grey relational coefficient

No.	Grey relational coefficient				
	MRR	KT	KB	ST	Angle
1	0.3333	0.5519	0.6112	0.3333	0.3787
2	0.3804	0.575	0.634	0.3831	0.3792
3	0.6603	1	1	0.7044	0.3932
4	0.7203	0.8012	0.6349	0.4864	0.7227
5	0.3497	0.7296	0.6186	0.5879	0.6196
6	0.4878	0.4906	0.5996	0.6452	0.3333
7	0.5597	0.8891	0.592	1	1
8	1	0.696	0.6846	0.71	0.4374
9	0.618	0.3333	0.3333	0.778	0.606

Table 6 Calculated grey relational grade and rank in optimization process

Exp. no.	MRR	KT	KB	ST	Angle	Grade	Rank
1	0.1167	0.0966	0.107	0.05	0.0568	0.427	9
2	0.1332	0.1006	0.1109	0.0575	0.0569	0.4591	8
3	0.2311	0.175	0.175	0.1057	0.059	0.7457	3
4	0.2521	0.1402	0.1111	0.073	0.1084	0.6848	4
5	0.1224	0.1277	0.1083	0.0882	0.0929	0.5395	6
6	0.1707	0.0859	0.1049	0.0968	0.05	0.5083	7
7	0.1959	0.1556	0.1036	0.15	0.15	0.7551	2
8	0.35	0.1218	0.1198	0.1065	0.0656	0.7637	1
9	0.2163	0.0583	0.0583	0.1167	0.0909	0.5406	5

where $\sum_{j=1}^{n} w_i = 1$ and $i = 1,2, \dots m$ and $j = 1, 2, 3, \dots n$.

Where n are the process responses. Greater grade shows closeness to optimum combination. Grey relational grades are listed in Table 6 with rank.

Plasma arc cutting is used as a primary cutting process to obtain rough dimension size for components. Cut edges should not have large taper increasing later machining. The MRR should be large for primary cutting process. The kerf width should be low to reduce metal loss. To apply grey analysis, equivalent importance is given to MRR, reduction of kerf width and straight parallel cut edges. Value of 0.35 is selected as weight for MRR and kerf width with nearly equal weight out of 0.35 for kerf width given to top and bottom kerf width. 0.3 is assigned to combination of straightness and bevel angle for straight and parallel edges. Table 7 lists the analysis results of grey relational grades for different levels of the parameters.

Table 7 Response table for the grey relational grade

Cutting parameter	Grey relational grade			Max-Min	Rank
	Level-1	Level-2	Level-3		
Current	0.5439	0.5775	0.6865	0.1425	2
Sod	0.6223	0.5874	0.5982	0.0349	4
Pressure	0.5663	0.5615	0.6801	0.1186	3
Speed	0.5024	0.5741	0.7314	0.2291	1

3.2 Factor Effects

Table 6 shows that the best combination of variables out of the variable range selected is the one implementing 40 A current, 5 bars pressure, 2.5 mm stand-off distance and 0.3044 m/min cutting speed. Analysing the grey relational grades for each level of each of the parameters also, it is identified that higher levels of speed, current, pressure and lower levels of stand-off distance are observed to be the optimum levels. The highest level of current gives sufficient energy for penetrating into the thickness of the metal. Heat energy generated is directly propositional to the square of the current. Hence increase in current will result in higher material removal. The higher cutting speed eliminates localized heating at a given point and helps reduce straightness and bevel angle due to lesser time available for cutting. Also, higher cutting speed reduces cutting time hence increases in MRR and reduces power cost for cutting. Lower stand-off distance gives narrow plasma cut which is more desirable. Narrow cut eliminates excess cutting of plate which is non-desirable and also lower stand-off distance reduces tapering cut effect which needs post cutting treatment like grinding. Higher level pressure selected helps in removing the heated molten metal immediately allowing a better and uniform heat transfer in thickness direction causing lesser kerf widths at top and bottom and smaller difference between kerf width at top and bottom too.

3.3 Analysis of Variance (ANOVA)

In the present study, ANOVA is performed using Minitab. Table 8 shows the ANOVA result for overall grey relational grade of MRR, kerf width, straightness and bevel angle. It is clear from Table 8 and Fig. 3 that cutting speed is most significant influence on responses optimization, which is about 55.80% followed by current, pressure and stand-off distance in that order. The trends seen in Fig. 3 and Table 8 also show the significance of the effects matching with the F-value and percentage contribution values.

Table 8 Results of ANOVA of grey relational grade

Source	DF	Seq SS	Adj SS	Adj MS	F-Value	P-Value	Contribution (%)
Current	1	0.0304	0.0304	0.03046	9.25	0.038	21.07
Stand-off distance	1	0.0008	0.0008	0.0008	0.26	0.634	0.60
Pressure	1	0.0194	0.0194	0.0194	5.89	0.072	13.42
Speed	1	0.0806	0.0806	0.0806	24.5	0.008	55.80
Error	4	0.0131	0.0131	0.0032			9.11
Total	8	0.1446					100.00

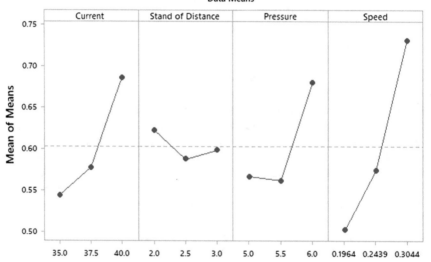

Fig. 3 Effect of PAC process parameters on multi-performance characteristics

4 Confirmation Test

Once the optimal level of process parameters is determined, a confirmation test is required to check the correctness analysis. Table 9 compares the estimated grey relational grade with grey relational grade actually obtained using the results of the confirmation test carried out at optimal settings. From the initial experiments and analysis, it is found that 40 A current, 5 bars pressure, 2.5 mm stand-off distance and 0.3044 m/min cutting speed have maximum grey relational grade, But the response table for grey shows that higher level of current (40 A), lower level of SOD (2 mm), higher level of pressure (6 bar) and higher level of speed (0.3044 m/min) will have higher grey relational grade. So by setting these parameters, another experiment was

Table 9 Comparison of initial condition and optimal condition grey relational grade

	Initial condition	Optimal condition
	A3B2C1D3	A3B1C3D3
MRR	3675.0000	3690.0000
Top kerf width	2.8046	2.6523
Bottom kerf width	1.5240	1.4556
Straightness	0.2421	0.2203
Bevel angle	12.054	11.2853
New grey relational grade	0.7553	0.8517

Here *A* current, *B* SOD, *C* gas pressure, *D* cutting speed

conducted and the results are found as shown in Table 9. It is found that grey relational grade is improved by 0.0964, which is 12.73% increase.

5 Conclusion

Application of GRA as optimization method and effect of machining parameters on MRR, kerf width, straightness and bevel angle in plasma cutting of SS-316L using Taguchi orthogonal array shows that the optimal combination is A3B1C3D3 (higher level of current, lower level of SOD, higher level of gas pressure, higher level of cutting speed). Based on ANOVA, most significantly influencing parameter is cutting speed, followed by current and gas pressure. Whereas stand-off distance is less effective parameter within the specific test range. To check the validity of the analysis, a confirmation test is also carried out. There is an improvement in grey relational grade from initial to optimal process parameter by about 12.73%.

References

1. Taguchi, G.: Introduction to Quality Engineering. Asian Productivity Organization, Tokyo (1990)
2. Ross, P.J.: Taguchi Techniques for Quality Engineering. McGraw-Hill, New York (1988)
3. Phadke, M.S., Dehnad, K.: Optimization of product and process design for quality and cost. Qual. Reliability Eng. Int. **4**(2), 105–112 (1988)
4. Mathiselvan, G., Sundaravel, S., Sindiri Chaitanya, Kaja Bantha Navas, R.: Application of taguchi method for optimization of process parameters in analyzing the cutting forces of lathe turning operation. Int. J. Pure Appl. Math. **109**(8), 129–136 (2016)
5. Arun, K.K., Ponnuswamy, D.: Optimization of dry end milling process parameters of Al-6063 alloy using Taguchi method. Int. J. Trend Sci. Res. Dev. **2**(3), 2058–2062 (2018)
6. Dubey, A.K., Yadava, V.: Multi-objective optimisation of laser beam cutting process. Opt. Laser Technol. **40**, 562–570 (2008)

7. Sharma, P., Verma, A., Sidhu, R.K., Pandey, O.P.: Process parameter selection for strontium ferrite sintered magnets using Taguchi L9 orthogonal design. J. Mater. Process. Technol. **178**, 147–151 (2005)
8. Yi, Q., Li, C., Tang, Y., Chen, X.: Multi-objective parameter optimization of CNC machining for low carbon manufacturing. J. Clean. Prod. **95**(15), 256–264 (2015)
9. Ghani, J.A., Choudhury, I.A., Hassan, H.H.: Application of Taguchi method in the optimization of end milling parameters. J. Mater. Process. Technol. **145**, 84–92 (2004)
10. Tarng, Y.S., Juang, S.C., Chang, C.H.: The use of grey-based Taguchi methods to determine submerged arc welding process parameters in hard facing. J. Mater. Process. Technol. **128**, 1–6 (2002)
11. Deng, J.L.: Introduction to grey system. J. Grey Syst. **1**, 1–24 (1989)
12. Deng, J.L.: Control problems of grey systems. Syst. Control Lett. **5**, 288–294 (1982)
13. Tsai, M.-J., Li, C.-H.: The use of grey relational analysis to determine laser cutting parameters for QFN packages with multiple performance characteristics. Elsevier J. Opt. Laser Technol. **41**, 914–921 (2009)
14. Das, M.K., Kumar, K., Barman, T.K., Sahoo, P.: Optimization of process parameters in plasma Arc cutting of EN 31 steel based on MRR and multiple roughness characteristics using grey relational analysis. Elsevier J. Procedia Mater. Sci. **5**, 1550–1555 (2014)
15. Pandian, P.P., Rout, I.S.: Parametric investigation of machining parameters in determining the machinability of Inconel 718 using taguchi technique and grey relational analysis. Procedia Comput. Sci. **133**, 786–792 (2018)
16. Kumar, S.S., Uthayakumar, M., Kumaran, S.T.: Parametric optimization of wire electrical discharge machining on aluminium based composites through grey relational analysis. J. Manuf. Proc. **20**(Part 1), 33–39 (2015)
17. Lin, J.L., Lin, C.L.: The use of the orthogonal array with grey relational analysis to optimize the electrical discharge machining process with multiple performance characteristics. Int. J. Mach. Tools Manuf **42**, 237–244 (2002)

Jaya Algorithm Based Intelligent Color Reduction

Raghu Vamshi Hemadri and **Ravi Kumar Jatoth**

Abstract The purpose of color quantization is to reduce colors in an image with the least parody. Clustering is a popularly used method for color quantization. Color image quantization is an essential action in several applications of computer graphics and image processing. Most of the quantization techniques are mainly based on data clustering algorithms. In this paper, a color reduction hybrid algorithm is proposed by applying Jaya algorithm for clustering. We examine the act of Jaya algorithm in the pre-clustering stage and K-means in the post-clustering phase, and the limitations of both the algorithms are overcome by their combination. The algorithms are compared by MSE and PSNR values of the four images. The MSE values are lower and PSNR values are higher for the proposed algorithm. The results explain that the proposed algorithm is surpassed both the K-means clustering and Jaya algorithm clustering for color reduction method.

1 Introduction

With the increase of technology, storage of information is not an issue. But storage of unnecessary information is a waste of resources. High-quality images contain a large amount of detailed information which increases transfer time and processing problems. So, preprocessing methods are introduced to overcome these problems. Color quantization is one of the image preprocessing methods to reduce the number of colors with the least possible distortion. In color quantization, a number of clusters or the number of colors is first selected, the pellet is formed, and the image is reproduced with the newly formed pellet. Color quantization algorithms [1] are divided into two types, splitting and clustering algorithms. Traditional splitting algorithms are median

R. V. Hemadri (✉) · R. K. Jatoth
Department of Electronics and Communication Engineering, National Institute of Technology, Warangal, India
e-mail: vamshi.hemadri@gmail.com

R. K. Jatoth
e-mail: ravikumar@nitw.ac.in

© Springer Nature Singapore Pte Ltd. 2020
R. Venkata Rao and J. Taler (eds.), *Advanced Engineering Optimization Through Intelligent Techniques*, Advances in Intelligent Systems and Computing 949,
https://doi.org/10.1007/978-981-13-8196-6_18

cut [2], center cut [3] and octree algorithms [4], etc., and clustering algorithms are
K-means clustering [5, 6], fuzzy c-means clustering, etc.

The clustering of colors can also be done by using nature-inspired or bio-inspired
algorithms [7–9] by taking the mean square error of a cluster as objective function.
In this paper, an overview of most used clustering algorithm, K-means algorithm
is given and a hybrid color quantization algorithm is proposed based on K-means
clustering and Jaya algorithm. The hybrid algorithm contains two stages of clustering,
pre-clustering and post-clustering. In the pre-clustering stage, Jaya algorithm is used
to find the cluster centers used in K-means clustering. K-means is applied using the
cluster centers from the pre-clustering stage. This paper is organized in seven sections,
first deals with introduction followed by K-means clustering, Jaya algorithm, the
proposed hybrid algorithm, results and simulations, conclusion and references.

2 K-MEANS Clustering

K-means is a clustering algorithm, type of unsupervised learning, and is used for
data having uncategorized or ungrouped data. K-means clustering iteratively groups
data into pre-defined K number of groups. In the first step of K-means clustering,
'k' (number of clusters) is user defined (taken as input). 'k' number of points of data
are randomly taken called its corresponding cluster center or centroid. Every point
of the data is assigned to the nearest cluster center. Mean of Euclidian distance of
all the points from the points and the cluster center in every cluster is calculated and
the cluster center is updated; i.e., the mean is the new cluster center, clusters are also
updated by assigning each data point to the nearest centroid. The process is repeated
for desired number of times to get perfect clusters.

There are both pros and cons for K-means clustering, it is very simple to implement, very fast clustering algorithm even for large values of 'k' and highly sensitive,
small changes in image changes the centroids. Whereas, initial centroids (randomly
generated), have strong impact on the final results (final centroids and clusters) [10].
The data distribution impacts final result (distribution of pixels in latent space). Normalization or standardization of data or datasets completely changes the final results.

3 Jaya Algorithm

The word 'Jaya' in Sanskrit means victory. The name itself speaks more about the
algorithm, this algorithm always toward victory; i.e., it always moves close to the
best and away from the worst solutions. It is proposed by Prof. Dr. Hab. Ravipudi
Venkata Rao in the year 2016 [11].

Let f be the function to be optimized, for instance, say minimum. Let f be a
function in m-variables, here $m = 3$ because we use RGB color format and each
pixel is a population member, has red, green, and blue intensity values for each

pixel. 'n' be the population size or the number or pixels taken in each iteration (i.e., $j = 1, 2, 3, \ldots, n$). $P_{best,i}$ be the pixel that gives the best solution (*minimum*) for f in ith iteration, $P_{worst,i}$ be pixel that gives the worst solution (*maximum*) for f in ith iteration. $P_{j,i}$ be the jth pixel of the population in ith iteration. The candidate solution is modified by using the below equation. Where $r_{j,i}, r'_{j,i}$ are two random values in the range [0, 1].

$$P'_{j,i} = P_{j,i} + r_{j,i}\left(P_{best,i} - |P_{j,i}|\right) - r'_{j,i}\left(P_{worst,i} - |P_{j,i}|\right) \qquad (1)$$

$r_{j,i}\left(P_{best,i} - |P_{j,i}|\right)$ drives the candidate pixel closer to global best solution and $r'_{j,i}\left(P_{worst,i} - |P_{j,i}|\right)$ drives the candidate pixel away from the global best solution. The flowchart of the Jaya algorithm is described below (Fig. 1).

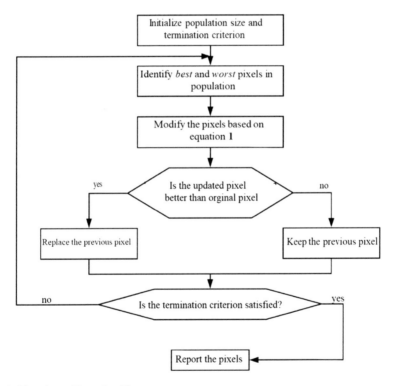

Fig. 1 Flowchart of Jaya algorithm

4 The Hybrid Jaya K-MEANS Algorithm

K-means algorithm struck up at local optima which leads to improper edges and variations of color is not properly observed in reduced image. Color quantization can also be done using nature-inspired algorithms like Jaya algorithm. Unlike standard clustering algorithms, nature-inspired algorithms escape local optima and reach global optima. But these algorithms take a long time to form clusters compared to standard algorithms. To avoid this, we limit clustering using nature-inspired algorithms to fixed number of iterations and this stage is called the pre-clustering stage. The centroids obtained are very close to correct cluster centers [8] and are corrected by few K-means iterations called post-clustering stage. This method of color reduction is called intelligent color reduction. In this paper, Jaya algorithm is used for pre-clustering stage. The hybrid algorithm gives centroids very close to correct centroids, so perfect edges and proper color variations can be observed in the reduced image which makes reduced image closer to original image.

The set of cluster centers is taken as a population member and number of such sets makes population.

$$p_i = \left[z_1^i, z_2^i, z_3^i, \ldots, z_k^i\right] \tag{2}$$

where p_i is a population member, $i = 1, 2, \ldots, n$. 'n' is the population size, z_j^i is the center of cluster j, $j = 1, 2, 3, \ldots, k$. 'k' is the number of clusters.

Each pixel of the image is assigned to its closest cluster center by calculating Euclidian distance; i.e., clusters are made. In quantization, the main objective is to reduce colors in the image. The reduced image should be visually similar to the original image; i.e., the distortion should be minimum. So, the objective function taken is (Table 1, Fig. 2).

Table 1 Description of parameters of Eq. (3)

Parameter	Description
np	Total number of pixels
k	Number of clusters
p_i	ith pixel
c_j	jth cluster
z_j	Center of jth cluster

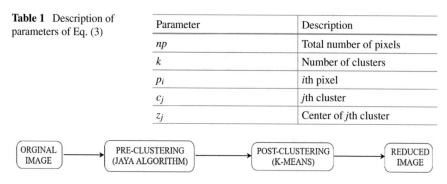

Fig. 2 Flowchart of the proposed hybrid algorithm

$$f = \frac{1}{np} \sum_{j=1}^{k} \sum_{p_i \in c_j} (p_i - z_j)^2 \tag{3}$$

The cluster centers obtained from the pre-clustering step are adjusted using the K-means clustering. K-means clustering depends on the initial population. To overcome this, the cluster centers obtained from the pre-clustering stage are given as initial population to the post-clustering stage (i.e., the K-means clustering stage).

5   Results and Simulations

In this paper, the proposed hybrid algorithm, K-means are compared by measuring peak signal-to-noise ratio (PSNR) and mean square error (MSE) [12]. For this comparison purpose, we use standard images like 'lena', 'peppers', 'mandrill', and 'jet'. All the images taken are of 512 * 512 size from USC-SIPI database. The code is implemented on Intel i5 7th generation processor and 4 GB RAM possessing computer on MATLAB 2016 version.

The algorithms are compared by MSE (mean square error) and PSNR (peak signal-to-noise ratio) values of the four images. MSE estimates how close the quantized image is to the original image. Smaller is the MSE closer or similar is the quantized image to the original image.

MSE can be computed by below formula:

$$\text{MSE} = \frac{1}{n} \sum_{i=1}^{n} (P_i - P_i')^2 \tag{4}$$

PSNR can be computed by below formula

$$\text{PSNR} = 10 \log_{10} \left(\frac{M^2}{\text{MSE}} \right) \tag{5}$$

where P_i' is the ith pixel of the quantized image, P_i is the ith pixel of the parent image, and n is the total number of pixels in the parent image. M is the maximum possible pixel value in the parent image.

As stated above, the algorithms are compared by MSE and PSNR values. The number of iterations for different algorithms is:

K-means: 20
Jaya + K-means: 40 Jaya iterations (10 population size) + 10 K-means iterations

From Tables 2 and 3 and Figs. 3 and 4, it is clear that the hybrid Jaya algorithm is better than K-means clustering, at the edges and the MSE values are smaller, PSNR values are higher for the proposed algorithm (Figs. 5 and 6).

Table 2 PSNR values for different images for different *K* values

	Image	K-Means	Proposed algorithm
$K = 8$	Lena	26.5512	26.8775
	Peppers	24.1389	24.3051
	Babbon	22.0893	22.4149
	Jet	25.9494	28.4493
$K = 16$	Lena	29.1908	29.6418
	Peppers	26.3484	26.6749
	Babbon	24.3348	24.8601
	Jet	28.7549	30.7882
$K = 32$	Lena	31.5594	32.1415
	Peppers	28.4675	29.2094
	Babbon	26.9549	27.08
	Jet	30.0618	32.6237
$K = 64$	Lena	33.8193	34.1466
	Peppers	30.9767	31.4305
	Babbon	28.9505	29.0871
	Jet	33.0366	36.2457

Table 3 MSE values for different images for different *K* values

	Image	K-Means	Proposed algorithm
$K = 8$	Lena	0.0022	0.002
	Peppers	0.0039	0.0037
	Babbon	0.0062	0.0057
	Jet	0.0025	0.0014
$K = 16$	Lena	0.0012	0.0011
	Peppers	0.0023	0.0021
	Babbon	0.0037	0.0033
	Jet	0.0013	8.3403e−04
$K = 32$	Lena	6.9833e−04	6.1073e−04
	Peppers	0.0014	0.0012
	Babbon	0.002	0.0019
	Jet	9.8588e−04	5.4655e−04
$K = 64$	Lena	4.1502e−04	3.8489e−04
	Peppers	7.9860e−04	7.1937e−04
	Babbon	0.0013	0.0012
	Jet	4.9699e−04	2.3737e−04

K	LENA	PEPPERS	MANDRILL	JET

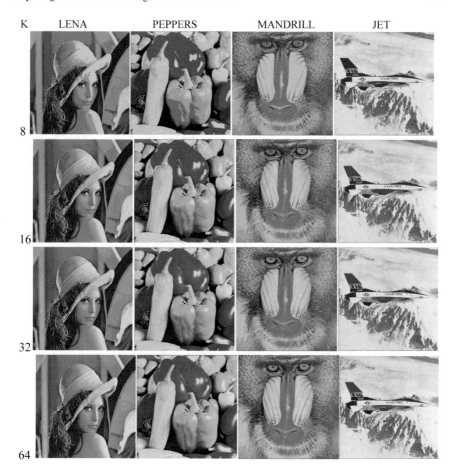

Fig. 3 Images of *K*-means algorithm

6   Conclusion

In this paper, a color reduction hybrid algorithm is proposed by applying Jaya algorithm for clustering in the pre-clustering stage and *K*-means in post-clustering stage. The limitations of both the algorithms are overcome by their combination. The algorithms are compared by MSE and PSNR values of the four images. The MSE values are lower and PSNR values are higher for the proposed algorithm, and perfect edges are obtained for the proposed algorithm.

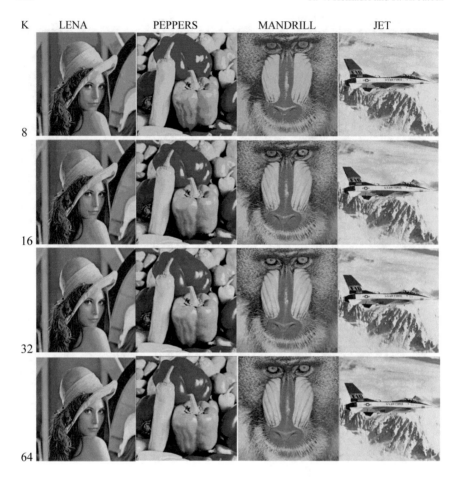

Fig. 4 Images of hybrid Jaya K-means algorithm

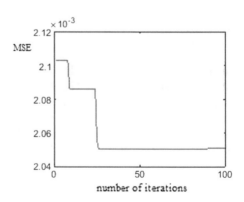

Fig. 5 Convergence plot of MSE of the proposed algorithm for $k = 8$ for Lena

Fig. 6 Convergence plot of
PSNR of the proposed
algorithm for $k = 8$ for Lena

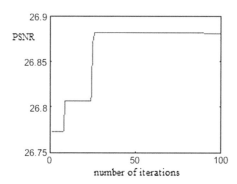

References

1. Scheunders, P.: A comparison of clustering algorithms applied to color image quantization. Pattern Recognit. Lett. **18**, 1379–1384 (1997). (New York). https://doi.org/10.1016/S0167-8655(97)00116-5
2. Heckbert, P.: Color image quantization for frame buffer display. In: Computer Graphics (Proceedings Siggraph), vol. 16, no. 3, pp. 297–307 (1982). https://doi.org/10.1145/800064.801294
3. Joy, G., Xiang, Z.: Center-cut for color image quantization. Vis. Comput. **10**, 62–66 (1993). (Berlin). https://doi.org/10.1007/BF01905532
4. Gervautz, M., Purgtathofer, W.: A simple method for color quantization: octree quantization. Academic, San Diego, CA (1990). https://doi.org/10.1007/978-3-642-83492-9_20
5. EmreCelebi, M.: Effective initialization of k-means for color quantization. In: Proceedings of IEEE International Conference on Image Processing, 2009 IEEE Press, Piscataway, NJ, USA, pp. 1629–1632. https://doi.org/10.1109/ICIP.2009.5413743
6. EmreCelebi, M.: Improving the performance of k-means for color quantization. Image Vis. Comput. **29**(4), 260–271 (2011). https://doi.org/10.1016/j.imavis.2010.10.002
7. Wang, Z., Sun, X., Zhang, D.: A swarm intelligence based color image quantization algorithm. In: International Conference on Bioinformatics and Biomedical Engineering: ICBBE 2007, pp. 592–595, July 2007
8. Yazdani, D., Nabizadeh, H., Kosari, E.M., Toosi, A.N.: Color quantization using modified artificial fish swarm algorithm. In: Proceedings of 24th International Conference on Advances in Artificial Intelligence (AI'11), Springer-Verlag, Berlin, Heidelberg, pp. 382–391 (2011). https://doi.org/10.1007/978-3-642-25832-9_39
9. Omran, M.G., Engelbrecht, A.P., Salman, A.: A color image quantization algorithm based on particle swarm optimization. Informatica **29**, 261–269 (2005)
10. Jain, A.K., Murty M.N., Flynn, P.J.: Data clustering: a review. ACM Comput. Surv. **31**(3), 256–323 (1999). https://doi.org/10.1145/331499.331504
11. Venkata Rao, R.: Jaya: a simple and new optimization algorithm for solving constrained and unconstrained optimization problems. Int. J. Ind. Eng. Comput. **7**, 19–34 (2016). https://doi.org/10.1080/0305215X.2016.1164855
12. Thung, K.-H.: A survey of image quality measures. In Proceedings of International Conference for Technical Postgraduates, pp. 1–4 (2009). https://doi.org/10.1109/TECHPOS.2009.5412098

Tooth Profile Optimization of Helical Gear with Balanced Specific Sliding Using TLBO Algorithm

Paridhi Rai and Asim Gopal Barman

Abstract Meshing performance of helical gears is affected by the sliding coefficients. In this paper, addendum modification is used to maximize the specific sliding coefficients of gears. Interference, undercutting, and strength of gears act as the design constraints for the design problem formulated. TLBO is used to perform design optimization of helical gears. The results achieved are compared with the results found in literature. It has been found that there is a significant improvement in the value of specific sliding. This will not only increase the wear resistance of the gear pair but also increase the service life and meshing performance of gears.

Keywords Specific sliding · Profile shift · Helical gears · Teaching–learning based optimization

Nomenclature

v_s	Sliding velocity
μ_{max}	Maximum sliding velocity
α_{at}	Tip transverse pressure angle
α_t	Transverse pressure angle
i	Transmission ratio
ε_α	Transverse contact ratio
m_t	Transverse module
h_{an}	Addendum modification coefficient
r_b	Base radius
S_{at}	Tranverse arc tooth thickness
σ_{Fcal}	Nominal bending stress
s_{Fall}	Allowable bending stress

P. Rai · A. G. Barman (✉)
Department of Mechanical Engineering, National Institute of Technology Patna, Patna 800005, India
e-mail: asim@nitp.ac.in

© Springer Nature Singapore Pte Ltd. 2020
R. Venkata Rao and J. Taler (eds.), *Advanced Engineering Optimization Through Intelligent Techniques*, Advances in Intelligent Systems and Computing 949, https://doi.org/10.1007/978-981-13-8196-6_19

σ_{Hcal} Nominal contact stress
σ_{Hall} Allowable contact stress
z_1 Number of pinion teeth
z_2 Number of gear teeth
z' Minimum number of teeth to avoid undercut
C Working center distance

1   Introduction

In order to improve the meshing performance and loading capacities of gears, using addendum modifications proves to be very advantageous in gear design optimization problem. Various design parameters related to gear design and its geometry are affected by addendum modification coefficients. This will help in optimization of gear design with special necessity.

Recent literature shows that there are various methods developed for achieving optimal values of profile shift coefficients for fulfilling various design criteria. Henriot [1] developed a graphical method to balance specific sliding coefficients of gear and pinion. A diagram for total profile shift is provided by DIN 3992 [2] which contributes to having teeth with balanced specific sliding velocity. It has been observed earlier that the values of addendum modifications obtained either from a graph or from approximate formulas are not exact and errors increase with interpolation method. To resolve the difficulties faced earlier, Pedero and Artes [3] presented approximate equations for profile shift so as to balance the specific sliding of both the gears. Diez Ibarbia et al. [4] reported effects of profile shift on efficiency of spur gears. Further, Chen et al. [5] demonstrated the effect of addendum modifications on the internal spur gear mesh efficiency. They conclude that there is an increase in the single-tooth mesh stiffness with an increase in profile shift, but decreases the multi-tooth mesh stiffness. Samo et al. [6] proposed adaptive grid refinement for optimization of tooth geometry. Atanasovska et al. [7] developed a new optimization technique called explicit parametric method (EPM) in order to calculate the optimal values of the profile shift coefficient, tooth root radius, and pressure angle. A multi-objective study with respect to transmission error is conducted by Bruyere and Velex [8]. Miler et al. [9] investigated the significance of profile shift on volume optimization of spur gears.

Many design optimization problems are very complex and non-linear with a lot of design variables and complicated constraints. Earlier, evolutionary optimization techniques are used to optimize the gear design [10] but a new population-based optimization technique, i.e., TLBO is used for minimization of specific sliding coefficients due to robustness and effectiveness of the algorithm. Low computational cost and high reliability are the features which led to the implementation of TLBO for the minimization design problem.

This paper presents the use of TLBO algorithm to balance the specific sliding coefficient for increasing the service life and gear efficiency. The design problem

considered by Abderazek et al. [10] is used for the implementation of TLBO algorithm. Various design constraints are formulated which should be satisfied for an
optimal design which includes constraints on bending strength, contact strength,
tooth thickness, involute interference, and contact ratio. The developed program of
TLBO is implemented on MATLAB platform. The results obtained are compared
with those obtained earlier [10]. There is a significant improvement in the results
obtained using TLBO algorithm.

The presented article is structured as follows. In the second section, mathematical
model is formulated which consists of design variables, objective function, and design
constraints. The next section describes the optimization technique used, i.e., TLBO
algorithm. Results and discussions about the article are made in section four. Finally,
the conclusion has been made in section five.

2   Mathematical Model

A general design optimization problem consists of design variables, objective function, and design constraints, which are briefly discussed below.

2.1   Design Variables

Profile shift coefficients of pinion and gear, x_1 and x_2, respectively, are the two design
variables considered in the formulation of design problem. The limits of these design
variables are mentioned in Table 1.

2.2   Objective Function

Relative sliding velocity is one of the important factors, which affects the meshing
[11]. When the sliding coefficients of two gears are balanced, the gear scoring is
reduced and its service life is improved. The proportion between sliding velocity and
rolling velocity of the mating tooth is known as sliding velocity. The mathematical
formula for sliding coefficient of pinion and gear μ_1 and μ_2 is defined as,

Table 1 Design variables for specific sliding calculation

Design variables	Lower limit	Upper limit
x_1	−0.5	0.8
x_2	−0.5	0.8

$$\mu_1 = \frac{v_{s_2} - v_{s_1}}{v_{s_1}} \tag{1}$$

and

$$\mu_2 = \frac{v_{s1} - v_{s2}}{v_{s2}} \tag{2}$$

where v_{s1} and v_{s2} are the pinion sliding velocity and gear sliding velocity.

The maximum sliding velocity coefficient arises when the addendum of one gear tooth mesh with another gear tooth. For attaining increased wear resistance and gear life, there is requirement to equalize the sliding coefficient of both pinion and gear. Hence, the objective function can be given by,

$$f(x_1, x_2) = |\mu_{2\,\mathrm{max}} - \mu_{1\,\mathrm{max}}| \tag{3}$$

where $\mu_{1\mathrm{max}}$ and $\mu_{2\mathrm{max}}$ are the maximum sliding velocity for pinion and gear, respectively, and can be expressed as

$$\mu_{1\,\mathrm{max}} = \frac{\tan \alpha_{\mathrm{at1}} - \tan \alpha_t'}{(1+i) \tan \alpha_t' - \tan \alpha_{\mathrm{at1}}} (1+i) \tag{4}$$

$$\mu_{2\,\mathrm{max}} = \frac{\tan \alpha_{\mathrm{at2}} - \tan \alpha_t'}{(1+1/i) \tan \alpha_t' - \tan \alpha_{\mathrm{at2}}} (1+1/i) \tag{5}$$

2.3 Design Constraints

The characteristics of the design problem define the type of constraints which must be imposed on the objective function. When the constraints are satisfied, acceptable solutions are obtained. The following design constraints are used in this design formulation.

- For achieving silent and steady gear operation with reduced vibration, the value of the transverse contact ratio should not be less than 1.2.

$$g_1(x_1, x_2) = \varepsilon_\alpha \geq 1.2 \tag{6}$$

- The tip diameter thickness should be more than $0.4m_t$ to have broader top land. For pinion,

$$g_2(x_1, x_2) = 0.4\,m_t - S_{\mathrm{at1}} \leq 0 \tag{7}$$

For gear wheel,

$$g_3(x_1, x_2) = 0.4\, m_t - S_{at2} \le 0 \tag{8}$$

Here, m_t denotes the transverse module. S_{at1} and S_{at2} are the transverse arc tooth thickness for pinion and gear, respectively.

- The maximum calculated bending and contact stress must not exceed the permissible values for both the stress. The expression can be expressed as [12]

$$g_4(x) = \sigma_{Fcal} - \sigma_{Fall} \le 0 \tag{9}$$

$$g_5(x) = \sigma_{Hcal} - \sigma_{Hall} \le 0 \tag{10}$$

- When the addendum of one gear mates with dedendum of other, interference occurs [13]. The reason for occurrence is the number of teeth on pinion is less than the prescribed minimum number of teeth.
Pinion interference

$$g_6(x) = \frac{m_t(z_2 + 2 + 2x_2)}{2} - \sqrt{r_{b2}^2 + (C \sin \alpha_t)^2} \le 0 \tag{11}$$

Gear interference

$$g_7(x) = \frac{m_t(z_1 + 2 + 2x_1)}{2} - \sqrt{r_{b2}^2 + (C \sin \alpha_t)^2} \le 0 \tag{12}$$

where r_b is the base radius and C is the working center distance.

- To prevent undercutting in pinion and gear, the following constraints must be satisfied.

$$g_8(x) = \frac{z' - z_1}{z'} h_{an} \le x_1 \tag{13}$$

$$g_9(x) = \frac{z' - z_2}{z'} h_{an} \le x_2 \tag{14}$$

where z' is a maximum number of teeth to prevent undercut and h_{an} is addendum coefficient. The main features of the helical gear used for design problem are reported in Table 2.

3 Teaching–Learning-Based Algorithm

TLBO is a population-based algorithm and is different from other evolutionary and swarm-based optimization algorithms [14]. It requires only commonly used design parameters such as generation number and population size rather than algorithm-based control design parameters. The teaching process used in the current scenario

Table 2 Input parameters for the design problem [10]

Input parameters	Value
Pinion teeth number	15
Gear teeth number	17
Face width	14
Module	5
Helix angle	16
Pressure angle	20

is the main basis of this algorithm. There are two basic modes of learning in this algorithm

1 Teacher phase—In this phase, the teachers teach to learners. The teacher attempts to improve the mean result of the class in a particular subject to his full ability. The output of this phase acts as the input for the learners phase. This means that learner's phase is dependent on the teachers phase.
2 Learner phase—In this phase, there is an increase in learner's knowledge because of the interaction of learners among themselves. A learner randomly discusses with other learners in order to enrich their knowledge, since other learners are assumed to have more knowledge.

This algorithm defines the group of learners as the population and various design variables are different subjects proposed to the learners. The results obtained by the learners refer to the fitness function of the optimization design problem. Teacher refers to the best solution obtained by the design algorithm. There is continuous interaction between different qualities of teacher and achievement of the learners.

There is a correlation between TLBO and the gear optimization which can be expressed as: Population is the class which consists of gear design, a learner in a class indicates gear design in the population, subjects taught in the class are the design variables, and the teacher is the gear design with minimum specific sliding.

4 Results and Discussions

The performance of TLBO algorithm can be verified by comparing the attained results with the previously published techniques. The value of profile shift coefficients attained is such that the specific sliding for pinion and gear are balanced. This means that the value of $\mu_{1\max}$ and $\mu_{2\max}$ are the same for both pinion and gear. Table 3 presents the comparison of results obtained using various optimization techniques.

Thus, it has been found that the values of profile shift achieved utilizing TLBO algorithm results in the perfect balancing of specific sliding. There is a decrease in the value of specific sliding for both pinion and gears compared to values obtained using other optimization techniques. This will lead to further improvement in the meshing performance of gears. Since both the values of specific sliding for pinion

Table 3 Comparison of results obtained

Parameters	AGr	DE	TLBO
x_1	0.3070	0.2842	0.3955
x_2	0.0460	0.2158	0.2189
$\mu_1 \max$	2.5449	2.4789	1.2257
$\mu_2 \max$	2.5072	2.4789	1.2257
$f_{\min}$	0.0377	0	0

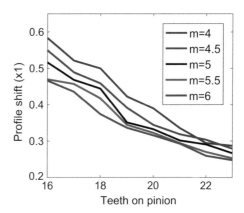

Fig. 1 Variation of pinion profile shift versus the number of teeth on pinion

and gear are same, this will further result in balancing of wear resistance property for the gear pair.

The summation of values of profile shift achieved using TLBO algorithm is also more than those obtained using DE and AGr. As the total addendum modifications increases, the tooth thickness also increases, thus decreasing the tooth fillet stress. There is also an increase in load carrying capacity of the gears. Thus, the bending strength of the gear pair will increase (Fig. 1).

A graph is plotted in between the pinion profile shift versus a number of teeth on pinion. The plot reveals that as a number of teeth increases, the profile shift decreases. This important prerequisite of the minimum number of teeth to avoid interference is also satisfied. Thus, it can be seen that TLBO algorithm proves to be more efficient and durable compared to other evolutionary algorithm such as DE. This will result in maximization of service life and wear resisting property of gears.

5 Conclusions

In this article, an advanced optimization technique is used to obtain optimal values of profile shift. Including profile shift coefficients in the design, procedure has many advantages such as avoiding undercutting, balancing specific sliding, and balancing flash temperature.

A constrained optimization mathematical model has been developed and an advanced population-based optimization algorithm is employed to perform optimization. The objective function is to maintain a balance between the specific sliding of pinion and gear. Bending strength, contact strength, undercutting, involute interference, and contact ratio are the design constraints imposed on the merit function. TLBO is employed to optimize the profile shift coefficients of both pinion and gears.

The results obtained using TLBO shows that there is a perfect balance between the specific sliding coefficients. The value of specific sliding coefficients is also lower compared to values obtained using other optimization techniques used in prior published articles. Even the sum of values of profile shift coefficients is more than earlier published result; this will further increase the tooth thickness at the critical section and increases the load carrying capacity. There is an increase in meshing performance of gears, service life, and scuffing resistance for gears. This proves the effectiveness of the optimization techniques proposed in this research article.

This mathematical model can be further extended to multi-objective optimization mathematical model using the optimization technique proposed in this article.

References

1. Henriot, G.: Engrenages: conception fabrication mise enoeuvre, 8th ed, Dunod, p 869. (2007)
2. DIN 3992: Addendum modification of external spur and helical gears (1964)
3. Pedrero, J.I., Artes, M., Garcia-Prada, J.C.: Determination of the addendum modification factors for gears with pre-established contact ratio. Mech. Mach. Theory **31**(7), 937–945 (1996)
4. Diez-Ibarbia, A., del Rincon, A.F., Iglesias, M., De-Juan, A., Garcia, P., Viadero, F.: Efficiency analysis of spur gears with a shifting profile. Meccanica **51**(3), 707–723 (2016)
5. Chen, Z., Zhai, W., Shao, Y., Wang, K.: Mesh stiffness evaluation of an internal spur gear pair with tooth profile shift. Sci. China Technol. Sci. **59**(9), 1328–1339 (2016)
6. Ulaga, S., Flasker, J., Zafosnik, B., Ciglaric, I.: Optimisation feature in gear design procedure. Strojniski Vestnik **47**(7), 286–299 (2001)
7. Atanasovska, I., Mitrovic, R., Momcilovic, D.: Explicit parametric method for optimal spur gear tooth profile definition. Adv. Mater. Res. **633**, 87–102 (2013)
8. Bruyère, J., Velex, P.: A simplified multi-objective analysis of optimum profile modifications in spur and helical gears. Mech. Mach. Theory **80**, 70–83 (2014)
9. Miler, D., Lončar, A., Žeželj, D., Domitran, Z.: Influence of profile shift on the spur gear pair optimization. Mech. Mach. Theory **117**, 189–197 (2017)
10. Abderazek, H., Ferhat, D., Atanasovska, I., Boualem, K.: A differential evolution algorithm for tooth profile optimization with respect to balancing specific sliding coefficients of involute cylindrical spur and helical gears. Adv. Mech. Eng. **7**(9), 1687814015605008 (2015)
11. ISO/TR 4467: Addendum modification of the teeth of cylindrical gears for speed-reducing and speed increasing gear pairs (1982)
12. ISO Standard 6336: Calculation of load capacity of spur and helical gears, international organization for standardization, Geneva, Switzerland (2006)
13. Rai, P., Barman, A.G.: Design optimization of spur gear using SA and RCGA. J Braz. Soc. Mech. Sci. Eng. **40**, 1–8 (2018)
14. Rao, R.V., Savsani, V.J., Vakharia, D.P.: Teaching–learning-based optimization: a novel method for constrained mechanical design optimization problems. Comput. Aided Des. **43**(3), 303–315 (2011)

Optimization of Nanofluid Minimum Quantity Lubrication (NanoMQL) Technique for Grinding Performance Using Jaya Algorithm

R. R. Chakule, S. S. Chaudhari and P. S. Talmale

Abstract The machining performance and the surface quality are the basic requirements of industries. At the same time, the machining process should be clean, economical and eco-friendly to sustain in globalized competitive environments. The wet
technique consumes large amount of cutting fluid to minimize temperature and friction generates during grinding process. The recent NanoMQL technique of cutting
fluid can substitute over wet grinding due to better cooling and lubrication obtained
using nanofluid and better penetration using compressed air at contact zone. The
experiments were conducted as per the design matrix using response surface methodology (RSM). The modeling and multi-objective optimization of NanoMQL process
are carried out for minimizing the surface roughness and cutting force using Jaya
algorithm. The study demonstrates the validity of regression models by comparing
the experimental test results conducted at optimized parameters value obtained from
Jaya algorithm with predicted values and is observed the close.

Keywords Grinding · Jaya algorithm · Modeling · NanoMQL · Optimization

1 Introduction

The grinding is a complex material removal process widely used for finishing the
machining surfaces. The high heat generation and friction occur due to the contact of
abrasive grits of wheel with workpiece surface for microseconds which significantly
affects on surface quality and wheel life [1]. The more amount of specific grinding

R. R. Chakule (✉) · S. S. Chaudhari
Yeshwantrao Chavan College of Engineering, Nagpur 441110, India
e-mail: r_chakule@rediffmail.com

S. S. Chaudhari
e-mail: sschaudhari@rediffmail.com

P. S. Talmale
Late G. N. Sapkal College of Engineering, Nashik 422213, India
e-mail: poonam.talmale05@gmail.com

© Springer Nature Singapore Pte Ltd. 2020
R. Venkata Rao and J. Taler (eds.), *Advanced Engineering Optimization
Through Intelligent Techniques*, Advances in Intelligent Systems and Computing 949,
https://doi.org/10.1007/978-981-13-8196-6_20

energy is required due to relatively more contact length of abrasive grits and work-piece surface and elastic–plastic deformation of material removal in comparison with other machining process. Hence, to improve the efficiency of grinding process, the effective lubrication at contact zone is the prime requirement. Despite using high amount of cutting fluid and widely used for manufacturing, the conventional/wet technique of cutting fluid has reported several drawbacks such as inefficient pen-etration into high hydrodynamic pressure-grinding zone, high cost, high pumping power and harmful to operator and environment [2–4]. To overcome these limita-tions, much effort has been taken by the researchers in the recent years to minimize the friction. The minimum quantity lubrication (MQL) and penetration of nanofluid by MQL (NanoMQL) are the better micro-lubrication methods recently focused for better machining results. In MQL, the cutting fluid is injected by pressurized air in spray form into the contact area of grits and workpiece [5]. The papers were reviewed on MQL for tribological study and process optimization to achieve the grinding performance [6–8]. In the last few years, nanofluids are used largely for machining application due to better thermal and tribological characteristics, high surface energy that gives effective cooling and lubrication effects. The thermophysi-cal properties of nanofluid and application of nanofluid using MQL (NanoMQL) for grinding performance are reviewed in papers [9–16].

The traditional methods of optimization can be used to find the optimal values of parameters, but the accurate values of parameters are not possible and may trap into local optima. The problem is more critical for NanoMQL grinding process where large number of parameters involved. To overcome these limitations, recently, Rao introduced Jaya algorithm which does not require any algorithm-specific param-eter to optimize the process parameters [17]. The optimization of multi-objective responses for optimal parameters values using Jaya algorithm and comparative study of machining results of NanoMQL process with wet and MQL grinding environments are studied. The derived mathematical model is considered in Jaya algorithm as fit-ness function for optimization.

2 Materials and Methods

The experiments were performed on oil hardening non-shrinking steel (OHNS) rect-angular plate of size 100*50*25 mm by horizontal surface grinding using vitrified Al_2O_3 grinding wheel having specification of A-46-3-L5-V8. The experimental runs have been determined by response surface methodology (RSM) using Minitab 17 statistical software. The levels of parameters are shown in Table 1. The homemade MQL setup used for experimentation is shown in Fig. 1. The nozzle position was set properly for effective penetration of cutting fluid. The concentration (0) means the experiments are conducted for MQL. The analytical grade of spherical γ-Al_2O_3 nanoparticle of 99.50% purity and mean diameter of 30–50 nm was used for synthe-sizing the stable nanofluid in distilled water of 0.15 and 0.30 vol. % concentrations. The grinding forces (tangential and normal force) were measured with the help of

Table 1 Levels of parameters

Level	Table speed (mm/min)	Depth of cut (μm)	Air pressure (bar)	Coolant flow rate (ml/hr)	Nanofluid concentration (vol. %)
Low	7000	20	2	250	0 (MQL)
Medium	10000	30	3	500	0.15
High	13000	40	4	750	0.30

Fig. 1 Experimental setup

strain gauge dynamometer, and surface roughness of workpiece surface along the grinding direction using surface roughness tester (SJ-210). The surface roughness is measured after 60th pass of wheel on workpiece and cutting forces at three locations on workpiece surface after 14, 32 and 46 passes of wheel. Figure 2 shows the characterizations of nanoparticles.

3 Modeling and Optimization

The regression models of surface roughness (R_a), cutting force (F_c) and their combine objective (CO) for NanoMQL grinding process in coded form are shown in Eqs. (1–3). The full quadratic models were formulated and models adequacy are checked by ANOVA based on coefficients of determination (R^2) values. The surface quality is the prime requirement of industries, and cutting force is more related with specific energy and wheel life. The input factors are coded as table speed (TS): x_1, depth of cut (DOC): x_2, coolant flow rate (CFR): x_3 and nanofluid concentration (NC): x_4.

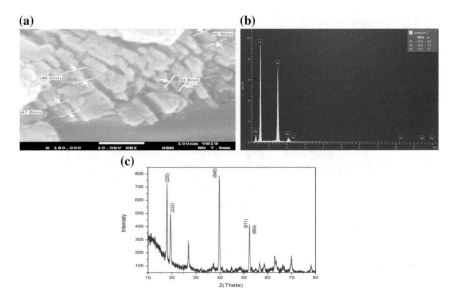

Fig. 2 Characterization techniques of nanoparticle: **a** SEM micrograph of Al_2O_3 **b** EDX image of Al_2O_3 **c** XRD image of Al_2O_3

$$R_a = 0.1726 + 0.000005x_1 - 0.00430x_2$$
$$- 0.000021x_3 - 0.0948x_4 + 0.000081x_2 * x_2 \qquad (1)$$

$$F_c = 26.87 + 0.004329x_1 + 0.743x_2$$
$$- 0.01740x_3 - 102.4x_4 + 235.3x_4 * x_4 \qquad (2)$$

$$CO = 0.8736 + 0.000053x_1 + 0.004794x_2 - 0.000317x_3$$
$$- 1.428x_4 + 3.025x_4 * x_4 - 0.000037x_1 * x_4 + 0.000487x_3 * x_4 \qquad (3)$$

The multiple regression coefficients (R^2) were obtained as 0.816, 0.832 and 0.959 for R_a, F_c and CO, respectively. The pred R-squared values are in close with adj R-squared values of R_a, F_c and CO. The significant parameters were determined using ANOVA at 95% confidence level and considered for modeling of responses. The residuals lie along and near to the straight line. The average predicted and actual experimental values of responses are within the range to make the model more adequate for accurate results. It shows that the errors are normally distributed. After adequacy of developed regression models, the optimizations of the process parameters were carried out using Jaya algorithm. The combine objective function was developed considering the normalized values of responses by assigning the weight of 0.75 and 0.25 for R_a and F_c, respectively, based on the importance of responses in industry. The combine objective function is generated using Eq. (4) which is the compromised outcome and different from the individual value [18]. The regression equation of combine objective was developed based on CO values of all experimental runs.

$$CO_{min} = w_1\left(R_a/R_{a_{min}}\right) + w_2\left(F_c/F_{c_{min}}\right) \tag{4}$$

where

w_1 weight assigned to surface roughness (R_a)
w_2 weight assigned to cutting force (F_c)

3.1 JAYA Algorithm

The recently developed Jaya algorithm is simple and free from algorithm-specific parameters. The optimal results also obtained in less value of function evaluations and memory requirement [19–24]. In the present research, the optimal values of NanoMQL process parameters were obtained by coding the Jaya algorithm to minimize the responses. The execution of Jaya algorithm is on the concept to obtain the best solution and avoid the worst solution according to Eq. (5). The objective functions were subject to significant parameters and range constraints such as 7000 $\leq$ TS $\leq$ 13000, 20 $\leq$ DOC $\leq$ 40, 250 $\leq$ CFR $\leq$ 750 and 0 $\leq$ NC $\leq$ 0.30.

$$X'_{m,n,i} = X_{m,n,i} + r_{1,m,i}\left(X_{m,best,i} - \left|X_{m,n,i}\right|\right)$$
$$- r_{2,m,i}\left(X_{m,worst,i} - \left|X_{m,n,i}\right|\right) \tag{5}$$

where $X_{m,best,i}$ is the value of variable m for the best candidate and $X_{m,worst,i}$ is the value of the variable m for the worst candidate. $X'_{m,n,i}$ is the updated value of $X_{m,n,i}$ and random numbers $r_{1,m,i}$ and $r_{2,m,i}$ are for the m variable during the ith iteration in the range [0, 1].

Figure 3 shows the convergence curve to optimize surface roughness, cutting force and combine objective function. It shows that the convergence graphs remain stable after it reaches to lower value of responses continuously. This shows that the Jaya algorithm does not get trapped in local optima. The elapsed time and convergence after number of generations are shown in Table 2. The number of population and generations for optimization of process parameters are considered 100 and 50, respectively.

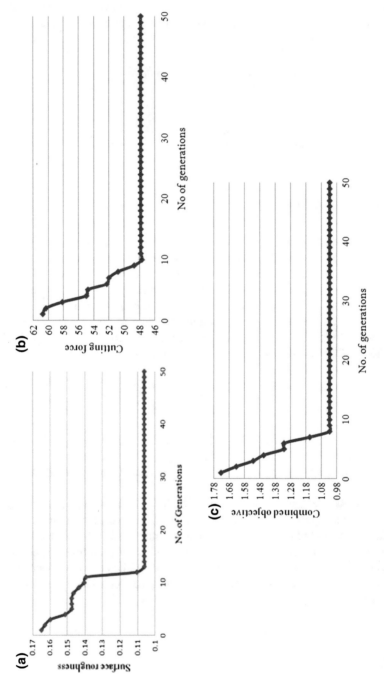

Fig. 3 Convergence curve for optimizing: **a** Surface roughness **b** cutting force **c** combine objective using Jaya algorithm

Table 2 Optimal values and confirmation experiments

Input/responses	TS	DOC	CFR	NC	R_a Pred (μm)	F_c Pred (N)	Experimental results	Elapsed time (sec)	Convergence
R_a	7000	26	750	0.30	0.106	–	0.112 μm	15.90	12
F_c	7000	20	750	0.22	–	47.8	–	15.09	9
CO	7000	23	750	0.30	0.107	51.8	0.115 μm 52.83 N	14.4	7

4 Results and Discussions

Figure 4 shows the response values under different machining environments such as wet, MQL and NanoMQL. The resultant of tangential force and normal force is treated as resultant force (F_c). The tangential force is measured along the grinding direction during reciprocating movement of worktable, whereas normal force mostly affected by workpiece hardness. The surface roughness and cutting force obtained in NanoMQL were 0.133 μm and 68.36 N at 0.30 vol. % concentration, whereas in wet grinding the values obtained as 0.2 μm and 87.1 N, respectively. The effective penetration of nanofluid using MQL approach in mist form into contact interface, and formation of tribofilm of nanoparticles on the surface of workpiece under air pressure are the main reasons for better workpiece surface finish. Thus, effective penetration of cutting fluid at contact area significantly reduces the friction coefficient and helps to maintain the grit sharpness. The better grinding results are obtained for NanoMQL process due to better lubricating and cooling effects compared to wet and MQL.

4.1 Confirmation Experiments

To validate the expected responses from Jaya algorithm, the experiments were conducted at optimal feasible values. The experimental test results at optimal parameters values obtained by Jaya algorithm and predicted values for responses are shown in Table 2. The % error 7.48 and 2.05% for surface roughness and cutting force were obtained by considering the combine objective function. This is supposed to be a good prediction of responses from the models, and obtained optimal grinding parameters are valid in the region of interest.

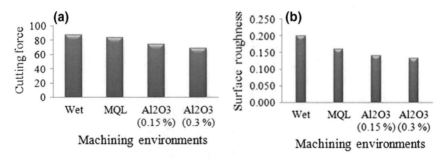

Fig. 4 Grinding responses under different machining environments: **a** Cutting force. **b** Surface roughness

5 Conclusions

1. The NanoMQL process significantly reduces the surface roughness and cutting force by 33.5 and 21.51% at 0.30 vol. % concentration over the wet grinding. The efficient lubrication and cooling due to slurry formation of nanofluid on workpiece surface using air pressure and better penetration of fluid in mist form at contact interface. The efficient lubrication maintains the grit sharpness better and results into less cutting force and surface roughness.

2. The modeling and optimization of NanoMQL process parameters were carried out for obtaining minimum surface roughness and cutting force at 95% confidence interval. From ANOVA, it was observed that the linear, square and interaction effects are statistically more significant. The parameter table speed found the most influencing parameter for minimizing the responses. The optimized values of parameters obtained from Jaya algorithm for combine objective function were TS (7000 mm/min), DOC (23 µm), CFR (750 ml/hr) and NC (0.30 vol. %) in selected ranges of process parameters.

3. The experimental trials were performed at optimal feasible values obtained from Jaya algorithm to validate the results. For combine objective, the percentage errors 7.48 and 2.05% for surface roughness and cutting force were obtained, respectively.

4. The convergence accuracy and speed of Jaya algorithm observed very high. The good optimal grinding results were obtained in less function evaluations. The results obtained from Jaya algorithm for NanoMQL process is beneficial to minimize the surface roughness and cutting for improving the grinding performance.

Acknowledgments The authors would like to thanks the Director, Visvesvaraya National Institute of Technology (VNIT) for providing facility to characterize the nanofluid and Sameeksha industry for extending the experimental facility.

References

1. Tawakoli, T., Hadad, M.J., Sadeghi, M.H., Daneshi, A., Stockert, S., Rasifard, A.: An experimental investigation of the effects of workpiece and grinding parameters on minimum quantity lubrication-MQL grinding. Int. J. of Machine Tools and Manufacture. **49** (12–13), 924–932 (2009). https://doi.org/10.1016/j.ijmachtools.2009.06.015
2. Sinha, M.K., Madarkar, R., Ghosh, S., Rao, P.V.: Application of eco-friendly nanofluids during grinding of Inconel 718 through small quantity lubrication. J. Cleaner Prod. **141**, 1359–1375 (2017). https://doi.org/10.1016/j.jclepro.2016.09.212
3. Kalita, P., Malshe, A.P., Arun Kumar, S., Yoganath, V.G., Gurumurthy, T.: Study of specific energy and friction coefficient in minimum quantity lubrication grinding using oil-based nanolubricants. J. Manuf. Proc. **14**(2), 160–166 (2012). https://doi.org/10.1016/j.jmapro.2012.01.001
4. Brinksmeier, E., Meyer, D., Huesmann-Cordes, A.G., Herrmann, C.: Metalworking fluids-mechanisms and performance. J. CIRP Ann. **64**(2), 605–628 (2015). https://doi.org/10.1016/j.cirp.2015.05.003

5. Kim, H.J., Seo, K.J., Kang, K.H., Kim, D.E.: Nano-lubrication: a review. Int. J. Precision Eng. Manuf. **17**(6), 829–841 (2016). https://doi.org/10.1007/s12541-016-0102-0
6. Chakule, R.R., Chaudhari, S.S., Talmale, P.S.: Evaluation of the effects of machining parameters on MQL based surface grinding process using response surface methodology. J. Mech. Sci. Technol. **31**(8), 3907–3916 (2017). https://doi.org/10.1007/s12206-017-0736-6
7. Tawakoli, T., Hadad, M.J., Sadeghi, M.H.: Influence of oil mist parameters on minimum quantity lubrication-MQL grinding process. Int. J. Mach. Tools Manuf. **50**(6), 521–531 (2010). https://doi.org/10.1016/j.ijmachtools.2010.03.005
8. Huang, X., Ren, Y., Jiang, W., He, Z., Deng, Z.: Investigation on grind-hardening annealed AISI5140 steel with minimal quantity lubrication. Int. J. Adv. Manuf. Technol. **89**(1–4), 1069–1077 (2017). https://doi.org/10.1007/s00170-016-9142-y
9. Lee, J., Yoon, Y.-J., Eaton, J.K., Goodson, K.E., Bai, S.J.: Analysis of oxide (Al_2O_3, CuO, and ZnO) and CNT nanoparticles disaggregation effect on the thermal conductivity and the viscosity of nanofluids. Int. J. Prec. Eng. Manuf. **15**(4), 703–710 (2014). https://doi.org/10.1007/s12541-014-0390-1
10. Mao, C., Zou, H., Zhou, X., Huang, Y., Gan, H., Zhou, Z.: Analysis of suspension stability for nanofluid applied in minimum quantity lubricant grinding. Int. J. Adv. Manuf. Technol. **71**(9–12), 2073–2081 (2014). https://doi.org/10.1007/s00170-014-5642-9
11. Chiam, H.W., Azmi, W.H., Usri, N.A., Mamat, R., Adam, N.M.: Thermal conductivity and viscosity of Al_2O_3 nanofluids for different based ratio of water and ethylene glycol mixture. Exp. Thermal Fluid Sci. **81**, 420–429 (2017). https://doi.org/10.1016/j.expthermflusci.2016.09.013
12. Zhang, D., Li, C., Jia, D., Zhang, Y., Zhang, X.: Specific grinding energy and surface roughness of nanoparticle jet minimum quantity lubrication in grinding. Chinese J. Aeronautics **28**(2), 570–581 (2015). https://doi.org/10.1016/j.cja.2014.12.035
13. Zhang, Y., Li, C., Jia, D., Li, B., Wang, Y., Yang, M., Hou, Y., Zhang, X.: Experimental study on the effect of nanoparticle concentration on the lubricating property of nanofluids for MQL grinding of Ni-based alloy. J. Mater. Proc. Technol. **232**, 100–115 (2016). https://doi.org/10.1016/j.jmatprotec.2016.01.031
14. Wang, Y., Li, C., Zhang, Y., Yang, M., Zhang, X., Zhang, N., Dai, J.: Experimental evaluation on tribological performance of the wheel/workpiece interface in minimum quantity lubrication grinding with different concentrations of Al_2O_3 nanofluids. J. Cleaner Production **142**(4), 3571–3583 (2017). https://doi.org/10.1016/j.jclepro.2016.10.110
15. Setti, D., Sinha, M.K., Ghosh, S., Rao, P.V.: Performance evaluation of Ti-6Al-4V grinding using chip formation and coefficient of friction under the influence of nanofluids. Int. J. Mach. Tools Manuf. **88**, 237–248 (2015). https://doi.org/10.1016/j.ijmachtools.2014.10.005
16. Mao, C., Zou, H., Huang, X., Zhang, J., Zhou, Z.: The influence of spraying parameters on grinding performance for nanofluid minimum quantity lubrication. Int. J. Adv. Manuf. Technol. **64**(9–12), 1791–1799 (2013). https://doi.org/10.1007/s00170-012-4143-y
17. Rao, R.V., Rai, D.P., Balic, J.: A new optimization algorithm for parameter optimization of nano-finishing processes. Int. J. Sci. Technol. **24**(2), 868–875 (2017). https://doi.org/10.24200/sci.2017.4068
18. Gupta, M.K., Sood, P.K., Sharma, V.S.: Optimization of machining parameters and cutting fluids during nano-fluid based minimum quantity lubrication turning of titanium alloy by using evolutionary techniques. J. Cleaner Production **135**, 1276–1288 (2016). https://doi.org/10.1016/j.jclepro.2016.06.184
19. Rao, R.V., Rai, D.P., Ramkumar, J., Balic, J.: A new multi-objective Jaya algorithm for optimization of modern machining processes. Adv. Prod. Eng. Manag. **11**(4), 271–286 (2016). https://doi.org/10.14743/apem2016.4.226
20. Rao, R.V.: Jaya: A simple and new optimization algorithm for solving constrained and unconstrained optimization problems. Int. J. Ind. Eng. Comput. **7**(1), 19–34 (2016). https://doi.org/10.5267/j.ijiec.2015.8.004
21. Rao, R.V., More, K.C.: Optimal design and analysis of mechanical draft cooling tower using improved Jaya algorithm. Int. J. Refriger. **82**, 312–324 (2017)

22. Rao, R.V., Rai, D.P.: Optimization of submerged arc welding process parameters using quasi-oppositional based Jaya algorithm. J. Mech. Sci. Technol. **31**(5), 2513–2522 (2017). https://doi.org/10.1007/s12206-017-0449-x
23. Rao, R.V., Kalyankar, V.D.: Parameter optimization of machining processes using a new optimization algorithm. Mater. Manuf. Proc. **27**, 978–985 (2012). https://doi.org/10.1080/10426914.2011.602792
24. Rao, R.V., More, K.C., Taler, J., Oclon, P.: Dimensional optimization of a micro-channel heat sink using Jaya algorithm. Appl. Therm. Eng. **103**, 572–582 (2016)

Performance Evaluation of Jaya Optimization Technique for the Production Planning in a Dairy Industry

Aparna Chaparala, Radhika Sajja, K. Karteeka Pavan and Sreelatha Moturi

Abstract In any manufacturing industry, for each production run, the manufacturer has to balance many variables, including the available quantity and quality of raw materials and associated components. Effectively balancing that ever-changing equation is the key to achieve optimum utilization of raw materials, maximum product profitability and adequate fulfilment of customer demand. The implementation of computer-based strategies can commendably balance such equation with minimum time. However, their efficiency would be enhanced, only when the applied algorithm is capable of providing solutions to real-world problems. Hence, to study the applicability of the recently proposed Jaya optimization method, it is used for finding the optimal master production schedule in a dairy industry. The performance of Jaya is also compared with the solution obtained when used teaching–learning-based optimization method (TLBO) for the same problem.

Keywords Process industries · Master production scheduling · Evolutionary algorithms · Jaya algorithm

1 Introduction

Over the last decade, the world has changed from a marketplace with several large, almost independent markets, to a greatly incorporated global market demanding an extensive multiplicity of products that comply with high quality, reliability and envi-

A. Chaparala (✉) · S. Moturi
Department of Computer Sceince and Engineering, RVR & JC College of Engineering (A), Guntur, AP, India
e-mail: chaparala36@gmail.com

R. Sajja
Department of Mechanical Engineering, RVR & JC College of Engineering (A), Guntur, AP, India
e-mail: sajjar99@gmail.com

K. Karteeka Pavan
Department of Computer Applications, RVR & JC College of Engineering (A), Guntur, AP, India

© Springer Nature Singapore Pte Ltd. 2020
R. Venkata Rao and J. Taler (eds.), *Advanced Engineering Optimization Through Intelligent Techniques*, Advances in Intelligent Systems and Computing 949,
https://doi.org/10.1007/978-981-13-8196-6_21

ronmental standards. Moreover, in the present scenario, fluctuating industry dynamics have influenced the objectives, operation and design of manufacturing systems by placing emphasis on (1) improved customer service, (2) reduced cycle time, (3) improved products and service quality, (4) reduced costs, (5) integrated information technology and process flow, (6) planned and managed movement and (7) flexible product customization to meet customer needs. Effective management of such production systems can be achieved by identifying customer service requirements, determining inventory levels and creating effective policies and procedures for the harmonization of production activities.

Outmoded techniques for solving master production scheduling (MPS) problems are narrow in application and also could yield a local optimal solution. Unlike the traditional optimization methods, metaheuristic algorithms own numerous characteristics that are desirable for solving problems of the said kind. Jaya algorithm is one such metaheuristic algorithm which is newly proposed and which relies on simple control parameters without much use of any algorithm-specific control parameters. This algorithm works on the idea that as the solution moves towards the best, it avoids the worst solution.

Master production scheduling has been extensively investigated since past few decades and it continues to fascinate the interests of both the industrial and academic sectors. The validity of tentative MPS must be checked at different levels before its realistic implementation in manufacturing system [1]. Several studies have been identified suggesting an authentication process to check the legitimacy of tentative MPS, few of which worth mentioning include the works of [2–4]. In addition to these, few researchers employed advanced optimization techniques to solve MPS problems for suitable environment with good robustness in data. Vieira et al. used simulated annealing [5], while Soares et al. [6] introduced new genetic algorithm structure for MPS problems. Later Vieira [7] has compared the solution quality of both genetic algorithms and simulated annealing for master production scheduling problem. Radhika et al. [8, 9] applied differential evolution for MPS problems in various fields, with multi-objectives of minimization of inventory level, maximization of service level and minimization of inventory level below safety stock. The experimental results from the literature have shown that Jaya is a powerful optimization method possessing better robustness [10]. Many researchers have started applying the Jaya algorithm to their research problems. Radhika applied Jaya methodology and have proved it to be better in comparison with GA [11]. A brief survey on the available literature about Jaya algorithm revealed that the algorithm is on par with other advanced algorithms like artificial bee colony algorithm (ABC), ant colony optimization (ACO), particle swarm optimization (PSO), etc. [12–14]. Since the Jaya algorithm does not need algorithm-specific parameters, its application to real-life optimization problems becomes easy and can be solved in much lesser time and with better robustness.

2   Problem Description

Dairy industry considered is a manufacturer, wholesaler and retailer of milk and milk products, manufacturing a variety of 21 products. This is the largest cooperative dairy farm in South India. The major products of their production line include sterilized flavoured milk (SFM), skim milk powder, doodh peda, table butter, ghee, curd, basundhi, malai laddu, kalakand, sweet lassi, etc. for which the chief constituent is milk. The products are accessible and are presented/offered in standard size packing, and consequently, the orders are taken in multiples of the standard size.

An important factor of the time window is that the products can be kept after production. The time window plays a crucial role in determining the level of inventory since most of the dairy products are prepared by milk and its associates, and they have a shelf life of two to three days with the exemption of a small number. The problem considered is typical since the demand is to be satisfied by adequate inventory levels and shortages are to be avoided as it directly influences the repute of the firm. The mathematical model and notations employed in the current paper are taken from the work of same author, Radhika et al. [8]. The subsequent section gives the details of the problem considered and the solution attained by using Jaya.

The scenario of dairy industry considered for the present work is with a planning horizon of twenty-four periods, twelve products and three production resources (which can be work centres, production lines or cells, machines). At the end of each period, it considered 100 units to be kept as safety inventory (safety stock). Table 1 gives the different production rates. Table 2 describes the different initial inventory quantity for each product and also presents the period-wise gross requirements for each product. Same case study was already solved by the same author using TLBO [15].

3   Results and Discussions

Relating the values obtained with the quantities being produced currently, it can be established that the percentage of unfilled demands is very little and at the same time maintaining very fewer inventory levels. The values in Table 3 infer that the Jaya methodology could successfully reduce the ending inventory (EI) levels which are of foremost concern to the dairy industry.

Table 4 advocates that the master schedules generated by the Jaya method are in close estimate with the input data, i.e. gross requirements. Figure 1 portrays the period-wise production for the product J3. A sample of period-wise utilization of resources for the product J3 is presented in Fig. 2. Figure 3 provides a comparision of total quantities produced for product J3.

The termination criterion is taken as the difference between any two successive fitness values to be less than 0.0001. The procedure is allowed to run for 50 times wherein the termination criteria are reached as presented in Fig. 4.

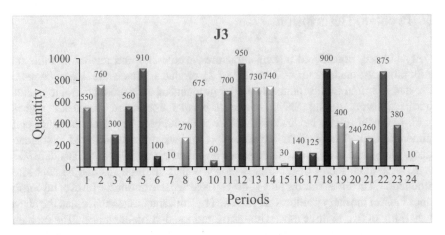

Fig. 1 Period wise production for J3

4 Conclusions

The complexity of parameter optimization problems increases with the increase in the number of parameters. Even with a rise in number of parameters, the proposed MPS Jaya model is shown to be further advanced and reliable for future research. The implementation of the proposed production scheduling system shows positive and encouraged improvement based on the data collected and is compared with the data before the implementation, where substantial quantity of time is saved by using

Table 1 Production rates

Products	Product code	Resources		
		Skimmer 1 (Sk1)	Skimmer 2 (Sk2)	Skimmer 3 (Sk3)
White butter	J1	120	120	90
Kalakand	J2	120	50	70
Paneer	J3	100	100	90
Doodh peda	J4	120	90	70
Basundhi	J5	40	100	130
Malai laddu	J6	120	100	130
Lassi	J7	50	70	110
Butter milk	J8	50	90	130
SF milk	J9	90	70	70
B ghee	J10	60	90	100
Milk cake	J11	70	50	100
Curd	J12	120	90	120

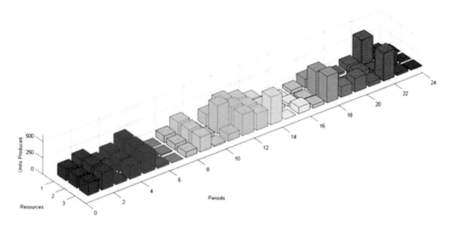

Fig. 2 Period wise utilization of resources for product J3

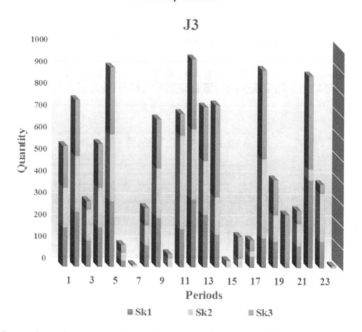

Fig. 3 Comparison of total quantities produced for product J3 using Jaya

Table 2 Period-wise gross requirements

Products	Initial inventory	Periods								
		1	2	3	4	20	21	22	23	24
J1	75	810	260	80	430	740	80	490	490	80
J2	40	690	170	260	190	910	500	890	370	590
J3	10	550	760	300	570	270	270	880	390	10
J4	20	540	590	30	490	280	620	650	720	390
J5	30	920	660	440	880	360	620	430	690	310
J6	50	680	980	160	150	150	500	720	130	890
J7	40	790	640	800	40	450	970	930	790	370
J8	60	290	800	280	830	590	550	950	850	920
J9	30	980	270	60	650	240	780	20	530	530
J10	20	20	700	520	360	270	920	480	960	470
J11	15	570	210	560	690	390	860	500	210	600
J12	10	810	930	770	360	820	130	210	340	240

the scheduling method in managing flow of processes and orders. It can also be inferred that Jaya methodology could efficiently solve the real-time case considered

Table 3 Performance Measures

Methodology	EI (units/h)	RNM (units/h)	BSS (units/h)
TLBO	41.61	91.78	16.51
Jaya	42.28	57.59	12.42

Table 4 Comparison of total quantities produced for each product

Product	Total obtained from Jaya (units)	Total gross requirement (units)
J1	11005	11150
J2	12445	12430
J3	100975	11050
J4	11990	11860
J5	13870	14020
J6	9130	9050
J7	13365	13380
J8	13435	13680
J9	12080	12270
J10	12825	12800
J11	11525	11540
J12	11510	11420

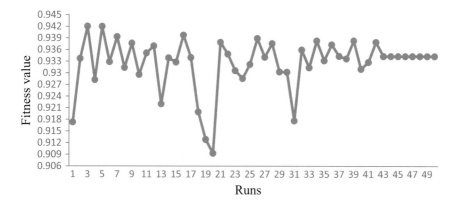

Fig. 4 Convergence of fitness values

on par with the TLBO, with a better convergence rate. The results obtained remained in line with the currently employed master schedules in the company. The same methodology could be applied to the rest of the products too.

References

1. Supriyanto, I.: Fuzzy multi-objective linear programming and simulation approach to the development of valid and realistic master production schedule; LJ_proc_supriyanto_de 201108_01, (2011)
2. Higgins, P., Browne, J.: Master production scheduling: a concurrent planning approach. Prod. Plan. Control **3**(1), 2–18 (1992)
3. Kochhar, A.K., Ma, X., Khan, M.N.: Knowledge-based systems approach to the development of accurate and realistic master production schedules. J. Eng. Manuf. **212**, 453–60 (1998)
4. Heizer, J.H., Render, B.: Operations management. Pearson Prentice Hall, Upper Saddle River, New York (2006)
5. Vieira, G.E., Ribas, C.P.: A new multi-objective optimization method for master production scheduling problems using simulated annealing. Int. J. Prod. Res. **42** (2004)
6. Soares, M.M., Vieira, G.E.: A new multi-objective optimization method for master production scheduling problems based on genetic algorithm. Int. J. Adv. Manuf. Technol. **41**, 549–567 (2009)
7. Vieira, G.E., Favaretto, F., Ribas, P.C.: Comparing genetic algorithms and simulated annealing in master production scheduling problems. In: Proceedings of 17th International Conference on Production Research, Blacksburg, Virginia, USA (2003)
8. Radhika, S., Rao, C.S., Pavan, K.K.: A differential evolution based optimization for Master production scheduling problems. Int. J. Hybrid Inf. Technol. **6**(5), 163–170 (2013)
9. Radhika, S., Rao, C.S.: A new multi-objective optimization of master production scheduling problems using differential evolution. Int. J. Appl. Sci. Eng. **12**(1), 75–86 (2014). ISSN 1727-2394
10. Abhishek, K., Kumar, V.R., Datta, S., Mahapatra, S.S.: Application of JAYA algorithm for the optimization of machining performance characteristics during the turning of CFRP (epoxy) composites: comparison with TLBO, GA, and ICA. Eng. Comput., 1–19 (2016)

11. Radhika, S., Srinivasa Rao, Ch., Neha Krishna, D., Karteeka Pavan, K.: Multi-objective opti-
    mization of master production scheduling problems using Jaya algorithm (2016)
12. Rao, R.V., Rai, D.P., Balic, J.: Surface grinding process optimization using Jaya algorithm. In:
    Computational Intelligence in Data Mining, vol. 2, pp. 487–495. Springer, India (2016)
13. Rao, R.V., More, K.C., Taler, J., Ocłoń, P.: Dimensional optimization of a micro-channel heat
    sink using Jaya algorithm. Appl. Therm. Eng. **103**, 572–582 (2016)
14. Pandey, H.M.: Jaya a novel optimization algorithm: what, how and why? In: Cloud System and
    Big Data Engineering (Confluence), 2016 6th International Conference, pp. 728–730. IEEE
    (2016)
15. Radhika, S., Srinivasa Rao, Ch., Neha Krishna, D., Swapna, D.: Master production scheduling
    for the production planning in a dairy industry using teaching learning based optimization
    method (2016)

Multi-objective Optimization in WEDM of Inconel 750 Alloy: Application of TOPSIS Embedded Grey Wolf Optimizer

G. Venkata Ajay Kumar and K. L. Narasimhamu

Abstract The current work focuses on multi-objective wire electrical discharge machining (WEDM) parameters optimization of Inconel 750 alloy. Taguchi L18 orthogonal array (OA) was used to carry the experiments in various WEDM parameters such as wire feed rate, pulse-on-time, pulse-off-time and water pressure. The output responses estimated are machining speed (cutting) and machined surface roughness of the part. Optimum machining parameters estimation is difficult in Taguchi process; a multi-objective optimization (MOO) technique known as TOPSIS is embedded with grey wolf optimizer (GWO). Initially, the multi-responses are converted to the relative closeness value, and then the heuristic approach is applied. Based on the optimal parametric setting value from GWO, a confirmation test has been conducted and compared with the fitness value.

Keywords Grey wolf optimizer · Wire electric discharge machining · Parametric optimization · TOPSIS

1 Introduction

Difficult-to-cut material with highly complex shape is a challenging job on the shop floor. Contour machining of the flat and curved surface is possible with non-conventional machining process, one such process is wire electrical discharge machining (WEDM) process perceived as wire cut EDM process [1]. In WEDM process, a wire moving slowly which travels along a set path cuts the workpiece with the

G. V. A. Kumar (✉)
Department of Mechanical Engineering, Annamacharya Institute of Technology & Sciences (Autonomous), Rajampet, Andhrapradesh, India
e-mail: ajay.ajay79@gmail.com

K. L. Narasimhamu
Department of Mechanical Engineering, Sree Vidyanikethan Engineering College (Autonomous), A. Rangampet, Andhrapradesh, India
e-mail: klsimha@gmail.com

© Springer Nature Singapore Pte Ltd. 2020
R. Venkata Rao and J. Taler (eds.), *Advanced Engineering Optimization Through Intelligent Techniques*, Advances in Intelligent Systems and Computing 949, https://doi.org/10.1007/978-981-13-8196-6_22

discharge spark. WEDM is commonly employed in the manufacturing of punches, dies and tools from high-hardness materials and intricate shapes for the electronics industry. Low strength-to-weight ratio alloys are used in turbo-machinery applications due to their thermo-mechanical loads [2]. Tool wear is one of the common problems in conventional machining. So, non-conventional machining techniques need to employ one such machining process WEDM. Apart from the tool wear, accuracy and surface finish of the WEDM components, stress-free and burr-free with straight edges are ideal as well as inexpensive [3]. WEDM founds economical, where instantaneous optimization of process parameters is still difficult, many methods have been addressed by various researchers. This work mainly focuses on the application of grey wolf optimizer (GWO) [4] in determining the optimum parameters (process) and closeness coefficient index forming multi-responses is evaluated with a technique for order preferences by similarity to ideal solution (TOPSIS) [5].

2 Literature Review

In the recent studies, GWO has applied in machining process optimization of the abrasive water-jet (AWJM) process [6] to estimate the material removal rate (MRR), the surface roughness (SR), overcut and taper. Somvir investigated the effect of various machining parameters in WEDM of Udimet-L605 superalloy using particle swarm optimization (PSO) [7] and developed the valid solution for the wire wear ratio, cutting speed and dimensional deviation. Yuvaraj [8] applied TOPSIS in the multi-objective optimization of abrasive water-jet cutting process of AA5083-H32. Further, the literature on the use of multi-criteria decision making (MCDM) [9] technique and multi-objective decision making (MODM) techniques is presented in Table 1.

The literature review shows the Inconel is difficult to machine compared to other alloys. Very few researchers have optimized the various WEDM process parameters in machining Inconel- and Nickel-based superalloys.

3 Methodology

This study attempts a new method by combining the MCDM technique and MODM technique in finding the optimum parametric setting in WEDM [19]. The MCDM technique used here is TOPSIS, where the multi-responses are converted into the relative closeness index value, and then a regression equation is developed. The parametric optimization is estimated from the GWO. The convergence rate for GWO algorithm for achieving the optimal solution is faster, as well as GWO [4] maintains an excellent stability between exploitation and exploration. TOPSIS has the option to compute the closet to the ideal solution comparing to other MADM methods and it has less mathematical computation [20] and simple to apply.

Table 1 Optimization in WEDM

Author	Work material	Input parameters	Output parameters	Experimental/mathematical models	Multi-objective optimization technique
Garg [10]	Inconel 625	P_{ON}, P_{OFF}, SGV, WF	CS, GC, SR	Response surface methodology	Desirability approach
Selvam [11]	Hastelloy C-276	P_{ON}, P_{OFF}, WF, C, SV	SR, KW	Taguchi L27 Orthogonal array (OA)	Genetic algorithm (single-objective optimization)
Ashok [12]	Aluminium alloy6061 (matrix) and nano-SiC and nano-B_4C	P_{ON}, P_{OFF}, C, SV	MRR	Taguchi L18 OA	Artificial Neural Network (ANN) and Fuzzy logic
Pawan Kumar [13]	Inconel 825	P_{ON}, P_{OFF}, SGV, PC, WT, WF	MRR, SR, WWR	Central composite design of RSM	Grey relational analysis
Huang [14]	YG15 tool steel	P_{ON}, P_{OFF}, P, CF, WT, WS, WP	SR, MRR	Taguchi L18 OA	Taguchi technique
Majumder [15]	Shape memory alloy Nitinol	P_{ON}, DC, WF, WT, FP	SR, MH	Taguchi L27 OA	Fuzzy logic, multi-objective optimization on the basis of ratio analysis (MOORA)
Rajyalakshmi [16]	Inconel 825	P_{ON}, P_{OFF}, CSE, FP, WF, WT, SGV, SF	MRR, SR, SG	Taguchi L36 OA	GRA
Varun [17]	EN 353	P_{ON}, P_{OFF}, PC, SV	MRR, SR, KW	RSM	GRA coupled with genetic algorithm (GA)

(continued)

Table 1 (continued)

Author	Work material	Input parameters	Output parameters	Experimental/mathematical models	Multi-objective optimization technique
Rao [18]	Oil hardened and nitrided steel (OHNS)	P_{ON}, P_{OFF}, PC, SF	SR, MS	RSM	Artificial bee colony (ABC)
Somvir [7]	Udimet-L605	P_{ON}, P_{OFF}, PC, WT, SGV, WF	CS, DD, WWR	Taguchi L27 OA	PSO

P_{ON}—Pulse-on-time, P_{OFF}—Pulse-off-time, SGV—Spark gap voltage, SF—Servo feed, WF—Wire feed rate, DC—Discharge current, C—Current, SV—Servo voltage, P—Power, CF—Cutting feed rate, WS—Wire speed, WP/FP—Water/flushing pressure, PC—Peak current, CSE—Cutting servo, WT—Wire tension; CS—Cutting speed, GC—Gap current, SR—Surface roughness, GC—Gap current, KW—Kerf width; MRR—Material removal rate; WWR—Wire wear ratio; MH—Micro-hardness; SG—Spark gap; MS—Machining speed; DD—Dimensional deviation

3.1 Technique for Order Preference by Similarity to Ideal Solution (TOPSIS) Method

The TOPSIS method was established by Hwang and Yoon (1981) [20]. The finest alternative was selected by the precise Euclidean distance from the positive ideal solution (PIS) and the farthest to the negative ideal solution (NIS).

Step 1: Analysing the objective and evaluation of the attributes.

Step 2: Obtain the normalized decision matrix by Eq. 1.

$$R_{ij} = \frac{m_{ij}}{\sqrt{\sum_{j=1}^{M} m_{ij}^2}} \tag{1}$$

Step 3: Find the relative importance weights for the distinct attributes which correspond to the objective, and summation of all weights should be equal to one.

Step 4: Obtain the weighted normalized decision matrix from Eq. 2.

$$V_{ij} = w_j R_{ij} \tag{2}$$

Step 5: Obtain the PIS and NIS from Eq. 3.

$$A^+ = \{V_{ij}^+ 1, \ldots \ldots V_n^+\} = \{(\max V_{ij} | i \in I'), ('\min V_{ij} | i \in I'')'$$
$$A^- = \{V_{ij}^- 1, \ldots \ldots V_n^-\} = \{(\max V_{ij} | i \in I'), ('\min V_{ij} | i \in I'')' \tag{3}$$

Step 6: Enumerate the separation measures by (4) and (5)

$$D_j^+ = \sqrt{\sum_{i=1}^{n} \left(V_{ij} - V_j^+\right)^2} \tag{4}$$

$$D_j^- = \sqrt{\sum_{i=1}^{n} \left(V_{ij} - V_j^-\right)^2} \tag{5}$$

Step 7: Enumerate the closeness value (relative) to the ideal solution by Eq. 6.

$$P_i^* = \frac{D_j^-}{D_j^+ + D_j^-} \tag{6}$$

Step 8: The alternatives are now ranked based on the value obtained, P_i^* in the decreasing order.

3.2   Grey Wolf Optimizer

GWO mimics the hunting behaviour of grey wolves developed by Ali et al. (2014) [4]. Grey wolf, called western wolf, is extremely found in areas of Eurasia and North America and are treated as zenith predators, apex of the food chain. Grey wolves are the group of 5–12 on average, alpha (α), beta (β), delta (δ) and omega (ω) wolves in the hierarchy. Alpha wolves are responsible for decision making about hunting, where to sleep (place), when to wake (time) and so on. Next are beta wolves, subordinate to alpha wolves and helps in decision making and other movements in the hunting process. Delta wolves need to acknowledge alphas and betas, caretakers, elders belong to his category. Last in group are omega wolves which are a scapegoat, allowed to eat the prey and little importance individual in the group. The mimicing nature of GWO, was applied to the parametric optimization of WEDM process, GWO technique applied to AWJM process [6] in parametric optimization and its performance is also validated. The following steps give a brief about GWO.

- Initialize the population and coefficient vectors
- Hunting process starts by attacking their prey; wolves are approaching towards the prey
- Evaluate the fitness function value (each search agent)
- Identifying the best search agent for each set of α, β and δ wolves
- Update the current positions of α, β and δ wolves
- Updating the coefficient vectors
- Updating the positions of α, β and δ wolves based on latest fitness function value
- Repeat the last two steps until an optimum solution is obtained.

4   Experimental Details

The experiments are carried out using the Ultracut 843, Electronica manufacturer machine. The four process parameters and mixed levels by using Taguchi design of experiments L18 orthogonal array were used; factors and levels are given in Table 2. A brass wire of 0.25 mm diameter was used and dielectric fluid as de-ionized water. The Inconel 750 alloy workpiece material has been taken as a flat having dimensions $270 \times 45 \times 7$ mm thickness. Experiments were conducted on Inconel 750 material and mounted on a WEDM machine tool, the cutting area of the specimens are $10 \times 5 \times 7$ mm thickness and are represented in Fig. 1.

The SR of the WEDM specimens was measured using Mitutoyo (SJ-210). The machining (cutting) speed was taken from the machine control unit digital display, taken the average of four values during the job cutting in WEDM in terms of mm/min. Here the surface roughness is lower-the-better and cutting speed is higher-the-better. The experimental results and relative closeness index are shown in Table 3.

Table 2 WEDM parameters with levels

Factors	Symbols	Level 1	Level 2	Level 3
Wire feed rate (mm/min)	A	3	5	–
P_{ON} (µs)	B	109	115	115
P_{OFF} (µs)	C	56	58	60
Water pressure (kg/cm^2)	D	8	10	12

Fig. 1 **a** Dimensions of the specimen **b** WEDM process in machining of Inconel 750 on Ultracut 843

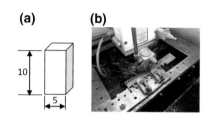

(a)   (b)

10

5

Table 3 Experimental results and relative closeness index

Experiment number	A	B	C	D	Surface roughness (SR) (µm)	Cutting speed (CS) (mm/min)	Relative closeness index (C_i)
1	3	109	56	8	1.8620	2.7625	0.6049
2	3	109	58	10	1.7953	2.8050	0.6419
3	3	109	60	12	1.9097	2.6925	0.5703
4	3	112	56	8	2.5567	3.0500	0.3048
5	3	112	58	10	2.3543	2.9375	0.3878
6	3	112	60	12	2.3467	3.1225	0.4271
7	3	115	56	10	2.9630	3.2850	0.2473
8	3	115	58	12	2.5770	3.3925	0.3785
9	3	115	60	8	2.6347	3.3225	0.3402
10	5	109	56	12	2.2433	2.7950	0.4249
11	5	109	58	8	1.4153	2.7475	0.7280
12	5	109	60	10	1.5175	2.6275	0.6799
13	5	112	56	10	2.6537	3.1200	0.2791
14	5	112	58	12	2.7090	3.3250	0.3154
15	5	112	60	8	2.3584	3.2875	0.4533
16	5	115	56	12	2.6623	3.1300	0.2783
17	5	115	58	8	2.9213	3.2575	0.2448
18	5	115	60	10	2.7877	3.4925	0.3362

5   Results and Discussions

Multi-objective optimization problem, the multiple responses are converted into single response (objective) problem. Here, TOPSIS a MCDM technique was used to convert into a single-objective problem. The surface roughness was lower-the-better and cutting speed was higher-the-better.

$$
\begin{aligned}
f(x) = {} & 121.9 + 0.522 * \text{Wirefeed} - 2.378 * \text{Pulse on} + 0.651 * \text{Pulse off} - 1.270 * \text{Waterpressure} \\
& + 0.010183 * \text{Pulse on}^2 - 0.00353 * \text{Pulse off}^2 - 0.00212 * \text{Waterpressure}^2 \\
& - 0.00227 * \text{Wirefeed} * \text{Pulse on} - 0.00435 * \text{Wirefeed} * \text{Pulse off} \\
& - 0.00293 * \text{Wirefeed} * \text{Waterpressure} - 0.001376 * \text{Pulse on} * \text{Pulse off} \\
& + 0.013450 * \text{Pulse on} * \text{Waterpresure} - 0.00340 * \text{Pulse off} * \text{Waterpressure}
\end{aligned} \tag{7}
$$

The relative closeness value (C_i) was calculated by using the TOPSIS method, treated as a single response in WEDM machining. A regression model was developed from the C_i values and parametric input values are expressed with the following objective function having R^2 as 99.84%. The objective function is shown in Eq. (7), this equation was treated as an objective function in the GWO algorithm. A MATLAB code has been developed for the GWO algorithm to get the optimal parametric values to the WEDM process parameters. The GWO parameters selected are 100 iterations. The convergence graph is shown in Fig. 2, and the optimal parametric setting from GWO was represented in Table 4.

The convergence curve was shown in Fig. 2, GWO converges rapidly towards the best solution. The GWO algorithm gives the optimal parametric setting of wire feed at 3 mm/min, P_{ON} at 109 μs, P_{OFF} at 60 μs and water pressure at 8 kg/cm^2, which is not available in the previous experimental run. Now, a confirmation experiment

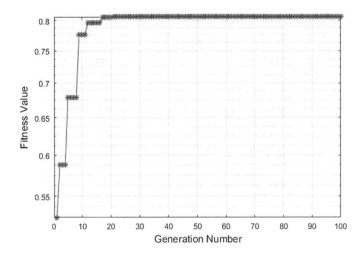

Fig. 2 Convergence diagram for relative closeness value

Table 4 Optimal parametric setting and fitness value by using GWO

Wire feed (mm/min)	Pulse on (μs)	Pulse off (μs)	Water pressure (kg/cm^2)	Fitness value
3	109	60	8	0.80629

was conducted to validate the robustness of the MOO technique. The confirmation test value converted into relative closeness value is 0.82484 which is 2.248% higher than the obtained fitness value.

6 Conclusions

Current study deals with the multi-objective optimization (MOO) of WEDM to obtain the parametric setting for maximum CS and minimum SR. The parametric setting (optimal) from the confirmation test is validated by conducting the confirmation test and again checked the adequacy of the technique. This analysis will help in finding the crucial process parameters which can reduce the surface roughness and maximize the cutting speed in WEDM of Inconel 750.

References

1. Williams, R.E., Rajurkar, K.P.: Study of wire electrical discharge machined surface characteristics. J. Mater. Process. Technol. **28**(1–2), 127–138 (1991)
2. Bewlay, BP., Weimer, M., Kelly, T., Suzuki, A., Subramanian, PR., Baker, I., Heilmaier, M., Kumer, S., Yashimi, K. (eds.) Inter Metallic Based Alloys-Science, Technology and Applications, Mrs Symposium Proceedings, vol. 1, no. 516, p. 49 (2013)
3. Anurag, S.: Wire-EDM: a potential manufacturing process for gamma titanium aluminides in future aero engines. Int. J. Adv. Manuf. Technol. **94**(1–4), 351–356 (2018)
4. Mirjalili, S., Mirjalili, S.M., Lewis, A.: Grey wolf optimizer. Adv. Eng. Softw. **69**, 46–61 (2014)
5. Lai, Y.J., Liu, T.Y., Hwang, C.L.: Topsis for MODM. Eur. J. Oper. Res. **76**(3), 486–500 (1994)
6. Chakraborty, S., Mitra, A.: Parametric optimization of abrasive water-jet machining processes using grey wolf optimizer. Mater. Manuf. Process. 1–12 (2018)
7. Nain, S.S., Garg, D., Kumar, S.: Investigation for obtaining the optimal solution for improving the performance of WEDM of super alloy Udimet-L605 using particle swarm optimization. Eng. Sci. Technol. Int. J. **21**(2), 261–273 (2018)
8. Yuvaraj, N., Pradeep Kumar, M.: Multiresponse optimization of abrasive water jet cutting process parameters using TOPSIS approach. Mater. Manuf. Process. **30**(7), 882–889 (2015)
9. Rao, R.V.: Advanced Modeling and Optimization of Manufacturing Processes : International Research and Development (2010)
10. Garg, M.P., Kumar, A., Sahu, C.K.: Mathematical modeling and analysis of WEDM machining parameters of nickel-based super alloy using response surface methodology. Sādhanā **42**(6), 981–1005 (2017)
11. Selvam, M.P., Kumar, P.R.: Optimization Kerf width and surface roughness in wire cut electrical discharge machining using brass wire. Mech Mech Eng **21**(1), 37–55 (2017)

12. Ashok, R., Poovazhagan, L., Srinath Ramkumar, S., Vignesh Kumar, S.: Optimization of material removal rate in wire-edm using fuzzy logic and artificial neural network. In: Applied Mechanics and Materials, vol. 867, pp. 73–80. Trans Tech Publications (2017)
13. Kumar, P., Meenu, M., Kumar, V.: Optimization of process parameters for WEDM of Inconel 825 using grey relational analysis. Decis. Sci. Lett. **7**(4), 405–416 (2018)
14. Huang, Y., Ming, W., Guo, J., Zhang, Z., Liu, G., Li, M., Zhang, G.: Optimization of cutting conditions of YG15 on rough and finish cutting in WEDM based on statistical analyses. Int. J. Adv. Manuf. Technol. **69**(5–8), 993–1008 (2013)
15. Majumder, H., Maity, K.: Prediction and optimization of surface roughness and micro-hardness using grnn and MOORA-fuzzy-a MCDM approach for nitinol in WEDM. Measurement **118**, 1–13 (2018)
16. Rajyalakshmi, G., Ramaiah, P.V.: Multiple process parameter optimization of wire electrical discharge machining on Inconel 825 using Taguchi grey relational analysis. Int. J. Adv. Manuf. Technol. **69**(5–8), 1249–1262 (2013)
17. Varun, A., Venkaiah, N.: Simultaneous optimization of WEDM responses using grey relational analysis coupled with genetic algorithm while machining EN 353. Int. J. Adv. Manuf. Technol. **76**(1–4), 675–690 (2015)
18. Rao, R.V., Pawar, P.J.: Modelling and optimization of process parameters of wire electrical discharge machining. Proc. Inst. Mech. Eng. Part B J Eng. Manuf. **223**(11), 1431–1440 (2009)
19. Nayak, B.B., Mahapatra, S.S., Chatterjee, S., Abhishek, K.: Parametric appraisal of WEDM using harmony search algorithm. Mater. Today Proc. **2**(4–5), 2562–2568 (2015)
20. Yoon, K.P., Hwang, C.L.: Multiple Attribute Decision Making: An Introduction, vol. 104. Sage publications (1995)

Experimental Investigations and Selection of Solid Lubricant Assisted Lubrication Strategy in Machining with the Use of PROMETHEE

M. A. Makhesana⑩ **and K. M. Patel**⑩

Abstract Manufacturing sector is always looking to find out an alternative in order to develop sustainable process. Machining plays a very important role in today's manufacturing sector. In order to avoid the problems caused by heat generated in machining, conventional coolant has been applied. The large quantity of these fluids causes environmental damage and also adding total production cost. In this context, the present work focuses on application of minimum quantity lubrication combined with the solid lubricants. Performance of calcium fluoride and molybdenum disulphide as solid lubricant is assessed with different concentration and particle size mixed with SAE 40 oil on response parameters like surface roughness, power consumption, chip-tool interface temperature and flank wear. In order to select best suitable lubricating combination, PROMETHEE a multiple attribute decision-making method is applied. Comparison of results revealed the effectiveness of calcium fluoride as solid lubricant over other machining conditions and minimum quantity lubrication (MQL). Ranking available from the PROMETHEE can be considered as feasible lubrication alternative in machining as compared to conventional fluid cooling. The results of the work will be useful to explore the possibility to consider the solid lubricants as an efficient alternative to cutting fluids.

Keywords Multiple attribute decision making · PROMETHEE · MQL · Solid lubricants

M. A. Makhesana (✉) · K. M. Patel
Mechanical Engineering Department, Institute of Technology, Nirma University, 382481 Ahmedabad, Gujarat, India
e-mail: mayur.makhesana@nirmauni.ac.in

K. M. Patel
e-mail: kaushik.patel@nirmauni.ac.in

241

1 Introduction

It has been the need of the present day to develop sustainable manufacturing process in view of requirements of customers. And hence it is very much required to upgrade the existing manufacturing practice to sustainable process [1]. In order to reduce the effects caused by heat generated in machining, conventional coolants and lubricants are used in large quantity. It can be noted that the total cost of these coolants are larger than of tooling cost for selected machining process. Minimum quantity lubrication concept was introduced and used [2] which supplies a very small quantity of cutting fluid to the machining area and thus reducing the problems and cost associated with conventional cooling. Many researchers has assessed the performance of MQL approach in various machining process like turning and grinding and analysed the effects on surface integrity, grinding power and wheel wear rate [3, 4].

It has been reported by the researchers that machining performance can be further improved with the application of solid lubricants. It can be applied in dry powder form or can be supplied by mixing with cutting oil and compressed air [5, 6]. Work has been reported with the application of solid lubricants in various machining processes, and the performance was assessed in terms of values of surface roughness, power consumption, process temperature and tool wear rate. Machining performance was improved with MQL and solid lubricant assisted lubrication [7, 8]. The researcher has also worked by combining MQL and application of solid lubricants in machining and studied various process response parameters. It is concluded that the solid lubricant has improved the machining performance as compared to that of dry machining and flood cooling [9, 10].

To solve various decision-making problems in industrial applications, use of various multiple attribute decision-making methods is reported in the literature [11]. Applications of Preference Ranking Organization Method for Enrichment Evaluations (PROMETHEE) were illustrated [12]. Analytic hierarchy process (AHP) is integrated with the fuzzy logic. It is concluded that the method is very effective in decision making in real-world selection problems. Applications of PROMETHEE are found in manufacturing, particularly for the planning of preventive maintenance activities and selection of lean manufacturing system [13, 14].

Looking to the status of research work done in the area of application of solid lubricants and MQL, very little work is available on the study of the effects of solid lubricant concentration and particle size during machining. Present work focused on the performance assessment of calcium fluoride and molybdenum disulphide as a solid lubricant with different concentration and particle size on response parameters. A decision-making methodology is used to find out best alternative among selected combination of lubricants.

2 Experimental Details

2.1 Experimental Set-Up

Minimum quantity lubrication (MQL) is a cooling method in which the small amount of high quality coolant (cutting fluid) mixed with compressed air forming an aerosol is supplied at cutting zone through nozzle. Minimum quantity lubrication (MQL) is also known as 'Near Dry Machining'. The compressed air and good quality lubricant oil in small quantity is mixed at mixing chamber of MQL system. The mixture is fed through nozzle in the form of aerosol exactly at interface of tool and workpiece. The lubricant applied to machining will create a thin film of lubricant between the tool-work interfaces and hence reduces the friction and heat generated, whereas compressed air acts as cooling medium by carrying away heat generated. The lubricant with MQL can be supplied as external or internal delivery. In case of internal supply, the mixture is supplied directly at cutting zone with rotary union, through spindle and coolant channel of the tool. A nozzle is used to direct the flow of lubricant into machining area in external delivery system. The minimum quantity lubrication system can have one or more nozzles as outlet. The experimental set-up contains lubricant reservoir, mixing block, nozzle and pressure regulator and mounted on the lathe machine. The same set-up is used to supply lubricant with minimum quantity and solid lubricant to the machining area. All the components are connected by hoses. Provision is made to supply compressed air from the compressor to carry the lubricant mixture and then to supply the same to the nozzle. The flow of lubricant can be controlled by the control knob provided with each nozzle. In order to ensure proper mixing of solid lubricant with oil, continuous stirring is ensured during experiments. The set-up with the machine is as shown in Fig. 1.

The turning tests were carried out with Kirloskar Turn Master lathe machine equipped with various feed and spindle rpm. EN31 bar with 200 mm length and

Fig. 1 MQSL set-up and stirrer mounted on machine

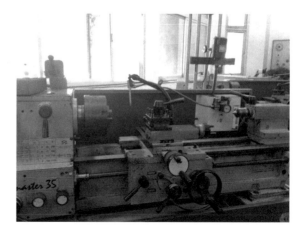

50 mm diameter used as a workpiece. The experiments were carried out by using coated carbide inserts with designation CNMG1204085 TN4000.

During the experiments, the values of process parameters were kept constant as cutting speed 130 m/min feed 0.2 mm/rev and depth of cut 0.5 mm. The experimental test was performed to measure surface roughness, power consumption, flank wear, and chip-tool interface temperature during various machining conditions. SAE 40 cutting oil was used as a lubricant and applied to machining area with the concept of minimum quantity lubrication. Two solid lubricants, calcium fluoride and molybdenum disulphide, were mixed with cutting fluid and applied with MQSL approach. Different concentration and particle size of lubricants were selected to check its effect on machining performance.

The surface roughness (R_a) was measured by Mitutoyo portable surface roughness tester, roughness values were measured at various locations and the average value was considered for final comparison. Power consumption during machining was measured by wattmeter placed between the machine tool and power source. The chip-tool interface temperature was measured by specially designed and calibrated tool-work thermocouple mounted on lathe machine. Tool flank wear was observed and measured under the microscope equipped with image analysis software after each machining test.

3   Results and Discussion

In order to optimize the machining process, it is essential to understand the effect of various machining conditions on the response variables. With this view, experiments were performed under various cutting conditions such as different solid lubricants with varying concentration and particle size by keeping fixed values of process parameters. Solid lubricants were added with 10 and 20% concentration with the particle size of 10 and 30 μm and mixed with SAE 40 oil. The average particle size of solid lubricants was confirmed by SEM analysis. The values of process parameters are selected based on the results obtained in form of optimized parameters by grey relational analysis. The results of the experiments are presented in Table 1.

As mentioned earlier, surface roughness value of machined workpiece was measured at three locations along the length and an average value is considered for comparison. Surface finish plays a very important role when it comes to the quality of machined components as it is affected by several factors during machining. It can be concluded from the comparison in Fig. 2 that lowest value of surface roughness is obtained when machining is done with 20% calcium fluoride of 10 μm particle size mixed with SAE 40 oil. For all other machining conditions with increased concentration and particle size of calcium fluoride and molybdenum disulphide, surface roughness value is larger as compared with MQL machining.

Improvement in Ra value with 10 μm particle size may be due to the ability of smaller particles to enter and retain between the tool-work interfaces and created a thin layer of lubrication which can be correlated with the least amount of chip-tool

Table 1 Measured values of output responses

S. No.	Machining condition	Surface roughness (μm)	Power consumption (W)	Chip-tool interface temperature (°C)	Flank wear (mm)
1	MQL	3.810	680	480	0.21
2	10% CaF$_2$ with 10 μm + SAE 40 oil	3.680	680	430	0.34
3	10% CaF$_2$ with 30 μm + SAE 40 oil	4.120	720	450	0.32
4	20% CaF$_2$ with 10 μm + SAE 40 oil	3.220	640	400	0.28
5	20% CaF$_2$ with 30 μm + SAE 40 oil	3.910	680	490	0.35
6	10% MoS$_2$ with 10 μm + SAE 40 oil	4.220	720	480	0.34
7	10% MoS$_2$ with 30 μm + SAE 40 oil	3.950	640	510	0.38
8	20% MoS$_2$ with 10 μm + SAE 40 oil	4.520	680	390	0.28
9	20% MoS$_2$ with 30 μm + SAE 40 oil	4.320	880	520	0.32

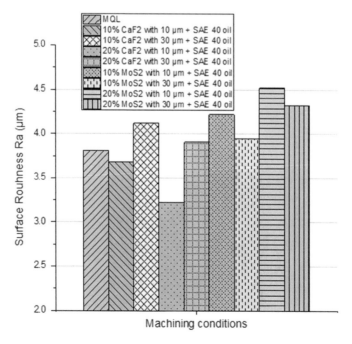

Fig. 2 Comparison of surface roughness in different machining conditions

temperature produced with 20% calcium fluoride of 10 μm particle size leading to lower tool flank wear and surface roughness.

As shown in Fig. 3, the highest chip-tool interface temperature is observed during machining with 20% molybdenum disulphide of 30 μm particle size added with SAE 40 oil. It may be due to increased concentration of solid lubricant has reduced the thermal conductivity of base oil which has resulted the larger amount of heat generation in the process.

Another reason may be the larger particle size of solid lubricant failed to maintain constant lubricating film between tool and workpiece resulted in larger flank wear of tool and surface roughness. However, it is observed in Fig. 4 that there is not much change in the values of measured cutting power. The values of cutting power in case of MQSL are almost in the same range than that of MQL machining which proves the effectiveness of solid lubricants to reduce the friction and provision of better lubrication.

The highest value of flank wear is recorded in case of machining with MoS_2 assisted machining with 10% concentration and 30 μm particle size. This is resulted as larger temperature produced with the same machining condition leading to the larger tool flank wear (Fig. 5).

However, smaller flank wear is observed with MQL machining as compared to the application of solid lubricants. Looking to the comparison and results available

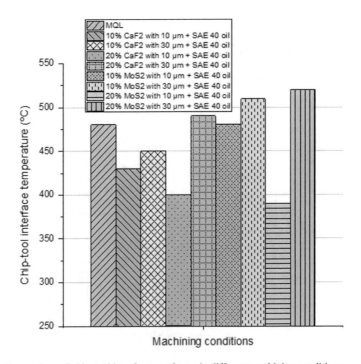

Fig. 3 Comparison of chip-tool interface roughness in different machining conditions

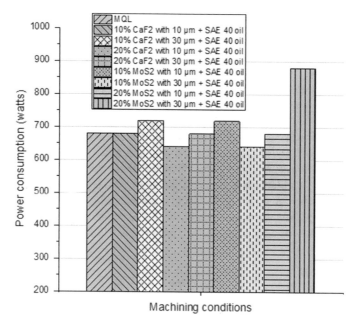

Fig. 4 Comparison of power consumption in different machining conditions

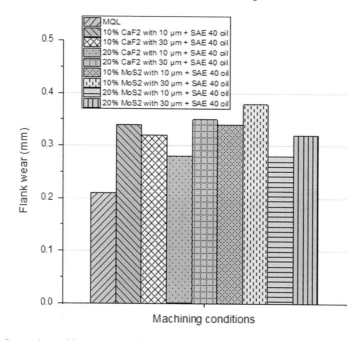

Fig. 5 Comparison of flank wear in different machining conditions

from experiments, it is difficult to select any one alternative as lubricant combination for machining under specific conditions. As calcium fluoride has performed better by lowering surface roughness and flank wear; however on the other side, MoS_2 performed by reducing chip-tool interface temperature. Looking to conflicting criteria for selection of best suitable lubricating conditions, multiple attribute decision-making methods can be used. In order to find the best alternative among all considered alternatives of lubrication, PROMETHEE (Preference Ranking Organization Method for Enrichment Evaluations) method introduced by Brans et al. [15] is applied. PROMETHEE is simple, systematic and logical method which can efficiently deal with the qualitative criteria. PROMETHEE is applied in different fields of decision-making problems as reported in literature [16]. However, few applications are found in production and manufacturing related decision-making problems.

The first step takes the normalization of the input matrix which is in the form of measured values of output responses. The normalized decision matrix, R_{ij} can be represented as,

$$R_{ij} = \frac{m_{ij}}{\sqrt{\sum\limits_{j=1}^{M} m_{ij}^2}} \tag{1}$$

Equal relative importance or weightage are considered for all response parameters. The multiple criteria preference index Π_{a1a2} is then defined as the weighted average of the preference functions Pi:

$$\Pi_{a1a2} = \sum\limits_{i=1}^{M} W_i P_{i,a1a2} \tag{2}$$

Π_{a1a2} represents the intensity of preference of the decision maker of alternative $a1$ over alternative $a2$ when considering simultaneously all the criteria. Its value ranges from 0 to 1.

For PROMETHEE outranking relations, the leaving flow, entering flow and the net flow for an alternative a belonging to a set of alternatives A are defined by the following equations:

$$\varphi^+(a) = \sum\limits_{x \in A} \Pi_{xa} \tag{3}$$

$$\varphi^-(a) = \sum\limits_{x \in A} \Pi_{ax} \tag{4}$$

$$\varphi(a) = \varphi^+(a) - \varphi^-(a) \tag{5}$$

$\varphi^+(a)$ is called the leaving flow, $\varphi^-(a)$ is called the entering flow and $\varphi\ (ai)$ is called the net flow. $\varphi^+(a)$ is the measure of the outranking character of a (i.e. dominance

Table 2 Resulting table

	1	2	3	4	5	6	7	8	9
1	–	0.50	0.75	1	1	1	0.75	0.75	1
2	0.75	–	0.75	0	1	1	0.75	0.50	0.75
3	0.25	0.25	–	0	0.50	1	0.50	0.25	1
4	0.75	1	1	–	1	1	1	0.75	1
5	0.25	0	0.50	0	–	0.50	0.75	0.50	0.75
6	0.25	0.25	0.25	0	0.50	–	0.50	0.25	0.75
7	0.25	0.50	0.50	0.25	0.25	0.25	–	0.50	0.75
8	0.50	0.75	0.75	0.50	0.75	0.75	0.50	–	0.75
9	0	0.25	0.25	0	0.25	0.25	0.25	0.25	–

Table 3 Leaving, entering and net flow

	φ^+	φ^-	φ	Ranking
1	6.75	3	3.75	2
2	5.5	3.5	2	3
3	3.75	4.75	−1.0	5
4	7.5	1.75	5.75	1
5	3.25	5.25	−2.0	6
6	2.75	5.75	−3.0	8
7	2.75	5	−2.2	7
8	5.25	3.75	1.5	4
9	1.5	6.75	−5.2	9

of alternative a over all other alternatives) and $\varphi^-(a)$ gives the outranked character of a (i.e. degree to which alternative $a1$ is dominated by all other alternatives). The net flow, $\varphi(a)$, represents a value function, whereby a higher value reflects a higher attractiveness of alternative a (Table 2).

From the results indicated in Table 3, alternative 4 which is 20% calcium fluoride and 10 μm particle size added with SAE 40 oil is the best alternative for selected machining parameters as compared to all other considered alternatives. The ranking is followed by MQL, machining with calcium fluoride and molybdenum disulphide with different concentration of solid lubricants and particle size.

4 Conclusion

Performance of solid lubricants with different concentration and particle size mixed with SAE 40 oil is assessed for turning operation. It has been found that machining performance is improved with the use of calcium fluoride with smaller particle size and 20% of concentration as a solid lubricant. Molybdenum disulphide has performed better by reducing chip-tool interface temperature. Lower values of surface roughness and flank wear are observed with the application of calcium fluoride due to its excellent lubricating properties. PROMETHEE as multiple attribute decision-making method is used to find the best lubricant combination from considered alternatives. Based on the ranking obtained by PROMETHEE, 20% calcium fluoride with 10 μm particle size added with SAE 40 oil is the best alternative. Results can be compared with experimental values where the smaller particle size of lubricant has improved lubrication by creating thin layer of lubricating film, whereas larger particle size of the solid lubricant leads to the increased value of surface roughness, flank wear and chip-tool interface temperature. Looking at the problems associated with conventional cutting fluid, the improved performance of solid lubricants can be considered as an alternative leading to a clean and sustainable machining process.

References

1. Pusavec, F., Krajnik, P., Kopac, J.: Transitioning to sustainable production- Part I: application on machining technologies. J. Clean. Prod. **18**, 174–184 (2010)
2. Klocke, F., Beck, T., Eisenblätter, G., Fritsch, R., Lung D. and Pöhls, M.: Applications of minimal quantity lubrication (MQL) in cutting and grinding. In: Proceedings of the 12th International Colloquium Industrial and Automotive Lubrification. Technische Akademie, Esslingen (2000)
3. Hafenbraedl, D., Malkin S.: Technology environmentally correct for intern cylindrical grinding. Mach. Metals Mag., 40–55 (2001)
4. da Silva, L.R., Bianchi, E.C., Catai, R.E., Fusse, R.Y., França, T.V., Aguiar, P.R.: Analysis of surface integrity for minimum quantity lubricant-MQL in grinding. Int. J. Mach. Tools Manuf. **47**, 412–418 (2007)
5. Reddy, N.S.K., Rao, P.V.: Experimental investigation to study the effect of solid lubricants on cutting forces and surface quality in end milling. Int. J. Mach. Tools Manuf. **46**, 189–198 (2006)
6. Vamsi Krishna, P., Srikant, R.R., Rao, D.N.: Experimental investigation to study the performance of solid lubricants in turning of AISI1040 steel. IMechE Part J: J. Eng. Tribol. **224**, 1273–1281 (2010)
7. Varadarajan, M.A.S., Philip, P.K., Ramamoorthy, B.: Investigations on hard turning with minimal cutting fluid application (HTMF) and its comparison with dry and wet turning. Int. J. Mach. Tools Manuf. **42**, 193–200 (2002)
8. Rahman, M.M., Senthil Kumar, A., Salam, M.U.: Experimental evaluation on the effect of minimal quantities of lubricant in milling. Int. J. Mach. Tools Manuf. **42**, 539–547 (2002)
9. Dilbagh Singh, M., Rao, P.V.: Performance improvement of hard turning with solid lubricants. Int. J. Adv. Manuf. Technol. **38**, 529–535 (2008)
10. Reddy, N.S.K., Rao, P.V.: Performance improvement of end milling using graphite as a solid lubricant. Mater. Manuf. Processes **20**, 673–686 (2005)
11. Rao, R.V.: Decision making in the manufacturing environment using graph theory and fuzzy multiple attribute decision making methods. Springer-Verlag, London (2007)
12. Rao, R.V., Patel, B.K.: Decision making in the manufacturing environment using an improved PROMETHEE method. Int. J. Prod. Res., 1–18 (2009). https://doi.org/10.1080/00207540903049415

13. Cavalcante C.A.V., De Almeida, A.T.: A multi-criteria decision-aiding model using PROMETHEE III for preventive maintenance planning under uncertain conditions. J. Qual. Maintenance Eng. **13**(4), 385–397 (2007)
14. Anand, G., Kodali, R.: Selection of lean manufacturing systems using the PROMETHEE. J. Model. Manag. **3**(1), 40–70 (2008)
15. Brans, J.P., Mareschal, B., Vincke, P.: PROMETHEE: a new family of outranking methods in multicriteria analysis. Proc. Oper. Res. **84**, 477–490 (1984)
16. Behzadian, M., Kazemzadeh, R.B., Albadvi, A., Aghdasi, M.: PROMETHEE: a comprehensive literature review on methodologies and applications. Eur. J. Oper. Res. **200**(1), 198–215 (2009)

BPSO-Based Feature Selection for Precise Class Labeling of Diabetic Retinopathy Images

Rahul Kumar Chaurasiya, Mohd Imroze Khan, Deeksha Karanjgaokar and B. Krishna Prasanna

Abstract Diabetic retinopathy (DR) is an eye disease caused by high levels of blood sugar in diabetic patients. A timely detection and diagnosis of DR with the aid of powerful algorithms applied to the eye fundus images may minimize the risk of complete vision loss. Several previously employed techniques concentrate on the extraction of the relevant retinal components characteristic to DR using texture and morphological features. In this study, we propose to estimate the severity of DR by classification of eye fundus images based on texture features into two classes using a support vector machine (SVM) classifier. We also introduce a reliable technique for optimal feature selection to improve the classification accuracy offered by an SVM classifier. We have applied the Wilcoxon signed rank test and observed the p-values to be 3.7380×10^{-5} for Set 1 and 7.7442×10^{-6} for Set 2. These values successfully reject the null-hypothesis against the p-value benchmark of 5%, indicating that the performance of SVM has significantly improved when used in combination with an optimization algorithm. Overall results suggest a reliable and accurate classification of diabetic retinopathy images that could be helpful in the quick and reliable diagnosis of DR.

Keywords Diabetic retinopathy · Texture features · Binary particle swarm optimization (BPSO) · Class labeling · Cross-validation

R. K. Chaurasiya (✉) · M. I. Khan · D. Karanjgaokar · B. K. Prasanna
Department of Electronics and Telecommunication, National Institute of Technology, Raipur 492010, India
e-mail: rkchaurasiya.39@gmail.com

M. I. Khan
e-mail: imroze786@gmail.com

D. Karanjgaokar
e-mail: kardeeksha@gmail.com

B. K. Prasanna
e-mail: krishnaprasanna1410@gmail.com

© Springer Nature Singapore Pte Ltd. 2020
R. Venkata Rao and J. Taler (eds.), *Advanced Engineering Optimization Through Intelligent Techniques*, Advances in Intelligent Systems and Computing 949,
https://doi.org/10.1007/978-981-13-8196-6_24

1 Introduction

A diabetic person often develops some ophthalmic complications, such as glaucoma, retinal and corneal abnormalities cataracts, retinopathies and neuropathies [1]. The most common among these are diabetic retinopathy [2, 3] which is a major cause of blindness [4]. Retinopathy is caused by acute or persistent damage to the retina of the eye. It is an ocular manifestation of systemic human disease and seen in diabetic condition. Diabetes is a worldwide problem that affects millions of people in the world [5, 6]. According to the research report, it is estimated that more than 21 crore people have diabetes mellitus and 20–28% of it might had developed retinopathy [7]. DR was found to be 18% prevalent in an urban population with diabetes in India [8] and is becoming an issue of increasing concern as the world's population ages [9, 10]. Hence, in order to prevent diabetic retinopathy, its early detection and diagnosis are very important. Early detection can ensure reduced visual impairment due to the timely and prompt medical care.

Annual re-examination of a normal retinal patient is necessary because 5–10% of normal patients can develop DR with time and this condition is known as mild. Repeated examination of patients with retinal micro-aneurysms and hard exudates or other abnormalities within 6–12 months is recommended and if condition persists, it is known as moderate. Category beyond moderate is severe and very severe diabetic retinopathy. Both eyes are generally affected by DR. Often vision changes in the early stages of DR go unnoticed, but with time, irreversible vision loss conditions can be observed in many cases. The main aim of this study was to optimize the feature set for the improved classification accuracy and application of diabetic retinopathy. Retinal images were analyzed for the optimization and best feature selection, which can be further used for correct classification of DR images and better management of the disease.

2 Materials and Methods

Classification of images into various groups is achieved through stages of preprocessing, feature extraction, training and optimization. Images acquired from Messidor dataset are first preprocessed or enhanced to make them more suitable for further processing. A vector called the feature vector defines each preprocessed image. There are two types of features for every image:

a. *Texture Features*: Texture features represent the statistical properties of an image such as entropy, gray-level co-occurrence (GLC) of pixel pairs, gray-level run-length (GLR) of a particular intensity value, moment invariant (MI) and so on.
b. *Morphological Features*: Morphological features represent the distinct objects or components within an image.

Our project for classification of eye fundus images is based on the texture features only.

Extracted texture features are normalized in an interval [0,1] and then class labels are assigned to them based on the frequency of occurrence of "abnormal" entities in eye fundus images such as micro-aneurysms, hemorrhages, abnormal blood vessel growth and presence of hard exudates.

Support vector machine (SVM) classifier is used to train the dataset. It is then tested on a separate dataset to classify the images. Optimization techniques are used to improve classification accuracy achieved by the SVM classifier. The sequence of procedures adapted for accurate classification of diabetic retinopathy stages are depicted in Fig. 1.

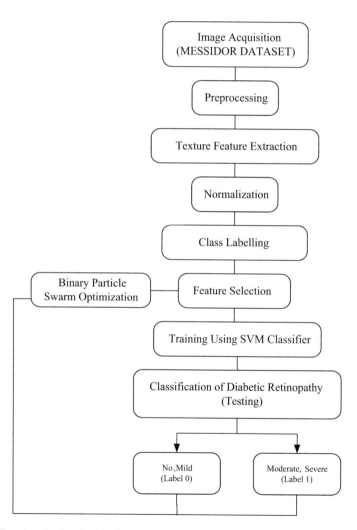

Fig. 1 Flowchart for the classification system of DR

2.1 Dataset

The Messidor Database [11] was used for the purpose of classification. The database consists of 1200 eye fundus color images obtained from three ophthalmologic departments. The 1200 images in TIFF format are packaged in three groups. Each group is divided into four groups containing 100 images accompanied with an Excel file enlisting medical diagnoses. Two sets of samples are constructed from each group. Set 1 consists of 100 images in which we have applied fivefold cross-validation for classification. Set 2 consisting of 400 images, we have applied fourfold cross-validation where we have three complete subgroups for training and one complete subgroup for testing.

2.2 Preprocessing

Preprocessing is applied to overcome the deficiencies in images such as lack of contrast, poor quality and distortion due to improper focusing and lighting to produce enhanced images that are more suitable for easier and accurate extraction of useful information. The retinal images are preprocessed using the following methods.

2.2.1 RGB to Gray Conversion

The texture features extracted from the image do not use any color information. RGB to gray conversion discards redundant or unnecessary information makes the file size smaller and convenient for further analysis.

2.2.2 Contrast Limited Adaptive Histogram Equalization (CLAHE) Algorithm

The algorithm divides an image into smaller regions and applies the histogram equalization to each of these regions [12]. This evenly distributes the intensity levels and enhances the hidden details in an image.

2.2.3 Green Channel Extraction

Channels in an image represent the proportion in which basic colors combine to create a digital image. Green channel extraction brings out the best contrast in images and plays a crucial role in medical image processing.

2.3   Texture Feature Extraction

The mutual and repetitive relationship among neighboring pixels over an area larger than the region of relationship can be defined as texture [13].

We have considered four sets of texture features, namely:

2.3.1   First-Order Measures: Intensity Histogram (6 Features)

A histogram is a graph that gives the frequency distribution of the pixels of an image against the corresponding pixel intensities, i.e., a plot of the number of pixels for each luminous intensity. These statistical properties derived from histogram are used for texture analysis.

2.3.2   Second-Order Measures: GLC Matrix of 22 Features

The GLC matrix functions calculate the frequency and spatial relationship among specific pixel pairs in an image, resulting in a GLCM. The number of gray levels G in the image defines the size of this matrix.

2.3.3   Third-Order Measures: GLR Matrix of 11 Features

A gray level run length matrix (GLRLM) is a two-dimensional matrix that consists of sets of the same gray-leveled pixels which occur consecutively along particular direction called the gray-level run.

2.3.4   MI of Seven Features

MI is a set of algebraic parameters that remain invariant under scale change, translation and rotation. Hu moment invariants [14, 15] are the fundamental and the most popular invariant moments used for image processing and pattern recognition; they are specified in terms of normalized central moments. For each sample image, the aforementioned (total 46) features have been extracted. These features were concatenated together with their class labels as calculated by the procedure mentioned in Sect. 2.6.

2.4 Normalization

The texture features extracted above take extreme values that are not evenly distributed around the mean. Therefore, softmax scaling, a non-linear scaling method is used which limits the values into the interval [0,1] as given by the expression

$$\widehat{y_1} = \frac{1}{1 + \exp(-z)} \tag{1}$$

where $z = \frac{x_i - x_{min}}{x_{max} - x_{min}}$, $x_i = i$th feature vector for a feature x, $x_{min} = $ minimum value $x_{max} = $ maximum value for a feature x.

2.5 Class Label Assignment

In this work, the class labeling of images for supervised classification was performed based on two diagnoses provided in Messidor Database by the medical experts. Class labels were assigned according to the retinopathy and macular edema of the Messidor Database. The no and mild DR cases were assigned class 0, and moderate and severe cases were assigned class 1. This binary class labeling assigned by the medical experts was supposed to be achieved by application of classification methods.

2.6 Classification Using SVM

One of the widely accepted techniques, SVM classifier uses a statistical learning approach of structural risk minimization for classification. This classifier finds its usage in a vast range of applications in pattern recognition and image processing fields. SVM performs well on higher dimensional data spaces [16]. For two linearly separable classes, a binary SVM classifier produces an optimal hyperplane that separates vectors from both classes in feature space in the best possible manner and also maximizes the margin from vectors that are closest to the hyperplane in each class. The decision boundary is computed using a subset of the training examples known as the support vectors. The selected decision boundary generalizes well with the test samples with a small accuracy trade-off. The linear SVM works on minimizing the constraint $\|w^2\|$ and learns the segregating hyperplane

$$y_i = x W^T + W_0 \tag{2}$$

where $y_i \in [0, 1]$ is the output, $x = $ feature vector, $W = $ Weight vector, $W_0 = $ Bias. Figure 2 illustrates the selection of hyperplane by the SVM classifier to separate the two classes of data. The redline obtained by applying SVM algorithm proves out to be the best hyperplane for classification of data.

2.7 Optimized Feature Selection

An optimization algorithm is an algorithm that selects only those features from a given feature set of a sample that yield classification accuracy and eliminates the remaining *misleading* features that decrease the accuracy. These algorithms are therefore reset to make computation faster and which results in a better classification.

2.7.1 Binary Particle Swarm Optimization

PSO is extensively used as a powerful optimization technique. This algorithm mimics a discipline called swarm intelligence, which deals with the social interaction of animals with each other and their surrounding environment. PSO is inspired from the artificial simulation of the social behavior of bird flocking and fish schooling. This technique derives its roots from genetic algorithms and evolutionary programming and is used for optimization of continuous non-linear functions. It is computationally inexpensive and requires less memory [17]. A swarm consists of particles or agents that are defined by a feature vector in a multi-dimensional search space [18]. Each randomly initialized particle moves toward the best position with each successive iteration. This movement is achieved by adjustment of the "velocity" based on the particle's initial velocity (inertia), its own best experience (Personal Best) and the best global experience of other particles (Group Best).

Let the best position of the swarm be
$P_{g\text{Best}} = (p_{g1}, p_{g2}, \ldots, p_{gd})$. For ith particle in the swarm defined by a d-dimensional vector;
$X_i = (x_{i1}, x_{i2}, \ldots x_{id})$, the velocity vector;
$V_i = (v_{i1}, v_{i2}, \ldots v_{id})$ and the best position vector $P_{\text{Best}} = (p_{i1}, p_{i2}, \ldots, p_{id})$ is updated as

$$v_{id} = w * v_{id} + c_1 * \emptyset * (p_{id} - x_{id}) + c_2 * \emptyset * (p_{gd} - x_{id}) \tag{3}$$

$$x_{id} = x_{id} + v_{id} \tag{4}$$

where w = weight, c_1 and c_2 = acceleration coefficients and
$\emptyset$ = random variable in interval [0,1].

A feature vector consisting of 46 distinct features describes each image. The aim of a BPSO [19] algorithm is to select a subset of the feature vector that will maximize the classification accuracy. The BPSO algorithm applied in our paper uses a feature space of 46 dimensions, where each dimension corresponds to a particular texture

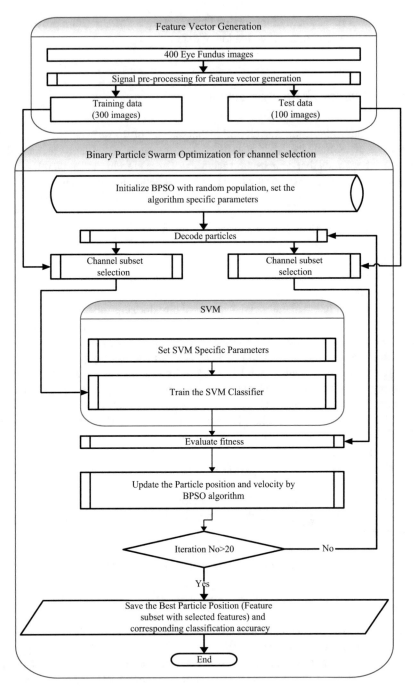

Fig. 2 Flowchart for the proposed methodology based on BPSO for feature selection and SVM for class label assignment

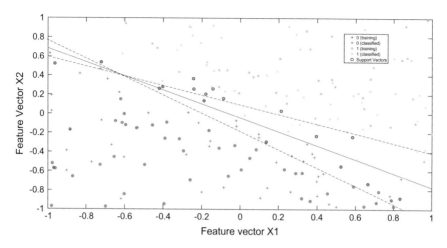

Fig. 3 Selecting the appropriate hyperplane for a representative dataset

feature defining the image. The values along these dimensions are binary, i.e., each dimension can take only two values, either 0 or 1, depending on whether a feature is rejected or selected, respectively. A selection vector of length 46 is constructed, and it is a string of 0s and 1s. The 100 "particles" in the BPSO algorithm in our paper are the 100 such randomly generated sets of selection vectors on the feature space.

Algorithmically, in every iteration, each "particle" or selection vector is multiplied with the entire feature set to obtain a subset of features to be trained and classified using SVM. Then a fitness rank is assigned to this particle based on the accuracy obtained from SVM. The best fitness of a particle will be its personal best (P_{Best}) score. Fitness rank is allotted for each particle in this way. With every new iteration, each of these particles tries to modify their selection vectors using update equations so that they may move toward achieving the G_{Best} score, i.e., the best fitness rank among all the particles. The convergence or the ability of the particles to achieve G_{Best} score depends on the number of iterations. At the end of the iterations, the selection vector corresponding to maximum accuracy is obtained at the output by the algorithm.

Figure 3 depicts a flowchart of the proposed method applied in Set 2, which includes BPSO and SVM for feature selection and classification, respectively. The algorithm is repeated multiple times to ensure that it reaches a global maximum.

3 Results

This section depicts the results obtained by applying the SVM classifier on Set 1 and Set 2 of eye fundus images. In addition, we have tabulated the results obtained by running BPSO together with SVM.

3.1   Performance of SVM for Classification

SVM classifier uses all the 46 texture features (22 GLC, 23 GLR, and 7 MI features) for classification. These features set also comprise some misleading features that lower the classification accuracy. The effect of misleading features is more prominent when SVM is applied on larger datasets that aggravate the classification accuracy as illustrated in the results of Table 1. The binary classification of the features by SVM is depicted in Fig. 3.

3.2   Performance of BPSO for Maximizing the Accuracy

The BPSO algorithm along with SVM was applied on both datasets. The number of particles was 100. As it is generally suggested that $c_1 + c_2$ should be less than or equal to 4, the acceleration coefficients c_1 and c_2 were both taken as 2. The initial value of inertia weight was taken as 0.9. Further, the weight was linearly decreased by a value 0.01 in each iteration. The decrease in inertia weight was motivated by the fact that initially, the swarms should explore more, and slowly particles should try to stabilize.

BPSO along with SVM reduces the effect of misleading features. As a result, the size of feature set was reduced and accuracy was improved, which can be observed from Table 1. When SVM is applied with BPSO on the dataset of 300 images, the time of convergence is large. This is due to the use of the approximated algorithm of SVM for larger datasets which increases the implementation complexity [20]. Hence, ensemble SVM was applied. It divides the 300 images into five sections (each of 60 images). As the number of data points in each section on which SVM was applied is fewer, so we faced with the curse of dimensionality. After a number of pilot studies, it was observed that the maximum number of times either BPSO was giving fixed accuracy or there was very less significant improvement in the accuracy values after around 12–16 iterations. Hence, in order to make the system fast and also giving sufficient number of iterations to BPSO, the maximum number of iterations was fixed to 20. The convergence curve of BPSO is depicted in Fig. 4.

Table 1 Tabulated classification results of the experiment

	SVM with all 46 features			SVM+BPSO with selected feature set			
	A (%)	S	P	#RF	A (%)	S	P
SET1	83.33	0.8	0.8667	29	96.43	94.38	98.48
SET2	63.19	0.388	0.875	26	73.94	71.66	76.22

A: Accuracy, S: Sensitivity, P: Specificity, #RF: Number of reduced features

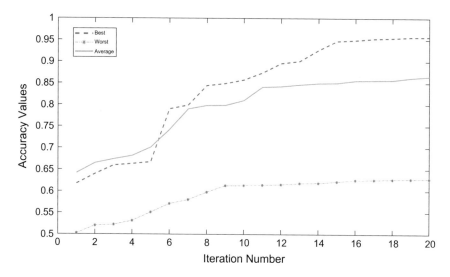

Fig. 4 Convergence curve of BPSO for selecting the most discriminative features to improve the classification accuracy

4 Conclusion

Diabetic retinopathy is a major cause of vision loss. Small aneurysms, hemorrhages, abnormal blood vessel growth, hard exudates have been found to be prevalent earliest clinical signs of retinopathy. Thus, computational identification and classification of retinal images into mild, moderate and severe condition is clinically significant and easy to diagnose. For best feature selection from retinal images, BPSO was employed in this work. The features that improve the accuracy of classification are selected and developed as an optimized feature set.

Using the proposed methodology of morphological features, optimal channel selection has improved the classification performance. The performance can be further improved using color-based features. The application of the most recent classification methods such as active learning, deep learning and random forest on color and morphology-based features is planned as future work.

References

1. Mishra, P.K., Sinha, A., Teja, K.R., Bhojwani, N., Sahu, S., Kumar, A.: A computational modeling for the detection of diabetic retinopathy severity. Bioinformation **10**, 556 (2014)
2. Cai, X., McGinnis, J.F.: Diabetic retinopathy: animal models, therapies, and perspectives. J. Diabetes Res. (2016)
3. Aiello, L.M., Cavallerano, J., Aiello, L.P., Bursell, S.E., Guyer, D.R., Yannuzzi, L.A., Chang, S.: Diabetic retinopathy. In: Retina Vitreous Macula, vol. 2 (1999)

4. Benson, W.E., Tasman, W., Duane, T.D.: Diabetes mellitus and the eye. In: Duane's Clinical Ophthalmology, vol. 3 (1994)
5. Shaw, J.E., Sicree, R.A., Zimmet, P.Z.: Global estimates of the prevalence of diabetes for 2010 and 2030. Diabetes Res. Clin. Pract. **87**, 4–14 (2010)
6. Chobanian, A.V.: Control of hypertension—an important national priority. Mass. Medical Soc. (2001)
7. Thomas, R.L., Dunstan, F., Luzio, S.D., Chowdury, S.R., Hale, S., North, R.V., Gibbins, R., Owens, D.R.: Incidence of diabetic retinopathy in people with type 2 diabetes mellitus attending the Diabetic Retinopathy Screening Service for Wales: retrospective analysis. BMJ **344**, e874 (2012)
8. Raman, R., Rani, P.K., ReddiRachepalle, S., Gnanamoorthy, P., Uthra, S., Kumaramanick-avel, G., Sharma, T.V.: Prevalence of diabetic retinopathy in India: Sankaranethralaya diabetic retinopathy epidemiology and molecular genetics study report 2. Ophthalmology **116** (2009)
9. Cedrone, C., Mancino, R., Cerulli, A., Cesareo, M., Nucci, C.: Epidemiology of primary glaucoma: prevalence, incidence, and blinding effects. Prog. Brain Res. **173**, 3–14 (2008)
10. George, R., Ramesh, S.V., Vijaya, L.: Glaucoma in India: estimated burden of disease. J. Glaucoma **19**, 391–397 (2010)
11. Decencière, E., Zhang, X., Cazuguel, G., Lay, B., Cochener, B., Trone, C., Gain, P., Ordonez, R., Massin, P., Erginay, A., Charton, B., Klein, J.-C.: Feedback on a publicly distributed image database: the Messidor database. Image Anal Stereol **33**(4) (2014)
12. Zhang, Y., Wu, X., Lu, S., Wang, H., Phillips, P., Wang, S.: Smart detection on abnormal breasts in digital mammography based on contrast-limited adaptive histogram equalization and chaotic adaptive real-coded biogeography-based optimization. Simulation **92**, 873–885 (2016)
13. Du, N., Li, Y.: Automated identification of diabetic retinopathy stages using support vector machine. In: Conference Automated identification of Diabetic Retinopathy Stages Using Support Vector Machine, pp. 3882–3886. IEEE (2013)
14. Zhang, Y., Yang, J., Wang, S., Dong, Z., Phillips, P.: Pathological brain detection in MRI scanning via Hu moment invariants and machine learning. J. Exp. Theor. Artif. Intell. **29**, 299–312 (2017)
15. Zhang, Y., Wang, S., Sun, P., Phillips, P.: Pathological brain detection based on wavelet entropy and Hu moment invariants. Bio-Med. Mater. Eng. **26**, S1283–S1290 (2015)
16. Xu, L., Luo, S.: Support vector machine based method for identifying hard exudates in retinal images. In: Conference Support Vector Machine Based Method for Identifying Hard Exudates in Retinal Images, pp. 138–141. IEEE (2009)
17. Kennedy, J.: Particle swarm optimization. In: Encyclopedia of Machine Learning, pp. 760–766. Springer (2011)
18. Chaurasiya, R.K., Londhe, N.D., Ghosh, S.: Binary DE-based channel selection and weighted ensemble of SVM classification for novel brain–computer interface using Devanagari Script-based P300 speller paradigm. Int. J. Hum.-Comput. Interact. **32**, 861–877 (2016)
19. Wang, S., Phillips, P., Yang, J., Sun, P., Zhang, Y.: Magnetic resonance brain classification by a novel binary particle swarm optimization with mutation and time-varying acceleration coefficients. Biomed. Eng./Biomedizinische Technik **61**, 431–441 (2016)
20. Blankertz, B., Muller, K.-R., Krusienski, D.J., Schalk, G., Wolpaw, J.R., Schlogl, A., Pfurtscheller, G., Millan, J.R., Schroder, M., Birbaumer, N.: The BCI competition III: validating alternative approaches to actual BCI problems. IEEE Trans. Neural Syst. Rehabil. Eng. **14**, 153–159 (2006)

Dynamic Analysis and Life Estimation of the Artificial Hip Joint Prosthesis

Akbar Basha Shaik and Debasish Sarkar

Abstract While discussing the hip joint failure, material selection and apposite dimension of the femoral head are a significant concern for the artificial hip replacement. In this context, an attempt was made to optimize both ball head and socket material from different combinations like femoral head (metal)—acetabular liner (Polyethylene) and femoral head (ceramic)—acetabular liner (ceramic) in consideration of a different set of femoral ball head size of 28, 30 and 32 mm. The material and femoral head size were optimized in the perspective of minimum stress that eventually enhances the prosthesis life and minimizes the wear of counter bodies. The hip joint prosthesis was designed in CATIA V5 R17 followed by finite element analysis (FEA) was performed in ANSYS 17.2. Dynamic FEA was performed when the 100 kg human in jogging. A theoretical optimization established the combination of ceramic–ceramic articulating body consists of 30.02 mm ID acetabular liner—30 mm OD femoral head made of zirconia toughened alumina (ZTA) experience less stress and deformation that eventually exhibit very low wear rate per cycle of jogging. This design exhibits 0.93 mm wear depth after 15 years of activity; however, similar theoretical analysis can be done under different degree of dynamic motions. The proposed material and design combination has excellent potential for the development of artificial hip joint prosthesis.

Keywords Total hip replacement (THR) · Prosthesis · Dynamic analysis · ZTA · Wear

A. B. Shaik · D. Sarkar (✉)
Department of Ceramic Engineering, National Institute of Technology, Rourkela 769008, Odisha, India
e-mail: dsarkar@nitrkl.ac.in

A. B. Shaik
e-mail: akbar3406@gmail.com

© Springer Nature Singapore Pte Ltd. 2020
R. Venkata Rao and J. Taler (eds.), *Advanced Engineering Optimization Through Intelligent Techniques*, Advances in Intelligent Systems and Computing 949,
https://doi.org/10.1007/978-981-13-8196-6_25

1 Introduction

The hip joint is ball and socket joint, where the ball and socket represent femoral head and acetabulum. Its main function at the hip joint is to support the weight of the human body both in static (standing) and dynamic (walking and running) positions and found as the most significant and most reliable joint in the body. Total hip replacement (THR) is a joint replacement surgery used to replace damaged, arthritic hip joint with an orthopaedic prosthesis. Instability and dislocation are the main problems for the failure of THR. Dislocation rates are 0.5–10% for primary THR [1] and 10–25% for revision THR [2]. Dislocation of the hip joint is mainly early dislocation and late dislocation. The reason for an early dislocation is due to the impingement of the femoral neck from the acetabular liner cup lip, and the late dislocation occurs due to wear rate between the femoral head and acetabular liner. The paper was mainly focused on generating the stresses in the hip joint and analysing the wear of the hip joint. To estimate the stresses, a finite element analysis was performed. In the present hip, joint prosthesis is made of components acetabular socket, an acetabular liner, femoral head and femoral stem. The femoral stem fixed in femoral bone F, and the acetabular socket fixed in acetabular bone A. The dynamic analysis was performed used a slow jogging load.

2 Finite Element Modelling of Hip Joint

Figure 1 represents hip joint CAD model developed using CATIA V5 R17 modelling software. The femoral stem, femoral head, an acetabular liner, acetabular socket, femur bone (bone F) and acetabular bone (bone A) separately built and assembled using CATIA V5 R17 assembly options. Three different hip joint models developed in CATIA V5 R17. The three models are 28-mm head THR model, 30-mm head THR model and 32-mm head THR model.

Three femoral head of size 28, 30 and 32 mm designed for analysis. To accommodate femoral head sizes of 28, 30 and 32 mm three acetabular liners were designed to match the femoral head sizes with a radial clearance of 100 μm in between femoral heads and acetabular liners are maintained. The acetabular liner inner and outer diameter are 28.02/38, 30.02/40, and 32.02/42 mm for 28, 30 and 32 mm femoral head, respectively was maintained. The inset distance of 28, 30 and 32 mm head is 2.93, 3.11 and 3.28 mm respectively. A 30 mm femoral head and accompanying acetabular liner with dimensions are shown in Fig. 2.

The femoral stem designed with stem length 81.73 mm, vertical height 24.65 mm, medial offset 38.58 mm, neck length 30.03 mm, taper 12/14 (Taper angle 3.814°) and neck shaft angle 130°. The acetabular socket inner and outer diameter are 38.02/48, 40.02/50, and 42.02/52 mm for 28, 30 and 32 mm femoral head, respectively, was maintained. All the three models exported into ANSYS 17.2. THR models finely mesh through ANSYS 17.2 with element size 5, 2.5 and 1 mm with tetrahedral

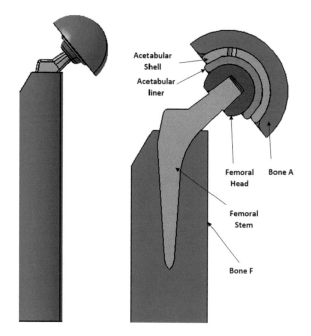

Fig. 1 THR CAD model system

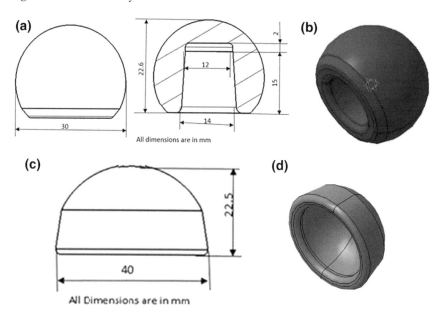

Fig. 2 a Femoral head dimension. **b** Femoral head CAD model. **c** Acetabular liner dimension. **d** Acetabular liner CAD model

elements. The von Mises stress values are reduced for 5 mm element size to 2.5 mm element size but further reduction in element size to 1 mm increases simulation time with negligible change in von Mises stress. Hence, the 2.5-mm element with tetrahedral elements is considered for analysis. After meshing the 28 mm THR model has 4,21,960 nodes and 2,90,078 elements, 30 mm THR model has 4,30,105 nodes and 2,95,434 elements, and 32 mm THR model has 4,39,268 nodes and 3,01,434 elements.

3 Material Properties of THR Model

Two bearings considered for analysis of each model, i.e., ceramic head/ceramic liner, metal head/Polyethylene liner. The material used for ceramic head and ceramic liner bearing is ZTA composite (95wt% Al_2O_3, 5wt% 3YSZ), and the materials used for metal head and polyethylene liner are Ti6Al4V and UHMWPE. All materials are assumed to be homogeneous, isotropic and linear elastic. By using the rule of mixture Young's modulus and Poisson's ratio of ZTA composite are calculated as 375 GPa and 0.3. The essential material properties required for the FEM analysis are tabulated in Table 1.

4 Loading and Boundary Conditions

An average human load 100 kg considered for analysis. The boundary conditions for the dynamic load applied in this work are normal to hip joint. The loads were determined from experimental observations of Bergmann et al. [4]. The high contact forces and also with the friction moments in the hip joints have the danger of fixation of the acetabular socket, and the failure of various hip implants is mainly due to cup

Table 1 Material properties used for FEM analysis [3]

S. No.	Material	Young's Modulus (GPa)	Poisson's ratio	Mass density (g/cm^3)	Yield strength (MPa)
1.	Alumina (Al_2O_3)	380	0.3	3.95	665
2.	Zirconia (ZrO_2)	210	0.3	6.05	711
3.	Ti-6Al-4V	110	0.3	4.5	800
4.	UHMWPE	1	0.4	0.93	23.56
5.	Simulative bone	20	0.3	1.932	–

loosening [5, 6]. It is vital to use moments to identify the effect of cup fixation. In addition to the contact forces and moments, the coefficient of friction between the components affects the loads in the THR assembly. Among the daily activities, the jogging considered for dynamic analysis.

The contact forces are maximum during slow jogging at 7 km/h. The 4839 N force acts on the femoral head was 68% higher than during walking and nearly five times the BW(100 kg). The jogging load pattern is shown in Fig. 3a, the maximum load of 4839 N and 0.8 Nm was applied at 2 s of load pattern. The abductor muscle force of 703 N is applied to the bone F to counter the femoral head force [7]. The base of bone F and the top surface of bone A fixed in all directions. The dynamic loading and boundary conditions of the THR model are shown in Fig. 3b.

The frictional coefficient of bearing interface varies with bearing materials. Fialho et al. [8] used the coefficient of friction 0.05 for ceramic on the ceramic bearing to calculate frictional heating. According to Rancourt et al. [9], the coefficient of friction between the interface between porous-surfaced metals and tibial cancellous bone was in between 0.3 and 1.3. The coefficient of friction between alumina on alumina bearing was 0.05–0.1 [10]. The coefficient of friction was 0.55 between Ti6Al4V and UHMWPE under bovine serum [11].

In the present study, the coefficient of friction between ceramic on the ceramic bearing is considered as 0.2, and the coefficient of friction between metal and Polyethylene bearing is considered as 0.15 respectively.

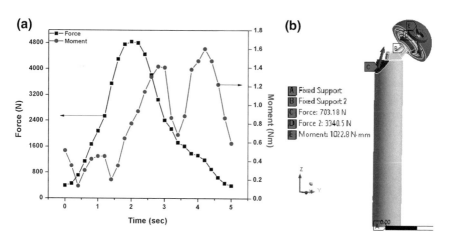

Fig. 3 **a** Variation of force and moment applied on the prosthesis during jogging. **b** Loads and boundary conditions on the THR model system

5 Results for Dynamic Analysis

The idea of analysis of two bearings and three head sizes is to analyse the stress and deformation on the dynamic load during jogging. The analysis of two bearings with three heads was tabulated in Table 2. The von Mises stress obtained from the dynamic analysis in ceramic/ceramic bearings is 182.33, 180.88 and 179.5 MPa for 28, 30 and 32 mm head, as the head size increased the stress values decreased. The same was observed in metal/polyethylene bearing.

The von Mises stress developed in both bearings is lower than the yield strength of prosthesis material used. The results of dynamic analysis show maximum stress concentrated on the upper neck portion of the femoral stem hip joint compared to femoral head and acetabular liner. For all THR model, Ti6Al4V material is used for the femoral stem, hence maximum stresses generated in the stem. The von Mises stress decreased with the increase of the head size due to an increase in the area of the bearing load. The stress generated in the bone for the two bearings are in the range of 7–20 MPa which is less than the yield strength of bone (104–121 MPa).

The Z-component load of jogging has a significant effect on the deformation of the femur bone. In case of metal on polyethylene bearing, the maximum deformation was observed in femoral head as the acetabular liner material was polyethylene and in case of ceramic and ceramic bearing the deformation observed in the metal acetabular socket as both femoral head and acetabular socket ceramic material used.

In a combination of different materials and head size, von Mises stress, total deformation and equivalent elastic strain are calculated and found ceramic–ceramic bearing surface is the best choice compared to metal/polyethylene bearing, as it has less stress and deformation.

Among the three heads, 30 mm ceramic is considered as optimum ceramic femoral head, as it has a high degree of freedom and lower size 28 mm results dislocation under continuous activity, but higher size 32 mm experience more deformation compare to other sizes.

6 Wear of 30 mm Ceramic/Ceramic Bearing Total Hip Replacement

Wear is the gradual removal of material from the surface of the body. The wear of hip prosthesis is due to contact of components, sliding distance and tribological properties of materials used. Archard's law was used to estimate the wear [12].

$Q = K_w * L * P_c$ where Q is the volume of wear debris produced, K_w wear coefficient, L is the sliding distance, P_c is the contact pressure.

$$dV = \Delta A dh = K_w * \sigma * A * ds$$
$$dh = K_w * \sigma * ds.$$

Table 2 Dynamic analysis: The von Mises stress and total deformation of three 28, 30 and 32 mm THR models with two bearings

THR model		28 mm	30 mm	32 mm
Metal/polyethylene bearing	von Mises stress (MPa)	 200.36 Max 178.24 156.11 133.98 111.85 89.724 67.596 45.468 23.34 1.2122 Min	 196.58 Max 174.88 153.18 131.48 109.78 88.085 66.386 44.688 22.989 1.2908 Min	 192.57 Max 171.32 150.08 128.83 107.58 86.337 65.09 43.844 22.598 1.3516 Min
	Total deformation (mm)	 0.35946 Max 0.33364 0.30783 0.28201 0.2562 0.23038 0.20457 0.17875 0.15294 0.12712 Min	 0.36496 Max 0.33708 0.30919 0.2813 0.25341 0.22552 0.19763 0.16975 0.14186 0.11397 Min	 0.37023 Max 0.34076 0.3113 0.28183 0.25237 0.2229 0.19344 0.16397 0.13451 0.10504 Min
	Simulation time (h)	13	12.5	12

(continued)

Table 2 (continued)

THR model		28 mm	30 mm	32 mm
Ceramic/ceramic bearing	von Mises stress (MPa)	182.33 Max 162.18 142.04 121.9 101.75 81.607 61.464 41.32 21.176 1.0319 Min	180.88 Max 160.92 140.97 121.01 101.05 81.091 61.133 41.174 21.216 1.2581 Min	179.5 Max 159.69 139.89 120.08 100.28 80.472 60.667 40.861 21.056 1.2508 Min
	Total deformation (mm)	0.3494 Max 0.31356 0.27772 0.24188 0.20604 0.1702 0.13437 0.098527 0.062688 0.026849 Min	0.36885 Max 0.33108 0.29331 0.25554 0.21777 0.17999 0.14222 0.10445 0.066678 0.028906 Min	0.38693 Max 0.34722 0.30751 0.2678 0.2281 0.18839 0.14868 0.10897 0.069258 0.029549 Min
	Simulation time (h)	20	18	17

Wear coefficient (K_w) is a function of roughness and material properties which can be obtained experimentally. Wear coefficient of Al_2O_3 and ZTA combination determined from the ball on disc experiment is 5.3E-8 mm^3/Nm [13]. The contact pressure obtained from analysis is (σ) = 56.75 MPa (see Fig. 4). Sliding distance (L) was estimated from wear depth (d_W).

The sliding distance was considered from walking cycle, during half walking cycle, the flexion and extension are 23° and is 17°, respectively. For the total walking cycle, hip joint rotates an angle of 80°. The radius of femoral head 15 mm is considered for analysis.

Angle of hip rotation during walking = 80° = 1.396 rad

$$\text{Sliding distance } (L) = \text{head radius} * \text{rotating angle}$$
$$= 15 * 1.396 = 20.94 \, \text{mm} = 20.94\text{E-3 m}$$

According to Archard's law,

$$\text{Wear depth } (d_W) = K_w * \sigma * L.$$
$$= 5.3\text{E-8} * 56.75 * 20.94\text{E-3}$$
$$= 6.298\text{E-8 mm/cycle.}$$

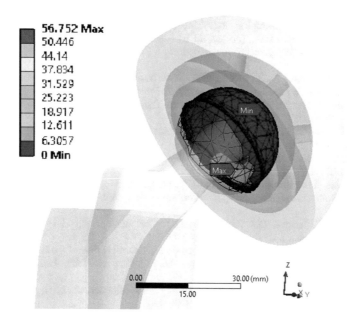

Fig. 4 Contact pressure of ceramic head and ceramic acetabular liner bearing

Average human takes one million steps in one year; the wear rate was estimated to be 0.062 mm/year. After 15 years, 0.93 mm of wear depth takes place at a constant point.

7 Conclusion

A typical 28, 30 and 32 mm THR model designed with components femoral stem, femoral head, acetabular liner and acetabular socket and the appropriate material was used for dynamic analysis. The load pattern of 100 kg human performing slow jogging at 7 km/h is considered for dynamic analysis. During dynamic analysis, the stress generated in the THR components is less than the yield stress of the prosthesis material used. Hence, the design is safe in the human body. The high stresses are observed in the femoral stem neck region. The wear is estimated by using Archard's law, and wear depth of 0.062 mm/year has calculated for the 30-mm femoral head of ZTA material. Based on the wear of femoral head, it is assumed that the hip joint is safe for 15 years.

References

1. Sariali, E., Leonard, P., Mamoudy, P.: Dislocation after total hip arthroplasty using Hueter anterior approach. J. Arthroplasty **23**(2), 266–272 (2008)
2. Alberton, G.M., High, W.A., Morrey, B.F.: Dislocation after revision total hip arthroplasty: an analysis of risk factors and treatment options. J. Bone Joint Surg. Am. **84**(10), 1788–1792 (2002)
3. Askari, Ehsan, Flores, Paulo, Dabirrahmani, Danè, Appleyard, Richard: A review of squeaking in ceramic total hip prostheses. Tribol. Int. **93**, 239–256 (2016)
4. Bergmann, G., Bender, A., Dymke, J., Duda, G., Damm, P.: Standardized loads acting in hip implants. PLoS ONE **11**(5), e0155612 (2016)
5. David, D., Graves, S., Tomkins, A.: Annual report 2013 Australian Orthopaedic Association National Joint Replacement Registry (2013)
6. Garellick, G., Karrholm, J., Rogmark, C., Rolfson, O., Herberts, P.: Annual report 2011. Swedish Hip Arthroplasty Register (2011)
7. Tsouknidas, A., Anagnostidis, K., Maliaris, G., Michailidis, N.: Fracture risk in the femoral hip region: a finite element analysis supported experimental approach. J. Biomech. **45**, 1959–1964 (2012)
8. Fialho, J.C., Fernandes, P.R., Eca, L., Folgado, J.: Computational hip joint simulator for wear and heat generation. J. Biomech. **40**, 2358–2366 (2007)
9. Rancourt D., Shirazi-Adl A., Drouin G., et al.: Friction properties of the interface between porous-surfaced metals and tibial cancellous bone. J. Biomed. Mater. Res. **24**, 1503–1519 (1990)
10. Khademhosseini, A.: Micro and nanoengineering of the cell microenvironment: technologies and applications. Artech House Publishers (2008)
11. Park, J.: Bioceramics: properties, characterizations, and applications. Springer (2008)
12. Wu, J.S., Hung, J., Shu, C., Chen, J.: The computer simulation of wear behavior appearing in total hip prosthesis. Comput. Methods Program. Biomed. **70**, 81–91 (2003)
13. Strey, N.F., Scandian, C.: Tribological transitions during sliding of zirconia against alumina and ZTA in water. Wear **376–377**, 343–351 (2017)

Modeling Elastic Constants of Keratin-Based Hair Fiber Composite Using Response Surface Method and Optimization Using Grey Taguchi Method

P. Divakara Rao⊙**, C. Udaya Kiran and K. Eshwara Prasad**

Abstract Tensile modulus is an important mechanical property of any material which is responsible for the stiffness of the material. In the present study, experimental testing is done to know Young's modulus and Poisson's ratio of human hair-fiber-reinforced polyester composite. Response surface methodology (RSM) technique is used to obtain empirical relations for Young's modulus and Poisson's ratio in terms of fiber weight fraction and fiber length. ANOVA is carried out to check the significance of the developed models. Results showed that predictions made by RSM models are in good concurrence with experimental results. Optimization of fiber weight fraction and length is carried out using RSM, and highest Young's modulus value of 4.062 GPa was observed at 19.95% fiber volume fraction and 29.32 mm fiber length. Multi-response optimization is carried using grey Taguchi method. Optimized factor levels using grey Taguchi method are obtained as 20% and 30 mm for fiber weight fraction and fiber length respectively.

Keywords ANOVA · Elastic constants · Human hair fiber · Response surface methodology · Multiple regression · Grey Taguchi

P. Divakara Rao (✉)
J.B. Institute of Engineering and Technology, Hyderabad, Telangana, India
e-mail: raodivakar.p@gmail.com

C. Udaya Kiran
Bhaskar Engineering College, Hyderabad, Telangana, India
e-mail: ukchavan@gmail.com

K. Eshwara Prasad
Siddhartha Institute of Engineering and Technology, Hyderabad, Telangana, India
e-mail: epkoorapati@gmail.com

© Springer Nature Singapore Pte Ltd. 2020
R. Venkata Rao and J. Taler (eds.), *Advanced Engineering Optimization Through Intelligent Techniques*, Advances in Intelligent Systems and Computing 949, https://doi.org/10.1007/978-981-13-8196-6_26

1 Introduction

Development of fiber-reinforced composites started with manmade fibers like glass, aramid, carbon, etc., due to their superior tensile properties [1]. Researchers shifted their focus onto natural fibers in the recent past with an aim to find alternative fibers because of increasing environmental constraints. Natural fibers have many advantages like easy renewability, good degradability, and availability across the globe. Moreover, natural fibers exhibit low specific weight and higher tensile modulus compared to glass fibers and aramid fibers, respectively [2, 3]. Manufacturing of short fiber polymer composites has gained acceleration due to low cost and easy processing of complex shapes [4].

Sizable research work has already been done on plant fibers, but use of animal fibers as reinforcing material is yet to be fully exploited. Keratin is the main constituent of animal fibers. Chicken feather and human hair are two main types of animal fibers which are available as waste by-products. Huge tonnage of waste feathers is produced every year that are creating a serious solid waste problem [5, 6]. Conventional methods of animal fiber disposal are restricted as they pose serious damage to the environment [7]. Some researchers threw some light on the use of chicken feather as reinforcing material in composites [8–11]. Waste hair thrown away into nature not only decomposes slowly but also leaves the constituent elements such as carbon, nitrogen, and sulfur. The best way to address the problems posed by the waste human hair is to invent new systems that use these wastes as a resource. This in turn contributes to the economy of any nation. India being second largest populated country, it can derive the benefits of this renewable, readily available human hair as a potential material resource for reinforcement in composites. Many potential uses of waste human hair are explained by Gupta [12].

Hair is naturally a fiber having diameter in few microns and strong in tension. These natural properties of hair can be exploited by way of using it as a reinforcing phase in composites. Fiber-reinforced composites can be designed for large specific moduli and specific strengths by taking suitable combinations of fiber and matrix in appropriate volume fractions.

Young's modulus signifies the stiffness of a material. It is extremely important to have reliable values of Young's modulus in engineering design. The study of effect of fiber weight ratio on Young's modulus of a material is a promising area for research. Different analytical models have been developed to correlate the relationship between various parameters of the filler materials. Simple tension test is a conventional method used for determining the tensile modulus. Many theoretical procedures based on rule of mixtures are also available to estimate Tensile modulus.

Mostly, conventional experimentation provides only the main effects of one parameter at a time keeping all other parameters at a constant value. A number of statistical tools and mathematical techniques form basis for developing response surface methodology (RSM) [13]. Along with main effects, RSM also provides detailed information about the interaction of parameters more effectively with minimum number of experiments. Further, this method provides optimization of various process parameter levels more accurately and completely [14].

Oleiwi et al. [15] have developed mathematical models using RSM for variety of tensile properties including Young's modulus of woven glass-reinforced unsaturated polyester composite. They found highest tensile modulus of 11.5 GPa at 11% fiber loading. Ahmed et al. [16] studied flexural strength of NWFA biocomposites using RSM. Ghasemi et al. [17] have developed models using response surface methodology for optimizing factors of talc, MAPP, and xGnP in polypropylene (PP) composites.

2 Experimental

2.1 Materials

2.1.1 Fiber Material

Major sources of hair in India are temples and barber shops. As most of women from India stay away from the chemical treatments, it is considered to be the highest graded hair. Short hair collected from the barbers in India is considered to be the lowest graded with less cost [18]. The cost of the hair depends on its length.

Hair collected from women in the age group of 35–40 years is used in the present work. Hair ranging from 16 to 22 inches in length is rinsed in detergent water and clean water and sun-dried. The hair fibers were then cut into 10, 20, 30, 40, 50 mm and stored.

2.1.2 Matrix Material

Polyester resin used as matrix material is obtained from M/s Anand composites, Hyderabad. Cobalt naphthenate and methyl ethyl ketone peroxide (MEKP) supplied by M/s. SP Engineering Ltd., Hyderabad, are used as accelerator and catalyst, respectively.

2.2 Preparation of Composite and Testing

Hand layup technique is used to prepare the human hair-fiber-reinforced polyester composite (HHRC). A layer of releasing agent (PVA) is applied on the cleaned and dried mild steel mold of 300X300X3 mm and dried it for 15 min. Hair fibers of 10 mm length are taken in required quantities to achieve a required fiber volume fraction that spread uniformly into the mold. Unsaturated polyester, catalyst, and

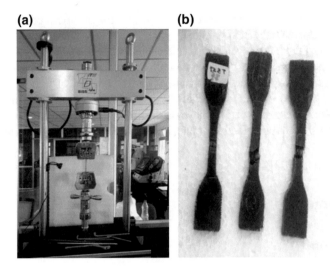

Fig. 1 **a** Specimen under test. **b** Broken specimens

accelerator in appropriate quantities are mixed, stirred, and poured into the mold. Proper care is exercised to achieve hair fiber uniformity in the mold. The mold is sealed tightly at room temperature with the help of bolt nut assembly for a period of 24 h. Similar procedure is adopted in making composite sheets of 5, 10, 15, 20, 25, 30, and 35% fiber weight fractions. Same number sheets are also made for 20, 30, 40, and 50 mm fiber lengths. Tensile test specimens are cut as per ASTM D 638 with over all specimen dimensions of 165 × 20 × 3 mm. Tensile tests are conducted on Plug N Play Nano Servo Hydraulic UTM of 25kN capacity. A specimen under test and broken specimens are shown in Fig. 1a and b, respectively. Average readings of five samples are presented in Table 2.

3   RSM Modeling of Elastic Constants

In the present study, two factors and five-level historical data models are chosen as appropriate RSM models for establishing the correlation between the elastic constants and the factors. Tensile modulus and Poisson's ratio of short human hair-fiber-reinforced polyester composite are considered as the two responses, and fiber weight ratio and fiber length are two input variables. Design Expert version 10, DX10-05-3 software is used for this purpose. Experimental matrix for historical data model consisting of two factors and five levels of each is shown in Tables 1 and 2.

Response–factor relationship is expressed in Eq. (1) below.

$$Y = F(x_1, x_2, x_3, \ldots, x_m) + \varepsilon \tag{1}$$

where $x_1, x_2, x_3, \ldots x_m$ are the factors, F is the unknown and complex true response function, ε represent variability crept into Y by other sources of variability which is not part of F. The actual response function must be approximated in terms of M as given in Eq. (2) below.

$$Y_P = M(x_1, x_2, x_3, \ldots, x_m) \tag{2}$$

where Y_P indicate predicted value of M.

Cubic RSM model is now written in Eq. (3)

$$Y_P = a_0 + a_1 x_1 + a_2 x_2 + a_3 x_1 x_2 + a_4 x_1^2 + a_5 x_2^2 \\ + a_6 x_1^2 x_2 + a_7 x_1 x_2^2 + a_8 x_1^3 + a_9 x_2^3 \tag{3}$$

where a_0 is the average of the responses and $a_1, a_2, \ldots a_9$ are regression coefficients.

Design Expert version 10, DX10-05-3 software, computes the regression coefficients of the suggested models.

Final equations in coded factors for both the elastic constants in terms of coded factors are given in Eqs. (4) and (5).

$$\text{Young's Modulus} = + 3.96 - 0.24 \times A + 0.40 \times B - 0.074 \times AB - 0.41 \times A^2 \\ - 0.26 \times B^2 - 0.012 \times A^2 B + 0.067 \times AB^2 \\ + 0.011 \times A^3 - 0.37 \times B^3 \tag{4}$$

$$\text{Poisson's ratio} = + 0.27 - .094 \times A - 0.057 \times B + 0.023 \times AB + 0.036 \times A^2 \\ + 0.014 \times B^2 - (4.571E - 003) \times A^2 B - 0.010 \times AB^2 \\ + 0.069 \times A^3 + 0.015 \times B^3 \tag{5}$$

Table 1 Process factor values with their levels

Factors	Parameters	Levels				
		-2	-1	0	$+1$	$+2$
A	Hair fiber length (mm)	10	20	30	40	50
B	Hair fiber weight ratio (%)	5	10	15	20	25

Table 2 Experimental design matrix

	Factor 1	Factor 2	Response 1	Response 2
Run	A: Fiber length (mm)	B: Fiber weight ratio (%)	Tensile modulus (GPa)	Poisson's ratio
1	10	5	3.38	0.43
2	10	10	3.45	0.36
3	10	15	3.88	0.34
4	10	20	3.95	0.31
5	10	25	3.5	0.28
6	20	5	3.58	0.41
7	20	10	3.67	0.39
8	20	15	3.95	0.32
9	20	20	3.95	0.30
10	20	25	3.72	0.29
11	30	5	3.73	0.30
12	30	10	3.82	0.30
13	30	15	4.02	0.25
14	30	20	4.13	0.22
15	30	25	3.89	0.25
16	40	5	3.53	0.30
17	40	10	3.56	0.28
18	40	15	3.58	0.25
19	40	20	3.75	0.25
20	40	25	3.45	0.22
21	50	5	3.13	0.31
22	50	10	3.25	0.30
23	50	15	3.32	0.28
24	50	20	3.49	0.27
25	50	25	3.05	0.26

4 Multi-Response Optimization Using Grey Taguchi Method

From Table 2, it can be observed that factor levels are different for optimum response values. Grey Taguchi method converts both elastic constants into grey relational grade for obtaining global optimum responses of random-oriented short human hair fiber composite material. Grey Taguchi method takes tensile modulus and Poisson's ratio as the inputs and calculates grey relational generation using Eq. (6) [19]

$$Z_{ij} = \frac{y_{ij} - \min(y_{ij}, i = 1, 2, \ldots, n)}{\max(y_{ij}, i = 1, 2, \ldots, n) - \min(y_{ij}, i = 1, 2, \ldots, n)} \qquad (6)$$

Equation (7) [19] calculates grey relation coefficient as given below

$$\gamma((k), yi(k)) = \frac{\Delta \min + \xi \, \Delta \max}{\Delta oj(k) + \xi \, \Delta \max} \qquad (7)$$

Equation (8) [19] calculates grey relational grade as given below

$$\bar{\gamma}_j = \frac{1}{K} \sum_{i=1}^{m} \gamma_{ij} \qquad (8)$$

5 Results and Discussion

5.1 RSM Model Adequacy and Verification of Elastic Constants

A software called Design Expert 10 was used to analyze the data collected from the experimentation. ANOVA was used to analyze both the responses and results are shown in Tables 3 and 4.

The ANOVA results indicate that the cubic regression models of both Young's modulus and Poisson's ratio are highly significant as the Fisher's F-test had a very low probability values. Also, the model F-values of 20.72 and 19.65 for Young's modulus and Poisson's ratio, respectively, imply that the models are significant. This low value of probability value means that there is only a 0.01% chance that an F-value of this large could happen due to noise. The goodness of fit of the models is further inspected using R^2 values. Results in Tables 3 and 4 showed that R^2 values of the models for Young's modulus and Poisson's ratio are 0.9256 and 0.9218, respectively. Adequate precision measures the signal-to-noise ratio. A ratio greater than 4 is desirable. Adequate values of 16.12 and 17.323 for Young's modulus and Poisson's ratio models indicate an adequate signal [20]. It is concluded from the above discussion that both the models can be used successfully to pilot the design space. Student's t-test and P-values establish the significance of each term in Eqs. (4) through (7) and are given in Tables 3 and 4 for Young's modulus and Poisson's ratio, respectively. All the model terms having "Prob > F" value less than 0.05 are significant [21]. For Young's modulus, the model terms A, B, A^2, B^2, B^3 are significant and for Poisson's ratio, A, B, AB, A^2, A^3 are noteworthy model terms. Predicted values of the model and investigational values for Young's modulus and Poisson's ratio are plotted in Fig. 2 to check the quality of the models. The correlation graph in Fig. 2 shows that the cubic model fits for both the responses is suitable. The R^2 values indicate that only variations of 7.44% of the Young's modulus and 7.82% of Poisson's ratio were not explained by these models.

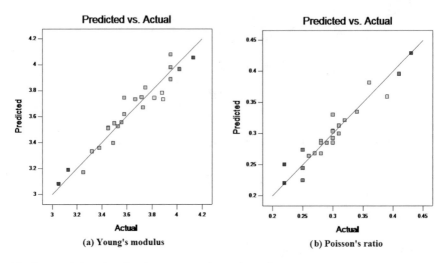

Fig. 2 Correlation of predicted responses and experimental responses

Table 3 ANOVA for response surface cubic model of Young's modulus

Source	Sum of squares	df	Mean square	F-value	P-value Prob > F	
Model	1.78	9	0.20	20.72	<0.0001	Significant
A-Fiber length	0.067	1	0.067	6.99	0.0184	
B-Fiber weight Ratio	0.19	1	0.19	20.34	0.0004	
AB	0.034	1	0.034	3.59	0.0775	
A^2	0.72	1	0.72	76.00	<0.0001	
B^2	0.30	1	0.30	31.72	<0.0001	
A^2B	3.150E–004	1	3.150E–004	0.033	0.8582	
AB^2	9.778E–003	1	9.778E–003	1.03	0.3271	
A^3	1.280E–004	1	1.280E–004	0.013	0.9093	
B^3	0.15	1	0.15	16.22	0.0011	
Residual	0.14	15	9.529E–003			
Cor total	1.92	24				
Std. Dev.	0.098		R-squared		0.9256	
Mean	3.63		Adj R-squared		0.8809	
C.V. %	2.69		Pred R-squared		0.7852	
PRESS	0.41		Adeq precision		16.120	

Table 4 ANOVA table of Poisson's ratio

Source	Sum of squares	df	Mean square	F-value	P-value Prob > F	
Model	0.064	9	7.115E–003	19.65	< 0.0001	Significant
A-Fiber length	0.011	1	0.011	29.30	< 0.0001	
B-Fiber weight ratio	3.936E–003	1	3.936E–003	10.87	0.005	
AB	3.364E–003	1	3.364E–003	9.29	0.008	
A^2	5.670E–003	1	5.670E–003	15.66	0.001	
B^2	8.229E–004	1	8.229E–004	2.27	0.152	
A^2B	4.571E–005	1	4.571E–005	0.13	0.727	
AB^2	2.314E–004	1	2.314E–004	0.64	0.436	
A^3	5.408E–003	1	5.408E–003	14.94	0.001	
B^3	2.420E–004	1	2.420E–004	0.67	0.426	
Residual	5.430E–003	15	3.620E–004			
Cor total	0.069	24				
Std. Dev.		0.019		R-squared		0.922
Mean		0.30		Adj R-squared		0.875
C.V. %		6.37		Pred R-squared		0.814
PRESS		0.013		Adeq precision		17.32

5.2 Influence of Fiber Length and Fiber Weight Ratio on Elastic Constants

The 3D response surface and contour plots for both Young's modulus and Poisson's ratio are shown in Figs. 3 and 4 for the developed cubic models explained by Eqs. (4) and (5). These 3D plots explain the combined effect of the fiber length and weight ratio on Young's modulus and Poisson's ratio. The contour plots of the responses given in Fig. 4 show increasing Young's modulus with increasing volume fraction of fibers and increased fiber length. Similarly, Poisson's ratio decreases with increasing fiber volume ratio of fibers.

It is observed from the Figs. 3a, 4a that the fiber weight fraction has maximum effect of the tensile modulus followed by fiber length. The tensile modulus increased with increase in fiber weight fraction. This is owing to the fact that increases in stiffer fiber content in less stiffer polyester increases the stiffness of the composite. Thus, it is obvious that the composite material becomes stiffer and gives high tensile modulus with increase in fiber weight fraction [22]. However, the tensile modulus decreases at higher fiber weight fractions. At low fiber weight fractions, fewer fibers are present in the composite and fibers fail even small loads. Hardly any load will be taken by these broken fibers and in turn present in the composite as holes. As a result, stiffness of the composite material sometimes falls below the stiffness value of the matrix material itself. Load-bearing capacity of fibers becomes effective when the fiber weight ratio is above the critical weight fraction [23]. A vinyl ester composite reinforced with short banana fibers, minimum critical fiber weight fractions is observed as 15%, and critical weight fraction is at 25% [24]. As suggested by Madsen et al. [25] in their research, the properties of the composite deteriorate beyond a maximum fiber weight fraction. In the present research, Young's modulus of the human hair fiber composite decreases beyond the fiber weight fraction of 20%. This is mainly because of improper impregnation of hair fiber in the polyester matrix. Further, fiber tensile modulus increases up to fiber length of 30 mm and with further increase in length results in decrease in the value. Reason for this is discontinuous chopped hair fibers of length more than 30 mm are not impregnated straight into the matrix.

From the surface and contour plots shown in Figs. 3b and 4b, Poisson's ratio is decreasing with increase in both fiber weight fraction and fiber length. Increase in hair fiber content increases stiffness of the composite and deformation of the composite decreases as the stiffness increases. Thus, declining trend of Poisson's ratio with increasing fiber weight fraction and length is witnessed. Similar results are reported by Manjusha et al. [26].

5.3 Optimization of Parameters of the Hair Composite

The response surface methodology is used to optimize fiber weight fraction and fiber length to maximize Young's modulus [16, 27]. Constraints chosen while maximizing the Young's modulus are shown in Table 5.

The apex of the response surface plots shown in Fig. 3 has the maximum achievable Young's modulus. A contour plot shown in Fig. 4 provides visual display of region

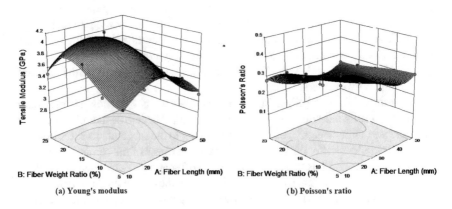

(a) Young's modulus   (b) Poisson's ratio

Fig. 3 Response surface plots

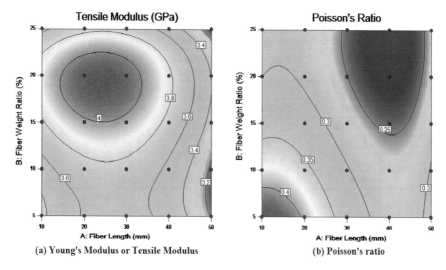

Fig. 4 Contour plots

of optimality of fiber length and fiber weight fraction. Contour plots take part in the examination of response surface. By analyzing the 3D response surface and contour plots given in Figs. 3 and 4, the maximum achievable Young's modulus is found to be 4.062 GPa and achievable Poisson's ratio corresponding to this value is 0.254. The optimal factors responsible for this maximum value are fiber weight fraction of 19.95% and fiber length of 29.32 mm. Contributions made by these factors can be ranked from their respective F ratio values provided in Tables 3 and 4. The terms having higher F ratio values are more significant.

5.4 Multi-Objective Optimization

Grey relational grade is calculated using Eq. (8) by considering 60% and 40% weightage for tensile modulus and Poisson's ratio, respectively. Distinguishing coefficient

Table 5 Optimal solution as suggested by the historical RSM Model

Constraints				Optimum values of Young's modulus
Name	Goal	Lower limit	Upper limit	
A: Fiber length	In range	10	50	29.32 mm
B: Fiber weight Ratio	In range	5	25	19.95%
Young's modulus	Maximize	3.05	4.13	4.062 GPa
Poisson's ratio	In range	0.22	0.43	0.254

Table 6 Grey relational grade calculations

Exp. No.	Grey relational generation		Grey relation coefficient		
	Tensile modulus	Poisson's ratio	Tensile modulus	Poisson's ratio	GRG
1	0.305556	1	0.418605	1	0.651163
2	0.37037	0.666667	0.442623	0.6	0.505574
3	0.768519	0.571429	0.683544	0.538462	0.625511
4	0.833333	0.428571	0.75	0.466667	0.636667
5	0.416667	0.285714	0.461538	0.411765	0.441629
6	0.490741	0.904762	0.495413	0.84	0.633248
7	0.574074	0.809524	0.54	0.724138	0.613655
8	0.833333	0.47619	0.75	0.488372	0.645349
9	0.833333	0.380952	0.75	0.446809	0.628723
10	0.62037	0.333333	0.568421	0.428571	0.512481
11	0.62963	0.380952	0.574468	0.446809	0.523404
12	0.712963	0.380952	0.635294	0.446809	0.5599
13	0.898148	0.142857	0.830769	0.368421	0.64583
14	1	0	1	0.333333	0.733333
15	0.777778	0.142857	0.692308	0.368421	0.562753
16	0.444444	0.380952	0.473684	0.446809	0.462934
17	0.472222	0.285714	0.486486	0.411765	0.456598
18	0.490741	0.142857	0.495413	0.368421	0.444616
19	0.648148	0.142857	0.586957	0.368421	0.499542
20	0.37037	0	0.442623	0.333333	0.398907
21	0.074074	0.428571	0.350649	0.466667	0.397056
22	0.185185	0.380952	0.380282	0.446809	0.406892
23	0.25	0.285714	0.4	0.411765	0.404706
24	0.407407	0.238095	0.457627	0.396226	0.433067
25	0	0.190476	0.333333	0.381818	0.352727

is chosen as 0.5 while calculating grey relational grade. Table 6 presents the results of grey Taguchi method. It can be seen from table that highest GRG is obtained at experiment no. 14.

Global optimum factor levels happen at experiment no. 14. Five experiments are conducted at these factor levels, and the average results are shown in Table 7.

Table 7 Global optimum factor levels and response values

Factors		Response 1	Response 2
A: Fiber length (mm)	B: Fiber weight ratio (%)	Tensile modulus (GPa)	Poisson's ratio
30	20	4.13	0.22

6 Conclusions

1. An empirical relationship between Young's modulus and fiber weight fraction and length is developed to forecast the Young's modulus of hair fiber polyester composite incorporating factors at 95% confidence level. Among the two factors, the variance analysis indicates that the fiber weight fraction is predominant factor that affects Young's modulus more, followed by fiber length. Similar empirical model is also developed for Poisson's ratio, where fiber length became the predominant factor followed by the fiber weight fraction.
2. The RSM optimization results indicate that a maximum tensile modulus value of 4.062 GPa is obtained under the conditions of 19.95% fiber weight fraction and 29.32 mm fiber length. Poisson's ratio at these conditions is found to be 0.254.
3. Multi-objective optimization is carried out using grey Taguchi method. Maximum GRG is found to be 0.7333 which occurred at optimum fiber length of 30 mm and fiber weight ratio of 20%. Optimum value of tensile modulus is 4.13 GPa, and Poisson's ratio at this condition is 0.22.

Acknowledgements My sincere gratitude to Dr. P. Ravinder Reddy, Professor and Principal, CBIT, Hyderabad, India, for providing laboratory support. Grateful acknowledgments are made to Stat-Ease for providing software support to carry out RSM.

References

1. Joly, C., Gauthier, R., Escouben, M.: Partial masking of cellulosic fiber hydrophilicity for composite applications. Water sorption by chemically modified fibers. J. Appl. Polym. Sci. **61**(1), 57–69 (1996). https://doi.org/10.1002/(sici)1097-4628(19960705)61:1
2. Sotton, M., Ferrari, M.: L'industrie Textile, vol. 1197, p. 58 (1989)
3. (a) Joseph, P.V., Joseph, K., Thomas, S.: Effect of processing variables on the mechanical properties of sisal-fiber-reinforced polypropylene composites. Compos. Sci. Technol. **59**(11), 1625–1640 (1999); (b) Cruz-Ramos C.A.: Natural fiber reinforced thermoplastics. In: Clegg D.W., Collyer A.A. (eds) Mechanical Properties of Reinforced Thermoplastics, pp. 65–81. Springer, Dordrecht (1986). PII:S0266-3538(99)00024-X
4. Parkinson, G.: Chementator: a higher use for lowly chicken feathers. Chem. Eng. **105**, 21 (1998)
5. Schmidt, W.F.: Innovative feather utilization strategies. In: Proceeding of the National Poultry Waste Management Symposium, Auburn University Printing Services, 19–22 October 1998, pp. 276–282 (1998)

6. Acda, M.N.: Waste chicken feather as reinforcement in cement-bonded composites. Philippine J. Sci. **139**(2), 161–166 (2010)
7. Winandy, J.E., Muehl, J.H., Micales-Glaeser, J.A., Schmidt W.F.M.: Chicken feather fiber as additives in MDF composites. J. Nat. Fiber **4**, 35–48 (2007). https://doi.org/10.1300/j395v04n01_04
8. Wool, R.P.: Bio-based composites from soy bean oil and chicken feather. In: Wool, R.P., Sun, X.S. (eds) Biobased Polymers and Composites, 411 p. Elsevier Academic Press (2005) https://doi.org/10.1016/b978-012763952-9/50013-7
9. Barone, J.R., Gregoire, N.T.: Characterization of fibre-polymer interactions and transcrystallity in short keratin fiber-polypropylene composites. Plast. Rubber Compos. **35**, 287–293 (2006) https://doi.org/10.1179/174328906x146478
10. Barone, JR., Schmidt, W.F.: Polyethylene reinforced with keratin fibers obtained from chicken feather. Compos. Sci. Technol. **65**,173–181 (2005) https://doi.org/10.1016/j.compscitech.2004.06.011
11. Hamoush, S.A., el-Hawary, M.M.: Feather fiber reinforced concrete. Concr. Int. **16**, 33–35 (1994)
12. Gupta, A.: Human Hair "Waste" and Its Utilization: Gaps and Possibilities. J. Waste Manage. **2014** (Article ID 498018), 1–17 (2014) https://doi.org/10.1155/2014/498018
13. Zare, Y., Garmabi, H., Sharif, F.: Optimization of mechanical properties of PP/nanoclay/CaCO3 ternary nanocomposite using response surface methodology. J. Appl. Polym. Sci. **122** (2011) https://doi.org/10.1002/app.34378
14. Chieng, B.W., Ibrahim, N.A., Wan Yunus, W.M.Z.: Optimization of tensile strength of poly (lactic acid)/graphene nanocomposites using response surface methodology. Polym. Plast. Technol. Eng. **51**, 791e799 (2012) https://doi.org/10.1080/03602559.2012.663043
15. Dr. Oleiwi, J.K., Al- Hassani, E.S., Mohammed, A.A.: Experimental investigation and mathematical modeling of tensile properties of unsaturated polyester reinforced by woven glass fibers. Engg. Technol. J, **32**, Part (A) (3) (2014)
16. Ahmad Rasyid, M.F., Salim, M.S., Akil, H.M., Ishak, Z.A.M.: Optimization of processing conditions via response surface methodology (RSM) of nonwoven flax fibre reinforced acrodur biocomposites. Procedia Chem. **19**, 469–476 (2016) https://doi.org/10.1016/j.proche.2016.03.040
17. Ashenai Ghasemi, F., Ghasemi, I., Menbari, S., Ayaz, M., Ashori, A.: Optimization of mechanical properties of polypropylene/talc/graphene composites using response surface methodology. Polym. Test. **53**(3) (2016) https://doi.org/10.1016/j.polymertesting.2016.06.012
18. Jagannathan, K., Panchanatham, N.: An overview of human hair business in Chennai. Int. J. Eng. Manage. Sci. **2**(4), 199–204 (2011)
19. Kuo, Yiyo, Yang, Taho, Huang, Guan-wei: The use of grey-based taguchi method to optimize multi response simulation problems. Eng. Optim. **40**(6), 517–528 (2008). https://doi.org/10.1080/03052150701857645
20. Velumani, S., et al.: Optimization of mechanical properties of non-woven short sisal fibre-reinforced vinyl ester composite using factorial design and GA method. Bull. Mater. Sci. **36**(4), 575–583 (2013)
21. Powell, B.C., Rogers, G.E.: The role of keratin proteins and their genes in the growth, structure and properties of hair. In: Jolles, P., Zahn, H., Hocker, E. (eds.) Formation and Structure of Human Hair, pp. 59–148. Birkhauser Verlag, Basel (1997)
22. Sudheer, M., Pradyoth, K. R., Somayaji, S.: Analytical and numerical validation of epoxy/glass structural composites for elastic models. Am. J. Mater. Sci. **5**(3C), 162–168 (2015) https://doi.org/10.5923/c.materials.201502.32
23. Harris, B.: Engineering composite materials. The Institute of Materials, London (1999)
24. Ghosh, R., Reena, G., Krishna, A.R., Raju, B.H.L.: Effect of fibre volume fraction on the tensile strength of Banana fibre reinforced vinyl ester resin composites. Int. J. Adv. Eng. Sci. Technol. **4**(1), 89–91 (2011)
25. Madsen, B., Thygesen, A., Liholt, H.: Plant fibre composites—porosity and stiffness. Compo. Sci. Technol. **69**(7–8), 1057–1069 (2009). https://doi.org/10.1016/j.compscitech.2009.01.016

26. Manjusha, K., Kondareddy, B., Pavan Kumar, D.: Effect of fiber length and weight on tensile response of natural fiber reinforced composite. Int. Journal of Eng. Res. Technol. **5**(04), 389–394 (2016)
27. Raji, N.A., et al.: Response surface methodology approach for transmission optimization of V-belt drive. Mod. Mechan. Eng. **6**, 32–43 (2016). https://doi.org/10.4236/mme.2016.61004

A Decennary Survey on Artificial Intelligence Methods for Image Segmentation

B. Vinoth Kumar, S. Sabareeswaran and G. Madumitha

Abstract The technique of breaking down an image into categorial regions containing each pixel with similar attributes is termed as image segmentation (IS). It is the preliminary step of image processing. This technique can be used for both grey-scale and colour images. This technique is applied everywhere, even in our personal Smartphone's camera while capturing pictures. And image segmentation is the most innovative problem under the computer vision domain. This paper provides various techniques that are available in the field of image segmentation and their pros and cons. A lot of research is being done by applying artificial intelligence techniques for the image segmentation problems. In this paper, an overview of artificial intelligence algorithm techniques such as machine learning, deep learning, meta-heuristics approaches that was used in the past decade has been discussed, and a comparative study about the same is carried out and the problems and recommendations for selection of appropriate method for image segmentation have been dealt with.

Keywords Artificial intelligence · Machine learning · Deep learning · Meta-heuristic techniques · Computer vision and image segmentation

B. Vinoth Kumar (✉) · S. Sabareeswaran · G. Madumitha
Department of Information Technology, PSG College of Technology, Coimbatore, Tamil Nadu, India
e-mail: bvk.it@psgtech.ac.in

S. Sabareeswaran
e-mail: ssabarish1997@gmail.com

G. Madumitha
e-mail: madhumithagopinath@gmail.com

© Springer Nature Singapore Pte Ltd. 2020
R. Venkata Rao and J. Taler (eds.), *Advanced Engineering Optimization Through Intelligent Techniques*, Advances in Intelligent Systems and Computing 949, https://doi.org/10.1007/978-981-13-8196-6_27

1   Introduction

Computer vision is a plot of computer science that enables the computers to visualize, identify and process the images in a similar way like humans do and also provides an efficient output. For example, this mechanism converts [1] the two-dimensional image into a three-dimensional image, for e.g., the part-picking robot that is used in industries, where it is easy for the robot to distinguish between the objects that are necessary and eliminate the rest. Either it may be the technique of digital image processing or computer vision, the pre-processing that is required for any such technique is image segmentation. So, it hits a typical role in the field of image processing. The main aim of IS [2] is to combine pixels in similar regions by making use of a kind of feature extraction. The entire image segmentation process is shown in Fig. 1 for easy visualization. The process of image segmentation can be done using traditional methods like clustering, region-based, edge-based, etc. or the modern techniques that uses artificial intelligence like CNN, ANN, genetic algorithm, etc.

One of the most noteworthy applications is Google's Pixel XL 2 mobile phone that has the best camera according to the recent statistics. The reason behind this is image segmentation using artificial intelligence which uses a deep learning mechanism known as convolution neural network (CNN). Although many surveys are available in the literature [3, 4], as far to our knowledge, the extensive survey has not been carried out for AI-based methods. So here, our aim is to explore the methods that are available for solving the problems in image segmentation using artificial intelligence.

Our contribution to this paper showcases the methods that are emerging in this field to solve the image segmentation problems. Part 2 discloses the categorization methods and gives a short note on traditional methods. Part 3 gives the Evident contributions of AI to image segmentation. Part 4 reveals the decennary survey that is done in this field for the past decade. Part 5 concludes the survey by quoting that AI-based methods provide better results than traditional methods when compared with an important performance measure like time complexity, accuracy, etc.

2   Categorization of Segmentation Methodologies

IS is broadly split into two main types that are the old traditional methods and rapidly emerging technologies that include machine learning techniques and deep learning techniques which come under artificial intelligence. In recent years, many researchers have showcased their works based upon the optimized algorithms so that

Fig. 1 Image segmentation process

it gives better results when compared to the traditional ones. This paper gives the categorization for the techniques that are used frequently by the research scholars in the past decade. The paper focuses to present the statistical survey about the status of improvisation in the field of image segmentation by giving a short note on the traditional methods and proceeding with the AI-based techniques and also predicting the future algorithms which is going to be based on the trending and modern technologies. This categorization that is proposed here is not based upon any specific criteria or any mechanism. The categorization is shown in Fig. 2.

Clustering is done by clubbing relevant regions with respect to segmentation boundaries that match the region edge. Mean shift is responsible for [5] the action of shifting each cluster data point to the mediocre of data points nearby. The region [6] level local knowledge of image is integrated into the clustering technique. Histogram threshold is constructed by splitting the range of the data into equal-sized bins [7]. A-IFS Histon [8] is an encrustation of the histogram. Adaptive histogram equalization (AHE) is a contrast enhancement method [9] that has demonstrated effectiveness. Region-based method is done by [10] splitting the image into regions by collecting homogenous pixels. Growing is a method [11] that compares the seed pixels with the nearby pixels with a threshold. Merging [10] is used to determine the important index of each cluster and then merge those cluster with the lower value of important index based upon the new merging rule. Hybrid [12, 13] is the combination of both region-based and edge-based. Energy-based is classified as Mumford Shah [14] and Markovian Bayesian fusion algorithm [15] which uses the mathematical function for segmentation. A graph cut [16] is used to perform an unsupervised image segmentation using hybrid graph model (HGM).

3 Evident Contribution of Artificial Intelligence to Image Segmentation

Many research scholars have contributed so much of their works in cutting edge technology. Artificial intelligence (AI) [17] is the process of creating intuitive machines that perform and behave like humans. Few of the process of AI include problem-solving, learning, planning, natural language processing, etc. Knowledge is the main core part of AI, where the machine tries to learn about itself and acts according to the situation and the surroundings. AI is further classified as machine learning (ML), meta-heuristic techniques and a sub-branch of ML is known as deep learning (DL).

3.1 Machine Learning

Machine learning [18] stands under the branch of AI that is driven towards the scientific advancement of human thinking. It grants the machine to tackle new scenarios

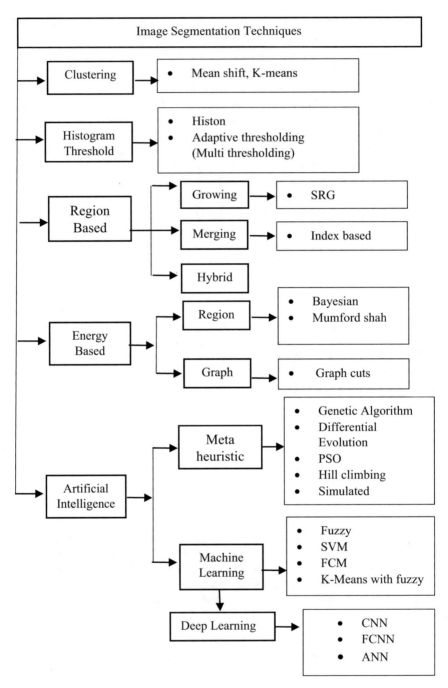

Fig. 2 Image segmentation methodologies

for analysing, self-determining, observation and experience. Some of machine learning algorithms are decision trees, naïve-based classification, support vector machine (SVM), fuzzy logic, etc. Different machine learning classifiers used in image segmentation technique are listed in Table 1.

Work Flow of ML. Initially, the image is captured and stored in the database and then loaded to the machine learning network. The second step is pre-processing, where the images at the lower level of abstraction are removed for better accuracy results. Next is the feature extraction process in which some of the techniques like histogram matching, grey-scale extraction, texture extraction and square of logarithm filtering are used. Since the input data are too large to the ML network, feature extraction or feature selection divides the large data into a reduced format (represented as pixel vectors of images) for better outcomes. After that, the feature extracted images are fed into the machine learning classifiers like SVM [19], random forest [20], LDA [21], K-means clustering with fuzzy logic [22], Bayesian, MAP-ML estimation, etc. In post-processing step, additional classification is done to get better result in image segmentation. The mean square root estimation (MSE), correct classification rate (CCR) and peak signal to noise ratio (PSNR) of the picture are used as a performance measure. The entire workflow of ML is shown in Fig. 3. Some of the Machine learning techniques proposed are listed below.

Linear Discriminant Analysis (LDA). The main principle of the algorithm is to preserve uttermost of the vector class knowledge without mislaying any metadata. It is widely applied in the latest applications like face detection, image recovery, etc. In this method, the algorithm classifies the image into positive and negative vector images and then takes the mean square root for better image segmentation results. There are some limitations in it where the two groups that are classified are not perfect in feature space and the training data set gets large minimum error rate which is also not desirable [21]. This can be overcome by applying SVM.

Support Vector Machine (SVM). SVM overcomes the pitfalls of LDA by calculating hyper-plane and scalar bias. It mostly uses quadratic programming to solve the problems. SVM is a binary classification algorithm. In this process [19], the 2D array of SVM model automatically learns the characteristics of the object. The proposed method calculates the superpixels initial scores that belong to foreground using the trained SVM models. It has some limitations like unable to store large feature space and the kernel selection is but tricky.

K-means. This comes under the clustering type of machine learning. K-means technique is influenced through initial cluster centroids and the number of clusters K [23] and does not guarantee unique clustering outputs all the time for the single input. Mignotte [22] proposed the very simple image segmentation technique with fusion method where it selects and extracts the image features and is handled as vectors. The set of inter-associated dots (pixels) corresponding to each of its particular class defines the various part of the image [22].

AdaBoost. It converts the weak learner into strong learner [24]. Some of the weak learners are decision trees, Naïve Bayes, LDA and SVM. When boosting the weak learner with AdaBoost, it helps in achieving better segmentation results.

Table 1 Different ML methods and its pros and cons

Refs.	Feature extraction	Machine learning method	Data set	Advantages/limitation
[21]	Content-based image retrieval, colour histogram is used	Speeded up robust feature (SURF), SVM and LDA	Medical image data set	• The image objects can be split up into various classifications in terms of the designed support vector classifier
[52]	Texture feature extraction using wavelet	SVM	CT images	• Hybrid method which combines SVM and texture feature extraction for liver parenchyma segmentation • Final integration of morphological operations delimits the noise
[24]	AdaBoost and EM-minimization	Conditional random fields (CRFs)	Brodatz set and a natural image (flowers)	• The proposed method is better than MAP and ML approaches
[53]	Threshold method	Fuzzy cellular neural network and advanced FCNN	CT images of liver	• It achieves better boundary and segmentation results • AFCNN gives better result when compared with FCNN. (fuzzy cellular NN)
[20]	Single histogram class model based on textons	Random forest	MSRC data set	• It achieves 67.7% accuracy

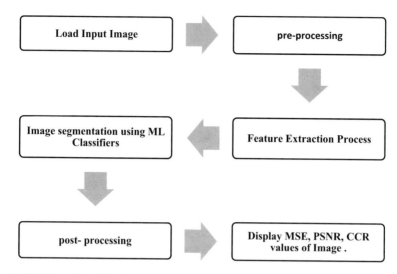

Fig. 3 Work flow of machine learning algorithms

3.2 Meta-Heuristic Techniques

Meta-heuristic optimization. It deals with optimization problems using meta-heuristic algorithms. Optimization can be considered as the maximization and minimization problems. To solve the maximization problem, efficient search or optimization algorithms are needed. The optimization algorithms can be categorized as gradient-based and derivative-free algorithms namely hill climbing and Nelder–Mead Downhill, respectively. On other perspective, it can be split as trajectory-based and population-based algorithms namely hill climbing and particle swarm optimization (PSO), respectively. In other way, it can also be split as deterministic and stochastic. GA and PSO are good instances of the stochastic model. Some of the noteworthy contributions are described below.

Simulated Annealing (SA). SA is based on metal annealing processing. SA can able to solve Euclidean distance problem using simulated annealing spectral clustering (SASC) algorithm proposed by Yang and Wang [25] and achieve better clustering compared with traditional methods.

Genetic Algorithm (GA). The essence of GA is to catch the most flawless solution in the chromosomal population, i.e. survival of the fittest. It uses contour a nonlinear and interpolation detector [26] method to get a global maximum which shows the percentage of image matching using optimized fitness function in CT scan and other medical-related visualization equipments. These results are better when compared to other clustering and region-based methods [5, 10].

Differential evolution (DE). It is an improved vector-based version of GA that comprises of the essential steps such as mutation, crossover and selection [27]. The main aim of DE is to find similarity between the cluster index and fuzzy index for segmentation of satellite images [28]. When compared with FCM [29], computation time decreases while using DE.

Particle Swarm Optimization (PSO). It is most frequently applied to all the areas of optimization and computational intelligence. Half-life constant particle swarm optimization (HCPSO) algorithm with Tsallis entropy threshold [30] has better result when compared to normal PSO [31]. It uses dynamic inertia weight (DW-PSO) algorithm [32] in medical image processing applications.

Meta-heuristic optimization with deformable models. This technique is better than machine learning approaches. Among various model-based methods, the deformable model has a robust image segmentation result. The term shape is known as deformable. The working process of this model initiates from the starting edge-shaped curve and its step-by-step process modification is done by applying energy functions. Deformable is of two types; their classification is shown in Fig. 4

Parametric approach. Snake model is the first proposed parametric deformable model by Terzopoulos and et al. [33]. This is also called as a active contour model. A snake is an energy boasting model, influenced by b-spline constraint and the external force moves it towards object edge detection and internal force resists the deformation [34–36]. Snake model is highly subjected to noise and very slow in computation. This pitfall is overcome by topological active nets [37, 38]. Essential analysis between active nets and snake is that active contour lacks resources to manage topology. And some other energy fitness functions like gaussian mixture, Tsallis entropy methods hybridized with meta-heuristic techniques are listed in Table 2.

Geometric approach. This method overcomes the primary limitation of parametric approach. The growth of curves and lines does not deny upon the peculiar way the curve has been organized. It is basically analysed that the procedure of deformable models changes the silhouette shape that relies only on geometrical magnitude. One of the examples of this type is the level set approach [39]. The essential benefit of the level set-based methods is the capability of a single image to split apart and merge

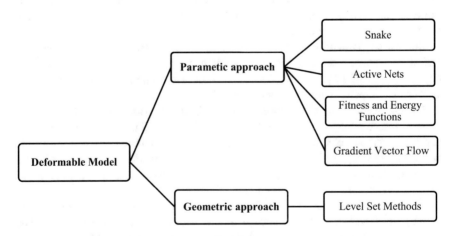

Fig. 4 Classification of deformable methods in image segmentation

Table 2 Different meta-heuristic techniques with deformable methods and its pros and cons

Refs.	Model type	Meta-heuristic method	Data set	Advantages/limitations
[28]	Xei-Beni (XB) index and type II fuzzy	DE	Satellite images	• Proposed method is compared with silhouette index of images, proposed DE gives better results
[30]	Tsallis entropy	Constant half-life particle swarm optimization	The standard Google's images like Lena, etc. images of (512 × 512) size	• When compared to normal GA, PSO and BF's proposed methods for multi-level thresholding gives a better quality of segmented images
[31]	FCM as fitness function	Particle swarm optimization	Berkeley data set	• It provides better results even when the image contains a huge noise ratio
[32]	Inertial weight	Dynamic inertia weight-PSO	Medical image segmentation	• Basic PSO execution time is less when compared to (DW-PSO) • Segmentation results are better in DW-PSO than basic PSO
[34, 35]	Snake	SA	Wood samples	• Works well on segmentation of knots with higher irregular-image samples
[36]	Snake	Active contour with GA	A large number of retinal images	• In medical image processing, it greatly affects the silhouette analysis (outlier analysis)
[54]	Gradient vector flow (GVF) snake	GA (adaptive balloon snake)	150 cardiac magnetic resonance images	• Automatic 150 heart image segmentation is proposed with less than 5 s with 8% error rate

(continued)

Table 2 (continued)

Refs.	Model type	Meta-heuristic method	Data set	Advantages/limitations
[55]	Poisson GVF	GA	Medical data set consists of MRI scan of liver (3×16)	• The computation time is robust when large data are given as input and automatic segmentation is also moderate when compared with state-of-the art method
[56]	Snake	PSO	CT scan and MRI scan images of brain	• Using PSO with canonical velocity update equation with snake kinematic makes to deliver results in an optimized way
[57]	Gaussian mixtures models	DE	83 images of a parasite named Trypanosoma	• Results are compared with Otsu's method provides minimum error rate of 5
[37, 38]	Topological active net (TAN)	Differential evolution with ANN	12 synthetic, 4 computed topographies	• This (TAN) approach combines the region and boundary segmentation techniques together • Trained ANN by DE provides segmentation results rapidly
[58]	Otsu's method	Chaotic-PSO	Six cement-based images	• CPSO has better search efficiency and accuracy than normal PSO
[59]	Active shape model-statistical approach	Differential evolution	320 real images, 20 synthetic images	• When compared with all six meta-heuristic techniques, it shows that DE is better that gives 90.9 and 93.0% accuracy of real and synthetic images, respectively

(continued)

Table 2 (continued)

Refs.	Model type	Meta-heuristic method	Data set	Advantages/limitations
[60]	Fuzzy active contour (FAC)	PSO	Nerve fibres contain 15 microscopic images	• This (FAC) model helps in boundary detection of nerve cells • PSO along with FAC helps in achieving 91% accuracy over other methods
[61]	Active shape model-statistical approach	Random forest + DE + Otsu's method	Histological images of mouse brains	• Average segmentation accuracy of the 92.25 and 92.11% that comprises of 15 real and 15 synthetic images, respectively.
[62]	Active shape model	Particle swarm optimization	Histological (hippocampus) brain	• 120 images yielded a perfect segmentation of about 89.2% accuracy
[63]	Active shape model	PSO + DE with NVIDI CUDA	Human video-sequences	• Robustness with respect to noise • When this method is compared with [62], DE provides better accuracy than PSO
[39]	Level set approach	GA for learning the level set and scatter search for post segmentation	CT scan and MRI scan medical images	• It uses a hybrid level set approach that gives better accuracy in medical image segmentation but slower than random forest

together without changing its characteristics. On the alternate side, there are two main demerits that reside in the complexity of the computation times which tends to be very slow.

3.3   Deep Learning

Deep learning (DL) is a sub-branch of ML that deals with algorithms influenced by the framework and skeleton of the neurons in the brain called artificial neural networks. Some of the DL algorithms are convolution neural networks (CNN), artificial neural networks (ANN), dynamic deep neural networks (DDNN), etc. Most widely preferred network for image segmentation is convolution neural network. Different DL methods and its pros and cons are discussed in Table 3.

Detailed View of Convolution Neural Networks (CNN). The convolutional neural networks (CNN) are described as a deflection of the standard neural networks. This deflection brings in a new peculiar network, which accommodates the existing convolution and pooling layers [41], rather than fully associated hidden layers. This was first popularized for overcoming the common obstacles of fully associated deep neural networks while approaching ambit magnitude structured input pictures for the process of segmentation [42]. CNN's [43] layers produce activation for input images that is used for segmentation. The above listed were some of the proposed works. CNN plays a vital role in computer vision's major tasks. This paper discusses more about the deep learning algorithms that work in the computer vision's major task as shown in Fig. 5.

Semantic Segmentation. In semantic segmentation, no objects only, pixels are detected. The problem of semantic segmentation is that input and output image will be a decision of a category of every pixel in that image. For example, a cat walking on the grass field so the output should be more number of pixels containing categories like grass, cat, trees, etc. This process is called semantic segmentation. One interesting thing about this type is that it does not distinguish between instances, for example, image of two cows are taken as an input then the output only provides the pixel representation for a single cow but the algorithm does not provide any distinguishing character between two cows. So, to overcome this, very basic technique implemented is sliding window approach. The idea is very inefficient because of the problem of overlapping patches. The next method approach [44, 45] is the fully convolutional network. Here, rather than extracting individual patches independently, imagine just having the network of whole giant stack of convolution layer. So, in this case, a bunch of convolution layers of $3 \times H \times W$ with zero padding exists and each convolution layer preserves the spatial size of the input. When the absorption picture is given to the proffered model, final output will be $C \times H \times W$, where C is the number of categories as shown in Fig. 6. That gives the classification score in every pixel in the absorption picture. After that training, this network is more or less similar to back-propagation network. One of the main drawbacks of FCNN is that it is highly expensive in computation. They reduce the spatial size of the image by performing

Table 3 Different DL methods and its pros and cons

Refs.	Activation function/classifiers	Deep learning method	Data set	Advantages/limitation
[44]	Jaccard distance	Convolution–deconvolution neural network (CDNN)	Online validation data set of 150 images training: 2000 dermoscopic images: 600	• They achieved the average 0.784 Jaccard index for the online validation data set • This is used for semantic segmentation • Disadvantage is that it is expensive
[45]	Alex Net, the VGG net and GoogLeN	Fully convolution neural network (FCNN)	PASCAL VOC and semantic segmentation data sets	• This gives better segmentation results than sliding window • It gives 30% improvement over the other • The main drawback of FCNN is highly expensive in computation
[47]	Non-maximal suppression (NMS)	Reginal proposal (which is a traditional computer vision technique)	PASCAL VOC data sets [40]	• This method overcomes the pitfall of FCNN, and this is highly inexpensive in computation • Drawback is that it is an old traditional approach
[48]	O-Net and T-Net	R-CNN (regions with CNN)	PASCAL VOC data set	• This works better than FCCN, in terms of computation and accuracy • It achieves mean average precision of about 53.3%, a 30% increase • This is very slow in computational
[49]	–	Faster R-CNN	PASCAL VOC data set	• This overcomes the drawback of R-CNN

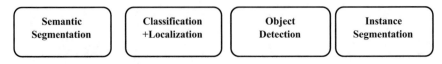

Fig. 5 Computer vision's major tasks

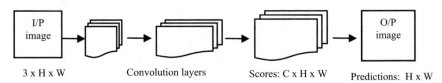

Fig. 6 Fully convolutional neutral network (FCNN)

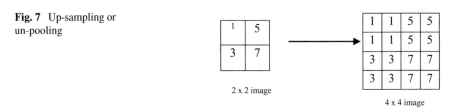

Fig. 7 Up-sampling or un-pooling

various types of pooling in the network like up-sampling or un-pooling and max un-pooling for better efficiency. Max Un-pooling is done to get a pixel perfect image, because inside the feature map from low-resolution layer, some feature may miss, by doing max un-pooling to find those missing features as shown in Fig. 7. Learning up-sampling by transpose convolution is another idea to learn about the feature map and weights of the image. Major drawback of transpose convolution is overlapping of pixels which may lead to decrease the efficiency of output image.

Classification and Localization. To learn more about the image in addition to protecting the category, for example, let us can take the cat, where is the object (cat) located in that image and also want to draw a bounding box on the region of cat in that image. This task is called classification and localization. Working module of this technique can use again the same machinery tool like a giant FCNN. The basic architecture is input image may feed into giant FCNN and output contains class scores and box coordinates. During the training time, there may be three types of losses like SoftMax loss, regression loss and L2 loss between the bounding box coordinates. This method [46] was proposed in human pose estimation.

Object Detection. It is a crucial topic and the core problem in computer vision. Moving on to the core object detection, consider the set of categories, for example, cats, dog, etc. Fix those categories. Now, the task of the algorithm is for the given input image every time one or more categories appear in the image. The algorithm will draw a box around it and it protects the category of that box, so this technique is different from classification and localization. This is because there may be a varying number of outputs for the input image and thus it is an interesting problem in core computer vision. Regional proposal is used instead of super slow sliding window approach for object detection. Regional proposal [47] is not a deep learning technique. It is a traditional computer vision technique. The working of this network is done by taking an input image and the output may be 2000 box around the image where the object might be present. It basically finds the blobby region in the image and it is relatively fast in its computation. The major drawback is that there will be a lot of noises in the image. All these were proposed in R-CNN network [48]. The drawbacks of R-CNN framework are training time, interference (detection) and the testing time is super slow. This problem of R-CNN is overcome by faster R-CNN [49], where the network itself proposes its own region proposals. Detection of objects without proposals is done by two methods which are SSD [50] and YOLO [51]. SSD is much faster than R-CNN but not accurate.

Instance Segmentation. Given an input image, the algorithm needs to predict one location and identify the object in an image similar to object detection. But rather than just predicting the boundary box for each the objects, it needs to predict the whole segmentation mask. And also, it predicts which pixel in the input image corresponds to each object instance. So, this is a hybrid technique between semantic and object detection. MASK R-CNN [40] is mostly proposed for instance object detection. It also works well for pose estimation. It is built upon the faster R-CNN framework and it runs five frames per sec under GPU. It really works well when compared to SSD, R-CNN, etc.

4 Decennary Survey

A decennary survey is a survey that has been done for the past ten years which means a decade. This paper explains the detailed review of the most frequently used techniques in this particular field so that it will be easy for the scientists in the future to further enhance their research in this field. The prime information that has inferred from this paper is that most of the researches are based on K-means clustering and CNN algorithms, when compared to rest.

This information has been taken from pie chart above in Fig. 8 for which the statistics have been taken for the past decade. So, this conveys that K-means clustering and CNN are the most frequently used techniques in AI-based methods of image segmentation. It is also evident from the graph Fig. 9 that during the past decade, traditional algorithms were used mostly in the former time and it gradually decreased by paving the way for AI-based algorithms in the latter period of time. Overall pros and cons of image segmentation techniques are discussed in above Table 4. There is a large number of algorithms available for image segmentation, but there is no perfect algorithm for all kind of image applications. To recommend the most efficient algorithm for image segmentation is a herculean task that demands a deep knowledge about the core problem and algorithms. Table 5 shows the problems and recommendation for selection of appropriate method for image segmentation.

Fig. 8 Pie chart
representation

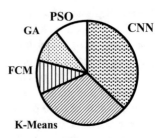

Fig. 9 Usage of traditional
versus AI techniques during
last decade

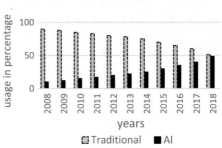

Table 4 Overall pros and cons of image segmentation techniques

Approaches	Advantages	Disadvantages
Clustering	• K-means is computationally fast when k is small • It produces more homogeneous regions	• Output is not desired with non-globular cluster • Expensive computational time
Histogram threshold	• Less computational time • Prior image information is not required for image segmentation	• Highly noise sensitive • Output is not efficient when applying this technique to glare, befog and blurred images
Region and edge detection	• Noise resilient • Proper selection of image seed gives better image segmentation results than other methods	• High computation time and memory • Segmentation with images having many edges produces over segmented results
Energy-based	• It achieves more reliable and accurate results	• Since it uses complex mathematical formula time complexity may high for such algorithms

(continued)

Table 4 (continued)

Approaches	Advantages	Disadvantages
Machine learning	• High Accuracy and speed compared to traditional methods	• Automatic segmentation is in developing stage using ML
Meta-heuristic	• Multiple object detection is moderate using deformable models	• Design made by user's decisions • Limited theoretical results
Deep learning	• Automatic image segmentation and most sensitive with real-world data can be easily handled	• In some cases, algorithm may fail which leads to background noise and may affect the accuracy

Table 5 Recommendation for selection of appropriate method for image segmentation problems

Problems	Technique used	Recommendation
Clustering	• K-means, Fuzzy-C-means (FCM), Mean shift, and manhattan distance	• K-means is better for clustering problems mainly because of its simplicity and of its accuracy
Object detection	• Meta-heuristic techniques like PSO, GA, DE, memetic algorithm and SA	• DE and memetic algorithm are better because of its high search capability, more optimization, accuracy and high convergence rate
Object Identification	• Evolutionary algorithm + classification algorithm like SVM, random forest, Naive Bayes, LDA, decision tree + deep leaning methods like CNN, ANN, FCNN, R-CNN, YOLO SSD and fuzzy cellular neural network	• CNN is better because of (pre-trained) automatic feature extraction for the given input, weight sharing and its accuracy

5 Conclusion and Future Work

In image processing, segmentation of image is an indispensable part of preliminary step so that it makes the image analysis and image compression more efficient and easier. This technique can be used for black–white and colour images. The time complexity and accuracy are the important performance measure in each and every algorithm. From the survey done, it is perceptible that AI-based methods provide better and optimal results than traditional methods based on the criteria of time complexity and accuracy. This survey also portrays the possible recommendation for selection of appropriate method for image segmentation problems. The AI-based algorithms that include machine learning, deep learning and meta-heuristic techniques are the trending and modern methods and it can also be enhanced in the future by learning the hyper-parameter, learning rate of the images in prior of constructing the neural network to deliver better and most optimal results.

References

1. Techopedia. http://www.techopedia/computer-vision
2. Liao, W., Rohr, K., Kang, C.-K., Cho, Z.-H., Stefan, W.: Automatic human brain vessel segmentation from 3D 7 tesla MRA images using fast marching with anisotropic directional prior (2012). https://doi.org/10.1109/isbi.2012.6235761
3. Kumar, V.B., Janani, K., Priya, M.N.: A survey on automatic detection of hard exudates in diabetic retinopathy (2017). https://doi.org/10.1109/icisc.2017.8068604
4. Kumar, V.B., Divya, S.: A study on optic disc localization methods in retinal images (In press)
5. Cheng, Y.: Mean shift, mode seeking, and clustering (1995). https://doi.org/10.1109/34.400568
6. Liu, G., Zhang, Y., Wang, A.: Incorporating adaptive local information into fuzzy clustering for image segmentation. IEEE Trans. Image Process. 24(11), 3990–4000 (2015). https://doi.org/10.1109/tip.2015.2456505
7. Raju, D.R.P., Neelima, G.: Image segmentation by using histogram thresholding (2012). doi: ijcset2012020103
8. Mushrif, M.M., Ray, A.K.: A-IFS Histon based multithresholding algorithm for color image segmentation. IEEE Signal Process. Lett. 16(3), 168–171 (2009). https://doi.org/10.1109/lsp.2008.2010820
9. Pizer, S.M., Amburn, P.E., Austin, J.D., Cromartie, R., Geselowitz, A., Greer, T., Romeny, B.H., Zimmerman, J.B., Zuiderveld.: Adaptive histogram equalization and its variations. Comput. Vis. Graph. Image Process. 39(3), 355–368 (1987). https://doi.org/10.1016/s0734-189x(87)80186-x
10. Kuan, Y.-H., Kuo, C.-M., Yang, N.-C.: Color-based image salient region segmentation using novel region merging strategy. IEEE Trans. Multimedia 10(5), 832–845 (2008). https://doi.org/10.1109/tmm.2008.922853
11. Preetha, M.M.S.J., Suresh, P.L., Bosco, J.M.: Image segmentation using seeded region growing. In: 2012 International Conference on Computing, Electronics and Electrical Technologies (2012). https://doi.org/10.1109/icceet.2012.6203897
12. Huang, Q., Dam, B., Steele, D., Ashley, J., Niblack, W.: Foreground/background segmentation of color images by integration of multiple cues. In: IEEE International Conference on Image Processing, pp. 246–249 (1995). https://doi.org/10.1109/icip.1995.529692
13. Haddon, J.F., Boyce, J.F.: Image segmentation by unifying region and boundary information. IEEE Trans. Pattern Anal. Mach. Intell. 12(10), 929–948 (1990). https://doi.org/10.1109/34.58867
14. Chan, T.F., Vese, L.A.: Active contours without edges. IEEE Trans. Image Process. 10(2), 266–277 (2001). https://doi.org/10.1109/83.902291
15. Mignotte, M.: A label field fusion bayesian model and its penalized maximum rand estimator for image segmentation. IEEE Trans. Image Process. 19(6), 1610–1624 (2010). https://doi.org/10.1109/tip.2010.2044965
16. Liu, G., Lin, Z., Yu, Y., Tang, X.: Unsupervised object segmentation with a hybrid graph model (HGM). IEEE Trans. Pattern Anal. Mach. Intell. 32(5), 910–924 (2010). https://doi.org/10.1109/tpami.2009.40
17. Techopedia-artificial intelligence. http://www.techopedia/artificialintelligence
18. Techopedia-machine learning. http://www.techopedia/machinelearning
19. Kim, K., Oh, C., Sohn, K.: Non-parametric human segmentation using support vector machine. In: 2016 IEEE International Conference on Consumer Electronics (ICCE) (2016). https://doi.org/10.1109/icce.2016.7430551

20. Barinova, O., Shapovalov, R., Sudakov, S., Velizhev, A.: Online random forest for interactive image segmentation (2012)
21. Hemjot, Sharma, A.: A refinement: better classification of images using LDA in contrast with SURF and SVM for CBIR system. Int. J. Comput. Appl. **117**(16), 31–33 (2015). https://doi.org/10.5120/20642-3349
22. Mignotte, M.: Segmentation by fusion of histogram-based K-means clusters in different color spaces. IEEE Trans. Image Process. **17**(5), 780–787 (2008). https://doi.org/10.1109/tip.2008.920761
23. Kumar, V.B., Karpagam, G.R., Rekha, V.N.: Performance analysis of deterministic centroid initialization method for partitional algorithms in image block clustering. Indian J. Sci. Technol. **8**(S7), 63 (2015). https://doi.org/10.17485/ijst/2015/v8is7/63376
24. Lee, S.H., Koo, H.I., Cho, N.I.: An unsupervised image segmentation algorithm based on the machine learning of appropriate features. In: 16th IEEE International Conference on Image Processing (ICIP). IEEE, Korea (2009). https://doi.org/10.1109/icip.2009.5413758
25. Yang, Y., Wang, Y.: Simulated Annealing Spectral Clustering algorithm for image segmentation. J. Syst. Eng. Electron. **25**(3), 514–522 (2014). https://doi.org/10.1109/jsee.2014.00059
26. Costin, H.: Elastic contour-based image segmentation using genetic algorithms. In: 2011 E-Health and Bioengineering Conference (EHB) (2011)
27. Kumar, V.B., Karpagam, G.R.: Evolutionary algorithm with memetic search for optic disc localization in retinal fundus images (In press)
28. Parihar, A.S.: Satellite image segmentation based on differential evolution. In: 2017 International Conference on Intelligent Sustainable Systems (ICISS) (2017). https://doi.org/10.1109/iss1.2017.8389245
29. Bezdek, J.C.: Pattern recognition with fuzzy objective function algorithms. In: Advanced Applications in Pattern Recognition book series (AAPR), pp. 43–93 (1981). https://doi.org/10.1007/978-1-4757-0450-1
30. Alva, A., Akash, R.S., Manikantan, K.: Optimal multilevel thresholding based on Tsallis entropy and half-life constant PSO for improved image segmentation. In: IEEE UP Section Conference on Electrical Computer and Electronics (2015). https://doi.org/10.1109/upcon.2015.7456685
31. Mirghasemi, S., Andreae, P., Zhang, M., Rayudu, R.: Severely noisy image segmentation via wavelet shrinkage using PSO and fuzzy C-means. In: IEEE Symposium Series on Computational Intelligence (2016). https://doi.org/10.1109/ssci.2016.7850051
32. Na, L., Yan, J., Shu, L.: Application of PSO algorithm with dynamic inertia weight in medical image thresholding segmentation. In: IEEE 19th International Conference on e-Health Networking, Applications and Services (Healthcom) (2017). https://doi.org/10.1109/healthcom.2017.8210769
33. Terzopoulos, D., Witkin, A., Kass, M.: Constraints on deformable models: recovering 3D shape and nonrigid motion. Artif. Intell. **36**(1), 91–123 (1988). https://doi.org/10.1016/0004-3702(88)90080-x
34. Cristhian, A.C., Sanchez, R., Baradit, E.: Detection of knots using X-ray tomographies and deformable contours with simulated annealing. Wood Res., 57–66 (2008)
35. Cristhian, A.C., Mario, A., Angel, D.: Simulated Annealing—A Novel Application of Image Processing in the Wood Area (2012). https://doi.org/10.5772/50635
36. Hussain, A.R.: Optic nerve head segmentation using Genetic Active Contours. In: International Conference on Computer and Communication Engineering, pp. 783–787 (2008). https://doi.org/10.1109/iccce.2008.4580712
37. Sierra, C.V, Novo, J., Santos, J., Penedo, M.G.: Frontiers in artificial intelligence and applications. In: Advances in Knowledge-Based and Intelligent Information and Engineering Systems, vol. 243, pp. 1380–1389 (2012)
38. Sierra, C.V, Novo, J., Santos, J., Penedo, M.G.: Emergent segmentation of topological active nets by means of evolutionary obtained artificial neural networks. In: Proceedings of the 5th International Conference on Agents and Artificial Intelligence (ICAART), vol. 2, pp. 44–50 (2013). https://doi.org/10.5220/0004195700440050

39. Mesejo, P., Valsecchi, A., Marrakchi-Kacem, L., Cagnoni, S., Damas, S.: Biomedical image segmentation using geometric deformable models and metaheuristics. Comput. Med. Imaging Graph. **43**, 167–178 (2015). https://doi.org/10.1016/j.compmedimag.2013.12.005

40. Everingham, M., Gool, L.V., Williams, C.K.I., Winn, J., Zisserman, A.: The Pascal visual object classes (VOC) challenge. Int. J. Comput. Vis. **88**(2), 303–338 (2010). https://doi.org/10.1007/s11263-009-0275-4

41. Hamid, O.A., Mohamed, A.R., Jiang, H., Deng, L., Penn, G., Yu, D.: Convolutional neural networks for speech recognition. IEEE/ACM Trans. Audio Speech Lang. Process. **22**(10), 1533–1545 (2015). https://doi.org/10.1109/taslp.2014.2339736

42. Lecun, Y., Bottou, L., Bengio, Y., Haffner.: Gradient-based learning applied to document recognition. Proc. IEEE **86**(11), 2278–2324 (1998). https://doi.org/10.1109/5.726791

43. Bardou, D., Zhang, K., Ahmed, S.M.: Classification of breast cancer based on histology images using convolutional neural networks. IEEE Access **6**, 24680–24693 (2018). https://doi.org/10.1109/access.2018.2831280

44. Yuan, Y.: Automatic skin lesion segmentation with fully convolutional-deconvolutional networks. In: IEEE Conference on Computer Vision and Pattern Recognition. IEEE, USA (2017). https://doi.org/10.1109/jbhi.2017.2787487

45. Shelhamer, E., Long, J., Darrell, T.: Fully convolutional networks for semantic segmentation. In: IEEE Conference on Computer Vision and Pattern Recognition (2015). https://doi.org/10.1109/cvpr.2015.7298965

46. Toshev, A., Szegedy, C.: DeepPose: human pose estimation via deep neural networks. In: IEEE Conference on Computer Vision and Pattern Recognition (2014). https://doi.org/10.1109/cvpr.2014.214

47. Zitnick, C.L, Dollar, P.: Edge boxes: locating object proposals from edges. In: European Conference on Computer Vision (2014). https://doi.org/10.1007/978-3-319-10602-1_26

48. Girshick, R., Donahue, J., Darrell, T., Malik, J.: Rich feature hierarchies for accurate object detection and semantic segmentation. In: IEEE Conference on Computer Vision and Pattern Recognition, IEEE, USA (2014). https://doi.org/10.1109/cvpr.2014.81

49. Ren, S., He, K., Girshick, R., Sun, J.: Faster R-CNN: towards real-time object detection with region proposal networks. IEEE Trans. Pattern Anal. Mach. Intell. **39**(6), 1137–1149 (2015). https://doi.org/10.1109/tpami.2016.2577031

50. Liu, W., Anguelov, D., Erhan, D., Szegedy, C., Reed, S., Fu, C.-Y., Berg, A.C.: SSD: single shot MultiBox detector. In: Lecture Notes in Computer Science, pp. 21–37 (2016). https://doi.org/10.1007/978-3-319-46448-0_2

51. Redmon, J., Divvala, S., Girshick, R., Farhadi, A.: You only look once: unified, real-time object detection. In: IEEE Conference on Computer Vision and Pattern Recognition (CVPR) (2016). https://doi.org/10.1109/cvpr.2016.91

52. Luo, S., Hu, Q., He, X., Li, J., Jin, J.S., Park, M.: Automatic liver parenchyma segmentation from abdominal CT images using support vector machines. In: ICME International Conference on Complex Medical Engineering (2009). https://doi.org/10.1109/iccme.2009.4906625

53. Wang, S., Fu, D., Xu, M., Hu, D.: Advanced fuzzy cellular neural network: application to CT liver images. Artif. Intell. Med. **39**(1), 65–77 (2007). https://doi.org/10.1016/j.artmed.2006.08.001

54. Teixeira, G.M., Pommeranzembaum, I.R., Oliveira, B.L.D., Lobosco, M., Santos, R.W.D.: Automatic segmentation of cardiac MRI using snakes and genetic algorithms. In: Bubak, M., Dongarra, J., VanAlbada, G.D., Sloot, P.M.A. (eds.), Computational Science, vol. 5103, pp. 168–177 (2008). https://doi.org/10.1007/978-3-540-69389-5_20

55. Hsu, C.-Y., Liu, C.-Y., Chen, C.-M.: Automatic segmentation of liver PET images, computerized medical imaging and graphics. Comput. Med. Imaging Graph. **32**(7), 601–610 (2008). https://doi.org/10.1016/j.compmedimag.2008.07.001

56. Shahamatnia, E., EbadzaSSdeh, M.M.: Application of particle swarm optimization and snake model hybrid on medical imaging. In: Proceeding of IEEE Third International Workshop on Computational Intelligence in Medical Imaging (2011). https://doi.org/10.1109/cimi.2011.5952043

57. Montiel, O.R.., Aguilar, C.M.A., López, S.C., Velasco, A.F.J., López, M.F.E., Pulido, F.L.: Images segmentation by using differential evolution with constraints handling. In: IEEE Latin American Conference on Computational Intelligence (LA-CCI) (2017). https://doi.org/10.1109/la-cci.2017.8285713

58. Liu, W., Deng, X., Shi, H.: Research on algorithm of PSO in image segmentation of cement-based. In: 7th International Conference on Cloud Computing and Big Data (CCBD) (2016). https://doi.org/10.1109/ccbd.2016.080

59. Mesejo, P., Ugolotti, R., Cunto, F.D., Giacobini, M., Cagnoni, S.: Automatic hippocampus localization in histological images using differential evolution-based deformable models. Pattern Recogn. Lett. **34**(3), 299–307 (2013). https://doi.org/10.1016/j.patrec.2012.10.012

60. Wang, Y.-Y., Sun, Y.-N., Lin, C.-C.K., Ju, M.-S.: Segmentation of nerve fibers using multi-level gradient watershed and fuzzy systems. Artif. Intell. Med. **54**(3), 189–200 (2012). https://doi.org/10.1016/j.artmed.2011.11.008

61. Mesejo, P., Ugolotti, R., Cagnoni, S., Cunto, F.D., Giacobini, M.: Automatic segmentation of hippocampus in histological images of mouse brains using deformable models and random forest. In: 25th IEEE International Symposium on Computer-Based Medical Systems (CBMS) (2012). https://doi.org/10.1109/cbms.2012.6266318

62. Ugolotti, R., Mesejo, P., Cagnoni, S., Giacobini, M., Cunto, F.D.: Automatic hippocampus localization in histological images using PSO-based deformable models. In: 13th Annual Conference Companion on Genetic and Evolutionary Computation (GECCO), pp. 487–494 (2011). https://doi.org/10.1145/2001858.2002038

63. Ugolotti, R., Nashed, Y.S.G., Mesejo, P., Ivekovic, S., Mussi, L., Cagnoni, S.: Particle swarm optimization and differential evolution for model-based object detection. Appl. Soft Comput. **13**(6), 3092–3105 (2013). https://doi.org/10.1016/j.asoc.2012.11.027

Air Quality Assessment Using Fuzzy Inference Systems

Seema A. Nihalani, Nandini Moondra, A. K. Khambete, R. A. Christian
and N. D. Jariwala

Abstract Air quality index (AQI) is a consistent pointer of the air quality for any locality. AQI may consider key air pollutants like sulphur dioxide, nitrogen dioxide, ground-level ozone, carbon monoxide and particulates. Several organisations worldwide calculate these indices, using different definitions. In India, Central Pollution Control Board, as well as State Pollution Control Boards, monitors the air quality level. A distinct sub-index is allocated to individual pollutant, and AQI is the maximum value of all these sub-indices. AQI is a numeral value that may be designated linguistically by stating that air quality is poor, fair and good [2]. When AQI upsurges, it is assumed that a substantial fraction of the residents is expected to face severe health effects. The present study is carried out to quantify air pollutant concentrations for four cities in Gujarat from 2011 to 2015. Conventional AQI is calculated by using a mathematical formula. Fuzzy logic system is used, and membership functions are fed as input to the Mamdani fuzzy inference system (FIS) to find out the fuzzy air quality index. Hence, the paper postulates a more consistent method for determining air quality index using the fuzzy logic system.

Keywords Air quality index · Fuzzy logic · Membership function · Rule editor

S. A. Nihalani (✉) · N. Moondra · A. K. Khambete · R. A. Christian · N. D. Jariwala
SVNIT, Surat, India
e-mail: seemanihalani@yahoo.com

N. Moondra
e-mail: nandini7808@yahoo.in

A. K. Khambete
e-mail: akk@ced.svnit.ac.in

R. A. Christian
e-mail: rac@ced.svnit.ac.in

N. D. Jariwala
e-mail: ndj@ced.svnit.ac.in

S. A. Nihalani
Parul University, Vadodara, India

© Springer Nature Singapore Pte Ltd. 2020
R. Venkata Rao and J. Taler (eds.), *Advanced Engineering Optimization
Through Intelligent Techniques*, Advances in Intelligent Systems and Computing 949,
https://doi.org/10.1007/978-981-13-8196-6_28

313

1 Introduction

Clean air by and large is considered as infinite and freely available natural resource. But at the present time, clean air cannot be taken for granted. Atmospheric pollution has seen a detrimental spin-off from human activities, apparently since first fire was lit by cavemen. The problem of air pollution is accelerated in magnitude with increasing urbanisation. Urbanisation, industrialisation and economic growth lead to substantial worsening of urban air quality. General air pollutants include total suspended particulates (TSP), respiratory suspended particulate matter (RSPM), sulphur dioxide (SO_2), nitrogen dioxide (NO_2) and carbon monoxide (CO), and particulates may include dust, smoke, pollen and other solid particles. Generally, these pollutants are present in low (background) concentrations naturally when they are principally not detrimental [3]. They become detrimental only when the concentration of these pollutants is substantially high as compared to the background value, and they begin to trigger off harmful effects [12].

AQI is an index utilised to narrate the complete environmental status based on the National Air Quality Standards. Individual AQI classification helps people to identify the amount of air pollution. Fuzzy logic is suitable to address real-life problems that generally contain a certain amount of ambiguity. Fuzzy logic utilises variants like low, medium or high instead of true or false and yes or no. Fuzzy sets are governed by input and output variables, membership functions and inference rules. The input variables are defined linguistically, and inference rules framed. The basic theories of fuzzy system are obtained from fuzzy logic given by Professor Zadeh. The two ordinarily used inference systems are Mamdani system and Sugeno system. Mamdani system assumes that the output membership functions shall be fuzzy data sets. After the combination process, each output variable has a fuzzy set that needs defuzzification. It includes converting the crisp values into linguistic terms of fuzzy sets. An individual linguistic term is assigned ranking using membership function. The most important input variables considered in this case are SO_2, NO_2 and PM_{10}. If the number of inputs is more, the number of rules is also more and hence the complexity rises.

2 Air Quality Index

AQI, as discussed earlier, is an index for recording everyday air quality. AQI emphasises the health effects that shall be witnessed by people within hours or days after inhaling polluted air. In the Indian context, AQI ranges from 0 to 500. The greater value of AQI implies a higher level of air pollution and the larger health alarms. The breakpoint concentrations for calculating AQI are shown in Table 1 [9].

Table 1 AQI values and breakpoint concentrations

AQI value	Category	SO_2	NO_2	CO	O_3	PM_{10}
0–100	Good	0–80	0–80	0–2	0–180	0–100
101–200	Moderate	81–367	81–100	2.1–12	180–225	101–150
201–300	Poor	368–786	181–564	12.1–17	225–300	151–350
301–400	Very poor	787–1572	565–1272	17.1–35	301–800	351–420
401–500	Severe	>1572	>1272	>35	>800	>420

2.1 Input Parameters

The input variables contributing to air quality differ extensively. Out of the various air pollution parametres, three parametres namely, SO_2, NO_2 and PM_{10} are considered for study purpose taking into account the combined effect of industries along with traffic movement for Ahmedabad, Vadodara, Surat, and Rajkot for five years from 2011 till 2015 [1]. The data is taken from Gujarat State Pollution Control Board website.

2.2 AQI Calculation

The formation of AQI should involve air quality data for a longer duration in addition to seasonal, diurnal and monthly meteorological parameters. Thus, air quality data of five years has been selected. AQI signifies overall quality trends in an improved manner as the combined results of all the pollutants and the associated standards can be taken into consideration. The result is an equation, which converts the input numerical values into a further simpler and accurate form. The sub-index of individual pollutant is obtained by physically measuring pollutants like SO_2, NO_2 and PM_{10}. For this study, AQI for each case is calculated by using a mathematical equation given below.

$$I_p = \frac{I_{Hi} - I_{Lo}}{BP_{Hi} - BP_{Lo}}(C_p - BP_{Lo}) + I_{Lo}$$

where

I_P sub-index for pollutant P
C_P measured concentration of pollutant P
BP_{Hi} higher value of breakpoint concentration of pollutant P
BP_{Lo} lower value of breakpoint concentration of pollutant P
I_{Hi} AQI value equivalent to BP_{Hi}
I_{Lo} AQI value equivalent to BP_{Lo}.

2.3 Air Quality in Ahmedabad, Rajkot, Surat and Vadodara

The SO_2, NO_2 and PM_{10} concentrations for Ahmedabad, Rajkot, Surat and Vadodara for the years 2011–2015 are shown in Figs. 1, 2, 3 and 4, respectively. Based on this data, the corresponding sub-index and AQI are shown in Table 2.

Fig. 1 Air quality in Ahmedabad

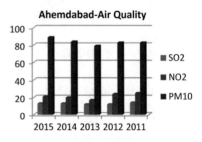

Fig. 2 Air quality in Rajkot

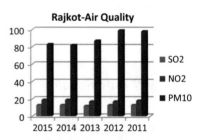

Fig. 3 Air quality in Surat

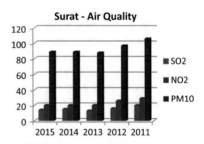

Fig. 4 Air quality in Vadodara

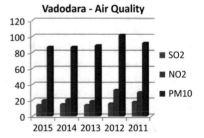

Table 2 Sub-indices and AQI values by equation

City	Year	SO_2	Ip for SO_2	NO_2	Ip for NO_2	PM_{10}	Ip for PM_{10}	AQI
Ahmedabad	2015	13	16.25	21	26.25	89	89	89
	2014	13	16.25	20	25	84	84	84
	2013	12	15	17	21.25	79	79	79
	2012	12	15	24	30	83	83	83
	2011	14	17.5	25	31.25	83	83	83
Rajkot	2015	13	16.25	19	23.75	83	83	83
	2014	13	16.25	19	23.75	82	82	82
	2013	12	15	17	21.25	87	87	87
	2012	13	16.25	17	21.25	99	99	99
	2011	13	16.25	18	22.5	98	98	98
Surat	2015	14	17.5	20	25	89	89	89
	2014	15	18.75	20	25	89	89	89
	2013	13	16.25	20	25	88	88	88
	2012	16	20	26	32.5	97	97	97
	2011	20	25	29	36.25	106	111.102	111.102
Vadodara	2015	14	17.5	20	25	87	87	87
	2014	15	18.75	21	26.25	87	87	87
	2013	14	17.5	19	23.75	89	89	89
	2012	16	20	33	41.25	102	103.02	103.02
	2011	18	22.5	30	37.5	92	92	92

3   Air Quality Analysis by Fuzzy Logic

The mathematical computation of air quality indices is operational; however, it has a severe shortcoming in that it does not take account of domain expert's know-how. AQI fuzzy logic control process consists of the following steps.

3.1   Defining Input and Output Variables' Input Parameters

In this case, we have considered three pollutants as input variables for determining AQI. The input variables are SO_2, NO_2 and PM_{10}, and the output variable is AQI. Linguistic variables are the input or output variants having values as words from English language, in place of numbers. A linguistic variable is usually disintegrated into a linguistic data set [4].

3.2 Membership Functions

Membership functions are utilised for fuzzification and defuzzification steps [5]. They relate the numerical input values (non-fuzzy) to fuzzy linguistic terms and vice versa. This includes utilising linguistic terms to convert crisp values into membership functions. The general membership functions used in FLS are trapezoidal, triangular, Gaussian, etc [8]. For our study, linguistic variables and membership functions mentioned below have been used as input and output variants.

(a) Linguistic variables for SO_2 are shown in Fig. 5.
(b) Linguistic variables for NO_2 are shown in Fig. 6.
(c) Linguistic variables for particulates (PM_{10}) are shown in Fig. 7.

3.3 Fuzzy Inference Rules

In fuzzy logic system, rules are formulated to regulate the output value. A fuzzy rule generally consists of a rule with an IF-THEN condition followed by a conclusion [10]. After fuzzification of input variables, the amount to which each antecedent is fulfilled for each rule is known. The assessment of the fuzzy inference rules and the amalgamation of the individual rules are implemented by applying fuzzy set operations [7]. After assessing each rule result, the results should be united to acquire

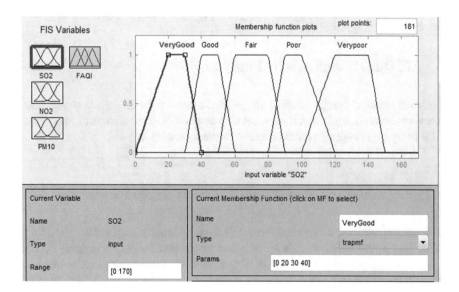

Fig. 5 Membership function for SO_2

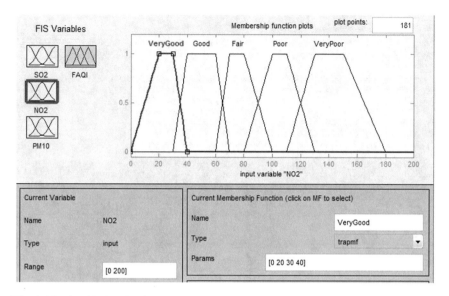

Fig. 6 Membership function for NO$_2$

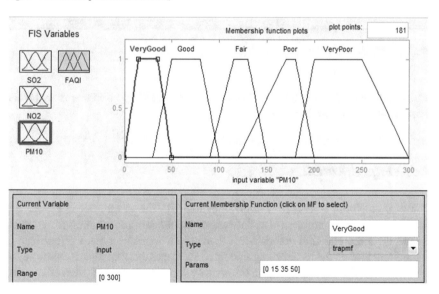

Fig. 7 Membership function for PM$_{10}$

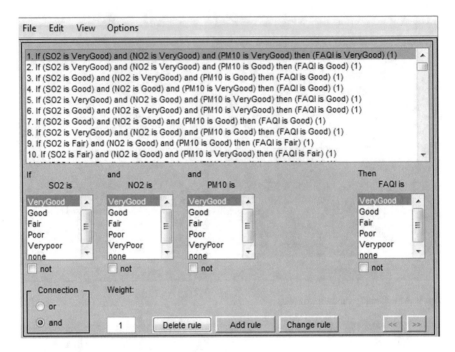

Fig. 8 Fuzzy inference rule

a final outcome. This process is known as inference. Fuzzy inference rules for the current case are shown in Fig. 8.

3.4 Defuzzification

The final step after inference is defuzzification [6]. The result after inference is a fuzzy value, which is required to be defuzzified for obtaining concluding crisp output. Hence, the requirement of defuzzifier is final output in crisp format. Defuzzification is accomplished on the basis of linguistic terms and membership functions employed for output variable. Defuzzification takes fuzzy set as the input and gives crisp value as the output [11]. Fuzzy logic supports the assessment of inference rules through the transitional steps, but the final result is typically a crisp output. Since fuzzy set in cumulation, includes an array of output values, defuzzification is necessary to get a single value from a data set. The defuzzified AQI and surface viewer are shown in Figs. 9 and 10.

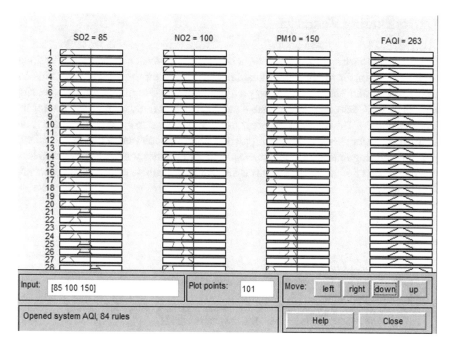

Fig. 9 Rule viewer for AQI

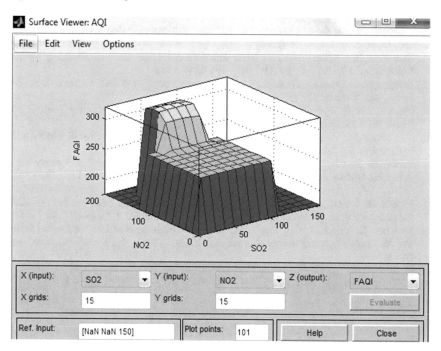

Fig. 10 Surface viewer for AQI

4 Concluding Remarks

The paper demonstrates the advantage of the fuzzy logic system over the conventional AQI method for air quality classification. One of the utmost reliable methods for policymakers and public at large to predict or study environmental trends is using various indices. Several countries are establishing their individual procedure for the AQI calculation. However, in the view of uncertainties surrounding air quality, the fuzzy logic system used in this paper can be a superior depiction of a dynamic system and thus offering entirely novel dimension to air quality monitoring. Nevertheless, the realisation of every fuzzy system is subject to the choice of domain experts and their observations.

References

1. Anand, A., Ankita, B.: Revised air quality standards for particle pollution and updates to the Air Quality Index, vol. 25, pp. 24–29 (2011)
2. Anand, K., Ashish, G., Upendu, P.: A study of ambient air quality status in Jaipur city (Rajasthan, India), using Air Quality Index. Nat. Sci. 9(6), 38–43 (2011)
3. Balashanmugan, P., Ramanathan, A., Elango, E., Nehru Kumar, V.: assessment of ambient air quality Chidambaram a South Indian Town. J. Eng. Sci. Technol. 7(3) (2012)
4. Daniel, D., Alexandra, A., Emil, L.: Fuzzy inference systems for estimation of Air Quality Index. Rom. Int. J. 7(2), 63–70 (2011)
5. Gopal, U., Nilesh, D.: Monitoring of air pollution by using fuzzy-logic. Int. J. Comput. Sci. Eng. IJCSE 2(7), 2282–2286 (2010)
6. Kumaravel, R., Vallinayagam, V.: Fuzzy inference system for air quality in using Matlab, Chennai, India. J. Environ. Res. Dev. 7(1), 181–184 (2012)
7. Lokeshappa, B., Kamath, G.: Feasibility analysis of air quality indices using fuzzy logic. Int. J. Eng. Res. Technol. 5(8) (2016)
8. Mckone, T.E., Deshpande, A.W.: Can fuzzy logic bring complex environmental problems into focus? IEEE 15, 364–368 (2010)
9. Nagendra, S.M., Venugopal, K., Jones, S.L.: Assessment of air quality near traffic intersections in Bangalore city using air quality indices. J. Elsevier, 1361–9209 (2007)
10. Pankaj, D., Suresh, J., Nilesh, D.: Fuzzy rule based meta graph model of Air Quality Index to suggest outdoor activities. Int. J. Comput. Sci. Eng. Technol. (IJCSET) 2(1), 2229–3345 (2010)
11. Saddek, B., Chahra, B., Chaouch Wafa, B., Souad, B.: Air Quality Index and public health: modelling using fuzzy inference system. Am. J. Environ. Eng. Sci. 1(4), 85–89 (2014)
12. Sadhana, C., Pragya, D., Ravindra, S., Anand, D.: Assessment of ambient air Quality Status and Air Quality Index of Bhopal city (Madhya Pradesh), India. Int. J. Curr. Sci. 9, 96–101 (2013)

Optimization of Water Releases from Ukai Reservoir Using Jaya Algorithm

Vijendra Kumar and S. M. Yadav

Abstract The scarcity of water resources is one of the most pervasive natural resource allocation problems faced by the water users and policymakers. Reservoir operation is the best solution to obtain its utmost possible performance. In the present study, the Jaya algorithm (JA) has been applied to optimize the water releases from Ukai reservoir at different dependable inflows. The model is optimized for four different dependable inflows namely 60, 65, 70, and 75%. The results from JA are compared with teaching–learning-based optimization (TLBO), particle swarm optimization (PSO), differential evolution (DE), and linear programming (LP). It was observed that JA performed better than TLBO, PSO, DE, and LP. The global optimum solution obtained using JA for 60, 65, 70, and 75% dependable inflow are 3224.620, 4023.200, 4672.800, and 5351.120, respectively in MCM. Based on the results, it is concluded that JA outperformed over TLBO, PSO, DE, and LP.

Keywords Optimization · Reservoir operation · Jaya algorithm · Ukai dam · Teaching–learning-based optimization

1 Introduction

Reservoir operations are complex engineering problem. With the growth in population, the demand for water in both consumptive and non-consumptive use has been increased drastically [1], resulting in increased competition and conflict among different water users. Additionally, due to the increasing demand, it is difficult to construct new water resources projects due to high investment and land acquisition problem [2]. Therefore, it is a big challenge for the policymaker is to address the

V. Kumar (✉) · S. M. Yadav
Civil Engineering Department, Sardar Vallabhbhai National Institute of Technology, Surat, Gujarat, India
e-mail: vij100000@gmail.com

S. M. Yadav
e-mail: shivnam27@gmail.com

© Springer Nature Singapore Pte Ltd. 2020
R. Venkata Rao and J. Taler (eds.), *Advanced Engineering Optimization Through Intelligent Techniques*, Advances in Intelligent Systems and Computing 949, https://doi.org/10.1007/978-981-13-8196-6_29

water management issues [3]. To overcome such problems, optimization would be an alternative solution for an existing projects. Often-substantial increases in benefits for relatively small improvements in operating policy. For decades, optimization techniques have been used to solve water resources problems [4].

Linear programming (LP), nonlinear programming (NLP), and dynamic programming (DP) have been broadly used to solve numerous water resource operation problems. These methods have its own advantages to solve the problems; however, it has its own drawbacks too [5]. Many times, it is observed that it stuck in the local optimal solution. To surmount the major drawbacks of conventional optimization techniques, various evolutionary algorithms (EA) have been developed. With recent advancements in computing power, EA has been a growing interest among researchers all over the world. Mostly because of its simplicity, flexibility, and gradient-free mechanism. Numerous researchers have applied different EA to solve complex water resources problems. For example, genetic algorithm (GA) [6], weed optimization algorithm [7], honey bee mating optimization [8], artificial neural network [9], cuckoo search algorithm [10], and shark algorithm [11] etc.

The existing evolutionary algorithms have few drawbacks. The main drawback, it is having internal parameters other than the common controlling parameters, i.e., a number of iterations and population size. For instance, GA needs tuning of reproduction parameters, mutation and crossover probability. Particle swarm optimization (PSO) requires cognitive and social parameters and inertia weight. Artificial bee colony (ABC) needs different numbers of bees such as scout bees, employee bees, and an onlooker's bees. Difference equation (DE) needs a crossover rate and scaling factor. These different internal parameters are called the algorithm-specific parameters. Similarly, the imperialist competitive algorithm (ICA), improve bat algorithm (IBA), weed optimization algorithm (WOA), krill herd algorithm, ant colony optimization (ACO), etc., used in reservoir operations also have internal parameters that is required to be tuned before executing the algorithm. If proper tuning is not done, it influences the overall performance of the algorithm and may result in a local optimum. In consideration of the above-stated problems, it becomes essential to look for algorithm free from internal parameters.

Jaya algorithm (JA) is a newly developed novel approach developed by Rao [12]. JA does not require any internal parameters. It works on the basic principle that it must move toward the best optimum solution and avoid the worst solution [13]. Various researchers have used JA to solve different optimization problems, such as micro-channel heat sink [14], traffic light scheduling problem [15], maximum power point tracking [16], facial emotion recognition [17], and multi-reservoir operation [18].

In the present study, JA has been used to optimize the reservoir operation to maximize the water allocation for various demand from Ukai reservoir at various dependable inflows. The results are compared with PSO, DE, TLBO and LP. The following section describes the methodology.

2 Methodology

2.1 Particle Swarm Optimization (PSO)

PSO is a swarm intelligence-based optimization technique proposed by Eberhart and Kennedy [19]. It works based on natural grouping and resembles to bird flocking. It is a population-based algorithm and similar to the genetic algorithm. It provides the optimal solution based on the individual and social behavior [20]. Following are the steps to run the PSO algorithm.

Step 1: Initialize a group of random particles. For each particle, generate positions X_i and velocity v_i.
Step 2: Evaluate the fitness and obtained Pbest_i, i.e.,

$$\text{Pbest}_i = \begin{cases} \text{Pbest}_i(k-1), & \text{if } F(\text{Pbest}_i(k)) \leq F(\text{Pbest}_i(k-1)) \\ \text{Pbest}_i(k), & \text{if } F(\text{Pbest}_i(k)) > F(\text{Pbest}_i(k-1)) \end{cases} \tag{1}$$

where Pbest_i represents the best positions of the individual, K represents the iteration $K = 1, 2, 3 \ldots k$.

Step 3: Find the global best position from the swarm, i.e.,

$$\text{Gbest}_i(K) = \max\{F(\text{Gbest}_1(k), F(\text{Gbest}_2(k), \ldots, F(\text{Gbest}_m(k))\} \tag{2}$$

Step 4: Update the velocity of each individual using Eq. (3),

if $v_i > v_{\max}$ then $v_i = v_{\max}$; if $v_i < v_{\min}$ then $v_i = v_{\min}$.

$$v_i(k+1) = w(k) * v_i(k) + c_1 * r_1(k) * (\text{Pbest}_i(k) - X_i(k)) \\ + c_2 * r_2(k) * (\text{Gbest}_i(k) - X_i(k)) \tag{3}$$

where v_i represent the velocity of ith individual, X_i is the position of ith individual, $w(k), c_1$ and c_2 are the internal parameters, term $w(k)$ is internal weight, $r_1(k)$ and $r_2(k)$ are the random numbers, c_1 and c_2 are the cognitive and social parameters.

Step 5: Update the position of each individual using Eq. (4).

$$X_i(k+1) = X_i(k) + v_i * (k+1) \tag{4}$$

Step 6: Once the maximum iterations are completed, it stops with an optimum solution else it reiterates from step 2.

2.2 Differential Evolution (DE)

DE is a heuristic technique used in the optimization. Storn and Price [21] introduced DE as an effective method for optimizing multi-model objective problems. Storn and Price [22] presented DE and tested with minimizing continuous space problems. The results were compared with adaptive simulated annealing (ASA), GA, annealed Nelder and Mead approach (ANM), and stochastic differential equations (SDE). The result shows that DE outperformed. The additional details for DE can be found in [22, 23]. Following are the steps to run DE.

Step 1: Decide the common controlling parameters and internal parameters.
Step 2: Generate the initial vector solutions randomly, i.e., $X_{i,G}^j = \{x_{i,G}^1, x_{i,G}^2, \ldots, x_{i,G}^D\}$. For a generation $G = 0$, it can be generated as per Eq. (5).

$$x_{i,0}^j = x_{\min}^j + R_{i,j} * \left(x_{\max}^j - x_{\min}^j\right) \tag{5}$$

where $x_{i,j,0}$ is the initial population at generation $G = 0$ at ith component of the jth vector, $i = 1, 2, \ldots, NP$, NP is the population size, $j = 1, 2, \ldots, D$, D is a dimensional parameter vector, $R_{i,j} = $ random number between[0, 1], $x_{\max}^j$ and $x_{\min}^j$ are the upper and lower bound.

Step 3: Next is the mutation operation to produce a mutation vector $V_{i,G}^j = \{V_{i,G}^1, V_{i,G}^2, \ldots, V_{i,G}^D\}$ for each target vector $X_{i,G}^j$.

$$V_{i,G}^j = X_{r_1^i,G} + F * \left(X_{r_2^i,G} - X_{r_3^i,G}\right) \tag{6}$$

where r_1^i, r_2^i, and r_3^i, are the randomly generated integers between [1, NP], F is the scaling factor varies between [0, 2].

Step 4: Next crossover operation is applied to each target vector $X_{i,G}^j$ and its corresponding mutation vector $V_{i,G}^j$ to obtain a trail vector $U_{i,G}^j = \{U_{i,G}^1, U_{i,G}^2, \ldots, U_{i,G}^D\}$. The basic binomial crossover is defined as Eq. (7).

$$U_{i,G}^j = \begin{cases} V_{i,G}^j, & \text{if (rand[0, 1]} \leq \text{CR) or } (j = j_{\text{rand}}) \\ X_{i,G}^j, & \text{otherwise} \end{cases} \tag{7}$$

where CR is the crossover rate within the range of [0, 1], j_{rand} is the randomly generated integers between [1, NP].

Step 5: Selection operation as per Eq. (8).

$$X_{i,G+1}^j = \begin{cases} U_{i,G}^j, & \text{if } f\left(U_{i,G}^j\right) \leq f\left(X_{i,G}^j\right) \\ X_{i,G}^j, & \text{otherwise} \end{cases} \tag{8}$$

Step 6: Increase the generation to $G = G + 1$. Once the maximum iterations are completed, it stops with an optimum solution else it reiterates.

2.3 Teaching–Learning-Based Optimization

TLBO is a metaheuristic optimization technique proposed by Rao et al. [24]. Its working principle is similar as a classroom, where teachers teach and students learn. In order to obtain good results, a strong bond between the teachers and the students is required. Apart, students try to interact with each other and further improve their results. Based on this teaching and learning concept, TLBO was developed. Following are the steps of TLBO algorithm.

Step 1: Decide the common controlling parameters.
Step 2: Generate the initial solutions randomly between the upper and lower bounds using the formula presented in Eq. (9).

$$\text{Randomly Generated Population} = [L + r * (U - L)] \tag{9}$$

where U is upper bound of the variables and L is lower bound of the variables, r is a random number generated between [0, 1].

Step 3: Identify the best solution from the randomly generated population.
Step 4: Next is the teaching phase, where the teacher tries to improve the results, using Eqs. (10) and (11).

$$\text{Difference mean} = r * (X_{\text{best}} - X_{\text{mean}}) \tag{10}$$

$$X_{\text{new}} = (X_{\text{old}} + \text{Difference mean}) \tag{11}$$

where $X_{\text{best}} = $ best solution(teacher), $X_{\text{mean}} = $ mean of all the students, $X_{\text{new}} = $ modified solution, $X_{\text{old}} = $ old solution. At the end of teaching phase, the new and old solutions are compared, and better solution is selected.

Step 5: Learning phase: In the second phase, any two random learners (i.e., X_i and X_j, where $i \neq j$), interact with each other to increase their knowledge. Here, two conditions can occur.

$$\text{if } f(X_i) < f(X_j); \quad \text{Then, } X_{\text{new},i} = X_{\text{old},i} + r_i * (X_j - X_i) \tag{12}$$

$$\text{if } f(X_i) > f(X_j); \quad \text{Then, } X_{\text{new},i} = X_{\text{old},i} + r_i * (X_i - X_j) \tag{13}$$

Step 6: Once the maximum iterations are completed, it stops with an optimum solution else it reiterates from the teaching phase. The flowchart of TLBO can be referred from [24].

2.4 Jaya Algorithm

JA works on the principle to get closer to the better solution by avoiding the failure. Following are the steps of Jaya algorithm

Step 1: Decide the common controlling parameters.
Step 2: Generate initial solutions between the upper and lower limits using Eq. (9).
Step 3: Identify the best solution and the worst solution.
Step4: Modified the solution as per Eq. (14).

$$X_{\text{new},j,k,i} = X_{j,k,i} + r_{1,j,i} * \left(X_{j,\text{best},i} - \left| X_{j,k,i} \right| \right) - r_{2,j,i} * \left(X_{j,\text{worst},i} - \left| X_{j,k,i} \right| \right)$$
$$(14)$$

where $X_{\text{new},j,k,i}$ is the new updated variable. The term "$r_{1,j,i}\left(X_{j,\text{best},i} - \left| X_{j,k,i} \right| \right)$" aids the solution to move faster toward the best solution and the term "$r_{2,j,i}\left(X_{j,\text{worst},i} - \left| X_{j,k,i} \right| \right)$" aids the solution to avoid the worst solution. $X_{j,\text{worst},i}$ and $X_{j,\text{best},i}$ are the worst and best solutions, respectively. $r_{1,j,i}$ and $r_{2,j,i}$ are random numbers generated between [0, 1].

Step 5: The better functional value is accepted obtained from step 3 and 4, and it becomes the input for the next loop. The process runs until the maximum iterations are reached. The flow chart of JA can be referred from [12].

The coding was done using MATLAB R2014b for JA, TLBO, PSO, and DE, and Lingo17 software was used for the LP model. The following section describes the study area and data collection.

3   Study Area and Data Collection

The Tapi River is the west flowing interstate river of India. It covers a vast area of Maharashtra state and other parts of Madhya Pradesh and Gujarat state. The total length of this west flowing river is 724 km. It originates from Multai in Betul district at an elevation of 752 m above MSL. The Tapi River drains an area of 65,145 km^2 out of which nearly 80% (51,504 km^2) lies in Maharashtra state, 9804 km^2 in Madhya Pradesh, and 3837 km^2 in Gujarat. The Ukai Dam is one of the biggest dams of Gujarat, having a capacity of 8510 million cubic meters (MCM). Out of which 7079.19 MCM as live storage and 1430.82 MCM as dead

Table 1 Salient features of Ukai reservoir

General item	Details
Mean annual rainfall in the watershed	785 mm
Maximum annual rainfall in the watershed	1191 mm
Catchment area	62,225 km^2
Top of dam	111.252 m
Area	65,145 km^2
State in the basin	Maharashtra, Madhya Pradesh, and Gujarat
Road width of the spillway	6.706 m
Type of spillway	Radial

storage. The dam is constructed across the Tapi River in 1972 and having its coordinates as 21° 14′ 53.52″N and 73° 35′ 21.84″E. The main purpose of the dam is for irrigation, power generation, and flood control. Table 1 shows the salient features of the Ukai reservoir. Figure 1 shows the index map of the study area prepared using ArcGis. Figure 1a represents the India map showing Tapi basin. Figure 1b represents the Tapi catchment area. Figure 1c shows the location of the dam and downstream of Ukai Dam; the irrigation water is delivered using Ukai left bank main canal, Kakrapar left bank main canal, Kakrapar right bank main canal, and Ukai right bank main canal, wherein it is represented using a different color combination. Major city downstream of Ukai dam is Surat, having a population of approximately 6.6 million. The domestic and industrial water is supplied from Ukai reservoir. Figure 2 depicts the line diagram of Ukai dam, wherein it shows the allocation of water for different uses and losses in the form of evaporation.

The required data had been collected from Surat Irrigation Circle (SIC) and Ukai Left Bank Division (ULBD) which is converted in useful form to carry out this study. The major data required for the analysis were water releases for irrigation, domestic and industrial purpose, storage and inflow of the reservoir, and evaporation losses.

4 Mathematical Models

The mathematical formulation of the objective function and constraints are as follows.

4.1 Objective Function

The objective function of the present study is to maximize the water release over a year for Ukai reservoir, and it is mathematically represented in Eq. (15).

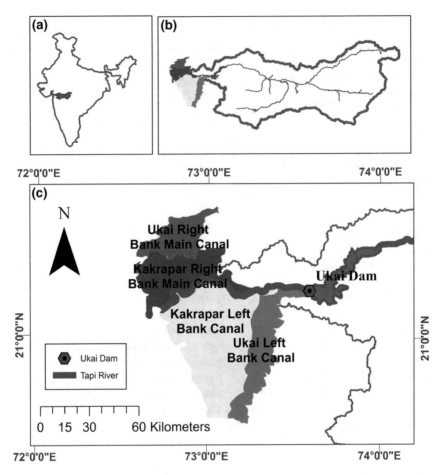

Fig. 1 Index map **a** India map showing Tapi basin, **b** Tapi basin, and **c** Cultivated land from Ukai reservoir

$$\text{Max } Z = \sum_{t=1}^{12} \text{RI} + \sum_{t=1}^{12} \text{RID} + \sum_{t=1}^{12} \text{RIND} \qquad (15)$$

where $Z =$ total releases from Ukai reservoir, including releases for irrigation, domestic purpose and industrial purpose. RI $=$ releases for irrigation, RID $=$ releases for domestic purpose, RIND $=$ releases for industrial purpose, $t =$ index for time period ($t = 1, 2, 3\ldots, 12$). In order to satisfy the reservoir storage constraint, the penalty is applied to the reservoir. Penalty parameter $g(m) = 5$, $p_{(m,t)(c)}$ is the penalty function, and c is the constraint. The penalty functions are expressed in Eqs. (16) and (17). The modified objective function is expressed as Eq. (18).

$$\text{If } f\left(p_{(m,t)(c)}\right) == 0, \text{ then penalty function} = 0 \qquad (16)$$

Fig. 2 Line diagram of Ukai Dam

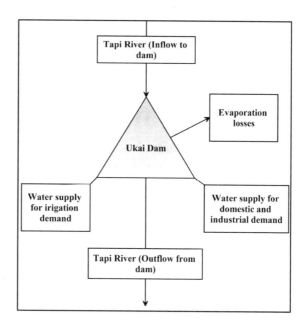

$$\text{elseif } f\left(p_{(m,t)(c)}\right) \neq 0, \text{ then penalty function} = \text{abs}\left(f\left(p_{(m,t)(c)}\right)\right)^2 \quad (17)$$

$$\text{Max } Z = \sum_{t=1}^{12} \text{RI} + \sum_{t=1}^{12} \text{RID} + \sum_{t=1}^{12} \text{RIND} - g(m) * \text{penalty function} \quad (18)$$

The above objective function is subjected to the following constraints.

4.2 Constraints

Irrigation demand constraint:

The releases for irrigation should be greater than or equal to minimum releases required for irrigation and less than or equal to maximum releases required for irrigation, for that particular time period.

$$\text{ID}_{t,\text{Max}} \geq \text{RI}_t \geq \text{ID}_{t,\text{Min}} \quad (19)$$

where $\text{ID}_{t,\text{Max}}$ = maximum irrigation demand for time period t. RI_t = release for irrigation for time period t. $\text{ID}_{t,\text{Min}}$ = minimum releases required for irrigation.

Domestic demand constraint:

The releases for domestic demand should be greater than or equal to minimum domestic demand and less than or equal to maximum domestic demand, for that particular time period.

$$DD_{t,\text{Max}} \geq RID_t \geq DD_{t,\text{Min}} \tag{20}$$

where $DD_{t,\text{Max}}$ = maximum domestic demand for time period t. RID_t = release for domestic for time period t. $DD_{t,\text{Min}}$ = minimum releases required for domestic use.

Industrial demand constraint:

The releases for industrial demand should be greater than or equal to minimum industrial demand and less than or equal to maximum industrial demand, for that particular time period.

$$INDD_{t,\text{Max}} \geq RIND_t \geq INDD_{t,\text{Min}} \tag{21}$$

where $INDD_{t,\text{Max}}$ = maximum industrial demand for time period t. $RIND_t$ = release for industrial for time period t. $INDD_{t,\text{Min}}$ = minimum releases required for industrial use.

Storage capacity constraint:

The reservoir storage should not be more than the maximum capacity of the reservoir and should not be less than the capacity at which hydropower generated, in any time period.

$$S_{t,\text{Max}} \geq S_t \geq S_{t,\text{Min}} \tag{22}$$

where $S_{t,\text{Max}}$ = maximum storage capacity of the reservoir in million cubic meter (MCM), $S_{t,\text{Min}}$ = minimum storage up to which hydropower generated in MCM.

Reservoir storage continuity constraint:

The constraint related to the continuity of storage, inflow, release, and evaporation of the reservoir at the different time period.

$$S_t + I_t - R_t - E_t = S_{t+1} \tag{23}$$

where S_t = active storage during period t in MCM, S_{t+1} = storage during $(t + 1)$, I_t = inflow during period t in MCM, R_t = release during period t in MCM, $R_t = RI + RID + RIND$, E_t = evaporation loss during period t in MCM.

Non-negativity constraint:

$$R_t > 0, RI > 0, RID > 0, RIND > 0 \tag{24}$$

5 Results and Discussions

The present study was carried out to maximize the releases from Ukai reservoir, at various dependable inflows. The dependable inflows of 60, 65, 70, and 75% were prepared using the Weibull method [25]. To maximize the release, Jaya algorithm (JA) was used and the results were compared with TLBO, PSO, DE, and LP.

Different population sizes were used in JA, TLBO, PSO, and DE, i.e., 10, 20, 40, 60, 80, 100 and 150, and the functional evaluation is taken as 1,000,000 for a fair comparison. No other internal parameters were required in JA and TLBO. In DE, the upper and lower scaling factors were taken as 0.8 and 0.2, respectively, and the crossover probability was taken as 0.2. In PSO, the inertia weight was taken as 1, the cognitive c_1 and social parameters c_2 were taken as 1.5 and 2.0, respectively, and the inertia weight damping ratio was taken as 0.99.

The results obtained using JA, TLBO, PSO, and DE were obtained over 15 independent runs, and Table 2 shows the best optimal solution obtained using different algorithms. It was observed that JA performed better as compared to other methods. The global optimum solution obtained using JA for 60, 65, 70, and 75% dependable inflow were 3224.620, 4023.200, 4672.800, and 5351.120, respectively in MCM. TLBO and PSO performed little less than the JA. The absolute solution obtained using LP for 60, 65, 70, and 75% were 3223.920, 4022.579, 4672.170, and 5350.420, respectively in MCM. DE and LP results were almost similar. Figure 3 shows the convergence curve of all the models. It was found that the JA convergence rate was faster as compared with other algorithms. Note, for clear view only iteration up to 500 have been shown.

For JA and DE, the best-optimized solutions were obtained at population size 10. And, for TLBO and PSO, the best solutions were obtained at population size 50, and the same results were obtained at population sizes 100 and 150.

The actual release of water from the dam is compared with the computed value of release from JA at different dependability level. Figure 4 represents the comparison between the actual and computed release using JA at different dependable inflow, during time period t. It can be observed that at 60% dependable inflow there is much more variation in comparison with the actual releases, but at 75% inflow, there is less variation as compared to actual releases of water.

Table 2 Computed maximum release from the Ukai reservoir at various dependable inflows

Algorithm	Maximum release in MCM			
	60%	65%	70%	75%
JA	**3224.620**	**4023.200**	**4672.800**	**5351.120**
TLBO	3224.520	4023.190	4672.770	5351.020
PSO	3224.520	4023.190	4672.770	5351.020
DE	3219.339	4022.210	4672.160	5350.420
LP	3223.920	4022.579	4672.170	5350.420

Bold values show the optimal solution

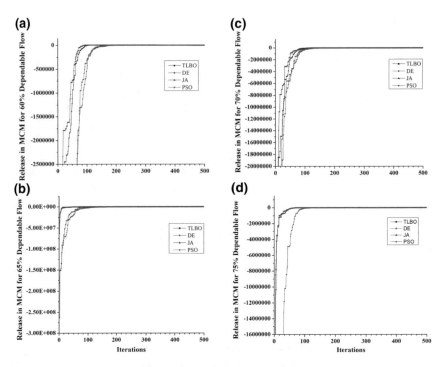

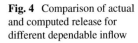

Fig. 3 Convergence curve **a** 60% **b** 65% **c** 70% **d** 75% dependable inflow

Fig. 4 Comparison of actual
and computed release for
different dependable inflow

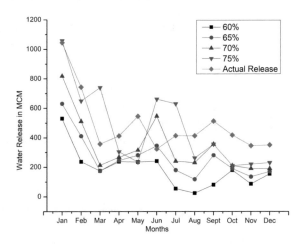

6   Conclusion

The present study was performed to obtain optimal reservoir operation to release the water from the Ukai dam for domestic, irrigation, and industrial purpose. Four different models have been developed, using different dependable inflow, i.e., 60, 65, 70, and 75%. To optimize the problem, Jaya algorithm has been used. The merit of the algorithm is that it does not require any algorithm-specific parameters and only required the common controlling parameters. The results from JA were compared with the TLBO, PSO, DE, and LP. It was found that JA outperformed over TLBO, PSO, DE, and LP. It is concluded that at 75% dependable flow variation to the actual release is less in comparison with other dependable flow. Thus, it is recommended to operate the reservoir as per 75% dependable inflow level model.

Acknowledgements   The authors like to thank Dr. R. V. Rao, Professor, Mechanical Engineering Department, SVNIT, and Mr. Ankit Saroj Research Scholar, SVNIT, for helping and guiding. The authors would also like to acknowledge with deep sense of gratitude the valuable help received from the authorities of Sardar Vallabhbhai National Institute of Technology (SVNIT) and the P. G. Section of Water Resources Engineering.

References

1. Mansouri, R., Torabi, H., Hoseini, M., Morshedzadeh, H.: Optimization of the water distribution networks with differential evolution (DE) and mixed integer linear programming (MILP). J. Water Resour. Prot. **07**, 715–729 (2015). https://doi.org/10.4236/jwarp.2015.79059
2. Adeyemo, J., Otieno, F.: Differential evolution algorithm for solving multi-objective crop planning model. Agric. Water Manag. **97**, 848–856 (2010). https://doi.org/10.1016/j.agwat.2010.01.013
3. Torabi Pudeh, H., Mansouri, R., Haghiabi, A.H., Yonesi, H.A.: Optimization of hydraulic-hydrologic complex system of reservoirs and connecting tunnel. Water Resour. Manag. **30**, 5177–5191 (2016). https://doi.org/10.1007/s11269-016-1477-5
4. Zeng, X., Kang, S., Li, F., Zhang, L., Guo, P.: Fuzzy multi-objective linear programming applying to crop area planning. Agric. Water Manag. **98**, 134–142 (2010). https://doi.org/10.1016/j.agwat.2010.08.010
5. Hosseini-Moghari, S.M., Morovati, R., Moghadas, M., Araghinejad, S.: Optimum operation of reservoir using two evolutionary algorithms: imperialist competitive algorithm (ICA) and cuckoo optimization algorithm (COA). Water Resour. Manag. **29**, 3749–3769 (2015). https://doi.org/10.1007/s11269-015-1027-6
6. Ravansalar, M., Rajaee, T., Kisi, O.: Wavelet-linear genetic programming: a new approach for modeling monthly streamflow. J. Hydrol. **549**, 461–475 (2017). https://doi.org/10.1016/j.jhydrol.2017.04.018
7. Azizipour, M., Ghalenoei, V., Afshar, M.H., Solis, S.S.: Optimal operation of hydropower reservoir systems using weed optimization algorithm. Water Resour. Manag. **30**, 3995–4009 (2016). https://doi.org/10.1007/s11269-016-1407-6
8. Afshar, A., Bozorg Haddad, O., Mariño, M.A.A., Adams, B.J.J.: Honey-bee mating optimization (HBMO) algorithm for optimal reservoir operation. J. Franklin Inst. **344**, 452–462 (2007). https://doi.org/10.1016/j.jfranklin.2006.06.001

9. Shamim, M.A., Hassan, M., Ahmad, S., Zeeshan, M.: A comparison of artificial neural networks (ANN) and local linear regression (LLR) techniques for predicting monthly reservoir levels. KSCE J. Civ. Eng. **20**, 971–977 (2016). https://doi.org/10.1007/s12205-015-0298-z

10. Ming, B., Chang, J.X., Huang, Q., Wang, Y.M., Huang, S.Z.: Optimal operation of multi-reservoir system based-on cuckoo search algorithm. Water Resour. Manag. **29**, 5671–5687 (2015). https://doi.org/10.1007/s11269-015-1140-6

11. Ehteram, M., Karami, H., Mousavi, S.F., El-Shafie, A., Amini, Z.: Optimizing dam and reservoirs operation based model utilizing shark algorithm approach. Knowl.-Based Syst. **122**, 26–38 (2017). https://doi.org/10.1016/j.knosys.2017.01.026

12. Venkata Rao, R.: Jaya: a simple and new optimization algorithm for solving constrained and unconstrained optimization problems. Int. J. Ind. Eng. Comput. **7**, 19–34 (2016). https://doi.org/10.5267/j.ijiec.2015.8.004

13. Rao, R.V., Waghmare, G.G.: A new optimization algorithm for solving complex constrained design optimization problems. Eng. Optim. **0273**, 1–24 (2016). https://doi.org/10.1080/0305215x.2016.1164855

14. Rao, R.V., More, K.C., Taler, J., Oclon, P.: Dimensional optimization of a micro-channel heat sink using Jaya algorithm. Appl. Therm. Eng. **103**, 572–582 (2016). https://doi.org/10.1016/j.applthermaleng.2016.04.135

15. Gao, K., Zhang, Y., Sadollah, A., Lentzakis, A., Su, R.: Jaya, harmony search and water cycle algorithms for solving large-scale real-life urban traffic light scheduling problem. Swarm Evol. Comput. **37**, 58–72 (2017). https://doi.org/10.1016/j.swevo.2017.05.002

16. Huang, C., Wang, L., Yeung, R.S.-C., Zhang, Z., Chung, H.S.-H., Bensoussan, A.: A prediction model-guided Jaya algorithm for the PV system maximum power point tracking. IEEE Trans. Sustain. Energy **9**, 45–55 (2018). https://doi.org/10.1109/TSTE.2017.2714705

17. Wang, S.-H., Phillips, P., Dong, Z.-C., Zhang, Y.-D.: Intelligent facial emotion recognition based on stationary wavelet entropy and Jaya algorithm. Neurocomputing **272**, 668–676 (2018). https://doi.org/10.1016/j.neucom.2017.08.015

18. Kumar, V., Yadav, S.M.: Optimization of reservoir operation with a new approach in evolutionary computation using TLBO algorithm and Jaya algorithm. Water Resour. Manag. **32**, 4375–4391 (2018). https://doi.org/10.1007/s11269-018-2067-5

19. Eberhart, R., Kennedy, J.: A new optimizer using particle swarm theory. In: Proceedings of the Sixth International Symposium on Micro Machine and Human Science. IEEE, pp. 39–43 (1995)

20. Baltar, A.M., Fontane, D.G.: Use of multiobjective particle swarm optimization in water resources management. J Water Resour. Plan Manag. **134**, 257–265 (2008). https://doi.org/10.1061/(ASCE)0733-9496(2008)134:3(257)

21. Storn, R., Price, K.: Minimizing the real functions of the ICEC'96 contest by differential evolution. In: Proceedings of IEEE International Conference on Evolutionary Computation. IEEE, pp. 842–844 (1996)

22. Storn, R., Price, K.: Differential evolution—a simple and efficient heuristic for global optimization over continuous spaces. J Glob. Optim. **11**, 341–359 (1997). https://doi.org/10.1023/A:1008202821328

23. Qin, A.K., Huang, V.L., Suganthan, P.N.: Differential evolution algorithm with strategy adaptation for global numerical optimization. IEEE Trans. Evol. Comput. **13**, 398–417 (2009). https://doi.org/10.1109/TEVC.2008.927706

24. Rao, R.V., Savsani, V.J., Vakharia, D.P.: Teaching–learning-based optimization: a novel method for constrained mechanical design optimization problems. Comput. Des. **43**, 303–315 (2011). https://doi.org/10.1016/j.cad.2010.12.015

25. Subramanya, K.: Engineering hydrology. Tata McGraw-Hill Educ, pp. 45–60 (2013)

Design of Antenna Arrays Using Chaotic Jaya Algorithm

R. Jaya Lakshmi and T. Mary Neebha

Abstract Antenna arrays are commonly used in communication systems to enhance desired radiation properties. The optimum design of antenna arrays is an important problem as it greatly affects the efficiency of the communication system. In this work, a variation of Jaya algorithm, i.e., Chaotic Jaya, is used to design planar antenna arrays to get minimized side lobe levels (SLL) thus improving the directivity. Two design examples of planar antenna arrays (concentric circular and hexagonal) are formulated by considering different inter-element spacings. Comparison of results between Chaotic Jaya algorithm and few other algorithms is done to demonstrate the algorithm's efficiency. The Chaotic Jaya algorithm was able to get much-reduced side lobe levels.

Keywords Optimization · Planar antenna array synthesis · Chaotic Jaya algorithm

1 Introduction

Antenna arrays find widespread applications in communications such as mobile, television transmissions in UHF and VHF frequency ranges, satellite, sonar, radar, etc. Antenna arrays are used commonly in satellite, mobile and radar communications systems to have improved performance in terms of directivity, signal-to-noise ratio (SNR), spectrum efficiency, etc. Therefore, the proper design of antenna arrays is a crucial factor in deciding the system's performance. Linear antenna arrays do not give efficient performance in all azimuth directions, despite having high directivity, whereas a circular array can perform all azimuth scan around its center. Hence,

R. Jaya Lakshmi (✉)
Department of Information Engineering and Computer Science, University of Trento, 381243 Trento, Italy
e-mail: jayalakshmi.ravipudi@studenti.unitn.it

T. Mary Neebha
Karunya Institute of Technology and Sciences, Coimbatore, India
e-mail: maryneebha@karunya.edu

© Springer Nature Singapore Pte Ltd. 2020
R. Venkata Rao and J. Taler (eds.), *Advanced Engineering Optimization Through Intelligent Techniques*, Advances in Intelligent Systems and Computing 949, https://doi.org/10.1007/978-981-13-8196-6_30

concentric circular antenna arrays (CCA) find applications in direction of arrival (DOA). Uniform circular arrays (UCCA) are the ones which have inter-element spacing of about half a wavelength.

An antenna array's radiation pattern is dependent on the array structure, number of ON and OFF elements, inter-element spacing and individual elements' amplitude and phase excitation, etc. One method of suppressing side lobe levels is array thinning. Thinning method for linear arrays is reported in the literature [7, 8] and can be very well used for circular and hexagonal arrays. Thinning means, at a time, only few elements of the array are ON condition while others are OFF. ON means the element is provided with feed, and OFF means either the element is not provided with feed or is connected to a dummy or matched load. This helps to reduce cost by decreasing the number of elements required and power consumption.

Design of arrays with large number of elements using thinning is a difficult task as solving the functions require efficient techniques. In recent years, researchers have applied various algorithms to get efficient design of antenna arrays, such as enhanced particle swarm optimization (PSO) [9], modified PSO [10], ant colony optimization (ACO), genetic algorithm (GA), differential evolution (DE), flower pollination algorithm and cuckoo optimization algorithm [13]. For concentric circular arrays, algorithms such as PSO, improved particle swarm optimization (IPSO), gravitational search algorithm and modified PSO [4], comprehensive learning PSO [5], decomposition-based evolutionary multi-objective optimization approach [3] and backtracking search algorithm [6] have been used. However, the above-mentioned optimization algorithms have parameters that are specific to them, i.e., those parameters are also required to be tuned apart from the common parameters. Keeping this in view, Rao [11] proposed Jaya algorithm that excludes the effort of tuning any specific parameters of the algorithm and requires the user to just tune the common parameters such as size of population and no. of generations. The algorithm is simple in implementation along with giving efficient results. The detailed working and the variants of Jaya algorithm are discussed in Rao [12].

The configuration of the concentric circular and hexagonal antenna arrays is presented in Sect. 2. A description of Jaya and Chaotic Jaya algorithm is given in Sect. 3. Section 4 defines the objective function used, and Sect. 5 includes the synthesis of concentric circular and hexagonal antenna array. Conclusions are given in Sect. 6.

2 Design Equations

2.1 Concentric Circular Antenna Array

The configuration of uniform two-ring concentric circular array with Np elements along the circle with radius r_p ($p = 1, 2$) is given in Fig. 1 [1]. Inter-element spacing is d_p which is half the operational wavelength. The radius of pth circle is given by

Fig. 1 Configuration of uniform two-ring concentric circular array antenna

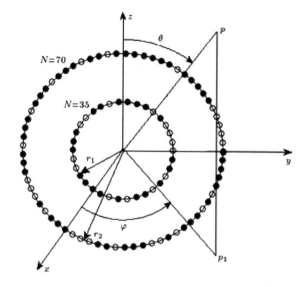

$$r_p = N_p \lambda / 4\pi \tag{1}$$

The angular position is given by:

$$\omega_{pn} = 2n\pi / N_p \tag{2}$$

The array factor (AF) is calculated as follows.

$$\text{AF}(\theta, \varphi) = \sum_{p=1}^{2} \sum_{n=1}^{N_p} I_{pn} e^{jkr_p \left[\sin\theta \cos(\emptyset - \emptyset_{pn}) - \sin\theta_0 \cos(\emptyset_0 - \emptyset_{pn}) \right]} \tag{3}$$

I_{pn} Excitation of nth element in pth ring
θ elevation angle
φ azimuthal angle
k $2\pi/\lambda$, wave number
$\theta_{0,0}$ direction of main beam's maximum.

$\theta_0 = 0$ and $\varphi_0 = 0$ are taken for the design problem and $= 0$ plane is considered. When the above-mentioned angles are substituted in Eq. (3), it gets reduced to:

$$\text{AF}(\theta, \varphi) = \sum_{p=1}^{2} \sum_{n=1}^{N_p} I_{pn} e^{\left[jkr_p \sin\theta \cos(\emptyset - \emptyset_{pn}) \right]} \tag{4}$$

2.2 Hexagonal Antenna Array

The configuration of uniform hexagonal antenna array is shown in Fig. 2. To visualize a hexagonal array, two concentric circles of radii r_1 and r_2 are considered. Each circle is having elements on it. The array consists of $2N$ elements: N on the mid-point of the sides of the hexagon and N on the vertices.

The array factor (AF) for this configuration is computed as given below:

$$\text{AF}(\theta, \varphi) = \sum_{n=1}^{N} A_n e^{jkr_1 \sin\theta(\cos\varphi_{1n}\cos\varphi + \sin\varphi_{1n}\sin\varphi)} + B_n e^{jkr_2 \sin\theta(\cos\varphi_{2n}\cos\varphi + \sin\varphi_{2n}\sin\varphi)} \quad (5)$$

$$\varphi_{1n} = 2\pi(n-1)/N \quad (6)$$

$$\varphi_{2n} = \varphi_{1n} + \pi/N \quad (7)$$

$$r_1 = d_e / \sin\left(\frac{\pi}{N}\right) \quad (8)$$

$$r_2 = r_1 \cos\left(\frac{\pi}{N}\right) \quad (9)$$

where A_n and B_n are relative amplitudes of nth element lying on the vertices and the mid-point, φ_{1n} is the angle between the nth element at the vertex point in XY plane and the x-axis, φ_{2n} is the angle between the nth element lying on the mid-point of the hexagon's side and x-axis, r_1 and r_2 are the radii of the two circles, and θ is the elevation angle, and φ is the azimuth angle.

Fig. 2 Configuration of uniform hexagonal array antenna

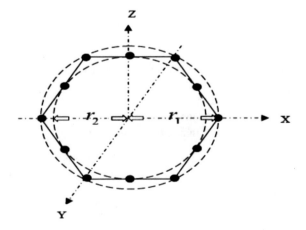

3   The Jaya Algorithm

The first step in the implementation of Jaya algorithm is to generate initial solutions. Therefore, P initial solutions are generated in random fashion by keeping in consideration the design variables' range values, i.e., upper and lower bounds. Then, every solution's variable values are updated using Eq. (10). Let the objective function be denoted with f. For demonstration, it is considered that the function is to be minimized (or maximized, according to the design problem). Say, the no. of design variables is 'nd.' best_f is used to denote the best solution and worst_f to denote the worst solution [12].

$$A(x + 1, y, z) = A(x, y, z) + r(x, y, 1)(A(x, y, b) - |A(x, y, z)|)$$
$$- r(x, y, 2)(A(x, y, w) - |A(x, y, z)|) \quad (10)$$

Among the population, the best and worst solutions are identified by their indices b and w. The indices to denote iteration, variable and the candidate solution x, y, z. $A(x, y, z)$ mean the yth variable of zth candidate solution in xth iteration. $r(x, y, 1)$ and $r(x, y, 2)$ are the randomly generated numbers in the range of [0, 1] and act as scaling factors and to ensure good diversification. Figure 3 depicts Jaya algorithm's flowchart.

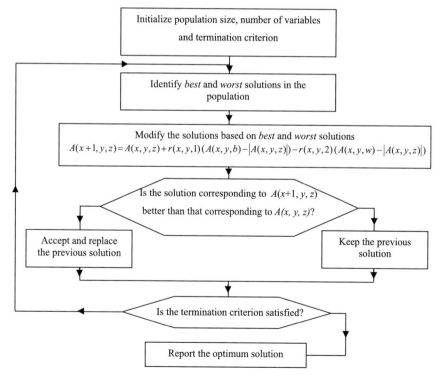

Fig. 3 Flowchart of Jaya algorithm

In this algorithm, the value of the objective function of every candidate solution of the population is improved. Values of variables of each solution are updated by the algorithm in order to move the corresponding function value closer to the best solution. After updating variable values, comparison between the new function value and corresponding old function value. Then, only the solutions having better objective function value (lower values for minimization and higher values for maximization problems) are included in the next generation.

3.1 Chaotic Jaya Algorithm

A variation of Jaya algorithm, i.e., Chaotic Jaya, is used in this paper. This variant helps in improving the speed of convergence and also improves the search space exploration without getting entrapped in the local optima. Chaos is a branch of mathematics that deals with systems that are sensitive to initial conditions. In chaos optimization algorithms, the chaotic maps are used to map the chaotic sequences to produce the design variables, instead of producing sequences that are simply random.

Different chaotic maps are available to introduce chaos in optimization, e.g., exponential map, circle map, logistic map and tent map. These maps are used to initialize population that is uniformly distributed so that the quality of initial population is improved. In these chaotic maps, an initial value is chosen in the range [0, 1] (or according to the range of chaotic map).

The only difference in the implementation Jaya algorithm and Chaotic Jaya algorithm is in the use of a chaotic map by the latter one. Chaotic Jaya uses the tent map chaotic function to generate random numbers, and the function is expressed by Eq. (11) [14].

$$x_{k+1} = \left\{ \begin{array}{ll} \frac{x_k}{0.7} & x_k < 0.7 \\ \frac{10}{3}(1 - x_k) & x_k \geq 0.7 \end{array} \right\} \tag{11}$$

where x_k is the previously generated chaotic random number and x_{k+1} is the newly generated chaotic random number.

4 Objective Function

The side lobe level is calculated by the following equation.

$$\text{SLL} = 20 \log_{10} \left(\frac{|AF(\theta, \varphi)|}{|AF(\theta, \varphi)|_{\max}} \right) \tag{12}$$

For two-ring concentric circular antenna array, the fitness function is given as follows [1].

$$\text{Fitness} = \max(\text{SLL}), \quad \text{FNBW}_o < \text{FNBW}_d$$
$$= 10^2, \quad \text{otherwise} \tag{13}$$

where FNBW is the first null beam width. The subscript o denotes the obtained FNBW and d denotes the desired FNBW. The desired FNBW for this work is $10°$.

For the case of hexagonal antenna array, the objective function is formulated as follows [2].

$$\text{Fitness} = \min(\max(20 \log|\text{AF}(\theta)|)) \tag{14}$$

$$\theta \in \left\{ \left[-90°, -|\text{FN}|° \right] \text{ and } \left[|\text{FN}|°, 90° \right] \right\}$$

where FN is first null in the antenna radiation pattern and $\text{AF}(\theta, \varphi)$ is given by Eq. (5). In all the cases, % thinning is calculated as the percentage of no. of OFF elements to total number of elements.

5 Case Studies

5.1 Synthesis of Concentric Circular Antenna Array

The inter-element spacing is considered as 0.5. Equation (4) is used to calculate the array factor, and Eq. (13) is the fitness function. This problem was worked out by [1] using firefly algorithm (FFA). In their paper, taking 50 as the population size and 500 as the maximum number of generation produced an optimal result for FFA algorithm. Here, the same case study is solved using Chaotic Jaya algorithm. To compare both algorithms fairly, the maximum number of function evaluations is kept same as that of FFA (i.e., 25,000). So population size is maintained as 50 and 500 is chosen as the maximum number of generations.

Table 1 shows the results of Chaotic Jaya algorithm for two-ring concentric circular array. By comparison, it is seen that minimum peak SLL by FFA is -18.36 dB whereas Chaotic Jaya algorithm resulted in a minimum peak SLL of -21.495287 dB which is an improvement over FFA results. Table 2 reports the amplitude distribution. The radiation pattern is shown in Fig. 4.

Table 1 Results of concentric circular

Algorithm	No. of ON elements	% Thinning	Max. SLL (dB)
Chaotic Jaya	53	49.52	−21.495287
[a]FFA	69	34.29	−18.36

[a]The results of FFA are reproduced from Basu and Mahanti [1]

Table 2 Amplitude distribution results of concentric circular array

Algorithm	Distribution of ON-OFF elements	
	Ring 1	Ring 2
Chaotic Jaya	0 0 0 1 1 0 1 1 1 1 0 1 1 0 1 1 0 1 0 1 0 1 0 1 1 1 1 0 1 1 1 0 0 1 0	0 0 0 0 0 1 0 0 1 1 1 1 1 0 1 1 0 0 1 1 1 0 0 1 1 0 0 0 1 0 0 0 0 1 0 0 1 0 0 1 0 0 1 1 0 1 0 1 0 1 1 1 1 1 1 1 0 0 1 0 1 0 1 0 0 0 0 0 0 1
[a]FFA	0 1 0 1 1 1 1 1 1 1 0 1 0 1 1 0 0 0 0 0 0 1 1 0 1 1 0 1 1 1 1 0 1 0	1 1 0 1 0 1 0 1 1 1 0 1 0 0 1 1 1 1 1 0 1 1 1 1 1 1 0 1 1 0 0 1 0 1 1 0 1 0 0 1 0 1 1 1 1 1 0 1 1 1 1 1 1 1 0 1 1 0 1 1 1 1 1 0 1 0 1 0 0 1

[a]The results of elements distribution using FFA are reproduced from Basu and Mahanti [1]

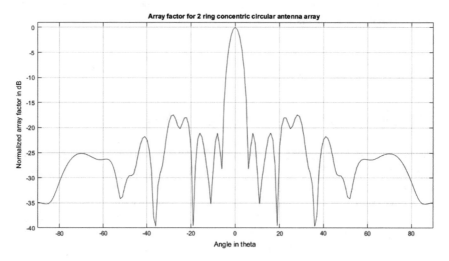

Fig. 4 Normalized array pattern of CCA

5.2 Synthesis of Hexagonal Antenna Array

For all the cases of hexagonal antenna array, Eq. (5) is used to calculate the array factor and Eq. (14) is the fitness function. The number of elements considered is $2N = 12$, i.e., six elements lie on the inner circle and the other six on the outer circle. This problem was worked out by [2] using improved particle swarm optimization (IPSO). In their paper, taking 50 as the population size and of 50 and 100 as the maximum number of generations gave the optimal result for IPSO algorithm. Here, the same case study is solved using Chaotic Jaya algorithm. To compare both algorithms fairly, the maximum number of function evaluations is kept same as that of IPSO (i.e., 5000). So population size is maintained as 50 and 100 is chosen as the maximum number of generations.

Case 1: 0.50λ spacing

The inter-element spacing is considered as 0.5λ. The optimized array patterns obtained by Chaotic Jaya algorithm for reducing SLL for 12-element array and of IPSO [2] is tabulated in Table 3. From Table 3, it is perceived that minimum peak SLL by IPSO is −23.85 dB whereas Chaotic Jaya algorithm resulted in a minimum peak SLL of −27.605378 which is an enhancement of IPSO results. The radiation pattern by Chaotic Jaya algorithm along with the pattern of a fully populated array is plotted in Fig. 5. Table 4 reports the amplitude distribution.

Case 2: 0.55λ spacing

Table 3 Results of hexagonal antenna array (0.50λ)

Algorithm	No. of ON elements	% Thinning	Max. SLL (dB)
Chaotic Jaya	6	50	−27.605378
[a]IPSO	6	50	−23.85

[a]The results of IPSO are reproduced from Bera et al. [2]

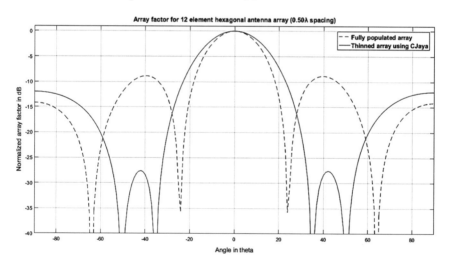

Fig. 5 Normalized array pattern of HA for 0.50λ spacing

Table 4 Hexagonal array's amplitude distribution for 0.50λ

Algorithm	Distribution of ON-OFF elements
Chaotic Jaya	[a]V: 1 0 1 1 1 0 [a]MP: 1 0 0 1 0 0
[b]IPSO	V: 1 0 0 1 0 1 MP: 0 1 0 0 1 1

[a]V vertices MP mid-point
[b]The results of elements distribution using IPSO are reproduced from Bera et al. [2]

The inter-element spacing is taken as 0.55λ. Table 5 reports the optimized array patterns obtained by Chaotic Jaya algorithm for reducing SLL for 12-element array and of IPSO [2]. From the table, it is clearly visible that the result of IPSO is −18.03 dB and Chaotic Jaya algorithm achieved a minimum peak SLL of −27.651426 dB which is an enhancement of IPSO results. Figure 6 depicts the radiation pattern achieved by Chaotic Jaya algorithm along with the pattern of a fully populated array. The amplitude distribution is given in Table 6.

Case 3: 0.60λ spacing

Table 5 Results of hexagonal antenna array (0.55λ)

Algorithm	No. of ON elements	% Thinning	Max. SLL (dB)
Chaotic Jaya	6	50	−27.651426
[a]IPSO	8	33.33	−18.03

[a]The results of IPSO are reproduced from Bera et al. [2]

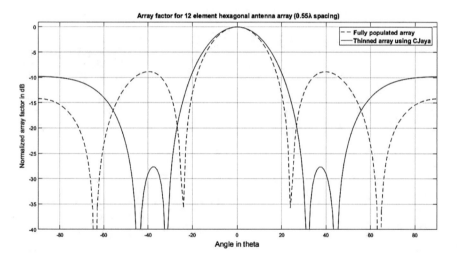

Fig. 6 Normalized array pattern of HA for 0.55λ spacing

Table 6 Hexagonal array's amplitude distribution for 0.55λ

Algorithm	Distribution of ON-OFF elements
Chaotic Jaya	[a]V: 1 0 1 1 0 1 [a]MP: 1 0 0 0 0 1
[b]IPSO	V: 1 0 0 0 0 0 MP: 1 1 0 0 1 1

[a]V vertices MP mid-point
[b]The results of elements distribution using IPSO are reproduced from Bera et al. [2]

The inter-element spacing is taken as 0.60λ. The optimized array patterns obtained by Chaotic Jaya algorithm for reducing SLL for 12-element array and of IPSO [2] is tabulated in Table 7. On inspection, it is easily perceivable that Chaotic Jaya algorithm resulted in a minimum peak SLL of −23.866137 dB which is clearly an enhancement over the SLL results of IPSO, i.e., −13.95 dB. The radiation pattern resulting from Chaotic Jaya algorithm align with the pattern of a fully populated array is plotted in Fig. 7. The amplitude distribution is given in Table 8.

Table 7 Results of hexagonal antenna array (0.60λ)

Algorithm	No. of ON elements	% Thinning	Max. SLL (dB)
Chaotic Jaya	6	50	−23.866137
[a]IPSO	9	25	−13.95

[a]The results of IPSO are reproduced from Bera et al. [2]

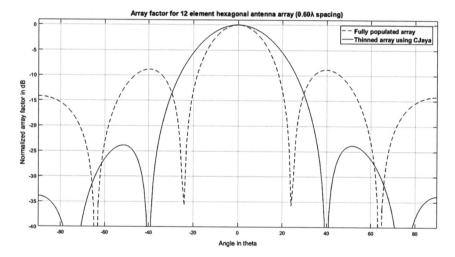

Fig. 7 Normalized array pattern of HA for 0.60λ spacing

Table 8 Hexagonal array's amplitude distribution for 0.60λ

Algorithm	Distribution of ON-OFF elements
Chaotic Jaya	[a]V: 1 0 0 1 0 0 [a]MP: 1 1 1 0 0 1
[b]IPSO	V: 1 0 1 0 1 0 MP: 1 0 0 0 1 1

[a]V vertices MP mid-point
[b]The results of elements distribution using IPSO are reproduced from Bera et al. [2]

6 Conclusions

In this work, synthesis of planar antenna array for suppression of SLL is considered. Concentric circular and hexagonal antenna arrays have been synthesized. Four optimization case studies have been formulated and the same are worked out successfully by Chaotic Jaya algorithm. Results' comparison is done between Chaotic Jaya algorithm, FFA and IPSO. It is noted that Chaotic Jaya algorithm accomplished the task by achieving enhanced results than the other above mentioned optimization algorithm with a high convergence speed.

First, the synthesis of two-ring concentric circular array is considered. The number of elements is 35 and 70 on the rings. It is perceived that Chaotic Jaya algorithm's array pattern attains reduced side lobe levels as compared to FFA algorithm's array pattern.

Under the synthesis of 12-element hexagonal antenna array, three case studies with spacings 0.5, 0.55 and 0.60λ are considered. In all the three, increased SLL suppression of SLL is evident from the array pattern of Chaotic Jaya algorithm than the SLL of conventional and IPSO optimized array pattern.

The results show that the Chaotic Jaya algorithm can effectively synthesize a planar antenna array while keeping the desired antenna parameters within acceptable limits. The implementation of Chaotic Jaya algorithm is relatively simple and excludes algorithm-specific parameters. In terms of having good solution search ability and high speed of convergence, the Jaya algorithm is complete in itself. However, with the incorporation of chaotic theory it becomes Chaotic Jaya algorithm and performance is enhanced. Therefore, the Chaotic Jaya algorithm may be further applied for solving optimization problems in the electromagnetic engineering field.

References

1. Basu, B., Mahanti, G.K.: Thinning of concentric two-ring circular antenna array using firefly algorithm. Trans. Comput. Sci. Eng. Electr. Eng. **19**(6), 1802–1809 (2012)
2. Bera, R., Lanjewar, R., Mandal, D., Kar, R., Ghoshal, S.P.: Comparative study of circular and hexagonal antenna array synthesis using Improved Particle Swarm Optimization. Procedia Comput. Sci. **45**, 651–660 (2015)
3. Biswas, S., Bose, D., Das, S., Kundu, S.: Decomposition based evolutionary multi- objective optimization approach to the design of concentric circular antenna arrays. Prog. Electromagnet. Res. **52**, 185–205 (2013)
4. Chattarjee, A., Mahanti, G.K.: Comparative performance of Gravitational Search Algorithm and modified Particle Swarm Optimization algorithm for synthesis of thinned scanned concentric ring array antenna. Prog. Electromagnet. Res. **25**, 331–348 (2010)
5. Elsaidy, S., Dessouky, M., Khamis, S., Albagory, Y.: Concentric circular antenna array synthesis using comprehensive learning particle swarm optimizer. Prog. Electromagnet. Res. **29**, 1–13 (2012)
6. Guney, K., Durmus, A., Basbug, S.: Backtracking search optimization algorithm for synthesis of concentric circular antenna arrays. Int. J. Antennas Propag. (2014)
7. Haupt, R.L.: Thinned arrays using genetic algorithms. IEEE Trans. Antennas Propag. **42**(7), 993–999 (1994)

8. Keizer, W.P.M.N.: Linear array thinning using iterative FFT techniques. IEEE Trans. Antennas Propag. **56**(8), 2757–2760 (2008)
9. Mangoud, M.A.A., Elragal, H.M.: Antenna array pattern synthesis and wide null control using Enhanced Particle Swarm Optimization. Prog. Electromagnet. Res. **17**, 1–14 (2009)
10. Pathak, N., Basu, B., Mahanti, G.K.: Combination of inverse fast Fourier transform and modified particle swarm optimization for synthesis of thinned mutually coupled linear array of parallel half-wave length dipoles antennas. Prog. Electromagnet. Res. **16**, 105–115 (2011)
11. Rao, R.V.: Jaya: a simple and new optimization algorithm for solving constrained and unconstrained optimization problems. Int. J. Ind. Eng. Comput. **7**, 19–34 (2016)
12. Rao, R.V.: Jaya: An Advanced Optimization Algorithm and Its Engineering Applications. Springer Nature, Switzerland (2018)
13. Singh, U., Rattan, M.: Design of linear and circular antenna arrays using cuckoo optimization algorithm. Prog. Electromagnet. Res. **46**, 1–11 (2014)
14. Jaya Lakshmi, R., Mary Neebha, T.: Synthesis of linear antenna arrays using Jaya, self-adaptive Jaya and Chaotic Jaya algorithms. AEU-Int. J. Electr. Commun. **92**, 54–63 (2018)

Optimization of Friction Drilling Process by Response Surface Methodology

Akshay Potdar◉ **and Sagar Sapkal**◉

Abstract Friction drilling is the process of producing hole in the sheet metal and hollow sections. Rotating conical tool is used to produce friction between tool and workpiece results in the formation of bushing around the hole. Different speed, feed and material thickness are used in the evaluation of the bushing height and bushing thickness. Coordinate measuring machine (CMM) is used to measure dimensions of the bushing height and width. The main objective of the experimentation is to optimize speed and feed to maximize bushing height and thickness for different material thicknesses (up to 2 mm). Furthermore, response surface methodology (RSM) technique is used for the optimization of speed and feed for different material thicknesses. It is found that as spindle speed of rotating tool increases, gradual increment in the bushing height and decrease in the bushing width.

Keywords Friction drilling · Bushing height · Bushing width · Response surface methodology

1 Introduction

Friction drilling is the process of making the hole by using the friction between the tool and workpiece. Friction drilling process mostly used for the thin sheet metals and hallow sections. Workpiece material can be steel, aluminium, copper and brass. Rotating conical tool is used to produce the friction between tool and workpiece. Tool used in the friction drilling is made up of tungsten carbide which has very high melting point (34,000 °C). As material gets softened, tool penetrates the material to form the bushing around the hole. This bushing is further used for forming thread. Once threads are formed bolt or screw can be easily inserted in the hole as shown

A. Potdar (✉) · S. Sapkal
Department of Mechanical Engineering, Walchand College of Engineering, Sangli, India
e-mail: akshaypotdar99@gmail.com

S. Sapkal
e-mail: sagar.sapkal@walchandsangli.ac.in

© Springer Nature Singapore Pte Ltd. 2020   351
R. Venkata Rao and J. Taler (eds.), *Advanced Engineering Optimization*
Through Intelligent Techniques, Advances in Intelligent Systems and Computing 949,
https://doi.org/10.1007/978-981-13-8196-6_31

Fig. 1 Friction drilling process [2]

in Fig. 1. This process can be used to eliminate the rivet nut, weld nut and insert nut. This process is very clean and fast to produce the hole and form the threads. Life of the drill is also very high up to 40,000 holes; therefore, it can beneficial for mass production. The added height of the bushing formed can lengthen the threaded portion of the hole and consequently increase the fastener clamp load for joining thin sheet metal. In Fig. 1, it can be seen that first three steps for the friction drilling and after that form tapping and inserting the bolt inside the hole. Material deformation around the tool causes the formation of collar around the hole [1].

2 Experimental Procedure

2.1 Experimental Setup

Tool. Tool used in the experimentation is the standard friction drill made up of the tungsten carbide (WC) material [1]. In the friction drilling process, temperature goes up to the 700–800 °C. Therefore, tool is made up of tungsten carbide that can withstand at high temperature. There are two types of the tool that can form the hole with collar and without collar. Tool of short type with the collar having diameter 7.3 mm is selected. It can form the threads of size 8 × 1.25 mm after form tapping process [2].

   Material. Friction drilling is normally used in the sheet metal and hollow sections. Material used in the experimentation is AISI 1015 mild steel pipe which is very commonly used in the industry. Steel pipe of AISI 1015 is commonly used in the industry.

   Machine Specification. Friction drilling performed on the column drilling machine, VMC machine or milling machine. Spindle speed should be 2000–4000 rpm with 1.5 kW motor power. In this research machine used is the Vertical Machining Center (VMC).

2.2 Experimental Details

Process Parameter Selection. Process parameters for the experimentation can be classified on the basis of tool, material and process parameters. Total input variables affecting the process are indicated in the fish-bone diagram (Fig. 2). Tool used in the experimentation is standard available for the steel made up of tungsten carbide having friction angle 36° [3]. Speed and feed are the parameters related to the machine process parameters that can vary bushing size. Parameters related to the material used in the experimentation are the variation in the thickness and the rest of the parameters are kept constant. Here for the experimentation purpose, speed, feed and material thickness are the factors used for affecting quality of the bushing. Therefore, these parameters are varied for the experimentation [4].

Response Surface Methodology (RSM). RSM technique is used for the design of experiments. There are two continuous variables (spindle speed and feed) and one categorical variable (material thickness) are used for the design of experiments in Minitab software. There are two types of design of experiments in RSM that are central composite and Box Behnken. central composite design (CCD) used for two continuous factors and Box Behnken design used for the three and above continuous factors, so here CCD is used. Range of the parameters is fixed to the axial points in the CCD. In this case, material thickness cannot be varied beyond the given levels, because it is a standard parameter, so material thickness included the categorical parameter in the software. Range of the parameters used is indicated in Table 1.

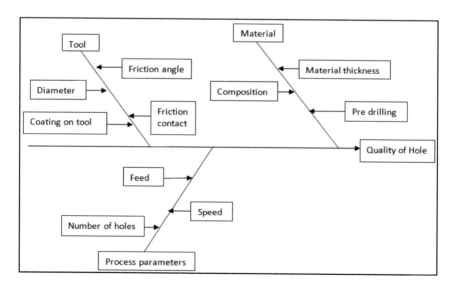

Fig. 2 Fish-bone diagram for total input parameters

Table 1 Range of parameters

Process parameters	Range		
	Low		High
Spindle speed (RPM)	2000		4000
Feed rate (mm/min)	70		150
Material thickness (mm)	1.2	1.6	2

3 Results and Discussion

3.1 Response Variables

Response variables related to the quality of bushing were decided through the practical application and literature review. It is found that bushing plays important role in the friction drilling process, because ultimately strength of the form threads depends on the bushing height and the bushing petal width (Fig. 3) [5].

As thickness varies, change in the bushing height and bushing width (petal width) will vary. So investigation and optimization of the bushing height and petal width is the main objective of this research. Table 2 denotes the target and weight of the parameters.

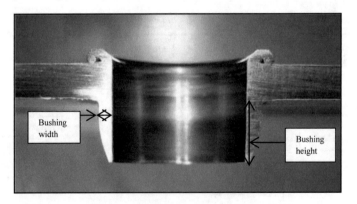

Fig. 3 Bushing specification [1]

Table 2 Target and weight of the parameters

Response variables	Goal	Lower (mm)	Target (mm)	Weight
Bushing thickness	Maximum	0.2996	0.61905	1
Bushing height	Maximum	3.48	4.5456	1

Fig. 4 Effect of input
variables on bushing height

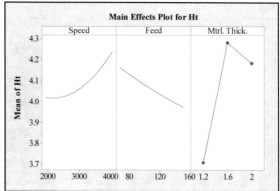

3.2 Analysis of Bushing Height

In this analysis, effect of input parameters on bushing height is investigated. Bushing height is measured with the help of CMM at four points. From the analysis, it is found that with increase in the speed, bushing height increases due to increase in friction heat with increased speed. Therefore, flow ability of the material increases with spindle speed, but as feed increases it causes decrease in bushing height due to low frictional heat generation [6]. There is also effect of material thickness. It has been seen that with increase in thickness from 1.2 to 1.6 mm bushing height increases but from 1.6 to 2 mm bushing height decreases (Fig. 4).

From the contour plot in Fig. 5, it is observed that maximum bushing height is in the speed range of 3500–4000 rpm for all material thickness. Feed for maximum height is in the range of 70–90 mm/min [7].

3.3 Analysis of Bushing Width

In this analysis, average bushing thickness is calculated. Bushing thickness is measured by the coordinate measuring machine (CMM) by taking the measurements at every 0.5 mm for the height of 2.5 mm. It is observed that with increase in the spindle speed, bushing thickness decreases due to increase in the flow ability. Increased flow ability causes the more material flow in the axial direction. Feed also affect the bushing thickness in positive way. As feed increases, bushing thickness increases as shown in main effects plot (Fig. 6). Material thickness is the major parameter affecting the bushing thickness. There is positive effect of thickness on bushing thickness that can be seen in Fig. 6. Material thickness to hole diameter (t/d) ratio is an important parameter, as t/d ratio increases bushing is more uniform with less cracks.

From the interaction plot in Fig. 7, it can be seen that maximum average thickness which is above 0.38 mm is obtained in the feed range of the 120–150. Interaction

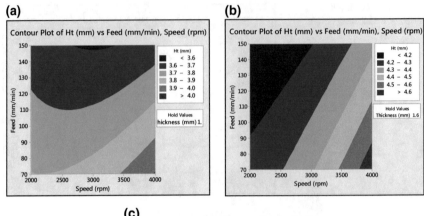

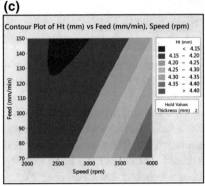

Fig. 5 Contour plot for bushing height **a** for 1.2 mm, **b** for 1.6 mm and **c** for 2 mm

Fig. 6 Effect of input variables on bushing thickness

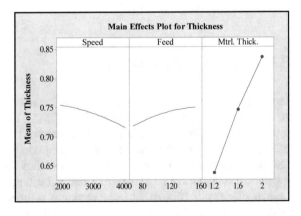

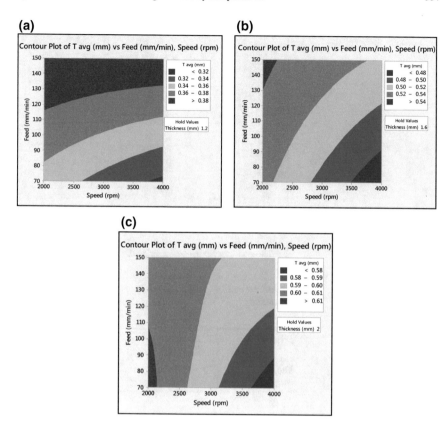

Fig. 7 Contour plot for bushing thickness **a** for 1.2 mm, **b** for 1.6 mm and **c** for 2 mm

graph indicates that for any speed range we can obtain the maximum average thickness by using feed rate above 130 mm/min.

3.4 Multi-objective Optimization

Multi-objective optimization is carried out in the Minitab-17 software by using RSM, for that purpose bushing height set to maximum and bushing thickness also set to maximum. This optimization gives speed 4000 rpm and feed 70 mm/min which is applicable to material thickness form 1.2 to 2 mm. In Fig. 8, maximum bushing height and thickness for 2 mm is given after multi-objective optimization. This optimized parameters can be used to get the uniform bushing which is the main objective of the research.

Fig. 8 Multi-objective
optimization for response
variables

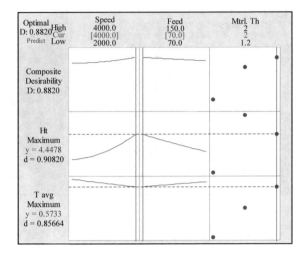

Table 3 Results of the
validation run

Optimized input parameters		Optimized results	Material thickness (mm)		
			1.2	1.6	2
Speed (rpm)	4000	Bushing height (mm)	3.77	4.54	4.45
Feed (mm/min)	70	Bushing thickness	0.36	0.48	0.56

3.5 Validation Run

Validation run is conducted to verify the results obtained by multi-objective optimization in the RSM. Table 3 indicates the results obtained after performing the experiments by using optimized input parameters.

After locating the results of validation run in the contour plot, it is found that results of bushing height and average bushing thickness are in the green zone that is in the optimum region of contour plot. Thus, results obtained after multi-objective optimization are confirmed in the validation test.

4 Conclusion

1. In the multi-objective optimization, spindle speed 4000 rpm and feed 70 mm/min is found to be optimum for maximum bushing height and thickness and applicable for material thickness up to 2 mm.
2. As speed increased bushing height also increased due to higher frictional heat and increased flow ability of material. Feed shows negative impact on bushing height.

There is decrease in the bushing height for 2 mm material thickness compared to 1.6 mm material thickness. It is due to increase in bushing width instead of bushing height.

References

1. Chow, H.M., Lee, S.M., Yang, L.D.: Machining characteristic of friction drilling on AISI 304 stainless steel. J. Mater. Process. Technol. **207**, 180–186 (2008)
2. Form Drill India, http://www.formdrill-india.com
3. Ku, W., Chow, H.M., Lin, Y., Wang, D., Yang, L.: Optimization of thermal friction drilling using grey relational analysis. Adv. Mater. Res. **154**, 1726–1738 (2011)
4. Ozek, C., Demir, Z.: Investigate the surface roughness and bushing shape in friction drilling on A7075-T651 and St 37 steel. Technol. Educ. Manag. J. **2**, 170–180 (2013)
5. Ozler, L., Dogru, N.: An experimental investigation of hole geometry in friction drilling. Mater. Manufact. Process. **28**, 470–475 (2013)
6. Pantawane, P.D., Ahuja, B.B.: Experimental investigation and multi-objective optimization of friction drilling on AISI 1015. Int. J. Appl. Eng. Res. **2**, 448–461 (2011)
7. Liang, L.D., Lee, S.M., Chow, H.M.: Optimization in friction drilling for SUS 304 stainless steel. Int. J. Adv. Manufac. Technol. **53**, 935–944 (2011)

Optimization of Parameters in Friction Stir Welding of AA6101-T6 by Taguchi Approach

Deepak Kumar and Jatinder Kumar

Abstract In this experimental analysis, mechanical and metallurgical properties of friction stir welding of aluminium alloy (AA6101-T6) are examined. A cylindrical threaded tool of high-speed steel (H13) is used for friction stir welding of working material AA6101-T6. For optimization of parameters like feed, rotational speed and tilt angle, the design of experiment considered is Taguchi method with three levels and three parameters. ANOVA statistical tool is used for optimization of parameters for tensile strength and microhardness of the weld. The microstructure of the weld surface and fractography of the fractured weld surface are studied by scanning electron microscope.

Keywords FSW · Tensile strength · Microhardness · Taguchi · ANOVA · SEM

1 Introduction

Friction stir welding is a modern fabrication technique used for aluminium alloys. It was invented by The Welding Institute (TWI), in the UK in 1991 and patented by W. M Thomas. It is a relatively new joining method and is also called as a solid-state fabrication process [1]. It is an environment-friendly and single-phase process and is suitable for the joining of almost all types of aluminium alloys which are widely used in the aviation industry. In this process, a specially designed rotating tool is traversed along the line of the joint which is inserted into the abutting edges of sheets [2]. The parts are clamped rigidly on the fixture on a backing plate to control the abutting joint surfaces from being forced apart. When the tool shoulder starts to touch the surface of the workpiece, then the plunging action is stopped. Shoulder of the tool should touch the working plates. The main aim of the rotating tool is to heat and soften

D. Kumar (✉) · J. Kumar
Mechanical Engineering Department, N.I.T. Kurukshetra, Haryana, India
e-mail: deepak.mmb@gmail.com

J. Kumar
e-mail: jatin.tiet@nitkkr.ac.in

© Springer Nature Singapore Pte Ltd. 2020
R. Venkata Rao and J. Taler (eds.), *Advanced Engineering Optimization Through Intelligent Techniques*, Advances in Intelligent Systems and Computing 949,
https://doi.org/10.1007/978-981-13-8196-6_32

the workpiece and produce the welding joint due to the movement of material. A sufficient amount of heat can be obtained from this process which is used for welding of joint, and the temperature of the pairing point can be raised to a level at which surfaces susceptible to friction can be welded together. Due to the rotational and transverse speed of the tool, there occurs localized plastic heating which softens the working material around the tool pin [3]. The tool then travels along the joint until it reaches the end of the workpiece, and by this process, both the plates are welded by frictional heat. A schematic diagram of the FSW process shown in Fig. 1 [4].

Numerous researches around the world investigated on FSW with different wide series of materials. Mishra et al. [2] published a series of books on FSW and FSP which provides an informative and comprehensive study of the process. Sarsilmaz and Caydas [5] performed FSW on dissimilar aluminium alloy (AA1050 & AA5083) using full factorial design approach. In their investigation, they studied the effect of parameters on mechanical properties (UTS and hardness) of the weld. ANOVA statistical tool was used for this investigation which helped in reporting the significant factors and predicted the optimum result of the experiment. Elanchezian et al. [6] carried out optimization of FSW parameters on aluminium alloys (AA8011 & AA6062) by using the Taguchi approach. Responses such as tensile, impact, microhardness were calculated with ANOVA tool and Taguchi L9 orthogonal array and also evaluated the effects of parameters on responses. SEM was manipulated to test the quality of the surface texture. Elatharasan et al. [7] carried out the optimization of FSW parameters on AA6061-T6 by using RSM. In their investigation, they build the relationship between the parameters and responses (elongation, yield stress and tensile strength) with the help of RSM mathematical model with three parameters and three levels. They performed 20 runs and analysed the results with RSM methodology. Sergey et al. [8] investigated on optimization of processing microstructural properties on weld of FSW 6061-T6. They considered parameters and analysed their effects. They reported that the microhardness lowered at stir zone from base metal and tool travel speed 760 mm/min had yield joint efficiency 93% and localization of strain in HAZ.

There is no study reported on statistical optimization of the FSW parameters with three levels and three parameters at a time for the selected aluminium alloy (AA6101-T6). Taguchi approach is used for investigation with three levels and three parameters at a time.

2   Experimentation

2.1   Set-up of Friction Stir Welding

We performed the experiment on a vertical milling machine with some modifications on the machine. As per the requirement of our work, we designed the fixture which can easily fit on the bed of the machine. A threaded cylindrical tool of high-speed

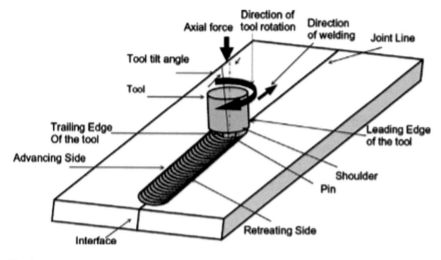

Fig. 1 A schematic diagram of the FSW process [4]

Fig. 2 FSW set-up on a vertical milling machine and fixture

steel with a shank diameter of 20 mm was fixed into cullet. A schematic pictorial diagram of machine set-up is shown in Fig. 2.

2.2 Material Used

Aluminium alloy AA6101-T6 (dimensions—150 × 60 × 6 mm) was used in this experiment. Alloy 6101-T6 is an extruded product as a heat-treatable alloy and offers good weldability, bendability, formability and electrical conductivity as well as thermal conductivity [9]. Temper—T6 means artificially aged and heat treated (Tables 1 and 2).

Table 1 (%) Chemical composition of plate AA6101-T6 [9]

Chemical composition	Al grade	Cu	Mg	Si	Fe	Mn	Others	Al
wt%	6101-T6	0–0.05	0.40–0.90	0.30–0.70	0.50	0.03	0.1	Rest

Table 2 Mechanical properties of aluminium alloy AA6101-T6 [9]

Density gm/cc	Melting point	UTS MPa	Yield strength MPa	Young's modulus (GPa)	% Elongation	Specific heat	Conductivity W/MK
2.7	635	221	172	75	19	$10^{-4}/°C$	219

2.3 Tool Used

The tool made from those types of materials could withstand in the process and offer enough frictional heat. The tool must be sufficiently stronger and have high melting temperature and higher wear resistance than the base material. It was designed in CATIA software and then manufactured by specific design. The hardness of the tool was less, so we hardened the tool till 60 HRC by tempering process [10] (Table 3; Fig. 3).

2.4 Design of Experiment

In this experiment, the Taguchi approach is applied for the optimization of parameters. It is more systematic and efficient approach and relates to inadequate knowledge of statistics. We selected L9 orthogonal array for optimizing the parameters. All three parameters have DOF = 6 and total DOF = $N-1$ = 8 where N = no. of trials [11].

3 Results

Friction stir welding was conducted for the investigation of parametric effect on responses like tensile strength and microhardness which are tabulated below. The

Table 3 Tool description

Material	H13 (mm)
Shoulder dia.	20
Pin length	5.8
Pin base dia.	6.3

Fig. 3 Design of tool

tensile test was performed on digital UTM machine which gave a graph between stress and strain. In microhardness test where indentation is the form used for calculation, we took an average of response values (Table 4).

3.1 Effect on Tensile Strength

The values of tensile strength at levels of 1, 2 and 3 for each parameter were plotted which is shown below in the figure. It demonstrates that the tensile strength increases with the increase in rotational speed but decreases after reaching a certain maximum

Table 4 Results of testing in L9 representations

Experiments	rpm (R)	Feed (S)	Tilt angle	Av. tensile strength (MPa)	Av. microhardness (HV) at weld centre
Exp-1	710	16	0	135	68.96
Exp-2	710	25	1	139	72.93
Exp-3	710	40	2	140	75.82
Exp-4	1000	16	1	141	69.98
Exp-5	1000	25	2	154	80.19
Exp-6	1000	40	0	144	72.86
Exp-7	1400	16	2	140	75.52
Exp-8	1400	25	0	147	78.26
Exp-9	1400	40	1	134	74.25

Table 5 ANOVA for tensile strength (*S/N* data)

Source	DOF	Adj SS	Adj MS	F-Value	P-Value
R	2	110.889	55.444	19.96	0.048
S	2	118.22	59.111	21.28	0.045
Θ	2	67.556	33.778	12.16	0.076
Error	2	5.556	2.778		
Total	8	302.22			

MS—mean of squares, SS—squares sum, *F*- [*Vs/Ve*], *P* < 0.05—95% confidence level for significance

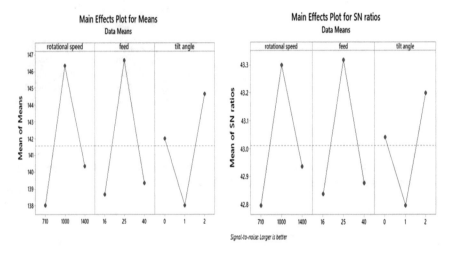

Fig. 4 Parametric effect on the response (tensile strength)

value; same is for feed, but for tilt angle, it decreased with increases in tilt angle; after reaching a minimum value, the tensile strength increases with the increase in the tilt angle (Table 5; Fig. 4).

3.2   Selection of Optimum Levels

Taguchi's signal to noise ratio is the logarithmic function of desired output. *S/N* ratio is the ratio of the mean to standard deviation. Here mean refers to signal and standard deviation refers to noise. The ratio depends on the quality characteristic of the process to be optimized. The significance of the process variables towards tensile strength, Analysis of variance (ANOVA) was performed. It was noticed that some factors more effective for the response of tensile strength. Rank and delta assign to each factors indicates the importance of each factor on each response. The response Table 5 show the average of each response characteristic (*S/N* and means) for each level factor. The Tables include ranks based on delta statistics, which compare the relative magnitude of effects. MINITAB 17 was used for plotting table and calculating the delta. For tensile strength, the larger is better type quantity [11].

3.3 Estimation of Optimum Response (Tensile Strength)

Significant variables are A_2 and B_2.

$$\mu_T = \bar{T} + (\bar{A}_2 - \bar{T}) + (\bar{B}_2 - \bar{T})$$

The overall mean of tensile strength $\bar{T} = 141.556$ MPa

Predicted optimal value $\mu_T = 151.444$ MPa

CI_{CE} and CI_{POP} are calculated by Eqs. (i) and (ii) [11, 12].

$$(i)\ CI_{POP} = \sqrt{\frac{F_\alpha(1, f_e)V_e}{n_{eff}}} \quad (ii)\ CI_{CE} = \sqrt{F_\alpha(1, f_e)V_e\left[\frac{1}{n_{eff}} + \frac{1}{R}\right]}$$

$F\alpha\ (1, f_e)$ = F-ratio level of confidence $(1–\alpha)$ aligned with DF 1 and error DOF (f_e)

$n_{eff} = N/1 + $ (DF estimated response) $= 1.2857$, $F\ 0.05\ (1, 2) = 18.5120$ (chart F-value) [12].

Ninety-five per cent confidence intervals of (CI_{CE}) and (CI_{POP}) are calculated from the below equations [11, 12].

$$CI_{POP} = \pm 6.324, \quad CI_{CE} = \pm 7.5591$$

The optimal values of parameters are: Rpm $(A_2) = 1000$ rpm and feed $(B_2) = 25$ mm/min,

3.4 Effect on Microhardness

$$\mu_{MH} = \bar{T} + \left(\bar{B}_2 - \bar{T}\right) + \left(\bar{C}_3 - \bar{T}\right)$$

Predicted optimal value $\mu_{MH} = 79.991$ HV

Ninety-five per cent confidence intervals of $(CI_{CE}) = \pm 3.7599$ and $(CI_{POP}) = \pm 3.146$

The optimal values of process variables: feed $(B_2) = 25$ mm/min and tilt angle $(C_3) = 2°$

3.5 Confirmation Experiment

The values of tensile strength and microhardness calculated from confirmation experiments are tabulated in Table 6.

3.6 Microstructural Analysis of Surface Using Scanning Electron Microscopy (SEM)

Scanning electron microscope was used to characterize the micro fracture surfaces of the tensile tested specimens to understand the failure patterns. A scanning electron microscope develops images of the surface by scanning it with a focused beam of electrons. Special attention is required for analysis of fractography; firstly, we prepare a sample for which we conduct our study. The tensile tested sample which was fractured during the tensile test at UTM machine selected for the study. A small sample was cut from the welded portion into 5×10 mm. It was examined by SEM at a different magnification at different points of the weld. Material flow (lump material deposits) observed in the weld zone of FSW is shown in Fig. 5a.

Failures consist of different size and shape of dimples present in welded portion, some times voids form prior to necking. In Fig. 5c, we can observe local plastic deformation at the deepest portion, and in Fig. 5b at the side, upper portion of the

Table 6 Predicted optimal values, confidence intervals and results of CE

Response	Optimal parameters	Predicted optimal value	Predicted interval with 95% confidence level	Av. of CE value
Tensile strength	$A_2 B_2$	151.444 MPa	$CI_{POP} =$ $145.44 < \mu_T$ < 157.768 $CI_{CE} =$ $143.882 < \mu_T$ < 158.99	149 MPa
Microhardness	$B_2 C_3$	79.991 HV	$CI_{CE} =$ $76.2311 < \mu_{MH}$ < 83.75 $CI_{POP} =$ $76.84 < \mu_{MH}$ < 83.134	78.23 HV

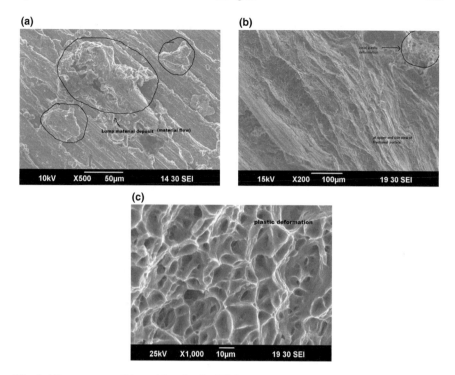

Fig. 5 Microstructure of the weld surface by SEM

surface, we can observe partially plastic and partially ductile failure. Due to the presence of other particles such as transgranular particles in the fracture area, failure due to local plastic deformation was observed during a tensile test.

4 Conclusions

- Tool rotational speed was found as the most significant parameter for tensile strength and feed for microhardness.
- The tensile strength of joints increases with an increase in rotational speed in the range of 710–1000 rpm and reaches a maximum at 1000 rpm. After that, it decreases with an increase in rotational speed.
- Microhardness increases with an increase in welding speed from 16 to 25 mm/min and reaches maximum at 25 mm/min and then decreases with any further increases in welding speed from 25 to 40 mm/min.
- Tilt angle was found to be insignificant for tensile strength. However, it was found to be significant for other response.

- The optimized value of tensile stress was found to be 151.444 MPa, and microhardness was 79.991 HV which indicates a significant improvement in both responses from the initial setting of the first experiment. Moreover, the optimized results have been validated through a confirmation experiment.
- From the microstructure analysis of the welded samples, localized as well as widespread plastic deformation was visible for few samples. Lumped material deposits were observed resulting in partially plastic and partially ductile failure of weld during a tensile test. Plastic brittle failure at the deepest portion of fracture was also observed.

References

1. Thomas, W.M., Nicholas, E.D., Needham, J.C., Murch, M.G., Temple-Smith, P., Dawes, C.J.: Friction stir butt welding, International Patent Application No. PCT/GB92/02203; US Patent No. 5460317 (1995)
2. Mishra R.S., Ma, Z.Y.: Friction stir welding, and processing. J. Mater. Sci. Eng. **50**, 1–78 (2005)
3. Boz, M., Kurt, A.: The influence of stirrer geometry on bonding and mechanical properties in the friction stir welding process. Mater. Des. **25**, 343–347 (2004)
4. Kumar K and Kailash Satish V.: On the role of axial load and effect of interface position on the tensile strength of a friction stir welded aluminium alloy. Materials Des. **29**, 791–797 (2008)
5. Sarsilmaz, F., Caydas F.: Statistical analysis of mechanical properties of friction-stir-welded AA 1050/AA 5083 couples. Int. J. Adv. Manufact. Technol. **43**(3–4), 248–255 (2009)
6. Elanchezhiana, C., Vijaya Ramnath, B., Venkatesan, P., Sathish, S., Vignesh, T., Siddharth, R.V., Vinay, B., Gopinath, K.: Parameter optimization of friction stir welding of AA8011-6062 using mathematical method. Procedia Eng. **97**, 775–782 (2014)
7. Elatharasan, G., Senthil Kumar, V.S.: An experimental analysis and optimization of process parameter on friction stir welding of AA 6061-T6 aluminium alloy using RSM. Procedia Eng. 1227–1234 (2012)
8. Sergey Malopheyev, A.N., Igor Vysotskiy, A., Vladislav Kulitskiy, A., Sergey Mironov, B., Rustam Kaibyshev: Optimization of processing-microstructure-properties relationship in friction-stir welded 6061-T6 aluminium alloy. Mater. Sci. Eng., A **662**, 136–143 (2016)
9. Hamilton-Carter, Dymek, S., Blicharski, M.: Mechanical properties of AA 6101-T6 welds by friction stir welding and metal inert gas welding. Arch. Metall. Mater. **52**, 67–72 (2007)
10. Lockwood, Reynolds: Digital Image Correlation for Determination of Weld and Base Metal Constitutive Behaviour. USC, Department of Mechanical Engineering, USA
11. Ross, P.J.: Taguchi techniques for quality engineering. McGraw-Hill Book Company, New York (1988)
12. Roy, R.K.: Design of experiments using the Taguchi approach. Wiley (2001)

Evolutionary Optimization in Master Production Scheduling: A Case Study

Radhika Sajja, K. Karteeka Pavan, Ch. Srinivasa Rao and Swapna Dhulipalla

Abstract Over the last decade, the world has transformed from a marketplace with several large, almost independent market, to a highly integrated global market demanding a wide variety of products that comply with high quality, reliability, and environmental standards. Production scheduling system is one among the management tools that is widely used in manufacturing industries proving its capabilities and effectiveness through many success stories. Effective management of such production systems can be achieved by identifying customer service requirements, determining inventory levels, creating effective policies and procedures for the coordination of production activities. The aim of the current work is to benefit engineers, researchers and schedulers understand the factual nature of production scheduling in vibrant manufacturing systems and to reassure them to ponder how manufacture scheduling structures can be improved. A real world case of a large scale steel manufacturing industry is considered for this purpose and an evolutionary based meta-heuristic paradigm viz Teaching Learning Based Optimization method (TLBO) is used for creation of an effective Master Production Schedule (MPS) by the selection of optimum process parameters. The investigation on various data sets proved that the suggested algorithm is robust, automatic and efficient in finding an optimal master production schedule when compared to the currently followed conventional approach.

R. Sajja (✉) · S. Dhulipalla
Department of Mechanical Engineering, RVR & JC College of Engineering (A), Guntur, Andhra Pradesh, India
e-mail: sajjar99@gmail.com

K. K. Pavan
Department of Computer Applications, RVR & JC College of Engineering (A), Guntur, Andhra Pradesh, India
e-mail: karteeka@yahoo.com

Ch. S. Rao
Department of Mechanical Engineering, AU College of Engineering (A), Visakhapatnam, Andhra Pradesh, India

© Springer Nature Singapore Pte Ltd. 2020
R. Venkata Rao and J. Taler (eds.), *Advanced Engineering Optimization Through Intelligent Techniques*, Advances in Intelligent Systems and Computing 949, https://doi.org/10.1007/978-981-13-8196-6_33

Keywords Master production scheduling · Multi-objective optimization · Discrete manufacturing · Evolutionary algorithms · Teaching-learning-based optimization

1 Introduction and Brief Survey of Literature

Numerous production related problems, comprising extreme work-in-process inventories and little machine consumption, can be attributed directly to poor scheduling. MPS is a function in Production Management System which provides mechanism for the dynamic interaction between marketing and manufacturing. Henceforth, incorrect choice in the MPS improvement may lead to infeasible implementation, which eventually causes meagre delivery performance. Since this type of scheduling problems enforce many other constraints that are not typically present in old-fashioned job shop scheduling, the generation of the optimized MPS has been more complex especially when objectives, like maximization of resource utilization, service level, and minimization of overtime,, chance of occurring stock outs and also minimization of inventory levels and setup times, are considered. The present work attempts to propose more efficient approach for MPS formation that is able to deal with the said obstacles.

Although MPS is not a new field, there is little amount of research on the application of heuristic and intelligent algorithms in MPS optimization problems [1]. Numerous investigators have applied the TLBO algorithm to their research problems. Few works related to scheduling using TLBO were also reported in the literature, which include the work of Kumar et al. [2] for the problem of short-term Hydro Thermal Scheduling (HTS) and the work of Keesari and Rao [3] for Job Shop Scheduling Problem (JSSP). Johnson and Kjellsdotter [4] have proved that sophisticated MPS methods directly affect the performance regardless the context. In the work of Radhika et al. [5], application of TLBO for developing an effective MPS is described and the suggested approach is compared with same data set done by Soares [6], using GA. The results demonstrated that the TLBO technique produced more optimal parameter values when compared to those obtained from GA.

Close examination of the literature discloses that ample work has not been described about the application of meta-heuristic techniques for solving MPS problems. Although evolutionary computation methods offer solutions that combine computational efficiency and good performance, evolutionary computational research has been criticized for many years about the consideration of non-natural test problems that are much meeker than real-life business cases. The current work makes a modest attempt to implement computer generated evolutionary computations for real-time complex production scheduling problems.

2 Problem Description

A real world case is considered in the present work. A huge scale steel manufacturing industry is selected for this purpose. The data is collected from an Indian based large scale steel industry, which was incorporated in the year 1999 as a public limited corporation. The company functions in different divisions few of which include the Rolling Division, Wire Drawing, Steel Melting Division, Trading Division, HC Wire Products division, Sponge Iron Division and Galvanized Wire Division. The present work considered the preparation of an efficient master production schedule using the proposed methodology for the products of wire drawing division. The major products of wire drawing units are high carbon wire products which include: pc wire, indented wire and spring steel wire; mild steel wire which include binding wire, gauge wire, straight rods. The problem of optimally allocating inventories at the production facility is a major issue. As these are nonperishable goods, they can be kept in the inventories for longer periods of time and hence the holding cost is quite high. The only concern is how much production should be there so that inventory holding cost is minimal while fulfilling the demands. All these conflicting objectives are grouped together and weightage factors namely c_1, c_2 and c_3 are assigned to each of the objectives, so that the choice of weight values lies in the hands of the decision maker, which will in turn depend on the priorities of the company. The objectives included are of conflicting type like maintaining more Ending Inventory (EI) levels at the completion of any period can always ensure of better service levels. These service levels are measured in the quantitative form of the amount of Requirements Not Met (RNM) i.e. to say a low value of RNM assures of better filled in requirements. The mathematical model and representations employed in the current paper are taken from the work of same author, Radhika et al. [7].

Minimize,

$$O = c_1 \text{EI} + c_2 \text{RNM} + c_3 \text{BSS} \tag{1}$$

Since the selected objective function is a minimization problem, a small O leads to a solution closer to the optimum. O cannot be directly used as a possible fitness function. One can, however, make a change to this function so that it can be considered as a measure of individual fitness, given as follows

$$\text{fitness} = \left[\frac{1}{1 + O_n} \right] \tag{2}$$

$$O_n = c_1 \frac{\text{EI}}{\text{EI}_{\text{max}}} + c_2 \frac{\text{RNM}}{\text{RNM}_{\text{max}}} + c_3 \frac{\text{BSS}}{\text{BSS}_{\text{max}}} \tag{3}$$

EI_{max}, RNM_{max} and BSS_{max} are the maximum values found during 'warm-up' from the initial population created. The subsequent sections give the details of the problem considered and the solution obtained using and TLBO. The steel industry under consideration has a planning horizon of twelve periods, six products and four resources (machines or production lines for instance).

As presented in Table 1, the industry considered different quantities (tonnes of units) of initial inventory quantity for each product and proposes to hold different quantities of safety stock in different periods. Table 1 also presents the gross require-

Table 1 Gross requirements for scenario

Product	Initial inventory	Periods											
		1	2	3	4	5	6	7	8	9	10	11	12
P 1	440	1470	1300	1710	1090	1440	2000	1860	2500	1560	1140	1450	1320
P 2	350	1320	1520	1560	1660	1530	1590	2080	2460	2200	1740	1200	1360
P 3	220	1360	1430	1120	1590	1740	1680	1950	2910	2810	1710	2120	2420
P 4	500	1450	1100	1480	1410	1130	1520	1950	2120	1580	1880	1890	2100
P 5	450	1270	1360	1680	2080	1430	1120	1280	2670	2080	2080	2000	1890
P 6	490	1420	1720	1760	1850	2000	1860	2820	2450	1950	1950	1740	2340
Safety Stock		200	180	300	250	170	220	250	200	180	290	210	230

Table 2 Product wise production rates on different resources

Product code	Product name	Production resources			
		Resource 1	Resource 2	Resource 3	Resource 4
P1	PC wire	400	100	150	200
P2	Indented wire	150	200	400	150
P3	Spring steel wire	200	350	150	100
P4	Binding wire	400	100	300	350
P5	Gauge wire	200	150	100	350
P6	Steel wire	300	200	250	200

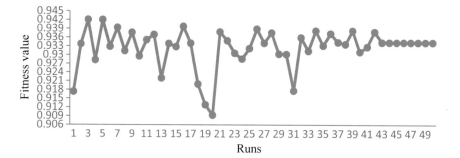

Fig. 1 Convergence of fitness values

ments of all the products for each period. The following Table 2 presents different production rates for each product on each different resource.

3 Results

Teaching Learning Based Optimization technique is applied to the above production scenario, using the objective function given in Eq. (2), with a teaching factor 1.8. Before the actual implementation of the algorithm we have run it several times and have noticed that the algorithm provides more optimal solutions when the teaching factor is 1.8. However according to Rao et al. [8], [9] the value can range from 1 to 2. The termination criterion is taken as difference between any two successive fitness values to be less than 0.00001. The procedure is allowed to run for 50 times where in the termination criteria is reached as presented in Fig. 1 and the best MPS thus obtained, corresponding to the largest fitness value is given in Table 3. For conciseness best MPS values for products P3–P5 are not given in table.

Figure 2 describes the way the four production resources are utilized to process the product P4 across all the twelve periods.

Table 3 Best MPS obtained

Products	Resources	Periods											
		1	2	3	4	5	6	7	8	9	10	11	12
P1	Res1	450	350	400	150	350	500	500	600	600	300	400	500
	Res2	450	200	400	50	350	500	500	600	300	300	300	100
	Res3	50	350	450	250	300	500	500	650	650	150	350	150
	Res4	500	400	450	600	400	500	350	650	0	350	400	550
	Total MPS	1450	1300	1700	1050	1400	2000	1850	2500	1550	1100	1450	1300
P2	Res1	350	350	500	500	350	650	350	650	550	250	350	300
	Res2	350	350	500	500	350	200	750	450	550	150	100	350
	Res3	200	400	350	300	400	700	150	650	500	1250	350	350
	Res4	400	400	200	350	400	0	800	700	600	50	400	350
	Total MPS	1300	1500	1550	1650	1500	1550	2050	2450	2200	1700	1200	1350
P6	Res1	400	400	400	450	550	50	700	250	200	600	400	550
	Res2	350	400	450	450	550	300	750	700	850	150	400	600
	Res3	400	450	450	450	550	650	600	750	0	600	450	600
	Res4	250	450	450	500	350	850	750	750	900	600	450	550
	Total MPS	1400	1700	1750	1850	2000	1850	2800	2450	1950	1950	1700	2300

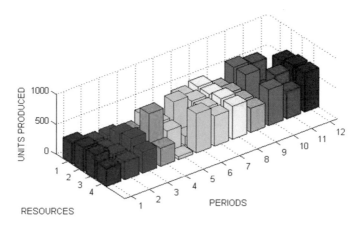

Fig. 2 Period wise utilization of resources for product P4

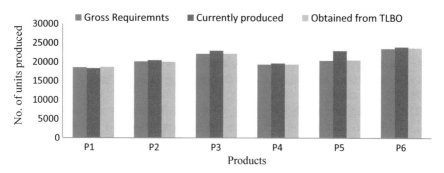

Fig. 3 Comparison of products produced

Table 4 Performance Measures

	EI (units/h)	RNM (units/h)	BSS (units/h)
Obtained from current production	157.8	2.8	3.94
Obtained from TLBO	111.6	3.04	3.06
% Reduced	29.2	8.5 (increased)	22.3

As seen from Fig. 3 which gives a comparison of total quantities produced by the existing master schedule with those obtained when generated from TLBO. The figure clearly advocates that the master schedules produced by the TLBO method are in close estimate with the input data.

Table 4 reports the values of performance measures obtained in the proposed methodology. The MPS that was developed through the suggested method was compared with the industry data of quantities that are being produced currently, it can be established that the percentage of unfilled demands is very little and at the same time sustaining very fewer inventory levels. The values in Table 4 also conclude that the proposed TLBO methodology could effectively reduce the Ending Inventory (EI) levels by almost 29% which is of major concern to the steel industry. The results given in Table 4 report the huge decrease in the primary objective (reducing EI) of the industry, along with a remarkable decrease in the Below Safety Stock units. However there is marginal increase in RNM, which cannot be considered as a drawback of the proposed solution. The choice is left to the decision maker to choose on his preferences among the objectives.

4 Conclusions

Manufacturing industry is always complex and a challenging environment because of the unpredictable uncertainties and complications that might occur anytime throughout the production operation. To preserve a profitable growth, a company must develop different approaches to reduce its operational cost. These approaches raise the demand for new and evolved management systems. Production scheduling system is one of the management tools that is widely used in manufacturing industries proving its capabilities and effectiveness through many success stories. In the present work, an evolutionary based meta-heuristic approach namely, Teaching Learning Based Optimization algorithm is used in optimization of parameters of a Master Production Schedule.

The implementation of the proposed production scheduling system shows positive and encouraged improvement based on the data collected and is compared with the data before the implementation, where significant amount of time is saved by using the scheduling method in managing flow of processes and orders. The results obtained were found to be in line with the currently employed master schedules in the company. The same methodology could be applied to the rest of the products too. The output from the tested evolutionary methodology, TLBO is found to be in close approximation with the gross requirements, suggesting that the proposed method could be used for optimizing the raw material inputs for the production of different products under study.

The TLBO algorithm is being applied in many fields, a review of all such applications can be found in the publication of Rao [9].

References

1. Wu, Z., Zhang, C., Zhu, X.: An ant colony algorithm for master production scheduling optimization. In: Proceedings of the 2012 IEEE 16th International Conference on Computer Supported Cooperative Work in Design (2012)
2. Kumar Roy, P., Sur, A., Pradhan, D.K.: Optimal short-term hydro-thermal scheduling using quasi-oppositional teaching learning based optimization. Eng. Appl. Artif. Intell. 26(10), 2516–2524 (2013)
3. Keesari, H.S., Rao, R.V: Optimization of job shop scheduling problems using teaching-learning-based optimization algorithm. OPSEARCH, 1–17 (2013)

4. Jonsson, P., Kjellsdotter Ivert, L.: Improving performance with sophisticated master production scheduling. Int. J. Prod. Econ. **168**, 118–130 (2015)
5. Radhika, S., Rao, C.S., Swapna, D., Sekhar, B.P.C: Teaching learning based optimization approach for the development of an effective master production schedule. Int. J. **7**(5) (2017)
6. Soares, M.M., Vieira, G.E.: A new multi-objective optimization method for master production scheduling problems based on genetic algorithm. Int. J. Adv. Manufact. Technol. **41**(5), 549–567 (2009)
7. Radhika, S., Rao, C.S.: A new multi-objective optimization of master production scheduling problems using differential evolution. Int. J. Appl. Sci. Eng. **12**(1), 75–86 (2014)
8. Rao, R.V., Kalyankar, V.D.: Parameter optimization of modern machining processes using teaching–learning-based optimization algorithm. Eng. Appl. Artif. Intell. **26**(1), 524–531 (2013)
9. Rao, R.: Review of applications of TLBO algorithm and a tutorial for beginners to solve the unconstrained and constrained optimization problems. Decis. Sci. Lett. **5**(1), 1–30 (2016)

Optimal Irrigation Planning in the Eastern Gandak Project Using Linear Programming—A Case Study

M. Bhushan and L. B. Roy

Abstract Linear programming was used for conjunctive use of canal water and groundwater to develop a cropping pattern for optimal net saving to the farmers in Bhagwanpur Distributary of the Eastern Gandak project. Constraints like total land availability, total water availability, fertilizer availability, labour availability and crop affinity were used. Average hour of operation in a year and average discharge of tube wells were estimated to calculate the total groundwater availability in a year. Six different combinations of canal water and groundwater were used. The optimum result gave the maximum net saving of Rs. 26,346/ha for the conjunctive use of 90% availability of both canal water and groundwater.

Keywords Linear programming · Conjunctive use · Cropping pattern

1 Introduction

There is an urgent need to produce more grains to feed the growing population on the globe. For this, either the area under the agriculture should be increased or yield per unit area should be increased. The present cropping pattern might not be satisfactory in fulfilling the need. Therefore, there is need to best utilize the scarce water resources and accordingly plan the cropping pattern. Frizzone et al. [1] developed a separable linear programming model and considered a set of factors which influence the irrigation profit. They also made a model that maximizes the net income in specific range of water availability. Raju and Kumar [2] have developed a fuzzy linear programming model for the evaluation of management strategies for the case study of Sri Ram Sagar project. They considered three conflicting objectives, i.e. net benefit, crop production and labour employment for irrigation planning. They concluded that

M. Bhushan · L. B. Roy (✉)
NIT Patna, Patna, India
e-mail: lbroy@nitp.ac.in

M. Bhushan
e-mail: mani.tuntun@gmail.com

© Springer Nature Singapore Pte Ltd. 2020
R. Venkata Rao and J. Taler (eds.), *Advanced Engineering Optimization Through Intelligent Techniques*, Advances in Intelligent Systems and Computing 949,
https://doi.org/10.1007/978-981-13-8196-6_34

381

there have been decreases in net benefit, crop production and labour employment by 2.38, 9.6 and 7.22%, respectively, in fuzzy linear programming as compared to the ideal value in the crisp linear programming model. Sethi et al. [3] developed two models to determine optimum cropping pattern and allocation of groundwater from government and private tube wells, according to soil type, agriculture and season for a coastal river basin in Odisha. Gajja et al. [4] studied the effect of land irrigability classes on the productivity of the crops. In the poor irrigability land classes, the growing of high-water requiring crops has led to the twin problem of waterlogging and secondary salinization. They concluded that crops selection should be according to land irrigability class to increase the production and minimize the input cost. Khare et al. [5] studied feasibility of conjunctive use planning in a link canal. An economic engineering optimization model was used to explore conjunctive use of groundwater and surface water resources using linear programming. With conjunctive use, they saved a considerable quantity of surface water for use in other needy area. Gore and Panda [6] used single objective linear programming allocation model and multi-objective fuzzy technique to allocate land under selected crops in the Aundha Minor Irrigation Scheme in Maharashtra, India, so as to maximize the net benefit and production. However, they advised farmers to advocate the optimum cropping pattern obtained from multi-objective allocation model for better return. Regulwar and Gurav [7] developed irrigation planning model for crop planning in irrigated agriculture. The multi-objective fuzzy linear programming was used by them to maximize various objectives, i.e. net benefit, yield production, employment generation and manure utilization. These objectives were maximized simultaneously by considering the decision-makers' satisfaction level. The model was found to be helpful to take decision under conflicting situation. Mirajkar and Patel [8] used crisp linear programming to get solution of maximization of three conflicting objectives with both maximization and minimization. The cost of cultivation was minimized, and employment generation and net benefit were maximized. The applicability of the method was demonstrated through a case study in Kakrapar Right Bank Main Canal (KRBMC) under Ukai command area in India.

Srivastava and Singh [9] studied fuzzy multi-objective goal programming-based optimal crop planning in Soraon Canal Command Area in U.P., India. They found that the scenario, with 50% of canal water availability, achieved the maximum satisfaction level because of its lower limit in all cases.

Therefore, in the present paper effort has been made to optimize the net saving using linear programming for the command area of Bhagwanpur Distributary of Eastern Gandak project. The Bhagwanpur Distributary is between latitude $25°\ 52'\ 30''$N to $26°\ 3'\ 0''$N and longitude $85°7'\ 30''$E to $85°15'\ 0''$E of Vaishali Branch Canal (VBC), which takes off from Tirhut Main Canal at 553.89 RD (1 RD = 1000 ft) as given in Fig. 1. VBC runs up to 48 km from its head regulator and thereafter it is known as Bhagwanpur Distributary which is 33 km in length. Its total cultivable command area is 1841 ha, and its gross command area is 2250 ha. The main crops are rice and wheat acquiring nearly 80% of the crop area. These are also the most water requiring crops practised in the area. The cropping intensity during kharif, rabi and hot weather are 59.71, 36.7 and 5%, respectively. The temperature during

Fig. 1 Location map of the study area

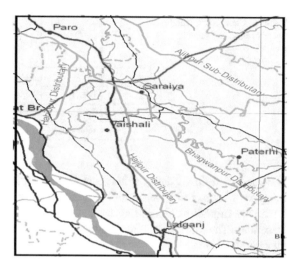

summers is nearly 35–42 and 18–29 °C during winters. Heavy rain is witnessed during monsoons seasons. Little rains are observed sometimes during summers and winters. The maximum rain occurs during July to September. Crops like rice, wheat and green gram are favoured by the climatic conditions and banana is the perennial crop in the area.

2 Input Data and Methodology

2.1 Input Data

Canal Water. The field canal efficiency and application efficiency were assumed to be 70 and 59%, respectively, from the research carried out in similar soil condition [10]. The availability of canal water during kharif season for the period from 2012 to 2014 is given in Table 1. The charges for canal water are as given in Table 2.

Table 1 Monthly average discharges and water availability in the study area

Month	2012 (m^3)	2013 (m^3)	2014 (m^3)	Average (ha-cm)	Net volume (ha-cm)
July	5123	823,052	1,143,774	6573.01	1932.5
August	18,254	3,206,445	2,953,017	20,592.38	6054.15
September	30,963	3,439,127	1,563,362	16,778.17	4932.77
October	20,078	583,752.9	2,152,987	2730.43	802.74

Table 2 Charges for canal water in Eastern Gandak project

S. no.	Crop	Canal water charge (Rs./Acre)		
		(1983–1995)	(1995–2001)	After rabi 2001
1.	Rice	36.20	70	88
2.	Wheat	20.70	60	75
3.	Sugarcane	63.80	120	150
4.	Vegetables	63.80	120	150

Rainfall: Rainfall data for 36 years (1981–2016) are taken from Pusa meteorological station. Effective rainfall was calculated using USDA SCS method. The maximum rainfall occurs in the month of July and the minimum in the month of December. The four months from June to September have the most of the rainfall in the command area.

Groundwater: The total number of tube wells in the command area was found to be 971. The average discharge of the tube wells has been estimated as 54 m³/h, and the average operation of tube well has been estimated as 87 h/year. Based on this the total available groundwater was estimated to be 45,617.58 ha cm.

Agriculture: Crop coefficient (K_c) data were taken from the FAO-56 [11]. The other data were taken from the detailed project report of Gandak Command Area Development Agency.

Climatological Data: The 30-year data of the nearby station Pusa were used to determine reference crop evapotranspiration using CROPWAT 8.0 software.

Labour Requirements: Table 3 shows the labour requirements, and these data are taken from the detailed project report of Gandak Command Area Development Agency.

Cost of Production: Various inputs by the farmer for the production of crops are seed cost, fertilizer cost, land preparation cost, weeding, harvesting and thrashing cost. Table 3 shows the requirement of seed per hectare and the unit cost of seed.

Net Income: Farmers get income from the both yield of product and by-product. For some of the crops like potato, mustard and green gram, there is no by-product. Yield of the products and by-products together with their unit costs is shown in Table 4.

2.2 Methodology

Objective Function: For linear programming model, objective function is given as

$$\text{Max}(f) = \sum_{j=1}^{3}\sum_{k=1}^{7} \text{NI}_{ij} * \text{LA}_{ij} - \text{CWC}\sum_{j=1}^{3}\text{CWA}_j - \text{GWC}\sum_{i=1}^{7}\text{TOP}_i \quad (1)$$

Table 3 Labour, seed and fertilizer requirements for crops

S. no.	Crop	Labour requirement (man-days)/ha	Seed requirement kg/ha	Unit cost of seed (Rs./kg)	Fertilizer requirement		
					N (kg/ha)	P (kg/ha)	K (kg/ha)
1.	Rice	162	50	30	100	50	50
2.	Wheat	73	100	45	100	50	50
3.	Mustard	30	5	80	70	60	35
4.	Lentil	60	50	80	40	60	30
5.	Potato	85	2000	25	120	60	60
6.	Maize	84	20	50	120	60	60
7.	Green gram	50	25	100	40	60	30

Table 4 Yield and unit cost of product and by-product

Crop	Yield of product (Kg. per ha)	Unit cost of product (Rs. per kg)	Yield of by-product (Kg per ha)	Unit cost of by-product (Rs. per kg)
Rice	3000	15	3000	0.42
Wheat	3000	15	2500	4
Mustard	1250	50	–	–
Lentil	900	65	1000	5
Potato	28,000	7	–	–
Green gram	750	35	–	–
Maize	4000	15	–	–

where

$$NI_{ij} = \left[\left(SP_{ij} * Y_{ij} \right) + \left(SPB_{ij} * YB_{ij} \right) - CP_{ij} \right] \tag{2}$$

where j = season index (1 = kharif, 2 = rabi and 3 = hot weather); i = crop index (rice, wheat, mustard, lentil, potato, maize, green gram); f = net Income from the crop; NI_{ij} = net income from the crop excluding the cost of water; LA_{ij} = land area under the crop I in the jth season; CWC = cost of canal water; CWA = canal water area under different crop irrigated In kharif season; GWC = Groundwater pumping cost; TOP_{ij} = time of pumping (in hours) for ith crop; NI_{ij} = net income from the ith crop excluding the cost of water; SP_{ij} = minimum support price of the main product of the crop; Y_{ij} = yield of the main product; SPB_{ij} = selling price of the by-product of the crop; YB_{ij} = yield of by-product of the crop; and CP_{ij} = cost of production of crop excluding the cost of water.

System Constraints: The maximization of the objective function is subjected to the constraints as per the resource limitations as discussed below.

Land Area Constraint. Land area allocated to different crops during different season must not exceed the available area in that season having the irrigation facility in the command.

$$\sum_{k=1}^{3} \sum_{i=1}^{7} LA_{ij} \leq LA_t \tag{3}$$

where LA_t = total available land with irrigation facility in different season.

Food Productivity Constraint. For fulfilling the demand of basic crops such as rice and wheat to meet the demand of the population of the area, this constraint has been imposed.

$$\sum_{k=1}^{3} \sum_{i=1}^{7} Y_i * LA_i \leq F_i \tag{4}$$

where F_i = average consumption per year of that crop * population of the area.

Crop Area Constraint: The crop permissible variation in the crop area is taken into consideration by this constraint. The variation is taken from 0.75 to 1.25 times the existing crop area.

$$Min(i) \leq LA_i \leq Max(i) \tag{5}$$

Labour Availability Constraint: There is limited availability of the human resource in the command area for the agricultural production which also affects the production of the crops. The average working day for each labour has been assumed to be 100 days per year.

$$\sum_{i=1}^{7} LR_i * LA_i \leq TAL \tag{6}$$

where LR_i = labour requirement for the ith crop; TAL = total available labour in the area (in man-days)

Canal Water Supply Constraint: There is limited availability of canal water and it is supplied only during kharif season. Therefore, kharif season water requirement must not exceed canal water supplied.

$$CWV_i \leq ACVW \tag{7}$$

where ACVW = total available canal water

Groundwater Supply Constraint: Total available tube wells in the area are 1100. The total hour of operation of each tube well is 24 h per year. The average discharge of the tube well is 12 l/s.

$$GWV_i \leq AGVW \tag{8}$$

where AGVW = (No. of tube well) × (average discharge of tube well in cumecs) × (hour of operation per year of the tube well).

Non-Negativity Constraint: The land area allocated to different crop must be positive.

$$LA_i \geq 0 \tag{9}$$

3 Results and Discussion

3.1 Total Canal Water Availability

The canal runs only during the kharif season, i.e. from 15th of July to 15th of October due to construction and extension of the irrigation scheme during the last five years. The average of total available canal water and net available canal water in the kharif season is 3,500,506 ha-cm and 13,722.16 ha-cm, respectively. The largest available volume is in the month of September, and the lowest is in the month of October as given in Table 5.

3.2 Total Groundwater Availability

The total available groundwater for a year was estimated to be 45,617.58 ha-cm.

3.3 Gross and Net Irrigation Requirements

The gross irrigation requirement (GIR) and net irrigation requirement (NIR) were determined using CROPWAT 8.0, and the results are given in Table 6.

Table 5 Total and net volume of canal water availability in different months

Month	Total volume of water (ha-cm)	Net volume of water (ha-cm)
July	4929.87	2070.54
August	15,444.28	6486.59
September	12,583.62	5285.12
October	2047.82	860.08

Table 6 Crop water requirement

Crop	Rice	Wheat	Mustard	Lentil	Maize	Green gram	Potato
GIR (mm)	982.02	267.46	82.00	101.91	336.25	211.39	193.02
NIR (mm)	497.92	222.26	42.00	80.21	132.2	79.19	126.31

3.4 Net Saving from Different Crops

In the command area, for some of the crops, labour is paid by giving them the yield, whereas in some other cases of crops like mustard and potato, labours are paid in cash. For cash payment, labour cost was considered to be Rs. 200 per day. The net savings per hectare for different crops were estimated from the field study of different cost input data, and the result is given in Table 7. Potato gives the maximum net return of Rs. 81,500 per hectare followed by lentil. Being the perishable crop, the area allocation of potato is dependent on the decision of the farmers.

Study was also carried out for six different combinations of availability of canal water and groundwater for conjunctive use. The combinations considered are as shown in Table 8, which includes availability of canal water as 100% for case 1 and 2, 90% for case 3 and 4 and 85% for case 5 and 6. This table also includes availability of groundwater as 100% for case 1, 85% for case 2, 100% for case 3, 90% for case 4, 100% for case 5 and 90% for case 6. These combinations have been considered so as to encourage conjunctive use groundwater and canal water in the study area.

3.5 Optimal Area Allocation to Different Crops Under Different Case

Total seven crops, i.e. rice, wheat, mustard, lentil, maize, green gram and potato, were taken under consideration for the allocation using linear programming model. The optimum land area allocated to the different crops and the total area to be cultivated for different cases are as shown in Table 9. Similarly, the net saving per unit area

Table 7 Net saving from different crops

Crops	Rice	Wheat	Mustard	Lentil	Potato	Green gram	Maize
Net saving (Rs./ha)	19,710	22,100	36,250	43,860	81,500	13,188	61,940

Table 8 Different cases of water availability

Scenarios	Canal water (in ha cm)	Groundwater (in ha cm)	Total water (in ha cm)	% of currently used (canal water, groundwater)
Case I	14,702 (24%)	45,617 (76%)	60,319	(100, 100)
Case II	14,702 (27%)	38,774 (73%)	53,476	(100, 85)
Case III	13,231.8 (22%)	45,617 (78%)	58,848.8	(90, 100)
Case IV	13,231.8 (24%)	41,055 (76%)	54,286.8	(90, 90)
Case V	12,496.7 (21%)	45,617 (79%)	58,113.7	(85, 100)
Case VI	12,496.7(23%)	41,055 (77%)	53,551.7	(85, 90)

Table 9 Optimal land allocation in hectares to different crops for different cases

Crop	Case I	Case II	Case III	Case IV	Case V	Case VI
Rice	629.64	621.9	621.9	621.9	621.9	621.9
Wheat	577.5	346.5	535.50	346.5	502.56	346.5
Mustard	88.2	88.2	88.2	88.2	88.2	88.2
Lentil	68.64	54.89	68.64	68.64	68.64	64.22
Potato	52.5	52.5	52.5	52.5	52.5	52.5
Maize	77.77	25.92	77.77	77.77	77.77	25.92
Moong	118.21	70.92	118.21	118.21	118.21	70.92
Total area	1612.46	1260.83	1562.72	1373.2	1529.78	1270.16

Table 10 Net saving per unit area for different scenarios

Scenarios	Case I	Case II	Case III	Case IV	Case V	Case VI
Total net saving (in Million Rs.)	42.03	32.33	40.95	**36.17**	40.22	32.75
Net saving (Rs./ha)	26,069	25,647	26,209	**26,346**	26,295	25,786

Table 11 Comparison of net saving and total water available

Decreasing order of net saving per unit area		Decreasing order of water availability		
Scenario	Net saving/ha (Rs./ha)	Scenarios	Total water available (ha cm)	Total area cultivated (ha)
Case IV	26,346	Case I	60,319	60,319
Case V	26,295	Case III	58,848.8	58,848.8
Case III	26,209	Case V	58,113	58,113
Case I	26,069	Case IV	54,286.8	54,286.8
Case VI	25,786	Case VI	53,551.7	53,551.7
Case II	25,647	Case II	53,476	53,476

for different cases is given in Table 10, from which it is observed that the maximum net saving is for case 4, i.e. Rs. 26,346 per hectare. Case 4 has given the best result among all the six cases due to availability of canal water and groundwater as 90%.

It is always seen that as the water availability increases, the land cultivated increases, but it is not necessary that increase in area of cultivation gives increase in net saving per unit area. The cases arranged in the order of decreasing net saving per unit area and the cases arranged in the order of decreasing water availability are as given in Table 11. This order is so because land area cultivation is linearly dependent on water availability till all the area gets allocated to maximum profit and crops get their water demand. However, the net saving per unit area depends on various other constraints due to which the relation between the net saving per unit area and total water available is not linear.

4 Conclusions

From the above results and discussion, the following conclusions are drawn.

1. From this study, it is concluded that LP can be used as an efficient tool in the optimal allocation of resources when we deal with the function which depends on a large number of known and unknown factors.
2. Land area cultivation is linearly dependent on water availability till all the area gets allocated to maximum profit and crops get their water demand. However, the net saving per unit area depends on various other constraints due to which the relation between the net saving per unit area and total water available is not linear.
3. Six different scenarios for combinations of canal water and groundwater were developed. The best obtained condition was for the fourth case, which gave maximum net saving of Rs. 26,346/ha per unit area.

References

1. Frizzone, J.A., Coelho, R.D., Dourado-Neto, D., Soliani, R.: Linear Programming Model to Optimize the Water Resource Use in Irrigation Projects: An Application to the Senator Nilo Coelho Project (1997)
2. Raju, K.S., Kumar, D.N.: Irrigation planning of Sri Ram Sagar project using multi objective fuzzy linear programming. ISH J. Hydraul. Eng. 6(1), 55–63 (2000)
3. Sethi, L.N., Kumar Nagesh, D., Panda, S.N., Mal, B.C.: Optimal crop planning and conjunctive use of water resources in a Coastal River Basin. J. Water Resour. Manag. (2002)
4. Gajja, B.L., Chand, K., Singh, Y.V.: Impact of land irrigability classes on crop productivity in canal command area of Gujrat: an economic analysis. Agric. Econ. Res. Rev. 19, 83–94 (2006)
5. Khare, D., Jat, M.K., Deva, Sunder J.: Assessment of water resources allocation options: conjunctive use planning in a Link Canal command. Resour. Conserv. Recycl. 51, 487–506 (2007)
6. Gore, K.P., Panda, R.K.: Development of multi objective plan using fuzzy technique for optimal cropping pattern in command area of Aundha minor irrigation project in Maharashtra State (India). In: International Conference on Computer and Computing Technologies in Agriculture, pp. 735–741. Springer, Boston, MA. (2008)
7. Regulwar, D.G., Gurav, J.B.: Fuzzy approach based management model for irrigation planning. J. Water Resour. Prot 2(6), 545–554 (2010)
8. Mirajkar, A.B., Patel, P.L.: Optimal irrigation planning using multi-objective fuzzy linear programming models. ISH J. Hydraul. Eng. 18(3), 232–240 (2012)
9. Srivastava, P., Singh, R.M.: agricultural land allocation for crop planning in a canal command area using fuzzy multi-objective goal programming. J. Irrig. Drainage Eng. 143(6), 04017007 (2017)
10. Singh, A.: Optimal allocation of resources for the maximization of net agricultural return. J. Irrig. Drainage Eng. 138, 830–836 (2012)
11. Allen, R.G., Pereira, L.S., Raes, D., Smith, M.: Guideline for Computing Crop Water Requirement. Irrigation and Drainage paper 56, FAO Rome (1998)

Optimization of Squeeze Casting Process Parameters to Investigate the Mechanical Properties of AA6061/Al$_2$O$_3$/SiC Hybrid Metal Matrix Composites by Taguchi and Anova Approach

L. Natrayan, M. Senthil Kumar and Mukesh Chaudhari

Abstract Hybrid metal matrix composites (HMMC) have gained importance in industrial applications owing to their tailorable properties. The properties of HMMC are influenced by the casting process parameters. In this work, AA6061-reinforced Al$_2$O$_3$/SiC HMMC were fabricated by squeeze casting technique. The effect of four levels and four input parameters are squeeze pressure, pressure holding time, die material, and melt temperature is investigated. L16 orthogonal array was selected to perform the experiments. Taguchi technique identified best process parameters that exhibited highest hardness, ultimate tensile strength (UTS), and yield strength. ANOVA demonstrated the contribution ratio. The process parameters 100 MPa squeeze pressures, 20-s holding time, die steel material, and 750 °C melt temperature showed optimal value for hardness, UTS, and yield strength.

Keywords Squeeze casting · Taguchi technique · Anova · Hardness · UTS · Yield strength · Process parameters · HMMC

1 Introduction

Aluminum reinforced with ceramic particulate called as aluminum metal matrix composite (AMC's). Aluminum alloy find it difficult to meet the increasing scope for heavy and modern application. AMCs improved mechanical as well as microstructural properties such as strength, wear resistance, low thermal expansion, stiffness, high strength-to-weight ratio, and tensile strength [1]. Mechanical properties of AMCs are affected by the fabrication process and reinforcement type [2]. AMCs serves as a suitable alternate for automobile, defense, aeronautical, and some other application [3]. The most commonly used reinforcements are Al$_2$O$_3$ and SiC. The reinforcement Al$_2$O$_3$ offered good wear resistance, tensile strength, hardness, and density [3]. Wear resistance increase with the addition of SiC in to Al alloy

L. Natrayan (✉) · M. Senthil Kumar · M. Chaudhari
School of Mechanical and Building Sciences, VIT, Chennai 600127, Tamil Nadu, India
e-mail: natrayanphd@gmail.com

© Springer Nature Singapore Pte Ltd. 2020
R. Venkata Rao and J. Taler (eds.), *Advanced Engineering Optimization Through Intelligent Techniques*, Advances in Intelligent Systems and Computing 949,
https://doi.org/10.1007/978-981-13-8196-6_35

Table 1 Chemical composition of Al 6061

Element	Al	Mg	Si	Fe	Cu	Mn	Cr	Zn	Ti
Weight %	97.82	0.94	0.53	0.23	0.17	0.14	0.08	0.07	0.02

[4]. From the literature review, mechanical properties of AMC's fabricated by stir casting and powder metallurgy technique reported agglomeration and porosity [5]. In order overcome the defect, the squeeze casting technique has been selected [6]. The effect of process parameters on mechanical properties of Al 6061/Al$_2$O$_3$/SiC is studied by taguchi analysis. Design of experiments (DOE) method were carried out to improve the progress on overall performances of the fabricated AMC by using taguchi technique to optimizing the process parameters such as squeeze pressure, pressure holding time, die material, and melt temperature. DOE technique is used for exploring the potential conditions implicate in investigates such as variables, parameters, and multiple factor. Analysis of variance (ANOVA) was applied to evaluate the contribution percentage of process parameter on hardness, UTS, and yield strength.

2 Experimental Method

2.1 Materials and Methods

Al6061-T6 was purchased from Bharat aerospace metals-Mumbai and chemical compositions listed in Table 1. Al$_2$O$_3$/ SiC/Gr particles were purchased from scientific research laboratory (mean particle size is 10 μm).

2.2 Composites Preparation

Al 6061-T6 is selected as the base material, Al$_2$O$_3$ (5wt%)/ SiC (5wt%) is selected as reinforcement. The average size of the reinforcement is 10 μm. Reinforced matrix improved mechanical behavior up to 10 wt% and decreased the mechanical properties with increasing reinforcement wt% due to the agglomeration of the hard reinforcement particles that leads to porosity [5, 6]. The squeeze casting setup used for HMMC fabrication is shown in Fig. 1. A hydraulic stainless steel plunger controls the squeeze pressure during solidification. It consists of base die and two steel speacer. The cylinder-molded die is considered to distribute continuous pressure, ensure discharge of the cast, and minimize the leakage. The cylinder die cavity measures 50 mm diameter and 250 mm length [7]. Casted specimens were machined according to the ASTM standards for evaluating the mechanical properties.

Fig. 1 Squeeze casting
machine arrangement

3 Materials Characterizations

3.1 Hardness

Vickers hardness test is taken to measure the hardness on fabricated composite as per ASTM E10 standard. Fabricated composite samples were cut to the dimension $15 \times 25 \times 10$ mm and polished with abrasive paper. The diamond ball indenter is exposed to the load of 0.5 kgf for 10 s. Hardness measurement was carried out at two altered positions towards challenge the potential result of indenter on the firmer elements, finally taken the average of all the two measured results [8]. It has been noted that fabricated composite hardness was high compared to Al6061 base alloy. This shows the perfect bonding between the ceramic particle and the matrix phase. The result of the hardness test represents that the matrix containing reinforcement showed high hardness due to the presence of hard reinforcement in the matrix phase that enabled good resistance to indentation [9].

3.2 UTS and Yield Strength

ASTM E8M standard was considered for preparing tensile test specimens. Figure 2 shows the machined tensile test specimens. The UTE100 FIE model universal tester was utilized for studying the tensile and yield strength of the hybrid samples under ambient temperature (29 °C). The results indicate that the UTS and yield strength

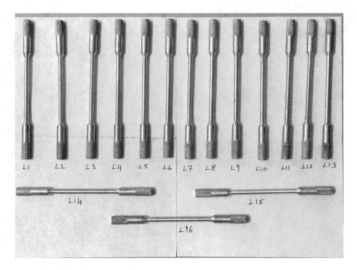

Fig. 2 Tensile testing specimens

increase concerning the modified squeeze casting process parameters. Al_2O_3/SiC particles reinforcement restricts the dislocation motion; this results in the improvement of yield and ultimate tensile strengths. The phenomenon shows that also there is better interfacial bonding between the matrix and the reinforcement [10].

4 Results and Discussion

4.1 Statistical Analysis

Taguchi's technique indicates the experimental and analytical perceptions in order to determine the most significant parameter on the results [11]. Taguchi technique employed reduced the number of experiments. Taguchi robust design is a powerful tool for the design of a high-quality system.

$$\frac{S}{N} = -10\,log\left(\frac{1}{n}\sum_{i=1}^{n}\frac{1}{y^2}\right) \quad \text{(larger is better)} \tag{1}$$

The process variables and their corresponding ranges considered is shown in Table 2. L16 orthogonal arrays with sixteen experimental runs were selected. Sixteen specimens were fabricated according to the experimental design shown in Table 3. Hardness, UTS, and yield strength of the prepared composite tested were recorded. Obtained response value were taken.

Table 2 Squeeze casting factors and their levels

Factors	Symbols	Units	Level 1	Level 2	Level 3	Level 4
Squeeze pressure	A	MPa	60	80	100	120
Holding time	B	Sec	10	20	30	40
Die material	C	–	DS	CS	SGI	SS
Melt temperature	C	°C	700	750	800	850

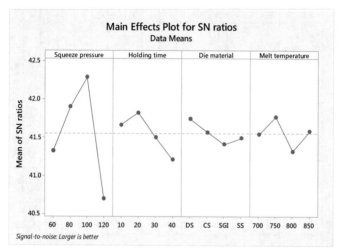

Fig. 3 Response curve (S/N ratio) of hardness

Die materials such as die steel (DS), carbon steel (CS), SG iron (SGI), and stainless steel acts as four level in this optimization, degree of freedom (DOF) $= N - 1$, Where, N = number of level. For each factor, DOF equal to: For squeeze pressure (A): DOF $= 4 - 1 = 3$, holding time (B): DOF $= 4 - 1 = 3$, die material $(C) = 4 - 1 = 3$, melt temperature (D): DOF $= 4 - 1 = 3$. DOF for the experiment is $= 16 - 1 = 15$, by varying four factors at four levels optimization is carried out.

Table 3 provides the final response and S/N ratios calculated for the important parameters yield, ultimate tensile strength, and hardness. Tables 4 and 5 represent adjusted YS and UTS. Similarly, Figs. 3, 4, and 5 depict the effect of S/N ratio curves.

By using Taguchi's method for optimization, the response variables like the hardness, UTS, and yield strength can be determined by the means of S/N ratio values. The experimental results of prepared composites were measured as the quality distinguishing with the perception of "the large the better" [10–14].

Figure 3 shows that the response curve (S/N ratio) for hardness. It shows squeeze pressure that increase from 60 to 100 MPa and decreases from 100 to 120 MPa, and here, that optimum squeeze pressure 100 MPa shows maximum hardness. Pressure holding time increases from 10 to 20 s and decreases beyond 20 s. Pressure holding time at 20 s shows maximum optimum value. The die material Die steel shows better

Table 3 Orthogonal array (L16) with S/N ratio for hardness, UTS, and yield strength

Squeeze pressure (MPa)	Holding time (sec)	Die material (°C)	Melting temp (°C)	Hardness (HV)	UTS (MPa)	Yield strength (MPa)	S/N ratio for hardness	S/N ratio for UTS	S/N ratio for yield strength
60	10	SS	700	120	328	280	41.584	50.317	48.943
60	20	SGI	750	118	324	286	41.438	50.211	49.127
60	30	CS	800	113	320	279	41.062	50.103	48.912
60	40	DS	850	115	317	270	41.214	50.021	48.627
80	10	SGI	800	121	322	278	41.656	50.157	48.881
80	20	SS	850	128	318	285	42.144	50.049	49.097
80	30	DS	700	123	312	280	41.798	49.883	48.943
80	40	CS	750	126	308	286	42.007	49.771	49.127
100	10	CS	850	129	320	289	42.212	50.103	49.218
100	20	DS	800	137	337	298	42.734	50.553	49.484
100	30	SS	750	132	329	291	42.411	50.344	49.278
100	40	SGI	700	123	322	295	41.798	50.157	49.396
120	10	DS	750	115	320	292	41.214	50.103	49.308
120	20	CS	700	112	315	287	40.984	49.966	49.158
120	30	SGI	850	109	306	280	40.749	49.714	48.943
120	40	SS	800	98	297	276	39.825	49.455	48.818

Table 4 Mean of S/N ratio for process parameters using larger is better (hardness)

Level	Squeeze pressure	Holding time	Die material	Melt temperature
1	41.32	41.67	41.74	41.54
2	41.90	41.83	41.57	41.77
3	42.29	41.50	41.41	41.32
4	40.69	41.21	41.49	41.58
Delta	1.60	0.61	0.33	0.45
Rank	1	2	4	3

Table 5 Mean of S/N ratio for process parameters using larger is better (UTS)

Level	Squeeze pressure	Holding time	Die material	Melt temperature
1	50.16	50.17	50.14	50.08
2	49.96	50.19	49.99	50.11
3	50.29	50.01	50.06	50.07
4	49.81	49.85	50.04	49.97
Delta	0.48	0.34	0.15	0.14
Rank	1	2	3	4

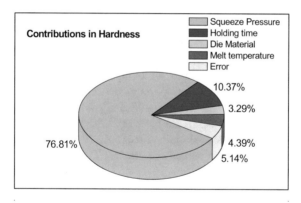

Fig. 4 Percentage contribution in hardness

result compare to carbon steel, SG Iron and stainless steel. S/N ratio of melting temperature increased 750 °C and decreased with above this temperature. Thus the optimum melting temperature is 750 °C. Hardness of the squeeze casting varies significantly for the modified level of individual process parameters. The hardness increased with increase in the squeeze pressure (upto 100 Mpa) and pressure holding time (upto 20 s). It significantly decreased with the increasing melting temperature and changing die material.

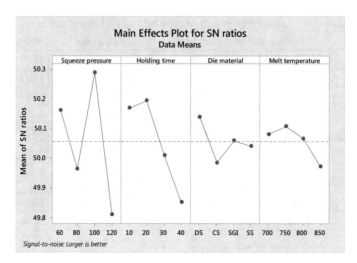

Fig. 5 S/N ratio response curve of UTS

Figure 5 shows that the S/N ratio response curve for UTS shows squeeze pressure decreased from 60 to 80 MPa and increased with decreased at 100–120 MPa, hence the optimum squeeze pressure is 100 MPa. Pressure holding time smoothly increased the level 10–20 s and decreased. Figure 5 shows pressure holding time 20 s that shows maximum optimum value. The die martial Die steel shows better result compare to other consider die materials. S/N ratio of melting temperature increased 750 °C and randomly decreased 850 °C, thus the optimum melting temperature are 750 °C.

UTS of squeeze casting results varies significantly through the modified individual process parameters and levels. UTS has increased with increase in the squeeze pressure (upto 100 Mpa) and pressure holding time (upto 20 s). It significantly decreases with increasing the melting temperature (more than 750 °C) and changing die material.

Figure 7 exposed the response curve (S/N ratio) of yield strength, squeeze pressure has increased at 60–100 MPa and decreased at 120 MPa, here that optimum squeeze pressure 100 MPa shows maximum rate.

Pressure holding times smoothly increased at 10–20 s and suddenly decreased at 20–40-s time range. Pressure holding time 20 s shows maximum optimum value in the S/N ratios graph. Die steel material shows better result. S/N ratios of melting temperature increased at 700–750 °C and suddenly decreased at 850 °C; thus, the optimum melting temperature is 750 °C. The graph shows a significant difference in yield strength with changing the process parameters and their levels. The yield strength increased with increase in the squeeze pressure (100 Mpa) and holding time (20 s). When increasing the melting temperature and changing the die material, yield strength gets decreased.

Table 6 Response table for signal to noise ratios larger is better (yield strength)

Level	Squeeze pressure	Holding time	Die material	Melt temperature
1	48.90	49.09	49.19	49.11
2	49.01	49.22	49.10	49.21
3	49.34	49.02	49.09	49.02
4	49.06	48.99	49.03	48.97
Delta	0.44	0.22	0.07	0.24
Rank	1	3	4	2

Table 7 Hardness ANOVA table

Source	DF	Seq. SS	Contribution (%)	Adj. SS	Adj. MS	F-value
Squeeze pressure	3	1079.19	76.81	1079.19	359.73	14.95
Holding time	3	145.69	10.37	145.69	48.56	2.02
Die material	3	46.19	3.29	46.19	15.40	0.64
Melt temperature	3	61.69	4.39	61.69	20.56	0.85
Error	3	72.19	5.14	72.19	24.06	
Total	15	1404.94	100.00			

4.2 Analysis of S/N Ratio

The various individual factors of process responses can be effectively identified by using statistical ANOVA method [12]. ANOVA result depicts the influence of various factors on ultimate tensile, yield strengths, and hardness. Tables 5, 6, and 7 expose both the S/N ratio and mean value of the samples. The process, a parameter that possesses a significant influence on tensile strengths was predicted by the F-test. The remarkable changes in performance characteristics are an indication of larger F-value [13, 14].

The variation in reduction of relative power is represented by percent contribution. Performance of material is greatly influenced by a small variation of reinforcement percentage. The contribution percentage "f" can be calculated by the squares of each specific item of a material [14]. In Table 7, F-value and contribution percentage act as a controllable process parameters to attain the maximum hardness, and uncontrollable parameters probability is indicated by the P-value. Figure 4 and Table 7 reported that the contribution of squeeze pressure was 76.81%, pressure holding time 10.37%, followed by the die material, melting temperature with 3.29, 4.39% and 5.14% error. Squeeze pressure exposed major contribution to reach maximum hardness. The ratio of explained variation to total variation is called the coefficient of determination (R^2). R^2 for the above model is 94.86%. It is demonstrated a higher degree of correlation between the experimental and the predicted values.

Figure 6 and Table 8 describe the contribution percentage of UTS. The squeeze pressure contribution 51.67%, holding time contribution is 28.66%, followed by

Table 8 UTS ANOVA table

Source	DF	Seq. SS	Contribution (%)	Adj. SS	Adj. MS	F-value
Squeeze pressure	3	718.19	51.67	718.19	239.40	4.93
Holding time	3	401.19	28.86	401.19	133.73	2.75
Die material	3	67.19	4.83	67.19	22.40	0.46
Melt temperature	3	57.69	4.15	57.69	19.23	0.40
Error	3	145.69	10.48	145.69	48.56	
Total	15	1389.94	100.00 •			

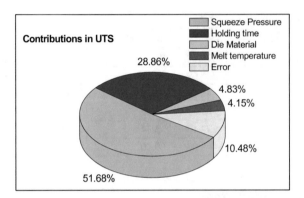

Fig. 6 Percentage contribution in UTS

the die material, melting temperature with 4.83, 4.15% contribution. Finally, error contributed to 10.48%. Squeeze pressure exposed major contribution to achieve maximum ultimate tensile strength and pressure holding time acts as a secondary major process parameter in this optimization. R^2 for the above models is 89.52%.

Figure 8 and Table 9 describe the contribution percentage of yield strength, squeeze pressure contributes 55.01%, the pressure holding time contributes 15.21%, followed by the die material, melt temperature with 11.87, 16.41% contribution in the process, and finally, error has contributed on 1.49%. Squeeze pressure exposed major contribution to reach maximum yield strength and pressure holding time acts as a secondary major process parameter in this optimization. R^2 for above models is 88.13%. It exposed a high rate of correlation that exists between the experimental values and predicted values.

4.3 Regression Equation

A complete model of regression equation was developed using Minitab 18 software. The conventional least square method which developed the equations by reducing the sum of the square residuals [15, 16]. Four operating parameters and four levels

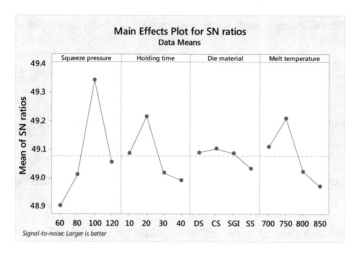

Fig. 7 S/N ratio response curve of yield strength

Fig. 8 Contribution in yield strength

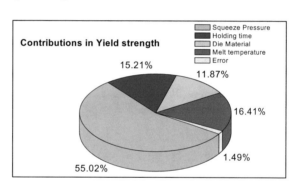

Table 9 Yield strength ANOVA table

Source	DF	Seq. SS	Contribution (%)	Adj. SS	Adj. MS	F-value
Squeeze pressure	3	461.00	55.01	461.00	153.667	4.63
Holding time	3	127.50	15.21	127.50	42.500	1.28
Die material	3	99.50	11.87	99.50	33.167	0.13
Melt temperature	3	137.50	16.41	137.50	45.833	1.38
Error	3	12.50	1.49	12.50	4.167	
Total	15	838.00	100.00			

were considered in this present study. The regression model developed is shown in the Eqs. 2, 3, 4.

$$\text{Hardness} = 119.94 - 0.091\,\text{Squeeze pressure} - 0.218\,\text{Holding time}$$
$$+ 1.12\,\text{Die material} - 0.0065\,\text{Melt temperature} \qquad (2)$$

$$\text{UTS} = 318.44 - 0.1312\,\text{Squeeze pressure} - 0.413\,\text{Holding time}$$
$$+ 0.77\,\text{Die material} - 0.0265\,\text{Melt temperature} \qquad (3)$$

$$\text{Yield strength} = 284.55 + 0.1300\,\text{Squeeze pressure} - 0.155\,\text{Holding time}$$
$$+ 0.65\,\text{Die material} - 0.0390\,\text{Melt temperature} \qquad (4)$$

The process parameter optimization equations created using Taguchi's design (hardness):

$$\eta_{opt} = m + (mA_3 - m) + (mB_2 - m) + (mC_1 - m) + (mD_2 - m)\ldots\ldots \quad (5)$$

Here, m is considered as over all mean of S/N response value, the mean of S/N value for squeeze pressure at level 3 takes mA_3, mB_2 is mean S/N data for holding time at level 2, the mean S/N response data takes as mC_1 for die material level 1 and mD_2 is the mean of S/N response of melt temperature at level 2. Overall mean for S/N value for hardness taken m $= 41.551$ from Table 3.

Therefore,

$$\eta_{opt} = 41.551 + (42.29 - 41.551) + (41.83 - 41.551) + (41.74 - 41.551)$$
$$+ (41.77 - 41.551) = 42.977$$

$$\text{For larger is better}: y_{opt}^2 = 10^{\eta_{opt}/10}$$
$$y_{opt}^2 = 10^{42.977/10} \rightarrow y_{opt(\text{Hardness})} = 137$$

optimum value of hardness is 137 HV.
S/N mean value of UTS is $m = 50.056$,

$$\eta_{opt} = 50.056 + (50.29 - 50.056) + (50.19 - 50.056) + (50.14 - 50.056)$$
$$+ (50.11 - 50.056) = 50.562$$

$$\text{For larger is better}: y_{opt}^2 = 10^{\eta_{opt}/10}$$
$$y_{opt}^2 = 10^{50.562/10} \rightarrow y_{opt\,(\text{UTS})} = 337$$

Therefore, optimum value of UTS 337 MPa.
The S/N mean value of yield strength $m = 49.078$

$$\eta_{opt} = 49.078 + (49.34 - 49.078) + (49.22 - 49.078) + (49.19 - 49.078)$$
$$+ (49.21 - 49.078) = 49.49$$

For larger is better : $y_{opt}^2 = 10^{\eta_{opt}/10}$

$$y_{opt}^2 = 10^{49.49/10} \rightarrow y_{opt \, (\text{Yield strength})} = 298$$

Optimum value of yield strength is 298 MPa.

From the above results, obtained were same with experimental response. Process parameters such as squeeze pressure (100 MPa), holding time (20 s), die steel material, and melting temperature (750 °C) showed optimum results.

5 Conclusion

In this proposed research work, Al6061/Al$_2$O$_3$/SiC composites specified the optimum squeeze casting parameters using Taguchi's method. It improved the quality of the product. Experiments were conducted using L16 orthogonal array considering the four factors and four levels. The parameters considered were squeeze pressure, holding time, die material, and melting temperature. The following were the conclusions drawn:

- Level $A_3B_2C_1D_2$ parameters such as squeeze pressure 100 MPa (level 3), holding time 20 s (level 2), die steel material (level 1), and melting temperature 750 °C (level 2) exhibited optimum results.
- Squeeze pressure 120 MPa, holding time 40 s, stainless steel die material, and melting temperature 800 °C results showed minimum hardness and UTS.
- Squeeze pressure 60 MPa, holding time 40 s, die steel material, and melting temperature 850 °C evaluation show minimum yield strength.
- Squeeze pressure demonstrated the major contributions to achieving maximum hardness, UTS, and yield strength. Pressure holding time acts as a secondary major process parameter in this optimization. Regression equation defined the statistical relationship between the response variable and the process parameter.

References

1. Prasad, S.V., Asthana, R.: Aluminum metal-matrix composites for automotive applications: tribological considerations. Tribol. Lett. 17(3), 445–453 (2004)
2. Hashim, J., Looney, L., Hashmi, M.S.J.: Particle distribution in cast metal matrix composites—Part I. J. Mater. Process. tech. 123(2), 251–257 (2002)

3. Jahangiri, A., et al.: The effect of pressure and pouring temperature on the porosity, microstructure, hardness and yield stress of AA2024 aluminum alloy during the squeeze casting process. J. Mater. Process. Tech. **245**, 1–6 (2017)
4. Kok, Metin: Production and mechanical properties of Al_2O_3 particle-reinforced 2024 aluminium alloy composites. J. Mater. Process. Tech. **161**(3), 381–387 (2005)
5. Natrayan, L., Senthil Kumar, M.: Study on Squeeze Casting of Aluminum Matrix Composites—A Review. Advanced Manufacturing and Materials Science, pp. 75–83. Springer, Cham (2018)
6. Senthil, P., Amirthagadeswaran, K.S.: Experimental study and squeeze casting process optimization for high quality AC2A aluminium alloy castings. Arab. J. Sci. Eng. **39**(3), 2215–2225 (2014)
7. Natrayan L., et al.: An experimental investigation on mechanical behaviour of SiCp reinforced Al 6061 MMC using squeeze casting process. Inter. J. Mech. Prod. Eng. Res. Develop. **7**(6), 663–668 (2017)
8. Dulyapraphant P et al.: Applications of the Horizontal Squeeze Casting Process for Automotive Parts Manufacturing. Light Metals 2013. Springer, Cham, pp 425–429 (2016)
9. Ravindran, P., et al.: Tribological properties of powder metallurgy–processed aluminium self-lubricating hybrid composites with SiC additions. Mater. Des. **45**, 561–570 (2013)
10. Sukumaran, K., et al.: Studies on squeeze casting of Al 2124 alloy and 2124-10% SiCp metal matrix composite. Mater. Sci. Eng., A **490**(1–2), 235–241 (2008)
11. Teng, Liu, et al.: An investigation into aluminum–aluminum bimetal fabrication by squeeze casting. Mater. Des. **68**, 8–17 (2015)
12. Vijian, P., Arunachalam, V.P.: Optimization of squeeze cast parameters of LM6 aluminium alloy for surface roughness using Taguchi method. J. Mater. Process. Tech. **180**(1–3), 161–166 (2006)
13. Natrayan, L., Senthil kumar, M., Palani kumar, K.: Optimization of squeeze cast process parameters on mechanical properties of Al_2O_3/SiC reinforced hybrid metal matrix composites using taguchi technique. Mater. Res. Express. 5, 066516 (2018)
14. Maleki, A., Niroumand, B., Shafyei, A.: Effects of squeeze casting parameters on density, macrostructure and hardness of LM13 alloy. Mater. Sci. Eng., A **428**(1–2), 135–140 (2006)
15. Manjunath Patel, G.C., Krishna, Prasad, Parappagoudar, Mahesh B.: Modelling and multi-objective optimization of squeeze casting process using regression analysis and genetic algorithm. Aust. J. Mech. Eng. **14**(3), 182–198 (2016)
16. Eazhil, K.M., et al.: Optimization of the process parameter to maximize the tensile strength in 6063 aluminum alloy using Grey based Taguchi method. Adv. Nat. Appl. Sci. **11**(4), 235–242 (2017)

Parametric Optimization of Electrochemical Machining Process Using Taguchi Method and Super Ranking Concept While Machining on Inconel 825

Partha Protim Das and Shankar Chakraborty

Abstract To exploit the fullest machining potential of electrochemical machining (ECM) process while machining on Inconel 825, it is recommended to operate the machine with the optimal combination of machining process parameters. Past researchers have already applied grey relational analysis (GRA) as an optimization tool so as to obtain the optimal parametric combination of ECM process. In this paper, based on the experimental data obtained by the past researchers, Taguchi method and super ranking concept is applied to analyze the efficacy of the proposed approach in obtaining the optimal parametric combination of ECM process. The derived parametric combination is validated with respect to the predicted response values, obtained from the developed regression equations which show that the proposed approach results in improved response values than that obtained by past researchers. Finally, Analysis of variance (ANOVA) is applied to identify the influence of each process parameters for the considered ECM process.

Keywords Taguchi method · Super ranking · ECM process · Optimization · Process parameter · Response · ANOVA

1 Introduction

Inconel 825 is a nickel-based super alloy that provides excellent resistance to corrosive environment. It finds its application in aerospace, oil and gas well piping, nuclear plant, and many chemical industries. Electrochemical machining (ECM) process is a well-known machining process because of its ability to generate intri-

P. P. Das (✉)
Department of Mechanical Engineering, Sikkim Manipal Institute of Technology, Sikkim Manipal University, Majitar, Sikkim, India
e-mail: parthaprotimdas@ymail.com

S. Chakraborty
Department of Production Engineering, Jadavpur University, Kolkata, West Bengal, India
e-mail: s_chakraborty00@yahoo.co.in

© Springer Nature Singapore Pte Ltd. 2020
R. Venkata Rao and J. Taler (eds.), *Advanced Engineering Optimization Through Intelligent Techniques*, Advances in Intelligent Systems and Computing 949,
https://doi.org/10.1007/978-981-13-8196-6_36

cate shapes and geometries on various advanced engineering materials [1]. In ECM process, the materials are eroded from the workpiece due to the electrochemical dissolution at their atomic level. Precession machining along with quality surface finish, provides an upper hand to ECM process, mainly in automobile and aerospace industries. In ECM process, it is often recommended to operate the machine at the optimal parametric combination, as a slight change in the process parameters may affect the responses adversely. Selection of optimal combination of process parameters is often considered to be challenging with the increased number of process parameters and responses, thus making it a multi-objective optimization problem. Several optimization tools are available nowadays that can be effectively deployed to obtain the optimal parametric combination such as Taguchi methodology, preference ranking organization method for enrichment evaluation (PROMETHEE), grey relational analysis (GRA), combinative distance-based assessment (CODAS), and VIKOR method. They have been observed to be quite promising in obtaining the optimal parametric combination resulting in enhanced machining performance.

Optimization of various machining process parameters using different optimization approaches being the topic of research interest since the past few years has been explored by many researchers. Jain and Jain [2] used real-coded genetic algorithm to obtain the optimal combination of three important ECM process parameters namely tool feed rate, electrolyte flow velocity, and applied voltage in order to minimize the geometrical inaccuracies subject to temperature, choking, and passivity constraints. Considering current, voltage, flow rate, and gap as the controllable process parameters, Asokan et al. [3] have applied GRA as a multi-response optimization tool for an ECM process while taking into account material removal rate (MRR) and surface roughness (Ra) as the responses. Rao and Yadava [4] proposed a hybrid approach combining Taguchi method with GRA for optimization of Nd:YAG laser cutting process in order to minimize responses such as kerf width, kerf taper, and kerf deviation. Mukherjee and Chakraborty [5] used biogeography-based optimization (BBO) algorithm so as to obtain the optimal parametric combination for an ECM process and also for a wire electrochemical turning process. Tang and Yang [6] considered voltage, cathode feed rate, electrolyte pressure, and electrolyte concentration as ECM process parameters, and applied GRA so as to improve MRR, Ra, and side gap while machining on 00Cr12Ni9Mo4Cu2 material. Considering input process parameters as pulse-on-time, pulse current, pulse-off-time and the speed of the wire drum, Lal et al. [7] applied Taguchi-based GRA approach for a wire electro-discharge machining (WEDM) process so as to improve surface quality characteristics such as Ra and kerf width. Bose [8] presented the application of Taguchi methodology aided with fuzzy logic as a multi-criteria decision-making (MCDM) technique to obtain the optimal combination of process parameter for an electrochemical grinding process. Rao and Padmanabhan [9] optimized the input process parameters of ECM process while integrating Taguchi method with utility concept. They considered applied voltage, electrolyte concentration, electrode feed rate, and percentage of reinforcement content as the process parameters while considering MRR, Ra and radial over cut as the responses. Mehrvar et al. [10] proposed a hybrid technique combining differential evolution algorithm and response surface methodology to a MCDM problem considering voltage, electrolyte flow rate, tool feed rate, and electrolyte concentration as the process parameters in a ECM process to optimize two machining criteria, i.e., MRR and Ra. Selecting compact load, current, and pulse-on-time as process parameters Rahang and Patowari [11] applied Taguchi method to optimize the performance measures such as TWR, MRR, Ra, and edge deviation. Dhuria et al. [12] proposed a hybrid Taguchi-entropy weight-based GRA method to optimize MRR and TWR of ultrasonic machining (USM) process while considering slurry type, tool type, power rating, grit size, tool treatment, and workpiece treatment as the process parameters. Antil et al. [13] selected voltage, electrolyte concentration, inter-electrode gap, and duty factor as the control parameters in an electrochemical discharge drilling of SiC reinforced polymer matrix composite and applied Taguchi's grey relational analysis to obtain the optimal parametric mix. Cole et al. [14] considered electrolyte flow rate, duty cycle, and pulsed voltage range as process parameters in an ECM process and applied Taguchi method to optimize with respect to the aspect ratio, surface roughness, and the rate of machining. Manikandan et al. [15] integrated GRA with Taguchi method to obtain the optimal parametric combination of feed rate, flow rate of electrolyte and concentration of electrolyte as an ECM process parameters while optimizing MRR, Ra, overcut, circularity error, and perpendicularity error as responses. Chakraborty et al. [16] applied GRA-based fuzzy logic approach for obtaining the optimal parametric settings of three NTM processes such as abrasive water-jet machining (AWJM) process, ECM process, and ultrasonic machining (USM) processes. Taddese [17] investigated the effect of electric voltage, electrolyte concentration, electrolyte pressure, and motor rotational speed on surface roughness while performing electrochemical honing for surface generation on cast iron of ASTM 35 using Taguchi method. Chakraborty and Das [18] adopted multivariate loss function approach to obtain the optimal parametric combination of ECM, electro-discharge machining (EDM), and WEDM processes. From the extensive review of the above literatures, it has been noticed that several optimization tools, such Taguchi method, GRA, PCA, and VIKOR has been used extensively to obtain the optimal parametric combination of ECM as well as other machining processes. As these tools are often conventional in nature and can often lead to a near optimal solution. In this paper, Taguchi method and super ranking concept being a new and efficient approach is applied to check its effectiveness in obtaining the optimal parametric combination of an ECM process based on data obtained from past researchers. Regression equations are also developed for each response representing the relationship between the process parameters and the responses. The obtained optimal parametric combination is thus verified with those obtained by past researchers based on the developed regression equation. Finally, ANOVA is also applied to identify the most influencing process parameter for the considered ECM process.

2   Taguchi Method and Super Ranking Concept

Taguchi method has emerged as an effective tool that deals with responses influenced by multi-variables. It was first introduced by Taguchi [19, 20]. But it has a drawback, as the method can only optimize a single response at a time without considering the effect of it on other responses. A simple and easy to apprehend version of this method is proposed by Besseris [21] that can be effectively deployed to solve difficult problem with multi-responses simultaneously. Taguchi method and super ranking concept starts with the identification of controlling process parameters (signal factors) and performance measures (noise factors). The working ranges for each of these control parameters are also identified. As per Taguchi's orthogonal design of experimental, a suitable array is then selected considering all control parameters along with their ranges. These designs are so selected to understand of variation of each process parameters with a minimum number of experiments. Based on each trial run, the performance measures (responses) were measured and noted. The measured responses are then converted to signal-to-noise (SN) ratio using Taguchi's three classes. For responses with higher-the-better characteristics Eq. (1) is used, for smaller-the-better, Eq. (2) is used and for responses where in-between values are desired, i.e., nominal-the-better, Eq. (3) is used.

$$SN = -10 \log_{10}\left[\frac{1}{n}\sum \frac{1}{x_{ij}^2}\right] \quad i = 1, 2, \ldots, m \text{ and } j = 1, 2, \ldots, n \quad (1)$$

$$SN = -10 \log_{10}\left[\sum \frac{x_{ij}^2}{n}\right] \quad (2)$$

$$SN = 10 \log_{10}\frac{\mu^2}{\sigma^2} \quad (3)$$

where n is the total number of responses, x_{ij} is the responses for the ith alternative and jth criterion, μ and σ is the mean and the standard deviation of the responses.

The calculated SN ratios are then converted to rank variables, such that the highest SN ratio for each response will receive, following the second highest will receive rank 2 and so on. If there is a tie between two or more SN ratios, then the average rank will be assigned. These ranks are then squared and added with respect to each experimental trial to obtain a single response column known as sum of squared rank (SSR). These SSR values further receive a ranking such that the smallest SSR value receives rank 1, followed by the second smallest value receives rank 2 and so on. These ranks are called as super ranks (SR). A smaller SSR value indicates that the particular experimental trial is the best among all the trials.

3 Analysis of Taguchi Method and Super Ranking Concept

Based on Taguchi's L_9 orthogonal array, Nayak et al. [22] conducted nine set of experiments with an ECM process on Inconel 825 as workpiece using a 'I' shaped tool. They considered concentration (A) (in g/l), voltage (B) (in volts), and feed rate (C) (in mm/min) as the machining process parameters, along with three level variation as shown in Table 1. The inter-electrode gap is kept constant at 0.5 mm. For performance measures material removal rate (MRR) (in mm³/min), surface roughness (R_a) (in μm) and longitudinal overcut (overcut) (in mm) were chosen as the responses. The detailed experimental design plan along with the measured responses is exhibited in Table 2, respectively.

Among the three responses, MRR is the only response with higher-the-better characteristic, and R_a and overcut are of smaller-the-better characteristic. Equations (1)–(3) are used to convert these response values to SN ratios depending upon the type of quality characteristics. Now as explained in Sect. 2, these SN ratios are then assigned ranks such that the higher SN ratios for each rank will receive rank 1 and the lowest value will receive rank 9 for this nine set of experimental design plan. The calculated SN ratios and the assigned ranks are presented in Table 3. These ranks are then squared and added for each experimental trial to obtain the final SSR value as exhibited in Table 3, respectively. From the table, it can be seen that experimental trial number 8 with parametric combination as $A_3B_2C_1$ has the smallest SSR value indicating it to be the best experimental trial among the nine trials.

Table 1 Process parameters with levels for ECM process [22]

Factors with symbol	Unit	Levels		
		Level 1	Level 2	Level 3
Concentration (A)	g/l	80	95	111.11
Voltage (B)	volts	5	10	15
Feed rate (C)	mm/min	0.2	0.4	0.6

Table 2 Experimental details for ECM process [22]

Exp. no.	A	B	C	MRR	R_a	Overcut
1	80	5	0.2	26.74759	0.923	0.09
2	80	10	0.4	27.30646	0.771	0.1105
3	80	15	0.6	20.83473	0.536	0.114
4	95	5	0.4	21.23712	0.842	0.1005
5	95	10	0.6	31.7439	0.643	0.1215
6	95	15	0.2	23.39436	0.92	0.1055
7	111.11	5	0.6	25.27217	0.83	0.107
8	111.11	10	0.2	32.93989	0.872	0.096
9	111.11	15	0.4	12.00456	0.68	0.124

Table 3 SN ratios, rank, squared rank, SSR, and SR for ECM process

Exp. No.	S/N ratios			Rank			Squared rank			SSR	SR
	MRR	R_a	Overcut	MRR	R_a	Overcut	MRR	R_a	Overcut		
1	28.5457	0.69597	20.9151	5	9	1	25	81	1	107	6
2	28.7253	2.25891	19.1328	3	4	6	9	16	36	61	2
3	26.3758	5.4167	18.8619	8	1	7	64	1	49	114	7
4	26.5419	1.49376	19.9567	7	6	3	49	36	9	94	5
5	30.0332	3.83578	18.3085	2	2	8	4	4	64	72	3
6	27.3822	0.72424	19.535	6	8	4	36	64	16	116	8
7	28.0529	1.61844	19.4123	5	5	5	25	25	25	75	4
8	30.3544	1.18967	20.3546	1	7	2	1	49	4	54	1
9	21.5869	3.34982	18.1316	9	3	9	81	9	81	171	9

Table 4 Response table for SSR

Factors	Levels			Max–min	Rank
	Level 1	Level 2	Level 3		
Concentration	**92**	94.3333	100	6	3
Voltage	92	**62.3333**	133.6667	71.3334	1
Feed rate	92.3333	108.6667	**87**	21.6667	2

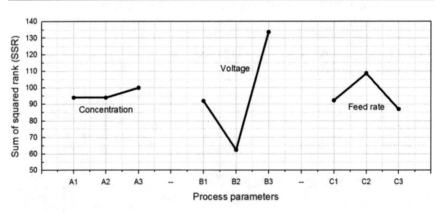

Fig. 1 Response graph for SSR

Table 4 and Fig. 1 represent the response table and corresponding response graph for the SSR values. These values are obtained by averaging the SSR values at their corresponding level of experimental trial. The max–min column in the table indicates that voltage is the most influencing factor among the three considered process parameters which is also supported by the sleep slope of voltage as seen from response graph. It can be seen from the response table and graph that in order to obtain the best responses, the optimal combination of the process parameters must be set as

Table 5 ANOVA results

Source	DoF	Adj SS	Adj MS	f-value	% contribution
Concentration	2	72.0	36.00	0.04	0.69
Voltage	2	7704.7	3852.33	3.97	73.52
Feed rate	2	764.7	382.33	0.39	7.29
Error	2	1938.7	969.33		18.50
Total	8	10480.0			100

concentration = 80 g/l, voltage = 10 V, and feed rate = 0.6 mm/min, respectively, can also be identified as $A_1B_2C_3$. Finally, to identify the influence of each process parameters ANOVA is applied and the results were presented in Table 5. From the table, it can be seen that with 73.52% contribution, voltage is the most influencing process parameter followed by feed rate, for the considered ECM process.

From the above analysis, it is seen that experiment number 8 with parametric combination as $A_3B_2C_1$ is the most preferred trial among the nine experimental trials. But from the response table, it is found to be $A_1B_2C_3$, based on the average SSR values at their corresponding level. It is mainly because the average values of SSR are found to be lower with setting $A_1B_2C_3$, then with setting $A_3B_2C_1$ which indicates that the chances of getting lower SSR are with $A_1B_2C_3$ parametric combination. But Nayak et al. [22] found it as $A_1B_2C_1$ employing GRA. The only difference is with feed rate, where using the proposed approach the optimal parametric combination found with feed rate at level 3, whereas the past researchers found at level 1. Thus, in order to show the superiority of the proposed method as a multi-response optimization tool, the obtained optimal parametric combination using Taguchi method and super ranking concept is compared to that obtained by past researchers. Using Eq. (4), the SSR values for both the parametric combinations can be predicted, which shows that the predicted SSR value with $A_1B_2C_3$ parametric combination is improved by 9.75%, i.e., from 54.663 to 49.333 with compared to $A_1B_2C_1$.

$$S_p = S_m + \sum_{i=1}^{n} (\overline{S_i} - S_m) \tag{4}$$

where S_p is the predicted SSR, n is the total number of process parameters, S_m corresponding to the mean SSR value for the nine experimental trials, and $\overline{S_i}$ is the mean SSR value for the ith level of process parameters.

In order to fully justify the dominance of the derived parametric combination, the regression equations are also developed representing the relationship between the measured responses and the considered process parameters. Equations (5)–(7) show the developed regression equations for MRR, R_a, and overcut, respectively. Based on these regression equations, a comparison between the derived parametric combinations with that of the past researcher is shown in Table 6. From the table it is interesting to see that with the proposed parametric combination, there is significant

Table 6 Comparison table for ECM process

Optimization approach	MRR (mm³/min)	R_a (μm)	Overcut (mm)
Taguchi and super ranking method ($A_1B_2C_3$)	35.685	0.6372	0.0816
GRA (By past researchers) ($A_1B_2C_1$)	33.4817	0.84	0.091
Improvement (%)	6.58	24.14	10.33

improvement of 6.58, 24.14, and 10.33% for MRR, R_a, and overcut, respectively. Thus, it can be concluded that the proposed approach outranks the optimization approach adopted by past researchers resulting in significant improvement of the machining performance.

$$MRR = 36.9 - 0.0051 \times A - 0.0567 \times B - 0.4 \times C \tag{5}$$

$$R_a = 1.015 + 0.00160 \times A - 0.01530 \times B - 0.588 \times C \tag{6}$$

$$Overcut = 0.0627 + 0.0000132 \times A + 0.0001533 \times B + 0.00425 \times C \tag{7}$$

4 Conclusions

This paper presents the application Taguchi method and super ranking concept in obtaining the optimal combination process parameters so as to improve the electro-chemical machining performance while machining on Inconel 825 workpiece. From the above analysis, it can be clearly observed that the proposed approach outperforms the approach adopted by the past researchers showing an improvement of 6.58, 24.14, and 10.33% for MRR, R_a, and overcut, respectively. Moreover, an ANOVA result identifies voltage as the most influencing process parameter for the said application. Since, the entire analysis is based on experimental data obtained from the past literatures, thus there is no scope of performing any validation experiments in order to justify the derived results. The proposed approach can be further applied to various machining processes so as to obtain the optimal parametric combination for enhanced machining performance.

References

1. El-Hofy, H.: Advanced Machining Processes: Nontraditional and Hybrid Machining Processes. McGraw-Hill, New York (2005)
2. Jain, N.K., Jain, V.K.: Optimization of electro-chemical machining process parameters using genetic algorithms. Mach. Sci. Technol. **11**(2), 235–258 (2007)

3. Asokan, P., Kumar, R.R., Jeyapaul, R., Santhi, M.: Development of multi-objective optimization models for electrochemical machining process. Int. J. Adv. Manuf. Technol. **39**(1–2), 55–63 (2008)
4. Rao, R., Yadava, V.: Multi-objective optimization of Nd: YAG laser cutting of thin superalloy sheet using grey relational analysis with entropy measurement. Opt. Laser Technol. **41**(8), 922–930 (2009)
5. Mukherjee, R., Chakraborty, S.: Selection of the optimal electrochemical machining process parameters using biogeography-based optimization algorithm. Int. J. Adv. Manuf. Technol. **64**(5–8), 781–791 (2013)
6. Tang, L., Yang, S.: Experimental investigation on the electrochemical machining of 00Cr12Ni9Mo4Cu2 material and multi-objective parameters optimization. Int. J. Adv. Manuf. Technol. **67**(9–12), 2909–2916 (2013)
7. Lal, S., Kumar, S., Khan, Z.A., Siddiquee, A.N.: Multi-response optimization of wire electrical discharge machining process parameters for Al7075/Al$_2$O$_3$/SiC hybrid composite using Taguchi-based grey relational analysis. Proc. Inst. Mech. Eng. B: J. Eng. Manuf. **229**(2), 229–237 (2015)
8. Bose, G.K.: Multi objective optimization of ECG process under fuzzy environment. Multidiscip. Model. Mater. Struct. **11**(3), 350–371 (2015)
9. Rao, S.R., Padmanabhan, G.: Parametric optimization in electrochemical machining using utility based Taguchi method. Int. J. Eng. Sci. Technol. **10**(1), 81–96 (2015)
10. Mehrvar, A., Basti, A., Jamali, A.: Optimization of electrochemical machining process parameters: combining response surface methodology and differential evolution algorithm. Proc. Inst. Mech. Eng. E: J. Proc. Mech. Eng. **231**(6), 1114–1126 (2017)
11. Rahang, M., Patowari, P.K.: Parametric optimization for selective surface modification in EDM using Taguchi analysis. Mater. Manuf. Processes **31**(4), 422–431 (2016)
12. Dhuria, G.K., Singh, R., Batish, A.: Application of a hybrid Taguchi-entropy weight-based GRA method to optimize and neural network approach to predict the machining responses in ultrasonic machining of Ti–6Al–4V. J. Braz. Soc. Mech. Sci. Eng. **39**(7), 2619–2634 (2017)
13. Antil, P., Singh, S., Manna, A.: Electrochemical discharge drilling of SiC reinforced polymer matrix composite using Taguchi's grey relational analysis. Arab. J. Sci. Eng. **43**(3), 1257–1266 (2018)
14. Cole, K.M., Kirk, D.W., Singh, C.V., Thorpe, S.J.: Optimizing electrochemical micromachining parameters for Zr-based bulk metallic glass. J. Manuf. Processes. **25**, 227–234 (2017)
15. Manikandan, N., Kumanan, S., Sathiyanarayanan, C.: Multiple performance optimization of electrochemical drilling of Inconel 625 using Taguchi based grey relational analysis. Eng. Sci. Technol. Int. J. **20**(2), 662–671 (2017)
16. Chakraborty, S., Das, P.P., Kumar, V.: Application of grey-fuzzy logic technique for parametric optimization of non-traditional machining processes. Grey Syst.: Theory Appl. **8**(1), 46–68 (2018)
17. Taddese, F.: Parameter optimization of ECH process for surface finish on ASTM-35 cast iron. Int. J. Adv. Manuf. Technol. 1–12 (2018)
18. Chakraborty, S., Das, P.P.: A multivariate quality loss function approach for parametric optimization of non-traditional machining processes. Manag. Sci. Lett. **8**(8), 873–884 (2018)
19. Taguchi, G.: Introduction to quality engineering: designing quality into products and processes (No. 658.562 T3) (1986)
20. Taguchi, G., Konishi, S.: Taguchi methods: Orthogonal arrays and linear graphs: tools for quality engineering, Amer. Suppl. Inst. (1987)
21. Besseris, G.J.: Multi-response optimisation using Taguchi method and super ranking concept. J. Manuf. Technol. Manag. **19**(8), 1015–1029 (2008)
22. Nayak, A.A., Gangopadhyay, S., Sahoo, D.K.: Modelling, simulation and experimental investigation for generating 'I'shaped contour on Inconel 825 using electro chemical machining. J. Manuf. Processes. **23**, 269–277 (2016)

Dependency of Bead Geometry Formation During Weld Deposition of 316 Stainless Steel Over Constructional Steel Plate

M. K. Saha, S. Sadhu, P. Ghosh, A. Mondal, R. Hazra and S. Das

Abstract Weld bead geometry influences mechanical properties, microstructure of the weld joint or weld overlay. It is much biased by heat input of a particular welding technique. In current work, weld bead of 316 austenitic stainless steel is produced on E250 low alloy steel by gas metal arc welding process using 100% carbon dioxide as shielding gas. Nine sets of welding current and welding voltage combinations were chosen for producing nine weld beads, keeping travel speed constant throughout the experiment. Two identical set of experiments were repeated. Experimental results depicted that the width of weld bead, PSF, RFF extended with increment in heat input, while height of reinforcement and depth of penetration declined slightly for the identical condition. Quadratic equations are generated successfully between different bead geometry parameters and heat input by means of polynomial regression analysis which agree with the real data.

Keywords Gas metal arc welding · Weld bead geometry · Heat input ·
Polynomial regression analysis

M. K. Saha · S. Sadhu · P. Ghosh · A. Mondal · R. Hazra · S. Das (✉)
Department of Mechanical Engineering, Kalyani Government Engineering College, Kalyani, Nadia 741235, West Bengal, India
e-mail: sdas.me@gmail.com

M. K. Saha
e-mail: manassaha71@gmail.com

S. Sadhu
e-mail: souvik.kgecme@gmail.com

P. Ghosh
e-mail: pritamghosh762@gmail.com

A. Mondal
e-mail: mondalajit830@gmail.com

R. Hazra
e-mail: hazra.ritesh2013@gmail.com

© Springer Nature Singapore Pte Ltd. 2020
R. Venkata Rao and J. Taler (eds.), *Advanced Engineering Optimization Through Intelligent Techniques*, Advances in Intelligent Systems and Computing 949,
https://doi.org/10.1007/978-981-13-8196-6_37

1 Introduction

The performance of the weld joint may be revealed by the components of geometry of weld bead such as weld bead width (W), reinforcement height (R) and penetration depth (P) and certain shape factor such as RFF (reinforcement form factor) and PSF (penetration shape factor). Figure 1 shows a pictorial view of weld bead and its parameters. Good quality welding demands more deposition with full penetration. Cladding is one of the surfacing techniques in which few millimetre deposition of corrosion-resistant material is being done over low-grade material to improve mechanical strength and corrosion resistance of the base material. Cladding by welding is thus welding of two dissimilar materials. Unlike welding a joint, cladding by welding demands sufficient deposition with lowest possible penetration for avoiding dilution. At the same time, joining strength of dissimilar welding in case of cladding must be sufficient enough so that it can be intact at working atmosphere even with such low penetration [1, 2]. Figure 1 shows the schematic diagram of one typical weld bead macrograph and its components.

Heat input gives the energy for melting and coalescence of base metal and consumable electrode/filler material at the weld joint made by fusion welding. Weld bead shape thus can be influenced by controlling heat input. Heat input can be controlled by proper selection of process parameters having optimum values. Heat input is influenced by three process parameters like welding current, motion of electrode holder and welding voltage. Nasir et al. observed in an experiment that microstructure of carbon steel showed coarse grains and reduced impact strength at higher heat input. At higher heat input, weld parts are developed having higher penetration but higher distortion [3].

GMAW is one of the cost effective, easy-to-work for various position, semi-automatic and user-friendly process that has spread over a wide spectrum of engineering field for the last few decade. When CO_2 gas is used as shielding gas in gas metal arc welding, it is called metal active gas welding. Quality of weld joint can be improved by proper selection of process parameters within a favourable range in case of GMAW. Metal active gas welding is cheaper that yields good quality welded

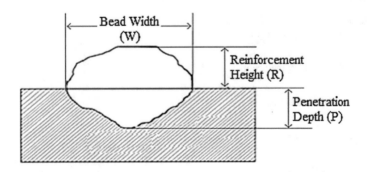

Fig. 1 Weld bead geometry and its components

part having a considerable amount of spatter. Shenet al. carried out an experiment on submerged arc welding of high strength, structural steelplates with varying heat input by means of using single and double electrodes. From the experimental results, they concluded that weld bead width, penetration depth, reinforcement height, penetration area and deposition area, size of heat affected zone increased with increasing heat input; whereas, as the heat input increased, the bead contact angle decreased [4]. Frei et al. observed in a GTAW hot wire and cold wire method that higher heat input produced coarse grain whereas lower heat input produced fine grain structure [5]. They observed also that heat affected zone was narrow where compositional gradient was formed from the outer surface of the fusion zone towards welding [6]. Kah et al. reviewed GMAW processes and showed that the proper design of experiment as well as algorithm control of prime process parameters could be achieved in GMAW process [7]. These would be beneficial for welding of low heat sensitive materials.

Several works were done to explore different numerical models to forecast weld bead components at different levels of process parameters for a particular set of materials. Weld beads formed are very sensitive to the welding process, composition of electrode and base materials set, heat input and different process parameters of corresponding welding processes. Several mathematical models were generated by using neural network [8], RSM [9], ANN [10–12], regression analysis [13–16], SA [15], GA [17–19, 21] and some other statistical tools [20] for predicting weld bead components produced and/or optimizing process parameters by different welding processes. Sarma et al. developed a new indexing technique that may optimize multi-input multi-output system like weld process parameters and weld bead components [18]. Goutam and Vipin generated mathematical model to predict the width of weld bead (W), penetration depth (P) and deposition height (R) with the change of current, voltage, travel speed and tip-to-nozzle distance by means of central composite design in submerged arc welding process [22].

Desirable weld beads are still difficult to achieve as the process is very hazardous, and it is dependent on the skill of the operator. To get the feedback about the dimension regarding weld bead, Soares et al. introduced geometric weld bead analysis system continuously monitored by robot [23].

In the current investigation, deposition of 316 austenitic stainless steel beads are deposited over (E250) low alloy constructional steel with the help of gas metal arc welding process using only carbon dioxide (CO_2) as shielding gas at a constant gas flow rate of 16 litre/min. Process parameters such as welding current and welding voltage are selected in nine different values so that nine number of weld beads for every set of experiment are formed by nine different heat input, keeping travel speed of electrode constant. The entire set of experiment is performed twice. With the real data, relationship between weld bead geometry components and heat input are calculated by means of quadratic regression method.

2 Experimental Method

2.1 Base Material

Base material used for bead-on-plate experiment is E250 low alloy steel, which is generally used for constructional work. The base material possesses chemical composition that is shown in Table 1.

2.2 Consumable Electrode Used

Filler electrode used for the experiment is (316) austenitic stainless steel. The alloying constituents of wire electrode measured by chemical analysis are given in Table 2. It may be noted that 316 steel contains about 15% Cr, 9.9% Ni, 2.1% Mo and 1.1% Mn. Presence of these alloying elements is expected to impart resistance to atmospheric corrosion.

2.3 Experimental Set up

Weld bead was deposited by a MIG/MAG welding machine (model No. Auto K 400, ESAB India Ltd.) having 60% duty cycle. The range of current and voltage of used

Table 1 Configuration of low alloy constructional steel (E250) base metal

% wt. of C	0.201	% wt. of Mo	<0.0017	% wt. of Nb	0.0053
% wt. of Si	0.1509	% wt. of Ni	<0.0024	% wt. of Ti	0.001
% wt. of Mn	0.533	% wt. of B	<0.001	% wt. of V	0.0038
% wt. of P	0.0859	% wt. of Al	0.0031	% wt. of Sn	<0.0014
% wt. of S	0.0389	% wt. of Co	0.0071	% wt. of Ce	0.0101
% wt. of Cr	<0.0011	% wt. of Cu	0.0231	% wt. of Fe	98.94

Table 2 Configuration of 316 austenitic stainless steel filler material

% wt. of C	0.076	% wt. of Mo	2.09	% wt. of Ti	0.013
% wt. of Si	0.182	% wt. of Ni	9.94	% wt. of V	0.047
% wt. of Mn	1.102	% wt. of Nb	0.04	% wt. of W	0.026
% wt. of P	0.029	% wt. of Al	0.0105	% wt. of Sn	0.010
% wt. of S	0.008	% wt. of Co	0.074	% wt. of Ce	0.010
% wt. of Cr	15.04	% wt. of Cu	0.342	% wt. of B	<0.001
% wt of Fe	<70.95				

Fig. 2 Experimental set up

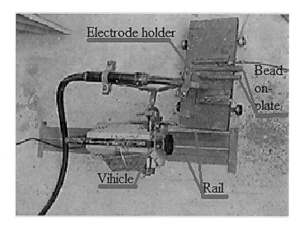

welding machine is 0–400 A and 0–75 V, respectively. The welding gun is mounted on a motor operated vehicle with a facility of speed variation and movement in a straight line path along a guided rail. The shielding gas used is 100% CO_2. Gas flow rate was constant and kept at a value of 16 (litre/min). Figure 2 shows the setup of bead-on-plate experiment.

2.4 Parameter Selection

Several beads were generated by the same set of material combination with different values of three process parameters that influence heat input greatly, namely welding current, welding voltage and motion of electrode torch. Results of trial run suggested favourable zone of process parameters, and those yielded better weld bead shape. From the outcome of trial runs, nine values of welding voltage and welding currents were selected within 24–30 V and 140–215 A, respectively, keeping torch travel speed constant at 420 mm/min so that nine weld beads were generated. The whole experiment was replicated twice. Table 3 shows the combination of process parameters for the present work.

Heat input is the energy required to produce weld bead per unit length of weld. It is treated as a significant attribute because it influences the cooling rate. The more the heat input, the more time required to release the heat outside the atmosphere. The effect will be slow rate of cooling. The mechanical properties and metallurgical structure of weld zone as well as heat affected zone (HAZ) may be affected by cooling rate. Heat input is obtained by the following equation.

$$Q = \frac{(V \times I)60}{(S) \times 1000}\eta \tag{1}$$

Table 3 Process parameters for the present work

Sl. No.	Weld voltage (V) [V]	Weld current (I) [A]	Welding speed (S) [mm/min]	Heat Input (Q) [kJ/mm]
1	24	160	420	0.548
2	26	150	420	0.557
3	28	140	420	0.56
4	26	180	420	0.668
5	28	170	420	0.68
6	30	160	420	0.685
7	28	200	420	0.798
8	26	215	420	0.8
9	30	190	420	0.814

Here, Q signifies heat input in kJ/mm, S expresses welding speed (mm/min), V represents voltage (V), I is equal to current (A) and η refers efficiency of GMAW and is taken 0.8 for the experiment.

2.5   Visual Inspection

Beads-on-plate were observed visually to evaluate the major welding defects and nature of weld beads.

2.6   Macroscopic Examination of Weld Bead

Each weld bead on base plate was cut in transverse direction. Fresh cut surface was ground by pedestal grinder and then polished by a series of emery papers in belt grinder and rotary discs. Finally, the surface was mirror finished by buffing (using rotary disc, velvet cloth and using a mixture of alumina powder and water). Prepared sample surface was etched by 2% nital solution and observed by Tool Makers Microscope with 40 magnification. Height, depth and width of weld bead were measured by vertical and horizontal vernier scales attached with the microscope.

3   Results and Discussion

Visual inspection shows bead formed in each case to be continuous. No weld defect was observed for the weld bead samples produced. However, spatters were formed in medium or high quantity in each case. Results obtained for two sets of replicated

Table 4 Spatter obtained in visual inspection of two sets of repeated experimentations

Sample No.	Voltage (V) [V]	Current (I) [A]	Weld speed (S) [mm/min]	Heat input (Q) [kJ/mm]	Spatter 1st rep.	Spatter 2nd rep.
1	24	160	420	0.548	High	Medium
2	26	150	420	0.557	High	Medium
3	28	140	420	0.56	High	Low
4	26	180	420	0.668	Medium	Medium
5	28	170	420	0.68	High	Medium
6	30	160	420	0.685	Medium	Medium
7	28	200	420	0.798	High	Low
8	26	215	420	0.8	High	Medium
9	30	190	420	0.814	High	Medium

samples are shown in Table 4. The number of spatter par weld bead is quiet more in the first set of experiment than the second set, whereas the train of spatter formation is almost same in both cases.

Table 5 shows weld bead components obtained from two replicated experiments at different heat input conditions. Table 6 represents two shape factors like RFF and PSF from two set of experiments at different heat input values. Average values obtained from both set of experiments corresponding to the width of weld bead, reinforcement or deposition height, penetration depth, different shape factor like RFF and PSF against different heat input. Graphical representation of change in weld bead width, reinforcement height, depth of penetration, RFF and PSF with the increase in heat input are shown in Figs. 3, 4, 5, 6 and 7 for two sets of experiments, respectively. Figure 3 shows that the increment of width of weld bead is happening with the increment in heat input. Maximum portion of heat applied for welding may be used to widen the bead rather than to make it high or deep within the experimental domain. Deviations occurred in the results with respect to train value also raises with enlarging in heat input.

Figure 4 depicts that the height of reinforcement does not change appreciably with an increment in heat input. The trendline is slightly decreasing within the experimental territory. Deviations occur in results with respect to trend value, and it expands with a hike in heat input. Most of the heat may have been spent for melting of electrode over the base material so that the filler material spreads wider. The more the voltage and current, the metal transfer mode will be changed from globular to spray mode and finer droplets spread over native metal.

Figure 5 reveals that the penetration depth remains unaltered appreciably with enlargement in heat input within experimental zone. Deviations of experimental results with respect to trend value increase in the first phase and then decrease with the heat input increment. However, trend line has a slight decreasing tendency with larger heat input.

Table 5 Elements of weld bead geometry measured from two set of experiments at different heat input

Sample No.	Heat input (kJ/mm)	Bead width (W) 1st rep	Bead width (W) 2nd rep	Bead height (R) 1st rep	Bead height (R) 2nd rep	Penetration (P) 1st rep (mm)	Penetration (P) 1st rep (mm)
1	0.548	6.8	4.75	2.1	2.3	2.87	1.85
2	0.557	7.6	5.65	1.62	2.4	2.42	2.6
3	0.56	7.4	6.64	2.2	2.5	1.82	2.6
4	0.668	7.9	8.96	2.35	1.47	1.8	3.63
5	0.68	6.45	9.84	2.18	2.1	1.87	2.4
6	0.685	8	10.85	2.68	1.45	1.596	2.75
8	0.798	8.8	11.1	2.16	1.88	1.49	2.72
7	0.8	8.7	10.4	2.2	1.3	1.94	2.7
9	0.814	11.4	12.3	1.55	2.8	2.6	1.4

Table 6 RFF and PSF obtained from two set of experiments at different heat input

Sample No.	Heat input (kJ/mm)	RFF 1st set	RFF 2nd set	RFF average	PSF 1st set	PSF 2nd set	PSF average
1	0.548	3.23809	2.06522	2.65165	2.36934	2.56757	2.46845
2	0.557	4.69136	2.35417	3.52276	3.14049	2.17308	2.65678
3	0.56	3.36364	2.656	3.00982	4.06593	2.55385	3.30989
4	0.668	3.36170	6.09524	4.72847	4.38889	2.46832	3.42860
5	0.68	2.95872	4.68571	3.82221	3.44919	4.1	3.77459
6	0.685	2.98508	7.48276	5.23392	5.01253	3.94545	4.47899
8	0.798	3.95454	8	5.97727	4.48454	3.85185	4.16819
7	0.8	4.07407	5.90425	4.98916	5.94595	4.08088	5.01341
9	0.814	7.35484	4.39286	5.87385	4.38462	8.78571	6.58516

RFF or reinforcement form factor is the fraction of weld bead width (W) to reinforcement height (R). W increases with increase in Q, whereas R remains almost the same with increase in Q, and thus, the fraction gets increased with increase in Q. Deviation of the test results with respect to trend value increases with an increase in heat input.

PSF or penetration shape factor is the ratio of bead width (W) to the depth of penetration (P). At increasing condition of heat input, bead width (W) increases appreciably; whereas, depth of penetration (P) decreases slightly and thus the ratio of those two increases along with heat input.

Fig. 3 Plot of bead width
(W) against heat input (Q)

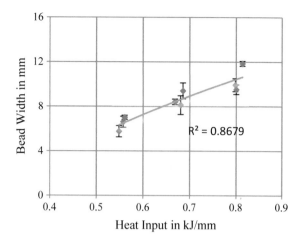

Fig. 4 Plot of reinforcement
height (R) against heat input
(Q)

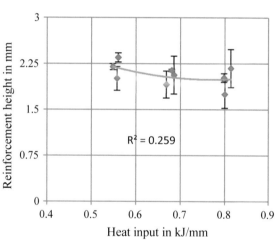

Fig. 5 Plot of depth of
penetration (P) against heat
input (Q)

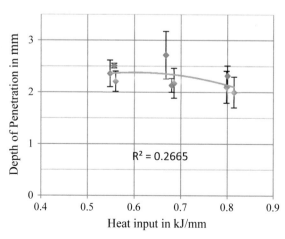

Fig. 6 Plot of reinforcement form factor (RFF) against heat input (Q)

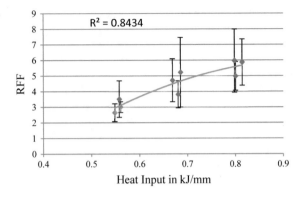

Fig. 7 Plot of penetration shape factor (PSF) against heat input (Q)

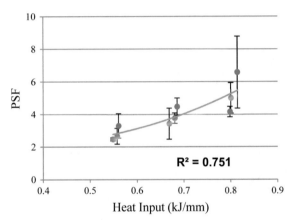

4 Regression Analysis

Regression analysis was used to obtain the relationship of input variables with output results. Among different types of relationships, polynomial quadratic equation was evaluated. ANOVA table prepared for each quadratic equation obtained from regression analysis is found becoming significant even at 95% confidence level. Relation between weld bead width, reinforcement height, depth of penetration, RFF and PSF with heat input are expressed by Eqs. (2–6), respectively.

$$W = -6.46 + 28.15\, Q - 8.74\, Q^2 \tag{2}$$

$$R = 4.459 - 6.39\, Q + 4.136\, Q^2 \tag{3}$$

$$P = 0.126 + 7.49\, Q - 6.218\, Q^2 \tag{4}$$

$$\mathrm{RFF_{Average}} = -10.65 + 34.63\, Q - 17.90\, Q^2 \tag{5}$$

Table 7 Analysis of variance: weld bead width versus heat input

Source	DF	SS	MS	F	P
Regression	2	24.7581	12.3791	19.71	0.002
Error	6	3.7685	0.6281		
Total	8	28.5266			

Table 8 Analysis of variance: reinforcement height versus heat input

Source	DF	SS	MS	F	P
Regression	2	0.063540	0.0317700	1.05	0.407
Error	6	0.181799	0.0302998		
Total	8	0.245339			

Table 9 Analysis of variance: depth of penetration versus heat input

Source	DF	SS	MS	F	P
Regression	2	0.104783	0.0523916	1.08	0.398
Error	6	0.291734	0.0486223		
Total	8	0.396517			

Table 10 Analysis of variance: reinforcement form factor (RFF) versus heat input

Source	DF	SS	MS	F	P
Regression	2	10.0323	5.01615	16.13	0.004
Error	6	1.8660	0.31100		
Total	8	11.8983			

$$PSF_{Average} = 4.09 - 10.88\,Q + 15.46\,Q^2 \tag{6}$$

Each table among Tables 7, 8, 9, 10 and 11 represents the analysis of variance table for each regression analysis. Each ANOVA table gives value of P and F of each regression equation at 95% confidence level which is shown in Table 12. On the basis of higher values of F, R^2 along with lower value of P, ANOVA table suggests strong relationship between weld bead width, PSF and RFF with heat input while there are feeble relations between height of reinforcement and penetration depth with heat input.

5 Conclusion

From the results obtained from current experiments conducted, following conclusions may be drawn:

Table 11 Analysis of variance: penetration shape factor (PSF) versus heat input

Source	DF	SS	MS	F	P
Regression	2	9.7246	4.86228	8.99	0.016
Error	6	3.2452	0.54086		
Total	8	12.9697			

Table 12 Values of P and F obtained using ANOVA

Sl. no.	Relation between	P value	F value	R^2 value
1	Weld bead width and heat input	0.002	19.71	0.8679
2	Reinforcement height and heat input	0.407	1.05	0.259
3	Depth of penetration and heat input	0.395	1.09	0.2665
4	PSF and heat input	0.016	8.99	0.8434
5	RFF and heat input	0.004	16.13	0.761

- Heat input is one important controlling factor for weld bead geometry and shape factors on the whole.
- As heat input increases, weld bead width, reinforcement form factor and penetration shape factor increase. On the other hand, as heat input increases, reinforcement height decreases slightly. Depth of penetration decreases with increase in heat input within the experimental territory.
- The second degree equation between weld bead widths, reinforcement height, penetration depth, penetration shape factor and reinforcement form factor with heat input is generated successfully by means of polynomial regression analysis which agrees with real data.
- ANOVA suggests regression equations are significant at very high (95%) confidence level. P and F values obtained from ANOVA table signify strong relationship between heat input and weld bead width, RFF, PSF rather than the relationship between heat input and depth of penetration, reinforcement height.
- Most suitable bead geometry achieved is at 30 V welding voltage, 190 A weld current and 420 mm/min electrode torch travel speed. This condition may be suggested for weld cladding for this work-electrode combination within the experimental domain.

References

1. Saha, M.K., Das S.: A Review on different cladding techniques employed to resist corrosion. J. Assoc. Eng. India **56**(1&2) (2016)
2. Saha, M.K., Das, S.: Gas metal arc welding and its anti-corrosive performance—a brief review. Athens J. Techno. Engg. **5**(2), 154–174 (2018)
3. Nasir, N.S.M., Razab, M.K.A.A., Ahmad, M.I., Mamat, S.: Influence of heat input on carbon steel microstructure. ARPN J. Eng. Appl. Sci. **12**(8), 2689–2697 (2017)

4. Shen, S., Oguocha, I.N.A., Yannacopoulos, S.: Effect of heat input on weld bead geometry of submerged arc welded ASTM A709 Grade 50 steel joints. J. Mater. Proc. Techno. **212**(1), 286–294 (2012)
5. Frei, J., Alexandrov, B.T., Rethmeier, M.: Low heat input gas metal arc welding for dissimilar metal weld overlays part II: the transition zone. Weld. World **62**(2), 317–324 (2018)
6. Frei, J., Alexandrov, B.T., Rethmeier, M.: Low heat input gas metal arc welding for dissimilar metal weld overlays part I: the heat-affected zone. Weld. World. **60**(3), 459–473 (2016)
7. Kah, P., Mvola, B., Suoranta, R., Martikainen, J.: Modified GMAW processes: control of heat input. J. Comput. Theor. Nanosci. **19**(3), 710–718 (2013)
8. Sreeraj, P., Kannan, T., Maji, S.: Simulation and parameter optimization of GMAW process using neural networks and particle swarm optimization algorithm. Int. J. Mech. Eng. Robot. Res. **2**, 130–146 (2013)
9. Palani, P.K., Murugan, N.: Development of mathematical models for prediction of weld bead geometry in cladding by flux cored arc welding. Int J. Adv. Manu. Tech. **30**, 669–676 (2006)
10. Campbell, S., Galloway, A., McPherson, N.: Artificial neural network prediction of weld geometry performed using GMAW with alternating shielding gases. Weld. J. **91**(6), 174S–181S (2012)
11. Nagesh, D.S., Datta, G.L.: Genetic Algorithm for optimization of welding variables for height to width ratio and application of ANN for Prediction of bead geometry for TIG welding process. Appl. Soft Com. **10**, 897–907 (2010)
12. Sreeraj, P., Kannan, T.: Modelling and prediction of stainless steel clad bead geometry deposited by GMAW using regression and artificial neural network models. Adv. Mech. Eng. **2012**, 1–12 (2012)
13. Mondal, A., Saha, M.K., Hazra, R., Das, S.: Influence of heat input on weld bead geometry using duplex stainless steel wire electrode on low alloy steel specimens. Cogent. Eng. **3**, 1143598 (2016)
14. Kannan, T., Yoganandh, J.: Effect of process parameters on clad bead geometry and its shape relationships of stainless steel claddings deposited by GMAW. Int. J. Adv. Manu. Tech. **47**, 1083–1095 (2010)
15. Kolahan, F., Heidari, M.: Modeling and optimization of MAG welding for gas pipelines using regression analysis and simulated annealing algorithm. J. Sci. Ind. Res. **69**(4), 177–183 (2010)
16. Saha, M.K., Dhara, L.N., Das, S.: Variation of Bead Geometry of 316 Austenitic Stainless Steel Weld with Varying Heat Input Using Metal Active Gas Welding. In: National Conference on Leveraging Simulation & Optimisation Techniques for Productivity Enhancement and Manufacturing Excellence at SNTI, Jamshedpur, Jharkhand, India (2018)
17. Saha, M.K., Hazra, R., Mondal, A., Das, S.: Effect of heat input on geometry of austenite stainless steel weld bead on low alloy steel. J. Inst. Eng. (India): Ser C. (2018). https://doi.org/10.1007/s40032-018-0461-7
18. Sharma, A., Verma, D.K., Arora, N.: A scheme of comprehensive assessment of weld bead geometry. Int. J. Adv. Manuf. Techno. **82**(9–12), 1507–1515 (2016)
19. Senthilkumar, B., Kannan, T., Madesh, R.: Optimization of flux-cored arc welding process parameters by using genetic algorithm. Int. J. Adv. Manuf. Tech. **93**(1–4), 35–41 (2017)
20. Sreeraj, P., Kannan, T., Maji, S.: Optimization of process parameters of stainless steel clad bead geometry deposited by GMAW using integrated SA-GA. Int. J. Res. Aeronaut. Mech. Eng. **1**(1), 26–52 (2013)
21. Rodrigues, L.O., Paiva, A.P., Costa, S.C.: Optimization of the FCAW process by weld bead geometry analysis. Weld. Int. **23**(2), 261–269 (2009)
22. Gautam, U., Vipin.: Weld bead geometry prediction model by design of experiments for mild steel. Int. J. Mech. Eng. Rob. Res. **3**(3), 517–527 (2014)
23. Soares, L.B., Weis, A.A., Rodrigues, R.N., Drews, P. Jr., Guterres, B., Botelho, S.S.C., Filho, N.D.: Seam Tracking and Welding Bead Geometry Analysis for Autonomous Welding Robot. In: Proceedings of Conference IEEE Latin American Robotics Symposium, at Curitiba, Brazil (2017)

Performance Assessment of Inter-Digital Capacitive Structure-Finite Element Approach

Sheeja P. George[ID]**, Johney Isaac**[ID] **and Jacob Philip**[ID]

Abstract Incorporation of modern sensors and software functionalities has transformed traditional measuring instruments into smart ones. Even though capacitors are passive components, they find extensive applications as sensors to measure parameters such as pressure, acceleration, proximity, electric/magnetic fields and material properties. Capacitor-based systems also find wide applications in touch-screen systems. In all these systems, the measured capacitance is the input to an automated system. Response of the system to variations in the input capacitance needs to be understood and analyzed for the design of any system for optimum performance. Currently inter-digital capacitive sensors with different configurations are under development for various applications. Finite element methods provide an approximate solution to modeling problems by numerical tools. The computed values can be compared with measured ones for validation. Modeling helps to vary the parameters like area of the electrode, gap between the electrodes and the dielectric constant and to investigate the effect of each on the effective capacitance value of a multi-element inter-digital capacitor configuration. Such a configuration has been modeled and analyzed using the finite element analysis tool ANSYS 15.0. The results of the analysis will be presented and discussed.

Keywords Inter-digital capacitive sensor · Finite element method · ANSYS

S. P. George (✉)
Department of Electronics, College of Engineering, Chengannur, Kerala, India
e-mail: sheejapgmanoj@yahoo.com

J. Isaac
Department of Instrumentation, CUSAT, Kochi, Kerala, India

J. Philip
Amaljyothi College of Engineering, Kanjirappally, Kottayam, Kerala, India

© Springer Nature Singapore Pte Ltd. 2020
R. Venkata Rao and J. Taler (eds.), *Advanced Engineering Optimization Through Intelligent Techniques*, Advances in Intelligent Systems and Computing 949, https://doi.org/10.1007/978-981-13-8196-6_38

1 Introduction

Instrumentation systems implemented with capacitive sensors measure the capacitance between the conductors separated by a dielectric medium. This measured value becomes the input to the control circuitry which analyzes the value and accordingly controls the operation of the system. The capacitance between the two conductors of a capacitor is given by

$$C = (\varepsilon_0 \varepsilon_r A)/d \tag{1}$$

where C is the capacitance, ε_0 is the permittivity of free space, ε_r is the relative permittivity, A is the area and d is the distance between the electrodes. The capacitance value can be varied by changing either the area or the distance between the plates. Physical quantity is allowed to vary these parameters and the change in capacitance is proportional to the physical quantity. Accordingly, displacement and pressure are the mostly traceable physical quantity with capacitive sensors which make them useful as proximity sensors, pressure sensors and to measure flow, liquid level and as switches. As the capacitance is dependent on the dielectric constant of the medium, capacitive sensors are also ideal for material identification.

By increasing the effective area of the electrodes, the capacitance can be increased. Inter-digital structure-coplanar periodic finger-like pattern of parallel electrodes— incorporates this principle and has the advantage of providing more capacitance per unit chip area than a two-plate parallel configuration [1]. The design of inter-digital electrode started as early as in 1891, found in the patent of Tesla [2]. The formation of coplanar structure from parallel plate is as shown in the Fig. 1. A review article discussing the various aspects of inter-digital sensors and transducers can be found in literature [3] which elaborates its versatile applications like non-destructive testing, micro electro mechanical systems, microwave filters, and as sensors such as acoustic sensors, pressure sensors, humidity sensors, biosensors, level sensors and food testing.

The structure of inter-digital capacitive structure is shown in the Fig. 2 [4]. Fingers are of width W and G is the gap between the fingers. $\lambda = 2(W + G)$ is the spatial wavelength, and L the length of the fingers. The comb group is connected either to

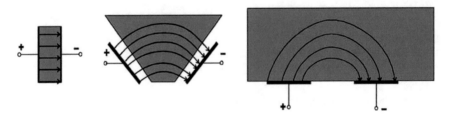

Fig. 1 Coplanar structure from a parallel plate capacitor [3]

Fig. 2 Inter-digital
capacitor [4]

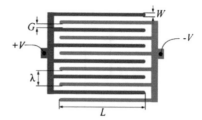

positive or negative terminal. One of the key features of the inter-digital capacitor is simplified modeling.

The most important parameter of the sensor, sensitivity and selectivity has a strong link with sensor geometry. The effect of electrode geometry on the performance of inter-digital structure has appeared in literature. The capacitance varies depending on the material dielectric constant and the electrode geometry [5, 6]. The outcome of the electrode thickness, length and the number of electrode fingers on the change of capacitance are been investigated [7].

The finite element analysis helps to handle complex geometries and is used extensively for sensor modeling, optimization and performance evaluation, especially for structures that are difficult to model analytically. All the modeling reported in the case of inter-digital structures so far considered a particular geometry for a particular application. A comparative study on the performance of inter-digital structures using finite element method for a total number of fingers of thirteen is presented. For the analysis, change in the number of positive and negative terminals, length of the fingers and width of the fingers is presented. The analysis of inter-digital capacitive structures and comparison of capacitance value obtained by direct measurement and simulation through finite element analysis tool is also incorporated.

2 Configuration of Inter-digital Structures and Measurements

The basic comb structure considered is with $W = 15$ mils, $G = 50$ mils and $L = 350$ mils. W and L are altered to achieve the structural variations. In the present illustration, the structural variation is accomplished by fixing the total number of fingers as 13, and then a change in the number of positive and negative electrodes are made. Finger arrangement of 1_2_1, 1_3_1, 1_5_1 & 1_11_1 can be attained with the total number of fingers of 13 as given in Fig. 3. Table 1 gives the terminal distribution in different configurations. In each case, the width of the finger and the length of fingers are altered to find the dependence on capacitance.

The inter-digital capacitors are of planar in structure and are comparatively simpler to fabricate using the usual PCB fabrication techniques. The comb structure was designed using Proteus and was printed onto FR4 board using the screen print-

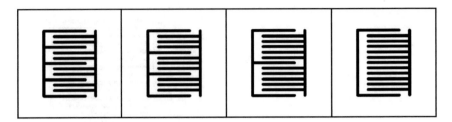

Fig. 3 Inter-digital structure configuration with 13 fingers [1_2_1, 1_3_1, 1_5_1 & 1_11_1]

Table 1 Terminal distribution

Total number of fingers	Configuration	Number of positive terminals	Number of negative terminals
13	1_2_1	5	8
	1_3_1	4	9
	1_5_1	3	10
	1_11_1	2	11

ing technique. The unwanted portion in the board is removed through etching, and traces of inter-digital structure remain in the board. Inter-digital structures with varying finger width (W) and varying finger length (L) were fabricated for all the four categories.

Direct measurement of capacitance is obtained using the LCR measurement feature of the Hioki Impedance Analyzer IM3570. The instrument allows test signal of DC or a band of frequency in the range 4 Hz to 5 MHz with levels of 5 mv to 5 V. The four terminal probes provide improved accuracy. Simulation is done using finite element analysis tool ANSYS.

3   Simulation of Inter-digital Structures

Finite element analysis (FEA) is a tool of numerical method for solving engineering problems. In FEA, an area is been divided into several elements. Then elements are reconnected at nodes as though nodes are drops of glue that hold elements together. These processes result in polynomial, and the coefficients represent the boundary conditions of the neighbor ones. The literature reveals that FEA error is less than 0.18% when compared to analytic solutions. Several commercial packages are available for the same, and for the present work, the tool ANSYS 15.0 is used since it helps to perform the electrostatic analysis. With the electrostatic field analysis, electric field distribution and electric scalar potential due to charge distribution or applied potential can be determined. The load we are applying in this analysis is voltage. The elements that are used in the charge-based electrostatic analysis are PLANE121,

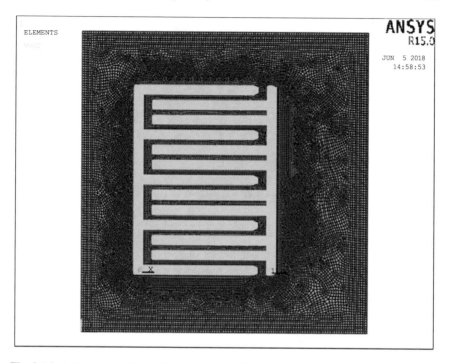

Fig. 4 Meshed geometry of inter-digital structure with 1_2_1 configuration

SOLID122 and SOLID123. In the present analysis, the element PLANE121 is considered. PLANE121 is a 2D element of eight nodes with the degree of freedom-voltage, at each node.

The electrostatic analysis starts with building the model by selecting the appropriate element which is made available with setting the preference to electric. The element type and the material property (permittivity) are to be defined before the required geometry is created. The loads for electrostatic analysis can be applied on the solid model or on the finite element model. The solution is obtained through the SOLVE command and the general post processor menu gives a review of the results of the analysis (Figs. 4, 5).

The capacitance calculation of an electrostatic system is based on the energy principle and is computed using the CMATRIX macro [8]. Within the macro, the structure is considered as a three electrode system, third being the ground and the capacitance is calculated using the Eq. (2).

$$W = (1/2) * \left(C V^2 \right) \tag{2}$$

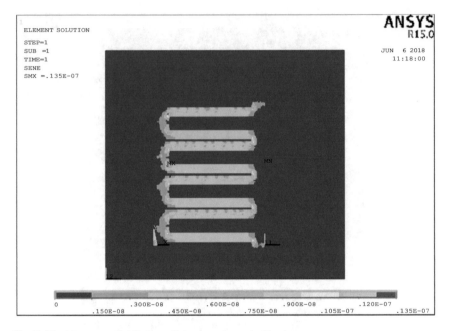

Fig. 5 Electric energy distribution of structure shown in Fig. 4

4 Results and Discussion

The direct measurement was obtained through LCR meter and simulation values using FEA tool ANSYS. Measurement and simulation are done on all the fabricated inter-digital structures.

4.1 Number of Negative Electrodes

The capacitance measured for the all the fabricated inter-digital structures, and an analysis on the basis of number of negative electrodes is performed. For this, a particular geometry is considered with a finite W and L. The analysis reveals that the capacitance decreases as the number of negative terminals increases. The result is as shown in Fig. 6.

4.2 Change in Finger Width (W)

The structural variation in width is obtained by changing the finger width (W) to the values 20 mils, 25 mils and 30 mils keeping the finger length constant. As the

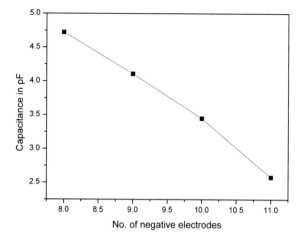

Fig. 6 Capacitance according to the number of negative electrodes

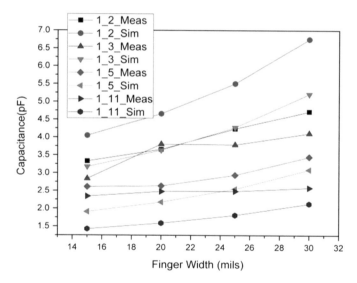

Fig. 7 Capacitance obtained through measurement and simulation for change in finger width

finger width increases, the effective area increases and the capacitance increases. The result is shown in Fig. 7. Among the different configurations in 13 finger, the structure 1_5_1 has the close relation between the measured value and the simulated value. This agrees with the performance evaluated in [9].

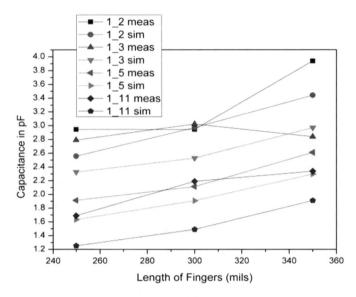

Fig. 8 Capacitance obtained through measurement and simulation for change in length of the fingers

4.3 Change in Length of the Fingers (L)

Fabrication of inter-digital capacitors was carried out with the length of fingers as 250 mils, 300 mils and 350 mils with width of the finger and the gap between the fingers kept constant for all the structures. As the length of the finger is increased, the distance between the electrodes decreases and the capacitance increases. The result is shown in Fig. 8.

5 Conclusion

The knowledge of capacitance is essential in the design of electrostatic devices, MEMS, etc. This paper verifies the validity of the electrostatic analysis of finite element method with that of the measured values of inter-digitated structures with a total number of fingers 13. The good agreement of experimental values with simulated values demonstrates the feasibility of finite element analysis for the design of inter-digital structures. Further analysis can be done on these models to find the effect of the medium and substrate type and thickness.

Acknowledgements The first author would like to acknowledge Mr. Akash Rajan, Research Scholar, College of Engineering, Thiruvananthapuram, Kerala, India, for his valuable help.

References

1. Zhu, L., et al.: Adjustable bandwidth filter design based on inter-digital capacitors. IEEE Microwave Wirel Compon Lett **18**(1) (2008)
2. Tesla, N.: Electric condenser, U.S. Patent 464 667 (1891)
3. Mamishev, A.V., et al.: Interdigital Sensors and Transducers. In: Proceedings of the IEEE, vol. 92, pp. 808–845 (2004)
4. Igreja, R., Dias, C.J.: Analytical evaluation of the interdigital electrodes capacitance for a multi-layered structure. Sens Actuators A **112**, 291–301 (2004)
5. Igreja, R., Dias, C.J.: Dielectric response of interdigital chemocapacitors: the role of the sensitive layer thickness. Sens Actuators B **115**, 69–78 (2006)
6. Mamishev, A.V., Takahashi, A.R., Du, Y., Lesieutre, B.C., Zahn, M.: Assessment of Performance of Fringing Electric Field Sensor Arrays. In: Proceedings of IEEE Conference on Electrical Insulation and Dielectric Phenomena, pp. 918–921 (2002)
7. Lim, Y.C., et al.: A Micromechanical Biosensor with Interdigitated Capacitor Readout. In: IEEE/ICME international Conference on Complex Medical Engineering (2011)
8. ANSYS Mechanical APDL Theory Reference Release 15.0
9. Mohd Syaifudin, A.R., et al.: Measurements and performance evaluation of novel inter-digital sensors for different chemicals related to food poisoning. IEEE Sens. J. **11**(11), 2957–2965 (2011)

MADM-Based Approach for Selection of Welding Process for Aluminum Tube Manufacturing

Ravindra Singh Saluja⊙ **and Varinder Singh**⊙

Abstract Traditionally, some of the alternative processes for manufacturing welded tubes have been high-frequency (HF) alternating current resistance welding and tungsten inert gas (TIG) welding. Recently, the laser welding (LW) is also emerging as a potential process owing to reduction in its cost. The HF electric resistance welding gained wider popularity and is almost always used without any second thoughts for manufacturing aluminum tube industry. A scientific examination of the decision criteria may make a way for systematic consideration and justification of newer technological developments to reap competitive advantages. This paper presents the use of analytic hierarchy process (AHP) and modified AHP to justify the selection of suitable welding technique for welded tube manufacturing.

Keywords Welded tubes · AHP · Welding processes · Multiple attributes

1 Introduction

Tube manufacturing is one of the major industries supporting many major sectors of industry dealing with engineering products. Most of the tubes are manufactured presently by technique involving bending a sheet or strip to give it a tube-like shape and welding it forming a seam. This type of manufacturing technique is preferred one in comparison to seamless tube manufacturing as it offers significant advantages on economics and production front. Tubes manufactured by this technique are commonly called welded tubes and are able to satisfactorily meet all general quality requirements. By subjecting such products to post-processing operations such as cold drawing and some other operations, even stringent quality requirements are met easily. The choice of welding technique plays a major role in defining the cost and

R. S. Saluja (✉) · V. Singh
Department of Mechanical Engineering, K. K. Birla Goa Campus, BITS Pilani, Pilani, India
e-mail: rsaluja@goa.bits-pilani.ac.in

V. Singh
e-mail: vsingh@goa.bits-pilani.ac.in

© Springer Nature Singapore Pte Ltd. 2020
R. Venkata Rao and J. Taler (eds.), *Advanced Engineering Optimization Through Intelligent Techniques*, Advances in Intelligent Systems and Computing 949, https://doi.org/10.1007/978-981-13-8196-6_39

quality of such tubes. In general, the techniques of manufacturing different tubes are decided purely on technical considerations. Recent research has pointed out the inadequacies of such selection process and have suggested the consideration of multiple criteria in a systematic manner for selecting appropriate welding processes for applications such as hardfacing of carbon steels, for the trivial butt joints, and so on [1–4]. However, for deciding welding process for this commercially important tube manufacturing industry, no systematic treatment has been found in the literature to deal with multiple criteria. In the modern times, a vast amount of technological research is resulting in the advent of newer techniques in welding technology. A frequent reconsideration of such decisions by systematically considering newer alternatives may also yield newer insights and may result in fruitful innovations. Thus, a systematic decision-making process is very much needed that can take the multiple criteria as well as multiple alternatives into consideration for selection of the most appropriate welding technique for welded tube manufacturing. Different multi-attribute decision-making (MADM) tools may be used for analyzing this decision. Analytic hierarchy process (AHP) is one of the popular ones used by a number of researchers for decision making in different applications [5]. Technique for Order Preference by Similarity to Ideal Solution (TOPSIS) is another method wherein the comparative value of the Euclidean distance from the positive and negative ideal hypothetical alternatives in a multi-dimensional Euclidean space is used to solve multi-attribute decision problems [1, 6, 7]. A hybrid AHP-TOPSIS approach has also been in use to provide decision support in many applications. Similarly, there are many more MADM tools, all with the aim of resolving the multi-attribute decision situation in a systematic manner while following different algorithms. Researchers [6] have also discussed that the use of different MADM tools does not create significant difference in results and the choice of the tool may be based on minimizing the computational load. On the other hand, the fuzzy scales have also been adopted by a number of researchers within AHP, TOPSIS, and different other MADM methodologies [8, 9–11]. However, AHP in its basic form also holds significant promise and is widely used to address decision situations involving subjective judgments in pairwise comparison of attributes as well as alternatives in the domain of the chosen attributes of decision making.

The present paper exploits AHP for proposing a systematic decision analysis for selecting welding processes in tube manufacturing. As AHP offers a simple way to systematically arrive at the user preference through subjective pairwise comparisons of alternatives as well as criteria and is found as easy to adapt by the experts contributing to the preference scores, it has been chosen for the current work. The present paper also uses combined AHP and TOPSIS for comparison of the ranking obtained for the available options of welding process for tube manufacturing. Particularly, in a high-volume production industry like welded tube manufacturing, if emerging technologies may fit as a better choice, it may result in a huge impact in terms of enhanced competitiveness at the business level. Thus, a systematic decision analysis for the existing choices is much desirable. The paper presents the methodology in brief followed by implementation to the specific case. Thereafter, the discussion of results and conclusions as well as future scope follows.

2 Methodology and Implementation

This section discusses the steps in the proposed methodology of AHP and AHP-TOPSIS model.

2.1 AHP Model

The general steps in AHP are described below based on established literature sources [1, 3–6, 10, 11].

(a) Index the welding processes that engineers can lead into consideration for potential pick.
(b) Find the criteria and measurable value for criteria that could be used.
(c) Next, the pairwise comparison matrix of all set criteria is completed by capturing relative importance as per expert views.
(d) Once the pairwise comparison has been captured in a matrix for a criterion, the normalized values are estimated. The values of each column are normalized by dividing each value in column by the sum of the values in the same column. Then the average of each row is calculated to provide priority weights (PW) of the criteria.
(e) Similar to steps (c and d), the pairwise comparisons of all the alternatives are used to obtain priority weights (PW) of each alternative with respect to different criteria.
(f) Also, before weights are calculated in steps (c, d, and e), it is necessary to check the pairwise comparisons are consistent. For that purpose, a measure known as consistency ratio (CR) is obtained as below.

$$CR = CI/RI,$$

where consistency index $(CI) = (\lambda_{max} - n)/(n - 1)$,
$\lambda_{max} = $ maximum eigenvalue,
$n = $ dimension of matrix,
and RI is a random index based on a table in reference [5].
So as long as CR < 0.1, the evaluations arrived at are considered consistent.

(g) For AHP-based selection, the weighted normalized decision matrix is calculated by summations of the products of criteria weights and alternative weights to obtain the final rankings.
(h) Compute the weighted priority scores for each alternative based on all criteria.
(i) In the end, the ranks are assigned to alternatives having the highest weighted priority score.

2.2 *AHP-TOPSIS Model*

(a) Compute the weighted normalized decision matrix by using procedure as given in AHP model up to step (g).
(b) Determine the hypothetical best and worst alternatives in the domain of alternatives considered by identifying the set of best and worst weighted criteria values. These are labeled as positive ideal and negative ideal solutions.
(c) Determine the positive ideal and negative ideal distances.
(d) The relative closeness to the ideal solution for each alternatives is calculated through the closeness coefficient.
(e) Alternatives are ranked based on the value of closeness coefficient.

3 Implementation of Methodology

The case of manufacturing a welded tube with the following design specifications has been considered for this study. In this work, mainly the problem of selection of welding process for this kind of tube manufacturing has been particularly addressed.

Design Specifications

   Material: Aluminum,
   Diameter: 30 mm,
   Thickness: 2 mm.

Many welding processes are available for joining metals in the tube welding. However, based on the design specifications and the guidelines of Society of Manufacturing Engineering [12, 13], three welding processes have been shortlisted for this case for further study, namely high-frequency (HF) electric resistance welding, tungsten inert gas (TIG) welding, and laser welding (LW). Many research papers [1, 3, 4] dealing with selection of welding processes for different applications use many different attributes. In the present study, five most important attributes relevant for the case under consideration were used as described in Table 1. Welding selection methdology is graphically presented in the Fig. 1 and Fig. 2 presents the hierarchy of the decision problem in graphical form for quick glance. The first level shows the overall goal which is to select the best welding process available for tube manufacturing. At the second level, attributes affecting the decision, such as setup requirement (SR), worker skill requirement (WSR), operator fatigue (OF), running cost (COST) and quality (QY), will lead to the systematic pairwise comparison of these criteria among each other to obtain the priority weights. At the third level, the alternatives, namely high-frequency (HF) electric resistance welding, tungsten inert gas (TIG) welding, and laser welding (LW), that are to be compared on the criteria at level 2 are presented.

Table 1 Description of attributes considered [3, 4]

S. no.	Attributes	Description
1	Setup requirement (SR)	Welding parameters setting, cleaning the base metal, clamping joint in fixtures
2	Worker skill requirement (WSR)	Refer to the skill level of worker required
3	Occupational factor (OF)	Arc glare, smoke, and fumes
4	Running cost (COST)	Running cost for mass production
5	Weld quality (QY)	Weld bead appearance, surface finish

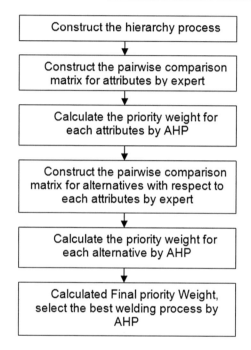

Fig. 1 Welding process selection methodology for tube manufacturing

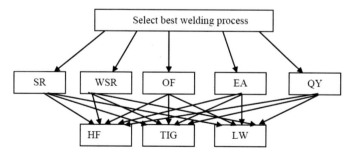

Fig. 2 Schematic diagram of analytic hierarchy process

Table 2 Strength of preference for pairwise comparison [3–5]

Strength of importance	Verbal judgment
1	Equally important
3	Moderately preferable
5	Strongly preferable
7	Very strongly preferable
9	Extremely strongly preferable
2, 4, 6, 8	Intermediate values
1/2, 1/3, 1/4, etc.	Reciprocal of above values

Table 3 Comparison of attributes

Attributes	SR	WSR	OF	COST	QY	PW
SR	1.00	0.33	0.20	0.14	0.11	0.03
WSR	3.00	1.00	0.33	0.20	0.14	0.07
OF	5.00	3.00	1.00	0.33	0.20	0.13
COST	7.00	5.00	3.00	1.00	0.33	0.26
QY	9.00	7.00	5.00	3.00	1.00	0.50
CR = 0.054						

Table 4 Pairwise comparison of alternatives with respect to the setup requirement

Process	TIG	HF	LW	PW
TIG	1.00	0.14	0.33	0.083
HF	7.00	1.00	5.00	0.723
LW	3.00	0.20	1.00	0.193
CR = 0.056				

The strength of preference for pairwise comparison is indicated using numbers in Table 2. It is used for writing the pairwise comparison matrices in the AHP procedure implementation.

The pairwise comparison of criteria for tube manufacturing is depicted in Table 3. It may be observed that the cost of the welding process is considered as the most important attributes followed by operator fatigue and so on. The pairwise comparison of alternatives with regard to each of these attributes is depicted in Tables 4, 5, 6, 7, and 8. In all these tables, the priority weights (PW) are calculated according to step (d) for AHP. Also, the calculated values of consistency ratios according to step (f) are listed, which were all found to be less than 0.1 to fulfill the important prerequisite for proceeding further with the analysis. The results of Tables 4, 5, 6, 7, and 8 are summarized in Table 9. Then, overall priority weights of alternatives were calculated according to step (g) and step (h) in section AHP model and are reported in Table 10.

Table 5 Pairwise comparison of alternatives with respect to the worker skill requirement

Process	TIG	HF	LW	PW
TIG	1.00	0.20	3.00	0.193
HF	5.00	1.00	7.00	0.723
LW	3.00	0.14	1.00	0.083
CR = 0.056				

Table 6 Pairwise comparison of alternatives with respect to the occupational factor

Process	TIG	HF	LW	PW
TIG	1.00	0.25	6.00	0.25
HF	4.00	1.00	9.00	0.69
LW	0.17	0.11	1.00	0.058
CR = 0.094				

Table 7 Pairwise comparison of alternatives with respect to the running cost

Process	TIG	HF	LW	PW
TIG	1.00	0.14	0.13	0.060
HF	7.00	1.00	0.33	0.302
LW	8.00	3.00	1.00	0.636
CR = 0.056				

Table 8 Pairwise comparison of alternatives with respect to the weld quality

Process	TIG	HF	LW	PW
TIG	1.00	2.00	0.20	0.189
HF	0.5	1.00	0.17	0.111
LW	3.00	7.00	1.00	0.699
CR.= 0.047				

Table 9 Composite weight of welding processes

Process	SR	WSR	OF	COST	QY
TIG	0.083	0.193	0.25	0.060	0.189
HF	0.723	0.723	0.69	0.302	0.111
LW	0.193	0.083	0.058	0.636	0.699

Table 10 Composite ranking of welding processes by AHP model

Process	SR	WSR	OF	COST	QY	PW	Ranking
TIG	0.002	0.013	0.025	0.0158	0.095	0.152	3
HF	0.025	0.049	0.1	0.078	0.055	0.309	2
LW	0.006	0.005	0.008	0.165	0.351	0.537	1

Table 11 Composite ranking of welding processes by AHP-TOPSIS model

Processes	PID	NID	RC	Ranking
TIG	0.309	0.043	0.124	3
HF	0.308	0.122	0.283	2
LW	0.104	0.331	0.761	1

In AHP-TOPSIS method, normalized matrix obtained in AHP model is used, to obtain hypothetical positive and negative ideal distances (PID and NID) as well as relative closeness (RC) values as according to steps in Sect. 2.2.

Positive ideal (P^+) and negative ideal (N^-) alternatives are identified with values from normalized matrix in Table 10 as follows.

$$P_{SR}^+ = 0.025, \quad P_{WSR}^+ = 0.049, \quad P_{OF}^+ = 0.1, \quad P_{COST}^+ = 0.165, \quad P_{QY}^+ = 0.351$$
$$N_{SR}^- = 0.002, \quad N_{WSR}^- = 0.005, \quad N_{OF}^- = 0.008, \quad N_{COST}^- = 0.0158, \quad N_{QY}^- = 0.055$$

The PID, NID, and RC calculations of alternative TIG welding are as shown below

$$PID = ((0.025 - 0.002)^2 + (0.049 - 0.013)^2 + (0.1 - 0.025)^2 + (0.165 - 0.0158)^2$$
$$+ (0.351 - 0.095)^2)^{1/2}$$
$$PID = 0.309$$
$$NID = ((0.002 - 0.002)^2 + (0.005 - 0.013)^2 + (0.008 - 0.025)^2$$
$$+ (0.0158 - 0.0158)^2 + (0.055 - 0.095)^2)^{1/2}$$
$$NID = 0.043$$

$$\text{Relative closeness (RC)} = \frac{NID}{NID + PID}$$

Relative closeness value for TIG welding is 0.124.

Similar way distances (PID and NID) from hypothetical positive and negative ideal alternatives, relative closeness values (RC), and rankings of other alternative are listed in Table 11.

4 Results and Discussion

In Table 10, the ranking of welding processes by using AHP model is shown, and in Table 11, the ranking of welding processes by using AHP-TOPSIS model is shown. In both methods, ranking of welding processes is found to be consistent and this shows robustness of ranking. The utility of such decision analysis is that it considers different practical aspects and uses the provided information systematically, which help the design engineer and manufacturing engineer to choose the most appropriate process considering several criteria simultaneously. Moreover, AHP and AHP-TOPSIS models take care of subjective information, scale of judgments, and pairwise comparison through established mechanism of evaluation through consistency ratio and consistency index.

Based on the value of overall priority weights in Table 10, the highest value stands for laser welding (LW). It shows that it ranks as the highest preferred option followed by HF and TIG. LW is standing high in the order of preference mainly on the basis of running cost and quality as compared to other processes. In high-frequency process, initial parameter setting, base metal preparation and smoke is less as compared to other processes. And in Table 5, processes are compared with respect to work skill requirement (WSR) and found that WSR is less in HF process because less equipment is required in handling during the process. In Table 7, processes are compared with respect to running cost and the analysis found that laser welding score is high compared to other processes, because laser welding running cost is less followed by HF and TIG. In Table 8, processes are compared with respect to quality and the analysis found that laser welding score is high in comparison to other processes because in laser welding, heat-affected zone and weld surface finish are better as compared to other processes.

The present work is a pioneering attempt to systematically address the multi-attribute selection issue of welding process selection for the seam weld-based tube manufacturing.

5 Conclusions and Future Scope

In this work, the AHP and AHP-TOPSIS models have been used for ranking the welding processes for tube manufacturing. The conclusions drawn from the study, and the future scope is presented below:

(i)   The case of manufacturing a welded tube with the following design specifications has been considered for this study: material—aluminum, diameter—30 mm, and thickness—2 mm.

(ii) AHP method offers a simple way to systematically arrive at the user preference through subjective pairwise comparisons of alternatives as well as criteria and is found as easy to adapt by the experts.

(iii) In both methods, AHP and AHP-TOPSIS, ranking is found to be consistent. This shows the robustness of ranking.

(iv) It has been found that laser welding (LW) is the most appropriate process for meeting the requirements of present case, followed by high-frequency (HF) welding, then tungsten inert gas (TIG) welding.

(v)   The analysis presented in this paper provides decision-making assistance for the selection of a suitable welding process by considering several attributes.

(vi) In future, there is scope to compare and rank the existing available options with other recent developments processes like hybrid welding.

References

1. Mostafa, J., Vahdat, S.E.: A fuzzy multi-attribute approach to select the welding process at high pressure vessel manufacturing. J. Manuf. Process. **14**, 250–256 (2012). https://doi.org/10.1016/j.jmapro.2011.10.006
2. Darwish, S.M., Tamimi, A.A., Habdan, S.A.: A knowledge base for metal welding process selection. Int. J. Mach. Tools Manuf. **3**, 1007–1023 (1997). https://doi.org/10.1016/S0890-6955(96),00073-9
3. Ravisankar, V., Balasubramanian, V., Muralidharan, C.: Selection of welding process to fabricate butt joints of high strength aluminum alloys using analytic hierarchic process. Mater. Des. **27**, 373–380 (2006). https://doi.org/10.1016/j.matdes.2004.11.008
4. Balasubramanian, V., Varahamoorthy, R., Ramachandran, C.S., Muralidharan, C.: Selection of welding process for hardfacing on carbon steels based on quantitative and qualitative factors. Int. J. Adv. Manuf. Technol. **40**, 887–897 (2009). https://doi.org/10.1007/s00170-008-1406-8
5. Saaty, T.L.: Analytic Hierarchy Process. McGraw Hill Publications, New York (1980)
6. Zamani- Sabzi, H., King, J.P., Gard C. C., Abudu, S.: Statistical and analytical comparison of multi-criteria decision-making techniques under fuzzy environment. Oper. Res. Perspect. **3**, 92–117 (2016). https://doi.org/10.1016/j.orp.2016.11.001
7. Singh, V., Agrawal, V.P., Deb, P.: A Decision Making Method for Selection of Finish Process. In: Proceedings of IEEE International Conference on Industrial Engineering and Engineering Management), Macao, China. 7th–10th December (IEEM), pp. 38–42 (2010). https://doi.org/10.1109/ieem.2010.5674417
8. Chang, D.Y.: Applications of the extent analysis method on fuzzy AHP. Eur. J. Oper. Res. **95**(3), 649–655 (1996). https://doi.org/10.1016/0377-2217(95),00300-2
9. Deng, H.: Multi-criteria analysis with fuzzy pair-wise comparison. Int. J. Approximate Reasoning **21**(3), 215–231 (1999). https://doi.org/10.1109/FUZZY.1999.793038
10. Rao, R.V.: Machinability evaluation of work materials using a combined multiple attributes decision-making method. Int. J. Adv. Manuf. Technol. **28**, 221–227 (2006). https://doi.org/10.1007/s00170-004-2348-4
11. Rao, R.V., Davim, J.P.: A decision-making framework model for material selection using a combined multiple attribute decision-making method. Int. J. Adv. Manuf. Technol. **35**, 751–760 (2008). https://doi.org/10.1007/s00170-006-0752-7
12. Baheti, M.: Everything about welded tube technology. DeeTee Industries Ltd., Indore, India (2016)
13. Veilleux, R.F., Petro, L.W.: Tool and Manufacturing Engineer's Handbook. Manufacturing Management, vol. 5, 4th edn. Society of Manufacturing Engineers (1998) ISBN 978-0-872633-06-3

Screening of Organic Brake Pad Materials Using MADM Technique

K. N. Hendre and B. D. Bachchhav

Abstract Asbestos friction materials are found to be harmful to human health and are being replaced. A research is being carried out in developing non-asbestos-based brake pad materials. Considerable efforts are being made in testing and evaluation of various tribological, metallurgical, mechanical and thermal properties of such materials. However, selection of novel brake pad material that meets all requirements becomes a complex task. In this paper, six novel brake pad materials are compared with asbestos friction material and ranked based on nine different attributes using multi-attribute decision-making (MADM) method. The subjective as well as the objective weights are considered. Subjective weights are calculated using analytical hierarchy process (AHP). The databases were collected through standard published research work. As per the order of ranking based on objective, subjective and integrated weights, Periwinkle shell-based brake pad material is found to be the best suitable choice among seven possible alternatives. This work will further lead to the development of asbestos-free eco-friendly friction materials for the manufacturing of automotive brake pads with an objective of enhanced longevity. However, in order to have a commercially viable product, tribo-mechanical behavior of a newly developed materials are to be assessed for a suitable application.

Keywords MADM · Brake pad materials · Subjective and objective importance

1 Introduction

The utilization of asbestos friction material is being outlawed because of various health risks associated with the fumes generated during manufacturing and oper-

K. N. Hendre · B. D. Bachchhav (✉)
Department of Mechanical Engineering, All India Shri Shivaji Memorial Society's College of Engineering, Pune, Maharashtra 411001, India
e-mail: bdbachchhav@aissmscoe.com

K. N. Hendre
e-mail: kishorhendre77@gmail.com

© Springer Nature Singapore Pte Ltd. 2020
R. Venkata Rao and J. Taler (eds.), *Advanced Engineering Optimization Through Intelligent Techniques*, Advances in Intelligent Systems and Computing 949, https://doi.org/10.1007/978-981-13-8196-6_40

ations [1–9]. A lot of research is carried out in the development of asbestos-free organic brake pads such as palm kernel shell (PKS) [1], Bagasse [2], egg shell [3], Periwinkles shell [4–6], cocoa bean shell [7], banana peels [8], maize husk [9] for their potential use. These materials are yet to prove their ability as a commercial efficiency. Due to the diversity of applications and varieties, selection of an appropriate non-asbestos brake pad material becomes a very complex task. It depends on number of diverse properties such as physical, chemical, mechanical, metallurgical and tribological. Also, in order to replace asbestos friction material, considerable efforts are needed in testing and evaluation of various tribological, metallurgical and thermal properties. Brake pad material selection problem considering both qualitative as well as quantitative attributes has not been addressed so far in open literature. The method, as developed by Rao and Patel [10], considered an objective as well as subjective weightage of importance to decide the integrated values of decisiveness. Several researchers have also used AHP as a decision-making tool [11–13] for prior utilization of material properties. Recently, Bachchhav et al. [13, 14] evaluated the critical property of a novel brake pad material for automotive applications and lubricant selection for metal forming processes by using AHP. The subjective importance of characteristics was patterned utilizing analytical hierarchy process (AHP) [15]. It is a multi-criteria alternative approach and gave a powerful method for appropriately evaluating the applicable information, utilizing a couple insightful examinations between the parameters. In this paper, authors made an attempt to solve a complex organic brake pad material selection problem by a multi-attribute decision-making (MADM) method.

2 Multi-attributes Decision-Making Methodology

In order to solve the complex selection problems due to large number of diversified attributes and alternatives, MADM method was invented by Rao and Patel [10]. The decision table contains attributes, alternatives, weight or relative importance of each attributes and the performance measures of alternatives. The stepwise procedure to solve this problem is discussed below.

2.1 Preparation of Decision Table

To meet the specific braking conditions, physical, chemical, tribological and metallurgical properties of brake pads are to be within specified limits. The attributes considered are: thickness swell in oil, wear rate, flame resistance, thermal conductivity, friction coefficient, specific gravity, thickness swell in water, compressive strength and hardness. Quantitative values for above attributes are referred from [8, 9]; however, qualitative value is expert opinion in the field of brake pads as shown in Table 1.

Table 1 Brake pad material attributes

S. no.	Attributes alternatives	TS(O) (%)	WR (mg/m)	FR	TC (W/mK)	FC	SG (g/cm³)	TS(W) (%)	CS (MPa)	H (MPa)
1	Asbestos-based brake pad	0.3	3.8	ExL	0.5	0.4	1.8	0.9	110.0	101.0
2	Palm kernel shell-based brake pad	0.4	4.4	VL	1.4	0.4	1.6	5.0	103.5	92.0
3	Bagasse-based brake pad	1.1	4.2	BA	4.8	0.4	1.4	3.4	105.6	100.5
4	Egg shell-based brake pad	1.1	4.0	AA	3.5	0.3	1.6	3.2	103.0	99.1
5	Periwinkle shell-based brake pad	0.3	3.0	H	1.9	0.4	1.0	0.3	147.0	116.7
6	PKS + CBS + MH-based brake pad	0.5	2.1	EH	0.3	0.3	0.8	0.9	10.3	127.8
7	Banana peels-based brake pad	1.1	4.1	EL	4.2	0.4	1.2	3.2	95.6	98.8

Abbreviations used for attributes: TS(O): Thickness swell in oil (%), WR: Wear rate (mg/m), FR: Flame resistance, TC: Thermal conductivity (W/mK), FC: Friction coefficient, SG: Specific gravity (g/cm³), TS(W): Thickness swell in water (%), CS: Compressive strength (MPa), H: Hardness (MPa). Exceptionally low (ExL); Very low (VL); Below average (BA); Above average (AA); High (H); Exceptionally high (EH); Extremely low (EL)

Table 2 Values of selection attributes

Qualitative measures of selection attributes	Fuzzy number	Assigned crisp score
Exceptionally low	M_1	0.0455
Extremely low	M_2	0.1364
Very low	M_3	0.2273
Low	M_4	0.3182
Below average	M_5	0.4091
Average	M_6	0.5000
Above average	M_7	0.5909
High	M_8	0.6818
Very high	M_9	0.7721
Extremely high	M_{10}	0.8636
Exceptionally high	M_{11}	0.9545

FR is considered as qualitative attribute and SG, WR, TS(W), TC, TS(O), CS, H, FC are quantitative attributes. WR, TS(W), TS(O), FR, H, are non-beneficial attributes (lower values are advantageous) and SG, TC, CS, FC, are beneficial attributes (higher values are advantageous). The quantitative values were referred from the research work carried out by [8, 9]. Qualitative values are assigned to the identified attributes through fuzzy logic analysis. A fuzzy conversion scale, proposed by Rao and Patel [10] is used here to present qualitative attributes in quantitative form. The qualitative attributes are converted to quantitative attributes based on fuzzy conversion scale (11-point scale is shown in Table 2). Table 3 presents the data for all nine attributes after appropriate conversion to quantitative terms.

2.2 Normalization of Data

The values associated with the attributes (x_{ij}) are in different units such as specific gravity in (g/cm^3), wear rate is in (mg/m) and compressive strength is in (MPa), etc. Hence, normalization is required for each element of the decision table. The values of attributes can be normalized using the following equation:

$$x'_{ij} = x_{ij} \bigg/ \sum\nolimits_{i=1}^{n} x_{ij} \tag{1}$$

where x'_{ij} is the normalized value of x_{ij}, and $\sum_{i=1}^{n} x_{ij}$ is the summation of all the values of jth attribute for the 'n' number of alternatives. The normalized data values for attributes of different alternatives are given in Table 4.

Determinations of weights of relative importance of the attributes have been done based on objective weights or subjective weights or integrated weights.

Table 3 Quantitative values for brake pad material attribute

S. no.	Attributes alternatives	TS(O) (%)	WR (mg/m)	FR	TC (W/mK)	FC	SG (g/cm³)	TS(W) (%)	CS (MPa)	H (MPa)
1	Asbestos-based brake pad	0.3	3.8	0.04	0.5	0.4	1.8	0.9	110.0	101.0
2	Palm kernel shell-based brake pad	0.4	4.4	0.22	1.4	0.4	1.6	5.0	103.5	92.0
3	Bagasse-based brake pad	1.1	4.2	0.40	0.2	0.4	1.4	3.4	105.6	100.5
4	Egg shell-based brake pad	1.1	4.0	0.59	1.1	0.3	1.6	3.2	103.0	99.1
5	Periwinkle shell-based brake pad	0.3	3.0	0.68	1.9	0.4	1.0	0.3	147.0	116.7
6	PKS + CBS + MH-based brake pad	0.5	2.1	0.95	0.5	0.4	0.8	0.9	10.3	127.8
7	Banana peels-based brake pad	1.1	4.1	0.13	0.4	0.4	1.2	3.2	95.6	98.8

Table 4 Normalized data for calculating the weights of attributes

S no.	Attributes alternatives	TS(O) (%)	WR (mg/m)	FR	TC (W/mK)	FC	SG (g/cm³)	TS(W) (%)	CS (MPa)	H (MPa)
1	Asbestos-based brake pad	0.05	0.14	0.01	0.08	0.14	0.19	0.05	0.16	0.13
2	Palm kernel shell-based brake pad	0.08	0.17	0.07	0.23	0.15	0.16	0.29	0.15	0.12
3	Bagasse-based brake pad	0.21	0.16	0.13	0.04	0.15	0.14	0.20	0.15	0.13
4	Egg shell-based brake pad	0.22	0.15	0.19	0.17	0.10	0.16	0.18	0.15	0.13
5	Periwinkle shell-based brake pad	0.07	0.11	0.22	0.30	0.14	0.10	0.02	0.21	0.15
6	PKS + CBS + MH-based brake pad	0.11	0.08	0.31	0.08	0.14	0.08	0.05	0.01	0.17
7	Banana peels-based brake pad	0.22	0.16	0.04	0.07	0.14	0.12	0.18	0.14	0.13

2.3 The Objective Weights of Importance of Attributes

The objective weights of importance for the attributes are calculated by the following equation:

$$V_j = (1/n) \sum_{i=1}^{n} \left(x'_{ij} - \left(x'_{ij} \right)_{\text{mean}} \right)^2 \tag{2}$$

where V_j is the statistical variance of the data corresponding to the jth attribute and $\left(x'_{ij} \right)_{\text{mean}}$ is the mean value of x'_{ij}.

The objective weights of the jth attributes (W_j^o) are calculated by the following equation:

$$W_j^o = V_j \Big/ \sum_{i=1}^{m} V_j \tag{3}$$

$W_{TS(O)}^o = 0.133$, $W_{WR}^o = 0.022$, $W_{FR}^o = 0.261$, $W_{TC}^o = 0.220$, $W_{FC}^o = 0.005$, $W_{SG}^o = 0.033$, $W_{TS(W)}^o = 0.230$, $W_{CS}^o = 0.085$, $W_H^o = 0.006$.

2.4 The Subjective Weights (SW) of Importance

The subjective weights (SW) of the attributes were computed using analytical hierarchy process (AHP) [13] and are shown below:

$W_{TS(O)}^S = 0.05$, $W_{WR}^S = 0.38$, $W_{FR}^S = 0.25$, $W_{TC}^S = 0.15$, $W_{FC}^S = 0.02$, $W_{SG}^S = 0.02$, $W_{TS(W)}^S = 0.04$, $W_{CS}^S = 0.02$, $W_H^S = 0.03$.

2.5 The Integrated Weights of Importance of Attributes

An integrated weight of the attributes can be computed by the following equation:

$$W_j^i = W_{\text{obj}} W_j^o + W_{\text{Sub}} W_j^s \tag{4}$$

2.6 Computation of Preference Index

Preference index for purely objective, purely subjective and integrated is calculated using the following equations:

$$P_i^O = \sum_{j=1}^{m} W_j^o x_{ij}'' \tag{5}$$

$$P_i^S = \sum_{j=1}^{m} W_j^s x_{ij}'' \tag{6}$$

$$P_i^I = \sum_{j=1}^{m} W_j^i x_{ij}'' \tag{7}$$

where P_i^O preference index is based on purely objective weights of attributes, P_i^S is the preference index based on subjective weights of attributes, and P_i^I is the preference index based on integrated weights of attributes. Integrated weights are calculated giving 20% subjective and 80% objectives, 40% subjective and 60% objectives, equal weightages and vice-a-versa. x_{ij}'' for beneficial and for non-beneficial attributes were calculated as Eqs. (8) and (9), respectively.

$$x_{ij}'' = x_{ij}'^{b} / \left(x_{ij}'^{b}\right)_{max} \tag{8}$$

$$x_{ij}'' = \left(x_{ij}'^{nb}\right)_{min} / x_{ij}'^{nb} \tag{9}$$

The preference index for alternative brake pads is computed using the above equations and are listed in the Table 5.

Table 5 Preference index for alternatives based on subjective and objective weights

S. no.	Alternatives	P_i^O	P_i^I $W_{obj}=0.2$ $W_{sub}=0.8$	P_i^I $W_{obj}=0.4$ $W_{sub}=0.6$	P_i^I $W_{obj}=0.5$ $W_{sub}=0.5$	P_i^I $W_{obj}=0.6$ $W_{sub}=0.4$	P_i^I $W_{obj}=0.8$ $W_{sub}=0.2$	P_i^S
1	Asbestos-based brake pad	0.69	0.66	0.63	0.62	0.60	0.57	0.54
2	Palm kernel shell-based brake pad	0.49	0.49	0.49	0.48	0.48	0.48	0.48
3	Bagasse-based brake pad	0.23	0.26	0.28	0.29	0.31	0.33	0.36
4	Egg shell-based brake pad	0.36	0.37	0.39	0.40	0.41	0.42	0.44
5	Periwinkle shell-based brake pad	0.77	0.78	0.79	0.80	0.81	0.82	0.83

(continued)

Table 5 (continued)

S. no.	Alternatives	P_i^O	P_i^I $W_{obj} = 0.2$ $W_{sub} = 0.8$	P_i^I $W_{obj} = 0.4$ $W_{sub} = 0.6$	P_i^I $W_{obj} = 0.5$ $W_{sub} = 0.5$	P_i^I $W_{obj} = 0.6$ $W_{sub} = 0.4$	P_i^I $W_{obj} = 0.8$ $W_{sub} = 0.2$	P_i^S
6	PKS + CBS + MH-based brake pad	0.30	0.37	0.43	0.46	0.49	0.55	0.62
7	Banana peels-based brake pad	0.32	0.33	0.34	0.35	0.36	0.37	0.39

Table 6 Ranking of brake pad materials

S. no.	Alternatives	P_i^O	P_i^I $W_{obj} = 0.2$ $W_{sub} = 0.8$	P_i^I $W_{obj} = 0.4$ $W_{sub} = 0.6$	P_i^I $W_{obj} = 0.5$ $W_{sub} = 0.5$	P_i^I $W_{obj} = 0.6$ $W_{sub} = 0.4$	P_i^I $W_{obj} = 0.8$ $W_{sub} = 0.2$	P_i^S
1	Asbestos-based brake pad	II	II	II	II	II	II	III
2	Palm kernel shell-based brake pad	III	III	III	III	IV	IV	IV
3	Bagasse-based brake pad	VII	VII	VII	VII	VII	VII	VII
4	Egg shell-based brake pad	IV	V	V	V	V	V	V
5	Periwinkle shell-based brake pad	I	I	I	I	I	I	I
6	PKS + CBS + MH-based brake pad	VI	IV	IV	IV	III	III	II
7	Banana peels-based brake pad	V	VI	VI	VI	VI	VI	VI

3 Rankings of Brake Pad Materials

The alternatives are managed in the descending order of the preference index. The alternative for which the value of preference index is greater is the best choice as a non-asbestos brake pad material. The seven alternative brake pad materials are ranked based on the computed values of the preference index and the ranks are mentioned in Table 6.

4 Conclusion

In this paper, a multi-attribute decision-making (MADM) method is used for a complex problem of a novel brake pad material selection. Ranking of alternatives is done based on the preference system, i.e., purely objective, subjective and combination of them. Order of ranking based on objective is (5-1-2-4-7-6-3) and for subjective (5-6-1-2-4-7-3) for integrated preference is (5-1-2-6-4-7-3). Based on the results, an alternative number 5, i.e., Periwinkle shell-based brake pad material is the best suitable choice among all preferential schemes. Here, Periwinkle shell-based brake pad materials are very good alternatives to asbestos-based commercial brake pad materials. However, an alternative number 3, i.e., Bagasse-based brake pad is a least preferred choice as a brake pad material. Furthermore, research can be extended in order to study tribo-mechanical performance evaluation of Periwinkle shell-based brake pad material as inorganic materials for its commercial viability into the automotive market. Also, application-based formulation of brake pad material is possible by varying additives and their volume fraction. A database needs to be generated in this direction by the researchers.

References

1. Ikpambese, K.K., Gundu, D.T., Tuleun, L.T.: Evaluation of palm kernel fibers (PKFs) for production of asbestos-free automotive brake pads. J. King Saud Univ. Eng. Sci. **28**, 110–118 (2016). https://doi.org/10.1016/j.jksues.2014.02.001
2. Aigbodion, V.S., Agunsoye, J.O.: Bagasse (Sugarcane Waste): Non-asbestos Free Brake Pad Materials. LAP Lambert Academic Publishing, Germany. ISBN 978-3-8433-8194-9 (2010)
3. Edokpia, R.O., Aigbodion, V.S., Obiorah, O.B., Atuanya, C.U.: Evaluation of the properties of eco-friendly brake pad using egg shell particles-gum arabic. Res. Phys. https://doi.org/10.1016/j.rinp.2014.06.003
4. Aku, S.Y., Yawas, D.S., Madakson, P.B., Amaren, S.G.: Characterization of periwinkle shell as asbestos-free brake pad materials. Pac. J. Sci. Technol. **13**(2), 57–63 (2012)
5. Amaren, S.G., Yawas, D.S., Aku, S.Y.: Effect of periwinkles shell particle size on the wear behaviour of asbestos free brake pad. Res. Phys. **3**, 109–114 (2013)
6. Yawas, D.S., Aku, S.Y., Amaren, S.G.: Morphology and properties of periwinkle shell asbestos-free brake pad. J. King Saud Univ. Eng. Sci. **28**, 103–109 (2016)
7. Olabisi, A.I., Ademoh, N.A., Boye, T.E.: Development of asbestos-free automotive brake pad using ternary agro-waste fillers. J. Multi. Eng. Sci. Technol. (JMEST) **3**(7), 5307–5327 (2016)
8. Idris, U.D., Aigbodion, V.S., Abubakar, I.J., Nwoye, C.I.: Eco-friendly asbestos free brake-pad: using banana peels. Eng. Sci. **27**, 185–192 (2015)
9. Ademoh, N.A., Olabisi, A.I.: Development and evaluation of maize husks asbestos-free based brake pad. Ind. Eng. Lett. **5**(2), 67–80 (2015)
10. Rao, R.V., Patel, B.K.: A subjective and objective integrated multiple attribute decision making method for material selection. Mater. Des. **31**, 4738–4747 (2010)
11. Ishizaka, A., Labib, A.: Review of the main development in the analytic hierarchy process. Expert Syst. Appl. **38**, 14336–14345 (2011)
12. Jaybhaye, M.D., Lathkar, G.S., Basu, S.K.: Critically Assessment of Clean Metal Cutting Parameters Using Diagraph and AHP. In: National Conference on Advances in Manufacturing Systems, Jadhavpur University, India, pp. 378–383 (2003)

13. Hendre, K.N., Bachchhav, B.D.: Critical property assessment of novel brake pad materials by AHP. J. Manuf. Eng. **13**(3), 148–151 (2018)
14. Bachchhav, B.D., Lathkar, G.S., Bagchi, H.: Criticality assessment in lubricant selection for metal forming process By AHP. Int. J. Manuf. Technol. Ind. Eng. **3**(1), 31–35 (2012)
15. Saaty, T.L.: The Analytical Hierarchy Process. McGraw-Hill, New York, NY (1980)

Selection of Coating and Nitriding Process for AISI 4140 Steel Material to Enhance Tribological Properties

Pathan Firojkhan, Nikhil Kadam and S. G. Dambhare

Abstract The Multi Criteria Decision method is used for selection of coating and nitriding process is VIKOR (VIšekriterijumsko Rangiranje) method. Analytical hierarchy process (AHP) and entropy methods are used to find out weightages for the material criteria like Young's modulus (E), hardness (H), H/E, and H^3/E^2. The VIKOR is applied to the five alternatives, namely AISI 4140 alloy steel, nitrided AISI 4140 alloy steel, nitrided and TiN-coated AISI 4140 alloy steel, nitrided and TiAlN-coated AISI 4140 alloy steel, and nitrided and WCC-coated AISI 4140 alloy steel. The experiment is performed on the universal testing machine and hardness testing machine which is then used for analysis. Nitrided and WCC-coated AISI 4140 alloy steel is predicted to be the best suitable material for the bearing materials.

Keywords AHP · Entropy method · VIKOR method results

1 Introduction

In the recent century, the tremendous amount of development is carried out in the field of tribology and surface engineering. Most of the research is focused on improving the wear resistance of industrial materials and applications. Metal deposition and coating techniques are used for improving the friction loss in the industrial application, thereby increasing the lifespan of materials. Hence, to improve wear resistance, surface roughness, and texture research, extensive research is carried out which includes the addition of a coating or metal deposition on the surface of materials [1, 2]. As far as the coating on the material is concerned, it depends on the properties, viz. modulus of elasticity (E), hardness (H), adhesion between coating and substrate,

P. Firojkhan (✉) · S. G. Dambhare
Department of Mechanical, Dr. D. Y. Patil Institute of Engineering, Management, and Research, Pune, India
e-mail: fzpathan@gmail.com

N. Kadam
Bits Pilani K K Birla Goa Campus, Goa, India

© Springer Nature Singapore Pte Ltd. 2020
R. Venkata Rao and J. Taler (eds.), *Advanced Engineering Optimization Through Intelligent Techniques*, Advances in Intelligent Systems and Computing 949, https://doi.org/10.1007/978-981-13-8196-6_41

elastic recovery, wear resistance, and coefficient of friction (COF) [3]. Each coating material will add different performances, some properties will improve one aspect of the material, and at the same time, it can have an adverse effect on the other. It is essential to select the best coating material which gives the best performance by considering different material properties and parameters. Multi-criteria decision-making methods (MCDMs) are utilized for the selection of coating depending upon the material properties, parameters, and applications [4]. MCDM methods are evolved during the last century for various varieties of applications. As per the applications, researchers had developed different MCDM methods. Some of MCDM methods are based on a graphical approach, for example, Ashby method [5], and on mathematical approaches, viz. the linear assignment method [6], complex proportional assessment (COPRAS) [7], preference ranking organization method for enrichment evaluation (PROMETHEE) [8], elimination and choice expressing the reality (ELECTRE) [9], *technique for order performance by similarity to ideal solution* (TOPSIS) [10, 11], *VlseKriterijumska Optimizacija Kompromisno Resenje* (VIKOR) [12], and grey relation method [13]. The selection of coating materials is carried out on the basis of properties like modulus of elasticity (E), hardness (H), COF, and bonding between coating substrate and materials. Moreover, some researcher found factors like H/E and H^3/E^2 are effective in the selection process [14–16].

In the present work, five materials with a different coating and nitriding are considered. Following are the materials used to identify the best-suited coating materials: AISI 4140 alloy steel, nitrided AISI 4140 alloy steel, nitrided and TiN-coated AISI 4140 alloy steel, nitride and TiAlN-coated AISI 4140 alloy steel, and nitrided and WCC-coated AISI 4140 alloy steel which are selected to compare the mechanical properties. The experimental testing is carried out to find mechanical properties. The selection of best materials is based on four factors, viz. Young's modulus, hardness, H/E, and H^3/E^2. The weightages of each factor are calculated with the help of the analytical hierarchy process (AHP) [17], entropy method [18, 19], and compromised weighting method. The VIKOR method results is used to identify and decide the best coating material among the five different coating materials.

2 Methods and Experiments

The elasticity modulus and hardness of five material alternatives, viz. (1) AISI 4140 alloy steel, (2) nitrided AISI 4140 alloy steel, (3) nitrided and TiN-coated AISI 4140 alloy steel, (4) nitrided and TiAlN-coated AISI 4140 alloy steel, and (5) nitrided and WCC-coated AISI 4140 alloy steel, are measured by standard ASTM tests. The tensile specimens are prepared as per ASTM E8 04 by machining and finishing with length 300 and 12.5 mm in diameter. The gauge length of the testing is 50 mm. The length of the reduced cross section is found to be 60 mm with a radius of fillet 10 mm. The prepared specimens are then nitrated and PVD-coated with TiN, TiAlN, and WCC with the thickness of 4 μm as per the particular materials. The measured value of the elasticity modulus is illustrated in Table 1.

Table 1 Tribological and mechanical characteristics of coatings

Sr. no.	Coating material	E (GPa)	H (GPa)	H/E	H^3/E^2
1	AISI 4140 alloy steel	210.00	7.0632	0.0336	0.0080
2	Nitrided AISI 4140 alloy steel	225.00	12.5274	0.0557	0.0388
3	Nitrided and TiN-coated AISI 4140 alloy steel	250.00	22.5630	0.0903	0.1838
4	Nitrided and TiAlN-coated AISI 4140 alloy steel	280.00	29.4300	0.1051	0.3251
5	Nitrided and WCC-coated AISI 4140 alloy steel	280.00	29.4300	0.1051	0.3251

The Tensile Specimens are prepared as per ASTM E8 04 by machining and finishing with length 300 mm and diameter 12.5 mm with a gauge length of 50 mm and length of reduced cross section 60 mm with a radius of fillet 10 mm. The prepared specimens are nitrated and PVD coated with TiN, TiAlN and WCC with the thickness of 4 μ. The tensile specimen for each alternative is shown in Fig. 1 (i) also the dimension of the tensile is shown Fig. 1 (ii). The measured value of elasticity modulus for each specimen is illustrated in Table 1.

The compression test specimens are prepared as per ASTM E9-89a by machining to a length of 38 mm and a diameter of 13 mm with the ratio of L/D equal to 3. Specimens are nitrated and PVD-coated with TiN, TiAlN, and WCC with the thickness of 4 μ. The compression test specimens are shown in Fig. 2.

The specimen is kept between two faceplates of compression testing machine; the upper faceplate is moved downward and upward by the wheel present above the upper faceplate. The force is applied by the servo motor and the system. After certain compression, the specimen fails and does not take more compressive force and the corresponding compressive force is displayed in the dial indicator system.

The hardness test is performed on Rockwell hardness tester used for measuring hardness of the specimen, respectively.

3 Decision-Making Methods

Many applications of AISI 4140 alloy steel have a high frictional force which leads to abrasive wear. Hence, it is essential to improve the wear resistance. Therefore, AISI 4140 alloy steel is generally being coated with different coating materials. However, selecting an appropriate coating for particular tribological application is still complex and difficult because the tribological response of coating depends on the number of factors. Young's modulus, hardness, H/E, and H^3/E^2 ratio properties are considered for the selection of coating materials as per the literature available. Five alternative AISI 4140 steels with a different combination of nitriding deposition and coatings

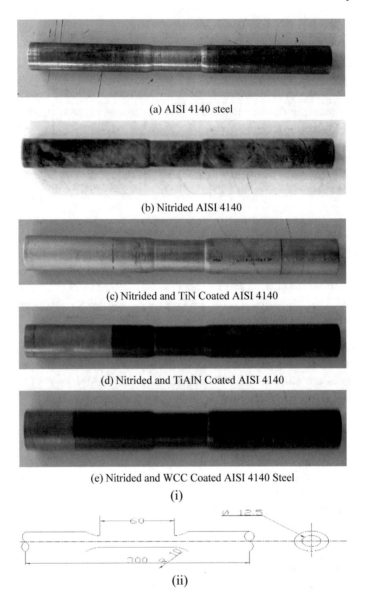

(a) AISI 4140 steel

(b) Nitrided AISI 4140

(c) Nitrided and TiN Coated AISI 4140

(d) Nitrided and TiAlN Coated AISI 4140

(e) Nitrided and WCC Coated AISI 4140 Steel

(i)

(ii)

Fig. 1 (i) Tensile test specimen. (ii) Specification of tensile test

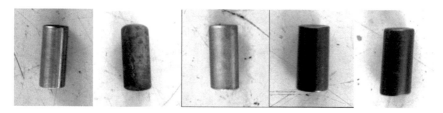

Fig. 2 Compression test specimen AISI 4140 steel, nitrided AISI 4140, nitrided and TiN-coated AISI 4140, nitrided and TiAlN-coated AISI 4140, nitrided and WCC-coated AISI 4140 steel

are used for the study. The properties are illustrated in Table 1. E, H, H/E, and H^3/E^2 are the criteria selected for this study. To improve the low coefficient of friction and high wear resistance, Young's modulus should be reduced (cost criteria) and other factors H, H/E, and H^3/E^2 should be improved (benefit criteria).

3.1 Determination of Weights

3.1.1 Compromised Weighting Method

The compromised weighting method is used to determine the final weight of the criteria. The weightages of the respective criteria are determined by the combination of weightages from the AHP method and entropy method as described in Eq. (1)

$$w_j = \frac{A_j \times En_j}{\sum_{j=1}^{m} A_j \times E_j} \quad j = 1, 2, \ldots n \quad (1)$$

A_j and E_j are the weights of jth criteria by AHP method and entropy method, respectively.

3.1.2 Analytical Hierarchy Process (AHP) Method [20]

AHP method is based on the hierarchy of the criteria, and AHP method is developed by Saaty to find the weight of all criteria by using a pairwise comparison between them. The AHP method works on three steps: (i) structure of system, (ii) the pairwise comparison matrix of the respective criteria over another by using relative importance grade as described by Saaty [20], and (iii) determination of the weight of the respective criteria based on the eigenvector method. The detailed procedure of the AHP method is as given below:

(a) Formulate the comparative pairwise matrix for each criterion according to relative importance. The pairwise comparison matrix (P) for n number of criteria is described in Eq. (2).

$$P = \begin{vmatrix} 1 & a_{12} & - & a_{1n} \\ a_{21} & 1 & - & a_{2n} \\ - & - & - & - \\ a_{n1} & a_{n2} & - & 1 \end{vmatrix} ; a_{ij} = \frac{1}{a_{ji}} \quad ; a_{ij} \neq 0 \qquad (2)$$

(b) The geometric mean is calculated for each jth criterion by Eq. (3); then, the normalized weight (α_j) of the respective criteria is determined to help normalization of the geometric mean as described in Eq. (4).

$$GM_j = \left(a_{j1} \times a_{j2} \times a_{j3} \times \cdots \times a_{jn} \right)^{1/n} \qquad (3)$$

$$A_j = \frac{GM_j}{\sum_{j=1}^{n} GM_j} \qquad (4)$$

(c) The next steps are performed to check whether the weights are under the prescribed limit. The multiplication of a pairwise comparison matrix with the weight of each jth criterion gives us the $n \times 1$ column matrix which is used to obtain the constancy values (CVs).

$$c = P \times \alpha = \begin{vmatrix} 1 & a_{12} & 1 & a_{1n} \\ a_{21} & 1 & - & a_{2n} \\ - & - & - & - \\ a_{n1} & a_{n2} & - & 1 \end{vmatrix} \begin{vmatrix} \alpha_1 \\ \alpha_2 \\ - \\ \alpha_n \end{vmatrix} = \begin{vmatrix} c_1 \\ c_2 \\ - \\ c_n \end{vmatrix} ; \quad CV_j = \frac{c_j}{A_j}$$

(d) The average of all consistency values obtained after the equation is called as eigenvalue $\lambda_{\max}$, and consistency index (CI) is determined by using the following equation $CI = (\lambda_{\max} - n) / (n - 1)$.

(e) The relative index (RI) is used to determine consistency ratio, and RI is calculated based on the number of criteria [20].

(f) The consistency ratio (CR) is determined to check the acceptability of the weights $CR = CI/RI$.

(g) The accepted upper limit of consistency ratio is 0.1. If it is not achieved, the whole process should be repeated with different comparative values.

3.1.3 Entropy Method

Entropy method is an objective weighting method which uses probability theory which measures the uncertainty in output responses of the respective criteria to determine the weightages of the respective criteria.

(a) To determine the weights by using entropy method, first output responses of the respective criteria are normalized and stored in a decision matrix by using Eq. (6).

$$N_{ij} = \frac{x_{ij}}{\sqrt{\sum_{i=1}^{m} x_{ij}^2}}; \quad i = 1, 2, \ldots, m; \, j = 1, 2, .., n \tag{6}$$

The entropy value T_j of jth criteria is calculated with the help of Eq. (7). The K value is calculated based on the number of alternatives present. K is a constant which normalizes entropy values between 0 and 1, $k = 1/\ln m$

$$T_j = -k \sum_{i=1}^{m} N_{ij} \ln N_{ij}; \quad i = 1, 2, , \ldots, m; \, j = 1, 2, \ldots, n. \tag{7}$$

(b) The degree of divergence (d_j) of the jth criteria is calculated by using Eq. (8)

$$d_j = \left|1 - T_j\right| \tag{8}$$

(c) The normalization of a degree of divergence for all criteria gives us the weight

$$En_j = d_j / \sum_{j=1}^{n} d_j. \tag{9}$$

3.2 VIKOR Method Results

The algorithm of VIKOR is explained in the following steps.

(a) Normalization of the decision matrix. The normalization process is performed with Eq. (10).

$$N_{ij} = \frac{x_{ij}}{\sqrt{\sum_{i=1}^{m} x_{ij}^2}} \tag{10}$$

(b) Determination of best and worst values of the respective criteria function has depended on their maximization (benefit) or minimization (cost) conditions.

$$N_j^* = \begin{cases} \underset{j}{\text{Max }} N_{ij} \ldots\ldots\ldots\ldots\ldots\ldots\ldots\ldots\ldots\ldots.\text{for benefit criteria} \\ \underset{j}{\text{Min }} N_{ij} \ldots\ldots\ldots\ldots\ldots\ldots\ldots\ldots\ldots.\text{for cost criteria} \end{cases} \tag{11}$$

$$
N_j^* = \begin{cases} \underset{j}{\text{Min }} N_{ij} \dots\dots\dots\dots\dots\dots\dots\dots\dots\dots\dots\dots\text{for benefit criteria} \\ \underset{j}{\text{Max }} N_{ij} \dots\dots\dots\dots\dots\dots\dots\dots\dots\dots\dots\dots\text{for cost criteria} \end{cases} \tag{12}
$$

(c) Calculate utility and regret measure using Eqs. (13) and (14) for alternative material $j = 1, 2, \dots, m$.

$$
S_i = \sum_{i=1}^{n} \frac{w_i \left(N_j^* - N_{ij} \right)}{\left(N_j^* - N_j^- \right)} \tag{13}
$$

and

$$
R_i \underset{i}{\text{max}} \left[\frac{w_i \left(N_i^* - N_{ij} \right)}{\left(N_j^* - N_j^- \right)} \right] \tag{14}
$$

where S_i and R_i are used for formulation ranking to measure Q_i. Compute the value of Q_j by using Eq. (15),

$$
Q_i = \frac{\vartheta \left(S_i - S^* \right)}{\left(S^- - S^* \right)} + \frac{(1 - \vartheta)(R_i - R^*)}{(R^- - R^*)} \tag{15}
$$

$$
S^* = \underset{i}{\text{min}} S_i \ S^- = \underset{i}{\text{max}} S_i; \ R^* = \underset{i}{\text{min}} R_i \ R^- = \underset{i}{\text{max}} R_i
$$

(d) The range of the value of ϑ is from 0 to 1. Three values of ϑ are used: 0.1 which shows "veto", 0.5 shows accepted solution "consensus", and 0.9 shows consideration of all factors "majority". Most of the time, ϑ is taken as 0.5. So, VIKOR ranking is calculated for all values if the solution of each value shows a higher correlation coefficient. Ranking of alternatives is carried as per the ascending value of Q_i.

4 Results and Discussion

4.1 Determination of Weightages

In this study, hardness is considered to be more important than other criteria like Young's modulus, H/E, and H^3/E^2. H/E and H^3/E^2 ratios are given same importance for the selection of the coating with the low friction coefficient and high wear resistance. Table 4 gives us the pairwise comparison matrix (Table 2).

Table 2 Pairwise comparison matrix of criteria

	E	H	H/E	H^3/E^2
E	1	0.33	0.5	0.5
H	3	1	0.5	0.5
H/E	2	2	1	1
H^3/E^2	2	2	1	1

Table 3 Weights of criteria

	E (GPA)	H	H/E	H^3/E^2
An (AHP method)	0.1251	0.2166	0.3292	0.3292
En (entropy method)	0.2679	0.2532	0.2583	0.2206
W (compromised weighting)	0.1362	0.2230	0.3456	0.2952

Table 4 VIKOR analysis

Sr. no	S_j	R_j	Q_j	Rank
1	0.8638	0.3456	1.0000	5
2	0.7030	0.2666	0.7049	4
3	0.3494	0.1316	0.1465	3
4	0.1362	0.1362	0.0109	2
5	0.1362	0.1362	0.0109	1

Table 3 shows the weights of the respective criteria obtained after all three methods. Weights obtained after using the AHP method are given α_j for all criteria's are shown in the first row. AHP methods give the highest weight to H^3/E^2 and H/E and the lowest to Young's modulus (E), while hardness (H) is the intermittent importance. The weightages for the pairwise matrix are calculated with the help of Eqs. (3) and (4) as shown in Table 3. Moreover, to check the reliability of the AHP method, λ_{max} was calculated whose value found to be 4.293 and C.I. value—$0.02117 < 0.1$ and C.R. value—$0.01890 < 0$.

The compromised weights of criteria (w_j) are determined using Eq. (1). As seen in Table 4, H/E is the most important criteria followed by H^3/E^2, H, and E.

4.2 VIKOR Method Results

According to the steps given in Sect. 3.2, normalizing of a decision matrix for the respective criteria is carried out by Eq. (10) where x_{ij} values are taken from Table 1. Best and worst values for the respective criteria are calculated with Eqs. (11)–(12). Utility and regret measure (S_j, R_j) are determined by Eqs. (13)–(14). VIKOR index Q_j is then determined to help by Eq. (15) which is given in Table 4. Then, Q is arranged in ascending order and best rank is given to the least value of Q. The rankings with respective alternatives are shown in Table 4.

5 Conclusion

This study proposes a better way to select material for tribological applications with the help of different mechanical properties by using MCDM methods. The weights of the respective criteria for material selection are conducted out through a compromised weighting method which yields the most important measures to be a ratio of hardness and modulus of elasticity (H/E). Nitride-deposited and WCC-coated AISI 4140 is selected.

References

1. Shtansky, D.V., Sheveiko, A.N., Petrzhik, M.I., Kiryukhantsev-Korneev, F.V., Levashov, E.A., Leyland, A.: Hard tribological Ti–B–N, Ti–Cr–B–N, Ti–Si–B–N and Ti–Al–Si–B–N coatings. Surf. Coat. Technol. **200**, 208–212 (2005)
2. Musil, J.: Hard and super hard nanocomposite coatings. Surf. Coat. Technol. **125**, 322–330 (2000)
3. Edwards, K.L.: Materials influence on design: a decade of development. Mater. Des. **32**, 1073–1080 (2011)
4. Caliskan, H.: Selection of boron-based tribological hard coatings using multi-criteria decision-making methods. Mater. Des. **50**, 742–749 (2013)
5. Rathod, M.K., Kanzaria, H.V.: A methodological concept for phase change material selection based on multiple criteria decision analysis with and without fuzzy environment. Mater. Des. **50**, 742–749 (2011)
6. Jahan, A., Ismail, M.Y., Mustapha, F., Sapuan, S.M.: Material selection based on ordinal data. Mater. Des. **31**, 3180–3187 (2010)
7. Maity, S.R., Chatterjee, P., Chakraborty, S.: Cutting tool material selection using grey complex proportional assessment method. Mater. Des. Des **36**, 372–378 (2012)
8. Chauhan, A., Vaish, R.: A comparative study on material selection for micro-electromechanical systems. Mater. Des. **41**, 177–181 (2012)
9. Chatterjee, P., Chakraborty, S.: Material selection using preferential ranking methods. Mater. Des. **35**, 384–393 (2012)
10. Shanian, A., Savadogo, O.: A material selection model based on the concept of multiple attribute decision making. Mater. Des. **27**, 329–337 (2006)
11. Gupta, N.: Material selection for thin-film solar cells using multiple attribute decision-making approach. Mater. Des. **32**, 1667–1671 (2011)
12. Dağdeviren, M., Yavuz, S., Kılınç, N.: Weapon selection using the AHP and TOPSIS methods under fuzzy environment. Expert Syst. Appl. **36**, 8143–8151 (2009)
13. Pathan, F., Gurav, H., Gujrathi, S.: Optimization for tribological properties of glass fiber-reinforced PTFE composites with grey relational analysis. J. Mater. (2016)
14. Jahan, A., Edwards, K.L.: VIKOR method results for material selection problems with interval numbers and target-based criteria. Mater. Des. **47**, 759–765 (2013)
15. Leyland, A., Matthews, A.: On the significance of the H/E ratio in wear control: a nanocomposite coating approach to optimized tribological behavior. Wear **246**, 1–11 (2000)

16. Tsui, T.Y., Pharr, G.M., Oliver, W.C., Bhatia, C.S., White, R.L., Anders, S., et al.: Nanoindentation and Nano Scratching of Hard Carbon Coatings for Magnetic Disks. In: Proceedings of Materials Research Society Symposium, San Francisco, pp. 447–52 (1995)
17. Chauhan, A., Vaish, R.: Hard coating material selection using multi-criteria decision making. Mater. Des. **44**, 240–245 (2013)
18. Kao, P.S., Hocheng, H.: Optimization of electrochemical polishing of stainless steel by grey relational analysis. J. Mater. Process. Technol. **140**, 255–259 (2003)
19. Lin, C.L.: Use of the Taguchi Method and grey relational analysis to optimize turning operations with multiple performance characteristics. Mater. Manuf. Processes **19**(2), 209–220 (2004)
20. Saaty, T.L.: How to make a decision: the analytic hierarchy process. Eur. J. Oper. Res. **48**, 9–26 (1990)

Risk-Pooling Approach in Inventory Control Model for Multi-products in a Distribution Network Under Uncertainty

Vats Peeyush, Soni Gunjan and A. P. S. Rathore

Abstract Risk-pooling is a substantial approach for controlling inventory in supply chain under uncertainty. In this approach, independent demands of two or more regions are aggregated; as a result, the demand uncertainty is reduced. This approach is also very much useful in reducing the safety stocks. In this study, risk-pooling approach is used in inventory optimization for three different types of products in a two-stage supply chain network under service-level constraint. This study represents a mathematical approach for inventory optimization using risk-pooling concept for different types of the product, and it provides optimum inventory policy (reorder point, ordering quantity, and total cost) along with expected shortages per cycle (ESC), fill rate (FR), safety stock (SS), and average inventory (AI).

Keywords Risk-pooling · Inventory optimization · Multi-products · Uncertainty

1 Introduction

Risk-pooling is the most effective way for reducing the safety stocks in supply chain [1]. This approach is also very helpful in reducing the total expected inventory costs and in improving customer service in a supply chain system [2]. In a risk-pooling system, it is allowed to an over-supplied region to provide the supply to the insufficient supplied region by adding the transportation and the inventory holding cost [3]. Various levels of risk-pooling were explored by Bernstein et al. [4] which can affect

V. Peeyush (✉) · S. Gunjan · A. P. S. Rathore
Malaviya National Institute of Technology, Jaipur, India
e-mail: peeyushvatsscholar@gmail.com

S. Gunjan
e-mail: gunjan1980@gmail.com

A. P. S. Rathore
e-mail: apsr100@yahoo.co.in

© Springer Nature Singapore Pte Ltd. 2020
R. Venkata Rao and J. Taler (eds.), *Advanced Engineering Optimization Through Intelligent Techniques*, Advances in Intelligent Systems and Computing 949,
https://doi.org/10.1007/978-981-13-8196-6_42

the profit, sales of product, and decisions related to the component capacity in the system. The greater level of pooling can increase the profit and can support the unbalanced capacities. Risk-pooling can reduce the demand uncertainty and this approach can be utilized in the effective inventory management [5], and in this approach, independent demands of two or more regions are aggregated; as a result, the demand uncertainty is reduced [6]. Risk-pooling is more dominating in supply chain when inventory costs have a bigger percentage in supply chain costs. The major impacts of risk-pooling on supply chain metrics are as follows [8]:

1. Increasing with the level of pooling, the safety stock decreases.
2. Overhead costs for maintaining facilities decreases with the risk-pooling.
3. Under risk-pooling, the outbound transportation costs increase due to increased distances between the facilities and retailers.

2   What Is Risk-Pooling

For understanding the concept of risk-pooling, let us assumed that, there are n retailers, and a central warehouse or distributor. These retailers lie in the proximity regions of central warehouse, and the demands of these retailers are independent and follow the normal distribution [7]. Let the demands and the standard deviations of demands for these n retailers are $(D_1, \sigma_1), (D_2, \sigma_2), (D_3, \sigma_3),\dots (D_n, \sigma_n)$. Let (D_{agg}, σ_{agg}) are the demand and standard deviation of central warehouse (Fig. 1).

According to the risk-pooling approach, the demand of the central warehouse is equal to the summation of all demands of retailers, and mathematically, it can be expressed as

$$D_{agg} = D_1 + D_2 + D_3 + \cdots + D_n$$

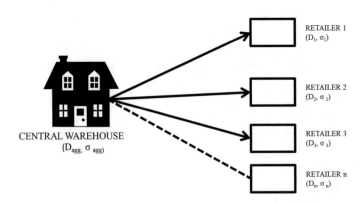

Fig. 1 Concept of risk-pooling

$$D_{agg} = \sum_{i=1}^{i=n} D_n$$

where D_{agg} is the aggregated demand of central warehouse.

And the standard deviation of demand for the central ware house may be expressed as

$$\sigma_{agg} = \sqrt{\sigma_1^2 + \sigma_2^2 + \sigma_3^2 + \cdots + \sigma_n^2}$$

$$\sigma_{agg} = \sqrt{\sum_{i=1}^{i=n} \sigma_i^2}$$

where σ_{agg} = aggregated standard deviation of demand for central warehouse.

The demand and the standard deviation of demand of central warehouse is denoted as (D_{agg}, σ_{agg}), and this central warehouse is capable to fulfil the demands of all n retailers.

If a distributor is shipping more than one retailer and the demand of all retailers are aggregated, then the demand during lead may be represented as

$$\mu_{xj} = D_{agg} \times L$$

where D_{agg} is average aggregated demand for distributor and L is the lead time for replenishment.

The expression for standard deviation during lead time is given below

$$\sigma_{xj} = \left(\sigma_j^2 \times L + \sigma_L^2 \times D_{agg}^2 \right)^{1/2}$$

where σ_j is the aggregated demand's standard deviation for distributor j, σ_L is the standard deviation for lead time

3 A Mathematical Approach for Multi-products Inventory Optimization Under Risk-Pooling

In this model, the risk-pooling approach is used for finding optimized inventory policies and optimized cost under service-level constraint. For this, it is considered that multi-products are shipping in a two-stage supply chain network from distributors to retailers on the demand of retailers. The objective function of this model is to minimize the total expected cost of the system which includes ordering cost, inventory carrying costs, cost for setting a distributor, operating cost of distributor, and the cost of shipment from distributor to retailer [8]. Here, a single-objective nonlinear optimization problem is developed. The outcome of this problem provides the optimum inventory policy (reorder point and ordering quantity) and the optimum expected cost. For developing this mathematical approach, the following assumptions are considered:

1. Continuous review inventory system is adopted.
2. Demand and lead time are random variable and are normally distributed.
3. The retailer's demands are independent.
4. The fill rate for all the distributors is the same.
5. Single distributor can ship more than one retailer, but a single retailer cannot be shipped by more than a single distributor.
6. A fixed distributor can fulfill the demand of some specific retailers only.
7. Transportation cost for each item is the same.
8. Partial fulfillment is not allowed for retailers.
9. Pipeline inventory is not considered.
10. Shortages and backorders are not allowed.

3.1 Symbols, Notation, and Sets

Distributors, j
Retailer, k
Products, p.

3.2 Parameters

Cost of establishing of a distributor (C_j)
Operating cost of central warehouse (O_j)
Inventory holding cost (H_j)
Fixed ordering cost (F_j)
Aggregated demand for distributor (D_{agg})
Demand of retailer (D_k)
Standard deviation of demand of retailer (σ_k)
Transportation cost from distributor to retailer (T)
Distance from distributor to retailer (d_{j-k})
Lead time (L)
Mean demand during lead time $(\mu_j = D_{agg}*L)$.

3.3   Variables

Ordering quantity (Q_j)
Reorder point (r_j).

3.4   Objective Function

The concept of risk-pooling is incorporated as mean demand during lead time in the function of total cost of the system especially in the mathematical expression of inventory carrying cost. A single-objective nonlinear optimization problem is developed which includes the ordering cost, inventory carrying costs, cost for setting a distributor, operating cost of distributor, and the cost of transportation from distributor to retailer for the formulation of this mathematical approach. The solution of this problem provides the optimum inventory policy along with minimum cost.

Total cost = ordering cost + inventory holding cost + operating cost + facility cost + transportation cost, total cost (TC)

$$Z_1 = \min(\sum_p \sum_j F_j(D_{agg}/Q_j) + \sum_p \sum_j \{Q_j/2 + (r_j - \mu)\}^* H_j^* v_p +$$
$$\sum_p \sum_j \left(\sum_k D_{agg}^* T^* d_{j-k}\right) + \sum_p \sum_j C_j + \sum_p \sum_j O_j^* D_{agg})$$

3.5   Constraints

Ordering quantity constraint, $Q_j \geq 1$
Reorder point, $r_j \geq \mu_j$ $(\mu_j = D_{agg}^*LT)$
Service-level constraints, Cycle service level, CSL ≥ 0.975
Where CSL is defined as CSL = NORMDIST $(r_j, \mu_j, \sigma_L, 1)$.

3.6   Product miles

The product miles [8] may be defined as the product of demand and distance. If X_{j-k} is the distance from distributor j to retailer k, then

$$\text{Product miles (PM)} = \sum_k \sum_j X_{j-k}^* D$$

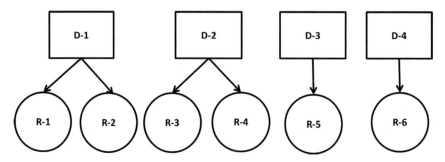

Fig. 2 Scenario considered in two-stage distribution network

4 Numerical Illustration

4.1 Scenario Considered for Risk-Pooling in Two-Stage Supply Chain

For solving the above problem, a two-stage supply chain network is considered between distributors and retailers. For the sake of simplicity, only four distributors and six retailers are considered, assuming that, these four distributors are sufficient for fulfilling the demand of all six retailers. Let the four distributors are denoted by D-1, D-2, D-3, and D-4, and the six retailers are denoted by R-1, R-2, R-3, R-4, R-5, and R-6. In this problem, it is also considered that three different types of product are shipping from distributor to retailer and presented by P-1, P-2, and P-3. İt is assumed that that R-1 and R-2 lie in proximity region of D-1, and R-3 and R-4 are situated in proximity region of D-2. D-3 fulfills the demand of R-5, and D-4 fulfills the demand of R-6. The demands of R-1 and R-2 are aggregated for D-1, which is known as risk-pooling. The same approach is applied in R-3 and R-4 for D-2. There is no need to apply risk-pooling approach for R-5 and R-6, as these retailers are standing alone in the system. This system can be illustrated in Fig. 2.

4.2 Parameters for Numerical Experimentation

For solving the above problem, four distributors and six retailers are considered in a two-stage supply chain network assuming that all distributors are capable to satisfy the demands of all retailers for all products. The numerical parameters are as follows: (Tables 1, 2, 3, 4, 5, and 6).

Table 1 Operating costs (O_j) and facility cost (C_j) for all types of product

	Operating costs/unit(O_j) (Rs)			Facility cost (C_j) (Rs)		
	P-1	P-2	P-3	P-1	P-2	P-3
D-1	4	9	11	2200	2300	2500
D-2	5	8	10.5	2200	2300	2500
D-3	5.5	8.5	12	2200	2300	2500
D-4	7	9.5	11.5	2200	2300	2500

Table 2 Fixed ordering costs (F_j) and inventory holding cost (H_j) for all types of product

	Fixed ordering cost/unit (F_j) (Rs)			Inventory holding cost (H_j) (Rs)		
	P-1	P-2	P-3	P-1	P-2	P-3
D-1	450	550	700	1.5	1.8	2.5
D-2	450	550	700	1.6	1.9	3
D-3	450	550	700	1.5	1.7	3.5
D-4	450	550	700	1.4	1.8	3

Table 3 Distance from distributors to retailers ($d_{j\text{-}k}$) in kilometers

	R-1	R-2	R-3	R-4	R-5	R-6
D-1	20	22				
D-2			30	20		
D-3					31	
D-4						59

Table 4 Demand (D_k) (in number of units) and standard deviations (σ_k) of retailers (in number of units)

	P-1 (D_k, σ_k)	P-2 (D_k, σ_k)	P-3 (D_k, σ_k)
Retailer-1	(2500, 125)	(2700, 135)	(2400, 120)
Retailer-2	(2300, 115)	(2900, 145)	(2780, 139)
Retailer-3	(2200, 110)	(2800, 140)	(2300, 115)
Retailer-4	(2400, 120)	(2200, 110)	(2900, 145)
Retailer-5	(2100, 105)	(2700, 135)	(2700, 135)
Retailer-6	(2300, 115)	(2600, 130)	(2850, 143)

Table 5 Lead time (L) in days and transportation cost (per unit in Rs)

Lead time for all types of products (L)	0.15 days
Standard deviation of lead time (SDLT)	0.05 days
Transportation cost for all types of products (T)	0.45 Rs/Unit

Table 6 Product value (v_p) in rupees/unit for all types of products for all retailers

P-1	50
P-2	55
P-3	60

4.3   Solution Procedure and Optimization Results

For solving this problem, the numerical testing is done in AIMMS© language and solved with CONPOT solver on an Intel (R) Core (TM)2 Duo CPU T 6570 with memory 4 GB. After receiving the solver results, some other performance parameters are also evaluated with the help of Excel sheet. These performance parameters are expected shortages per cycle (ESC), fill rate (FR), safety stock (SS), and average inventory [9, p. 337–344].

$$\text{ESC} = -\text{SS}[1 - \text{NORMDIST}(\text{SS}/\sigma_L, 0, \ 1, \ 1) + S_L \text{NORMDIST}(\text{SS}/\sigma_L, 0, \ 1, \ 0)]$$

Fill rate (FR)

$$\text{FR} = \frac{Q - \text{ESC}}{Q}$$

Safety stock (SS)

$$\text{SS} = r - \mu$$

Average inventory (AI)

$$\text{AI} = Q/2 + \text{SS}$$

Inventory Positions (IP)

$$\text{IP} = Q + r$$

Solver results and the results of Excel sheet are depicted simultaneously in the following tables for all three types of products.

5   Result Analysis

There are four distributors and six retailers in the distribution network. Three different types of products are shipping from distributors to retailers. As discussed earlier that these distributors are denoted by D-1 to D-4, retailers are represented by R-1 to R-6 and the three different types of products are shown by P-1 to P-3. In the given study, it is assumed that retailers R-1 and R-2 are pooled for distributor D-1, retailers R-3 and R-4 are pooled for distributor D-2, while retailers R-5 and R-6 are standing alone with the distributors D-3 and D-4. For this scenario, the results are obtained through nonlinear optimization which is depicted in Tables 7, 8, and 9.

Table 7 Results for product 1

	Q	r	μ	SS	ESC	FR	AI	IP	PM	Total cost
D-1	831	849	720	129	0.621929	0.999252	545	1680	385,000	165,453
D-2	788	814	690	124	0.596321	0.999243	518	1602	375,000	183,247
D-3	550	395	315	80	0.384133	0.999301	355	945	445,000	107,168
D-4	596	432	345	87	0.420717	0.999294	385	1028	605,000	182,698

Table 8 Results for product 2

	Q	r	μ	SS	ESC	FR	AI	IP	PM	Total cost
D-1	864	990	840	150	0.72436	0.99916	582	1854	400,400	231,717
D-2	795	885	750	135	0.65,119	0.99918	532	1680	390,000	229,404
D-3	617	507	405	102	0.49388	0.99920	411	1125	462,800	149,800
D-4	589	489	390	99	0.47559	0.99919	393	1077	629,200	221,743

Table 9 Results for product 3

	Q	r	μ	SS	ESC	FR	AI	IP	PM	Total cost
D-1	762	917	777	140	0.673147	0.99911	521	1678	369,600	251,478
D-2	697	917	780	137	0.658514	0.99905	485	1613	360,000	276,619
D-3	465	507	405	102	0.493885	0.99893	335	972	427,200	192,986
D-4	516	536	427	108	0.521323	0.99898	366	1051	580,800	273,181

As discussed earlier, four different uncertainty levels are taken. These uncertainty levels are 0.2, 0.4, 0.6 and 0.6. The optimization results show that all the performance parameters of the given system i.e. reorder point, safety stocks, expected shortages per cycle, average inventory, inventory positions, and the total cost increase, as the uncertainty level increases from 0.2 to 0.8 for all the three different types of products. Figure 3, 4, 5, 7 and 8 also show that reorder point, safety stocks, expected shortages per cycle, average inventory, inventory positions, and the total cost increase, as the uncertainty level increases from 0.2 to 0.8 for all the three different types of products. Only the fill rate decreases as the uncertainty level increases from 0.2 to 0.8 for all types of products (Fig. 6). It is observed that there is a very minute decrement in the fill rate, as the uncertainty level increase, still the system is providing more than 99% fill rates. From Fig. 9, it is clear that there is a very little increment in the total cost for all three different types of products. There is a little increment in the total cost due to the increment in the inventory holding cost component. Therefore, it can be proved that risk-pooling is an effective strategy for reducing safety stocks and inventory carrying cost under different uncertainty levels. This approach is also very much effective for providing improved customer service.

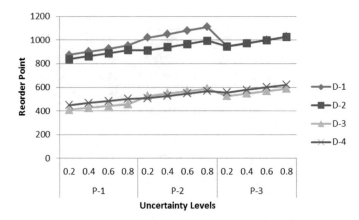

Fig. 3 Reorder point versus uncertainty levels

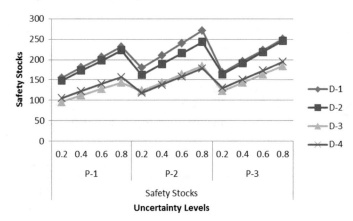

Fig. 4 Safety stocks versus uncertainty levels

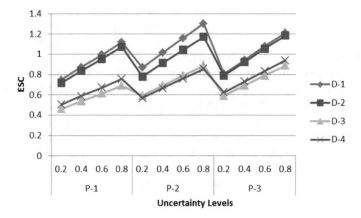

Fig. 5 ESC versus uncertainty levels

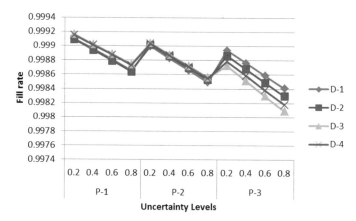

Fig. 6 Fill rate versus uncertainty levels

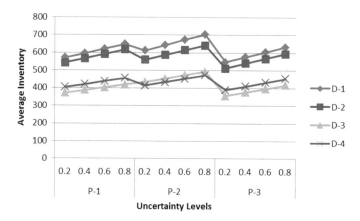

Fig. 7 Average inventory versus uncertainty levels

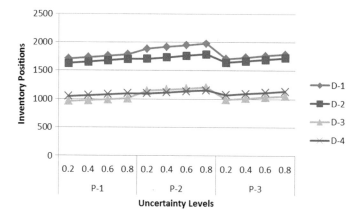

Fig. 8 Inventory positions versus uncertainty levels

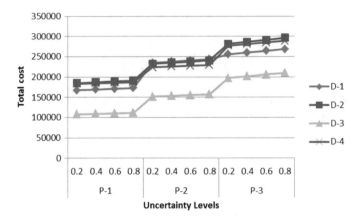

Fig. 9 Total cost versus uncertainty levels

6 Conclusion and Future Research Directions

In this study, a mathematical approach is developed for multi-products under risk-pooling for inventory optimization in a two-stage supply chain network. For solving this problem, only four distributors, six retailers, and three different types of products are considered. The first two retailers are aggregated for distributor 1 (risk-pooling is applied); the next two retailers are aggregated for distributor 2 (risk-pooling is applied). Retailers 5 and 6 are standing alone in the system. It is important to note that the fill rate is greater than 99% under fixed service-level (0.975) constraint for all type products. Some other performance parameters are also calculated for evaluating the performance of the system under fixed service-level constraint. These performance parameters are expected shortages per cycle (ESC), fill rate (FR), safety stock (SS), average inventory (AI), and inventory position (IP). The effect of uncertainties is evaluated on different performance parameters on reorder point (r), shortages per cycle (ESC), fill rate (FR), safety stock (SS), average inventory (AI), inventory position (IP), and total cost (TC). Some improved computational algorithms can be used for obtaining better results. More complex network may be taken for future study. The results prove that except fill rate, all the performance parameters increase, as the uncertainty level increases. Therefore, it can be concluded that risk-pooling is an effective strategy for providing improved customer service under various levels of uncertainties.

References

1. Tagaras, G.: Pooling in multi-location periodic inventory distribution systems. Omega, Int. J. Manage. Sci. **27**, 39–59 (1999)
2. Tagaras, G.: Effects of pooling on the optimization and service levels of two-location inventory systems. IIE Trans. **21**(3), 250–257 (1989)

3. Chang, P.L., Lin, C-T.: On the effect of centralization on expected costs in multi-location newsboy problem. J. Oper. Res. Soc. 42(11), 1025–1030 (1991)
4. Bernstein, F., Decroix, G.A., Wang, Y.: The impact of demand aggregation through delayed component allocation in an assemble-to-order system. Manage. Sci. **57**(6), 1154–1171 (2011)
5. Eyan, A., Fouque, T.: Benefiting from risk pooling effect: internal (component commonality) vs external (demand reshape) efforts. Int. j. Serv. Oper. Manag. **1**(1), 90–99 (2005)
6. Gerchak, Y., Mossan, D.: On the effect of demand randomness on inventories and costs. IIE Trans **40**(4), 804–807 (1992)
7. Sobel, M.J.: Risk pooling. In: Building Intuition, pp. 155–174. Springer, Boston. MA. (2008)
8. Gaur, S., Ravindran, A.R.: A bi-criteria model for inventory aggregation problem under risk pooling. Comput. Ind. Eng. **51**, 482–501 (2006)
9. Chopra, S., Meindl, P., Kalra, D.V.: Supply chain management: strategy, planning and operations, 5th Edn. Pearson Education Inc. pp. 337–344 (2013)

Optimization of Magnetic Abrasive Finishing Process Using Principal Component Analysis

S. B. Gunjal and P. J. Pawar

Abstract Study of process parameter and their selection is very important in the process performance point of view; this issue is attempted mostly by Taguchi method. However, Taguchi approach can be applied only to single-objective problem, and in case of multi-objective problem, it gives different levels and it becomes difficult to interpret these results. Principal component analysis (PCA) transforms the set of uncorrelated components to get the optimum level of combination for all the responses. In this paper, PCA is applied to a case study having three responses, i.e., change in surface roughness, tangential cutting force, and normal magnetic force. The levels of parameters obtained by PCA show the improved results for responses than those obtained by Taguchi method.

Keywords Magnetic abrasive finishing · Principal component analysis · Taguchi · Multi-objective

1 Introduction

Traditional finishing processes such as honing, lapping uses the rigid tool which may develop the substantial stresses in the workpiece due to which it is not possible to finish the delicate parts with these processes. Also, it may develop the cracks and temperature defects, affecting the performance of machined component. These limitations of traditional finishing processes are overcome by advanced finishing processes (AFP) [1]. Among the various AFP, magnetic abrasive finishing (MAF) is gaining importance due to its capability to achieve the higher surface finish. Also this process gives the low-cost option for finishing as the setup for this process can be

S. B. Gunjal (✉) · P. J. Pawar
Department of Production Engineering, K. K. Wagh Institute of Engineering Education and Research, Nashik, Maharashtra, India
e-mail: gunjsandip@rediffmail.com

P. J. Pawar
e-mail: pjpawar1@rediffmail.com

© Springer Nature Singapore Pte Ltd. 2020
R. Venkata Rao and J. Taler (eds.), *Advanced Engineering Optimization Through Intelligent Techniques*, Advances in Intelligent Systems and Computing 949, https://doi.org/10.1007/978-981-13-8196-6_43

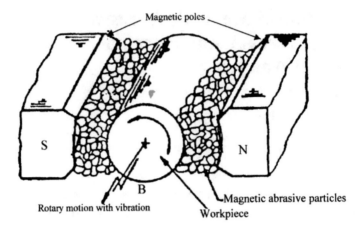

Fig. 1 Schematic of MAF *Source* Wear 250, p. 385

mounted on the existing machine tool. In MAF, magnetic abrasive particles (MAPs) are used between the tool and workpiece gap. In MAP, abrasive particles and iron particles are mixed in proper proportion and binder is added to hold these particles together. The magnetic field applied in the working gap between tool and workpiece will hold this mixture. Finishing operation is performed by pressing these MAP against the surface of workpiece, and working principle of MAF process is shown in Fig. 1 [2]. Relevant literature in the related area is as discussed below.

2   Literature Review

Judal et al. [3] used the response surface methodology for material removal rate (MRR) and surface roughness; in this work, attempt is made to find the optimum combination of input variables for the response parameters. Taguchi design of experiment is applied by Singh et al. [4] to know the effect of various input parameters on surface roughness, cutting force, and normal magnetic force during MAF. They found voltage and working gap of MAF. Yang et al. [5] determined the optimal parameter conditions for surface roughness and MRR, using Taguchi parameter design. They conducted the study to optimize the parameters to reduce operation cost, and different levels of input parameters are obtained for the two responses. The improvement in surface roughness and MRR by applying a magnetic field to abrasive flow machining is discussed by Singh et al. [6], and they concluded that surface roughness and MRR are significantly affected by magnetic.

From the literature reviewed, it is observed that for most of the cases Taguchi method is applied to address the multi-response problem. However, it can be applied only to single-objective problem, and in case of conflicting multi-objective optimization problem, it gives different levels, so it becomes difficult to interpret these results to achieve the best performance. Some researchers considered the response surface methodology for multi-objective problem. Moreover, it is the extended version of design of experiment and the output will be yet in terms of levels only.

Therefore, in this work principal component analysis (PCA) is applied for the multi-objective problem considered by Singh et al. [4]. PCA transforms the set of uncorrelated variables to get the optimum level of combination for all the responses.

3 Principal Component Analysis (PCA)

PCA is a popular technique for reduction of dimensionality. The PCA will minimize the dimensionality of a data set having a larger number of interrelated variables. It maintains the permissible variation in the initial data set. This can be done by altering to a revised set of variables, called as the principal components (PCs). They are ordered and uncorrelated; thus, first few will keep most of the initial data variation. It selects a minimum number of components for the variance of basic multi-response. The different steps in PCA are as follows:

Step (1): For each quality characteristics, S/N ratios are obtained. It is normalized as

$$X_l^*(m) = \frac{X_l(m) - X(m)^-}{X(m)^+ - X(m)^-} \tag{1}$$

where $X_l^*(m)$ is the normalized S/N ratio for mth quality characteristic in lth experimental run, $X_l(m)$ is the S/N ratio for mth quality characteristic in mth experimental run, $X_{(m)}^-$ and $X_{(m)}^+$ is the minimum and maximum of S/N ratios, respectively, for mth quality characteristic in all experimental runs.

Step (2): Multi-response is represented in matrix form.

Step (3): For normalized multi-response matrix, correlation coefficient is calculated as below.

$$R_{lm} = \frac{\text{cov}\left(X_{l(m)}^*, X_{l(k)}^*\right)}{\sigma X_{l(m)}^* \times \sigma X_{l(k)}^*} \tag{2}$$

where $\text{cov}\left(X_{l(m)}^*, X_{l(k)}^*\right)$ is the covariance of sequences $X_l^*(m)$ and $X_{l(k)}^*$; $\sigma X_{l(k)}^*$ represents the standard deviation for sequence $X_{l(k)}^*$

Step (4): Calculate the eigenvectors and the eigenvalues of the correlation matrix.

Step (5): The principal components are to be calculated as follows.

$$p(l)^h = \sum_{m=1}^{n} X_l^*(m) \times v_h(m) \tag{3}$$

Table 1 Input parameters and various levels [4]

Parameter	Level 1	Level 2	Level 3
Voltage, X_1 (V)	7.5	9.5	11.5
Working gap, X_2 (mm)	1.25	1.50	1.75
Rotational speed, X_3 (rpm)	90	125	180
Grain size, X_4 (mesh number)	400	800	1200

where $p(l)^h$ is the kth principal component corresponding to lth experimental run, and $v_h(m)$ is the mth element of hth eigenvector. The total principal component index (TPCI) for lth experimental run (P_l) is calculated as below.

$$P_l = \sum_{h=1}^{n} (p_i h) \times e(h) \tag{4}$$

where

$$e(h) = \frac{\text{eig}(h)}{\sum_{h=1}^{n} \text{eig}(h)} \tag{5}$$

where $\text{eig}(h)$ is the hth eigen value.

Step (6): For each experimental run, the average factor effect of each level is calculated from the TPCI. From the maximum value of TPCI, the optimum parameter level is predicted [7].

4 Application of PCA

PCA is applied for the multi-response problem considered by Singh et al. [4]. They used the Taguchi method to find out critical parameters influencing the quality of the surface. Different input parameters and their levels considered in the paper are as shown in Table 1. The responses considered are change in surface roughness (ΔR_a), normal magnetic force (F_{mn}), and tangential cutting force (F_c). For a three-parameter level, they considered L_9 orthogonal array (OA), and Table 2 shows the experimental design and the experimental results summarized by them.

For change in surface roughness (ΔR_a), tangential cutting force (F_c), and normal magnetic force (F_{mn}), linear regression models are as given below [4].

$$\Delta R_a = 0.137 + 0.0317X_1 - 0.0367X_2 + 0.0100X_3 + 0.0167X_4 \tag{6}$$

$$F_{mn} = 34.1 + 36.3X_1 - 32.3X_2 - 7.83X_3 + 25.2X_4 \tag{7}$$

$$F_c = 29.3 + 8.68X_1 - 6.50X_2 - 2.85X_3 + 2.64X_4 \tag{8}$$

Table 2 Experimental design (L₉ OA) and results [4]

S. No.	A	B	C	D	Average ΔR_a	Average F_{mn} (N)	Average F_c (N)
1	7.5	1.25	90	400	0.16	55	32
2	7.5	1.50	125	800	0.14	35	22
3	7.5	1.75	180	1200	0.14	25	19
4	9.5	1.25	125	1200	0.24	135	42
5	9.5	1.50	180	400	0.17	37	29
6	9.5	1.75	90	800	0.14	71	29
7	11.5	1.25	180	800	0.26	150	45
8	11.5	1.50	90	1200	0.21	133	48
9	11.5	1.75	125	400	0.18	50	32

Table 3 S/N ratio for various responses

S. No.	S/N ratio		
	Average ΔR_a	Average F_{mn}	Average F_c
1	−15.92	34.81	30.10
2	−17.08	30.88	26.85
3	−17.08	27.96	25.58
4	−12.40	42.61	32.46
5	−15.39	31.36	29.25
6	−17.08	37.03	29.25
7	−11.70	43.52	33.06
8	−13.56	42.48	33.62
9	−14.89	33.98	30.10

Signal-to-noise (S/N) ratio is calculated by using Eq. 9 as the objective of the experiment is to maximize the responses. Variation in response related to desired value under various noise conditions is measured by signal-to-noise ratio. Table 3 shows the S/N ratio for various responses considered in the experiment.

$$\frac{S}{N} = -10\log \sum \left(\frac{\frac{1}{y^2}}{n} \right) \tag{9}$$

Table 4 shows the normalized S/N ratio; it is calculated by using Eq. 3.

Correlation coefficient of normalized multi-response matrix is calculated by using Eq. 4 which is shown in Table 5.

Eigenvectors and eigenvalues for correlation coefficient matrix are calculated by using the MATLAB 2010, as shown in Table 6.

The principal component (PC) is calculated by using Eq. 5, and total principal component index (TPCI) is calculated by using Eq. 6 as shown in Table 7.

Table 4 S/N ratio (normalized)

S. No.	S/N ratio(normalized)		
	Average ΔR_a	Average F_{mn}	Average F_c
1	0.22	0.44	0.56
2	0.00	0.19	0.16
3	0.00	0.00	0.00
4	0.87	0.94	0.86
5	0.31	0.22	0.46
6	0.00	0.58	0.46
7	1.00	1.00	0.93
8	0.66	0.93	1.00
9	0.41	0.39	0.56

Table 5 Correlation coefficient between the responses

	Average ΔR_a	Average F_{mn}	Average F_c
Average ΔR_a	1.00	0.85	0.88
Average F_{mn}	0.85	1.00	0.94
Average F_c	0.88	0.94	1.00

Table 6 Eigenvalues and eigenvectors

S. No.	Eigenvalues			Eigenvectors
	X_1	X_2	X_3	λ
1	−0.7618	−0.2763	0.5859	2.7805
2	0.6323	−0.5140	0.5797	0.1625
3	0.1410	0.8121	0.5663	0.0570

Table 7 Principal component and total principal component index

S. No.	Principal components			TPCI
	PC1	PC2	PC3	
1	0.04	0.24	0.71	0.07
2	0.04	0.00	0.24	0.04
3	0.00	0.00	0.00	0.00
4	−0.42	0.56	1.37	−0.33
5	−0.03	0.35	0.48	0.00
6	0.11	−0.04	0.73	0.11
7	−0.49	0.66	1.48	−0.39
8	−0.17	0.51	1.42	−0.10
9	−0.09	0.38	0.69	−0.05

Table 8 Effect of TPCI on each level

Level	Voltage, A	Working gap, B	Rotational speed, C	Grain size, D
1	0.110	0.660	0.070	0.020
2	0.230	0.060	0.340	0.240
3	0.540	0.060	0.400	0.440

Table 9 Calculation of responses by using linear regression model

S. No.	A	B	C	D	Average ΔR_a,	Average F_{mn}	Average F_c
1	1.00	1.00	1.00	1.00	0.16	55.47	31.27
2	1.00	2.00	2.00	2.00	0.15	40.54	24.56
3	1.00	3.00	3.00	3.00	0.14	25.61	17.85
4	2.00	1.00	2.00	3.00	0.23	134.34	42.38
5	2.00	2.00	3.00	1.00	0.17	43.81	27.75
6	2.00	3.00	1.00	2.00	0.13	52.37	29.59
7	3.00	1.00	3.00	2.00	0.26	137.61	45.57
8	3.00	2.00	1.00	1.00	0.19	95.77	42.13
9	3.00	3.00	2.00	3.00	0.19	106.04	38.06
PCA levels	3.00	1.00	3.00	3.00	0.28	162.81	48.21

For every experimental run, the effect of average factor for each level is calculated from the TPCI as shown in Table 8. Maximum TPCI value indicates the optimum parameter level. Thus, it is level 3 for voltage, level 1 for working gap, level 1 for rotational speed, and level 3 for grain size. Table 9 shows the average ΔR_a, average F_{mn}, and average F_c which are calculated by using the linear regression models of Eqs. 6, 7, and 8. Table 9 shows that for the levels of PCA, all the responses are having the highest value as compared to levels used in the Taguchi design of experiment.

5 Conclusions

1. From the literature reviewed, it is observed that in most of the cases, Taguchi method is applied to address the multi-response problem. Taguchi approach in real sense is not optimization a technique. Moreover, it can be applied only to single-objective problem, and in case of conflicting multi-objective optimization problem, it gives different levels, and thus, it becomes difficult to interpret these results to achieve the best performance. Response surface methodology is the extended version of design of experiment, and the output will be yet in terms of levels only.
2. Principal component analysis converts a set of correlated variables into a set of uncorrelated variables. Principal components decrease the complexity of the

multi-response problems and reduce the number of dimensions. Accordingly, using these uncorrelated variables, in parameter design stage the optimal conditions are easily selected in an objective manner.

3. PCA is successfully applied for the multi-objective problem, and it is observed that for the three responses considered, the optimal levels of parameters are $A3$ (Voltage: 11.5 V), $B1$ (Working gap: 1.25 mm), $C3$ (Rotational speed: 180 rpm), and $D3$ (Grain size: 1200 mesh number).

4. Optimum values of responses by Taguchi experimental design are (i) change in surface roughness $(\Delta R_a) = 0.26$, (ii) normal magnetic force $(F_{mn}) = 150$ N, and (iii) tangential cutting force $(F_c) = 45$ N. By using the proposed linear regression model given in the paper, the optimal values obtained by PCA are (i) change in surface roughness $(\Delta R_a) = 0.28$, (ii) normal magnetic force $(F_{mn}) = 162.81$ N, and (iii) tangential cutting force $(F_c) = 48.21$ N.

References

1. Heng, L., Kim, Y.J., Mun, S.D.: Review of super finishing by the magnetic abrasive finishing process. High Speed Mach. **3**, 42–55 (2017). https://doi.org/10.1515/hsm-2017-0004
2. Jain, V.K., Kumar, P., Behera, P.K., Jayswal, S.C.: Effect of working gap and circumferential speed on the performance of magnetic abrasive finishing process. Wear **250**, 384–390 (2001). https://doi.org/10.1016/S0043-1648(01)00642-1
3. Judal, K.B., Yadava, V.: Cylindrical electrochemical magnetic abrasive machining of AISI-304 stainless steel. Mater. Manuf. Process. **28**, 449–456 (2013). https://doi.org/10.1080/10426914.2012.736653
4. Singh, D.K., Jain, V.K., Raghuram, V.: Parametric study of magnetic abrasive finishing process. J. Mater. Process. Technol. **149**, 22–29 (2004). https://doi.org/10.1016/j.jmatprotec.2003.10.030
5. Yang, L., Lin, C., Chow, H.: Optimization in MAF operations using Taguchi parameter design for AISI304 stainless steel. Int J. Adv. Manuf. Technol. **42**, 595–605 (2009). https://doi.org/10.1007/s00170-008-1612-4
6. Singh, S., Shan, H. S.: Development of magneto abrasive flow machining process. Int. J. Mach. Tools Manuf. **42**, 953–959 (2002). https://doi.org/10.1016/s0890-6955(02)00021-4
7. Su C., Tong, L.: Multi-response robust design by principal component analysis. Total Quality Manag. **8**, 409–416 (1997). https://doi.org/10.1080/0954412979415

Dynamic Distribution Network Expansion Planning Under Energy Storage Integration Using PSO with Controlled Particle Movement

Santosh Kumari, Prerna Jain, Dipti Saxena and Rohit Bhakar

Abstract Distribution network expansion planning (DNEP) is a multiobjective problem fulfilling demand growth, ensuring reliable supply and minimizing total expansion cost. It is a sequential methodology to plan for the reinforcement of existing or installation of new feeders and substations. This paper handles dynamic DNEP incorporating energy storage systems (ESSs), used to shave the peak demand and valley filling. Monte Carlo simulation method is used to introduce uncertainty in load demand of the system A.C. power flow, annual and daily load duration curve are used. Reliability index is modeled for composite system of lines and ESSs. Planning is formulated as constrained, mixed integer nonlinear programming problem (MINLP) and solved using inertia PSO with controlled particle movement to avoid randomness and premature convergence. An 11 kV, 30-bus radial distribution network is considered as a case study. Simulation results validate the consideration of ESSs in DNEP by a significant improvement in the voltage profile, losses, total expansion cost and reliability of the distribution system. The proposed PSO proves efficient with improved convergence and reduced planning cost.

Keywords Distribution network expansion planning · Energy storage systems · Energy not supplied · PSO

S. Kumari (✉) · P. Jain · D. Saxena · R. Bhakar
Department of Electrical Engineering, Malaviya National Institute of Technology Jaipur, Jaipur, India
e-mail: 2015pes5106@mnit.ac.in

P. Jain
e-mail: pjain.ee@mnit.ac.in

D. Saxena
e-mail: dsaxena.ee@mnit.ac.in

R. Bhakar
e-mail: rbhakar.ee@mnit.ac.in

© Springer Nature Singapore Pte Ltd. 2020
R. Venkata Rao and J. Taler (eds.), *Advanced Engineering Optimization Through Intelligent Techniques*, Advances in Intelligent Systems and Computing 949,
https://doi.org/10.1007/978-981-13-8196-6_44

497

1 Introduction

In order to meet the incessantly increasing peak load demand, distribution network expansion planning (DNEP) is carried out to supply reliable energy to the customers at minimum investment and operational cost while satisfying various technical constraints, over a planning horizon. DNEP is done either by the reinforcement of existing or installing new feeders and/or capacity expansion of substation and installing new ones [1, 2]. It can be broadly classified into two categories as static and dynamic based on the number of planning horizon. In static type, single planning horizon is considered, whereas in dynamic (multistage), planning horizon is distributed in several stages of time frame [1–6]. In static planning, to consider a number of planning horizons, DNEP is done separately for each and the total cost is not optimal.

Energy storage systems (ESSs) are being widely adopted by distribution network operators due to their applications in mitigating renewable resources uncertainties [7], microgrid [8], frequency and voltage control, peak shaving, valley filling, stability enhancement, congestion management, power quality, and reliability improvement [9, 10]. In context with leveling the load of the system, the load profile gets flattened and consequently the load factor is improved [7]. Their integration also contributes toward the clean environment. Accordingly, it is important to investigate their impact on the DNEP problem.

DNEP has been largely dealt through different objective functions, variables, and constraints, for a variety of one load level, multiload level, and probabilistic and fuzzy load levels [11–13]. DNEP problem is considered as multiobjective optimization problem in [12]. DGs are incorporated with DNEP, and their siting and sizing have been done in [13–15], but it is observed that incorporation of DGs result in reverse power flow [16]. The estimation of reliability with DGs is done in [17, 18]. To the best of author's knowledge, few papers have included ESSs in DNEP. Storage units and DGs both are considered in expansion planning in [19], but it deals partially with storage units. In [20], multistage DNEP is performed with main focus on storage units, but it does not comprise the reliability assessment of the composite system of lines and ESSs and has not analyzed the reliability of the planned system. Reliability assessment is an integral part of system planning [21]. Also, [19] and [20] perform planning with the forecasted peak load without modeling its uncertainty which is necessary for more accurate results. The use of dispersed ESS as an alternative for new wiring, new substation, substation expansion, and DG installation has been considered in [20] but needs extension with reliability and uncertainty considerations.

Based on the literature, this paper aims to incorporate ESS in dynamic expansion planning along with new lines as an expansion option to satisfy the expected rise in peak load demand while modeling its uncertainty. Monte Carlo technique is used for handling uncertainty in simulations, and thereafter, planned distribution system is evaluated for voltage, power flow through lines, line losses, and reliability. Reliability is assessed by energy not supplied (ENS) index, modeled for distribution lines integrated with ESS at different load buses. DNEP problem is a constrained mixed integer nonlinear programming model and has been solved using various metaheuristics techniques like GA, Tabu search, and artificial bee colony method [1, 2, 5, 6, 9, 14, 17]. One of the well-stabilized evolutionary algorithms is particle swarm optimization (PSO). DNEP problem is solved using PSO in [15, 19]. PSO requires tuning of the coefficients to regulate the movement patterns of the particles in order to avoid randomness and premature convergence [22]. Movement patterns of particle in a PSO can be characterized by two aspects. The first is the correlation between its consecutive positions, modeled by its base frequency, and the second is its range of movement, modeled by its variance of movement. Suitable values of PSO coefficients can be found using a set of equations to guarantee achieving a given base frequency and variance of movement, i.e., controlling the movement pattern of particles [23]. The proposed expansion planning work is solved using a novel inertia PSO having controlled particle movement where a smooth movement of particles between consecutive positions is adopted. An 11 kV, 30-bus radial distribution network with five years' timeframe for planning is considered as a test case.

This paper is framed as follows: Sect. 2 presents the problem formulation and discusses the reliability, in case of radial distribution system in the presence of both ESS and newly installed lines, Sect. 3 discusses the proposed methodology, and Sect. 4 presents the simulation and results of the considered radial distribution system.

2 Problem Formulation

DNEP is formulated as minimization of total expansion cost of newly installed lines and ESSs. This cost includes the investment cost of lines and ESSs and operational cost of ESSs. The above-defined problem is dynamic, so it is solved for different stages.

2.1 Investment Cost of Newly Installed ESSs

Investment cost of ESSs depends on the number of ESSs and their installed capacity. If installed capacity increases, cost increases significantly. Let F_1 represents the investment cost of newly installed ESSs at any stage.

$$F_1 = E_s \times IC_s \times VE_s \times \frac{1}{(1+k)^{s-1}} \quad \forall s \in S \quad (\$/stage) \tag{1}$$

Here, S is the total number of stages considered, s is a particular stage, E_s is the number of newly installed ESSs in the system at stage s, IC_s is the investment cost of E_s ($/kWh), and VE_s (kWh) is a vector of the capacity of ESSs (which are installed at stage s). k is the discount rate and hence $\frac{1}{(1+k)^{s-1}}$ converts investment cost into present value.

2.2   Investment Cost of Newly Installed Lines

Investment cost of lines depends upon the number of lines and their location in the system. This cost at any stage s is defined as:

$$F_2 = L_s \times IV_s \times \frac{1}{(1+k)^{s-1}} \quad \forall s \in S \quad (\$/\text{stage}) \tag{2}$$

Here, L_s is the number of new lines which are installed at stage s, and IV_s is a vector which is investment cost of $L_s(\$)$.

2.3   Operational Cost of ESSs

Operational cost of ESSs is formulated as:

$$F_3 = CU_s \times S_s \times VE_s \times OP_s \times \frac{1}{(1+k)^{s-1}} \quad \forall s \in S \quad (\$/\text{stage}) \tag{3}$$

Here, OP_s is the operational cost of ESS ($\$/KWh$) at stage s, S_s is defined as time duration of stage s, and CU_s shows the cumulative value of E_s.

2.4   Cost of Energy Purchased from Upstream Network

It includes the cost of energy which is purchased from the upstream network. This energy purchased is varying in nature as the load demand is also changing with time; i.e., to satisfy the varying demand of load, energy purchased from grid varies which leads to change in its purchased cost [15].

$$F_4 = \sum_{s \in S} T_s (EC_s \times P_s) \forall \quad s \in S \quad (\$/\text{stage}) \tag{4}$$

Here, T_s denotes the time period of stage s, EC_s (KWh) denotes the energy required to purchase from the grid, and P_s ($\$/KWh$) denotes the cost of the energy purchased at any stage s.

2.5   Final Objective Function

It includes the minimization of overall cost over the complete planning horizon.

$$\min F = \sum_{s=1}^{S} (F_1 + F_2 + F_3 + F_4) \tag{5}$$

2.6 Problem Constraints

The proposed DNEP incorporating ESSs is subjected to the following constraints.

$$V_n^{\min} \leq V_n \leq V_n^{\max} \quad \forall n \in rn \tag{6}$$

$$S_b \leq S_b^{\max} \quad \forall b \in rb \tag{7}$$

$$P_n = P\text{node}_{n+1} + P\text{load}_{n+1} + P\text{loss}_n \tag{8}$$

$$Q_n = Q\text{node}_{n+1} + Q\text{load}_{n+1} + Q\text{loss}_n \tag{9}$$

$$CU_s = CU_{s-1} + E_s \quad \forall s \in S \tag{10}$$

$$L_s \leq L_s^{\max} \quad \forall s \in S \tag{11}$$

$$E_s \leq E_s^{\max} \quad \forall s \in S \tag{12}$$

$$P_{\text{ch}} \leq P_{\text{ch}}^{\text{rate}} \tag{13}$$

$$P_{\text{disch}} \leq P_{\text{disch}}^{\text{rate}} \tag{14}$$

$$E_{\text{disch}} \leq \eta_{\text{ESS}} \times E_{\text{ch}} \tag{15}$$

$$E_{\text{ch}}^d = P_{\text{ch}}^d \times T_{\text{ch}}^d + (E_{\text{disch}}^{d-1} - \eta_{\text{ESS}} \times E_{\text{ch}}^{d-1}) \quad \forall d \in nd \tag{16}$$

Here, (6) is voltage constraint, V_n is the bus voltage of any node n, and $V_n^{\max}$ and $V_n^{\min}$ are its maximum and minimum limits, respectively. rn is the set of all buses. In the proposed problem considered, voltage limits are 1.1 and 0.9 at any of the bus of the considered distribution system. Constraint (7) defines transferred apparent power flow in a bth branch limited by its maximum value $S_b^{\max}$. Equations (8) and (9) are AC power flow equations. Backward–forward method is used here. P_n and Q_n are the active and reactive power flowing out of bus n. $P\text{node}_{n+1}$ and $Q\text{node}_{n+1}$ are the active and reactive power at node $n + 1$, and $P\text{loss}_n$ and $Q\text{loss}_n$ are power loss in the line connecting node n to $n + 1$. Constraint (10) calculates the cumulative value of E_s Constraint (11)–(12) denotes that number of new lines installed between two nodes and the number of new ESSs installed at each nodes should be less than their maximum defined limits $L_s^{\max}$ and $E_s^{\max}$, respectively. In this work, the maximum number of lines to be installed is restricted to two in each corridor (route between two nodes). Equations (13)–(14) define charging and discharging power of storage units to be less than or equal to their rated values. Here, P_{ch} and P_{ch}^{rate} denote the charging power and its rated value; P_{disch} and P_{disch}^{rate} denote the discharging power and its rated value. Constraint (15) defines discharged energy E_{disch} to be less than or equal to charged energy E_{ch}, where η_{ESS} defines ESSs efficiency. Constraint (16) defines the battery SOC equation.

2.7 Estimation of Reliability in Radial Distribution System

For estimating the reliability, the principle of series components is applied and ENS is computed as reliability index, given by (17) [21]

$$\text{ENS} = \sum_i L_a(i) \times U_i \tag{17}$$

Here, i is the load point/bus and U_i is its annual outage time in hours/year, calculated using failure rate λ_i (failures/year) and outage time r_i (hours) of load point i as in (18).

$$U_i = \lambda_i r_i \tag{18}$$

λ_i and r_i depend on the λ and r of the lines connected to the ith bus. For a radial system, they are the sum of the respective values of the connected lines [21]. If in a corridor, there are parallel lines of unique λ and r, then the composite values can be calculated as (19), where two lines are assumed to be in parallel of failure rates and outage times as λ_1, λ_2 and r_1, r_2, respectively

$$\lambda_{composite} = \frac{\lambda_1 \lambda_2 (r_1 + r_2)}{1 + \lambda_1 r_1 + \lambda_2 r_2}, \quad r_{composite} = \frac{r_1 r_2}{r_1 + r_2} \tag{19}$$

$L_a(i)$ is the average load at load point i and can be calculated using load curve as (20):

$$L_a(i) = \frac{(\text{Total Energy Demand in period of interest})}{(\text{period of interest in years})} \tag{20}$$

2.8 Estimation of ENS in the Presence of ESSs

Due to the installation of ESSs in the system, the required energy demand at load points E_d reduces at the time of their discharging.

$$E_d(i) = \begin{cases} E_{\text{demand}}(i) - E_{\text{disch}}(i) \, t \in t_{\text{disch}} \\ E_{\text{demand}}(i) \, t \notin t_{\text{disch}} \end{cases} \tag{21}$$

Here, E_{demand} denotes demand of energy without ESSs, E_{disch} denotes energy supplied by ESS, and t_{disch} denotes time period of discharging.

3 Proposed Methodology

In this work, DNEP is formulated as dynamic problem and models uncertainties of forecasted peak load. It is a mixed integer nonlinear programming (MINLP) problem. Different entities like ESSs and lines are installed optimally in order to satisfy the increased demand, while satisfying constraints. The problem is solved by novel inertia PSO having controlled particle movement problem.

3.1 Integration of ESSs in DNEP

Steps required for evaluating the considered DNEP problem with integration of ESSs modeling are as follows:

a. Initially, planning horizon of five years is set on the first stage.
b. Annual load profile is chosen according to historical data given in [18]. Annual load profile is set on the first level.
c. Setting daily load profile on the first level [18].
d. Modeling of ESSs is done based on daily load profile. So, whenever the load demand is high, ESSs are connected in system in discharge mode, and when load demand is low, ESSs are connected in charge mode, and at medium load demand, they are disconnected from the system.
e. Population of proposed PSO is generated with design variables as the number of new installed lines and ESSs along with their location.
f. Then, AC power flow is carried out to check the problem constraints for each particle using Eqs. (6)–(16).
g. If there is at least one constraint violated, current particle is removed and new is evaluated.
h. Then, planning objective (5) is evaluated for each particle, and personal best and global best are chosen. Population is updated. If PSO convergence is reached, go to next step; otherwise, go to step (f) and repeat.
i. All daily and annual load assessments are checked.
j. All stages of planning horizon are checked and evaluated. This process is repeated until all stages are not verified.

3.2   Considering Integration of Lines as Well ESSs

Steps required for evaluating the considered DNEP problem with integration of lines as well as ESSs are given in the flowchart of Fig. 1. Modeling of ESSs is done based on daily load profile. So, as shown in the flowchart, whenever the load demand is high, ESSs are connected in system in discharge mode, and when load demand is low, ESSs are connected in charge mode, and at medium load demand, they are disconnected from the system.

3.3   Considering Uncertainty in Forecasted Peak Load

In the earlier works on DNEP including ESS and line integration on distribution lines, it has been assumed that the actual peak load would differ from the forecast value with zero probability. This is extremely unlikely in actual practice as the forecast is normally predicted on past experience. The uncertainty in such a value can be described by a probability distribution functions (PDFs) whose parameters can be determined from past experience, future load modeling, and possible subjective evaluation. This then can be modeled in the DNEP by appropriate uncertainty simulations. Published data has suggested that the load uncertainty can be reasonably described by a normal distribution with a mean and standard deviation. Forecast peak load is the mean. Figure 2 depicts the daily peak load levels considered over a period of time, with associated uncertainty as normally distributed.

Monte Carlo simulations are an important technique for uncertainty handling and are used in this work. Following the technique, at first a large number of probable peak load values are generated based on the PDF. Then, for each load scenario, DNEP is evaluated. The expectation of the variables over the DNEP solutions for all load scenarios is the desired result. The algorithm for DNEP considering lines and ESS integration along with modeling of load uncertainty is shown in Fig. 3.

3.4   PSO with Controlled Particle Movement

Movement patterns of particle in a PSO can be characterized by two factors. One is the correlation between its consecutive positions which can be modeled by its base frequency (f). The second is its range of movement which can be modeled by its variance of movement (V_c). If the base frequency is small, mid-range, or large, then the particle movement is smooth, chaotic, or jumping at each iteration, respectively. Experimental tests have shown that small frequencies are more effective with a large number of PSO iterations. High value of variance is beneficial for any number of iterations. For Inertia PSO, the particle velocity is defined as:

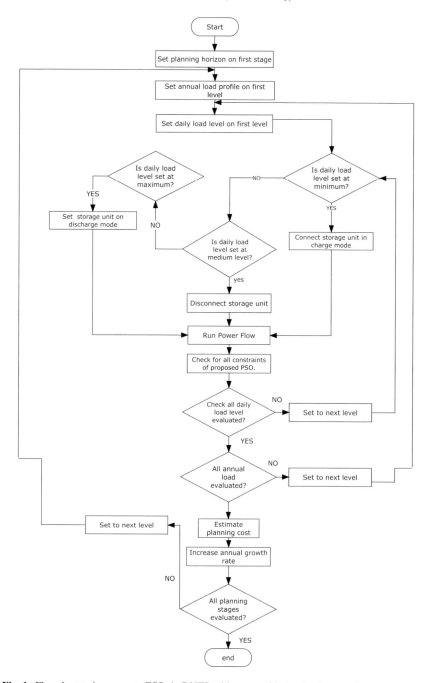

Fig. 1 Flowchart to incorporate ESSs in DNEP without considering load uncertainty

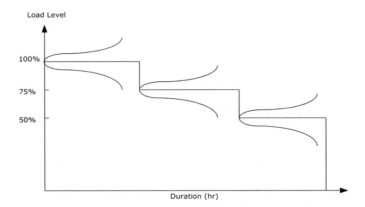

Fig. 2 Load duration curve with uncertainty

$$v_{t+1}^i = \omega v_t^i + c_1 R_{1t}(p_t^i - x_t^i) + c_2 R_{2t}(g_t - x_t^i) \tag{22}$$

Here, v_t^i is the previous iteration value multiplied by inertia weight ω; c_1 and c_2 are real numbers as acceleration coefficients, which are positive constants; for ith particle x_t^i is position in tth iteration and p_t^i is the best position till tth iteration; R_{1t} and R_{2t} are random numbers; g_t is the global best position; here, the inertia weight parameter is used to get balance between the global and local search; if its value is large, then it prefers global search, and if low, it facilitates local search. Most frequently used coefficient values ($c = 2$) impose mid-range resulting chaotic movement. A set of equations have been developed that enable to find coefficients values to guarantee achieving a given f and V_c, i.e., controlling the movement pattern of particles [21]. These are as (23) and can be solved to get c and ω. The number of iterations is less for short-run problem as compared to the long-run problem. Experimental values of V_c and F have been found as for short-run ($V_c = 25.6$, $F = 0.2$) and for long-run ($V_c = 6.4$, $F = 0.2$). Overall, the combination of $c = 1.711897$ and $\omega = 0.711897$ ($F = 0.25$ and $V_c = 25.6$) has been identified to perform better in all cases and gives smooth particle movement [21]. The proposed work uses the inertia PSO with these coefficient values, obtained using set of Eq. (23) (Figs. 4 and 5).

$$\begin{cases} c = 1 + \omega - 2\cos(2\pi F)\sqrt{\omega} \\[2mm] c = \dfrac{-48 V_c \omega^2 + 48 V_c}{28 V_c + \omega - 20 V_c \omega + 1} \end{cases} \tag{23}$$

4 Simulation Result

A sample test system of 11 kV, 10 MVA, 30-bus radial distribution system, having annual load growth of 9%, is used. The number of candidate lines present is twenty-

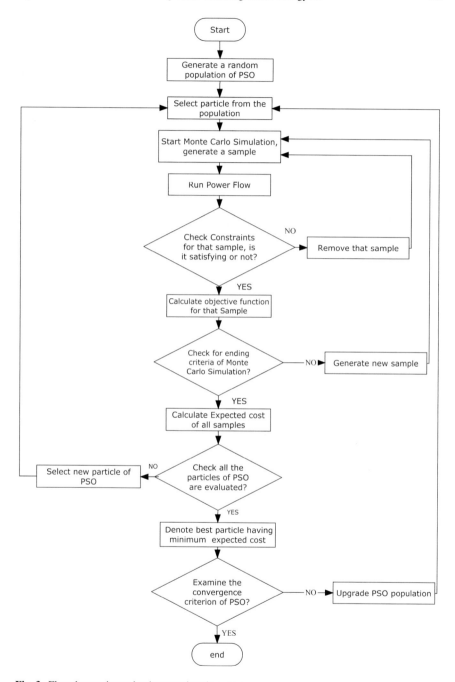

Fig. 3 Flowchart to insert load uncertainty in system

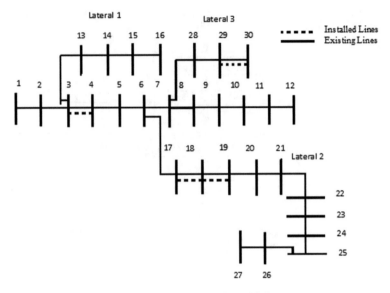

Fig. 4 Distribution system after the first stage of planning with lines

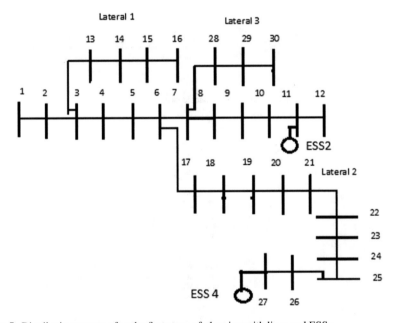

Fig. 5 Distribution system after the first stage of planning with lines and ESSs

Table 1 Market price with different load levels

Loading and cost of energy			
Load level	100	75	50
Time duration	1500	5000	2260
Market price	0.30 ($/year)	0.25 ($/year)	0.15 ($/year)

Table 2 Annual load profile having four levels

Load level (%)	100	83	68	54
Time duration (days)	29	144	131	56

Table 3 Daily load profile having three levels

Load level (%)	100	75	50
Time duration (h)	8	8	8

nine. New installed lines are not more than two in each corridor between two buses. To estimate the reliability, failure rate of each line is considered as 2 (failures/year) and outage time is 194.66 h/year [10]. Four ESSs are to be installed, as ESS1, ESS2, ESS3, and ESS4 with different capacities and charging rates.

Relevant data for the considered problem has been obtained from [20]. Tables 1, 2, and 3 give different data. Forecasted peak load level of Table 2 has 15% standard deviation as per normal PDF.

In the proposed DNEP, two cases are considered: (a) planning, when objective function does not include energy purchased from network; (b) when it includes. Further in case (a), the comparison of two plans is performed: (i) planning with lines; (ii) planning with ESSs and lines. The main objective of the formulated problem is to reduce the overall planning cost. It is cleared from the Table 4 that in the second plan, seven new ESSs are installed in the system, while in the first plan, ten new lines are installed to fulfill the load growth. It can be observed that no new lines are installed in the second plan. Thus, the results in Table 4 show that ESSs are at the first priority to be installed as compared to lines. In case (b), the results are compared with respect to load uncertainty. Further in this case, two schemes are considered. In the first scheme, load uncertainty is incorporated. While in the second one, uncertainty is not incorporated in the planning. The results of both schemes are shown in Table 5. The comparison of both these schemes depicts that total planning cost increases when uncertainty is considered, compared to the second scheme but provides more realistic solutions.

Table 4 DNEP with and without ESS

	DNEP with ESS and lines		DNEP with lines	
	Installed lines	Installed ESSs	Installed lines	Installed ESSs
Year 1	–	ESS 2-Bus 11 ESS 4-Bus 27	Line 3, 17, 18, 29	–
Year 2	–	ESS 3-Bus 27	Line 16, 19	–
Year 3	–	ESS 3-Bus 11	Line 4	–
Year 4	–	ESS 3-Bus 23	Line 1, 23	–
Year 5		ESS 1-Bus 27 ESS 3-Bus 11	Line 16	
Overall planning cost		172,430 $	196,760 $	

Table 5 DNEP with ESS and lines including energy purchased from network in the presence of with and without load uncertainty

	DNEP with ESS and lines in the presence of uncertainty		DNEP with lines without uncertainty	
	Installed lines	Installed ESSs	Installed lines	Installed ESSs
Year 1	–	ESS 1-Bus 13 ESS 3-Bus 23 ESS 3-Bus 27 ESS 3 Bus 23	–	ESS 1-Bus 13 ESS 3-Bus 23 ESS 3-Bus 27
Year 2	–	ESS 1-Bus 27 ESS 4-Bus 23	–	ESS 1-Bus 21 ESS 2-Bus 27
Year 3	–	ESS 1-Bus 23	–	ESS 3-Bus 27
Year 4	–	ESS 3-Bus 23	–	ESS 1-Bus 29 ESS 2-Bus 27 -
Year 5		ESS 2-Bus 29 ESS 4-Bus 11	–	ESS 4-Bus 23
Overall planning cost (M$)		7.685		6.811

4.1 Planning Cost Reduction

It can be observed from Table 1 that in the first plan, the overall planning cost is $172,430, and in the second plan, the planning cost is $196,760. Hence, when the planning is performed with ESSs and lines, the overall planning cost is reduced by 12.36%.

Table 6 Comparison of cost using the conventional PSO and proposed PSO

	Planning with ESSs and lines using proposed PSO	Planning with ESSs and lines using conventional PSO [20]
Planning cost	172,430 $	174,790 $

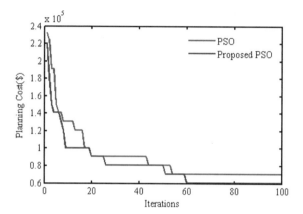

Fig. 6 Convergence curve for PSO and the proposed PSO in the first stage of planning

4.2 Planning Cost with Conventional PSO and Proposed PSO

It can be observed from Table 6 that in the first plan, the overall planning cost is $172,430, and in the second plan, the planning cost is $196,760. Hence, when planning is performed with ESSs and lines, the overall planning cost is reduced by 12.36%. As observed from Table 5, when planning is performed considering the energy price from the network and load uncertainty evaluated as schemes in case b. The planning cost in the second scheme is 6.811 M$ which increases to 7.685 M$ in the first scheme, which includes load uncertainty (Fig. 6).

4.3 Improvement in Voltage Profile

As storage units are installed in system, voltage profile is improved. Voltage of the buses is observed and compared for three cases: (a) without planning (base case), (b) planning with lines, and (c) planning with ESSs and lines. From Fig. 7, it can be observed that in the first case, voltage profile of the system is not in its limits. In second case, voltage at the buses comes under their limits but is at the verge of instability. In last case, voltage at buses improves significantly and is above their lower limits, which demonstrate the stability of the system.

Fig. 7 Voltage profile of the system

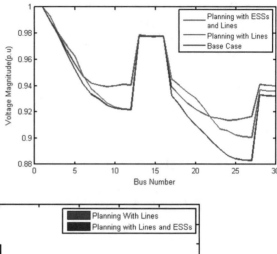

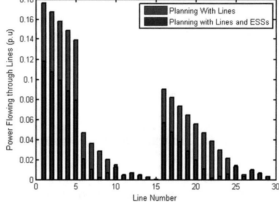

Fig. 8 Transmitted power through the lines with both plans

4.4   Reduction in Power Flow Through Lines

Due to installing ESSs in the system, the transmitted power flow through the lines reduces significantly compared to the second plan (as defined above), observed from Fig. 8.

4.5   Comparison of ENS and Losses

Reliability index ENS and losses are compared for both plans in Table 7. It is observed that supply reaching the customers becomes more reliable as energy not supplied to the system reduces by 297,227 (kWh/year) in case of planning with lines and ESSs. As shown in Table 8, ENS is compared when energy purchase from grid and uncertainty of load is taken, In this case, two schemes are considered as described above. In the

Table 7 Comparison of losses and energy not supplied (ENS) for planning with and without ESSs

	Planning with ESSs	Planning without ESSs
ENS	1,432,761 (KWh/year)	1,729,988 (KWh/year)
Losses	81.099 KW	98.4735

Table 8 Comparison of ENS considering energy purchased from grid with and without load uncertainty

	Planning with uncertainty	Planning without uncertainty
ENS	1,338,677 (KWh/year)	1,416,541 (KWh/year)

first scheme, ENS is reduced compared to the second scheme by 77,864 (kWh/year). From Table 7, in the second plan, losses are 98.4735 KW which are reduced up to 81.099 KW in the first plan. Hence, the installation of ESS assists to enhance the reliability of the system with the improvement in network losses.

5 Conclusion

In this work, a framework has been designed to perform expansion planning by optimally placing ESS and lines, including uncertainty related to load demand in the radial distribution system for minimizing the overall planning cost and operational cost. Controlled particle movement is used in inertia PSO to reduce its randomness and premature convergence and applied as a solution technique. The results show remarkable improvement in the convergence of PSO and final cost of planning. The installation of ESSs has shown positive impacts by improving network losses, transmitted power flow through lines along with improvement in voltage profile and reliability. It significantly improves congestion of the lines while enhancing life span of the distribution system

Acknowledgements The authors acknowledge the financial support provided by DST Rajasthan Grant No. 7(3) DST/Rand D/2016/3286 titled "Distribution Network Planning Mechanisms for state utilized under large integration of EV's and DERs in Smart Grid Framework."

References

1. Neimane, V.: Distribution network planning based on statistical load modeling applying genetic algorithms and Monte-Carlo simulations. In: IEEE Porto Power Tech Proceedings, vol. 3 (2001)
2. Miranda, V., Ranito, J.V., Proenca, L.M.: Genetic algorithms in optimal multistage distribution network planning. IEEE Trans. Power Syst. **9**(4), 1927–1933 (1994)
3. Mazhari, S.M., Monsef, H: Dynamic sub-transmission substation expansion planning using learning automata. Electr. Power Syst. Res. **96**, 255–266 (2013)

4. Haffner, S., Pereira, L.F.A., Pereira, L.A., Barreto, L.S.: Multistage model for distribution expansion planning with distributed generation part II: numerical results. IEEE Trans. Power Deliv. **23**(2), 924–929 (2008)
5. Zonkoly, A El.: Multistage expansion planning for distribution networks including unit commitment. IET Gener. Transmiss. Distrib. **7**(7), 766–78 (2013)
6. Shivaie, M., Ameli, M.T., Sepasian, M.S., Weinsier, P.D.: A multistage framework for reliability-based distribution expansion planning considering distributed generations by a self-adaptive global-based harmony search algorithm. Reliab. Eng. Syst. Saf. **139**, 68–81 (2015)
7. Chen, H., Ngoc, T., Yang, W., Tan, C., Li. Y.: Progress in electrical energy storage system : A critical review. Prog. Nat. Sci. **19**, 291–312 (2009)
8. Bortolini, M., Gamberi, M., Graziani, A.: Technical and economic design of photovoltaic and battery energy storage systems. Energy Convers. Manag. **86**, 81–92 (2014)
9. Cossi, A.M., Da Silva, L.G.W., Lazaro, R.A.R., Mantovani, J.R.S.: Primary power distribution systems planning taking into account reliability, operation and expansion costs. IET Gener. Transm. Distrib. **6**(2), 274–284, (2012)
10. Saboori, H., Hemmati, R., Jirdehi, M.A.: Reliability improvement in radial electrical distribution network by optimal planning of energy storage systems. Energy **93**, 2299–2312 (2015)
11. Georgilakis, P.S., Hatziargyriou, N.D.: A review of modern power distribution planning in the modern power systems era: models, methods and future research. Elect. Power Syst. Res. **121**, 89–100 (2015)
12. Ramirez-Rosado, I.J., Dominguez-Navarro, J.A.: Possibilistic model based on fuzzy sets for the multiobjective optimal planning of electric power distribution networks. IEEE Trans. Power Syst. **19**(4), 1801–1810 (2004)
13. El, khattam., W., Hegazy,Y. G., Salama, M. A.: An integrated distributed generation optimization model for distribution system planning. IEEE Trans. Power Syst. **20**(2), 1158–1165, (2005)
14. Ouyang, W., Cheng, H., Zhang, X., Yao, L.: Distribution network planning method considering distributed generation for peak cutting. Energy Convers. Manag. **51**(12), 2394–2401 (2010)
15. Hemmati, R., Hooshmand, R., Taheri, N.,: Electrical power and energy systems distribution network expansion planning and DG placement in the presence of uncertainties. Int. J. Electr. Power Energy Syst. **73**, 665–673 (2015)
16. Hatta, H., Asari, M., Kobayashi, H.: Study of energy management for decreasing reverse power flow from photovoltaic power systems. In: IEEE PES/IAS Conference Sustainable. Alternative Energy, 1–5 (2009)
17. Ziari, I., Ledwich, G., Ghosh, A., Platt, G.: Integrated distribution systems planning to improve reliability under load growth. IEEE Trans. Power Del. **27**(2), 757–765 (2012)
18. Ram, J.: Reliability and costs optimization for distribution networks expansion using an evolutionary algorithm. IEEE Trans. Power Syst. **16**(1), 111–118 (2001)
19. Sedghi, M., Aliakbar-golkar, M., Haghifam, M.: Electrical power and energy systems distribution network expansion considering distributed generation and storage units using modified PSO algorithm. Int. J. Electr. Power Energy Syst. **52**, 221–230 (2013)
20. Saboori, H., Hemmati, R., Abbasi, V.: Multistage distribution network expansion planning considering the emerging energy storage systems. Energy Convers. Manag. **105**, 938–945 (2015)
21. Billinton, R., Allan, R.N.: Reliability Evaluation of Power Systems. Plenum Press, New York (1984)
22. Kennedy, J., Eberhart, R.: Particle swarm optimization. In: IEEE International Conference on Neural Networks, 1942–1948
23. Bonyadi, M.R., Michalewicz, Z.: Impacts of coefficients on movement patterns in the particle swarm optimization algorithm. IEEE Trans. Evol. Comput. (2016)

Development of Plant Layout for Improving Organizational Effectiveness by Hybridizing GT, TOPSIS and SLP

Biswanath Chakraborty⊙ **and Santanu Das**⊙

Abstract A well-designed and interconnected spatial arrangement is the objective of developing a plant layout which consists mainly of processing units, warehouses and administrative section. A good layout would improve processing of a manufacturing unit. There are numerous constraints like safety, construction design, maintenance, operations of machines, retrofit and pallet load which must be balanced economically towards optimizing organizational effectiveness. The present work is concerned about developing a plant layout which will be effective both economically and functionally. Amalgamating three techniques like group technology (GT), TOPSIS and systematic layout planning (SLP), one model has been developed to create an effective plant layout.

Keywords Plant layout · Spatial arrangement · Retrofit · Pallet load · GT · TOPSIS · SLP

1 Introduction

In the night of 2 and 3 December 1984, the Union Carbide India Limited (UCIL), Bhopal, was shattered by the tragic accident on the leak of toxic methyl isocyanate (MIC) and the court verdict on this issue was announced on 13 May 2010 where C. Sahay, the CBI Counsel, had argued that the tragedy occurred due to defective design of the UCIL's factory and its poor maintenance [1]. It was considered as the world's worst industrial disaster [2]. Usually, the plant layout design had been ignored previously. But increased competition and government regulations have forced the contractors and manufacturers to implement potential strategy at every stage of layout

B. Chakraborty · S. Das (✉)
Department of Mechanical Engineering, Kalyani Government Engineering College, Kalyani 741235, West Bengal, India
e-mail: sdas.me@gmail.com

B. Chakraborty
e-mail: bchakraborti85@gmail.com

© Springer Nature Singapore Pte Ltd. 2020
R. Venkata Rao and J. Taler (eds.), *Advanced Engineering Optimization Through Intelligent Techniques*, Advances in Intelligent Systems and Computing 949, https://doi.org/10.1007/978-981-13-8196-6_45

design so that the optimization of the production process through improvement of layout design can be achieved as well as organizational effectiveness can be improved.

Plant layout designing basically involves the spatial allocation of workstation area with equipment and with proper connectivity among them. A lot of studies were carried out for many decades in order to find out the best possible layout which could be proved as safe as well as optimized from the point of views of construction, connectivity, retrofit and production facilities. Most of the research works on layout design can be divided into three major categories: algorithmic approaches, procedural approaches and a combination of both of them.

The techniques under the algorithmic approach include graph theoretical approach, mixed integer nonlinear programming (MINLP) approach and mixed integer linear programming (MILP) approach. A graph theoretical approach was considered by Jayakumar and Reklaitis [3], where they could establish a relationship between layouts of units of a chemical plant having a single floor with the graph partitioning problem. In 1996, they made an improvement on their previous investigation for multiple floors [4]. In this later work, they focused on the optimal allocation of unit to floors for achieving optimization in general plant layout design. Some researchers explored the safety aspect of the plant layout design. Penteado and Ciric, in 1996, worked on a chemical plant to improve the safety issues by considering some basic ways. They advocated that the safety measure could be confirmed in two ways. They argued that some protection devices can be installed which will be able to reduce the chance of severity of an accident. Otherwise, a barrier wall can be erected by increasing the in-between space of the units which will reduce the probability of spreading an accident from one unit to another [5]. They considered an algorithmic-based mixed integer nonlinear programming (MINLP) approach for making a trade-off between the financial risks associated with an accident. When the units were to be separated for lowering the risk, there would be an increase in the cost of piping and land as well as the cost of installing the protection devices for reducing the risk of an accident would also go up. Georgiadis et al. [6, 7] conducted their research work for determining the exact number of floor of a plant by using the mixed integer linear programming (MILP) problem. Papageorgiou and Rotsein made an improvement for determining the allocation of equipment items to floors through MILP and could provide the layout of each floor in detail [8]. Considering both process plant layout and safety, Patsiatzis et al. conducted a research work using mathematical programming [9–11].

In a procedural approach, research works were done mainly on two techniques. Some of the research works were conducted following different multi-criteria decision-making (MCDM) and multi-attribute decision-making (MADM) procedures. Other group of researchers used another technique like systematic layout planning (SLP) for improving plant layout design.

The third group combined procedural technique (SLP) with the algorithmic approach (e.g. graph-based theory—GBT and pairwise exchange method—PEM) for developing optimized production layout of food industries having small and medium scale. Analytic hierarchy process (AHP) was used as a multi-attribute decision-making methodology to select and sort the manufacturing plant layouts on the basis of priority weights [12]. Out of different layout alternatives, Yang and Kuo made an important contribution in the selection of plant layout by combining a procedural approach with the algorithmic approach for layout design [13]. In 2007, Yang and Hung made an improvement over their study of 2003 by further incorporating other multi-attribute decision-making (MADM) like AHP, TOPSIS and fuzzy TOPSIS to find out a group of best-sorted layout alternative that had already been generated through software Spiral® [14]. In 2008, Azadeh et al. [15] further made an improvement in the work of Yang et al. and used AHP and principal component analysis (PCA) for layout planning. Wiyaratn and Watanapa tried to increase the productivity by improving the plant layout using SLP procedure [16, 17]. Zhou et al. conducted their research work for making improvements in plant layout based on process analysis [18]. Ojaghi et al. in 2015 combined both the procedures like systematic layout processing (SLP), graph-based theory (GBT) and pairwise exchange method (PEM) for optimizing the production layout of food industries of small and medium scale [19]. In 2016, Barnwal and Dharmadhikari conducted their research work for optimizing plant layout using SLP method only [20] for a specific application.

It is true that through the algorithmic approaches like Spiral® and LayOPT®, the layout design of the industries can be made and applied. But, at the same time, it is also argued that these approaches normally oversimplify the constraints related to the layout design issues. These approaches also simplify the objectives to reach a surrogate objective function for achieving the probable solution. On the other hand, through procedural approaches, both qualitative and quantitative objectives can be incorporated in the layout design process [21]. In 2016, Fahad et al. [22] discussed on productivity improvement through SLP which was proposed by Muther in 1961.

Layout designing and its review for determining its effectiveness sometimes become challenging as well as time consuming and error-prone because of its inherent multi-criteria dependence. The main focus of the algorithmic approaches was the minimization of flow distance in order to minimize the material handling cost within the units. On the contrary, procedural approaches relied on experts' opinion and experience on subjective issues. Any single approach may not be sufficient to select an optimized plant layout. However, in 1969, Burbidge presented a paper [23] on the production flow analysis at a seminar where he advocated the concept of dividing the plant of a factory into units on the basis of three methods, namely line layout, group layout and functional layout.

The present research work provides a theoretical approach of optimizing the plant layout by suitably hybridizing GT, TOPSIS and SLP techniques. These techniques are to be used sequentially to arrive at a layout design which can certainly improve the productivity by reducing capital investment in layout design. The investment, thus saved from layout design, can be used in future at the time of retrofit, and this way the overall improvement of the organizational effectiveness can be achieved. There is hardly any initiation found where those three techniques have been combined for the development of plant layout.

The present work suggests an optimized *process plant layout* design where the entire job scenario has to be grouped first on the basis of processes and functionalities into modules or workstations through GT [24]. This layout will be considered as group layout. The workstations generated in this way are to be ranked to find out the relative necessity of closeness of different workstations on the basis of six criteria used by the industries. This ranking will be provided by one of the MCDM techniques, TOPSIS. Finally, on the basis of this relative closeness requirement of each workstation, the layout has been generated using SLP method.

2 Methodology

The process plant layout design is basically based on the following characteristics like construction cost, connectivity cost, retrofit and production facilities [6, 9–12] in the following ways:

i. **Construction Cost**: The cost of occupied land area and the covered square feet area is to be balanced in such a way so that the layout can be made more compact.
ii. **Connectivity Cost**: These costs are related to establish a connection among different groups or modules through piping, pumps, LAN for data transfer, etc., which are to be made simpler for reducing capital investment.
iii. **Retrofit**: Many a time, it requires to add equipment or a new accessory to the existing system or to install a new machine in the existing plant which was either not required or not available at the time of designing the plant layout. Proper allocation of funds is to be made available, and the provision of space must be provided in the layout.
iv. **Production Facilities**: Layout should be made in such a way so that there may be facilities for the movement of goods and operators around the plant machineries. Normally, 40% of the total covered area may be considered as extra aisle space which is to be managed properly.

In order to satisfy the above-mentioned requirements of the industries, one of the main technologies for making the layout effective is to adopt the group layout which is based on group technology. It is concerned with the division of the plant into groups or workstations on the basis of similarities of processes and functions. All the manufacturing units consist of at least three types of departments or sections:

i. Administrative section
ii. Processing section and
iii. Non-production activity section other than administrative section like storage (warehouse and other), restrooms (men and women), locker rooms (men and women), tool crib, and receiving and shipping area.

It is considered that the major gains and the success of the group technology (GT) arise from a simplification of the material flow system. Therefore, the primary objective of implementation of GT is to make the material flow as simple as possible inside each workstation. So, capital may be released through proper implementation of GT in a plant and that released capital can be utilized to upgrade technology in the

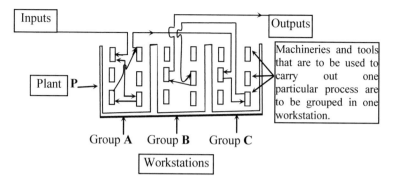

Fig. 1 Group layout based on functions of different families of components on each workstation

long run at retrofit. The following strategies are to be adopted in order to simplify the material flow process and subsequent implementation of GT:

i. Each part should be manufactured/processed, as far as possible, in one group only (Fig. 1). In Fig. 1, the entire plant P is divided into three groups/workstations, namely A, B and C. Each group is created with machineries which are of similar type.

ii. Each type of machine should be installed in one workstation only, as far as practicable.

iii. No restriction on machine usage sequence will be followed.

iv. Processes of incompatible type, in any way, should be processed in different workstations.

v. Any other process, which is not compatible with others, should be treated as exception and should be processed separately. These types of exceptional cases should be avoided, as far as possible.

Suppose one special purpose lathe has been installed at workstation A. So, all the jobs which require processing on a special purpose lathe, be at whatever stage of operation it is, will come at workstation A, get processed through it and go away. This strategy will certainly reduce the probability of duplication of machineries, and thus, better utilization of funds and machines will take place. At the same time, if the number of machines gets reduced, risk of portfolio investment [25] will get reduced. As a result, depreciation and insurance charges, inter alia, accrued on them will also get reduced. This will certainly provide a relief at capital investment front.

Since the entire process is modularized into workstations, it will be easier to replace and maintain at retrofit at a minimum cost without hampering the rest of the production process. Modularization of the entire plant into groups/workstations, thus, will expedite the production process.

3 Implementation of Group Layout Using GT

Group layout can be implemented by adhering to the following strategies:

 i. It will be better if the route cards are renumbered on the basis of rescheduling of operations. Route cards are created to record processing method of every component produced in the factory. It also records all operations and the corresponding machines that will be used from the material issue to final component completion. The routes/path along the networks and the time of processing along that route/path are to be recorded and numbered on the route card. At the time of group layout, the route cards are to be renumbered suitably on the basis of changed workstations' positions.
 ii. The routes may be needed to sort into packs.
 iii. Packs are required to be allocated into groups/workstations.
 iv. The pallet loads of the machines are to be calculated and checked, and machines are to be allocated into workstations accordingly.
 v. To eliminate the exceptional processes by using either of these following methods:

 a. Re-routing the operations of the exceptional processes to other machines in other workstations, if exists.
 b. Changing the method of processing.
 c. The designs of the components may also be modified.
 d. Components may be purchased instead of manufacturing.

 vi. Finally, the flow of system network of the changed layout is to be developed.

4 Implementation of Technique for Order Preference Using Similarity to Ideal Solution (TOPSIS)

GT helps to split up the plant into different workstations and to group those workstations on the basis of their similarities. As soon as the creation and grouping of the workstations get over, the next issue will be to position and place those workstations on the basis of their relative priorities, i.e. which workstation is to be positioned much close to other workstation and which workstation is not to be placed too close. So, a comparative study is required to be carried out for ranking the workstations with respect to a group of criteria that are used as the major constraints during production. In this work, it is proposed to carry out one of the MCDM techniques, TOPSIS, to prioritize the modules or workstations on the basis of certain weights which are used by the industries. The criteria for the selection may be divided into two categories:

 i. Quantitative: Material handling distance denoted by (C_1), adjacency score denoted by (C_2), shape ratio denoted by (C_3)

ii. Qualitative: Flexibility denoted by (C_4), accessibility denoted by (C_5), ease of maintenance denoted by (C_6).

It may be presumed that there are ten different modules or groups or workstations which are considered as ten different alternatives for ranking and are denoted by A_1, A_2, A_3, …, A_{10}. Converting the qualitative/subjective criteria to numbers by using proper scale, the alternatives can be sorted on the basis of weights where weights are to be collected through the unanimous decisions of the industry experts. These six criteria will be like the different parameters through which the objective function can be developed and the problem can be optimized.

At this stage, through TOPSIS, it will be finally possible to rank the workstations on the basis of six above-mentioned criteria.

5 Implementation of Systematic Layout Planning (SLP)

The entire processing unit has been grouped into separate unique modules or workstations through GT, and each of the workstations has been ranked through TOPSIS on the basis of the six above-mentioned criteria which are normally followed in the industries. On the basis of these ranked workstations, the layout can be planned using the activity relationship diagram proposed in SLP by Richard Muther in 1961. The relative closeness of the workstations is to be plotted following a rating scale like this (Table 1):

The advantage of this proposed model is that instead of considering only the material handling effort, which was the general trend of consideration for designing the layout through SLP, six criteria, including the material handling effort in terms of distance, have been considered and on the basis of these six criteria, the workstations have been ranked; e.g. workstation A_3 is unimportant with respect to A_8, and hence, they are not required to be placed closely. The activity relationship diagram (Fig. 2) represents the degree of closeness through alphabets, the reasons of closeness are represented through numbers, and the block diagram of the layout of process plant

Table 1 Closeness value and code representing the reasons behind the closeness value

Value	Closeness	Code	Reason
A	Absolutely essential	1	Face-to-face personal contact
E	Essential but not absolutely essential	2	Meetings, teamwork
I	Important	3	Shared staffing
O	Ordinary closeness	4	Convenience
U	Not important	5	Common type of space
X	Not desirable	6	Reception of visitors

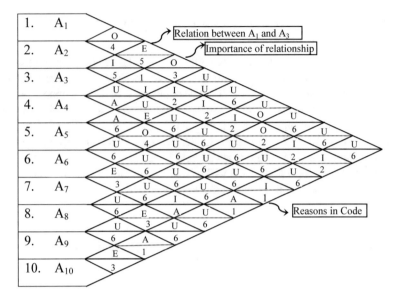

Fig. 2 Diagrammatic representation of the closeness of workstations generated through TOPSIS

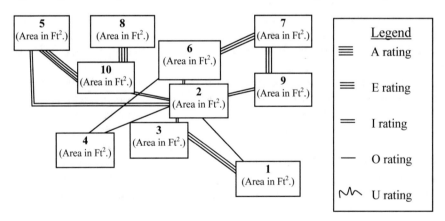

Fig. 3 Space relationship diagram

(Fig. 3) could be drawn on the basis of it. Each numeral in Fig. 3 represents one workstation or group. Generally, 40% area (of the total covered space required by all the workstations) is to be added with this as aisle space.

The diagrammatic representation of Fig. 2 provides the closeness requirements of different workstations. On the basis of this diagram, the space relationship diagram can be developed (Fig. 3). For example, the representation ☐ 1 ▤ 3 ☐ represents the "essential (E)" space relationship between workstations 1 and 3.

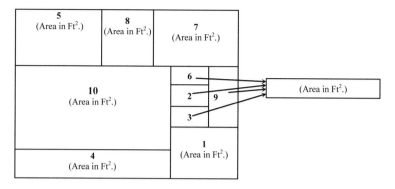

Fig. 4 Block diagram of the process layout of the plant

The space relationship diagram (Fig. 3) provides data to generate the final diagram of the process layout as depicted in Fig. 4. For example, workstations 1 and 3 are placed closed to each other as their relationship has been depicted as "essential".

6 Conclusion

In the present work, an integrated approach of GT, TOPSIS and SLP has been introduced for the first time by the authors to design a cost-effective and flexible process layout of a manufacturing plant. Through group layout, the entire plant can be grouped into several workstations, and machines are allocated to each workstation in such a way that all the similar processes can be done on machine(s) completely which is (are) located in one workstation only. Data collected through this phase will be subsequently supplied to TOPSIS. TOPSIS helps to rank those groups/workstations to ascertain their relative closeness on the basis of 6 (six) criteria used by the industries (constraints). Finally, data related to the ranked workstations will be provided to SLP model. Through SLP model, the activity relationship chart can be developed. Following this chart, the block design of the space relationship diagram and the plant layout can be generated.

The present approach of developing the process plant layout on the basis of six criteria will certainly improve the present scenario on these following issues:

i. Partitioning the plant into workstations will reduce the probability of an accident and, thus, improve the safety issue.
ii. Better utilization of machines can be attained, and thus, fewer machines will be required.
iii. Greater flexibility on both equipment and manpower can be attained for specific tasks.
iv. Reduce the risk of capital investment or portfolio investment.
v. Greater job satisfaction can be achieved by the operators.
vi. Specialized supervision will be possible.

Thus, this new hybrid method can be suitably applied to obtain an optimized plant layout to improve organizational effectiveness.

References

1. Bhopal Gas Tragedy Verdict in June after 26-Year Wait. https://www.thehindu.com/news/national/Bhopal-gas-tragedy-verdict-in-June-after-26-year-wait/article 16301057.ece
2. Bhopal Disaster. https://en.wikipedia.org/wiki/Bhopal_disaster
3. Jayakumar, S., Reklaitis, G.V.: Chemical plant layout via graph partitioning-1 single level. Comput. Chem. Eng. **18**, 441–458 (1994). https://doi.org/10.1016/0098-1354(94)88022-0
4. Jayakumar, S., Reklaitis, G.V.: Chemical plant layout via graph partitioning—II multiple levels. Comput. Chem. Eng. **20**, 563–578 (1996). https://doi.org/10.1016/0098-1354(95)00208-1
5. Penteado, F.D., Ciric, A.R.: An MINLP approach for safe process plant layout. Ind. Eng. Chem. Res. **35**(4), 1354–1361 (1996). https://doi.org/10.1021/ie9502547
6. Georgiadis, M.C., Rotstein, G.E., Macchietto, S.: Optimal layout design in multipurpose batch plants. Ind. Eng. Chem. Res. **36**(11), 4852–4863 (1997). https://doi.org/10.1021/ie9702845
7. Georgiadis, M.C., Schilling, G., Rotstein, G.E., Macchietto, S.: A general mathematical programming approach for process plant layout. Ind. Eng. Chem. Res. **23**(7), 823–840 (1999). https://doi.org/10.1016/S0098-1354(99)00005-8
8. Papageorgiou, L.G., Rotstein, G.E.: Continuous-domain mathematical models for optimal process plant layout. Ind. Eng. Chem. Res. **37**(9), 3631–3639 (1998). https://doi.org/10.1021/ie980146v
9. Patsiatzis, D.I., Papageorgiou, L.G.: Safe Process Plant Layout Using Mathematical Programming. Eur. Symp. Comput. Aided Process Eng. **12**, 295–300 (2002)
10. Patsiatzis, D.I., Papageorgiou, L.G.: Optimal multi-floor process plant layout. Comput. Aided Chem. Eng. **9**, 475–480 (2001). https://doi.org/10.1016/S1570-7946(01)80074-2
11. Patsiatzis, D.I., Knight, G., Papageorgiou, L.G.: An MILP approach to safe process plant layout. Chem. Eng. Res. Des. **82**(5), 579–586 (2004). https://doi.org/10.1205/026387604323142612
12. Abdul-Hamid, Y.T., Kochhar, A.K., Khan, M.K.: An analytic hierarchy process approach to the choice of manufacturing plant layout. Proc. Inst. Mech. Eng, Part B: J. Eng. Manuf. **213**(4), 397–406 (1999). https://doi.org/10.1243/0954405991516868
13. Yang, T., Kuo, C.: A hierarchical AHP/DEA methodology for the facilities layout design problem. Eur. J. Oper. Res. **147**(1), 128–136 (2003). https://doi.org/10.1016/S0377-2217(02)00251-5
14. Yang, T., Hung, C.C.: Multiple-attribute decision making methods for plant layout design problem. Robot. Comput. Integr. Manuf. **23**(1), 126–137 (2007). https://doi.org/10.1016/j.rcim.2005.12.002
15. Azadeh, A., Izadbakhsh, H.R.: A multi-variate/multi-attribute approach for plant layout design. Int. J. Ind. Eng. **15**(2), 143–154 (2008)
16. Wiyaratn, W., Watanapa, A.: Improvement plant layout using systematic layout planning (SLP) for increased productivity. Int. J. Ind. Manuf. Eng. **4**(12), 1382–1386 (2010)
17. Tak, C.S., Yadav, L.: Improvement in Layout Design using SLP of a small size manufacturing unit: a case study. IOSR J. Eng. **2**(10), 1–7 (2012)
18. Zhou, D.W., Wang, L.T., Feng, L.Y., Zhao, D.Z., Guan, T.S.: Research on Improvement of Plant Layout Based on Process Analysis. In: Proceedings of the 21st International Conference on Industrial Engineering and Engineering Management, pp. 253–256 (2014). https://doi.org/10.2991/978-94-6239-102-4_53
19. Ojaghi, Y., Khademi, A., Yusof, N.M., Renani, N.G., bin Syed Hassan, S.A.: Production layout optimization for small and medium scale food industry. Procedia CIRP **26**, 247–251 (2015). https://doi.org/10.1016/j.procir.2014.07.050

20. Barnwal, S., Dharmadhikari, P.: Optimization of plant layout using SLP method. Int. J. Innov. Res. Sci. Eng. Technol. **5**(3), 3008–3015 (2016). https://doi.org/10.15680/IJIRSET. 2016.0503046

21. Muther, R., Hales, L.: Systematic Layout Planning: Management and Industrial Research Publications, 4th edn, Marietta, GA 30067, USA (2015)

22. Naqvi, S.A.A., Fahad, M., Atir, M., Zubair, M., Shehzad, M.M.: Productivity improvement of a manufacturing facility using systematic layout planning. Cogent Eng. **3**, 1–13 (2016). https:// doi.org/10.1080/23311916.2016.1207296

23. Burbidge, J.L.: Production flow analysis. IEEE Xplore Digital Library **50**(4.5):139–152 (1971). https://doi.org/10.1049/tpe:19710022

24. Chakraborty, B., Das, S.: Optimization of production scheduling using critical path method and line of balance in group technology. Ind. Eng. J. **8**(8), 35–42 (2015)

25. Chakraborty, B., Das, S.: Evaluation criteria of project risk and decision making through beta analysis and TOPSIS towards achieving organizational effectiveness. In: Proceedings of the Computational Intelligence, Communications, and Business Analytics, Kalyani, India, p. 71 (2018)

Optimization of Process Parameters Using Hybrid Taguchi and VIKOR Method in Electrical Discharge Machining Process

B. Singaravel⑩, **S. Deva Prasad**⑩, **K. Chandra Shekar**⑩,
K. Mangapathi Rao⑩ **and G. Gowtham Reddy**⑩

Abstract Electrical discharge machining (EDM), based on thermal erosion principle, is used to machine materials in the area of aerospace, surface texturing, die,
and mold manufacturing. In this work, an attempt is made to use oil extracted
from vegetable as dielectric fluid and to carry out optimization study of EDM using
Taguchi-based hybrid optimization technique. Taguchi optimization method is a statistical approach to determine optimum process parameters, but it is appropriate
for solving single-objective problem. Most of the manufacturing industries problems are involved with multiple objectives. Hence, Taguchi method must unite with
some other techniques such as gray relational approach, desirability function, and
utility concept to multiple objectives. Taguchi method is combined with VlseKriterijumska Optimizacija I Kompromisno Resenje in Serbian (VIKOR) for solving
multiple objectives simultaneously. The performance of commercial dielectric fluid
(kerosene) is compared with two different types of vegetable oils such as Jatropha oil
and cottonseed oil. Copper is selected as electrode material which is cryogenically
treated and tempered. The result showed that the proposed approach has easy computational steps for solving multiple objectives simultaneously and the same can be
applicable to other machining process parameter optimizations.

Keywords EDM · Vegetable oil · Cryogenic · Hybrid Taguchi and VIKOR

B. Singaravel (✉) · S. Deva Prasad · K. Chandra Shekar · K. Mangapathi Rao ·
G. Gowtham Reddy
Department of Mechanical Engineering, Vignan Institute of Technology and Science, Hyderabad,
TS, India
e-mail: singnitt@gmail.com

S. Deva Prasad
e-mail: s.devaprasad@gmail.com

K. Chandra Shekar
e-mail: kcschandra2003@gmail.com

K. Mangapathi Rao
e-mail: kmprao63@gmail.com

G. Gowtham Reddy
e-mail: gowthamanreddy@gmail.com

© Springer Nature Singapore Pte Ltd. 2020
R. Venkata Rao and J. Taler (eds.), *Advanced Engineering Optimization
Through Intelligent Techniques*, Advances in Intelligent Systems and Computing 949,
https://doi.org/10.1007/978-981-13-8196-6_46

1 Introduction

EDM is used to machine difficult-to-cut materials in the area of making dies, mold, and tools. In EDM, dielectric fluid plays a major role for producing ionization followed by decomposition, flushing out debris, generation of plasma, cooling the electrode and workpiece. It is known, EDM processes using hydrocarbon-based dielectric fluids generate harmful elements and affecting environment and operator health. Recently, vegetable oil-based dielectric is tried to overcome the above issues in EDM process [1]. In EDM process, minimum electrode wear with superior material removal plays the most important role because of productivity, geometrical dimensions, and cost of the electrode. Cryogenic treatment is a type of heat treatment which is used to enhance the material properties such as wear resistance, reduced residual stress, refinement fine grain size, better electrical properties, toughness and making this process an eco-friendly process [2].

Taguchi method is a unique optimization method which uses orthogonal array and signal-to-noise (S/N) ratio in optimization study of design parameters and improves the quality of the solution. Generally, it is preferable for solving single-objective problems effectively, but real-world problems are multiple objectives. Multi-criteria decision making (MCDM), a statistical approach, is widely used in finding a best solution from the complex solution space of the problem in study and to support the decision making. VIKOR, one of the MCDM methods, also known as a compromise ranking method, is used for ranking and selecting a set of alternatives with conflicting criteria using multi-criteria ranking index. This indexing measures the closeness of a solution with respect to the positive ideal solution. VIKOR, a multi-objective optimization approach to solve multiple criteria problems, provides a near-optimal solution in terms of maximum high group utility for the opponent with its majority and minimum of individual regret [3].

2 Literature Review

Valaki et al.'s [1] study includes sustainability analysis in EDM using biodiesel (Jatropha-based) as dielectric and compared with kerosene. The effect of input parameters such as gap voltage, current, pulse on and pulse off times on process outputs e.g., Material Removal Rate (MRR), surface hardness and Surface Roughness (SR) were investigated. The result concluded that enhanced machining performance was observed using vegetable oil-based dielectric fluid. Ng et al. [4] investigated vegetable-based dielectric in EDM process for sustainability in the process. The effects of canola and sunflower-based biodiesel dielectric in conventional dielectric EDM process are investigated. The result concluded that MRR value was increased with canola and sunflower oils as dielectric at both low- and high-energy settings. Also, the result indicated that EDM process will be more cleaner and sustainable if vegetable oil as dielectric. Also, they stressed about research work is required in the area of different vegetable oils as dielectric and large amount of dielectric must present during machining process.

Kumar et al. [2] conducted the performance analysis of cryogenically treated tool on machinability of Inconel 718 in powder-mixed EDM process. The result revealed that there was a reduction in electrode wear rate and wear ratio. Sundaram et al. [5] studied the effect of cryogenically treated copper electrode on EDM process. Results show that significant in rate of material removal while using treated electrodes.

Majumdar et al. [6] used VIKOR method for optimization of process parameters in coating process. In their study, the effect of Ni-ion concentration, Cu-ion and reducing agent concentration of the chemical bath on an average mass deposition and SR were optimized using VIKOR. The result showed that compromise ranking of the different parametric values was obtained and an optimum combination of the parameters is suggested. Rao [7] optimized wire EDM process parameters using VIKOR method. The influence of pulse-on and pulse-off times along with wire tension and wire feed studied on MRR and SR. The result revealed that pulse-on time was influenced more than other parameters based on their VIKOR index value. Biswas et al. [8] used VIKOR method for welding process parameter optimization, and the result shown by optimum process parameters was achieved for a welded joint. Initially, S/N ratios were calculated based on output values. A decision matrix was generated using calculated S/N ratios and VIKOR method in converting a multi-objective problem into a single-objective problem.

From the literature review, few researchers have studied vegetable oil as dielectric during EDM process. Limited effort has been done in process parameters' optimization using hybrid Taguchi and VIKOR method. Hence, an attempt is made to optimize process parameters and vegetable oil-based dielectric in machining of die steel using EDM process.

3 Experimental Setup

The experiments in the work are conducted on die-sinking EDM machine (Fig. 1). In this work, die steel with grade AISI D2 is selected as workpiece material, which is used for die making and it can be categorized into difficult-to-machine materials. Copper is used as electrode (diameter of 12 mm), and this is cryogenically treated subsequently tempered with the temperature of -196 and 150 °C, respectively. Figure 2 shows the cryogenic and tempering cycle on copper electrode. Process parameters selected to work are types of dielectric (kerosene, sunflower oil, and canola oil), current, and pulse-on time. Table 1 shows the information on selected process parameters and their levels in this work. Taguchi L_9 array is used to decide the number of experiments to be conducted. These values are selected based on the preliminary experiments and literature. Output parameters considered in this study are MRR, TWR, and SR. The values of depth of cut as 1.5 mm and pulse-on time as 100 μs are kept constant in experiments. Table 2 shows the outcomes of the experiments that are conducted.

Fig. 1 EDM machine

Fig. 2 Cryogenic and
tempering cycle of copper
electrode

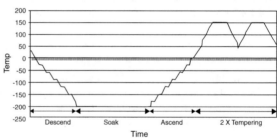

Table 1 Process parameters
in the study and their
respective values at different
levels

Parameters	Units	Level 1	Level 2	Level 3
Dielectric fluid	–	Kerosene	Jatropha oil	Cottonseed oil
Current	A	6	8	10
Pulse-on time	μs	200	300	400

Weight difference between before machining and after machining of workpiece and electrodes is observed using digital weighing machine; Eq. (1) is used to calculate the MRR, and Eq. (2) is used to calculate TWR

$$\text{MRR} = \frac{W_{\text{wbm}} - W_{\text{wam}}}{t \times \rho_{\text{w}}} \text{ mm}^3/\text{min} \tag{1}$$

$$\text{TWR} = \frac{W_{\text{ebm}} - W_{\text{eam}}}{t \times \rho_{\text{e}}} \text{ mm}^3/\text{min} \tag{2}$$

Table 2 Experimental results

Ex. No	Dielectric fluid	Current (A)	Pulse-on time (μs)	MRR (mm^3/min)	TWR (mm^3/min)	SR (μm)
1	Kerosene	6	200	0.2160	0.0314	3.014
2		8	300	0.1801	0.0187	2.778
3		10	400	0.2220	0.0221	2.939
4	Jatropha oil	6	300	0.0921	0.0190	2.883
5		8	400	0.1677	0.0224	2.730
6		10	200	0.1751	0.0272	2.903
7	Cottonseed oil	6	400	0.0994	0.0059	3.081
8		8	200	0.1433	0.0213	3.324
9		10	300	0.1792	0.0237	2.645

where

W_{wbm} = workpiece weight before machining (g),

W_{wam} = workpiece weight after machining (g),

W_{ebm} = electrode weight before machining,

W_{eam} = electrode weight after machining,

t = machining time (min),

ρ_w = density of workpiece material (g/cm^3),

ρ_e = density of electrode material (g/cm^3).

Surface roughness of the machined components is calculated using Talysurf surface roughness tester. The measurements are repeated at three locations on a machined sample, and the average values are considered for understanding the machining performance.

4 Methodology

VIKOR, a MCDM method, is used to obtain near-optimal solutions for a multiple criteria problem. The following procedure is used in hybrid Taguchi and VIKOR method [8]

Step 1 Problem statement to establish objectives of study, in this work, non-beneficial attributes, i.e., minimization objectives, includes SR and TWR while MRR is considered as a beneficial attribute, i.e., a maximization objective.

Step 2 Calculate Signal-to-Noise (*S/N*) ratio. In Taguchi's approach, *S/N* ratio is used to find optimum process parameters for the objective function in the problem of investigation. The ratio between signal mean value and noise standard deviation gives *S/N* ratio in this study. The highest level of *S/N* ratio presents optimum combination of process parameters.

If the objective function is to maximize, larger-the-better characteristics are considered:

$$\frac{S}{N} = -10 * \text{Log}_{10}\left(\frac{1}{n}\sum_{i-1}^{n}\frac{1}{y^2}\right) \tag{3}$$

If the objective function is to minimize, smaller-the-better characteristics are considered:

$$\frac{S}{N} = -10 * \text{Log}_{10}\left(\frac{1}{n}\sum_{i-1}^{n}y^2\right) \tag{4}$$

Table 3 shown below lists the value of S/N ratio in this investigation.

Step 3 Construct decision matrix, which is used to indicate the output value obtained in the experiments. In this work, S/N ratio values are considered as values of decision matrix.

Step 4 Determine the normalized decision matrix, and the purpose of normalization is to convert the original score into a comparable score while considering different nature of parameters. In this work, Eqs. (5) and (6) are considered for normalization, and Table 3 presents the value of normalized matrix. The calculated S/N ratio of selected objectives is normalized using Eqs. (5) and (6).

If the output value is maximization type, then the normalization equation is

$$x_i^*(k) = \frac{x_i^0(k) - \min x_i^0(k)}{\max x_i^0(k) - \min x_i^0(k)} \tag{5}$$

If the output value is minimization type, then the normalization equation is

Table 3 Result of *S/N* ratio and normalized matrix

Sl. No.	S/N ratio			Normalized matrix		
1	−13.310	30.061	−9.582	0.9688	0.0311	0.5715
2	−14.894	34.563	−8.874	0.7619	0.2380	0.2147
3	−13.072	33.112	−9.363	1.0000	0.0000	0.4612
4	−20.724	34.392	−9.196	0.0000	1.0000	0.3770
5	−15.508	32.995	−8.723	0.6817	0.3182	0.1384
6	−15.131	31.302	−9.256	0.7309	0.2690	0.4073
7	−20.047	64.569	−9.773	0.0883	0.9116	0.6677
8	−16.874	33.399	−10.4	0.5031	0.4968	1.0000
9	−14.933	32.503	−8.448	0.7568	0.2431	0.0000

$$x_i^*(k) = \frac{\max x_i^0(k) - x_i^0(k)}{\max x_i^0(k) - \min x_i^0(k)} \tag{6}$$

where $x_i^*(k)$ is a normalized value of the kth element in ith sequence, $x_i^0(k)$ is original sequence of S/N ratio, where $i = 1, 2, 3, \ldots, m$ and $k = 1, 2, \ldots, n$ with $m = 9$ and $n = 3$: $\max x_i^0(k)$ is highest value of $x_i^0(k)$; $\min x_i^0(k)$ is the lowest value of $x_i^0(k)$.

Step 5 Calculate a positive ideal solution (A^*) and a negative ideal solution (A^{**}). The following Eqs. (7) and (8) are used in calculating a A^* and negative A^{**}.

$$\begin{aligned} A^* &= \{(\max N_{ij} | j \in J), (\min N_{ij} | j \in J')\} \\ &= \{N_1^*, N_2^*, \ldots, N_n^*\} \end{aligned} \tag{7}$$

$$\begin{aligned} A^{**} &= \{(\min N_{ij} | j \in J), (\max N_{ij} | j \in J')\} \\ &= \{N_1^{**}, N_2^{**}, \ldots N_n^{**}\} \end{aligned} \tag{8}$$

J and $J' = 1, 2, 3, \ldots, n$.
where J is associated with the benefit criterion and J' is associated with the cost criterion.

Step 6 Determine utility and regret measures using Eqs. (9) and (10). The expression used to find utility measure (S_i) and regret measure (R_i) is,

$$S_i = \sum_{j=1}^n w_j \frac{\left(N_j^* - N_{ij}\right)}{\left(N_j^* - N_j^-\right)} \tag{9}$$

$$R_i = \text{Max}_j \left[w_j \frac{\left(N_j^* - N_{ij}\right)}{\left(N_j^* - N_j^-\right)} \right] \tag{10}$$

where w_j refers to weight of the jth criterion. The values of S_i and R_i are presented in Table 4.

Step 7 Determine the VIKOR index and find by the following expression

$$Q_i = v \left[\frac{(S_i - S^*)}{(S^- - S^*)} \right] + (1 - v) \left[\frac{(R_i - R^*)}{(R^- - R^*)} \right] \tag{11}$$

where Q_i is the VIKOR value of ith alternative, $i = 1, \ldots, m$; $S^* = \text{Min}(S_i)$, $S^- = \text{Max}(S_i)$, $R^* = \text{Min}(R_i)$, $R^- = \text{Max}(R_i)$, and v is the weight of the maximum group utility (and is usually set to be 0.5). Table 4 has VIKOR index values that are calculated and the best (near-optimal) solution is determined among the various alternative having smaller VIKOR index value to help in decision of decision maker.

SI. No.	Utility measure	Regret measure	VIKOR index	Rank
1	0.5233	0.3226	0.7369	8
2	0.4044	0.253	0.2333	3
3	0.4866	0.3330	0.7306	7
4	0.4585	0.3330	0.6885	5
5	0.3790	0.2270	0.0692	1
6	0.468	0.2433	0.2809	4
7	0.5553	0.3035	0.6950	6
8	0.6660	0.3330	1.0000	9
9	0.3330	0.2520	0.1180	2

Table 4 Result of utility measure, regret measure, VIKOR index, and respective rank

Step 8 Rank the alternatives using VIKOR index value

5 Results and Discussions

In this investigation, multiple output parameters considered include MRR, TWR, and SR. Table 2 shows the experimental results obtained. Hybrid Taguchi and VIKOR method are combined to investigate the experimental data with the help of VIKOR index, and its minimum value is used to find out optimum process parameters' combination. Based on the minimum value of VIKOR index, the optimum parameters observed are current as 8 A and pulse-on time as 400 μs with cottonseed oil as dielectric fluid.

MRR can be increased by higher spark energy density which is induced due to lower breakdown voltage at higher current level. In this study, medium level of current value is obtained as optimum, i.e., 8 A; hence, medium value of MRR is observed. Higher TWR is observed due to higher oxygen content of vegetable oil leading highly conductive discharge channel with number of ions striking on the electrode surface [1]. Cryogenically treated electrode gives less tool wear due to the formation of hard carbide compounds on the surface of electrode, refinement of fine grain size than non-treated electrodes; hence, wear resistance and thermal properties of treated electrodes are high. The electric field strength, of cryogenically cooled tool electrode, generates an electrode emission current density which is higher than that is required by conventional electrode. Hence, it leads to minimum tool wear [2]. Vegetable oil has higher thermal conductivity than hydrocarbon oil, hence shallow craters that can be minimized due to better heat transfer. It is used to minimize surface roughness of the machined surface. Vegetable oils do not contain any harmful organics; due to this, the process leads to become sustainable. Results revealed that those vegetable oils as dielectric provide equal or more performance than conventional dielectric. It is providing clean, safe, and environmental free machining environment.

It is noticed that the VIKOR approach is easy to understand and execute when compared with other optimization methods such as genetic algorithm and gray relational analysis. This method uses a simple ratio analysis and fewer mathematical computations to support the decision maker in reaching a near-optimal solution. For this simplicity, VIKOR is widely accepted for various decision-making problems' investigation.

6 Conclusion

The optimization of machining parameters in EDM process of D2 grade die steel by the application of hybrid Taguchi and VIKOR method is considered in this study. The major observations obtained from this study include:

- Hybrid Taguchi and VIKOR method are successfully employed for the estimation of optimum process parameters, and VIKOR index value is used to find out optimum parameters.
- The optimized process parameters obtained current as 8 A, pulse-on time as 400 μs with cottonseed oil as dielectric fluid.
- Cryogenic treatment on electrodes is used to enhance the properties of electrodes, and reduction tool wear is observed.
- In any manufacturing practice, environmental impact, personal health, and operator safety are a prime concern. In this work, it is imposed the sustainability in EDM process through vegetable oil as dielectric fluid.
- The proposed experimental and statistical approach was simple, useful, and reliable methodology in optimizing process parameters. Also this method could be extended to study other machining process parameters.

Acknowledgements The authors would like to thank IEI R&D cell for supporting and funding this research work. Ref.: RDPG2017006.

References

1. Valaki, J.B., Rathod, P.P., Sankhavara, C.D.: Investigations on technical feasibility of *Jatropha curcas* oil based bio dielectric fluid for sustainable electric discharge machining (EDM). J. Manuf. Process. **22**, 151–160 (2016)
2. Kumar, A., Maheshwari, S., Sharma, C., Beri, N.: Machining efficiency evaluation of cryogenically treated copper electrode in additive mixed EDM. Mater. Manuf. Process. **27**, 1051–1058 (2012)
3. Tong, L.I., Chen, C.C., Wang, C.H.: Optimization of multi-response processes using the VIKOR method. Int. J. Adv. Manuf. Technol. **31**, 1049–1057 (2007)
4. Ng, P.S., Kong, S.A., Yeo, S.H.: Investigation of biodiesel dielectric in sustainable electrical discharge machining. Int. J. Adv. Manuf. Technol. **90**, 2549–2556 (2017)
5. Sundaram, M.M., Yildiz, Y., Rajurkar, K.P.: Experimental study of the effect of cryogenic treatment on the performance of electro discharge machining. In: ASME International Manufacturing

Science and Engineering Conference. American Society of Mechanical Engineers, pp. 215–222 (2009)

6. Majumdar, G., Biswas, N., Pramanik, A., Sen, R.S., Chakraborty, M.: Optimization of mass deposition and surface roughness for ternary Ni-Cu-P electroless coating using VIKOR method. Adv. Mater. Process. Technol. **3**, 186–195 (2017)

7. Rao, C.M.: Effect of WEDM process parameters on the multiple responses using VIKOR analysis. J. Recent Act. Prod. 3 (2018)

8. Biswas, S.A., Datta, S., Bhaumik, S., Majumdar, G.: Application of VIKOR Based Taguchi Method for Multi Response Optimization: a case study in submerged arc welding (SAW). In: Proceedings of the International Conference on Mechanical Engineering. pp. 26–28 (2009)

Multi-response Optimization of Burnishing of Friction-Welded AA6082-T6 Using Principal Component Analysis

R. S. Tajane and P. J. Pawar

Abstract Ball burnishing is employed as post-welding treatment for AA6082-T6 friction-welded part to enhance surface and surface properties. In this paper, the principal component analysis was employed as a tool of multi-response optimization, to investigate the effect of control parameters on multiple responses of burnishing process. Four controllable factors such as burnishing speed, burnishing feed, burnishing force, and number of passes at five levels each and three responses such as surface roughness, surface hardness, and tensile strength were studied. The optimum combination of control parameters and their levels for multiple responses based on the total principal component was determined. The analysis of variance was used to find out the most influential burnishing parameter for the multiple responses problems. The overall performance index of optimal level of parameters (0.9562) is calculated; it reveals that the principal component analysis can effectively acquire the optimal combination of burnishing parameters.

Keywords Friction welding · Ball burnishing · Multi-response optimization · Principal component analysis · Analysis of variance

1 Introduction

Aluminum alloys of 6000 series and their composites are used in oil refineries, automobile, novel and space application [1]. Gas tungsten arc welding of AA6082 structure reduces the strength and forms solidification cracking in their fusion and heat-affected zone. It also produces distortion and poor surface properties that restrict

R. S. Tajane (✉) · P. J. Pawar
Department of Production Engineering, K. K. Wagh Institute of Engineering Education and Research, Nasik, Maharashtra, India
e-mail: rstajane@yahoo.co.in

P. J. Pawar
e-mail: pjpawar1@rediffmail.com

Savitribai Phule Pune University, Pune, Maharashtra, India

© Springer Nature Singapore Pte Ltd. 2020
R. Venkata Rao and J. Taler (eds.), *Advanced Engineering Optimization Through Intelligent Techniques*, Advances in Intelligent Systems and Computing 949, https://doi.org/10.1007/978-981-13-8196-6_47

its application in various fields [1, 2]. However, these problems can be reduced to enhance its durability and reliability in their applications by using friction welding for joining and burnishing as a post-welding treatment to improve the strength of the weld and surface properties. Friction welding though induced low tensile stress 100 MPa at surface and 200 MPa at middle as compared to conventional welding is prone to corrosion and fatigue failure. It also reduces surface hardness at heat-affected zone (HAZ) [3]. Furthermore, Al and Al alloy cannot be finished by grinding or any chip produced processes because of clogging and poor machinability. Burnishing, however, can be used to overcome these problems. It improves not only the surface finish but also the surface mechanical properties. Burnishing improves hardness and replaces the tensile stress induced during friction welding by deep and high magnitude compressive stress (−450 MPa) with low amount of cold work at low cost [4]. Abrasive machining methods such as grinding, lapping, and honing are generally used to improve the surface finish and not the surface properties. Moreover, these methods induce residual tensile stresses at the surface that deteriorates the fatigue performance of the part [5, 6]. On the other hand, shot peening and laser peening improve surface hardness and fatigue life by inducing residual compressive stresses at the surface but with a large amount of cold work. Furthermore, the peening deteriorates the surface finish that is susceptible to corrosion damage [7, 8]. However, burnishing can find a solution to the above problem. It is easy and low-cost process to enhance the surface characteristics by plastically deforming picks into valleys. Besides producing the better surface finish, burnishing also improves surface hardness, corrosion and wear resistance, immunity to stress corrosion cracking, and fatigue life by inducing residual compressive stresses [9, 10]. Burnishing is applied to enhance surface finish and produces deep compressive residual stress to improve fatigue and corrosion fatigue performance of FSW aluminum alloy [3, 4]. Burnishing can be easily opted to friction stir welding and friction stir process (FSP) components, either simultaneously or as a post-weld process on conventional machines, CNC machine, or with robot. [11]. Yu and Wang [12] analyzed the effect of burnishing parameters on the surface roughness of aluminum alloy by using a polycrystalline diamond ball. The effect of burnishing parameters on roughness, hardness, and fatigue life of the AA6061-T6 component was investigated by El-Axir [13]. Loh and Tam [14] conducted experiment and identified burnishing speed, burnishing feed, burnishing force, number of passes, ball material and ball size, workpiece material, and lubricant as the key parameters that affect the surface properties of burnished part. Taguchi technique [15], a single-response optimization tool, is used widely by the researcher in the recent year for improving the process performance. However, today's engineering scenario focuses on multi-response optimization rather than single-response optimization; therefore, many researchers have tried to combine Taguchi method with other methods to optimize several responses simultaneously [16–20]. Zhang et al. [16] used a regression method to optimize all process parameters and levels simultaneously. El-Khabeery and El-Axir [17] used response surface methodology (RSM) with group method of data handling technique (GMDH). Sagbas [18] implemented desirability function approach together with RSM to minimize surface roughness in ball burnishing. Esme [19] employed gray relation analysis in which weight of responses was determined by AHP method. El-Axir et al. [20] used multifactor experimental design that has been proposed by Box and Hunter. Researchers [21–23] used the principal component analysis (PCA) with Taguchi method as a tool to solve the multi-response problem. PCA statistically converts original performances into principal components (eigenvectors). In those case studies, the first principal component having eigenvalue greater than one is considered for optimization using Taguchi method. However, it is unsure whether the variable level found by the first principal component is optimal or not.

In this paper, the total principal component (TPC) is extracted from all principal component and their weights are statistically determined from data itself. TPC converts multi-response problem into single-response problem, and Taguchi tool is applied to determine the factor levels. The analysis of variance (ANOVA) was conducted to find out the most important burnishing parameter for the problem of multiple responses.

2 Application Example

2.1 Material

The aluminum alloy AA6082-T6 has chemical composition as given in Table 1. The material in the form of rolled bar of 15 mm diameter was used for the experimentation.

2.2 Process Parameters

In this study, burnishing speed, burnishing feed, burnishing force, and number of passes are used as process parameters at five levels each to investigate its effects. The process parameters and their levels are given in Table 2. Taguchi suggested 25 number of experiments for this combination of factors and levels [24]. Therefore, L_{25} orthogonal array is sufficient to investigate the complete burnishing experimental design space.

Table 1 Chemical composition of aluminum AA6082-T6 alloy (% weight)

	Si	Fe	Cu	Mn	Mg	Cr	Zn	Ti	Al
Min	0.70	–	–	0.40	0.60	0.04	–	–	Reminder
Max	1.30	0.50	0.10	1.00	1.20	0.15	0.20	0.20	

Table 2 Process parameters and their levels

Code	Parameter	Unit	Level 1	Level 2	Level 3	Level 4	Level 5
A	Speed (v),	m/min	13.0	20.0	30.0	47.0	75.0
B	Feed (f),	mm/rev	0.045	0.071	0.112	0.18	0.28
C	Force (F),	N	60	118	175	235	300
D	No. of passes	–	1	2	3	4	5

2.3 Process Performance Parameters

In this research, the attempt was made to study the effect of burnishing parameter on surface roughness (Ra), hardness (Hv), and tensile strength (TS) of friction-welded AA6082-T6 part. The surface roughness indicates the quality of the burnished surface, whereas hardness and tensile strength are indicators of surface topographical changes caused by the burnishing process.

3 Determination of Optimal Process Parameter Using PCA

PCA, a statistical tool of multi-response optimization, is used in the paper to find out the best control burnishing parameters' level combination to optimize surface roughness, hardness, and tensile strength of friction-welded AA6082-T6 part.

3.1 Detail of Experiment

AA6082-T6 parts are cut to the size and joined by friction welding process. Impurities at face of joint come out in the form of flash. Flash is removed, and job is turned on lathe, with the same setting burnishing applied as a post-weld treatment. The 25 experiment was conducted as per L_{25} orthogonal array, and PCA was applied to investigate the multi-response problem. All principal components are integrated into one total principal component using weights of each principal component.

3.1.1 Friction Welding

Friction welding was performed on friction welding technology (FWT) machine (model T40). In the friction welding, one of the parts rotates at a constant speed ($s = 1300$ rpm), while other is stationary. The stationary part is axially advanced against the rotating part with friction pressure ($P_f = 11$ bar) for a predetermine friction time ($t_f = 2.5$ s). This much time is sufficient to bring the end of both the parts in plastic state. Then, the rotary component is brought to stop within the braking time, and the axial pressure on the stationary part is increased to an upset pressure ($P_u = 16$ bar) for a calculated upset time ($t_u = 3$ s). Upset pressure forces out the impurities at the contact surface into flash to produce good quality weld. Rotational speed, friction pressure, friction time, forging pressure, and forging time are the most significant parameters in the friction welding operation.

3.1.2 Burnishing

A burnishing tool shown in Fig. 1 has two balls that exert a force on workpiece by using spring of 6 kg/mm stiffness. It has been specially designed and developed for two passes at a time to reduce the burnishing time and can be easily attached on conventional lathe [18]. In the present work, a conventional lathe (*Kirloskar make, model: Turnmaster 40*) was used which shows simplicity and adaptability of the burnishing technology for the commercial application in rural and small-scale industries. The lathe has maximum speed of 1600 rpm and 3.5 kW spindle power. Initially, AA6082-T6 bars were turned with cutting speed of 175 m/min and feed rate of 0.045 mm/rev using kerosene as lubricant. A CNMG 120408 carbide insert was used for turning to have better finish because the burnishing response depends on the initial surface roughness [25]. The preburnished average roughness in the range 0.532–0.715μm was measured with Mitutoyo HM-200 Series 'Surfcom 130A' surface tester. The surface hardness is measured on micro Vickers hardness testing machine, the hardness obtained was around 100.53 Hv, and tensile strength was around 238 N/mm². In this research, ball burnishing experiments are performed as per plan shown in Table 3 using kerosene.

The workpiece to be burnished was held in chuck; cutting tool was mounted on tool post and burnishing tool on cross-slide as shown in Fig. 1. Workpiece was first turned with set values of parameters, then the burnishing is carried out without

Fig. 1 Combined burnishing and turning process on lathe

Table 3 L_{25} orthogonal array, measured responses

Run no.	Process parameters				Experimental results			S/N ratio		
	A	B	C	D	Ra	Hv	TS	Ra	Hv	TS
					(μm)	(Hv)	(N/mm^2)			
1	1	1	1	1	0.149	131.84	320.58	16.5363	42.4008	50.1187
2	1	2	2	2	0.096	119.80	321.00	20.3245	41.5694	50.1301
3	1	3	3	3	0.084	123.47	316.58	21.5144	41.8316	50.0097
4	1	4	4	4	0.110	131.34	317.88	19.1459	42.3681	50.0453
5	1	5	5	5	0.160	139.26	288.33	15.8995	42.8768	49.1978
6	2	1	2	3	0.168	116.57	299.82	15.4938	41.3316	49.5372
7	2	2	3	4	0.102	120.09	293.30	19.7997	41.5902	49.3462
8	2	3	4	5	0.115	130.52	285.58	18.7609	42.3137	49.1146
9	2	4	5	1	0.104	132.34	291.57	19.6593	42.4336	49.2949
10	2	5	1	2	0.139	119.23	313.92	17.1189	41.5278	49.9364
11	3	1	3	5	0.112	128.75	303.23	18.9898	42.1946	49.6354
12	3	2	4	1	0.080	128.75	283.48	21.9745	42.1946	49.0504
13	3	3	5	2	0.127	124.69	326.21	17.9012	41.9163	50.2699
14	3	4	1	3	0.082	116.84	324.90	21.7591	41.3522	50.2350
15	3	5	2	4	0.109	125.61	281.20	19.2515	41.9801	48.9803
16	4	1	4	2	0.118	109.47	289.28	18.5624	40.7859	49.2264
17	4	2	5	3	0.133	127.79	283.88	17.5230	42.1300	49.0627
18	4	3	1	4	0.089	123.93	314.20	21.0448	41.8633	49.9441
19	4	4	2	5	0.116	132.84	312.73	18.7108	42.4665	49.9034
20	4	5	3	1	0.178	131.51	300.98	14.9916	42.3790	49.5708
21	5	1	5	4	0.192	125.76	296.07	14.3189	41.9908	49.4279
22	5	2	1	5	0.124	134.87	320.53	18.1316	42.5985	50.1174
23	5	3	2	1	0.132	123.03	249.04	17.5885	41.7999	47.9254
24	5	4	3	2	0.151	136.43	271.63	16.4013	42.6982	48.6796
25	5	5	4	3	0.125	134.87	280.30	18.0850	42.5985	48.9525

removing the workpiece from the lathe chuck to retain the same turning alignment and to avoid the roundness error. Literature reveals that surface finish improves with large diameter ball, whereas small diameter balls are effective in augmenting the surface hardness [8, 26–28]. A carbon chromium steel ball of 12 mm diameter freely available from ball bearing is used successfully, which has the effect on both surface finish and hardness.

3.2 Application of PCA

Taguchi divides the process responses into three characteristics: nominal the best, larger the better, and smaller the better. If the surface roughness value measured in this study increases then it become worse, such a quality characteristics is said to be smaller the better. Hardness and tensile strength become better as the value of the measure increases; these responses are said to have larger-the-better characteristics. The problems of multi-response optimization were thus investigated using PCA and Taguchi smaller-the-better and larger-the-better methodologies as follows. The PCA technique was first presented by Pearson (1901) and further developed by Hotelling (1933). The PCA and Taguchi method are combined in this study to deal with multi-response problems. First, the multi-response observations were converted into the *S/N* ratios using Eq. (1) for lower-the-better and Eq. (2) for higher-the-better characteristics.

$$\eta_{ij} = -10 \log_{10} \left(\frac{1}{n} \sum_{j=1}^{n} y_{ij}^2 \right) \tag{1}$$

$$\eta_{ij} = -10 \log_{10} \left(\frac{1}{n} \sum_{j=1}^{n} y_{ij}^{-2} \right) \tag{2}$$

where y_{ij} is the *i*th experiment at the *j*th test and n is the total number of the tests/reading for the given (*j*th) response. The *S/N* ratios for all responses in each trial run of L_{25} are calculated and given in Table 3.

Then, *S/N* ratio is normalized using Eq. (3) to overcome the problem of the difference between units and summarized in Table 5.

$$x_i^*(j) = \frac{x_i(j) - x_i(j)^-}{x_i(j)^+ - x_i(j)^-} \tag{3}$$

where $x_i^*(j)$ is the normalized response, $x_i(j)^+$ is the maximum of $x_i(j)$, and $x_i(j)^-$ is the minimum of $x_i(j)$.

Then, the correlation coefficient matrix was calculated by Eq. (4) from the normalized response matrix and listed in Table 4.

$$R_{jl} = \left(\frac{\text{Cov}\left[x_i^*(j), x_i^*(l)\right]}{\sigma_{x_i^*(j)} \times \sigma_{x_i^*(l)}} \right), \quad j = 1, 2, \ldots, m; \quad i = 1, 2, \ldots, n \tag{4}$$

where $\text{Cov}\left[x_i^*(j), x_i^*(l)\right]$ is the covariance of sequences $x_i^*(j)$ and $x_i^*(l)$; $\sigma_{x_i^*(j)}$ and $\sigma_{x_i^*(l)}$ are the standard deviation of sequence $x_i^*(j)$ and $x_i^*(l)$, respectively.

Table 4 Correlation coefficient matrix, weights, and eigenvectors

	Correlation coefficient matrix			Eigenvalue	Weights	Eigenvector V_k		
	Ra	Hv	TS			PC1	PC2	PC3
Ra	1	−0.2438	0.2184	$\lambda_1 =$ 1.4024	0.4675	0.7728	−0.0693	0.6309
Hv	−0.2438	1	−0.1365	$\lambda_2 =$ 0.8648	0.2886	0.5176	0.6441	−0.5633
TS	0.2184	−0.1365	1	$\lambda_3 =$ 0.7329	0.2443	−0.3673	0.7618	0.5335

Eigenvalue and eigenvectors of correlation matrix are evaluated by using MAT-LAB and Eq. (5) and listed in Table 4. The three elements of the eigenvector are the weights of the three responses.

$$(R - \lambda_k I_m)V_{ik} = 0 \qquad (5)$$

where λ_k is eigenvalues, $\sum_{k=1}^{n} \lambda_k = n$, $k = 1, 2, \ldots, n$; $V_{ik} = [a_{k1}a_{k2}\cdots a_{kn}]^T$ is eigenvectors corresponding to the eigenvalue λ_k.

The first principal component (PC) is the sum of the products of the normalized value of three responses multiplied by the elements of the eigenvector, which related to the biggest eigenvalue. The other PCs were obtained in the same way using Eq. (6). The results are all listed in Table 5.

$$PC_{mk} = \sum_{i=1}^{n} X_m^*(i) \cdot V_{ik} \qquad (6)$$

The principal components are produced in order to reduce variance, and hence, the first principal component, PC_{m1}, accounts for maximum variance in the data. The first principal component for fourth run (PC_{41}) is calculated as:

$$PC_{41} = \sum_{i=1}^{n} X_4^*(i).V_{i1}$$
$$= 0.7728 \times 0.6305 + 0.5176 \times 0.7567 + (-0.3673) \times 0.9042 = \mathbf{0.5468}$$

Coefficients of determination (weights) of the principal components are evaluated from eigenvalue using Eq. (7) and listed in Table 4.

$$W_k = \frac{\lambda_k}{\sum_{k=1}^{n} \lambda_k} = \frac{\lambda_k}{n}, \quad k = 1, 2, \ldots, n \qquad (7)$$

where, W_k, represents the weight of the principal component, PC_{mk}. Weight of the first principal component is calculated as:

Table 5 Normalized value of process characteristics, principal components, and TPC

Run no.	Normalized response			Principal component			Total principal component
	Ra	Hv	TS	PC1	PC2	PC3	TPC
1	0.2896	0.7724	0.9355	0.2800	1.1901	0.2468	0.5342
2	0.7845	0.3747	0.9404	0.4548	0.9034	0.7855	0.6649
3	0.9399	0.5001	0.8890	0.6587	0.9342	0.7855	0.7691
4	0.6305	0.7567	0.9042	0.5468	1.1325	0.4539	0.6930[a]
5	0.2065	1.0000	0.5427	0.4778	1.0432	−0.1435	0.4890
6	0.1535	0.2610	0.6875	0.0012	0.6812	0.3166	0.2743
7	0.7159	0.3847	0.6060	0.5298	0.6598	0.5583	0.5742
8	0.5802	0.7307	0.5072	0.6403	0.8168	0.2251	0.5897
9	0.6976	0.7880	0.5841	0.7324	0.9042	0.3078	0.6782
10	0.3657	0.3548	0.8577	0.1513	0.8566	0.4885	0.4370
11	0.6101	0.6737	0.7294	0.5523	0.9473	0.3945	0.6276
12	1.0000	0.6737	0.4799	0.9453	0.7302	0.5074	0.7763
13	0.4679	0.5406	1.0000	0.2741	1.0776	0.5242	0.5668
14	0.9719	0.2708	0.9851	0.5294	0.8575	0.9861	0.7356
15	0.6443	0.5712	0.4499	0.6283	0.6660	0.3248	0.5650
16	0.5543	0.0000	0.5549	0.2245	0.3843	0.6457	0.3735
17	0.4185	0.6428	0.4851	0.4780	0.7546	0.1607	0.4802
18	0.8786	0.5153	0.8610	0.6294	0.9269	0.7234	0.7381
19	0.5737	0.8038	0.8437	0.5495	1.1206	0.3593	0.6677
20	0.0879	0.7619	0.7018	0.2045	1.0193	0.0006	0.3896
21	0.0000	0.5763	0.6408	0.0629	0.8594	0.0173	0.2813
22	0.4980	0.8669	0.9349	0.4902	1.2361	0.3247	0.6648
23	0.4271	0.4850	0.0000	0.5811	0.2828	−0.0037	0.3522
24	0.2720	0.9146	0.3217	0.5654	0.8153	−0.1720	0.4573
25	0.4919	0.8669	0.4381	0.6680	0.8580	0.0558	0.5732

[a]Sample calculations are given below

$$W_1 = \frac{\lambda_1}{\sum_{k=1}^{n} \lambda_k} = \frac{\lambda_1}{n} = \frac{1.4024}{1.4024 + 0.8648 + 0.7329} = \frac{1.4024}{3} = 0.4675$$

Antony [29] used the first principal component having eigenvalue larger than one to convert multi-response problem to single-response. Taguchi method was applied to this first principal component to obtain optimal level of the control parameter. In this study, more than one principal component have an eigenvalue larger than one which is a case of today's complex manufacturing processes. To deal with this situation, more than one principal component rather all are integrated into total principal component (TPC) by using the coefficient of determination (weights).

The total principal component (TPC) based on all principal components and their weight is determined by Eq. (8). It is the sum of products of the principal component and their weights. TPC are selected to replace the original responses for further analysis and recorded in Table 5. The total principal component index (TPC_i) corresponding to ith experimental run is computed as follows.

$$TPC_i = \sum_{k=1}^{n} PC_i(k) \times W(k) \tag{8}$$

where $PC_i(k)$ is the principal component of ith run and kth characteristic; $W(k)$ is weight of kth characteristic. TPC for fourth run is calculated as:

$$TPC_4 = \sum_{k=1}^{n=3} PC_i(k) \times W(k)$$
$$= (0.5468) \times 0.4675 + (1.1325) \times 0.2883 + (0.4539) \times 0.2443 = 0.6930$$

4   Result and Discussion

4.1   Analysis of Variance

ANOVA is conducted using statistical software to determine which control variable significantly affects the process responses. The results of ANOVA for the total principal component are listed in Table 6. Based on ANOVA of total principal component, the recommended level of burnishing speed, burnishing feed, burnishing force, and number of passes was A_3, B_4, C_1, and D_5, respectively. Also, the two parameters, burnishing feed and burnishing speed, are found to be the main factors with the selected multiple performance characteristics.

4.2   Confirmation Tests

Once the optimal burnishing parameters are identified, experiment was conducted at optimal level to verify the improvement of the process responses using this optimal combination. Optimal level test result is shown in Table 7.

Validation experiment of optimal solution ($A_3B_4C_1D_5$) was conducted, the results are included in the orthogonal array, and the overall performance index of optimal condition is compared with experiments conducted.

Table 6 Response table and ANOVA of total principal component

Code	Levels					Optimal level	Degree of freedom	Sum of squares	Mean square	F	%
	1	2	3	4	5						
A	0.630	0.510	**0.654**	0.529	0.465	3	4	0.1300	0.0325	8.23	24.61
B	0.418	0.632	0.603	**0.646**	0.490	4	4	0.1970	0.0493	12.47	37.30
C	**0.621**	0.504	0.563	0.601	0.499	1	4	0.0614	0.0153	3.88	11.62
D	0.546	0.499	0.566	0.570	**0.607**	5	4	0.0311	0.0078	1.97	5.89
Error							8	0.1087	0.0136	3.44	20.58
Total							24	0.5282			100.00

Table 7 Test reading of optimal condition $A_3B_4C_1D_5$

	Process parameters				Experimental results		
	A	B	C	D	Surface roughness (μm)	Surface hardness (Hv)	Tensile strength (N/mm²)
Optimal level	3	4	1	5	0.084	130.34	324.58

Table 8 Overall performance index (Z)

Run no.	Process parameters				Experimental results			Normalized value			Z index
	A	B	C	D	Ra	Hv	TS	Ra	Hv	TS	
5	1	5	5	5	0.160	139.26	288.33	0.4969	1.0000	0.8839	0.7364
12	3	2	4	1	0.080	128.75	283.48	1.0000	0.9245	0.8690	0.9462
13	3	3	5	2	0.127	124.69	326.21	0.6257	0.8953	1.0000	0.7948
Optimal level	3	4	1	5	0.084	130.34	324.58	0.9484	0.9359	0.9950	**0.9562**

4.3 Overall Performance Index

The overall performance index (Z) is calculated for all experiment runs. It is the sum of the product of weights and normalized value of quality characteristics. Experimental run no. 12 ($A_3B_2C_4D_1$) among all 25 run has largest Z index as follow:

$$Z_{12} = 0.4675 \times 1.000 + 0.2883 \times 0.9245 + 0.2443 \times 0.8690 = 0.9462$$

It is also observed from Table 8 that individual quality characteristics achieved in optimal level are inferior as compared to experimental run no. 5, 12, and 13. For example, run no. 12 has 0.080 μm as a lowest roughness as related to the 0.084 μm attained in optimal condition. Similarly, experimental run no. 5 and 13 have 139.26 Hv, 326.21 N/mm² largest value for hardness and tensile strength respectively as compared to 130.34 Hv and 324 N/mm² in optimal level combination. However, the overall performance index at optimal level is 0.9507, and it is largest among all experiment run. This concludes the multi-responses are optimized by $A_3B_4C_1D_5$ level of control parameters. It confirms and validates the result.

5 Conclusions

It is very difficult to improve all these characteristics, namely surface roughness, hardness, and tensile strength, simultaneously due to their conflicting nature. Hence, an attempt is made in research work to apply principal component analysis for multi-response optimization of burnishing process. The effect of four parameters, namely burnishing speed, burnishing feed, burnishing force, and number of passes, is investigated. The result of multi-response optimization shows that the optimal level of parameter should be $A_3B_4C_1D_5$. It can be understood from the above level combination that:

1. Smaller value for burnishing force (60 N) is recommended with large number of passes to avoid the surface cracking because of excessive strain hardening at surface. This may increase the surface roughness.
2. Smaller force at medium speed and feed will not provide sufficient time to deform material to harden it; hence, five burnishing passes are recommended to deform it successively to increase the strain hardening and tensile strength.
3. Moderate value of speed (30 m/min) and feed (0.18 mm/rev.) does not allow the sufficient time to deform material plastically. However, the number of passes at low force will successively deform the material and increase the hardness and tensile strength.
4. The surface hardness, surfaces roughness and tensile strength of experiment number 5, 12 and 13 is observed to be better as compared to PCA optimal level. However PCA performance index (0.9562) is better as shown in Table 8.
5. It observed from ANOVA that burnishing feed and speed have highest contribution, and the number of passes has lowest contribution in multi-response optimization.

References

1. Wang, B., Xue, S., Ma, C., Han, Y., Lin, Z.: Effect of combinative addition of Ti and Sr on modification of AA4043 welding wire and mechanical properties of AA6082 welded by TIG welding. Trans. Nonferrous Met. Soc. China **27**(2), 272–281 (2017). https://doi.org/10.1016/S1003-6326(17)60031-1
2. Idris, M., Ismail, S., Bahari, M.J., Shuib, N.: Thermal deformation of gas metal arc welding on aluminum alloy T-joints. Eng. Solid Mech. 21–26 (2018). https://doi.org/10.5267/j.esm.2017.11.003
3. Hornbach, D., Mason, P.: Improving corrosion fatigue performance of AA2219 friction stir welds with low plasticity burnishing. In: International Surface Engineering Conference. pp. 1–7 (2002)
4. Prevéy, P., Mahoney, M.: Improved fatigue performance of friction stir welds with low plasticity burnishing. Residual stress design and fatigue performance assessment. In: Proceedings Thermec. pp. 1–7 (2003)
5. Loh, N.H., Tam, S.C.: Effects of ball burnishing parameters on surface finish—a literature survey and discussion. Precis. Eng. **10**(4), 215–220 (1988). https://doi.org/10.1016/0141-6359(88)90056-6
6. Yeldose, B.C., Ramamoorthy, B.: An investigation into the high performance of TiN-coated rollers in burnishing process. J. Mater. Process. Technol. **207**(1–3), 350–355 (2008). https://doi.org/10.1016/j.jmatprotec.2008.06.058
7. Scheel, J., Prevéy, P., Hornbach, D.: Safe life conversion of aircraft aluminum structures via low plasticity burnishing for mitigation of corrosion related failures. In: Presented at Department of Defense Corrosion Conference, Washington DC. pp. 1–16 (2009)

8. Hassan, A.M., Momani, A.M.: Further improvements in some properties of shot peened components using the burnishing process. Int. J. Mach. Tools Manuf. **40**(12), 1775–1786 (2000). https://doi.org/10.1016/S0890-6955(00)00018-3

9. Hassan, A.M.: Improvements in some properties of non-ferrous metals by the application of the ball burnishing process. J. Mater. Process. Technol. **59**, 250–256 (1996)

10. Hassan, A.M.: The effects of ball- and roller-burnishing on the surface roughness and hardness of some non-ferrous metals. J. Mater. Process. Technol. **72**, 385–391 (1997)

11. Prevey, P.S., Hornbach, D.J., Jayaraman, N.: Controlled plasticity burnishing to improve the performance of friction stir processed Ni-Al Bronze. In: Proceedings Thermec. Canada. pp. 2–7 (2006). www.lambdatechs.com

12. Yu, X., Wang, L.: Effect of various parameters on the surface roughness of an aluminium alloy burnished with a spherical surfaced polycrystalline diamond tool. Int. J. Mach. Tools Manuf. **39**, 459–469 (1999)

13. El-Axir, M.H.: Investigation into roller burnishing. Int. J. Mach. Tools Manuf. **40**(11), 1603–1617 (2000). https://doi.org/10.1016/S0890-6955(00)00019-5

14. Loh, N.H., Tam, S.C.: A study of the effects of ball-burnishing parameters on surface roughness using factorial design. J. Mech. Work. Technol. **18**, 53–61 (1989)

15. Ross, P.J.: Taguchi Techniques for Quality Engineering. Mc Graw-Hill, New York (1996)

16. Zhang, T., Bugtai, N., Marinescu, I.D.: Burnishing of aerospace alloy: a theoretical-experimental approach. J. Manuf. Syst. 2–8 (2014). https://doi.org/10.1016/j.jmsy.2014.11.004

17. El-Khabeery, M.M., El-Axir, M.H.: Experimental techniques for studying the effects of milling roller-burnishing parameters on surface integrity. Int. J. Mach. Tools Manuf. **41**(12), 1705–1719 (2001). https://doi.org/10.1016/S0890-6955(01)00036-0

18. Sagbas, A.: Analysis and optimization of surface roughness in the ball burnishing process using response surface methodology and desirabilty function. Adv. Eng. Softw. **42**(11), 992–998 (2011). https://doi.org/10.1016/j.advengsoft.2011.05.021

19. Esme, U.: Use of grey based Taguchi method in ball burnishing process for the optimization of surface roughness and microhardness of AA 7075 aluminum alloy. Mater. Technol. **44**(3), 129–135 (2010). http://www.scopus.com/inward/record.url?eid=2-s2.0-79951519669&partnerID=40&md5=9a1d486851f2b34944b21d1864eacdb1

20. El-Axir, M.H., Othman, O.M., Abodiena, A.M.: Improvements in out-of-roundness and microhardness of inner surfaces by internal ball burnishing process. J. Mater. Process. Technol. **196**(1–3), 120–128 (2008). https://doi.org/10.1016/j.jmatprotec.2007.05.028

21. Fung, C., Kang, P.: Multi-response optimization in friction properties of PBT composites using Taguchi method and principle component analysis. J. Mater. Process. Technol. **170**, 602–610 (2005). https://doi.org/10.1016/j.jmatprotec.2005.06.040

22. Lu, H.S., Chang, C.K., Hwang, N.C., Chung, C.T.: Grey relational analysis coupled with principal component analysis for optimization design of the cutting parameters in high-speed end milling. J. Mater. Process. Technol. **209**, 3808–3817 (2008). https://doi.org/10.1016/j.jmatprotec.2008.08.030

23. Ancio, F., Gámez, A.J., Marcos, M.: Study of turned surfaces by principal component analysis. Precis. Eng. (2015). https://doi.org/10.1016/j.precisioneng.2015.09.006

24. Shiou, F.J., Chen, C.H.: Freeform surface finish of plastic injection mold by using ball-burnishing process. J. Mater. Process. Technol. **140**, 248–254 (2003). https://doi.org/10.1016/S0924-0136(03)00750-7

25. Li, L.F., Xia, W., Yao, Z.Z., Zhao, J., Tang, Z.Q.: Analytical prediction and experimental verification of surface roughness during the burnishing process. Int. J. Mach. Tools Manuf. **62**, 67–75 (2012). https://doi.org/10.1016/j.ijmachtools.2012.06.001

26. Seemikeri, C.Y., Brahmankar, P.K., Mahagaonkar, S.B.: Investigations on surface integrity of AISI 1045 using LPB tool. Tribol. Int. **41**(8), 724–734 (2008). https://doi.org/10.1016/j.triboint.2008.01.003

27. Low, K.O., Wong, K.J.: Influence of ball burnishing on surface quality and tribological characteristics of polymers under dry sliding conditions. Tribol. Int. **44**(2), 144–153 (2011). https://doi.org/10.1016/j.triboint.2010.10.005

28. Hassan, A.M., Maqableh, A.M.: The effects of initial burnishing parameters on non-ferrous components. J. Mater. Process. Technol. **102**(1), 115–121 (2000). https://doi.org/10.1016/S0924-0136(00)00464-7

29. Antony, J.: Multi-response optimization in industrial experiments using Taguchis quality loss function and principal component analysis. Qual. Reliab. Eng. Int. **16**, 3–8 (2000)

Design Optimization of Helicopter Rotor with Trailing-Edge Flaps Using Genetic Algorithm

Saijal Kizhakke Kodakkattu

Abstract Helicopters are useful in many peculiar applications such as search and rescue, law enforcement, medical transport, tourism and military use as it can take-off and land vertically. Also it can hover in the air. But the high vibration levels in the helicopters in comparison with the fixed-wing aircraft is a major concern. Trailing-edge flap approach is a successful technique in alleviating rotor hub vibration in the helicopter. But the actuation of trailing-edge flaps is the major challenge. Optimization studies are carried out to reduce rotor vibration along with flap control power with flap locations and torsional stiffness as design variables. Several optimization studies are carried out in the past employing response surface method, neural network and various other metamodelling techniques. The present work endeavours to achieve optimal locations of the flaps and rotor blade torsional stiffness using genetic algorithm (GA). Results show that optimization using genetic algorithm provides some very good design solution with lesser computational effort in comparison with previous works.

Keywords Helicopter rotor · Optimization · Genetic algorithm

1 Introduction

Helicopters are the peculiar class of flying vehicles with their ability to hover, fly backward, forward and laterally as well as vertical take-off and landing. The major concern in the helicopter flight is the high vibration compared to other fixed-wing aircrafts due to the unsteady aerodynamic environment surrounding the rotor blade. Excessive vibration in helicopters has many adverse effects such as damage to structural components, errors in the instrument reading, inaccurate weapon delivery and high maintenance cost. Commonly used passive methods such as isolation mountings and dynamic vibration absorbers for helicopter vibration control are very simple and

S. Kizhakke Kodakkattu (✉)
Government Engineering College Kozhikode, West Hill P.O., Kozhikode 673008, Kerala, India
e-mail: saijalkk@gmail.com

© Springer Nature Singapore Pte Ltd. 2020
R. Venkata Rao and J. Taler (eds.), *Advanced Engineering Optimization Through Intelligent Techniques*, Advances in Intelligent Systems and Computing 949, https://doi.org/10.1007/978-981-13-8196-6_48

easy to implement. But there is a huge weight penalty associated with them. With the improvements in the field of smart materials technology, actively controlled trailing-edge flaps (TEF) with piezoelectric stack actuators are a major choice for the reduction of rotor hub vibration with the benefit of localized low-power actuation devices. Dual trailing-edge flaps are more effectual in comparison with single flap in reducing vibration in the helicopter rotor.

In the TEF approach, each flap is actuated using servo-stack actuators, which can be accurately phased to neutralize most of the undesired vibrations in the rotor blade. This technique provides a significant reduction in vibration in contrast with other passive and active control devices with minimal control power requirement. For an efficient system, the positioning of the flaps is to optimize along with rotor properties such as torsional stiffness. Viswamurthy and Ganguli carried out an optimization study to reduce vibration and flap power requirement with positions of the twin flaps as the design variables [1]. Polynomial response surfaces are formulated to approximate both vibration and flap power. Later, Kodakkattu et al. included torsional stiffness of the blade in the design variables [2]. They used polynomial response surface and neural network metamodels to represent the objective functions in the optimization. The inclusion of torsional stiffness of the rotor blade in the design variables made the two objective functions extremely nonlinear, and polynomial response surface was incapable of approximating the objective functions adequately. In another study, Kodakkattu et al. used the orthogonal array-based response surface and kriging metamodels to represent these highly nonlinear objective functions in the optimization with torsional stiffness and flap locations as the design variables [3, 4]. Orthogonal array-based response surface approximates these highly nonlinear objective functions appositely in the optimization. This optimization study is carried out at different flap length and flying conditions as well.

Many studies in the past demonstrated the capability of a genetic algorithm in the optimization. Liu et al. carried out a theoretical optimization to develop a plate-fin heat exchanger for the hydraulic retarder [5]. Pathan et al. developed a real-coded constrained genetic algorithm (GA) and carried out the performance assessment study of some selected classical optimization problems [6]. In the present work, the global optimization toolbox in MATLAB is used in the single-objective and multiobjective optimization using a genetic algorithm (GA). The fitness function, number of variables, lower and upper limits of the variables and number of iteration are the data required for the optimization, and then select a solver, define optimization problem and set the optimization options. Visualization of intermediate and final results is possible. The toolbox includes some plotting functions. Visualizations give feedback about optimization progress and help to make decisions to modify some solver options or stop the solver.

Studies in the past demonstrated the possibility of reducing vibration by optimally designing the helicopter rotor with TEF. But these studies were carried out a comprehensive search in the domain to find the optimum design solutions. In the present work, an efficient genetic algorithm is used in the optimization for improved design solution of rotor with TEF for better vibration reduction.

2 Optimization Problem

The inboard flap position, outboard flap position and torsional stiffness are denoted as x_{in}, x_{out} and GJ, respectively. Optimization is performed aiming the minimization of vibration and flap power requirement with twin trailing-edge flap locations (x_{in}, x_{out}) and rotor torsional stiffness (GJ) as variables. The rotor vibration and flap power requirement are normalized concerning to the values at the baseline and are represented as F_v and F_p, respectively. The rotor similar to a BO105 helicopter is taken for the aeroelastic analysis. The optimization problem can be presented as:

$$\text{Min.}\{F_v, F_p\}$$
$$\text{With constraints}: x_{in-lower} \le x_{in} \le x_{in-upper}$$
$$x_{out-lower} \le x_{out} \le x_{out-upper}$$
$$GJ_{lower} \le GJ \le GJ_{upper}$$

3 Results and Discussions

3.1 6% Flap Length

Initially, the length of TEF is taken as 6% of the helicopter rotor blade length. The baseline configuration and constraints on the variables are given in Table 1. From the Ref. [2], the vibration and flap control power functions formulated using response surface method (RSM) can be written as:

$$F_V = 0.9832 + 0.0125\,x_{in} - 0.065\,x_{out} + 0.1074\,GJ + 0.0191\,x_{in}^2 - 0.068\,x_{out}^2$$
$$- 0.0298\,GJ^2 - 0.0072\,x_{in}\,x_{out} + 0.011\,x_{in}\,GJ - 0.068\,x_{out}\,GJ$$

$$F_P = 0.9471 - 0.0183\,x_{in} + 0.2584\,x_{out} + 0.0346\,GJ + 0.0303\,x_{in}^2 + 0.0151\,x_{out}^2$$
$$+ 0.0304\,GJ^2 + 0.0294\,x_{in}\,x_{out} - 0.0245\,x_{in}\,GJ - 0.1466\,x_{out}\,GJ$$

Firstly, single-objective optimization studies are carried out minimizing F_v and F_p separately. These results are given Table 2. The GA tool in MATLAB is used for optimization. A population size of 50 is used for the study. The number of generations

Table 1 Baseline configuration and constraints on the variables

Variable	Lower limit	Baseline value	Upper limit
x_{in}	0.59 R	0.65 R	0.71 R
x_{out}	0.77 R	0.83 R	0.89 R
GJ	0.00461	0.00615	0.00769

Table 2 Results of single-objective optimization

Optimum value of the function	Coded value			Physical variable		
	x_{in}	x_{out}	GJ	x_{in}	x_{out}	GJ
$F_v = 0.4119$	−0.156	−1.732	−1.732	0.64 R	0.77 R	0.00461
$F_p = 0.1305$	0.442	−1.732	−1.732	0.67 R	0.77 R	0.00461

Table 3 Comparison of results obtained by RSM, GA and neural network analysis

Optimization method	Minimizing F_V			Minimizing F_P		
	x_{in}	x_{out}	GJ	x_{in}	x_{out}	GJ
Genetic algorithm (GA)	0.64 R	0.77 R	0.00461	0.67 R	0.77 R	0.00461
Response surface methodology (RSM)	0.64 R	0.77 R	0.00461	0.67 R	0.77 R	0.00461
Neural network analysis	0.66 R	0.89 R	0.00461	0.59 R	0.77 R	0.00461

in the genetic algorithm is fixed as 100. The fitness value of each generation is calculated by evaluating the objective functions. The selection for the next generation is based on the fitness value. The iterations are terminated once the generation satisfies the minimum criteria or the number of generations reaches the maximum prescribed.

Table 3 shows a comparison of single-objective optimization results obtained from an exhaustive search approach using response surface, neural network obtained from Ref. [2] and genetic algorithm.

The results show that optimization using exhaustive search and genetic algorithm using response surface functions give exactly same optimum design point. Neural network analysis results show some variation from the other results in the minimization of the helicopter rotor hub vibration alone, especially in the optimum outboard flap position and torsional stiffness. Single-objective optimization minimizing flap control power requirement provides exactly same optimum points except for the inboard flap position.

Multiobjective optimization minimizing both F_v and F_p is also carried out using a genetic algorithm, and the Pareto-optimal front obtained is given in Fig. 1. The GAMULTIOBJ tool in MATLAB is used for optimization with a population size of 50. Figure 1 shows three distinct points as the optimum points. The values of the two objective functions at this optimum are given in Table 4. The optimum value of hub vibration at these points is about 0.4510, which is about 55% reduction from baseline configuration. Similarly, the optimum value of flap power objective function is 0.1484, which is about 86% reductions compared to the values at the baseline configuration. The optimum design is $(x_{in}, x_{out}, GJ) = (0.66 R, 0.77 R, 0.00469)$. A comparison of the results of multiobjective optimization obtained from different optimization method is given in Table 5. The results show that genetic algorithm gives almost same optimum design as that obtained from an exhaustive search using response surface functions, but with much lesser computational effort.

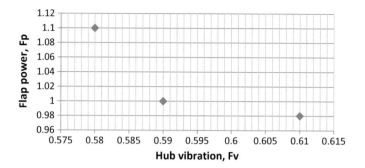

Fig. 1 Pareto-optimal front in the multiobjective optimization

Table 4 Optimum values of objective functions in the multiobjective optimization

Sl. No	F_v	F_p	x_{in}	x_{out}	GJ
1	0.4510893	0.1484554	0.39762	−1.7315	−1.6313
2	0.4510768	0.1484646	0.397645	−1.7315	−1.6314
3	0.4510765	0.14848983	0.39762	−1.7314	−1.63148

Table 5 Comparison of results obtained in the multiobjective optimization

Optimization method	Minimizing F_v and F_p		
	x_{in}	x_{out}	GJ
Genetic algorithm (GA)	0.66 R	0.77 R	0.00469
Response surface methodology (RSM)	0.65 R	0.77 R	0.00461
Neural network analysis	0.66 R	0.89. R	0.00461

Optimization using genetic algorithm is also done using the objective functions obtained from orthogonal array-based response surface as given in Ref. [3]. These orthogonal array-based response functions F_v and F_p are given by

$$F_v = 0.9906 + 0.0334\,x_{in} - 0.0957\,x_{out} + 0.1864\,GJ + 0.0416\,x_{in}^2 - 0.1856\,x_{out}^2$$
$$- 0.0683\,GJ^2 + 0.0090\,x_{in}\,x_{out} + 0.0237\,x_{in}\,GJ - 0.0798\,x_{out}\,GJ$$
$$- 0.0209\,x_{in}\,x_{out}^2 - 0.0248\,x_{in}\,GJ^2 + 0.0157\,x_{out}\,x_{in}^2 - 0.0467\,x_{out}\,GJ^2$$
$$+ 0.0402\,GJ\,x_{in}^2 - 0.1125\,GJ\,x_{out}^2$$

$$F_p = 1.0397 - 0.0721\,x_{in} + 0.1720\,x_{out} + 0.2460\,GJ - 0.0119\,x_{in}^2 - 0.1169\,x_{out}^2$$
$$- 0.0509\,GJ^2 + 0.0161\,x_{in}\,x_{out} - 0.0202\,x_{in}\,GJ + 0.0248\,x_{out}\,GJ$$
$$+ 0.0239\,x_{in}\,x_{out}^2 + 0.0419\,x_{in}\,GJ^2 - 0.0464\,x_{out}\,x_{in}^2 + 0.0802\,x_{out}\,GJ^2$$
$$- 0.0232\,GJ\,x_{in}^2 - 0.1460\,GJ\,x_{out}^2$$

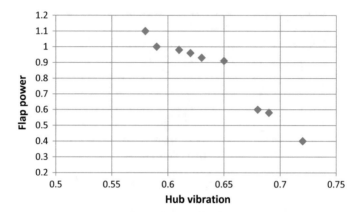

Fig. 2 Pareto front using genetic algorithm with orthogonal array response function at 6% flap length

Multiobjective optimization of the above objective functions is carried out using a genetic algorithm. The Pareto front obtained is given in Fig. 2. A comparison of these results with exhaustive search results using response surface is given in Table 6. The results show that the genetic algorithm gives almost same optimum point as that obtained using exhaustive search approach using RSM.

3.2 9% Flap Length

A similar study is conducted with a trailing-edge flap length of 9% of rotor blade length. The move limits of the flaps have been changed to avoid the overlapping of the flaps. The move limits and initial configuration of the design variables are given in Table 7. The functions, F_v and F_p, are formulated using RSM and are given as [3],

$$
\begin{aligned}
F_v = {} & 0.9906 + 0.0334\,x_{\text{in}} - 0.0957\,x_{\text{out}} + 0.1864\,\text{GJ} + 0.0416\,x_{\text{in}}^2 - 0.1856\,x_{\text{out}}^2 \\
& - 0.0683\,\text{GJ}^2 + 0.0090\,x_{\text{in}}\,x_{\text{out}} + 0.0237\,x_{\text{in}}\,\text{GJ} - 0.0798\,x_{\text{out}}\,\text{GJ} \\
& - 0.0209\,x_{\text{in}}\,x_{\text{out}}^2 - 0.0248\,x_{\text{in}}\,\text{GJ}^2 + 0.0157\,x_{\text{out}}\,x_{\text{in}}^2 - 0.0467\,x_{\text{out}}\,\text{GJ}^2 \\
& + 0.0402\,\text{GJ}\,x_{\text{in}}^2 - 0.1125\,\text{GJ}\,x_{\text{out}}^2
\end{aligned}
$$

$$
\begin{aligned}
F_p = {} & 1.0397 + 0.0721\,x_{\text{in}} - 0.1720\,x_{\text{out}} + 0.2460\,\text{GJ} + 0.0119\,x_{\text{in}}^2 - 0.1169\,x_{\text{out}}^2 \\
& - 0.0509\,\text{GJ}^2 + 0.0161\,x_{\text{in}}\,x_{\text{out}} + 0.0202\,x_{\text{in}}\,\text{GJ} - 0.0248\,x_{\text{out}}\,\text{GJ} \\
& - 0.0239\,x_{\text{in}}\,x_{\text{out}}^2 - 0.0419\,x_{\text{in}}\,\text{GJ}^2 + 0.0464\,x_{\text{out}}\,x_{\text{in}}^2 - 0.0802\,x_{\text{out}}\,\text{GJ}^2 \\
& + 0.0232\,\text{GJ}\,x_{\text{in}}^2 - 0.1460\,\text{GJ}\,x_{\text{out}}^2
\end{aligned}
$$

Table 6 Optimum design using genetic algorithm and orthogonal array response surface

RSM analysis						Genetic algorithm(GA)					
Coded value			Physical variable			Coded value			Physical variable		
x_{in}	x_{out}	GJ	x_{in}	x_{out}	GJ	x_{in}	x_{out}	GJ	x_{in}	x_{out}	GJ
1	−1	−1	0.71 R	0.77 R	0.00461	0.85994	−0.9831	−0.9915	0.70 R	0.77 R	0.00462

Table 7 Initial values and move limits of design variables for 9% flap length

Variable	Lower limit	Baseline value	Upper limit
x_{in}	0.61 R	0.65 R	0.69 R
x_{out}	0.79 R	0.83 R	0.87 R
GJ	0.00461	0.00615	0.00769

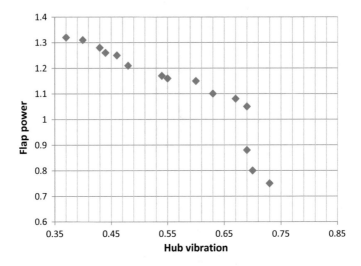

Fig. 3 Pareto-optimal front for 9% flap length

The Pareto-optimal front obtained in the multiobjective optimization using genetic algorithm is given in Fig. 3, and the comparison of the results is given in Table 8.

The results show that the optimum design points are similar, and hence, genetic algorithm can be used in the design optimization of the helicopter rotor blade with trailing-edge flaps.

4   Conclusions

From the results obtained in the present work, the following conclusion can be drawn:

- The genetic algorithm gives similar results as that of exhaustive search approach using response surface in both single- and multiobjective optimizations with much lesser computational effort.
- The multiobjective optimization of helicopter blade with twin trailing-edge flaps using genetic algorithm gives some very good design solution with about 55% reduction in hub vibration and 86% reduction in flap control power compared to the baseline configuration.

Table 8 Comparison of RSM analysis and genetic algorithm

RSM analysis						Genetic algorithm(GA)					
Coded value			Physical variable			Coded value			Physical variable		
x_{in}	x_{out}	GJ	x_{in}	x_{out}	GJ	x_{in}	x_{out}	GJ	x_{in}	x_{out}	GJ
1	−1	−1	0.71 R	0.77 R	0.00461	0.859	−0.983	−0.991	0.70 R	0.771 R	0.00462

- The computational cost in the design optimization of the helicopter rotor with trailing-edge flaps can be reduced to a great extent using genetic algorithm.

References

1. Viswamurthy, S.R., Ganguli, R.: Optimal placement of trailing-edge flaps for helicopter vibration reduction using response surface methods. Eng. Optim. **39**(2), 185–202 (2007)
2. Saijal, K.K., Ganguli, R., Viswamurthy, S.R.: Optimization of helicopter rotor using polynomial and neural network metamodels. J. Aircr. **48**(2), 553–566 (2011)
3. Kodakkattu, S.K., Joy, M.L., Prabhakaran Nair, K.: Vibration reduction of helicopter with trailing-edge flaps at various flying conditions. Proc. Inst. Mech. Eng., Part G: J. Aerosp. Eng. **231**(4) (2017)
4. Kodakkattu, S.K., Prabhakaran Nair, K., Joy, M.L.: Design optimization of helicopter rotor using kriging. Aircr. Eng. Aerosp. Technol. (2018). https://doi.org/10.1108/AEAT-12-2016-0250
5. Liu, C., Bu, W., Xu, D.: Multi-objective shape optimization of a plate-fin heat exchanger using CFD and multi-objective genetic algorithm. Int. J. Heat Mass Transf. **111**, 65–82 (2017)
6. Pathan, M.V., Patsias, S., Tagarielli, V.L.: A real-coded genetic algorithm for optimizing the damping response of composite laminates. Comput. Struct. **198**, 51–60 (2018)

Statistical Model for Spot Welding of Aluminum Alloy 6082T651 Using an Interlayer

Arick. M. Lakhani and **P. H. Darji**

Abstract Resistance spot welding of aluminum alloy 6082T651 is used to spot-weld the peripheral flange of the well member to the faceplate section of fuel rail in a diesel engine. The interlayer of SS304 having 0.5 mm thickness was introduced into the lap joint, an effect of which was the focus of the investigation. It is the need of modern automotive industries toward industrial fabrication with improved mechanical properties of thin aluminum sheets. This research paper summarizes work on the welding current, electrode force and welding time as input parameters, whereas failure load and nugget diameter were output parameters for this work. The range of electrode force was 1447–5653 N; the welding time was 0.6–7.5 cycles, and the welding current was 20–31 KA, during the performance of the experiment. A design matrix was generated by employing a central composite design (CCD) for experimental work. Effects of the welding process parameters on failure loads and nugget diameters were observed using ANOVA and then formulated by two separate equations.

Keywords Resistance spot welding · Interlayer · Central composite design · ANOVA

Arick. M. Lakhani (✉)
Faculty of Technology and Engineering, C.U. Shah University, Wadhwan 363020, Gujarat, India
e-mail: arick.lakhani@gmail.com; arick.lakhani.me@vvpedulink.ac.in

Department of Mechanical Engineering, V.V.P. Engineering College, Rajkot 360005, Gujarat, India

P. H. Darji
C.U. Shah College of Engineering and Technology, C.U. Shah University, Wadhwan 363020, Gujarat, India
e-mail: pranav_darji@rediffmail.com

© Springer Nature Singapore Pte Ltd. 2020
R. Venkata Rao and J. Taler (eds.), *Advanced Engineering Optimization Through Intelligent Techniques*, Advances in Intelligent Systems and Computing 949,
https://doi.org/10.1007/978-981-13-8196-6_49

1 Introduction

Steel and aluminum are widely used as construction materials for automotive industries. Hence, the hybrid structural body of aluminum and steel is required, as steel is less expensive than aluminum [1]. Automobile industry widely uses the two series of aluminum, namely 5xxx and 6xxx alloys [2]. Weight reduction of the automobile is the prime requirement in order to achieve better vehicle efficiency and less fuel combustion [3]. Conventional joining processes applied for joining of steel to aluminum are non-fusion joining methods such as adhesive bonding and mechanical fastening [4]. The high strength of spot weld is a must to sustain various loads. Resistance spot welding (RSW) process utilizes the concentrated Joule heat on sheets to be welded. These sheets are firmly held together through which current flows for a particular time duration and hence heat is produced [5]. Thus, a good quality spot weld joint is obtained by providing the optimum combinations of input parameters [6]. There are inherent differences in electrical, mechanical and thermal properties of aluminum and steel [7]. It is possible to weld aluminum alloy 5052 to stainless steel SUS304 with the help of cover plate by resistance spot welding [8]. Generation of heat in RSW is given by Eq. (1).

$$Q = I^2 R t \tag{1}$$

where Q is the generation of heat in joules, I is the welding current in amperes, R is the resistance in ohms, and t is the welding time in seconds. According to recent research, nanoparticles along with the cryorolling process in resistance spot welding process of aluminum alloy 6061 is being used [9]. Researchers have also studied properties and behavior of aluminum alloy 5754 under various cyclic loading conditions, using nickel as an interlayer [10]. The same material, 6082T651 is widely used in the application of friction stir welding; also, its finite element and microstructure were studied [11]. Also, numerical simulation of 5xxx series was investigated, namely for 5182 alloys [12]. Various researchers have contributed their work by applying design of experiment (DOE) in resistance spot welding, but no effort has been made to apply DOE in resistance spot welding of aluminum alloy 6082T651 using SS304 as an interlayer. A set of equations for the welding parameters was obtained from the analysis. Figure 1 shows the input and output parameters considered for this work.

2 Materials and Experimental Procedure

Design of experiments was performed, to investigate the effect of the SS304 interlayer (20 mm × 25 mm × 0.5 mm) with aluminum alloy 6082T651 (100 mm × 25 mm × 1 mm) in RSW process having a lap over of 20 mm. Both the materials were cut into specimens by machining.

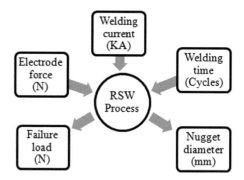

Fig. 1 Selected input and output parameters for the RSW process

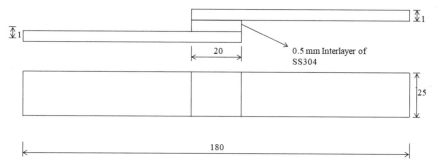

Fig. 2 Test coupon dimensions for weld with 0.5-mm interlayer (not to scale, dimensions are in mm)

Table 1 Chemical (wt%) and mechanical properties of aluminum alloy 6082T651

Tensile strength (MPa)	Yield strength (MPa)	Alloying elements					
		Si	Mn	Mg	Fe	Cu	Al
305.6	245.1	1.02	0.67	0.76	0.26	0.02	97.24

Figure 2 shows the arrangement of the experimental setup for 0.5 mm interlayer of SS304. Chemical and mechanical properties for both aluminum alloy 6082T651 and SS304 are shown in Tables 1 and 2, respectively. Surfaces of sheets were abraded randomly with silicon carbide P220 grade (emery paper) [13]. Further, it was cleaned by dry air jet, before welding. Test coupons were prepared according to the standards of ASME Section IX, before experiment [14]. Truncated cone copper electrode having a diameter of 6 mm was used.

Table 2 Chemical (mass%) properties of SS304

Tensile strength (MPa)	Yield strength (MPa)	Alloying elements							
		Si	Mn	C	P	S	Ni	Cr	Fe
515	205	0.85	1.25	0.06	0.04	0.02	8.0	18	Bal.

2.1 Measurements of Parameters

ARON 50/PR pneumatic resistance spot welding machine was used, manufactured by Kriton Weld Equipments Pvt. Ltd. Electrode force was measured manually from Eq. (2). Twenty experiments were conducted, based on designed parameters given by software Minitab 17. To measure failure load: the output, the tensile test was carried out using the TE-WDW-S1/2/5S universal testing machine, and failure load was measured in Newton (N). The tensile shear test was performed on the test coupons according to the standards of the Resistance Welders Manufacturer Association (RWMA) [15]. Nugget diameter: the output was measured by Mitutoyo Digital Vernier Caliper having least count 0.01 mm.

$$E = \frac{\pi D^2}{4} * P \tag{2}$$

where D is piston diameter, P is line pressure measured by the pressure gauge, and E is electrode force. While the welding current was measured by using Eq. (3),

$$\frac{V_s}{V_p} = \frac{I_p}{I_S} = \frac{N_s}{N_p} \tag{3}$$

where V_P is primary voltage, V_S is secondary voltage, I_P is primary current, I_S is secondary current, N_p is primary turns, and N_S is secondary turns. The primary current was measured by digital clamp meters, and the secondary current was calculated from the known turn ratio, while the welding time was displayed by the control panel of the welding machine.

2.2 Experimental Design

For constructing a second-order quadratic model for the response variable, an experimental design known as central composite design (CCD) is used. By applying CCD, one can avoid the use of complete three-level factorial experiment. Figure 3 shows CCD for $k = 3$ and $k = 2$ factors. For k number of factors, CCD includes: N_f factorial points (corners/cube points) each at two levels indicated by (1, +1) and N_c center points indicated by 0, $2k$ axial points (star points) (i) $k = 2$ and (ii) $k = 3$. The total

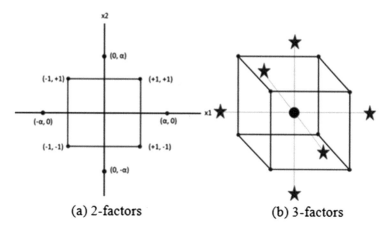

(a) 2-factors   (b) 3-factors

Fig. 3 Central composite design for **a** two factors and **b** three factors [16]

Table 3 Factors, levels, and values of welding parameters

Factors	Levels		
	−1	0	1
Electrode force (N)	2300	3550	4800
Welding time (Cycle)	2	4	6
Welding current (KA)	23	26	29

number of distinct design points that CCD should have must be equal to the number of experimental trials (N) for a composite design is shown in Eq. (4).

$$N = N_f + 2k + C \tag{4}$$

where C is the number of replications of center points. Its value should be at least $+$1 or $+2$, a minimum number of points desired in order to fit a second-order model. First, a $2k$ first-order model is used. If the model shows a lack of fit, then axial and center runs are added to incorporate the quadratic terms in the model. If the distance from the center of the design space to a factorial point is ± 1 unit for each factor, the distance from the center of the design space to a star point is $|\alpha| > 1$. It is important to select the value of α for the axial runs. Universally, the value of α is considered between 1 and k [16].

If $\alpha = 1$, the design is said to be face-centered. In order to get 20 numbers of runs in total for a CCD with 3 input parameters, ($C = 6$) center points are generally chosen. These 20 runs include ($N_f = 8$ cubes) points (cube corners) and ($2k = 6$) axial/star points as shown in Fig. 3b. In present work, central composite design is employed as shown in Table 3 [17]. Hence by using this method with less number of experiments, it is possible to measure the curved surface. The values and levels employed in CCD are listed in Table 3.

3   Result and Discussion

The values of the responses that is failure loads and nugget diameters as shown in Table 4 were achieved by performing twenty experiments. Inputs in the form of electrode force, welding time and welding current are given as shown in Table 4 for RSW process. These samples were then tested as discussed in the earlier section. Mean squares against experimental errors are compared with a specific confidence levels by ANOVA to see the significance of all factors [17, 18]. ANOVA is applied to observe the sensitivity of process and their significance on quality characteristics of failure load and nugget diameter of the welded joint. The regression model of the second order was taken into account as shown in Eq. (5),

$$y_k = \beta_0 + \sum_{i=1}^{3} \beta_i x_i + \sum_{i=1}^{3} \beta_{ii} x_i^2 + \sum_{j=2}^{3} \sum_{i=1}^{j-i} \beta_{ij} x_i x_j \tag{5}$$

Table 4 Experimental results for nugget diameter and failure load

Sample no.	Electrode force (N)	Welding time (cycle)	Welding current (KA)	Nugget diameter (mm)	Failure load (N)
1	1447.759	4	26	5.1	2750
2	3550	4	26	5.2	3465
3	3550	4	26	5.2	3652
4	3550	4	26	5.3	3585
5	4800	6	23	4.5	3452
6	3550	4	20.9546	3.6	3485
7	3550	0.6364	26	2.7	1890
8	4800	6	29	5.1	3615
9	5652.241	4	26	4.6	3030
10	2300	6	23	3.6	2451
11	2300	2	29	6.1	3380
12	2300	6	29	4.6	3489
13	3550	4	26	5.4	3511
14	2300	2	23	2.8	2030
15	3550	4	26	5.4	3515
16	3550	4	31.0453	6.5	4685
17	3550	4	26	5.5	3584
18	4800	2	23	3	2875
19	4800	2	29	4.9	3210
20	3550	7.3635	26	3.2	2775

Table 5 ANOVA for failure load

Source	DF	Adj. SS	Adj. MS	F-value	P-value
Model	9	7,448,049	827,561	74.72	0.000
Linear	3	2,798,529	932,843	84.23	0.000
E	1	378,278	378,278	34.15	0.000
W.T.	1	659,180	659,180	59.52	0.000
W.C.	1	1,761,072	1,761,072	159.00	0.000
Square	3	4,148,187	1,382,729	124.84	0.000
E * E	1	807,328	807,328	72.89	0.000
W.T. * W.T.	1	2,711,860	2,711,860	244.85	0.000
W.C. * W.C.	1	497,553	497,553	44.92	0.000
Two-way interaction	3	501,333	167,111	15.09	0.000
E * W.T.	1	25,538	25,538	2.31	0.160
E *W.C.	1	446,513	446,513	40.31	0.000
W.T. * W.C.	1	29,282	29,282	2.64	0.135
Error	10	110,756	11,076		
Lack of fit	5	88,024	17,605	3.87	0.082
Pure error	5	22,732	4546		
Total	19	7,558,805			

Model summary

S	R-sq	R-sq(adj.)	R-sq(pred.)
105.241	98.53%	97.22%	90.67%

where x_i denotes the codified entity of the input variables, namely: electrode force (E), welding current (W.C) and welding time (W.T), and y_k denotes the output variables, namely: nugget diameter (N.D) and failure load (F.L). β is the coefficient of regression obtained from the experimental results [19]. Equations (6) and (7) are the regression equations for output parameters failure load and nugget diameter, respectively.

Tables 5 and 6 show the ANOVA for failure load and nugget diameter, and for the weld joints, respectively. If the P-value is less than 0.0001, it means that the chances of model F-value (large value) could occur because noise is only a 0.01%. The model is insignificant if the P-value is greater than 0.1000. The determination coefficient R^2 (R-sq) in the ANOVA table is the ratio of explained variations to the total variations, which in turn is the measure of the degree of fit. If R^2 tends toward unity, the better response model fits the actual data [20]. It shows that there is less distinctness in predicted and actual values.

In order to obtain statistical model, F Model is used to disclose various possibility of differences by performing a single test, which is done by rejecting null hypothesis, thus multiple comparisons are avoided. The model F-value of 74.72 as shown in Table 5 for failure load indicates that the model is compelling. Moreover, there is only a 0.01% chance that this large value of F Model could occur due to noise.

Table 6 ANOVA for nugget diameter

Source	DF	Adj. SS	Adj. MS	F-value	P-value
Model	9	22.7586	2.52873	68.92	0.000
Linear	3	10.2469	3.41563	93.10	0.000
I	1	0.0142	0.01423	0.39	0.547
W.T.	1	0.2481	0.24815	6.76	0.026
W.C.	1	9.9845	9.98451	272.13	0.000
Square	3	9.7667	3.25557	88.73	0.000
E * E	1	0.3184	0.31845	8.68	0.015
W.T. * W.T.	1	9.6997	9.69966	264.37	0.000
W.C. * W.C.	1	0.0875	0.08754	2.39	0.153
Two-way interaction	3	2.7450	0.91500	24.94	0.000
E * W.T.	1	0.7200	0.72000	19.62	0.001
E * W.C.	1	0.4050	0.40500	11.04	0.008
W.T. * W.C.	1	1.6200	1.62000	44.15	0.000
Error	10	0.3669	0.03669		
Lack of fit	5	0.2936	0.05871	4.00	0.077
Pure error	5	0.0733	0.01467		
Total	19	23.1255			

Model summary

S	R-sq	R-sq(adj.)	R-sq(pred.)
0.191546	98.41%	96.99%	89.06%

Similarly, the model F-value of 68.92 as shown in Table 6 for the nugget diameter indicates that the model is compelling. Moreover, there is only 0.01% chance that this large value of F Model could occur due to noise.

$$\text{F.L.} = 3298 + 2.756E + 1159\text{W.T.} - 690\text{W.C.}$$
$$- 0.000151E * E - 108.45\text{W.T.} * \text{W.T.}$$
$$+ 20.65\text{W.C.} * \text{W.C.} + 0.0226E * \text{W.T.}$$
$$- 0.06300E * \text{W.C.} - 10.08\text{W.T.} * \text{W.C.} \qquad (6)$$

$$\text{N.D.} = -24.23 + 0.001730E + 3.232\text{W.T.} + 1.248\text{W.C.} - 0.000000E * E$$
$$- 0.2051\text{W.T.} * \text{W.T.} - 0.00866\text{W.C.} * \text{W.C.}$$
$$+ 0.000120E * \text{W.T.} - 0.000060E * \text{W.C.} - 0.0750\text{W.T.} * \text{W.C.} \qquad (7)$$

The discrepancy between the experimental value and the adapted response values can be presented by plotting a graph, and this plot is known as residuals versus the percentage frequency of respective cumulatives [21]. If the dispersion of

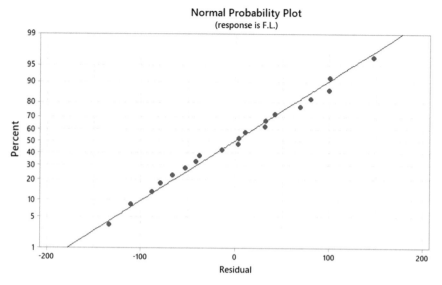

Fig. 4 Normal probability plot for failure load (F.L)

Table 7 Results of the confirmation test

Factors			Experimental output		Output from the empirical relation		Error %	
Electrode force (N)	Welding time (Cycle)	Welding current (kA)	N.D (mm)	F.L (N)	N.D (mm)	F.L (N)	N.D (mm)	F.L (N)
4800	6	23	4.5	3452	4.5	3447.99	0%	0.11%
3550	4	26	5.2	3585	5.2	3552.40	0%	0.99%
4800	6	29	5.1	3615	5	3572.70	1.9%	1.17%

residual is normal, the graph will be a straight line [22]. The normal probability plots of residuals for failure load and nugget diameter are shown in Figs. 4 and 5, respectively. It is observed from this graph that residuals are on the aligned line. The confirmation test was accomplished to authenticate the empirical relation as shown in Table 7. It has been observed that the statistical model has an error of less than 2% which is acceptable and indicates the best fit for the selection.

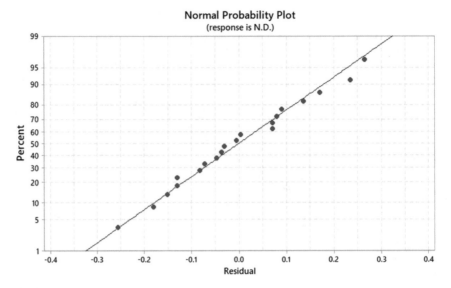

Fig. 5 Normal probability plot for nugget diameter (N.D)

4　Conclusion

– The statistical model provides the highest value of failure load (4685 N) and nugget diameter (6.5 mm) without expulsion using interlayer of 0.5 mm for a weld between the peripheral flange of well member and section of the faceplate in fuel rails of the diesel engine.
– Interlayer has the most significant influence in obtaining the highest value of failure load without an expulsion. Because by using interlayer at appropriate parameters weak intermetallic compounds (IMCs) were created at the faying surface, thus not disturbing weld quality and providing high strength.
– Regression equation obtained can be easily used to predict the values of nugget diameter and failure load, significantly.
– Normal probability plot justifies that errors are dispersed normally, and the experimental relationship is meticulously advanced.

References

1. Darwish, S.M., Soliman, M.S.: Variables of spot welding commercial aluminum sheets having different thickness. Int. J. Mater. Prod. Technol. **9**, 394–402 (1999). https://doi.org/10.1504/IJMPT.1994.036430
2. Zedan, M.J., Doos, Q.M.: New method of resistance spot welding for dissimilar 1008 low carbon steel-5052 aluminum alloy. Procedia Struct. Integrity **9**, 37–46 (2018). https://doi.org/10.1016/j.prostr.2018.06.008

3. Manladan, S.M., Yusof, F., Ramesh, S., Fadzil, M., Luo, Z., Ao, S.: A review on resistance spot welding of aluminum alloys. Int. J. Adv. Manuf. Technol. **90**, 605–634 (2017). https://doi.org/10.1007/s00170-016-9225-9

4. Oikawa, H., Ohmiya, S., Yoshimura, T., Saitoh, T.: Resistance spot welding of steel and aluminum sheet using an insert metal sheet. Sci. Technol. Weld. Joining **4**, 80–88 (1999). https://doi.org/10.1179/136217199101537608

5. Khanna, S.K., He, C.L., Agrawal, H.N.: Residual stress measurement in spot welds and the effect of fatigue loading on redistribution of stresses using high sensitivity Moiré interferometry. J. Eng. Mater. Technol. **123**, 132–138 (2001). https://doi.org/10.1115/1.1286218

6. Ghazali, F.A., Salle Z., Manurung, Y.H., Taib, Y.M., Hyie, K.M., Ahamat, M.A., Hamidi, S.A.: Three response optimization of spot-welded joint using Taguchi design and response surface methodology techniques. In The Advances in Joining Technology, pp. 85–95. Springer, Singapore (2018). https://doi.org/10.1007/978-981-10-9041-7_7

7. Satonaka, S., Iwamoto, C., Qui, R., Fujioka, T.: Trends and new applications of spot welding for aluminum alloy sheets. Weld. Int. **20**, 858–864 (2006). https://doi.org/10.1533/wint.2006.3677

8. Qiu, R., Iwamoto, C., Satonaka, S.: Interfacial microstructure and strength of steel/aluminum alloy joints welded by resistance spot welding with the cover plate. J. Mater. Process. Technol. **209**, 4186–4193 (2009). https://doi.org/10.1016/j.jmatprotec.2008.11.003

9. Zohoori-Shoar, V., Eslami, A., Karimzadeh, F., Abbasi-Baharanchi, M.: Resistance spot welding of ultrafine grained/nanostructured Al 6061 alloy produced by cryorolling process and evaluation of weldment properties. J. Manuf. Process. **26**, 84–93 (2017). https://doi.org/10.1016/j.jmapro.2017.02.003

10. Sun, M., Niknejad, S.T., Zhang, G., Lee, M.K., Wu, L., Zhou, Y.: Microstructure and mechanical properties of resistance spot welded AZ31/AA5754 using a nickel interlayer. Mater. Des. **87**, 905–913 (2015). https://doi.org/10.1016/j.matdes.2015.08.097

11. Gao, Z., Niu, J.T., Krumphals, F., Enzinger, N., Mitsche, S., Sommitsch, C.: FE modelling of microstructure evolution during friction stir spot welding in AA6082-T6. Weld. World **57**, 895–902 (2013). https://doi.org/10.1007/s40194-013-0083-x

12. Tao, J.F., Liang, G., Liu, C., Yang, Z.: Multi-field dynamic modeling and numerical simulation of aluminum alloy resistance spot welding. Trans. Nonferrous Met. Soc. China **22**, 3066–3072 (2012). https://doi.org/10.1016/s1003-6326(11)61572-0

13. Pereir, A.M., Bártoloa, P.J., Ferreira, J.M., Loureiro, A., Costa, J.D.M., Bártolo, P.J.: Effect of process parameters on the strength of resistance spot welds in 6082-T6 aluminum alloy. Mater. Des. **31**, 9 (2010). https://doi.org/10.1016/j.matdes.2009.11.052

14. Welding and brazing qualifications: ASME Sec 9. The American Society of Mechanical Engineers (2008). https://doi.org/10.1115/1.802892.ch9

15. Del Vecchio, E.J.: Resistance welding manual. Resistance welder manufacturers' association: 1 (1956)

16. Phadke, M.S.: Quality Engineering Using Robust Design, Prentice Hall PTR (1989). https://doi.org/10.1080/00401706.1991.10484810

17. Darwish, S.M., Al-Dekhial, S.D.: Statistical models for spot welding of commercial aluminum sheets. Int. J. Mach. Tools Manuf. **39**, 1589–1610 (1999). https://doi.org/10.1016/S0890-6955(99),00010-3

18. Juang, S.C., Tarng, Y.S.: Process parameter selection for optimizing the weld pool geometry in the tungsten inert gas welding of stainless steel. J. Mater. Process. Technol. **122**, 33–37 (2002). https://doi.org/10.1016/S0924-0136(02),00021-3

19. Tosun, N., Cogun, C., Tosun, G.: A study on kerf and material removal rate in wire electrical discharge machining based on Taguchi method. J. Mater. Process. Technol. **152**, 316–322 (2004). https://doi.org/10.1016/j.jmatprotec.2004.04.373

20. Kim, T., Park, H., Rhee, S.: Optimization of welding parameters for resistance spot welding of TRIP steel with response surface methodology. Int. J. Prod. Res. **43**, 4643–4657 (2005). https://doi.org/10.1080/00207540500137365

21. Kiaee, N., Aghaie-Khafri, M.: Optimization of gas tungsten arc welding process by response surface methodology. Mater. Des. **54**, 25–31 (2014). https://doi.org/10.1016/j.matdes.2013.08.032
22. Mushkudian, N.A., Einmahl, J.H.J.: Generalized probability-probability plots. J. Stat. Plann. Infer. **137**, 738–752 (2007). https://doi.org/10.2139/ssrn.607101

Fuzzy Programming Technique for Solving Uncertain Multi-objective, Multi-item Solid Transportation Problem with Linear Membership Function

Vandana Y. Kakran and Jayesh M. Dhodiya

Abstract The simple transportation problem (TP) is extended to the solid transportation problem (STP) by taking into consideration the mode of transportation as the third dimension along with the two dimensions of sources and destinations. A multi-objective, multi-item STP (MMISTP) with uncertain variables is known as uncertain MMISTP. Using the basic theory of uncertainty and the expected values of the uncertain parameters, the uncertain programming model is firstly converted into its deterministic form. By using the fuzzy programming technique, the transformed deterministic model is then solved using LINGO software to derive the optimal compromise solution of the problem. Finally, numerical examples are illustrated at the end to show the application of the uncertain programming model, and the obtained results are compared with results of other developed approaches.

Keywords Uncertain theory · STP · MMISTP · Fuzzy technique

1 Introduction

The simple TP was first developed by Hitchock [1] for determining the optimal solution so that the cost of distributing products from one place to another can be minimized. According to Hitchock [1], the transportation problem basically deals with two-dimensional properties, i.e. the number of sources (supply) and number of destinations (demand). Simple TP is valid only when there is one particular route available by which the transportation can be done. But, if one deals with the real-world situations, it is observed that there can be different choices available for the transportation. Selection of the mode of transportation is, therefore, a crucial parameter that affects the transportation cost. So, simple TP is not found to be suitable for

V. Y. Kakran (✉) · J. M. Dhodiya
S.V. National Institute of Technology, Surat, India
e-mail: vandana.kakran98@gmail.com

J. M. Dhodiya
e-mail: jdhodiya2002@yahoo.com

© Springer Nature Singapore Pte Ltd. 2020
R. Venkata Rao and J. Taler (eds.), *Advanced Engineering Optimization Through Intelligent Techniques*, Advances in Intelligent Systems and Computing 949,
https://doi.org/10.1007/978-981-13-8196-6_50

such practical situations, where one can choose different transportation modes and is modified further to STP. STP is an extension and generalization of the conventional TP where we consider the third-dimensional property as well, taking into account the transportation mode called conveyances along with the sources and destinations. It is called the generalization of the TP, because if we take only a single mode of transportation, then the extended STP is equivalent to the simple TP. TP has been extended to STP firstly by Schell [2]. The STP has also been studied by Ojha et al. [3] and Pramanik et al. [4] for an item with some fixed charge, price-discounted varying charge and vehicle cost. Bhatia et al. [5] aimed to minimize the shipment time of STP in his work.

In real-life applications, there exist situations where one needs to optimize more than one, i.e. two or more than two objectives at a particular time in STP. Such STP is classified as multi-objective STP (MOSTP). If the STP contains different number of items to be transported from one place to another, i.e. it does not deal with any single homogeneous commodity (or items), then it is known as multi-item STP (MISTP). On the other hand, if the STP contains more than one number of items as well as more than one number of objectives at a time, then it is called multi-objective, multi-item STP (MMISTP). The model for MMISTP was firstly suggested by Kundu et al. [6]. Moreover, the parameters involved in such problems may not be exact or sufficient due to incompleteness or lack of information. So, to represent such imprecise or uncertain data, Zadeh [7–9] used the concept of fuzzy sets, whereas Liu [10] used the concept of uncertainty theory.

In the literature, many researchers have worked upon TP with uncertain input parameters. Bit et al. [11] solved the MOSTP using Zimmerman's fuzzy approach [12]. Besides fuzzy approach, the uncertain theory [10] has been used by Sheng and Yao [13, 14]; Cui and Sheng [15]; Guo et al. [16] in the literature. Dalman [17] has also worked upon MMISTP with uncertain variables.

This paper presents a MMISTP with uncertain input parameters which is solved by using Zimmerman's fuzzy approach [12] with linear membership function. There are various types of membership functions (mfs) defined in fuzzy systems, e.g. linear, exponential or hyperbolic, and amongst these mfs, anyone of them can be used in this fuzzy approach. This paper utilizes linear mf for finding the solution to the problem.

2 Preliminaries

This section introduces some fundamental definitions of uncertainty theory.

Definition 1 [17] *Let ξ be an uncertain variable with regular uncertainty distribution $\Phi(x)$. If the expected value is available, then*

$$E[\xi] = \int\limits_{0}^{1} \Phi^{-1}(\alpha)d\alpha$$

where $\Phi^{-1}(\alpha)$ is the inverse uncertainty distribution of ξ.

Definition 2 [17] *An uncertain variable ξ is called normal if it has a normal uncertainty distribution*

$$\Phi(x) = \left(1 + \exp\left(\pi \frac{(e - x)}{\sqrt{3}\sigma}\right)\right), \quad -\infty < x < \infty, \quad \sigma > 0$$

denoted by $N(e, \sigma)$.

The inverse uncertainty distribution of a normal uncertain variable $N(e, \sigma)$ is

$$\Phi^{-1}(x) = e + \frac{\sigma\sqrt{3}}{\pi} \ln \frac{x}{1 - x}.$$

3 MMISTP Model

The formulation of MMISTP is stated as below in the following model (1):

Model 1:

$$\min f_t = \sum_{p=1}^{r}\sum_{i=1}^{m}\sum_{j=1}^{n}\sum_{k=1}^{l} c_{ijk}^{tp} x_{ijk}^{p}, \quad t = 1, 2, \ldots, S,$$

subject to the constraints:

$$\sum_{j=1}^{n}\sum_{k=1}^{l} x_{ijk}^{p} \leq a_{i}^{p}, i = 1, 2, \ldots, m; \ p = 1, 2, \ldots, r;$$

$$\sum_{i=1}^{m}\sum_{k=1}^{l} x_{ijk}^{p} \geq b_{j}^{p}, j = 1, 2, \ldots, n; \ p = 1, 2, \ldots, r;$$

$$\sum_{p=1}^{r}\sum_{i=1}^{m}\sum_{j=1}^{n} x_{ijk}^{p} \leq e_{k}, k = 1, 2, \ldots, l; \quad x_{ijk}^{p} \geq 0, \forall i, j, k, p.$$

where $p = 1, 2, \ldots, r$ represents the items to be delivered from m sources $O_i (i = 1, 2, \ldots, m)$ to n destinations $D_j (j = 1, 2, \ldots, n)$ by using $k = 1, 2, \ldots, l$ different modes of transportation (also called conveyances). There are $t = 1, 2, \ldots, S$ different objectives which need to be minimized in the MMISTP. c_{ijk}^{tp} indicates the transportation penalty assigned for transporting one unit of item from the origin (source) i to destination j by conveyance k for the item p. a_i^p and b_j^p denotes the total availability of items at ith source and total requirement of items at jth destination, respectively, corresponding to the pth item, whereas e_k represents the total transportation capacity of kth conveyance. The formulation of the above model (1) is done by considering the variables as the deterministic quantities. But uncertainty can be faced due to lack of information or knowledge, so we need to modify the model (1) to the uncertain programming model defined in the next section.

4   Uncertain Programming Model

Here, we consider the variables used for the cost per unit of item, availability of the source, requirement of the destination and conveyances as the variables with uncertain values. Let these uncertain variables be denoted by $\xi_{ijk}^{tp}, \tilde{a}_i^p, \tilde{b}_j^p, \tilde{e}_k$, respectively. In the uncertain model, the expected values of the objectives are optimized under the chance constraints instead of the given objective functions. So, this model is also known as the expected constrained programming model. The uncertain (expected) model for the above model (1) is introduced by Liu [10] as follows:

Model 2:

$$\min Z_t = E[f_t(x, \xi)] = E\left[\sum_{p=1}^r \sum_{i=1}^m \sum_{j=1}^n \sum_{k=1}^l \xi_{ijk}^{tp} x_{ijk}^p\right], t = 1, 2, \ldots, S,$$

subject to the constraints:

$$M\left\{\sum_{j=1}^n \sum_{k=1}^l x_{ijk}^p - \tilde{a}_i^p \leq 0\right\} \geq \Upsilon_i^p, i = 1, 2, \ldots, m; p = 1, 2, \ldots, r;$$

$$M\left\{\tilde{b}_j^p - \sum_{i=1}^m \sum_{k=1}^l x_{ijk}^p \leq 0\right\} \geq \beta_j^p, j = 1, 2, \ldots, n; p = 1, 2, \ldots, r;$$

$$M\left\{\sum_{p=1}^r \sum_{i=1}^m \sum_{j=1}^n x_{ijk}^p - \tilde{e}_k \leq 0\right\} \geq \delta_k, k = 1, 2, \ldots, l; \quad x_{ijk}^p \geq 0, \forall i, j, k, p.$$

where $\Upsilon_i^p, \beta_j^p, \delta_k$ are the constraint's specified confidence levels for i, j, k and p.

5   Deterministic Equivalence Model to Uncertain Model

Here, we will specify the deterministic model which is equivalent to the above model (2) defined using the uncertain parameters. Let us consider the uncertain variables $\xi_{ijk}^{tp}, \tilde{a}_i^p, \tilde{b}_j^p, \tilde{e}_k$ having the uncertainty distributions $\Phi_{\xi_{ijk}^{tp}}, \Phi_{\tilde{a}_i^p}, \Phi_{\tilde{b}_j^p}, \Phi_{\tilde{e}_k}$. Therefore, model (2) is transformed to the given below model (3) [17]:

Model 3:

$$\min Z_t = \sum_{p=1}^{r} \sum_{i=1}^{m} \sum_{j=1}^{n} \sum_{k=1}^{l} \int_{0}^{1} \Phi_{\xi_{ijk}^{tp}}^{-1}(\alpha) \, d\alpha, \ t = 1, 2, \ldots, S,$$

subject to the constraints:

$$\sum_{j=1}^{n} \sum_{k=1}^{l} x_{ijk}^{p} - \Phi_{\tilde{a}_i^{p}}^{-1}\left(1 - \Upsilon_i^{p}\right) \le 0, i = 1, 2, \ldots, m; \ p = 1, 2, \ldots, r,$$

$$\Phi_{\tilde{b}_j^{p}}^{-1}\left(\beta_j^{p}\right) - \sum_{i=1}^{m} \sum_{k=1}^{l} x_{ijk}^{p} \le 0, j = 1, 2, \ldots, n; \ p = 1, 2, \ldots, r,$$

$$\sum_{p=1}^{r} \sum_{i=1}^{m} \sum_{j=1}^{n} x_{ijk}^{p} - \Phi_{\tilde{e}_k}^{-1}(1 - \delta_k) \le 0, k = 1, 2, \ldots, l.$$

6 Solution Methodology

In this section, Zimmerman's fuzzy approach [12] is discussed to get the solution of the MMISTP problem. The sequential steps of this approach are described below:

Step 1. The deterministic model (3) of the given problem is solved as a single-objective problem subject to the given set of constraints, i.e. ignoring all the other objectives and considering only one objective at a time.

Step 2. Find the positive ideal solution (PIS), i.e. L_t, and negative ideal solution (NIS), i.e. U_t for every objective function. Calculate these values for each of the S objective functions.

Step 3. Now, we construct the linear membership function $\mu_t(Z_t)$ for the ith objective function as:

$$\mu_t(Z_t) = \begin{cases} 1, & \text{if } Z_t \le L_t \\ \frac{U_t - Z_t}{U_t - L_t}, & \text{if } L_t < Z_t < U_t \\ 0, & \text{if } Z_t \ge U_t, \forall t \end{cases}$$

Step 4. Now, we define the basic formulation of fuzzy approach by using the max-min operator as:

$$\max \lambda,$$

subject to the constraints:

$$\frac{U_t - Z_t}{U_t - L_t} \geq \lambda, t = 1, 2, \ldots, S$$

and the constraints given in model (3), with $\lambda \geq 0$ and $\lambda = \min\{\mu_t(Z_t)\}$.

Step 5.   Now this problem contains a single objective, which is solved by the LINGO software to get the compromise solution of the problem.

7   Numerical Illustrations

Numerical Illustration 1. Consider the MMISTP with three sources and two destinations having two different conveyances. This problem consists of two different objective functions. Find out the number of goods (products) to be supplied from given sources to the number of destinations using the provided two modes of transportation. The solution should be such that the total cost of transporting goods is optimized. In this problem, we consider normal uncertain variables only.

$$\xi_{ijk}^{tp} \sim N\left(e_{ijk}^{tp}, \sigma_{ijk}^{tp}\right), \qquad \tilde{a}_i^p \sim N\left(e_i^p, \sigma_i^p\right),$$

$$\tilde{b}_j^p \sim N\left(e_j'^p, \sigma_j'^p\right), \qquad \tilde{e}_k \sim N\left(e_k''^p, \sigma_k''^p\right),$$

$$i = 1, 2, 3, \quad j = 1, 2, \quad k = 1, 2, \quad t = 1, 2;$$

Solution:

$\Upsilon_i^p = 0.9$; $\beta_j^p = 0.9$; $\delta_k = 0.9$; are the confidence levels of the constraints for all i, j, k and p respectively. Using the data values given in above tables, the corresponding model (4) equivalent to model (3) is formulated as below:

Model 4:

$$\min Z_1 = E[f_1(x, \xi)] = \sum_{p=1}^{2}\sum_{i=1}^{3}\sum_{j=1}^{2}\sum_{k=1}^{2} e_{ijk}^{1p} x_{ijk}^p;$$

$$\min Z_2 = E[f_2(x, \xi)] = \sum_{p=1}^{2}\sum_{i=1}^{3}\sum_{j=1}^{2}\sum_{k=1}^{2} e_{ijk}^{2p} x_{ijk}^p;$$

subject to the constraints:

$$\sum_{j=1}^{2}\sum_{k=1}^{2} x_{ijk}^{p} - \left[e_{i}^{p} + \frac{\sigma_{i}^{p}\sqrt{3}}{\pi}\ln\frac{1-\gamma_{i}^{p}}{\gamma_{i}^{p}}\right] \le 0, \ i = 1, 2, 3, \ p = 1, 2.$$

$$\left[e_{j}^{\prime p} + \frac{\sigma_{j}^{\prime p}\sqrt{3}}{\pi}\ln\frac{\beta_{j}^{p}}{1-\beta_{j}^{p}}\right] - \sum_{i=1}^{3}\sum_{k=1}^{2} x_{ijk}^{p} \le 0, \ j = 1, 2, \ p = 1, 2.$$

$$\sum_{p=1}^{2}\sum_{i=1}^{3}\sum_{j=1}^{2} x_{ijk}^{p} - \left[e_{k}^{\prime\prime p} + \frac{\sigma_{k}^{\prime\prime p}\sqrt{3}}{\pi}\ln\frac{1-\delta_{k}}{\delta_{k}}\right] \le 0, \ k = 1, 2.$$

$$x_{ijk}^{p} \ge 0, i = 1, 2, 3, \ j = 1, 2, k = 1, 2, \ p = 1, 2.$$

Now substituting these data values from Tables 1, 2, 3, 4, 5, 6 and 7, we construct the following MMISTP model with the help of model (4):

Table 1 Transportation cost ξ_{ijk}^{11} for the first item corresponding to first objective

ξ_{ij1}^{11}	1	2	ξ_{ij2}^{11}	1	2
1	(10,2)	(9,1.5)	1	(7,2)	(5,1.5)
2	(8,1)	(9,1.5)	2	(5,1)	(6,1.5)
3	(8,1.5)	(17,1.5)	3	(4,1.5)	(7,1.5)

Table 2 Transportation cost ξ_{ijk}^{12} for the second item corresponding to first objective

ξ_{ij1}^{12}	1	2	ξ_{ij2}^{12}	1	2
1	(9,2)	(6,1.5)	1	(5,2)	(6,1.5)
2	(9,1)	(10,1.5)	2	(5,1)	(5,1.5)
3	(6,1.5)	(18,1.5)	3	(2,1.5)	(8,1.5)

Table 3 Transportation cost ξ_{ijk}^{21} for the first item corresponding to second objective

ξ_{ij1}^{21}	1	2	ξ_{ij2}^{21}	1	2
1	(20,2)	(19,1.5)	1	(30,2)	(25,1.5)
2	(23,1)	(22,1.5)	2	(26,1)	(26,1.5)
3	(16,1.5)	(17,1.5)	3	(33,1.5)	(29,1.5)

Table 4 Transportation cost ξ_{ijk}^{22} for the second item corresponding to second objective

ξ_{ij1}^{22}	1	2	ξ_{ij2}^{22}	1	2
1	(18,2)	(16,1.5)	1	(15,2)	(16,1.5)
2	(19,1)	(20,1.5)	2	(24,1)	(25,1.5)
3	(16,1.5)	(18,1.5)	3	(22,1.5)	(28,1.5)

Table 5 Capacity of the sources $\tilde{a}_i^p$

i	1	2	3	i	1	2	3
$\tilde{a}_i^1$	(32,1.5)	(35,1.5)	(30,3)	$\tilde{a}_i^2$	(22,2)	(25,1)	(20,1.5)

Table 6 Demands at destinations $\tilde{b}_j^p$

j	1	2	j	1	2
$\tilde{b}_j^1$	(10,1.5)	(12,1)	$\tilde{b}_j^2$	(5,2)	(5,1.5)

Table 7 Capacity of conveyances $\tilde{e}_k$

k	1	2
$\tilde{e}_k$	(80,1.5)	(110,2)

Model 5:

$$\min Z_1 = \left(10x_{111}^1 + 9x_{121}^1 + 8x_{211}^1 + 9x_{221}^1 + 8x_{311}^1 + 17x_{321}^1\right)$$
$$+ \left(7x_{112}^1 + 5x_{122}^1 + 5x_{212}^1 + 6x_{222}^1 + 4x_{312}^1 + 7x_{322}^1\right)$$
$$+ \left(9x_{111}^2 + 6x_{121}^2 + 9x_{211}^2 + 10x_{221}^2 + 6x_{311}^2 + 18x_{321}^2\right)$$
$$+ \left(5x_{112}^2 + 6x_{122}^2 + 5x_{212}^2 + 5x_{222}^2 + 2x_{312}^2 + 8x_{322}^2\right);$$

$$\min Z_2 = \left(20x_{111}^1 + 19x_{121}^1 + 23x_{211}^1 + 22x_{221}^1 + 16x_{311}^1 + 17x_{321}^1\right)$$
$$+ \left(30x_{112}^1 + 25x_{122}^1 + 26x_{212}^1 + 26x_{222}^1 + 33x_{312}^1 + 29x_{322}^1\right)$$
$$+ \left(18x_{111}^2 + 16x_{121}^2 + 19x_{211}^2 + 20x_{221}^2 + 16x_{311}^2 + 18x_{321}^2\right)$$
$$+ \left(15x_{112}^2 + 16x_{122}^2 + 24x_{212}^2 + 25x_{222}^2 + 22x_{312}^2 + 28x_{322}^2\right);$$

subject to the constraints:

$$x_{111}^1 + x_{112}^1 + x_{121}^1 + x_{122}^1 - \left(32 + \frac{1.5\sqrt{3}}{\pi}\right)\ln\frac{1-0.9}{0.9} \le 0,$$

$$x_{211}^1 + x_{212}^1 + x_{221}^1 + x_{222}^1 - \left(35 + \frac{1.5\sqrt{3}}{\pi}\right)\ln\frac{1-0.9}{0.9} \le 0,$$

$$x_{311}^1 + x_{312}^1 + x_{321}^1 + x_{322}^1 - \left(30 + \frac{3\sqrt{3}}{\pi}\right)\ln\frac{1-0.9}{0.9} \le 0,$$

$$x_{111}^2 + x_{112}^2 + x_{121}^2 + x_{122}^2 - \left(22 + \frac{2\sqrt{3}}{\pi}\right)\ln\frac{1-0.9}{0.9} \le 0,$$

$$x_{211}^2 + x_{212}^2 + x_{221}^2 + x_{222}^2 - \left(25 + \frac{1\sqrt{3}}{\pi}\right)\ln\frac{1-0.9}{0.9} \le 0,$$

$$x_{311}^2 + x_{312}^2 + x_{321}^2 + x_{322}^2 - \left(20 + \frac{1.5\sqrt{3}}{\pi}\right)\ln\frac{1-0.9}{0.9} \le 0,$$

$$\left(10 + \frac{1.5\sqrt{3}}{\pi}\ln\frac{0.9}{1-0.9}\right) - \left(x_{111}^1 + x_{112}^1 + x_{211}^1 + x_{212}^1 + x_{311}^1 + x_{312}^1\right) \le 0,$$

$$\left(12 + \frac{1\sqrt{3}}{\pi} \ln \frac{0.9}{1-0.9}\right) - \left(x_{121}^1 + x_{122}^1 + x_{221}^1 + x_{222}^1 + x_{321}^1 + x_{322}^1\right) \le 0,$$

$$\left(5 + \frac{2\sqrt{3}}{\pi} \ln \frac{0.9}{1-0.9}\right) - \left(x_{111}^2 + x_{112}^2 + x_{211}^2 + x_{212}^2 + x_{311}^2 + x_{312}^2\right) \le 0,$$

$$\left(5 + \frac{1.5\sqrt{3}}{\pi} \ln \frac{0.9}{1-0.9}\right) - \left(x_{121}^2 + x_{122}^2 + x_{221}^2 + x_{222}^2 + x_{321}^2 + x_{322}^2\right) \le 0,$$

$$x_{111}^1 + x_{121}^1 + x_{211}^1 + x_{221}^1 + x_{311}^1 + x_{321}^1$$

$$+ x_{111}^2 + x_{121}^1 + x_{211}^2 + x_{221}^2 + x_{311}^2 + x_{321}^2 - \left(80 + \frac{1.5\sqrt{3}}{\pi}\right) \ln \frac{1-0.9}{0.9} \le 0,$$

$$x_{112}^1 + x_{122}^1 + x_{212}^1 + x_{222}^1 + x_{312}^1 + x_{322}^1$$

$$+ x_{112}^2 + x_{122}^2 + x_{212}^2 + x_{222}^2 + x_{312}^2 + x_{322}^2 - \left(110 + \frac{2\sqrt{3}}{\pi}\right) \ln \frac{1-0.9}{0.9} \le 0.$$

Now, we apply fuzzy programming technique defined by Zimmerman [12] to get the Pareto-optimal solution of the above-mentioned problem with linear membership function. Model (5) is then transformed to the following problem with single-objective function using the steps described in Sect. 6:

Model 6:

$$\max \lambda$$

subject to the constraints:

$$\frac{U_1 - Z_1}{U_1 - L_1} \ge \lambda, \quad \frac{U_2 - Z_2}{U_2 - L_2} \ge \lambda,$$

and the constraints of the model (5).
where $U_1 = 1570.71$, $L_1 = 162.27$ and $U_2 = 4017.86$, $L_2 = 634.14$ are the maximum and minimum values of the individual objectives, respectively. Solving this model (6), we get the Pareto-optimal solution as:

$$\lambda = 0.957, x_{311}^1 = 11.82, x_{122}^1 = 13.21, x_{121}^2 = 6.82, x_{112}^2 = 1.95, x_{312}^2 = 5.47$$

and rest all variables are zero in the solution.

Here, λ represents the minimum of the above-formulated membership functions using the objective functions, i.e. $\lambda = \min\{\mu_t(Z_t)\}$. The individual membership function values for the objective functions Z_1 and Z_2 are obtained as $\mu_1(Z_1) = 0.957435267$ and $\mu_2(Z_2) = 0.957460981$, respectively. Considering the digits up to three decimal places, we obtain $\lambda = 0.957 = \min(0.957, 0.957)$.

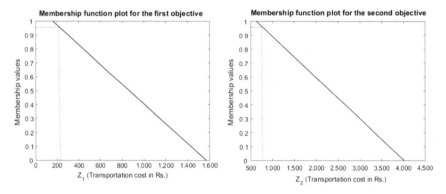

Fig. 1 Membership function value plots for the first and second objective functions

Table 8 Transportation cost ξ_{ijk}^{11} for the first item corresponding to first objective

ξ_{ij1}^{11}	1	2	3	4	ξ_{ij2}^{11}	1	2	3	4
1	(10,2)	(9,1.5)	(12,2)	(8,1.5)	1	(7,2)	(5,1.5)	(5,2)	(6,1.5)
2	(8,1)	(9,1.5)	(11,2)	(10,1)	2	(5,1)	(6,1.5)	(4,2)	(6,1)
3	(8,1.5)	(17,1.5)	(6,1.5)	(10,1.5)	3	(4,1.5)	(7,1.5)	(6,1.5)	(5,1.5)

The value of λ denotes that, both of these objective functions have at least 0.957 degree of satisfaction. The higher the value of degree of satisfaction λ, the better is the solution of the problem.

The objective function values corresponding to this Pareto-optimal solution are $Z_1^* = 222.22$ and $Z_2^* = 778.08$.

Here, we can observe from Fig. 1 that the dotted line in both the subplots for the first and second objective functions represents the degree of satisfaction on the Y-axis, and the Pareto-optimal value achieved at this membership function value is represented on the X-axis. In the first case, for the first objective function, the Pareto-optimal value attained is $\left(Z_1^* = 222.22\right)$ at $\lambda = 0.957$ degree of satisfaction. In the second case, for the second objective function, the Pareto-optimal value attained is $\left(Z_2^* = 778.08\right)$ at $\lambda = 0.957$ degree of satisfaction.

Numerical Illustration 2. In the previous example 1, consider four destinations instead of two destinations (i.e. $j = 1, 2, 3, 4$) along with the same number of origins (sources) and conveyances. The data values for this problem are given in the following Tables 8, 9, 10, 11, 12, 13 and 14.

Solution:

Here, $U_1 = 1665.50$, $L_1 = 368.26$ and $U_2 = 4255.92$, $L_2 = 1523.81$ are the maximum and minimum values of the individual objectives, respectively. In a similar way, we can obtain the Pareto-optimal solution for this numerical example 2 by using fuzzy programming approach with linear membership function which is:

Table 9 Transportation cost ξ_{ijk}^{12} for the second item corresponding to first objective

ξ_{ij1}^{12}	1	2	3	4	ξ_{ij2}^{12}	1	2	3	4
1	(9,2)	(6,1.5)	(3,2)	(7,1.5)	1	(5,2)	(6,1.5)	(7,2)	(9,1.5)
2	(9,1)	(10,1.5)	(11,2)	(10,1)	2	(5,1)	(5,1.5)	(3,2)	(3,1)
3	(6,1.5)	(18,1.5)	(8,1.5)	(12,1.5)	3	(2,1.5)	(8,1.5)	(7,1.5)	(3,1.5)

Table 10 Transportation cost ξ_{ijk}^{21} for the first item corresponding to second objective

ξ_{ij1}^{21}	1	2	3	4	ξ_{ij2}^{21}	1	2	3	4
1	(20,2)	(19,1.5)	(22,2)	(21,1.5)	1	(30,2)	(25,1.5)	(27,2)	(29,1.5)
2	(23,1)	(22,1.5)	(18,2)	(17,1)	2	(26,1)	(26,1.5)	(28,2)	(32,1)
3	(16,1.5)	(17,1.5)	(20,1.5)	(18,1.5)	3	(33,1.5)	(29,1.5)	(35,1.5)	(30,1.5)

Table 11 Transportation cost ξ_{ijk}^{22} for the second item corresponding to second objective

ξ_{ij1}^{22}	1	2	3	4	ξ_{ij2}^{22}	1	2	3	4
1	(18,2)	(16,1.5)	(23,2)	(17,1.5)	1	(15,2)	(16,1.5)	(17,2)	(19,1.5)
2	(19,1)	(20,1.5)	(21,2)	(20,1)	2	(24,1)	(25,1.5)	(23,2)	(23,1)
3	(16,1.5)	(18,1.5)	(18,1.5)	(12,1.5)	3	(22,1.5)	(28,1.5)	(27,1.5)	(23,1.5)

Table 12 Capacity of the sources $\tilde{a}_i^p$

i	1	2	3	i	1	2	3
$\tilde{a}_i^1$	(32,1.5)	(35,1.5)	(30,3)	$\tilde{a}_i^2$	(22,2)	(25,1)	(20,1.5)

Table 13 Demands at destinations $\tilde{b}_j^p$

j	1	2	3	4	j	1	2	3	4
$\tilde{b}_j^1$	(10,1.5)	(12,1)	(13,2)	(12,2)	$\tilde{b}_j^2$	(5,2)	(5,1.5)	(10,3)	(8,2)

Table 14 Capacity of conveyances $\tilde{e}_k$

k	1	2
$\tilde{e}_k$	(80,1.5)	(110,2)

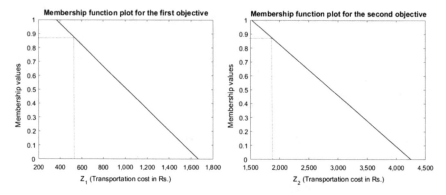

Fig. 2 Membership function value plots for the first and second objective functions

$$\lambda = 0.871, x_{141}^1 = 4.9, x_{241}^1 = 9.5, x_{311}^1 = 10.94, x_{331}^1 = 15.42, x_{122}^1 = 13.21,$$
$$x_{212}^1 = 0.88, x_{121}^2 = 6.82, x_{131}^2 = 5.33, x_{112}^2 = 7.42, x_{232}^2 = 8.30, x_{242}^2 = 10.42.$$

and rest all variables are zero in the solution.

Here, the individual membership function values for the objective functions Z_1 and Z_2 are obtained as $\mu_1(Z_1) = 0.871573494$ and $\mu_2(Z_2) = 0.871626691$, respectively. Considering the digits up to three decimal places, we obtain $\lambda = 0.871 = \min(0.871, 0.871)$. The value of λ denotes that both of these objective functions have at least 0.871 degree of satisfaction.

The objective function values corresponding to the obtained Pareto-optimal solution are $Z_1^* = 534.86$, $Z_2^* = 1874.54$, respectively.

Here, we can observe from Fig. 2 that, for the first objective function, the Pareto-optimal value attained is $\left(Z_1^* = 534.86\right)$ at $\lambda = 0.871$ degree of satisfaction. In the second case, for the second objective function, the Pareto-optimal value attained is $\left(Z_2^* = 1874.54\right)$ at $\lambda = 0.871$ degree of satisfaction.

In our both the numerical illustrations, the attained degree of satisfaction is nearly same when the objective functions are Pareto-optimal. But in other situations, the degree of satisfaction where objective functions become Pareto-optimal can have different membership function values for the individual objective functions.

8 Comparison and Results

The Pareto-optimal solution obtained for the numerical example 1 using linear membership function is $\lambda = 0.974, x_{311}^1 = 11.82, x_{122}^1 = 13.21, x_{121}^2 = 4.90, x_{112}^2 = 1.95, x_{122}^2 = 1.92, x_{312}^2 = 5.47$, and the objective function values corresponding to this optimal solution are $Z_1^* = 222.22$ and $Z_2^* = 778.08$.

The results obtained for numerical example 2 using linear membership function are compared with the results obtained from the convex combination and the minimizing distance function methods [17]. The comparison of the results is shown in the below table:

Convex combination method [17]	Minimizing distance method [17]	Fuzzy programming approach
$Z_1^* \in [336.964, 826.795]$	$Z_1^* = 551.148$	$Z_1^* = 534.86$
$Z_2^* \in [1408.9912, 2232.086]$	$Z_2^* = 1571.781$	$Z_2^* = 1874.54$

9 Conclusion

This paper presents a MMISTP with uncertain parameters which is transformed into deterministic model. The transformed model is then further solved using fuzzy programming technique in LINGO software, and the obtained results are compared with the results of other existing methods. It is concluded that the fuzzy programming technique gives an alternative approach to find the Pareto-optimal solution.

References

1. Hitchock, F.L.: The distributions of product from several sources to numerous localities. J. Math. Phys. **20**, 224–230 (1941)
2. Schell, E.D.: Distribution of a product by several properties. In: Direstorate of Management Analysis, Proceedings of the Second Symposium in Linear Programming 2, 615, DCS/comptroller HQUSAF, 615–664 (1955)
3. Ojha, A., Das, B., Mondal, S., Maity, M.: A solid transportation problem for an item with fixed charge, vehicle cost and price discounted varying charge using genetic algorithm. Appl. Soft Comput. **10**, 100–110 (2010)
4. Pramanik, S., Jana, D.K., Maity, K.: A multi objective solid transportation problem in fuzzy, bi-fuzzy environment via genetic algorithm. Int. J. Adv. Oper. Manag. **6**(1), 4–26 (2014)
5. Bhatia, H.L., Swarup, K., Puri, M.C.: Time minimizing solid transportation problem. Stat. J Theor. Appl. Stat **7**(3), 95–403 (1976)
6. Kundu, P., Kar, S., Maity, M.: Multi objective multi-item solid transportation problem in fuzzy environment. Appl. Math. Model. **37**(4), 2028–2038 (2013)
7. Zadeh, L.A.: Fuzzy sets. Inf. Control **8**, 338–353 (1965)
8. Zadeh, L.A.: The concept of a linguistic variable and its application to approximate reasoning-I. Inf. Sci. **8**(3), 199–249 (1975)
9. Zadeh, L.A.: The concept of a linguistic variable and its application to approximate reasoning-II. Inf. Sci. **8**(4), 301–357 (1975)
10. Liu, B.: Uncertainty theory, 2nd edn, pp. 205–234. Springer, Berlin Heidelberg (2007)
11. Bit, A.K., Biswal, M.P., Alam, S.S.: Fuzzy programming approach to multi objective solid transportation problem. Fuzzy Sets Syst. **57**, 183–194 (1993)

12. Zimmermann, H.-J.: Fuzzy programming and linear programming with several objective functions. Fuzzy Sets Syst. **1**, 45–55 (1978)
13. Sheng, Y., Yao, K.: Fixed charge transportation problem and its uncertain programming model. Ind. Eng. Manag. Syst. **11**(2), 183–187 (2012)
14. Sheng, Y., Yao, K.: A transportation model with uncertain costs and demands. Information **15**(8), 3179–3186 (2012)
15. Cui, Q., Sheng, Y.: Uncertain programming model for solid transportation problem. Information **15**(3), 342–348 (2013)
16. Guo, H., Wang, X., Zhou, S.: A transportation problem with uncertain costs and random supplies. Int. J. e-Navig. Marit. Econ. **2**, 1–11 (2015)
17. Dalman, H.: Uncertain programming model for multi-item solid transportation problem. Int. J. Mach. Learn. Cybernet. **9**(4), 559–567 (2016)

Investigation of Influence of Process Parameters in Deep Drawing of Square Cup

Chandra Pal Singh, Prachi Kanherkar, Lokesh Bajpai and Geeta Agnihotri

Abstract Aerodynamically designed machine parts are of very complicated shape and are used in automobile and aerospace industries. These parts are produced by forming process. The tendency of defects in formed part, viz. excessive thinning, wrinkling, and earing, makes the process complicated and is depended on many governing parameters. Holding forces, punch speed, and friction coefficient are the important process parameters, whereas die corner radius, clearance, punch nose radius, and blank thickness are important machine parameters. The present research work is focused to investigate the influence of process parameters in square cup. CATIA was used for modelling, and HyperWorks software was used for simulation, meshing, and parametric analysis. The investigation of the extent of process parameters' influence was estimated by Taguchi and ANOVA methods. Further, it is concluded that in a thin square cup made of 1 mm sheet thickness, friction coefficient dominates the process and it is the most influential parameter.

Keywords Process parameters · Taguchi · ANOVA

C. P. Singh (✉) · P. Kanherkar · L. Bajpai
Samrat Ashok Technological Institute Vidisha, Vidisha 464001, India
e-mail: chandraid@gmail.com

P. Kanherkar
e-mail: prachikanherkar0@gmail.com

L. Bajpai
e-mail: lokesh.bajpai.vds@gmail.com

G. Agnihotri
Mulana Azad National Institute Bhopal, Bhopal 462051, India
e-mail: dr.gagnihotri@gmail.com

© Springer Nature Singapore Pte Ltd. 2020
R. Venkata Rao and J. Taler (eds.), *Advanced Engineering Optimization Through Intelligent Techniques*, Advances in Intelligent Systems and Computing 949,
https://doi.org/10.1007/978-981-13-8196-6_51

1 Introduction

Geometrical shapes of complex nature can be produced with improved product strength and quality by forming process. Though many steps are required for achieving final product in forming process. Deep drawing is a forming process where a punch presses metal blank against die cavity, resulting in the formation of desired shape. The common defects such as wrinkling, excessive thinning, earing, and cracks at bending area are generally found in final product. Holding force, radius at die corner, punch movement speed, and radius at punch nose are the process parameters that affect the quality of product. Wrinkling tendency at flange is a major challenge to control, and the correct holding force can only avoid this tendency. Higher holding force results in low wrinkling tendency, but it leads to excessive stretching that may cause thinning at the bottom. Many research articles focused on the analysis of the effectiveness of parameters through finite element tools with the verified experimental results were studied. It is evident that holding force and friction coefficient have a great influence on wrinkling [1, 2]. Experimentally, the lubrication conditions were examined in axis-symmetric deep drawing of steel cup and the estimation of the effect of lubrication conditions between surfaces of blank and die was done [3]. Interrelationship between friction and distribution of strain was studied in the formation of steel cup [4]. Lubrication effect was studied to optimize the shape of the rectangular part [5]. Strain path depends on holding force and was studied in deep drawing [6]. Lubrication conditions were investigated to assess accurate friction between contact surfaces [7, 8]. Deep drawing of laminated sheet has been simulated, and the parametric study was carried [9]. Forming tool design is an important and complicated task in forming for defect-free products. Plastic deformation mechanism and interrelation and interaction between forming tools also control deep drawing along with material properties [10]. Wrinkling and springback phenomenon were studied in different boundary conditions [11]; for this, two models were created: One was with ¼ blank; and other was with full blank. Full model with higher computational time showed better and close result to the experimental result. A new property called hinge yield strength was estimated for simulation with paperboard to estimate springback and punch force, etc. [12]. Microforming process was studied taking wear as quality criterion to find the importance of lubrication conditions. It was concluded that surface roughness and wear at walls depend on lubrication [13]. Wrinkling depends on drawing depth; higher is the drawing depth, more will be wrinkling tendency. Further, the product quality depends on material type and required drawing force; higher is the drawing force, product quality will be poor. In the case of higher drawing depth, cushion and design of blank holder are of great importance [14]. Metal flow into die causes many defects. Nonuniformity of metal flow causes earing. Blank shape modification at various locations was done keeping yield strength and r-value in consideration, a earing defect-free cup was drawn [15].

2 Simulation and Modelling

Geometrical details were selected as per the literature [16], and CAD models of tooling have been developed in CAD software CATIA-V5-R12. For simulation purpose, HyperWorks 12.0 was used. All the CAD models were imported in IGES format in HyperWorks. After importing models in HyperMesh software, cleaning of geometries (if some surface has lost in importing) was done and meshed with shell elements. For the simulations of thin sheet, shell elements are proved to be suitable as shell elements can capture stretching and bending. Die was made as rigid by making all degrees of freedom null, and the punch was made to move in Z-direction. Blank was considered deformable part. Material properties as per the literature were created in HyperWorks environment and added to library of software. Finite element models were created in incremental HyperForm environment (with .hf extension). RADIOSS solver of HyperWorks was used for analysis. In post-processing, the simulation results in the form of thinning rate, von Mises stress, plastic strain, maximum stress point location, thickening tendency, and punch load were studied and compared with published article data [16]. On the verification of the simulation results, the model was verified. Once the models were verified, now these models were used to run simulations at different combinations of parameters. To study the importance of friction and its extent of influence, statistical method such as Taguchi's method and analysis of variance method were used. In Taguchi method, first design of experiments (DOE) was carried out. A range of three influential parameters (velocity of punch, blank holding force, and coefficient of friction) was decided. Simulations were run for planned scheme of parameters, and the results were studied. Statistical analysis was done to find the effectiveness of parameters. Machine parameters were kept as per the literature [16], and the values of parameters were changed to evaluate the effect of parameters.

Drawing depth for verifying model was kept as 20 mm, and holding force was kept 15 kN as mentioned in the literature [16]. Figure 1 shows EF model, and Table 1 depicts geometrical dominions of tooling. Friction coefficient between all contact surfaces was considered as 0.05 [16]. Material properties were selected as yield strength 276 Mpa, ultimate tensile strength 421 Mpa, Young's modulus 210 Gpa, strain hardening exponent 0.08, and Poisson's ratio 0.3 [16]. The punch velocity was selected as 5000 mm/s. Die, holder, and punch were constraint as rigid component, and these components were rigid meshed. So, these components are rigid meshed, and only blank or metal sheet was the deformable component. Mesh size was decided as 1 mm which was sufficient to capture all features of the square geometry. Further, there were no significant changes in the results observed at mesh size less than 1 mm. So, keeping geometrical details and simulation time in consideration, 1 mm mesh size was selected. After creating all similar boundary conditions, simulations were performed and the obtained results were compared with published article data. The model was verified with comparison of simulation results of plastic strain [16] as shown in Fig. 2. Variations in plastic strain were the same as mentioned in referred article [16], and maximum plastic strain was 0.35 as shown in Fig. 2. Percentage

Fig. 1 FE model of square cup formation showing die, punch, sheet, and blank holder

Table 1 Details of various parts [16]

Details of tooling for square cup	Dimensions (mm)
Blank size	124
Blank thickness	1
Punch size	70
Die size	74
Profile radius of die	5
Profile radius of punch	8
Radius at punch corner	10
Die corner radius	12
Blank size	124

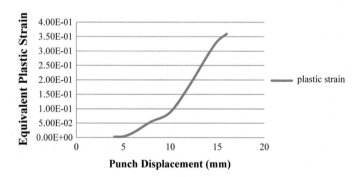

Fig. 2 Variation in equivalent plastic strain for an element in square

thickness variations at different sections of a square cup are shown in Fig. 3 Also, it can be seen that there is a thickening tendency at flange and thinning tendency at the bottom corner.

FLD curve was drawn at friction 0.05 which is the function of hardening exponent (n) and thickness (t). This diagram represents different zones' deformation such as failure zone, marginal zone, safe zone, compression zone, and loose metal zone.

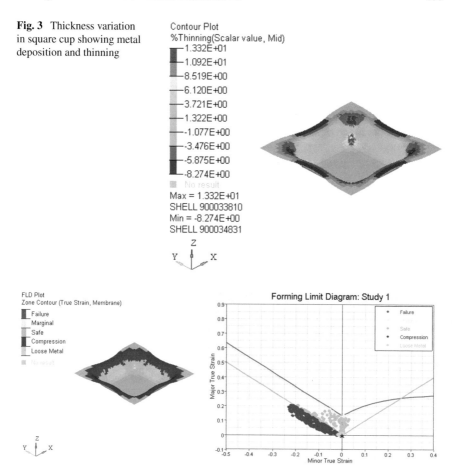

Fig. 3 Thickness variation in square cup showing metal deposition and thinning

Fig. 4 Forming limit diagram for square cup showing metal deposition and thinning

According to the strain value at element, particular element falls or locates in the particular zone. If excessive straining is there at element, then the element will be located in the failure zone; if element location is above or close to failure zone (red line), then it represents excessive straining or excessive thinning; if element is positioned between marginal and safe zone (between yellow and blue line), then it represents controlled or safe strain value; if the stain values at elements position element below safe zone, then it represents thickening of elements; if elements are located at outside of the safe zone, then it represents wrinkling. FLD in Fig. 4 is showing that there is thickening tendency at flange and thinning tendency at the bottom corner.

3 Statistical Analysis for Calculation of the Extent of the Influence of Parameters

Taguchi and ANOVA methods were employed to assess parameters' influence. In Taguchi's approach, the experiments are designed for individual factors' effect and the interaction between the factors (combination) can be studied. The experimental scheme design is an important step in Taguchi's approach. Taguchi used orthogonal arrays to design experiments. In orthogonal array, all factors are considered with all levels and are equally distributed. Signal-to-noise ratio, i.e. logarithmic transformation of mean square deviation, quantifies the influence of parameters. There are three approaches commonly used. In present study "larger is the best" approach was adopted. The function of this approach is given as $-10 \log\left(\frac{1}{n} \sum \frac{1}{y^2}\right)$, where y is the characteristic parameter. The thickness at various sections of formed part was selected as quality characteristics in the analysis. Minimum variation in thickness is considered as better quality of formed part. The combination of parameters, i.e. punch velocity, friction coefficient, and holder force, was studied. Table 2 shows the levels of parameters. The minimum value of holding force was selected on wrinkling threshold, and the value of coefficient of friction was selected from the literature and the lubrication conditions that generally prevailed in industrial applications. Table 3 shows Taguchi's orthogonal approach for designing experiments. Taguchi's orthogonal approach is based on design of the experiments or simulations such that all the three parameters will be correlated with each other with minimum numbers of experiments. The analysis of variances was used to find quantitative effect on the quality of product.

The following relation given below was used to perform analysis

$$\text{MS} = \left(\frac{1}{n} \sum \frac{1}{y^2}\right), (S/N)_i = -10 \times \log(\text{MS}), \text{ is } \overline{S/N} = \frac{1}{9} \sum_{i=1}^{i=9}(S/N)_i$$

$$\text{SS} = \sum_{i=1}^{i=9}\left((S/N)_i - \overline{(S/N)}\right)^2, (\text{SS})_i = \sum_{j=1}^{j=3}\left((S/N)_{ij} - \overline{(S/N)}\right)^2$$

where n is total elements considered, y corresponding thickness S/N ratio, $\overline{S/N}$ mean, SS sum of squares due to variation about mean, and $(\text{SS})_i$ the sum of squares due to variation about mean for ith parameter.

Table 2 Process parameters and their range for square cup

Level	Punch velocity	BHF (kN)	Coefficient of friction (μ)
1	4000	10	0.06
2	5000	15	0.1
3	6000	20	0.15

Table 3 Taguchi experimental design

Simulation	Punch speed	Blank holder force	Friction coefficient (μ)
1	1	1	1
2	1	2	2
3	1	3	3
4	2	1	2
5	2	2	3
6	2	3	1
7	3	1	3
8	3	2	1
9	3	3	2

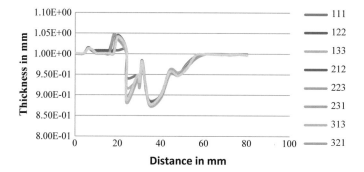

Fig. 5 Thickness distribution along OX-axis for square cup

Figure 5 shows thickness distribution along OX for square cup. To analyse the quality of product, thickness distribution throughout the product, as quality characteristics, was chosen. As material flow depends on the combination of parameters, these combinations result in the variation in thickness along the axis OX. Figure 5 shows, at low levels of parameters, material flow is smooth and it results in less thinning at the bottom of product. At higher levels of parameters, there is a large thinning at the side bottom corner due to the improper flow of material or overstretching of blank at corner and it is shown by curve 133. For statistical analysis, all simulations were conducted for drawing depth of 20 mm.

The analysis of variance was performed on forming process of square object. Square geometry can be considered as complex geometry. Tables 4 and 5 show the data of the analysis of variance and percentage contribution of each parameter in the case of the formation of square cup. The analysis of variance shows that friction coefficient is the most effective parameter, and its contribution is evaluated as 76.8%. The second effective parameter is punch velocity, and its contribution was evaluated as 15.8%. The influence of holding force is evaluated minimum, and its contribution is evaluated as 7.3%. From the above statistical analysis, it is clear that as complexity

Table 4 Data for ANOVA

Simulations	MS	$(S/N)_i$	$(S/N)_{ij}$
1	0.922495	0.70072	0.643635
2	0.927351	0.655117	
3	0.935937	0.575068	
4	0.92487	0.678386	0.666345
5	0.931058	0.620465	
6	0.922552	0.700183	
7	0.927351	0.655117	0.680794
8	0.924226	0.684436	
9	0.922271	0.702829	
1	0.922495	0.70072	0.678074
4	0.92487	0.678386	
7	0.927351	0.655117	
2	0.927351	0.655117	0.65334
5	0.931058	0.620465	
8	0.924226	0.684436	
3	0.935937	0.575068	0.65936
6	0.922552	0.700183	
9	0.924226	0.702829	
1	0.922495	0.70072	0.695113
6	0.922552	0.700183	
8	0.924226	0.684436	
2	0.927351	0.655117	0.678777
4	0.92487	0.678386	
9	0.922271	0.702829	
3	0.935937	0.575068	0.616883
5	0.931058	0.620465	
7	0.927351	0.655117	

of geometries increases, the contribution also gets changed, and it is evident that friction will be the most important process parameter in the complex geometries. In the case of simple geometry such as cylindrical cup, coefficient of friction has more influence on cup quality as compared to holding force [2], and in the present study, friction has maximum influence; thus, this validates the results of the present study. Further, in forming of thin sections of complex geometry, friction dominates the process and the contribution will be higher.

Table 5 Percentage contribution of each parameter (square cup)

Process parameters	Sum of squares (SS)$_i$	Percentage contribution $\left(\frac{SS_i}{SS} \times 100\right)$
Punch velocity	0.000704	15.8746
Blank holding pressure	0.000324	7.303427
Friction between contact surfaces	0.003408	76.82197

4 Conclusion

Many governing parameters are important for the execution of process. Holding forces, punch speed, and friction coefficient are the important process parameters, whereas die corner radius, punch nose radius, clearance, and thickness are the important machine parameters in drawing process. Wrinkling tendency at flange is a major challenge to control, and the optimum setting of parameters can control it. Square geometry was considered, and the analysis of variance was performed. The analysis of variance shows that the coefficient of friction has been the most influential and dominating parameter and its contribution was evaluated as 76.8% which is maximum among three parameters. The second influential parameter was velocity of the punch, and 15.8% contribution was evaluated. The contribution of holding force was minimum, and its contribution was evaluated as 7.3%. So it is evident that when the geometry of the object is complex and thin sheet is used for forming purpose, then friction coefficient between the surfaces plays the most vital role for defectless final product. In the case of simple geometry such as cylindrical cup, the coefficient of friction has more influence on cup quality as compared to holding force [2]. Thus, it can be concluded that deep drawing is a very complex process and many influential factors are to be set at optimum level. Further, the process parameters' influence also depends on the geometry type and thickness of sheet. In the case of thin sheet friction, the coefficient is the most influential parameter.

References

1. Amit, J., Narasimhan, K., Date, P.P., Maiti, S.K., Singh, U.P.: Sensitivity analysis of a deep drawing process for miniaturised products. J. Mater. Process. Technol. **147**, 321–327 (2004)
2. Padmanabhan, R., Oliveira, M.C., Alves, J.L., Menezes, L.F.: Influence of process parameters on the deep drawing of stainless steel. Finite Elem. Anal. Des. **43**(14), 1062–1067 (2007)
3. Allen, S.J., Mahdavian, S.M.: The effect of lubrication on die expansion during the deep drawing of axisymmetrical steel cups. J. Mater. Process. Technol. **199**(1), 102–107 (2008)
4. Yang, T.S.: Investigation of the strain distribution with lubrication during the deep drawing process. Tribol. Int. **43**(5), 1104–1112 (2010)
5. Hu, Z.: Realisation and application of size dependent FEM-simulation for deep drawing of rectangular work pieces. CIRP J. Manufact. Sci. Technol. **4**(1), 90–95 (2011)

6. Karupannasamy, D.K., Hol, J., de Rooij, M.B., Meinders, T., Schipper, D.J.: Modelling mixed lubrication for deep drawing processes. Wear **294**, 296–304 (2012)
7. Tommerup, Søren, Endelt, Benny: Experimental verification of a deep drawing tool system for adaptive blank holder pressure distribution. J. Mater. Process. Technol. **212**(11), 2529–2540 (2012)
8. Van, T.P., Jöchen, K., Böhlke, T.: Simulation of sheet metal forming incorporating EBSD data. J. Mater. Process. Technol. **212**(12), 2659–2668 (2012)
9. Amir, Atrian, Fereshteh-Saniee, Faramarz: Deep drawing process of steel/brass laminated sheets. Compos. B Eng. **47**, 75–81 (2013)
10. Tropp, M., Tomasikova, M., Bastovansky, R., Krzywonos, L., Brumercik, F.: Concept of deep drawing mechatronic system working in extreme conditions. Procedia Eng. **1**(192), 893–898 (2017)
11. Neto, D.M., Oliveira, M.C., Santos, A.D., Alves, J.L., Menezes, L.F.: Influence of boundary conditions on the prediction of springback and wrinkling in sheet metal forming. Int. J. Mech. Sci. **1**(122), 244–254 (2017)
12. Luo, L., Jiang, Z., Wei, D.: Reprint of Influences of micro-friction on surface finish in micro deep drawing of SUS304 cups. Wear **15**(376), 1147–1155 (2017)
13. Linvill, E., Wallmeier, M., Östlund, S.: A constitutive model for paperboard including wrinkle prediction and post-wrinkle behavior applied to deep drawing. Int. J. Solids Struct. **15**(117), 143–158 (2017)
14. Yang, S., McPhillimy, M., Supri, T.B., Qin, Y.: Influences of process and material parameters on quality of small-sized thin sheet-metal parts drawn with multipoint tooling. Procedia Manuf. **1**(15), 992–999 (2018)
15. Singh, A., Basak, S., Lin Prakas,P.S., Roy, G.G., Jha, M.N., Mascarenhas, M., Panda, S.K.: Prediction of earing defect and deep drawing behavior of commercially pure titanium sheets using CPB06 anisotropy yield theory. J. Manuf. Process **30**(33), 256–267 (2018)
16. Saxena, R.K., Dixit, P.M.: Numerical analysis of damage for prediction of fracture initiation in deep drawing. Finite Elem. Anal. Des. **47**(9), 1104–1117 (2011)

Selection of Machining Parameters in Ultrasonic Machining Process Using CART Algorithm

Shruti Dandge and Shankar Chakraborty

Abstract Data mining, also referred as 'knowledge mining from data', is the process in which useful information is extracted from a data and is transformed into an understandable pattern or structure. In this paper, one of the major data mining techniques known as classification method is applied to identify the most important machining parameters of ultrasonic machining of WC-Co composite material. The decision tree is developed by classification and regression tree algorithm (CART) to analyse the most predominant process parameters affecting the responses. The results of decision tree presented the effects and the major contribution of various machining parameters on material removal rate and tool wear rate. Analysis of variance is also applied to validate the outcomes of considered data mining technique.

Keywords Data mining · Classification · Decision tree induction · USM · CART algorithm

1 Introduction

Ultrasonic machining (USM) is a mechanical non-traditional machining process typically used for machining materials such as carbides, glass, ferrites, ceramics, quartz and germanium materials which possess high hardness/brittleness. In USM, a transducer is used to convert high-frequency electrical energy into mechanical vibrations which are then transmitted to an amplifying device known as horn (or booster). This causes the vibration of tool along its longitudinal axis at ultrasonic frequency approximately 20,000 Hz. Load is applied to the tool to provide feed in the vertical or orthogonal direction. Abrasive slurry continuously runs within the gap

S. Dandge (✉)
Department of Mechanical Engineering, Government Polytechnic, Murtizapur, Maharashtra, India
e-mail: shruti.dandge@gmail.com

S. Chakraborty
Department of Production Engineering, Jadavpur University, Kolkata, West Bengal, India
e-mail: s_chakraborty00@yahoo.co.in

© Springer Nature Singapore Pte Ltd. 2020
R. Venkata Rao and J. Taler (eds.), *Advanced Engineering Optimization Through Intelligent Techniques*, Advances in Intelligent Systems and Computing 949,
https://doi.org/10.1007/978-981-13-8196-6_52

between the tool and the workpiece. As the tool vibrates, the abrasive particles which are contained in the slurry strikes against the work surface which removes the metal in the form of micro-chip [1]. The variation of process parameters on machining characteristics of USM process had been reported by various investigators. It has been shown that power rating and grit size contributes maximally in controlling characteristics such as surface finish, tool wear, material removal and taper angle in USM operation. Chakravorty et al. [2] demonstrated and compared the effectiveness of four methods, i.e. weighted signal-to-noise (WSN) ratio method, grey relational analysis (GRA) method, multi-response signal-to-noise (MRSN) ratio method and utility theory (UT) method for optimization of multiple responses in USM process. Kataria et al. [3] applied analytical hierarchical process-based TOPSIS method for multi-objective optimization in ultrasonic drilling process and the results showed that power rating and grit size have a significant effect on cutting ratio, taper angle and a wear in USM. Bastos et al. [4] designed data mining-based prototype to develop the prediction system so as to forecast future failures on the basis of present records in manufacturing units which would estimate the probability of machine breakdown in maintenance activities of industrial units. Dey and Chakraborty [5] presented data mining approach to investigate the contribution of various process parameters of three non-traditional machining processes namely electrical discharge machining, electro discharge milling and laser beam machining for controlling the responses for the said application.

In this paper, data mining technique like decision tree is applied to study the influence of various machining characteristics on the considered responses. Several multi-objective techniques such as grey-Taguchi method, TOPSIS and Evolutionary algorithms have been employed in this direction. But, the application of data mining tools and techniques for analysing different USM characteristics on the considered responses is very limited. Thus the main aim of this paper is focussed on classification method, primarily in the form of classification and regression tree (CART) algorithm.

2 Decision Tree Induction for Classification Using CART

Classification is a form of data mining function in which classifier is constructed to predict the categorical class labels with good accuracy by analysing training data made of tuples and associated class labels. Induction of decision tree is one of the popular methods for classification. A decision tree is a classifier with a tree structure, in which each internal node (non-leaf node) represents a test on an attribute, each branch represents an outcome of the test, and each leaf node (or terminal node) shows a predicted class label. The initial state of a tree is the root node. Construction of decision tree is divided into two main steps: First, the growing stage in which the tree is actually built, while the second is the pruning stage to eliminate some tree nodes in order to avoid over-fitting, thus omitting the attributes that are not 'beneficial' for classification. The growing stage is the most important one, and it is based on selecting the order under which attributes are tested in each internal tree node, and the attribute value that splits to create a new tree node. Tree pruning identifies the noise or outliers in the training data, and thereby, eliminates the attributes which seems to be not useful for classification so as to improve the accuracy of data classification. Here, classification and regression trees (CART) algorithm available in STATISTICA software is used to construct the decision tress and develop the relationship between the predictor variables and the target variables. The decision tree starts with a single root node. During the construction of tree, the method first selects the predictor or attributes which best partitions the datasets into two different groups. Attribute selection function, called Gini index, is used in CART which measures the impurity of the node. The datasets which show minimum Gini index for a particular attribute, the same is chosen as the best splitting variable for the datasets at the time of classification [6]. At each stage, the two groups are further divided into subgroups on the basis of same or different predictor variables. In this manner, the tree continues to grow till all the decision nodes attain the state of terminal nodes which represents them in their pure form. At this point, the partitioning of attributes ends and all the elements in a node have the same class, i.e. a split is no longer possible. Further, decision rules can be extracted from the decision trees which can be easily understood especially if the constructed decision tree is large.

Thus, corresponding decision trees are developed by means of binary CART algorithm in STATISTICA software which helps to analyse the effects of USM parameters on responses (MRR and TWR). The tree induction CART algorithm procedure is summarized as follows:

Step1: The algorithm runs through a set of training samples, D.
Step2: If all datasets in D belong to class P, generate a node P and stop, otherwise decide an attribute F and produce a decision node.
Step3: Split the training samples in D into possible subsets in accordance with values of V by the attribute selection measure called Gini index.
Step4: Identify the optimum break points by selecting the predictor in such a manner that the dependent variable serves to be best split into two classes characterized by maximum internal uniqueness and maximum external differentiation.
Step5: Apply the algorithm recursively for all the subsets in D until all items or samples in a node have the same class, i.e. split is no longer possible.
Step6: Framing of decision rules, i.e. 'if... then...' rules. Rule: (Condition) → Y, where the condition is a combination of attributes and Y is the class label.

Kataria et al. conducted 36 experiments to study the influence of various experimental settings by variation in cobalt content, the thickness of work piece, abrasive grit size, tool profile, tool material and power rating on responses such as material removal rate (MRR) and tool wear rate (TWR) in the ultrasonic drilling of WC-Co composite material [7]. Each of the process parameters was set at different levels, i.e. cobalt content (6, 24%), thickness of work piece (3, 5 mm), profile of tool (solid, hollow), tool material (stainless steel, silver steel, nimonic-80A), grit size (200, 320, 500) and power rating (40, 60, 80%). Thus, the experimental data of this USM process is subjected to CART analysis (available in STATISTICA software) to identify the most influencing characteristics for achieving the enhanced machining response values by developing the corresponding decision trees with the following specifications:

(a) Split selection method—CART style exhaustive search for univariate splits
(b) Measure of goodness of fit-Gini index, Prior probabilities-estimated, misclassification cost-equal
(c) Stopping rule-Prune on misclassification error
(d) Stopping parameters—minimum $n = 5$
(e) Standard error rule—1.0
(f) Sampling—seed for random number generator $= 12$
(g) V-fold cross-validation, v-value $= 3$.

3 Analysis of Decision Tree

The data mining technique by means of CART algorithm is applied for analysing the influence of various USM parameters on responses MRR and TWR, which resulted in the generation of decision tree and induction of rules. The decision tree for MRR, as depicted in Fig. 1, is developed to analyse the variation of different machining characteristics (attributes) on MRR values. In this tree, 'Class 1' label comprise of MRR values ≤0.0150 g/min and 'Class 2' label denotes MRR values >0.0150 g/min where 0.0150 is the average value of MRR as obtained from experimental data. The developed tree has 6 numbers of split and 7 numbers of terminal nodes. Terminal nodes (leaves), shown by red outlines are those points on the tree where no further classification is possible, i.e. all elements in the node belongs to same class. While rest of the decision nodes are outlined with blue lines. The decision tree begins with a single first node, i.e. the top root node, which is represented as node 1. At first, all 36 datasets are allotted to root node and grouped as Class 1, as indicated by Class 1 label in the top extreme right of this node. At the initial classification stage, Class 1 is selected because there are more Class 1 datasets as represented by the histogram shown inside the node. The root node then generates two new child nodes with power rating as the first best split variable. The datasets with power rating values lower than or equal to 50% are routed to second node and assigned as Class 1 label and those with values greater than 50% are assigned to node number 3 and marked with Class 2 label. The attribute values 12 and 24 described above the nodes 2 and 3, respectively, signifies the number of cases sent to each of these two child nodes from their internal or (decision) node. At node 2, vertical histograms with same length above and below the reference line recognized it as a terminal node where no further partitioning or classification can be possible. On the other hand, node 3 acts as an impure node which can generate new child nodes due to unequal lengths of histograms displayed in the same. Thus, from node 3, 16 datasets with grit size values less than or equal to 410 (mesh) are classified as Class 2 at node number 4 whereas node number 5 is grouped as Class 1 classification with remaining 8 datasets having grit size greater than 410 (mesh). Based on the solid tool profile attribute, node 4 is further split into two nodes

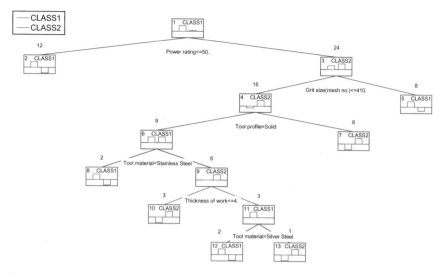

Fig. 1 Decision tree for material removal rate

with 8 datasets labelled as Class 2 for node number 7 which can be identified as a pure node, while 8 datasets are assigned to node 6 and categorized as Class 1. Thus, based on different machining characteristics such as tool material and thickness of work, the procedure of splitting of nodes continues further until the terminal nodes are reached and the nodes can no further be split. It can be clearly noticed, all the terminal nodes are as pure as possible, containing no misclassification error. It is observed that all six process parameters, i.e. cobalt content, the thickness of work piece, tool profile, tool material, abrasive grit size and power rating are responsible for performing the split operation. Hence these six parameters influence the response MRR, out of which power rating has the major contribution on MRR values.

From the developed decision tree, the following decision rules can be framed to aid the classification process.

Rule 1: If power rating ≤50%, then material removal rate ≤0.0150 g/min.
Rule 2: If power rating >50% and grit size ≤410 (mesh), then material removal rate >0.0150 g/min.
Rule 3: If power rating >50% and grit size >410 (mesh), then material removal rate ≤0.0150 g/min.
Rule 4: If power rating >50%, grit size ≤410 (mesh) and tool profile = solid, then material removal rate >0.0150 g/min.
Rule 5: If power rating >50%, grit size ≤410 (mesh), tool profile = solid and tool material = stainless steel, then material removal rate ≤0.0150 g/min.
Rule 6: If power rating >50%, grit size ≤410 (mesh), tool profile = solid, tool material = stainless steel and thickness of work ≤4 mm, then material removal rate >0.0150 g/min.

Rule 7: If power rating >50%, grit size ≤410 (mesh), tool profile = solid, tool material = stainless steel, thickness of work >4 mm and tool material = silver steel, then material removal rate >0.0150 g/min.

It is interesting to know that when the univariate split is performed, the predictor variables are rated on a 0–100 scale according to their potential importance for responses. In this case, it can be revealed from the Fig. 2 that power rating plays a vital role in controlling the response MRR. On the other hand, cobalt content and thickness of work is the least significant attribute in the classification process. Also, this can be reconfirmed from the ANOVA results of response variable MRR depicted in Table 1.

Similarly, for another response variable, i.e. tool wear rate, the corresponding decision tree is developed which is exhibited in Fig. 3 to determine the variation of six machining characteristics on tool wear rate. In this figure, datasets with tool wear rate less than or equal to 0.0037 g/min are termed as 'Class 1' label, whereas datasets with TWR values greater than 0.0037 g/min are grouped under 'Class 2' label (where 0.0037 is the calculated average value of TWR). Initially, at the root node, all the

Fig. 2 Relative importance of variables for MRR

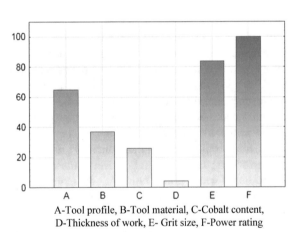

A-Tool profile, B-Tool material, C-Cobalt content, D-Thickness of work, E- Grit size, F-Power rating

Table 1 Analysis of variance for MRR

Source	DF	Adj SS	Adj MS	F-value	% contribution
Cobalt content	1	0.000138	0.000138	5.64	2.90
Thickness of work piece	1	0.000054	0.000054	2.19	1.13
Tool profile	1	0.000250	0.000250	10.22	5.25
Tool material	2	0.000370	0.000185	7.55	7.78
Grit size (mesh no.)	2	0.001036	0.000518	21.16	21.79
Power rating	2	0.002270	0.001135	46.35	47.74
Error	26	0.000637	0.000024		13.39
Total	35	0.004754			99.98

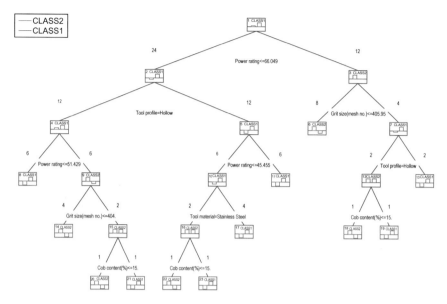

Fig. 3 Decision tree for tool wear rate

instances are primarily classified into Class 1 label, as there are more number of class 1 values. At the first split, datasets with power rating values of less than or equal to 66.049% are subjected to node number 2 and labelled as Class 1. There are 24 such datasets as shown in node 2. On the other hand, datasets with power rating greater than 66.049% are grouped as Class 2 label at node number 3. With grit size (mesh) as the splitting criteria, node 3 is further split in which, 8 items with grit size less than or equal to 405.95%, classified as Class 2 are sent to node 6 and with 4 items having grit size greater than 405.95 are merged in node number 7 and labelled as Class 1. At this point, node number 6 represents a terminal node. In this pattern, based on different splitting variables such as tool profile, tool material and cobalt content the process of classification continues until all the terminal nodes arrive where no further splitting is possible. In this tree, there are 11 split and 12 numbers of terminal nodes. From this tree, the following decision rules can be framed for TWR.

Rule 1: If power rating ≤66.049%, then tool wear rate ≤0.0037 g/min.
Rule 2: If power rating >66.049% and grit size ≤405.95, then tool wear rate >0.0037 g/min.
Rule 3: If power rating >66.049%, grit size >405.95 and tool profile = hollow, then tool wear rate ≤0.0037 g/min.
Rule 4: If power rating >66.049%, grit size >405.95, tool profile = hollow and cobalt content ≤15%, then tool wear rate >0.0037 g/min.
Rule 5: If power rating ≤66.049%, tool profile = hollow and tool material = stainless steel, then tool wear rate ≤0.0037 g/min.

Rule 6: If power rating ≤66.049%, tool profile = hollow, tool material = stainless steel and cobalt content ≤15%, then tool wear rate ≤0.0037 g/min.

Rule 7: If power rating ≤66.049%, tool profile = hollow and power rating ≤51.429, then tool wear rate ≤0.0037 g/min.

Rule 8: If power rating ≤66.049%, tool profile = hollow, power rating >51.429 and grit size ≤404, then tool wear rate >0.0037 g/min.

Rule 9: If power rating ≤66.049%, tool profile = hollow, power rating >51.429, grit size >404 and cobalt content ≤15%, then tool wear rate >0.0037 g/min.

Rule 10: If power rating ≤66.049%, tool profile = hollow, power rating >51.429, grit size >404 and cobalt content >15%, then tool wear rate ≤0.0037 g/min.

Figure 4 shows the importance of the considered machining parameters in influencing TWR, and it becomes quite apparent that power rating has the significant contribution for controlling TWR followed by grit size, tool profile, cobalt content and tool material. The role of power rating significantly affecting TWR values in USM process is also reassured from the ANOVA results of Table 2.

Fig. 4 Relative importance of variables for TWR

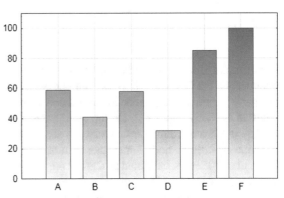

A-Tool profile, B-Tool material, C-Cobalt content, D-Thickness of work, E- Grit size, F-Power rating

Table 2 Analysis of variance for TWR

Source	DF	Adj SS	Adj MS	F-value	% contribution
Cobalt content	1	0.000003	0.000003	1.87	1.48
Thickness of work piece	1	0.000004	0.000004	2.32	1.98
Tool profile	1	0.000005	0.000005	3.00	2.47
Tool material	2	0.000020	0.000010	6.01	9.90
Grit size (mesh no.)	2	0.000054	0.000027	16.39	26.73
Power rating	2	0.000075	0.000037	22.83	37.12
Error	26	0.000043	0.000002		21.28
Total	35	0.000202			100

4 Conclusions

In this paper, a data mining approach employing CART algorithm is applied to USM process to identify the most important input parameters influencing the responses. From the developed decision trees, it is observed that power rating parameter has the maximum contribution in controlling the considered responses. The decision rules are framed for corresponding decision trees which shows the contribution of six process parameters on considered responses. The relative importance diagram for MRR and TWR also confirms the outcomes gained from the decision trees. At last, all these observations are validated using the ANOVA results. This classification method can also be applied to other machining processes to explore the effects of different operational characteristics on the responses.

References

1. Groover, M.P.: Fundamentals of Modern Manufacturing-Materials, Processes and Systems, 4th edn. Wiley, Hoboken (2010)
2. Chakravorty, R., Gauri, S.K., Chakraborty, S.: Optimization of multiple responses of ultrasonic machining (USM) process: a comparative study. Int. J. Ind. Eng. Comput. 4, 285–296 (2013)
3. Kataria, R., Kumar, J., Pabla, B.S.: Ultrasonic machining of WC-Co composite material: experimental investigation and optimization using statistical techniques. J. Eng. Manuf. 231(5), 867–880 (2017)
4. Bastos, P., Lopes, I., Pires, L.: Application of Data Mining in a Maintenance System for Failure Prediction. Taylor and Francis, Routledge, pp. 933–940 (2014)
5. Dey, S., Chakraborty, S.: Parameter selection in non-traditional machining processes using a data mining approach. Decis. Sci. Lett. 4, 211–226 (2015)
6. Han, J., Kamber, M., Pei, J.: Data Mining Concepts and Techniques, 3rd edn. Elsevier, Waltham, USA (2012)
7. Kataria, R., Kumar, J., Pabla, B.S.: Experimental investigation and optimization of machining of WC-Co composite using GRA method. Mater. Manuf. Processes 31(5), 685–693 (2015)

PSF-Based Spectrum Occupancy Prediction in Cognitive Radio

Jayesh Patil, Neeraj Bokde, Sudhir Kumar Mishra and Kishore Kulat

Abstract A primary user (PU) is the licensed owner of the spectrum allocated to it. Studies have shown that, for most of the time, spectrum allotted to the PU is unused. This creates an opportunity for the secondary user (SU) to access the spectrum allocated to PU when it is not in use. While opportunistically accessing the unused spectrum, the first step is spectrum sensing. A prior knowledge of spectrum occupancy helps in observing the used spectrum and also detecting the spectrum holes (unused spectrum). In cognitive radio network (CRN), due to the stochastic nature of the spectrum usage, it is difficult to obtain the prior information about the spectrum occupancy of PU. Real-time spectrum sensing and its assignment to SU take a significant amount of time. Hence, to reduce latency for SU, we can resort to the prior prediction of the occupancy. In this paper, we investigate the existing spectrum prediction methods and propose a pattern-sequence-based forecasting (PSF) method for spectrum allocation to SU. The occupancy study of global system for mobile communication (GSM) downlink band from 935 to 960 MHz is carried out and compared to the predicted occupancy.

Keywords Cognitive radio · Spectrum occupancy · Spectrum hole · Prediction · Time series · PSF

J. Patil (✉) · N. Bokde · S. K. Mishra · K. Kulat
Visvesvaraya National Institute of Technology, Nagpur, India
e-mail: jaypatil@students.vnit.ac.in

N. Bokde
e-mail: neeraj.bokde@students.vnit.ac.in

S. K. Mishra
e-mail: skmishra@students.vnit.ac.in

K. Kulat
e-mail: kdkulat@ece.vnit.ac.in

© Springer Nature Singapore Pte Ltd. 2020
R. Venkata Rao and J. Taler (eds.), *Advanced Engineering Optimization Through Intelligent Techniques*, Advances in Intelligent Systems and Computing 949,
https://doi.org/10.1007/978-981-13-8196-6_53

1 Introduction

There is an exponential growth in the popularity of new wireless application and devices for which the demand to access the radio spectrum is also increasing [1]. Radio spectrum is one of the most important assets; hence, it is very important that it should be managed properly. Efficient use of the existing spectrum is one of the major areas of research for the past many years. Studies have shown that the existing spectrum is inefficiently allocated [1]. A cognitive radio (CR) network consists of a primary user (PU) which has the legal rights to use the allocated spectrum, whereas a secondary user (SU) uses the spectrum in such a way that it causes no interference to the PU. An improper allocation of the existing spectrum gives rise to the spectrum hole, which can be opportunistically accessed by the SU. A spectrum hole provides an opportunity for a SU to access the spectrum of PU (licensed owner). This need can be fulfilled by dynamic spectrum access (DSA) which is a promising solution to deal with the growing demand for spectrum access.

The performance of a CR network depends on: spectrum sensing, spectrum decision, spectrum sharing, and spectrum mobility [2]. In spectrum sensing, the SU senses the presence (occupancy) of PU and thereafter recognizing the spectrum hole in the desired licensed band. In spectrum decision, the SU has to make a decision, whether to use the licensed band for CR purpose or not. If the band is faithful for CR application, then spectrum sharing is done but without any interference to the PU. As soon as the PU is detected, the evacuation from that channel is termed as spectrum mobility. Considering the fact that CR functions discussed above to introduce a significant amount of time delay which cannot be neglected, in order to minimize these delays one of the solutions is to predict the spectrum in advance.

Spectrum prediction is done to reduce the latency in the CR network. But spectrum prediction itself is a challenging task as it requires several processes to be considered such as PU activity prediction, channel status prediction, transmission rate prediction, and radio-environment prediction.

2 Literature Review

In this section, some of the states of art techniques used in cognitive spectrum predictions are introduced such as the hidden Markov model (HMM), neural network (NN), and the statistical method. A brief review of the HMM-based approach, followed by the artificial neural network, Bayesian inference, moving average, and autoregressive model-based prediction for spectrum sensing is provided.

2.1 Hidden Markov Model (HMM)-Based Prediction

Clancy and Walker [3] were first to use an HMM, Rabiner, and Juang [4] channel occupancy was modeled using HMM. In [5], the channel was modeled as Poisson distribution, and an HMM was used to predict the availability of the channel. To avoid the collision during transmission, an HMM trained using Baum-Welsh (BWA) was used. In [6], known-state sequence HMM (KSS-HMM) algorithm was used for channel activity prediction.

2.2 Artificial-Neural-Network-Based Prediction

Artificial neural networks (ANNs) are nonlinear mapping structures. These are based on the function of the human brain using multiple-layer perceptron (MLP). An MLP is a network of neurons also called as a perceptron. An MLP consists of one input layer and one output layer, whereas there is no restriction on the addition of hidden layers. It has been observed that an MLP-based prediction is slightly accurate compared to the HMM-based prediction [7]. This approach can be helpful in reducing the energy consumed by SU.

2.3 Bayesian-Interference-Based Prediction

Predicting a spectrum having stochastic but deterministic patterns has grabbed the attention of most researcher [8–10]. In [11], a spectrum occupancy prediction using Bayesian theorem approach is proposed, which ensembles Bayesian and exponential-weighted moving average (EWMA) to predict the spectrum hole. Using the Bayesian approach, the bit error rate is found to less at certain data distributions with less computational requirements, but at the cost of poor prediction performance. In [12], the distribution of intervals and the method to estimate the next transmission window were studied using Bayesian interference under various traffic loads. Also, Bayesian online learning (BOL) was used in [13] to predict the changes in the time series in advance and to track the activity of primary user efficiently.

2.4 Moving Average and Autoregressive-Model-Based Prediction

To predict a progression in a series of values moving-average-(MA)-based prediction is used. The autoregressive model (ARM) is used to predict the future channel states based on some previous observations over fading channels [14]. In ARM-based prediction, the model parameters are estimated using the Yule-Walker equation or some other approaches. Then, the observed sequences are used as input to the prediction rule and predict the future states. In [14], an ARM-based spectrum hole prediction model is used to predict the spectrum hole over fading channels. Each SU predicts the model parameters using the Yule-Walker equation which are further used as input to the prediction rule. Upon this model, Kalman filter is also used for the prediction of channels. The results show that the solution obtain is suitable for a Gaussian channel. For non-Gaussian noise, they proposed to use the particle filter. It is rooted in Monte Carlo simulation and Bayesian estimation. In [15], time series modeling (ARIMA) and machine learning (ML) techniques are discussed and compared. It is observed that when there is a periodic trend in the captured dataset, the time series model performs well with lesser computational complexity and better accuracy. Time series analysis is more helpful when the variations are modeled in the time-domain, whereas it fails when other parameters such as power threshold are incorporated in it [16]. The ML techniques are more accurate, but it comes at the cost of higher time consumption.

3   Pattern-Sequence-Based Forecasting Method

The pattern-sequence-based forecasting (PSF) method was proposed by [17, 18] to forecast a univariate time series with greater accuracy. The PSF has proved its superiority in time series prediction in various domains including electricity load forecasting [17], wind speed and power predictions [19, 20], and many others. In this method, any pattern present in the given time series can influence the results. In this method, marking (labeling) is used for patterns in the given data series. The normalization process is used in order to remove the redundancies in the raw captured data, which is as shown in expression 1:

$$X_j \leftarrow \frac{X_j}{\frac{1}{N} \sum_{i=1}^{N} X_j} \tag{1}$$

where N is its size in time units and X_j is the jth value of each cycle in time series. The marking used for different patterns obtain in the original input data time series is done by k-mean clustering. k-mean clustering is simple and consumes less time. Taking account of the above factors, the input series is transformed into series of marking (labels) which are further going to be used for the prediction purpose.

The further procedure of PSF method includes window-size selection, which is then followed by pattern sequence matching and estimation process. A marked series which is picked from the last position of given input data series of length W and then searched in the marked series which was obtained by clustering. A "window" is referred to sequence of markings of size W. If at all in investigating the "window" in the marked series if it is not found size is reduces by one unit. This is an iterative process. Iterating this process until the sequence is found confirms that some of the sequences are repeated in the labeled time series when the value of W is 1. Mean value of the closest obtained marking (labels) with expression 2 is used for the prediction of the next value.

$$\widehat{X}(t) = \frac{1}{size(ES)} \times \sum_{j=1}^{size(ES)} X(j) \tag{2}$$

where ES is vector containing the marked (labeled) sequence, X is the input data series, and size(ES) is the length of vector ES. The final step is denormalizing, essentially used to replace the labels with some suitable value in the real dataset. To keep the error low, the "window" size must be optimum. The optimum size selection of W is done by minimizing the expression 3,

$$\sum_{t \in T_s} \left\| \widehat{X}(t) - X(t) \right\| \tag{3}$$

where $\widehat{X}(t)$ is predicted value and $X(t)$ original values at time index t. The block diagram of the PSF method is given in Fig. 1. The further details of PSF methodology and its R package [21] are discussed in [22].

4 Measurement Systems

A careful selection of a site for occupancy measurement is an important aspect of the spectrum measurement survey. To identify or sense the frequency areas where the PU is not active is very crucial when it comes to spectrum occupancy measurement. Spectrum occupancy has a lot to do with the geographical location (area) where the sensing is done. As the geographical area changes, the occupancy of the frequency band also changes. Some factors which we should consider while spectrum measurement is: the site should have a limited number of the transmitter to prevent the

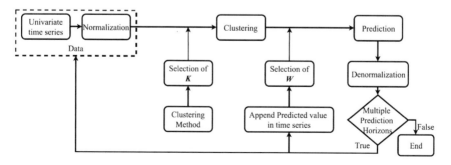

Fig. 1 Block diagram of PSF method. *Source* [20]

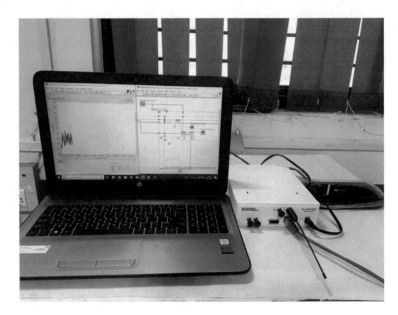

Fig. 2 Spectrum monitoring with USRP

intermodulation problem and man-made noise at measurement site must be maintained to lower level. One of the examples of man-made noise is impulsive noise from the ignition of auto-mobile vehicles [16].

The frequency spectrum for 935–960 MHz band was captured for measurement that has been performed at Department of Electronics in Visvesvaraya National Institute of Technology, Nagpur (India). The first step in determining the occupancy of the spectrum and further claim the availability of spectrum hole is spectrum sensing. Hence, it becomes very important than the measurement survey is carried out carefully. For any spectrum monitoring system, the key considerations are the type of equipment, the relevance of data captured, data processing and analyzing.

The system used for spectrum monitoring survey was configured as in Fig. 2. The system consists of USRP which is a software-defined radio (SDR) platform designed to implement software radio systems for various applications. The advantage of using the SDR platform is that it can configure the system multiple times as per the user requirement. The USRP is connected to a computer by Ethernet cable and linked with a software framework. Using the USRP N2922, a user can obtain a digital conversion rate of 100 MS/s on the reception chain and an analog conversion sample rate of 400 MS/s on transmission chain.

In the proposed study, the USRP was connected via a gigabit Ethernet interface to the computer running the spectrum-sensing application. A router switch can be used since the host computer is not having gigabyte port. But, it is completely optional. The spectrum sensing was programmed in the Lab-View software. Energy detection is a

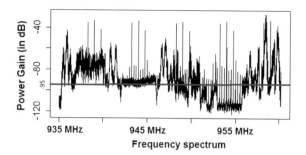

Fig. 3 Frequency spectrum captured with USRP for GSM band (red line represents the threshold at − 95 dBm)

comparison of the signal power to a computed threshold level, the step of calibrating the received signal power turned out to be an important step toward a more accurate decision.

5 Performance Comparison and Results

To evaluate the performance of prediction methods, there are three error performance comparison measures used as, mean absolute error, root mean square error and mean absolute percentage error, which are given as shown in expressions 5, 4, and 6, respectively.

$$\text{MAE} = \frac{1}{N} \sum_{i=1}^{N} \left| X_i - \widehat{X}_i \right| \tag{4}$$

$$\text{RMSE} = \sqrt{\frac{1}{N} \sum_{i=1}^{N} \left| X_i - \widehat{X}_i \right|^2} \tag{5}$$

$$\text{MAPE} = \frac{1}{N} \sum_{i=1}^{N} \frac{\left| X_i - \widehat{X}_i \right|}{X_i} \times 100\% \tag{6}$$

For prediction analysis, the frequency spectrum for 935–960 MHz is captured for five consecutive days (July 1 to 5, 2018) at an interval of an hour so that 5 days × 24 h = 120 total spectra are generated. The frequency spectrum of 1 h captured within the selected days is shown in Fig. 3.

The power gain at this spectrum band varies from −120 to −40 dBm. Then, the occupancy of each spectrum is evaluated for each hour with −95 dBm as a threshold, which is shown in the red line. The occupancy is measured as the percentage of power values of spectrum above the threshold value. Eventually, a spectrum occupancy time series is generated for six days. Further, this time series is predicted with distinct methods and then compared to the state-of-art methods. The PSF method is

Fig. 4 Spectrum occupancy
prediction with PSF method
(multiple-output strategy)

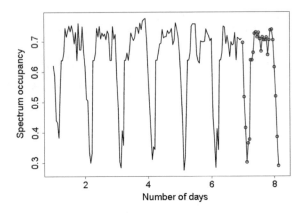

Table 1 Comparison of error
in spectrum occupancy
prediction

Methods	FFNN	LSVM	ARIMA	PSF
RMSE	0.065	0.107	0.083	0.026
MAE	0.049	0.088	0.062	0.020
MAPE	7.213	10.753	9.354	3.834

well known for its capability to capture the pattern present in a time series. Similarly, in the case of spectrum occupancy time series, the spectrum occupancy values
predicted for the next day (next 24 h). Using PSF with a multiple-output strategy is
shown in Fig. 4, whereas Fig. 5 shows the prediction results with ARIMA method.
(In Figs. 4 and 5, black lines represent the measure spectrum occupancy and blue
line are for the predicted data points) Here, "multiple-output strategy" means the
approach with which predicting entire prediction horizon (24 h) in a one-shot manner. Figures 4 and 5 are the good evidence to state that the PSF method is performing
better than ARIMA with multiple-output strategy. The PSF method captured the time
series pattern accurately for 2-h horizon in one shot (Fig. 4). This is of the inherent
superiority of PSF over other methods. Further, the study includes a comparison of
PSF method with continuous time series prediction method including ARIMA, feedforward neural network (FFNN), and Lagrangian support vector machine (LSVM).
The PSF method is operated in multiple-output strategy. Whereas, all other methods
are operated in direct multi-step forecast strategy, because these methods are not
able to perform efficiently in multiple output strategy. For the measured spectrum
occupancy time series, the comparison of methods is shown in Table 1. For all error
measure, the PSF method is having a minimum error. Also, the accuracy of prediction is exhibited in Fig. 6, which shows that the PSF has predicted the pattern more
accurately than ARIMA method.

Fig. 5 Spectrum occupancy prediction with ARIMA method (multiple-output strategy)

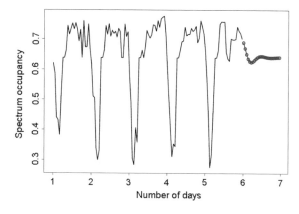

Fig. 6 One step ahead spectrum occupancy prediction with PSF and ARIMA method (multiple-output strategy)

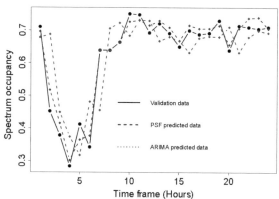

6 Conclusion

Spectrum occupancy prediction was performed for GSM downlink band. The data was analyzed for the band of 935–960 MHz keeping the threshold levels at −95 dBm. The data was analyzed using the PSF model. The predicted series was found to be well predicted as compared to ARIMA prediction. From the occupancy analysis, it was observed that most of the band is sparse for a maximum period of time which is well suited for a cognitive radio application. However, the prediction is a promising approach which helps to reduce the latency further helps in better realization of cognitive radio functions. The PSF method outperforms ARIMA when it comes to predicting time series for long-term accurately. Hence, the error accumulation problem is also minimized.

Bibliography

1. Ravi, S., West, D.M.: Spectrum policy in India. Center for Technology Innovation at Brookings, Washington, DC (2015)
2. Xing, X., Jing, T., Cheng, W., Huo, Y., Cheng, X.: Spectrum prediction in cognitive radio networks. IEEE Wirel. Commun. **20**(2), 90–96 (2013)
3. Clancy, T., Walker, B.: Predictive dynamic spectrum access. In: Proceedings of SDR Forum Technical Conference, vol. 1 (2006)
4. Rabiner, L.R., Juang, B.-H.: An introduction to hidden Markov models. ieee assp Mag. **3**(1), 4–16 (1986)
5. Akbar, I.A., Tranter, W.H.: Dynamic spectrum allocation in cognitive radio using hidden Markov models: Poisson distributed case. In: SoutheastCon, 2007, Proceedings. IEEE, pp. 196–201. IEEE (2007)
6. Devanarayana, C., Alfa, A.S.: Predictive channel access in cognitive radio networks based on variable order Markov models. In: Global Telecommunications Conference (GLOBECOM 2011), 2011 IEEE, pp. 1–6. IEEE (2011)
7. Tumuluru, V.K., Wang, P., Niyato, D.: Channel status prediction for cognitive radio networks. Wirel. Commun. Mob. Comput. **12**(10), 862–874 (2012)
8. Chen, H., Trajkovic, L.: Trunked radio systems: traffic prediction based on user clusters. In: 1st International Symposium on Wireless Communication Systems, 2004, pp. 76–80. IEEE (2004)
9. Moungnoul, P., Laipat, N., Hung, T.T., Paungma, T.: GSM traffic forecast by combining forecasting technique. In: 2005 Fifth International Conference on Information, Communications and Signal Processing, pp. 429–433. IEEE (2005)
10. Papadopouli, M., Shen, H., Raftopoulos, E., Ploumidis, M., Hernandez-Campos, F.: Short-term traffic forecasting in a campus-wide wireless network. In: IEEE 16th International Symposium on Personal, Indoor and Mobile Radio Communications, 2005, PIMRC 2005, vol. 3, pp. 1446–1452. IEEE (2005)
11. Jacob, J., Jose, B.R., Mathew, J.: Spectrum prediction in cognitive radio networks: a Bayesian approach. In: 2014 Eighth International Conference on Next Generation Mobile Apps, Services and Technologies (NGMAST), pp. 203–208. IEEE (2014)
12. Wen, Z., Fan, C., Zhang, X., Wu, Y., Zou, J., Liu, J.: A learning spectrum hole prediction model for cognitive radio systems. In: 2010 IEEE 10th International Conference on Computer and Information Technology (CIT), pp. 2089–2093. IEEE (2010)
13. Mikaeil, A.M.: Bayesian online learning-based spectrum occupancy prediction in cognitive radio networks. ITU J.: ICT Discov. (2017)
14. Wen, Z., Luo, T., Xiang, W., Majhi, S., Ma, Y.: Autoregressive spectrum hole prediction model for cognitive radio systems. In: IEEE International Conference on Communications Workshops, 2008, ICC Workshops' 08, pp. 154–157. IEEE (2008)
15. Agarwal, A., Sengar, A.S., Gangopadhyay, R.: Spectrum occupancy prediction for realistic traffic scenarios: time series versus learning-based models. J. Commun. Inf. Netw. 1–8 (2018)
16. Wang, Z., Salous, S.: Spectrum occupancy statistics and time series models for cognitive radio. J. Sig. Process. Syst. **62**(2), 145–155 (2011)
17. Alvarez, F.M., Troncoso, A., Riquelme, J.C., Ruiz, J.S.A.: Energy time series forecasting based on pattern sequence similarity. IEEE Trans. Knowl. Data Eng. **23**(8), 1230–1243 (2011)
18. Martínez-Álvarez, F., Troncoso, A., Riquelme, J.C., Aguilar-Ruiz, J.S.: LBF: a labeled-based forecasting algorithm and its application to electricity price time series. In: Eighth IEEE International Conference on Data Mining, 2008, ICDM'08, pp. 453–461. IEEE (2008)
19. Bokde, N., Troncoso, A., Asencio-Cortés, G., Kulat, K., Martínez-Álvarez, F.: Pattern sequence similarity based techniques for wind speed forecasting. In: International Work-Conference on Time Series (ITISE), vol. 2, pp. 786–794 (2017)
20. Bokde, N., Feijóo, A., Kulat, K.: Analysis of differencing and decomposition preprocessing methods for wind speed prediction. Appl. Soft Comput. **71**, 926–938 (2018)

21. Bokde, N., Asencio-Cortes, G., Martinez-Alvarez, F.: PSF: Forecasting of Univariate Time Series Using the Pattern Sequence-Based Forecasting (PSF) Algorithm, 2017. R package version 0.4
22. Bokde, N., Asencio-Cortés, G., Martínez-Álvarez, F., Kulat, K.: PSF: introduction to R package for pattern sequence based forecasting algorithm. R J. **9**(1), 324–333 (2017)

Hertzian Contact Stress Analysis in Roller Power Transmission One-Way Clutch by Using Finite Element Analysis

Karan A. Dutt, S. B. Soni and D. V. Patel⊙

Abstract A roller one-way clutch is used for transmitting torque in one direction while running free in the reverse. A special geometrical configuration of the inner race along with rollers has been used to transmit the torque from one race to another race in one direction only due to wedging action developed at two contact regions. During working of the clutch, contact stresses are developed in races. These contact stresses are considered to be very crucial in the design of the clutch in order to transmit the required torque without slippage or failure of any part. The main objective of the present work is to design and optimize various parameters of the clutch in such a way to limit these contact stresses within the permissible limit of the material and also verifying these designs by using finite element analysis. The verification of contact stress has been done by comparing the analytical values of contact stress with FEA values. The most effective parameters which can influence the functionality of the clutch have to be identified first. The optimization is done in sense of varying these parameters by staying in geometrical design range which results in improving working life and torque transmission capability of a roller clutch.

Keywords Roller clutch · One-way clutch · Hertzian contact stress · Finite element analysis

1 Introduction

The roller clutch is most commonly used as holdback device in which the outer race is fixed to machine structure. A unique geometrical shape has been provided to inner

K. A. Dutt (✉) · S. B. Soni · D. V. Patel
Mechanical Engineering Department, Institute of Technology, Nirma University, Ahmedabad 382481, India
e-mail: karandutt43me@gmail.com

S. B. Soni
e-mail: sbsoni1943@gmail.com

D. V. Patel
e-mail: dhaval.patel@nirmauni.ac.in

© Springer Nature Singapore Pte Ltd. 2020
R. Venkata Rao and J. Taler (eds.), *Advanced Engineering Optimization Through Intelligent Techniques*, Advances in Intelligent Systems and Computing 949,
https://doi.org/10.1007/978-981-13-8196-6_54

race in order to fulfil the requirement of one-way torque transmission. It is used to restrict or to stop the reverse motion or to backstop the machine for safety purpose when there is a power loss according to various applications, i.e. in agricultural process plants conveyors, stopping the reverse motion in roller coaster, etc. These freewheel one-way clutch provide the simplest and lowest cost-effective solutions for many applications as compared to clutches based on the pawl and ratchet type.

As the clutch having the main function to transmit the high magnitude of torque in one direction while running free in opposite direction, but by doing so there will be a high impact in between races and cylindrical rollers. This will generate the high stresses on races like high magnitude of contact stress between them and hoop stress in outer race subjected to burst.

2 Theory of Operation

Chesney [1] was the first who has elaborated two modes of operation in one-way clutch. The first mode is engaged mode. During the engaged mode, the clutch transmits a torque from inner race to outer race or vice versa due to the wedging effect through rollers. Significant contact forces are generated at the contact points between the rollers and races. Contact stresses, hoop stresses, and system deflections are the primary concerns in the engaged mode of operation. The second mode is freewheel mode. During freewheel mode, anyone race rotates freely without transmitting any torque to the other race. In roller clutch, rollers act as the wedging element at the contact surface, and it is accomplished through the utilization of inner race geometry and spring plunger arrangement.

2.1 Clutch Geometrical Specifications

The clutch assembly consists of various parts in order to perform its function which is as inner–outer race, spring plunger, and cylindrical rollers.

The contacts at inner race and outer race with rollers develop contact forces during operation which includes tangential and normal forces at inner and outer contact points as shown in Fig. 1. The suffix as 'o' indicates the forces developed at outer contact region, while 'i' indicates the forces at the inner contact region.

2.2 Strut Angle Limit

The contact forces at the inner and outer contact regions can be decomposed into normal force and tangential force as shown in Fig. 1. The basic relationship between tangential and normal forces with strut angle α can be shown by the following equations,

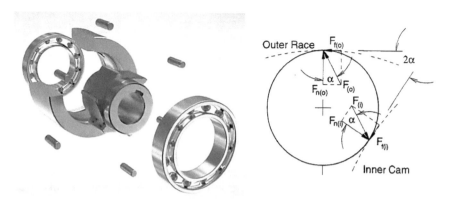

Fig. 1 Roller clutch exploded view with FBD (*Source* NMTG Mechtrans PVT LTD)

$$F_f \leq F \sin \alpha \tag{1}$$

$$F_n \leq F \cos \alpha \tag{2}$$

Also the total torque that can be transmitted from inner race to outer race through rollers is the product of the tangential force between the race and roller, the moment arm (radius of corresponding race at the contact), and the number of rollers (z) as,

$$M_t = F_f \cdot R \cdot z \tag{3}$$

The Strut angle limit depends on the magnitude of tangential force in order to transmit torque without slippage. The tangential force (F_f) is also the frictional force and its magnitude depends on the normal force (F_n) value, but it is limited by the coefficient of static friction value μ_s. From the free body diagram,

$$F_f \leq \mu_s \cdot F_n \tag{4}$$

The relation between μ_s and α can be derived by taking the ratio of (1) and (2),

$$F_f/F_n = \tan \alpha \tag{5}$$

From (4) and (5) the fundamental equation for strut angle limit can be defined as,

$$\tan \alpha \leq \mu_s \tag{6}$$

Equation (6) gives the range of strut angle and for case hardened steel as the value of coefficient of friction is in between 0.08 and 0.12 [2]. Considering the conservative value of 0.08, equation yields the maximum value of strut angle is 4.574°. If the value of strut angle falls above this maximum value, the slip phenomena of rollers will occur when the clutch engages and it will cost the functionality failure of the clutch.

3   Hertzian Contact Stress

When two separate surfaces touch each other in such way that they become mutually
tangent, they are said to be in contact. Contact stresses mainly occur when two bodies
come in contact by face to face, edge to face, or edge to edge under loading condition.
As the contact region is very small, the stresses generated will have high magnitudes.
The contact stress mainly depends on the type of contact surfaces and the material of
the bodies which are in contact. This plays a very important role in the application
like overrunning clutch for transmitting torques under high loading condition. The
contact stresses are theoretically determined from Hertz contact stress theory. This
theory mainly concerns about the stresses developed due to different types of contact
between two bodies. There are mainly two types of contacts one is contact between
two cylindrical bodies and the second one is contact between two spherical bodies. In
the case of one-way clutch, the contact between rollers and races is the type of contacts
between two cylindrical bodies. There are two possible forms of contact between a
race and roller-the concave/convex pair, as in between the outer race and roller and
convex/line contact pair, as in between the inner race and roller. The former is referred
to as conformal contact, whereas the latter is referred to as non-conformal contact
[3]. Non-conformal contact loads lead to higher contact stresses, and therefore, the
inner race/roller contact stresses should only be concerned. According to Hertz, due
to applied normal contact force (P) at contact regions, the maximum stress (σ_{max}) at
contacts in cylindrical connections can be determined from (7) (Fig. 2),

Fig. 2 Effective maximum
stress is the function of the
ratio of large to small
diameters (*Source* Orlov [3],
p 107)

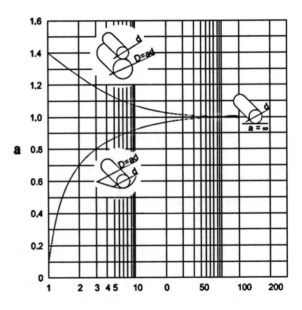

$$\sigma_{max} = 0.6 \sqrt{\left(\frac{P \cdot E}{l \cdot d}\right) \cdot \left(1 + \frac{1}{a}\right)} \, \text{MPa} \qquad (7)$$

4 Design of Roller Clutch

The clutch should have been designed against maximum contact stress at inner region only. This section includes the design calculations for existing model as NMTG NLR 40 and improved model by doing geometrical modifications to identified parameters. The material of the clutch assembly is alloy steel EN353 (Table 1).

4.1 Existing Design

Considering the service factor for repeated high impact loads, the design value of torque can be taken as $[Mt] = $ Service factor $\times$ Mt $= 2 \times 1005 = 2010$ Nm. Now from (1) and (2), the contact forces can be calculated as, $F_t = 5583.3$ N, $F_n = 89,649$ N, $F = P = 89,823$ N. Contact stress at inner contact region as per Hertz contact stress between inner race and roller from (7) will be 3964.7 MPa.

4.2 Improved Design

As the existing clutch having contact stress more than the permissible limits, it should be reduced by changing the roller dimensions by staying in geometrical design range. The contact stress can be reduced by increasing the number of rollers

Table 1 Comparison of design parameters

Parameters	Existing values	Improved values
Roller diameter, d	12 mm	10 mm
Roller length, l	36 mm	34 mm
Number of rollers, z	8	10
Strut angle, α	3.56°	4.574°
Inner diameter of outer race, D_1	90 mm	90 mm
Transmitted torque, T	1005 Nm	1005 Nm
Allowable contact stress (EN353)	3600 MPa	3600 MPa
Modulus of elasticity, E	210 GPa	210 GPa

Source NMTG Mechtrans PVT LTD

as well as reducing the length and diameter of the roller by 2 mm, respectively. The manufacturing of inner race should be in such a way to accommodate the rollers at new strut angle 4.574° keeping the same pitch. The outer race design should remain same.

The new values of contact stress have been calculated by the same equations as discussed earlier [4]. The maximum contact stress value will be at two convex surfaces. Hence considering the critical stress generated only at the at inner contact region, the new values for contact forces will be $F_t = 4467$ N, $F_n = 55{,}832$ N, $F = P = 56{,}011$ N, and contact stress at inner contact region from (7) will be 3528.8 MPa which is less than 3600 MPa.

5 Finite Element Analysis

The contact stress analysis is extremely complex process as the identification of actual contact region is very tedious. The contact analysis problem generally falls in category of nonlinearity due to inappropriate change in the stiffness of contact region also factors like surface roughness, material properties, temperature which also affect the actual result and make the analysis difficult to solve [5]. The following analysis is done in ANSYS as static structural FEA.

5.1 Geometric Modelling and Meshing

The geometric model of roller one-way clutch named as NMTG NLR 40 [6] has been prepared in one of the modelling software. The roller clutch assembly then imported in ANSYS for further analysis as shown in Fig. 3. The second step which is the most important in case of contact stress problem is meshing. Tetrahedron or hex-dominant is more suitable mesh element type for convergence. The contact element size will

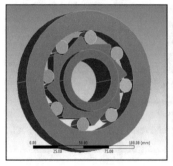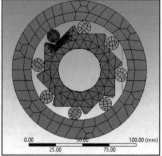

Fig. 3 Geometric model and meshing

be taken as 0.1–0.5 mm at contact regions for getting more accurate results [7]. In this case, two contact regions have been selected at one roller, one at the inner face and second at the outer face of races in order to avoid large computational time as developed contact stresses remain identical at all of the roller inner and outer contacts.

5.2 Boundary Conditions

There are mainly three boundary conditions to be concerned at the time of hold backing operation during which high contact stress has been developed in races.

1. Fixed support: The outer race is kept fixed as required for hold backing purpose.
2. Remote displacement: The relation between rollers and races has been given as rotation about the axis perpendicular to clutch front face.
3. Moment: The inner race has been given anticlockwise moment of magnitude 3800 Nm for hold backing purpose (Fig. 4).

5.3 Contact Stress

The contact stress analysis of one-way clutch has been performed on existing model. The FE analysis results show the maximum contact stress will be at the inner contact region as discussed earlier and the value is 3667.1 MPa. After modifying the clutch as per the new dimensions, the FE analysis results show the value of contact stress as 3471.9 MPa which is within the permissible limits of the material of the clutch (Fig. 5).

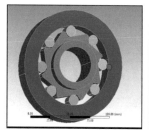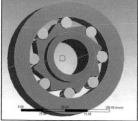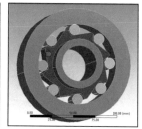

Fig. 4 Specified boundary conditions

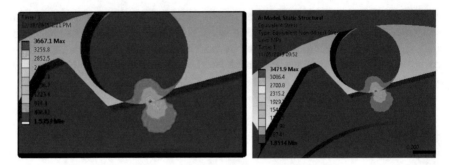

Fig. 5 Maximum contact stress analysis in existing and modified specimen

6 Results and Discussion

By modifying strut angle and increasing the number of rollers with reduced diameters and reduced lengths, the contact stress will be reduced and it will be verified by using FEA. The analysis shows the location of maximum contact stresses between roller and inner race contact region. The contact element size has been taken starting from 5 to 0.5 mm [8]. The result of the analysis includes some of the important factors like as decreasing the contact element mesh size or increasing the number of elements at contact regions, the value of contact stress increases accordingly [9]. The grid sensitivity graph indicates contact stress variations with increasing number of elements by reducing contact element mesh size (Fig. 6; Table 2).

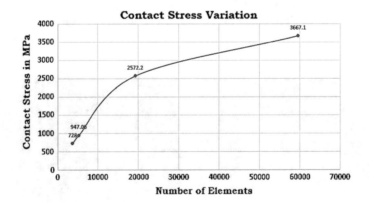

Fig. 6 Grid sensitivity graph for contact stress

Table 2 Contact stress analysis comparison

Model NLR 40	Maximum contact stress, N/mm^2		
Parameters	Theoretical values	FEA values	Safety
Existing design	3964.7	3667.1	>3600 Unsafe
Modified design	3528.8	3471.9	<3600 Safe

7 Conclusions

The aim of this work was to perform design, simulation, and optimization of roller one-way clutch followed by verification using finite element analysis. Regarding the same, the existing model of roller clutch has been analysed using theoretical approach based on solid mechanics. As the important design parameter for the clutch was contact stress induced because of geometrical constraint, the value of the same has been verified using FE-based CAE tools. As the generated stresses are more than safe limit values in existing model, various modifications are also suggested to optimize the performance of the product. This modification has been carried out by staying in geometrical constraint limit. The contact stress will be reduced by increasing the number of rollers with reduced diameter and length also the key role of the strut angle which has to be increased to some extent that the contact stress can be reduced to safe limits compared to existing design. The modified model has also been verified by FE analysis. The FE results showed close convergence with the theoretical results.

Acknowledgements The authors are sincerely grateful to Mr. Deepak Prajapati (Director, NMTG A'bad, India) for providing all facilities and details of the product.

References

1. Chesney, D.R., Kremer, J.M.: Generalized Equations for Roller One-Way Clutch Analysis and Design. SAE Paper 970682, SAE Technical Paper Series (1997)
2. Patil, S.: Friction tooth contact analysis along line of action of a spur gear using finite element analysis. In: International Conference on Advances in Manufacturing and Materials Engineering, AMME, Procedia Mater. Sci. **5**, 1801–1809. Elsevier, New York (2014)
3. Orlov P.: Fundamentals of Machine Design, vol. 1. MIR Publishers, Moscow
4. Design of Roller Type Unidirectional One Way Clutch, Design Data Book. PSG College of Technology, Coimbatore, India. ISBN 978-81-927355-0-4
5. Brezeanu, L.C.: Contact stresses: analysis by finite element method (FEA). In: The 7th International Conference Interdisciplinary in Engineering (INTER-ENG 2013), Procedia Technol. **12**, 401–410. Elsevier, New York (2014)
6. NMTG Product Design Parameter Selection Catalogue, NMTG India (2016)
7. Pipaniya, S.: Contact stress analysis of deep groove ball bearing 6210 using Hertzian contact theory. Int. J. Innovative Res. Eng. Sci. **7**, 8–16. ISSN 2319-5665 (2014)

8. Pandiyarajan, R., Starvin, M.S., Ganesh, K.C.: Contact stress distribution of large diameter ball bearing using Hertzian elliptical contact theory. Conference Paper. Procedia Eng. **38**, 264–269. Elsevier, New York (2012)
9. Xin, H.R., Zhu, L.: Contact stress FEM analysis of deep groove ball bearing based on ANSYS workbench. Appl. Mech. Mater. Res. **574**. Trans Tech, Switzerland (2014)

Numerical Simulation and Experimental Study on Lightweight Mechanical Member

G. Lakshmi Srinivas⏺ **and Arshad Javed**⏺

Abstract In the present work, a lightweight mechanical member is synthesized using topology optimization. For optimizing the initial material domain, minimum compliance is chosen as the objective function. A MATLAB code is developed for topology optimization. One end of the member is fixed, another end is subjected to an applied force, and the center of gravity exposed to gravitational as well as centrifugal forces with respect to the 50% weight reduction. In this way, a novel shape and size of the mechanical member are obtained. Along with the optimal topology, performance values such as compliance, maximum deflection, and Von-Mises stress are simulated at different grid size for convergence. These performance values are validated using ANSYS APDL 18.1 and COMSOL 5.3 simulations. To observe energy consumption and operational parameters like torque, an investigational study is also performed using the experimental setup for topologically optimized members. The current consumption and required torque values are compared among the different volume fraction topologies ranging from 0.2 to 0.5.

Keywords Topology optimization · Lightweight components · Mechanical member · Minimum compliance · Humanoid · Energy efficiency · Solid isotropic material with penalization · Volume fraction

1 Introduction

In the automation sector, the necessity of industrial robots is rising on account of their advanced superiority and productivity [1]. In the domain of energy-efficient robot, different attempts are made in past, such as reduction of components weight

G. Lakshmi Srinivas (✉) · A. Javed
Department of Mechanical Engineering, BITS-Pilani Hyderabad Campus, Hyderabad, Telangana 500078, India
e-mail: g.l.srinivas7@gmail.com

A. Javed
e-mail: arshad@hyderabad.bits-pilani.ac.in

© Springer Nature Singapore Pte Ltd. 2020
R. Venkata Rao and J. Taler (eds.), *Advanced Engineering Optimization Through Intelligent Techniques*, Advances in Intelligent Systems and Computing 949,
https://doi.org/10.1007/978-981-13-8196-6_55

by topology optimization, use of lightweight material, analysis of speed, operation scheduling, and identification of the least energy consuming trajectories. Among possible methodologies, topology optimizations is a promising method for reduction of energy consumption. It is an efficient method to reduce the mass of mechanical or structural components of machinery [2].

Significant work on the application topology optimization was initiated by Lohmeier et al. [3]. They reduced the weight of 22-DOF humanoid. Only a few structural parts were optimized by means of topology, and the power consumption was reduced by 35%. Later, Lohmeier et al. conducted an experimental work on 25-DOF humanoid robot, where the overall weight of the humanoid was reduced by 43.5% [4]. Similarly, Albers et al. applied a topology optimization method for ARMAR-III humanoid service robot [5]. The objective function was kept as stiffness maximization. In that work, the mass of the humanoid was reduced by 15%. Further, Albers et al. used a hybrid multibody dynamics system for topology optimization and achieved a weight reduction of 15% [6]. In most of the recent work, static loading condition is considered using the worst-case scenario. Huang and Zhang used solid isotropic material with penalization (SIMP) to optimize five-DOF MOTORMAN-HP20 [7]. For loading condition, the worst case was considered. Junk et al. and Yunfei et al. [8, 9]. Similarly, the dynamic analysis is used to compute the worst cases of loading and perform topology optimization. Chu et al. Briot and Goldsztejn [10, 11]. Few other methodologies were also introduced to model the dynamic behavior of the robot. A part-level meta-model was developed to capture the loading condition of four-DOF welding and painting robot by Kim et al. [12]. Liang et al. proposed a flexible multibody model for the robot considering joint flexibilities. In these recent attempts, the weight reduction was obtained in the range of 5–59% [13].

2 Methodology

Topology optimization is a gradient-based mathematical method help to optimize densities for given initial material domain subjected to boundary conditions and constraints. The objective function is selected as minimum compliance. The mathematical formulation of objective function reads as shown in Eq. (1). It represents the strain energy stored in the mechanical member.

$$
\left.
\begin{array}{c}
\underbrace{\min}_{x} : c(z) = D^{\mathrm{T}}SD = \sum_{e=1}^{N} S_{\mathrm{e}}(z_{\mathrm{e}}) d_{\mathrm{e}}^{\mathrm{T}} S_0 d_{\mathrm{e}} \\[2mm]
\frac{W(z)}{W_0} = r \\[2mm]
\text{subjected to}: \quad SD = L \\[2mm]
0 \le z \le 1
\end{array}
\right\}
\tag{1}
$$

where, '$c(z)$' is the compliance, 'r' is the volume fraction, 'D', 'L' and 'S' are the global displacement, force vectors, and global stiffness matrix, respectively, 'd_{e}' is

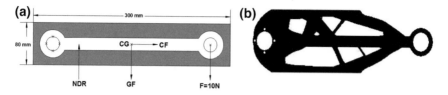

Fig. 1 **a** Boundary conditions: initial material domain, **b** optimized link

the element displacement vector, 'N' is the total number of elements, '$W(z)$' and 'W_O' are the material volume and design domain volume, respectively, 'z' is the design variable. S_e is the element stiffness matrix, S_0 is the stiffness matrix.

The mechanical member of one degree of freedom fixed at one end and another end is subjected to applied force, as shown in Fig. 1a. The length and height of the mechanical member are 300 and 80 mm, respectively. While it is rotating continuously, centrifugal forces (CF) come into picture and gravity (GF) pull it always vertically downwards at the centroid of the design domain.

The initial solid mechanical member is employed in the topology optimization problem, which means applying boundary conditions and constraints using gradient mathematical programming in MATLAB. In this mechanical member, some portion is defined as a non-design region (NDR). To do so, density parameters of these regions are defined with a fixed value, representing the solid material state. Thus, these regions are free from the optimization process. This provision is made to give a feasible optimal topology for assembly and other measurement purposes. In NDR, for assembly of the mechanical member with the motor shaft, a hole with a socket with constant dimensions is defined at the ends. At the central part along with the axis of the mechanical member a, solid rib is predefined in a rectangular shape.

In each iteration of the optimization process, the topology changes, thus the center of gravity (CG) also changes dynamically. In order to capture the effect of self-weight and centrifugal force, a subroutine is created which dynamically computes the CG. Also, the weight of the mechanical member also changes in each iteration until convergence. This change in the weight acting at the instance CG is also considered in the MATLAB code. The code is made to adapt to any grid size. In addition, mesh independency filter and grayscale removal filters are also included in the main routine [14]. The code is capable to compute deflections and stress for each node and element. Performance values are selected as the maximum deflection value and maximum Von-Mises stress acting in the mechanical member.

The member is considered to be made of mild steel (Young's modulus of elasticity: 200 GPa, Poisson's ratio: 0.33, density: 7700 kg/m³) with an external load of 10 N vertically downwards. It is desired to reduce the weight by 50%. The total weight of the initial material domain considering NDR is 1.0678 kg. The member is subject to a rotate with a maximum angular velocity of 4 rad/sec. Based on these boundary conditions and geometrical constraints, the optimal topology is synthesized, given in the next section.

2.1 Experimental Setup

An experimental setup is developed to observe the power consumption and torque exerted by the motor. The developed setup is shown in Fig. 2. The setup consists of alternate current servo motor (Model SINAMICS V90, 0.4 kW, rated torque 1.27 Nm. maximum 3000 rpm), which is the main component to drive the link. The motor was controlled through a motion controller (Model SIMOTICS S-1FL6, supply voltage 3 phase 400 V, Current 1.2 A). A programmable logic controller (PLC) is used to create an operational environment above motion controller (Model SIMATIC S7-1200, make Siemens). For easy and frequent operation at different rpm and position in terms of angular rotation span, a human–machine interface (HMI) on touch screen is developed using PLC (HMI: Model Simatic KTP700 basic, 7" TFT display).

Various links of different volume fraction are also fabricated. For this, the optimal topology is initially generated through the MATLAB platform, then the complicated contours of topology are refined through Adobe Illustrator software. It also helps to convert the .jpeg file of optimal topology into .dxf format. The .dxf file of optimal topology is imported into Mastercam software, which created the G-code for CNC application. A CNC milling machine (Bridgeport VMC GX600) is used to create the topology on a mild steel plate for different volume fraction values, as shown in Fig. 3.

Fig. 2 Experimental setup

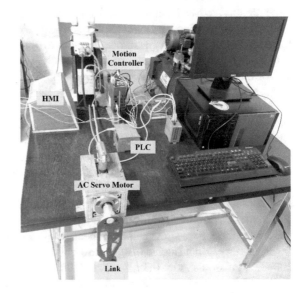

Fig. 3 Optimized link for volume fraction: **a** 1.0, **b** 0.5, **c** 0.4, **d** 0.3, **e** 0.2

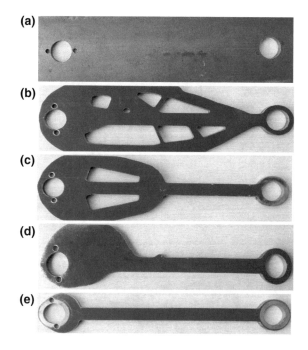

3 Results and Discussions

3.1 Optimal Topology

After implementation of topology optimization process through MATLAB code, the optimized link is obtained (Fig. 1b). The complete program and simulation executed in the Dell Precision Tower 5810 XCTO 825W, 32 GB 2400 MHz RAM, Intel Xeon Processor E5-1650 v3 (Six Core HT, 15 MB Cache, 3.5 GHz Turbo).

To check the mesh independence, the code is run for different element sizes. The MATLAB result for deflection, Von-Mises stress, and computational time is tabulated in Table 1. Topologies corresponding different element sizes is illustrated in Fig. 4.

It is observed from these topologies that the optimal topology is almost mesh independent. With increased mesh size, the smoothness of the boundaries also increases; however, large computational time is required. From these results, it is also observed that at element size 2400 × 640, the simulation converge with deflection of 0.00435 mm and Von-Mises stress of 8.16 MPa.

Table 1 Deflection, stress, and time for different element size

Element size	Deflection (mm)	Von-Mises stress (MPa)	Computational time (min)
300 × 80	0.00423	5.13	0.35
600 × 160	0.00411	6.07	1.55
900 × 240	0.00423	6.63	3.81
1200 × 320	0.00425	7.16	7.01
1500 × 400	0.00429	7.46	11.33
1800 × 480	0.00430	7.51	16.83
2100 × 560	0.00433	7.92	24.03
2400 × 640	**0.00435**	**8.16**	**32.46**
2700 × 720	0.00435	8.16	41.73
3000 × 800	0.00435	8.16	53.31

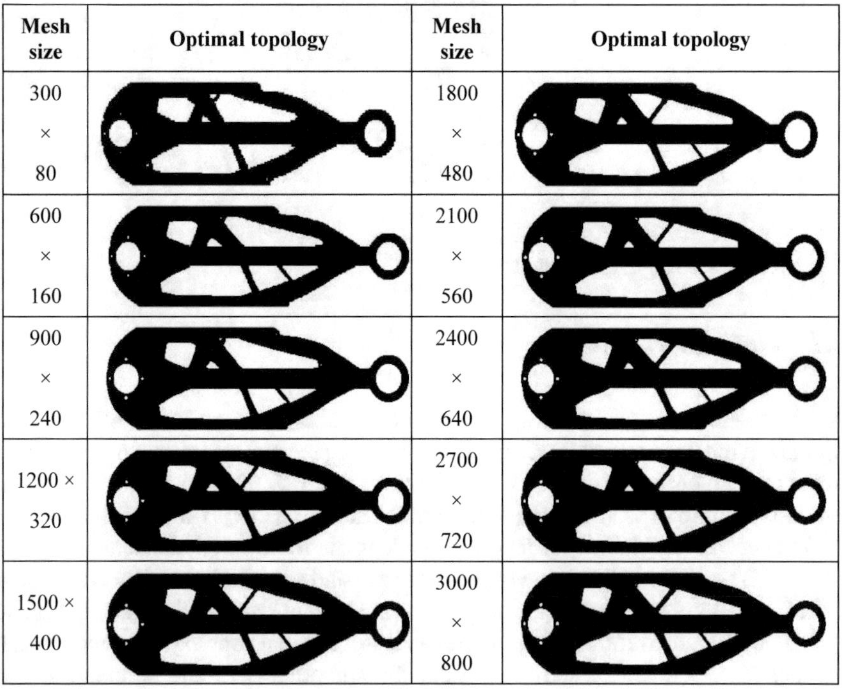

Mesh size	Optimal topology	Mesh size	Optimal topology
300 × 80		1800 × 480	
600 × 160		2100 × 560	
900 × 240		2400 × 640	
1200 × 320		2700 × 720	
1500 × 400		3000 × 800	

Fig. 4 Optimal topologies at a different mesh size

3.2 Performance Simulation Using ANSYS APDL

For validation of MATLAB results (shown above), the optimized topology at element size 2400 × 640 is imported to ANSYS APDL. The imported model extruded for 6 mm thickness to create volumes in the APDL work plane. Incorporate all material properties and define element type for the created model. Apply boundary conditions and constraints at different mesh tools to computing the deflection and Von-Mises stress. The simulation results are obtained at different mesh tools. Here, for illustration only fine and normal type of mesh tools are given in Table 2.

3.3 Performance Simulation Using COMSOL

The generated topologies of the mechanical member by MATLAB directly imported to COMSOL through the live link. A 3D geometry was created in the COMSOL work plane and physics (solid mechanics) is added. Apply the boundary conditions and constraints at different mesh tool to compute the results. The outcome of deflection and Von-Mises stress is shown in Table 3.

The performance result obtained using MATALB code, ANSYS, and COMSOL is compared. The convergence is within acceptable limits. The result from COMSOL slightly deviates. This difference is because of element type available in COMSOL. However, results are within permissible limits of safe design [15]. In order to prepare a concise manuscript, MATLAB 2017a routine is not presented in the document. The convergence values are simulated in ANSYS APDL and COMSOL for illustration and because of space constraint, only three mesh tools fine, normal and coarse are presented.

3.4 Experimental Analysis

As stated earlier, the experiments are performed for different optimal topologies of different volume fractions. From this experiment, the values of current consumption and torque supplied by the motor at different speeds are observed. During this observation, the motor ran through continuous rotation and the maximum value of the parameters (current and torque) is recorded. These parameters are shown in Fig. 5. It is evident that at lower volume fraction value, the current and torque requirement is lesser, and for higher values of volume fraction, these parameters attain higher values.

Also when the rpm is lesser, the variance between maximum and minimum numerical values of these parameters is larger, compared to the situation when the motor is running at a higher rpm. This is because of lower of viscous and frictional effect and high back current in the rotor coil of the motor at lower rpm. Hence, as long as

Table 2 Deflection and Von-Mises stress for different Mesh tool based on APDL

Mesh tool	Maximum deflection (mm) and maximumVon-Mises stress (MPa)
Fine Deflection: 0.00431 mm Von-Mises stress: 8.15 MPa	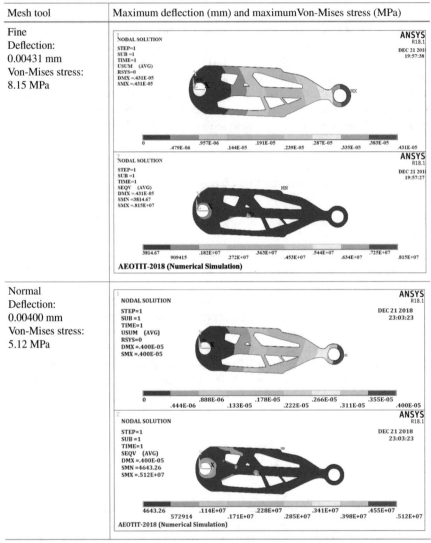
Normal Deflection: 0.00400 mm Von-Mises stress: 5.12 MPa	

the rigidity of link is maintained, the lowest possible volume fraction under the safe limits can be chosen for energy saving.

Table 3 Deflection and Von-Mises stress for different Mesh tool based on COMSOL

Mesh tool	Maximum deflection (mm) and maximum Von-Mises stress (MPa)
Fine Deflection 0.00405 mm Von-Mises stress: 8.20 MPa	
Normal Deflection 0.00405 mm Von-Mises stress: 5.9 MPa	
Coarse Deflection 0.00405 mm Von-Mises stress: 5.68 Mpa	

4 Conclusion

Topology optimization eliminates the needless densities according to the given constraints as volume fraction. Density is directly proportional to the mass if we decrease the mass of the mechanical member, the torque generated by the actuators subsequently decreases. The volume fraction is inversely proportional to the Von-Mises stress maximum deflection. Hence, the selection of the volume fraction should be

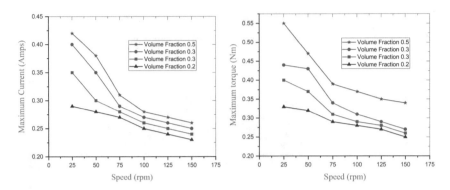

Fig. 5 Maximum current and maximum torque with respect to the speed

based on the factor of safety required. In the present case, the weight of the member is reduced by half, without sacrificing the strength. An experimental study was also performed to compare the different optimal topologies. The current consumption and torque supplied values were observed for different links at different rpm values. It was found that a maximum energy saving of 31% is obtained when a link of 0.2 volume fraction is used, compared to that of 0.5 volume fraction.

Acknowledgements The effort presented in this work is funded by the "Department of Science and Technology (DST)—Science and Engineering research board (SERB), India" (File Number: ECR/2017/000799), and Research Initiation Grant-BITS Pilani.

References

1. Statistics—IFR International Federation of RoboticsIfr.org. http://www.ifr.org/industrial-robots/statistics/ (2018). Retrieved 17 Sept 2018
2. Bendsøe, M., Sigmund, O.: Topology Optimization. Springer, Berlin (2003)
3. Lohmeier, S. et al.: Leg design for a humanoid walking robot: humanoid robots. In: 2006 6th IEEE-RAS International Conference on Humanoid Robots. https://doi.org/10.1109/ichr.2006.321325. IEEE (2006)
4. Lohmeier, S., Buschmann, T., Ulbrich, H.: Humanoid robot LOLA: Robotics and Automation. ICRA'09. In: (2009) IEEE International Conference on Robotics and Automation. https://doi.org/10.1109/robot.2009.5152578. IEEE (2009)
5. Albers, A. et al.: Methods for lightweight design of mechanical components in humanoid robots: humanoid robots. In: 2007 7th IEEE-RAS International Conference on Humanoid Robots. https://doi.org/10.1109/ichr.2007.4813934. IEEE (2007)
6. Albers, A., Ottnad, J.: System based topology optimization as development tools for lightweight components in humanoid robots: humanoid robots, 2008. In: Humanoids 2008-8th IEEE-RAS International Conference on Humanoid Robots. https://doi.org/10.1109/ichr.2008.4756024. IEEE (2008)
7. Huang, H.B., Zhang, G.: The topology optimization for l-shape arm of Motorman-HP20 robot. Appl. Mech. Mater. . 201. Trans Tech Publications (2012). https://doi.org/10.4028/www.scientific.net/AMM.201-202.871

8. Junk, S. et al.: Topology optimization for additive manufacturing using a component of a humanoid robot. Procedia CIRP **70**, 102–107. https://doi.org/10.1016/j.procir.2018.03.270. (2018)

9. Yunfei, B., Ming, C., Yongyao, L .: Structural Topology Optimization for a Robot Upper Arm Based on SIMP Method: Advances in Reconfigurable Mechanisms and Robots II, pp. 725–733. Springer, Cham (2016). https://doi.org/10.1007/978-3-319-23327-7_62

10. Chu, X. et al.: Multi-objective topology optimization for industrial robot: Information and Automation (ICIA). In: 2016 IEEE International Conference on Information and Automation (ICIA). https://doi.org/10.1109/icinfa.2016.7832132. IEEE (2016)

11. Briot, S., Goldsztejn, A.: Topology optimization of industrial robots: application to a five-bar mechanism. Mech. Mach. Theory **120**, 30–56 (2018). https://doi.org/10.1016/j.mechmachtheory.2017.09.011

12. Kim, B.J., et al.: Topology optimization of industrial robots for system-level stiffness maximization by using part-level metamodels. Struct. Multidisciplinary Optim. **54**(4), 1061–1071 (2016). https://doi.org/10.1007/s00158-016-1446-x

13. Liang, M., Wang, B., Yan, T.: Dynamic optimization of robot arm based on flexible multi-body model. J. Mech. Sci. Technol. **31**(8), 3747–3754 (2017). https://doi.org/10.1007/s12206-017-0717-9

14. Bourdin, B.: Filters in topology optimization. Int. J. Numer. Meth. Eng. **50**(9), 2143–2158 (2001)

15. Chalhoub, N.G., Ulsoy, A.G.: Control of a flexible robot arm experimental and theoretical results. J. Dyn. Syst. Meas. Control **109**(4), 299–309 (1987)

Robust Analysis of T-S Fuzzy Controller for Nonlinear System Using H-Infinity

Manivasagam Rajendran(ⓘ), P. Parthasarathy and R. Anbumozhi

Abstract Conventional controllers are not able to work well under the influence of delay and disturbance. Hence, it becomes necessary to design a robust fuzzy controller, which guarantees high stability, provides high disturbance rejection for any operating conditions and deals with model uncertainties. Here, the system response for the conventional PID controller, fuzzy logic controller and robust fuzzy controller is compared based on Sugeno model. Robust analysis of the system is carried out to guarantee the stability of the system. For given gamma values and weight values, we check the stability of the system. The results of robust fuzzy controller show that it works well under the influence of model uncertainty, delay and large disturbances.

Keywords Robust · Fuzzy · Stability · Disturbance · Sugeno

1 Introduction

1.1 Background

CSTR plays an important role in almost all chemical processes. PID controllers are most commonly used controllers in many industrial applications because it is simple. Conventional PID controllers are 1-DOF controller, which performs either servo-tracking or disturbance rejection problem at a time but not both problems concurrently. The main drawbacks of the PID controller are [1]:

M. Rajendran · P. Parthasarathy (✉) · R. Anbumozhi
K. Ramakrishnan College of Engineering, Trichy, India
e-mail: pspsarathy1@gmail.com

M. Rajendran
e-mail: manivasagammn3@gmail.com

R. Anbumozhi
e-mail: anbumozhieee@gmail.com

© Springer Nature Singapore Pte Ltd. 2020
R. Venkata Rao and J. Taler (eds.), *Advanced Engineering Optimization Through Intelligent Techniques*, Advances in Intelligent Systems and Computing 949,
https://doi.org/10.1007/978-981-13-8196-6_56

1. Deteriorate the system performance.
2. Produce large oscillations.

In a fuzzy system, first the number of inputs and outputs of the system is defined. Based on these inputs and outputs, we define rules for the system.

Extensive literature is available for fuzzy modeling using input–output data. For our work, we choose ISE as a performance index because the error is large and we try to minimize the ISE [2]. The rule base for fuzzy is a collection of IF-THEN statement [3].

For T-S fuzzy model, rule base is defined as:

If $Z_1(t)$ is M_{i1} and $Z_n(t)$ is M_{in} THEN
$$\dot{X} = A_i x(t) + B_i u(t) \qquad i = 1, 2, \ldots r$$
$$y(t) = C_i x(t) \qquad i = 1, 2, \ldots r$$

By using this rule, nonlinear system can be modeled.

1.2 H-Infinity Controller

H∞ control technique is found to guarantee good performance and robustness. First step in the design of the controller is to represent the system in the following standard configuration [4] (Fig. 1):

Let the state space representation of the system be given as

$$\dot{X} = Ax(t) + B_1 w(t) + B_2 u(t) \tag{1}$$

$$\dot{Z} = C_1 x(t) + D_{11} w(t) + D_{12} u(t) \tag{2}$$

$$\dot{Y} = C_2 x(t) + D_{21} w(t) \tag{3}$$

$P(s)$ may be further denoted as

$$\begin{bmatrix} Z \\ V \end{bmatrix} = P(S) \begin{bmatrix} w \\ u \end{bmatrix} = \begin{bmatrix} P11 & p12 \\ p21 & p22 \end{bmatrix} \begin{bmatrix} w \\ u \end{bmatrix}$$

Fig. 1 H-infinity controller

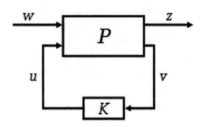

Fig. 2 Plant model for the synthesis of H-infinity controller

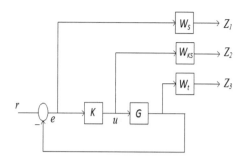

Here, $u = K(s)v$ (Fig. 2).

The generalized plant P is given as [5]

$$\begin{bmatrix} Z_1 \\ Z_2 \\ Z_3 \\ Z_4 \end{bmatrix} = \begin{bmatrix} W_s & -W_s G \\ 0 & W_{ks} \\ 0 & W_t G \\ I & -G \end{bmatrix} \begin{bmatrix} w \\ u \end{bmatrix} \qquad (4)$$

From Eqs. 1–4,

$$P = \begin{bmatrix} W_s & -W_s G \\ 0 & W_{ks} \\ 0 & W_t G \\ I & -G \end{bmatrix} = \begin{bmatrix} A & B_1 & B_2 \\ C_1 & D_{11} & D_{12} \\ C_2 & D_{21} & 0 \end{bmatrix} \qquad (5)$$

From Eqs. 4 and 5,

$$P = \begin{bmatrix} W_s S \\ W_{KS} K\, S \\ W_t S \end{bmatrix}$$

The H-infinity solution uses solutions of two algebraic Riccati equations (AREs). Dependency of z on w can be expressed as:

$$z = F_l(P, K)w \qquad (6)$$

$$F_l(P, K) = P11 + P12K(I - P22K)^{-1} P22 \qquad (7)$$

$$\|F_l(P, K)\|_\infty = \sup \sigma(F_l(P, K)(jw)) \qquad (8)$$

$$F_l(P, K)(jw)$$

The main objective of H∞ design is to find the controller K which minimizes the value of and tries to keep this within specified limits.

$$\min \| P \| = \min \begin{bmatrix} W_s S \\ W_{KS} K S \\ W_t S \end{bmatrix} = \gamma \qquad (9)$$

In the above equations,

$$W_{ks} = \text{constant} = 1$$

$$W_t = \frac{s + W_0/M}{As + W_0}$$

where M = sensitivity peak, A = max. allowable steady-state offset, and W_0 = desired bandwidth.

2 Hardware System: CSTR

2.1 Experimental Setup

DAC is used to interface CSTR with the personal computer (PC). Overall system consists of a tank, pump, rotameter, RTD. Figure 3 shows the block diagram of a CSTR tank interfaced with PC.

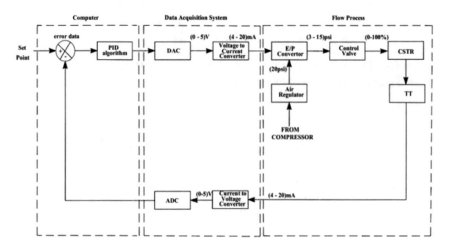

BLOCK DIAGRAM OF CONTINUOUS STIRRER TANK REACTOR

Fig. 3 Block diagram of CSTR

The pneumatic control valve uses air as an input and adjusts the flow of the water pumped to the CSTR jacket from a cold water tank. This flow maintains the temperature inside the tank at the desired value. The temperature of the liquid inside the tank is measured with the help of RTD and is transmitted in the form of (4–20) mA to the interfacing DAC module with the help of temperature transmitter to the personal computer (PC).

2.2 Calculation of TF

For calculating the transfer function of CSTR cooling process, the step response is taken into consideration. The transfer function is calculated by using the process reaction curve method [6]. The process has very large dead time and is highly damped. Therefore, the step response can be fitted into a simple first-order model with dead time.

3 Simulation and Results

From Table 2, it is clear that ISE for the optimized fuzzy is less than ISE for PID and fuzzy without optimization. In the simulation, we calculate ISE after a step change. Here, the initial value of the step input is kept at 50 and then step change given at time instant of 50. The final value of the step change is 40. This is done because our process is a cooling process, so we want to settle temperature from high value to low value (Figs. 4, 5, 6, 7, 8 and Table 1).

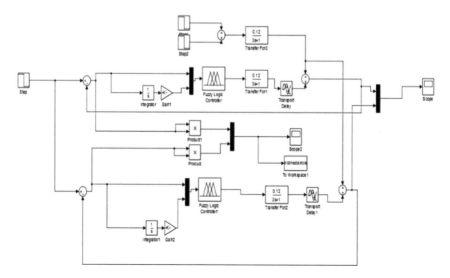

Fig. 4 Simulation for combined fuzzy

Table 1 Time domain specifications for fuzzy and PID

	PID	Fuzzy
Rise time (s)	6	5
Settling time (s)	17	12
Overshoot (%)	1.25	1.4

Fig. 5 Response for combined fuzzy PID

Table 2 Comparison of ISE

	ISE (with delay)
Fuzzy	3.590402
Optimized fuzzy	3.10073
PID	3.959913

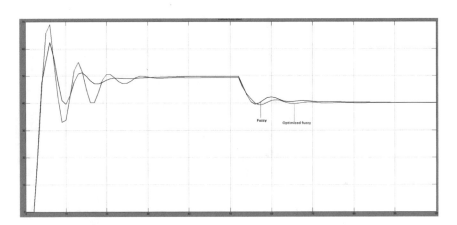

Fig. 6 Response for combined fuzzy

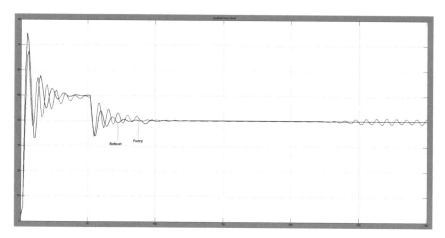

Fig. 7 Robust fuzzy with disturbance

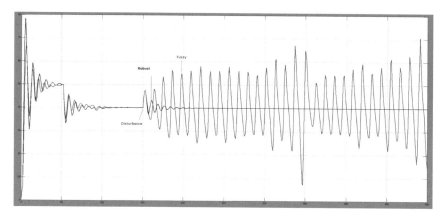

Fig. 8 Robust fuzzy with two disturbances and parameter change

3.1 Results with H-Infinity

The main aim of H-infinity design is to find controller K which minimizes the value of gamma [5]. For CSTR system, we take second-order transfer function and analyze the system with various values of the weight. The result for various values of weight function is given in Table 3.

With weight values obtained in the last row of the table, we perform H-infinity analysis of a system and this produces the following controller transfer function:

$$K = \frac{0.1132s^3 + 0.2673s^2 + 0.1926s + 0.03932}{s^4 + 3.324s^3 + 4.032s^2 + 2.1s + 0.3913}$$

Table 3 Gamma for different weight values

W1	W2	W3	γ(Max.)
$\frac{(s+1)}{(s+0.001)}$	1	$\frac{(s+1)}{(0.001s+1)}$	9.5507
$\frac{(s+1)}{(s+0.0005)}$	1	$\frac{(s+1)}{(0.0005s+5)}$	18.2228
$\frac{(s+1)}{(s+0.01)}$	1	$\frac{(s+1)}{(0.01s+1)}$	9.2146
$\frac{(s+1)}{(s+0.9)}$	1	$\frac{(s+1)}{(0.9s+1)}$	1.1041

Fig. 9 Largest singular value plot

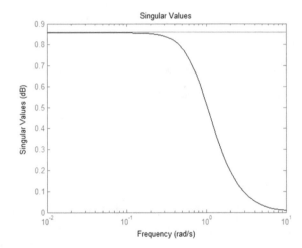

The plot of singular value is shown in Fig. 9.

4 Conclusion

From the response of the system for various controllers, it is clear that the fuzzy controller gives better performance compared to other controllers. Also, ISE (performance index) obtained with the optimized fuzzy is less than other controllers. Also, robust fuzzy controller is able to work well under the presence of large disturbances, delay and model uncertainties.

In this paper, a stepwise procedure for the design of H∞ controller has been presented in detail. It has been observed from analysis that H∞ controller guarantees robustness. Simple illustrative example of CSTR has been considered, and the loop shaping technique has been utilized to solve the problems. The simulation results presented here verify the validity of the loop shaping technique.

References

1. Manikandan, P., Geetha, M., Jerome, J.: Soft computing techniques based optimal tuning of virtual feedback PID controller for chemical tank reactor. IEEE Congr. Evol. Comput. (2014)
2. Mishra, C.K., Jebakumar, J.S., Mishra, B.K.: Controller selection and sensitivity check on the basis of performance index calculation. Int. J. Electr. Electron. Data Commun. 2(1) (2014)
3. Frank, P.M., Cao, Y.Y.: Analysis and synthesis of nonlinear time-delay systems via fuzzy control. IEEE Trans. Fuzzy Syst. 8(2) (2000)
4. Petkov, P. Hr., Gu, D.W., Konstantinov, M.: Robust Control Design with MATLAB. Springer
5. Sharma, V., Bansal, A.: Design and analysis of Robust H-infinity controller. In: National Conference on Emerging Trends in Electrical, Instrumentation & Communication Engineering, vol. 3, No. 2 (2013)
6. Stephanopoulos, G.: Chemical Process Control. Prentice Hall of India, New Delhi (2001)
7. Brijendra, kr. M., Saurabh, kr. B.: Fuzzy logic based temperature control of continuous stirred tank reactor. IJETT 12 (2014)
8. Kazuo, T., Wang, H.O.: Fuzzy Control Systems Design and Analysis (a linear matrix inequality approach). A Wiley Interscience Publication
9. Zhou, M., Li, R., Mo, H., Wang, F.Y., Xiao, Z.: Footprint of uncertainty for type-2 fuzzy sets. Inf. Sci. 272, 96–110 (2014)
10. Wu, D.: On the fundamental differences between type-1 and interval type-2 fuzzy logic controllers. IEEE Trans. Fuzzy Syst. 10(5), 832–848 (2012)
11. Jiang, X., Ge, X., Han, Q.-L.: Fuzzy sampled-data H∞ filtering for systems with time-varying delays and variable sampling periods. In: American Control Conference Fairmont Queen Elizabeth, Montréal, Canada (2012)
12. Ponnusamy, L., Jeeva, A., Sangapillai, S.: T-S fuzzy modeling for SISO and MIMO processes. In: Fifth International Conference on Advanced Computing (ICoAC) (2013)
13. Bharathi, M.,Selvakuma, C.: Fuzzy Based Temperature Controller For Continuous Stirred Tank Reactor. Int. J. Adv. Res. Electr. Electron. Instrum. Eng. (2018)

Adsorption of Crystal Violet Dye: Parameter Optimization Using Taguchi's Experimental Methodology

T. B. Gupta, D. H. Lataye and S. T. Kurwadkar

Abstract Adsorption of crystal violet dye (CV) was investigated using laboratory-developed activated carbon obtained from *Acacia nilotica* (AAC) sawdust. Taguchi's optimization methodology was used for optimizing various batch parameters for CV removal from liquid solutions by AAC. The effectiveness of three levels of Taguchi orthogonal array was examined to optimize adsorption parameters. The adsorption factors, viz AAC dosage, concentration of CV, temperature, and time, were used at three levels to study its overall effect on adsorption of CV by AAC. Taguchi's L_9 orthogonal array has been applied for experiment design and optimization, and the results were analyzed using ANOVA for raw and S/N data using 'bigger is better' characteristic. The ANOVA shows which CV concentration is the utmost significant with 55.60% contribution tracked by AAC dosage, temperature, and time of contact with 37.81, 3.95, and 2.11% contribution, respectively. The optimized combination of parameters for CV dye removal was found to be A_1, B_3, C_3, and D_3. The predicted (forecasted) values and confirmation respective values of total dye adsorbed onto AAC at optimization were observed to be 31.46 and 31.45 mg/g.

Keywords Adsorption · Crystal violet dye · Levels · Optimize · Taguchi

T. B. Gupta
Department of Civil Engineering, Ramdeobaba College of Engineering and Management, Nagpur 440013, India
e-mail: gupta_tripti3621@rediffmail.com

D. H. Lataye (✉)
Department of Civil Engineering, Visvesvaraya National Institute of Technology, Nagpur 440010, India
e-mail: diliplataye@rediffmail.com; dhlataye@civ.vnit.ac.in

S. T. Kurwadkar
Department of Civil and Environmental Engineering, California State University, Fullerton, 800 N. State College Blvd, Fullerton, CA 92831, USA
e-mail: skurwadkar@fullerton.edu

© Springer Nature Singapore Pte Ltd. 2020
R. Venkata Rao and J. Taler (eds.), *Advanced Engineering Optimization Through Intelligent Techniques*, Advances in Intelligent Systems and Computing 949,
https://doi.org/10.1007/978-981-13-8196-6_57

653

1   Introduction

The release of effluents from industrial sectors, like printing, textile, laundry, paper, and leather, into natural water bodies imparts toxicity to the marine life and living organisms. Besides being aesthetically unappealing, it causes an adverse effect on the environment. Hence, it is imperative that the wastewater containing dyes should be treated before its release to the environment. Currently, a variety of physical, biological, and chemical treatment approaches are commonly used for the elimination of dyes and colored impurities from wastewater. Among these various approaches, adsorption is found to be an efficient and economic remediation technology [1]. The commercially available adsorbents are often used for color, dyes, and dyestuffs reduction from wastewaters. Treatment by adsorption turns out to be more active if the adsorbent is economically sustainable and cost-effective over its operational life.

Optimization using Taguchi's method is internationally well accepted in designing experiments and forecasting an approach for control of quality of goods in manufacturing sectors [2]. Taguchi's approach optimizes designing and planning the quality into goods and processes. It is frequently used by various investigators in the optimization of adsorption of different adsorption factors [3–5].

In this research study, we present the results obtained from a series of batch adsorption experiments carried out using laboratory-developed adsorbent. Discarded sawdust of *Acacia nilotica* (AAC) was used as a raw material for developing the sorbent. Adsorption of crystal violet dye (CV) on to the newly developed sorbent material has been reported. Experimental parameters were optimized using Taguchi's optimization protocol.

Crystal violet is a water-soluble aromatic basic dye. It is a tri-aryl-methane dye and has anti-bacterial, antiseptic, and anti-fungal characteristics. The dye can be produced by condensing and cooling formaldehyde and di-methyl-aniline [6]. It is commonly used to color papers and also as a constituent of black and navy blue inks for coating, ballpoint nips, printing and pens, and inkjet and LaserJet printers. It is also used for coloring miscellaneous stuff such as manures, anti-freezers, soap cleansers, and leather. Despite several usages of CV, it is hazardous if ingested or inhaled and may cause problems like nausea, fatigue, dizziness, headache, skin redness, irritation, and cancer. It is harmful if swallowed and causes gastrointestinal tract irritation with vomiting, diarrhea, and abdominal pain.

2   Methods and Materials

2.1   Adsorbent

In this study, activated carbon (AC) obtained from sawdust (a waste product from the wood processing industry) of babul shrub (*A. nilotica*) that was readily accessible and perennially available. To prepare the adsorbent, sawdust was strained to get uniform particle size of 200–550 μm [7]. It was thoroughly washed to remove impurities that may be present. The cleaned sawdust was sun-dried for about 24 h followed by drying in oven at 105 °C for 2 h. To yield char, 25 mL of ortho-phosphoric (H_3PO_4) acid added with 50 g of dehydrated sawdust and dried in oven for about 24 h. The char thus formed was thermally activated in a temperature-controlled muffle furnace at 425 °C for 1 h. Post activation, the adsorbent was washed with DDW scrupulously. Afterward, it was dehydrated in the presence of air and, subsequently, oven dried at 105 °C for 120 min. This material was considered as an adsorbent for the present study.

2.2 Adsorbate and Supplementary Reagents

All the reagents used for conducting experiments were of analytical (AR) scale. The CV (chemical formula $C_{25}N_3H_3OCl$) was purchased from Loba Chem. Ltd., Mumbai. A solution of 1000 mg/L of CV made by adding 1 g CV powder mixed with 1 L of double distilled water. The solution was continuously stirred to make sure that the dye was homogeneously mixed. The solution was serially diluted with doubled distilled water for desired concentration for conducting adsorption experiments at various concentration levels. For analytical quantitation of dye concentration, the test samples were analyzed using Shimadzu ultraviolet and visible dual-beam spectrophotometer (Model No.: 2450). For CV, the wavelength conforming to maximum absorbance was found to be 590.5 nm.

2.3 The Rationale for Using Taguchi's Design Methodology

Taguchi's optimization protocol assists in identifying critical parameters in experiments, thereby obviating the need to conduct large experiments for studying the effect of all experimental parameters. The various factors that influence the adsorption process for adsorptive removal of CV were selected (Table 1). The orthogonal (OA) array, signal-to-noise (S/N) ratio, and variate analysis (ANOVA) conducted to better appreciate the importance of adsorption factors on the removal of CV.

2.3.1 Experimental Protocol

Series of batch adsorption experiments were conducted to identify critical factors that influence the sorptive removal of CV from the aqueous phase. Table 1 shows the parameters, descriptions, and levels.

Table 1 Different process parameters for sorption of CV on AAC by Taguchi's array	Factors	Parameters	Units	AAC levels		
				1	2	3
	A	AAC dosage (m)	g/50 mL	0.2	0.4	0.6
	B	CV concentration (C$_0$)	mg/L	50	100	200
	C	Temperature (T)	°C	293	303	313
	D	Contact time (t)	min	30	60	90

2.3.2 Orthogonal Array Selection and Parameter Consignment

Taguchi's experimental design protocol used in the study consists of various stages including identifying a specific array, number of process parameters along with its interactions, and the level numbers for parameters. The degrees of freedom of an array must be larger than or equal to the degrees of freedom essential for the experimentation. Based on this information, likely experimentation numbers (N) was determined. For AAC, $N = 3^4$, four parameters along with three levels were selected. Hence, for full factorial design, likely experimentation numbers (N) is 81.

To decrease the experiments in number, a minor likely set was nominated. For 81 full factorial designs (3^4), Taguchi used L_9 array where 9 specifies the number of trials needed. By four factors at three different levels, degrees of freedom essential is $= 4 * (3 - 1) = 8$; hence, a 3-level parameter has two degrees of freedom (no. of levels-1). Hereafter, an L_9 (3^4) OA has been chosen.

2.3.3 Experimental Approach

For adsorptive removal of CV by AAC, 50-mL ideal solution of the dye prepared in doubled distilled water at pH 6.56 having different dye concentrations was taken in a conical flask. The essential AAC dosage per experimental run was supplemented in each container. These containers were shaken, stirred, and mixed at 150 revolutions per minute in orbital shaker for requisite time and temperatures.

Adsorption capacity q_t was calculated by:

$$q_t = \frac{(C_0 - C_e)}{m} \tag{1}$$

C_o is the initial strength of CV (mg/L), C_e is the equilibrium strength (mg/L), m is the AAC dosage (g/L), and q_t is amount of CV accumulated on AAC at saturation (mg/g).

2.4 Evaluation of Experimentation Data

The acquired data from experimentations were evaluated using S/N ratio. It is the logarithmic function considered to variance optimization and method of ANOVA.

2.4.1 Signal-to-Noise (S/N) Ratio

Taguchi generated a transmutation for the function of loss, S/N ratio [2], looks at two features of distribution. The S/N ratio regulates the best set of situations. The outer array is rummage to allow the noise deviation into experimentations. It conglomerates the variance around this mean and the quality characteristic of mean level into a solitary metric structure [8]. The ratio S/N lessens the factors which are uncontrollable [9]. The following are three conditions of S/N ratio:

$$\text{Smaller is better,} \quad \frac{S}{N} = -10 \log \left[\frac{1}{n} \sum_{i=1}^{n} y_i^2 \right] \tag{2}$$

$$\text{Nominal is better,} \quad \frac{S}{N} = -10 \log \left[\frac{1}{n} \sum_{i=1}^{n} \frac{1}{y_i^2} \right] \tag{3}$$

$$\text{Bigger is better,} \quad \frac{S}{N} = 10 \log \frac{\mu^2}{\sigma^2} \tag{4}$$

where μ^2 = square of mean, σ^2 = observation variance, and y_i = response observation for n trials. Th bigger is better condition has been used for calculation of S/N ratio in the present work.

2.4.2 Analysis of Variance (ANOVA)

The raw data and ratio S/N data were afterward examined by analysis of variance (ANOVA). In ANOVA, several factors such as the degree of freedom, the sum of squares, variance, variance ratio, and percent contribution were deliberated. ANOVA applies to the experimentation outcomes to calculate the % contribution of each factor in contradiction of a specific confidence interval. Comprehensive procedure for the analysis of experiment data is addressed by different researchers [10]. The data were considered for outcome analysis and to compute the output of the sorption. After computation of the optimum conditions, the mean response (μ) was forecasted [11].

2.4.3 Forecast of the Mean

ANOVA analysis was done to evaluate the statistical significance of adsorption factors at optimum conditions and predetermined significance level. Once optimal process parameters were obtained, the mean response (μ) was estimated. The mean is predictable only from important factors [12].

2.4.4 Establishing Confidence Interval

Optimized parameters were checked to make sure that they are within the established confidence interval (CI) that represent the upper and lower bound values within these bounds the true mean is expected to lie. Both CI for the whole population (CI_{POP}) and the sample (CI_{CE}) were calculated [6]. The confidence intervals were calculated using:

$$CI_{POP} = \sqrt{\frac{F_\alpha (1, f_e) V_e}{n_{eff}}} \tag{5}$$

$$CI_{CE} = \sqrt{F_\alpha (1, f_e) V_e \left[\frac{1}{n_{eff}} + \frac{1}{R} \right]} \tag{6}$$

where $F_\alpha (1, f_e)$ are the F-ratio at a level of confidence $(1 - \alpha)$ at DOF one and DOF error (f_e).

The value of DOF error may be taken from the F-table. V_e designates error variance from ANOVA pooled. R designates the size of sample for confirmatory experimentation. It is seen from equations that the value $1/R$ reaches zero with R approaching infinity, and $CI_{CE} = CI_{POP}$. CI_{CE} becomes broader as R approaches 1 [5].

2.4.5 Confirmatory Experimentations

Once the critical parameters from Taguchi experimental design are identified, and their relative importance is recognized, a set of confirmatory experiments were conducted at using optimized parameters. For validation purposes, the difference Taguchi's optimized parameter values and the values obtained through the confirmatory experiments have to be within 5% confidence interval [6].

3 Results and Discussions

3.1 Taguchi's L_9 OA Experimental Methodology and Results

Table 1 shows the significant factors influencing the adsorptive elimination of CV, the array of values and their levels. Four factors (parameters) at three levels each are used in Taguchi's array. Set of nine experiments were conducted in triplicate. The experimental outcome of each set and S/N estimated for larger is a better situation for adsorption of CV on AAC which is given in Table 2.

From Table 2, it can be inferred that the S/N ratio value for every factor at levels 1, 2, and 3 is highest (i.e., 29.93) for set no. 7. Likewise, the adsorption capacity for every factor at levels 1, 2, and 3 is as well observed to be highest (i.e., 31.46 mg/g). However, the percent reduction is observed to be declined (i.e., 62.91%). Accordingly, it can be deduced that adsorption is directly reliant on the parameters (i.e., CV concentration, AAC dosage, time, and temperature).

3.2 Process Parameters' Effect

Factors like AAC dosage A (m), CV concentration B (C_0), temperature C (T), and time of contact D (t) at several levels considerably affect the values of response q_t (Table 2). For optimization of adsorption of CV using AAC, the parameter A, i.e., AAC dosage (m), revealed utmost effect at level 1; factor B, i.e., CV concentration (C_0), revealed utmost effect at level 3; factor C, i.e., temperature (T), revealed utmost effect at level 3; and factor D, i.e., time of contact (t), revealed utmost effect at level 3 (Fig. 2, Table 3).

The difference between levels 1 and 2 (i.e., $L_2 - L_1$) as well as levels 2 and 3 (i.e., $L_3 - L_2$) revealed the effect of one level over another. It also specifies that if the difference between levels is greater, then the effect of levels is also greater.

Figure 1 shows the relative influence of optimized process parameters (factors) on the adsorption of CV onto AAC at different levels. Adsorption capacity (q_t) tends to decrease as the process parameters raised from level to level. For example, raising

Table 2 Taguchi's array with values *of adsorption uptake capacity* q_t and S/N data for adsorption of CV on AAC

Run	Parameters				Experimental results						S/N ratio
	A	B	C	D		q_t (mg/g)		Percentage removal (%)			
	m	C_0	T	t	R_1	R_2	R_3	R_1	R_2	R_3	
1	0.4	200	293	60	19.84	19.87	19.78	79.35	79.47	79.11	25.95
2	0.4	100	313	30	12.2	12.24	12.18	97.58	97.95	97.46	21.73
3	0.6	50	313	60	4.16	4.16	4.16	99.79	99.78	99.82	12.38
4	0.2	100	303	60	18.1	15.99	17	72.4	63.95	68	24.59
5	0.6	100	293	90	8.32	8.32	8.32	99.83	99.82	99.84	18.4
6	0.6	200	303	30	13.48	13.51	13.2	80.89	81.04	79.23	22.54
7	0.2	200	313	90	30.59	33.53	30.25	61.17	67.06	60.51	29.93
8	0.4	50	303	90	6.23	6.24	6.24	99.76	99.87	99.82	15.9
9	0.2	50	293	30	12.21	12.11	12.11	97.71	96.86	96.86	21.69

Table 3 Main and Average effects of values *of* adsorption uptake capacity q_t for adsorption of CV on AAC: Raw and S/N data

Parameters	Raw data, average value			Main effects (Raw data)		S/N data, average value			Main effects (S/N Data)	
	L1	L2	L3	L2 − L1	L3 − L2	L1	L2	L3	L2 − L1	L3 − L2
A	20.21	12.76	8.62	−7.45	−4.13	25.4	21.19	17.77	−4.21	−3.42
B	7.51	12.52	21.56	5.01	9.04	16.66	21.57	26.14	4.92	4.56
C	13.43	12.22	15.94	−1.21	3.72	22.01	21.01	21.35	−1	0.34
D	12.58	13.67	15.34	1.09	1.67	21.99	20.97	21.41	−1.01	0.44

process parameter A (sorbent dose) leads to an overall decrease in adsorption capacity due to the abundance of adsorption sites and also coalescing tendency of the constituents at higher adsorption dose (m) [13]. Furthermore, at constant volume, owing to the locked connection, increasing sorbent dose (m) reduces the sorbate–sorbent ratio and also leads to the decline in adsorption capacity with increase AAC dosages. On the contrary, overall adsorption was found to increase as process parameter C (temperature) elevated from level 1 to level 3. At elevated temperature, the viscosity of the sorbate declines, leading to an increase in kinetic energy that facilitates the movement of the constituents and increases the overall diffusion rate. Similarly, raising parameter D [time of contact (t)] from level 1 to 3 also promotes adsorption due to easy accessibility of sorption sites. With the rise of the value of parameter B, i.e., AAC concentration (C_0) from level 1 to 3, the q_t also rises. The uptake capacity (q_t) rises with a rise in CV concentration (C_0) because of the reduced uptake among the ideal liquid and solid stage due to higher driving forces [13]. It is also observed that adsorption rises with the rise in factor C, i.e., temperature (T) from level 1 to 3 due to the decline in viscosity of the AAC solution, which upsurges the kinetic energy and movement of constituents, which results in a rise in the diffusion rate [14]. With the rise of parameter D, i.e., time of contact (t) from level 1 to 3, q_t too upsurges due to supplementary vacant situates accessible at first for adsorption of CV over AAC.

Figure 2 shows the parametric percentage contribution of different factors in complete adsorption of CV on AAC. It is seen from Fig. 2 that parameter B, i.e., CV concentration, (C_0) shows the maximum percentage contribution of 55.60%. These observations demonstrate that adsorption of CV gradually declines with a corresponding increase in the mass transfer forces [15, 16]. Following most persuading factor is parameter A, i.e., AAC dosage, with 37.81% share. After that, the parameter C, i.e., temperature, shows a 3.95% contribution, whereas latter parameter D, i.e., time of contact (t), displayed the slightest effect on adsorption progression with merely 2.11% contribution.

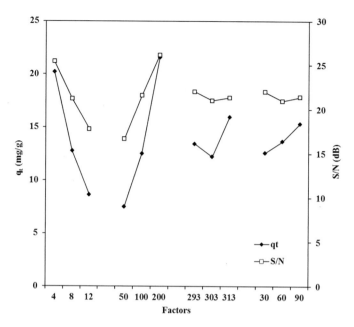

Fig. 1 Effect of parameters (factors) on adsorption of CV on AAC at different levels

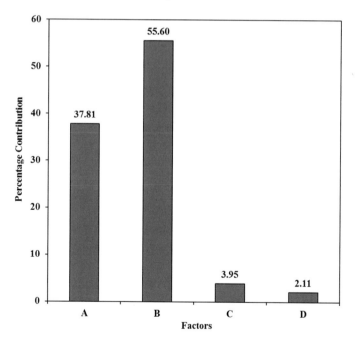

Fig. 2 Parametric percentage contribution of different factors in adsorption of CV onto AAC

3.3 Response Characteristics, Its Assessment, and Selection of Adsorption Parameters

Statistical measurements based on ANOVA were applied for evaluating the relative importance of each optimized parameters recognized through Taguchi optimization design. The ANOVA characteristics to evaluate the overall adsorption of CV are presented in Table 4. The ANOVA characteristic demonstrates that the total sum of the square of the standard deviations (S) is essentially a sum of the square of standard deviations for each of the optimized conditional factors A, B, C, and D and their respective error functions. The degree of freedom for error (f_e) is the alteration of DOF total (f_T) and DOF of controlled aspect (A, B, C, and D). All the standards are shown in Table 4. To augment the response value, i.e., q_t, the maximum values of q_t at a convinced level for a specific parameter are selected [2]. The response curves in Fig. 1 and Table 3 witnessed that for sorption of CV onto AAC, level 1 of parameter A (AAC dosage), level 3 of parameter B (CV concentration, C_0), level 3 of parameter C (temperature, T), and third level of parameter D (time of contact, t) have advanced q_t.

The average q_t (mg/g) at optimum levels of factors is as given:

$$\overline{A}_1 = 20.21, \ \overline{B}_3 = 21.56, \ \overline{C}_3 = 15.94, \ \text{and} \ \overline{D}_3 = 15.34 \ (\text{from Table 3}) \quad (7)$$

$$\text{Grand total } T = 1641.05 \ \text{mg/g with total runs } (N) = 27 \quad (8)$$

Therefore,

$$\overline{T} = T/N = 60.77 \quad (9)$$

$$\mu_{AAC} = \overline{T} + (\overline{A}_1 - \overline{T}) + (\overline{B}_3 - \overline{T}) + (\overline{C}_3 - \overline{T}) + (\overline{D}_3 - \overline{T}) = 31.46 \ \text{mg/g} \quad (10)$$

For the mean of population, 95% confidence interval and three confirmatory experimentations (CI_{POP} and CI_{CE}) for CV adsorption on AAC were obtained by taking N

Table 4 ANOVA of CV adsorption on AAC

Parameters	ANOVA qt					Pooled ANOVA qt				
	S	f	V	F	%	S	f	V	F	%
A	620.41	2	310.21	633.97	37.81	87.59	2	43.8	56.3	38.82
B	912.37	2	456.19	932.31	55.6	134.92	2	67.46	0.6	59.8
C	64.8	2	32.4	66.21	3.95	1.56	2	0.77	POOLED	0.68
D	34.66	2	17.33	35.42	2.11	1.55	2	0.78	1	0.69
Error	8.81	18	0.49	1	0.54	0	2	0	1	0
Totals	1641.05	26	816.61		100	225.62	0	112.81		100

= total numbers of set = $9 \times 3 = 27$ [12], $f_e = 26 - 8 = 18$, V_e (recalculated error variance after pooling) = 2.17 (from Table 4).

$$n_{\text{eff}} = \frac{N}{1 + [\text{total DOF}]} = 3 \tag{11}$$

From standard F-distribution table, $F\ 0.05\ (1, 18) = 4.4139$
Hence,

$$\text{CI}_{\text{POP}} = \sqrt{\frac{F_\alpha\ (1,\ f_e)\ V_e}{n_{\text{eff}}}} = \pm 1.788 \tag{12}$$

$$\text{CI}_{\text{CE}} = \sqrt{F_\alpha\ (1,\ f_e) V_e \left[\frac{1}{n_{\text{eff}}} + \frac{1}{R}\right]} = \pm 1.884 \tag{13}$$

The predicted and optimal values of process parameters, an average of capacity, and verification experimentation of q_t for adsorption of CV on AAC at 95% CI for CI_{POP} and CI_{CE} are shown in Table 5.

3.4 Confirmation Experiment

In current work, the amalgamation of the augmented adsorption capacity was found to be $A_1B_3C_3D_3$, i.e., factor A at first level and factors B, C, and D at third level. The confirmatory experiments were accomplished thrice with the optimized configurations. From confirmatory experiments, the average q_t was found 31.46 mg/g (shown in Table 5). To ensure that the experimental values were within the 95% confidence interval for CI_{CE}, a set of confirmatory experimental runs were conducted that supported our findings (Table 5).

Table 5 Predicted optimal q_t, CI and confirmation experiment results

Adsorption system	Optimal level process parameters	Predicted optimal value (mg/g)	Confidence interval 95%	Average confirmation capacity (mg/g)	Average of verification experiment (mg/g)
CV-AAC	A1B3C3D3	31.46	CIPOP: 29.67 < μAAC < 33.24	31.45	31.46
			CICE: 29.57 < μAAC < 32.34		

4 Conclusions

The results obtained through the application of Taguchi's design of experimentation method in adsorption and optimization of CV onto AAC show that it is a robust method that could be very serviceable from sources, time, and economic assessment. The significant parameters such as AAC dosage, CV concentration, temperature, and time of contact at three dissimilar levels are augmented in nine set of runs only. According to Taguchi's optimization technique, initial dye concentration plays a critical role in the overall adsorption of CV accounting for nearly 55.6% overall adsorption. The next critical parameter is the adsorbent dose accounting for 37.81% contribution to overall adsorption, followed by temperature with 3.95% contribution, followed by contact time with 2.11% contribution in the adsorption process. The optimal parameters at three different levels were observed to be A_1, B_3, C_3, and D_3. Taguchi's predicted optimized parameter values for the total adsorbed concentration of CV per gram of sorbent (q_t) dye 31.46 mg/g were found to be consistent with the average values obtained from the confirmatory experiments 31.45 mg/g. Both the values fall in the series of 95% CI_{CE}. The results demonstrate the utility of the optimization technique of Taguchi's design.

References

1. Wong, Y., Szeto. Y, Cheung, W., McKay, G.: Adsorption of acid dyes on chitosan equilibrium isotherm analyses. Process Biochem. **39**, 695–704 (2004)
2. Roy, R.: Design of Experiments using the Taguchi Approach, 16 Steps to Product and Process Improvement, pp. 110–136, issue 1. Wiley, New York (2001)
3. Lataye, D., Mishra, I., Mall, I.: Multicomponent sorptive removal of toxics-pyridine, 2-picoline, and 4-picoline from aqueous solution by bagasse fly ash: optimization of process parameters. Ind. Eng. Chem. Res. **47**, 5629–5635 (2008)
4. Lataye, D., Mishra, I., Mall, I.: Multicomponent sorption of pyridine and its derivatives from aqueous solution on to rice husk ash and granular activated carbon. Pract Periodical of Hazard. Toxic, and Radioactive Waste Manage (ASCE) **13**, 218–229 (2009)
5. Singh, K., Lataye, D., Wasewar, K.: Adsorption of fluoride onto sugarcane bagasse (Saccharum Officinarum): an application of Taguchi's design experimental methodology. J. Indian Water Works Assoc. IWWA **47**, 285–294 (2015)
6. Gessner, T., Mayer, U.: Triarylmethane and diarylmethanedyes, 6th edn. Ullmann's Encyclopedia of Industrial Chemistry, Weinheim (2002)
7. BIS—Bureau of Indian Standards Indian Standard methods of test for soil: Part 4—Grain size analysis, 2nd edn (IS 2720, New Delhi India) 4, pp. 6–8 (1985)
8. Barker, T.: Engineering Quality by Design, 1st edn, pp. 21–35. Marcel Dekker, New York (1990)
9. Silva, M., Carneiro, L., Silva, J., Oliveira, I., Filho, H., Almeida, C.: An application of Taguchi method (Robust Design) to environmental engineering: evaluating advanced oxidation process in polyester-resin wastewater treatment. Am. J. Anal. Chem. **5**, 828–837 (2014)
10. Ross, P.: Taguchi Techniques for Quality Engineering, 2nd edn, pp. 50–59. McGraw Hill, New York (1988)
11. Ross, P.: Taguchi Techniques for Quality Engineering, 2nd edn, pp. 102–179. McGraw Hill, New York (1996)

12. Taguchi, G., Wu, Y.: Introduction to the off-line quality control. Central Japan Quality Control Association Nagaya, Japan, 36–97 (1979)
13. Roy, R.: A Primer on the Taguchi Method, 2nd edn, pp. 83–88. Society of Manufacturing Engineers, Van Nostrand Reinhold, Michigan (1990)
14. Lataye, D., Mishra, I., Mall, I.: Pyridine sorption from aqueous solution by rice husk ash (RHA) and granular activated carbon (GAC)—parametric, kinetic, equilibrium and thermodynamic aspects. J. Hazard. Mater. **154**, 858–870 (2008)
15. Gupta, T., Lataye, D.: Adsorption of indigo carmine dye onto acacia nilotica (babool) sawdust activated carbon. J. Hazard. Toxic Radioact. Waste **21**(04017013), 1–11 (2017). https://doi.org/10.1061/(ASCE)HZ.2153-5515.0000365
16. Gupta, T., Lataye, D.: Adsorption of indigo carmine and methylene blue dye: Taguchi's design of experiment to optimize removal efficiency. SADHANA-Acad. Proceed. Eng. Sci. **43**, 170 (2018)

Minimization of Springback in Seamless Tube Cold Drawing Process Using Advanced Optimization Algorithms

D. B. Karanjule, S. S. Bhamare and T. H. Rao

Abstract Seamless tube manufacturing is mostly done using cold drawing process. Apart from scores on tube, chattering, eccentricity, springback is also one of the severe problems induced in seamless tube drawing process. It is because of the elastic energy stored in the tubes during forming process. This paper uses Taguchi method of experimentation for optimizing process parameters viz. die semi angle, land width and drawing speed to minimize springback. The regression model of Taguchi method using Minitab 17 is employed in Matlab as an input for further advanced optimization algorithms viz. particle swarm optimization (PSO), simulated annealing (SA) and genetic algorithm (GA). The results of these algorithms show that the process parameter values of 15° die semi angle, 10 mm land width and 8 m/min drawing speed give least springback with almost 10.5% improvement over Taguchi results.

Keywords Cold drawing · Springback · Particle swarm optimization · Simulated annealing · Genetic algorithm

D. B. Karanjule (✉)
Sinhgad College of Engineering, Vadgaon, Pune 411041, M.S, India
e-mail: karanjule.dada@gmail.com

S. S. Bhamare
Dr. Babasaheb, Ambedkar Technological University, Lonere 402103, M.S, India
e-mail: sunilsbhamare@gmail.com

T. H. Rao
Research and Development Department, Indian Seamless Metal Tubes Limited, Ahmednagar 414001, M.S, India
e-mail: hrthota@yahoo.co.in

© Springer Nature Singapore Pte Ltd. 2020
R. Venkata Rao and J. Taler (eds.), *Advanced Engineering Optimization Through Intelligent Techniques*, Advances in Intelligent Systems and Computing 949,
https://doi.org/10.1007/978-981-13-8196-6_58

667

1 Introduction

Geometrical and physical attributes of the product can be achieved by using different manufacturing techniques. Cold drawing is one of the important forming process for seamless tube manufacturing as it imparts closer dimensional tolerances, improved mechanical properties and better surface finish. This process consists of pulling the tube through a die and plug. Die controls the outer diameter, whereas plug controls the inner diameter of the drawn tube. One of the troublesome problems seamless tube production industry faces is springback in the tube making process. When a seamless tube is drawn from draw bench, due to residual stresses springback phenomenon occurs causing an undesired deformation in the final product [1]. This phenomenon is particularly severe in high strength steels, which are commonly used in the automotive industries for the performance improvement as well as light-weight design. Generally, the springback strain is an elastic strain which is measured with digital micrometer or coordinate measuring machine (CMM) 10 min after drawing from draw bench. The variation of outer diameter from targeted shape is recorded as a springback and is measured in milimeter.

2 Literature Review

Many research studies have been conducted over the last few decades for optimizing process parameters using Taguchi method [2–6] and optimization algorithms. For any objective function, the emphasis is on evaluating the global optimal values. Traditional optimization methods are found inefficient for many problems because of nonlinearity. This gives rise to advanced heuristic as well as meta-heuristic optimization algorithms which can handle any type of optimization problems like continuous or discontinuous, linear or nonlinear, stationary or nonstationary [7]. For any optimization problems, different heuristic algorithms are being used to obtain a solution by trial and error which may not be the best solution but it may be a good approximation to the exact solution. Due to less computational time, these methods are quite popular among the researchers. Meta-heuristics are treated as recent higher level techniques that include teaching–learning-based optimization (TLBO), genetic algorithm (GA), harmony search (HS), simulated annealing (SA), ant colony optimization (ACO), particle swarm optimization (PSO), imperialist competitive algorithm (ICA), bee algorithms (BA), firefly algorithms (FA), etc. A numbers of papers [8–13] could be found in the literature survey for deep understanding of all these algorithms. Particle swarm optimization (PSO) algorithm is a meta-heuristic technique related to social behavior of animals like flocking of birds. However, simulated annealing (SA) algorithm is related to process where an energy distribution of the substance is minimized by heating it above its melting point and then slowly cooling down. Whereas, genetic algorithm (GA) is an evolutionary method which is based on the principles of genetics and evolution. These modern meta-heuristics algorithms have found their usage in solving different optimization problems in various fields such as manufacturing, production systems, industrial planning, decision making, scheduling, pattern recognition, process parameter optimization in machining and many more.

Taking motivation from this, this research work explores three different algorithms viz. PSO, SA and GA in optimizing cold drawing process. The uniqueness of this work is on cold drawing process of seamless tube manufacturing for the selection of appropriate process parameters (combination of die semi angle, land width, drawing speed) toward minimizing springback.

3   Experimental Procedure

Seamless tubes are cold drawn on the draw bench machine, and the minimum variation in the drawn tube is measured. Springback is found less if the dimensional variation from targeted value is less. The ultimate aim of this study is to decide the best parameter setting of die semi angle, land width and drawing speed for a particular tube and die material combination. The cold drawn tubes are measured for outer diameter using Mitutoyo make digital micrometer of 1 μm accuracy. The springback is measured 10 min after drawing operation, once the stresses get settled.

3.1   Materials

AISI D3 die steel is taken for die and plug materials having capability of high wear, abrasion resistance and resistance to heavy pressure. This steel is widely used in industrial applications for different dies used in manufacturing processes like blanking, cold forming, stamping, punches, etc. Seamless tubes of C-45(EN-8D) material cold drawn from size of 33.40 mm outer diameter and 4.00 mm wall thickness are considered in this study. The chemical composition for tube material of C-45(EN-8D) consists 0.45% carbon, taken for study due to the fact that more the percentage of carbon, more will be the springback. The drawn size is 30 mm outer diameter.

3.2   Machine Tool

The experimentation is carried out on draw bench of Maruti make having capacity 50 Ton, maximum drawing speed 10 m/min and maximum width of drawn tube obtained is 30 mm. A draw bench for cold drawing seamless tubes consists of a die and plug control devices.

3.3 Taguchi Method

Taguchi method is utilized to design orthogonal array (OA) for the three parameters viz. die semi angle (DA), land width (LW) and drawing speed (DS) through two levels of DA viz. 10°, 15° and two levels of LW viz. 5, 10 mm whereas three levels of DS viz. 4, 6, 8 m/min. After discussions with industry experts and review of literature, it is found that these three geometrical parameters are crucial from springback point of view. L36 orthogonal array is used to conduct experiments. Drawn tubes are measured using digital micrometer of 1 μm accuracy for their outer dimensions. The variation from 30 mm is denoted as a springback and taken for further statistical analysis.

3.4 Experimentation

Table 1 shows the orthogonal array and the results of the experimentation carried on draw bench and measured with micrometer for EN 8 D(C-45) tube material and AISI D3 die and plug material.

4 Results and Discussions

4.1 Analysis of Variance (ANOVA)

The ANOVA is performed and is used to check the adequacy of developed model. The ANOVA Table 2 identifies the contribution of the process parameters that influence the response. If the P-value in the model is lower than 0.05 (for 95% confidence level), it indicates the statistical significance of the model. From Table 2, all the process parameters viz. die semi angle, land width and drawing speed are found significant. The springback is measured for various combinations of die angle, land width and drawing speed. The signal-to-noise (S/N) ratio values for springback are determined. Confirmatory tests are also conducted after finalizing the optimal parameter setting.

4.2 Development of Regression Model

From the results of experiments, an empirical equation is developed with linear regression technique for the springback. The regression equation is generated for mean values of springback under different conditions of input process parameters. Equation 1 indicates regression of springback (SB) in terms of die semi angle, land width and drawing speed.

Table 1 Experimental layout and results

S. No.	Die semi angle (°)	Land width (mm)	Drawing speed (m/min)	Actual measurement of OD (mm)	Springback (mm)
1	10	5	4	30.065	0.065
2	10	5	6	30.064	0.064
3	10	5	8	30.098	0.098
4	10	5	4	30.062	0.062
5	10	5	6	30.060	0.06
6	10	5	8	30.095	0.095
7	10	5	4	30.061	0.061
8	10	5	6	30.062	0.062
9	10	5	8	30.096	0.096
10	10	10	4	30.080	0.08
11	10	10	6	30.082	0.082
12	10	10	8	30.050	0.050
13	10	10	4	30.082	0.082
14	10	10	6	30.079	0.079
15	10	10	8	30.054	0.054
16	10	10	4	30.079	0.079
17	10	10	6	30.080	0.08
18	10	10	8	30.058	0.058
19	15	5	4	30.063	0.063
20	15	5	6	30.0248	0.0248
21	15	5	8	30.266	0.266
22	15	5	4	30.100	0.100
23	15	5	6	30.055	0.055
24	15	5	8	30.255	0.255
25	15	5	4	30.121	0.121
26	15	5	6	30.0258	0.0258
27	15	5	8	30.255	0.255
28	15	10	4	30.308	0.308
29	15	10	6	30.104	0.104
30	15	10	8	30.048	0.048
31	15	10	4	30.290	0.29
32	15	10	6	30.125	0.125
33	15	10	8	30.052	0.052
34	15	10	4	30.285	0.285
35	15	10	6	30.123	0.123
36	15	10	8	30.059	0.059

Table 2 Analysis of variance (ANOVA) of S/N ratio

Source	DF	Adj SS	Adj MS	F-value	P-value
DA	1	0.114695	0.114695	650.65	0.000
LW	1	0.005675	0.005675	32.19	0.000
DS	2	0.001600	0.000800	4.54	0.021
DA × LW	1	0.003600	0.003600	20.42	0.000
DA × DS	2	0.001448	0.000724	4.11	0.029
LW × DS	2	0.084801	0.042400	240.53	0.000
DA × LW × DS	2	0.055845	0.027922	158.40	0.000
Error	24	0.004231	0.000176		
Total	35	0.271895			

$$\begin{aligned}
\text{Springback(SB)} = {} & 0.871 - 0.0995 \text{X DA} - 0.1543 \text{X LW} \\
& - 0.1801 \text{X DS} - 0.00148 \text{DS X DS} \\
& + 0.01740 \text{ X DA X LW} + 0.02235 \text{ X DA X DS} \\
& + 0.02822 \text{ X LW X DS} - 0.003167 \text{X DA X LWXDS} \quad (1)
\end{aligned}$$

For better fitment of the model, the R squared value must be close to unity. For this data, the value is 0.9844. The fitness of the model is also judged by the good agreement of adjusted R squared value (0.9773) and the predicted R squared value (0.965).

The main effects of the factors on springback for each level are calculated, and the optimum level of each parameter is identified. From Figs. 1 and 2, it show that die semi angle of 15°, land width of 5 mm and drawing speed of 6 m/min give the least springback.

4.3 Optimization Using Advanced Optimization

An optimization algorithm is an iterative procedure to achieve optimum or a satisfactory solution. It begins with one or more design solutions and then checks new solutions iteratively to find required optimum solution. The objective function is formulated (Eq. 1) for applying different algorithms viz. particle swarm optimization (PSO), simulated annealing (SA) and genetic algorithm (GA).

Fig. 1 Main effects plot for S/N ratios

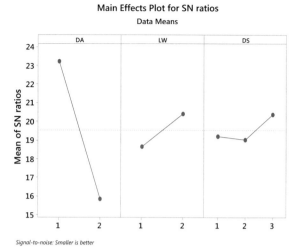

Fig. 2 Main effects plot for means

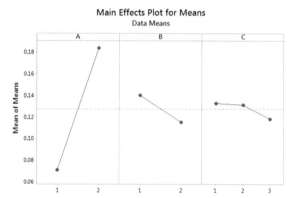

4.3.1 Optimization Using Particle Swarm Optimization (PSO)

PSO algorithm is a technique where particles are considered as a conceptual entities flying through the multi-dimensional search space. Each particle is assigned with a position and a velocity at any instant. The position vector of a particle with respect to the origin is considered as a trial solution in the search space.

The objective function can be written as

$$\begin{aligned}
\text{Minimize SB} = {} & 0.871 - 0.0995 \text{X DA} - 0.1543 \text{X LW} \\
& - 0.1801 \text{X DS} - 0.00148 \text{X DS X DS} \\
& + 0.01740 \text{X DA X LW} + 0.02235 \text{X DA X DS} \\
& + 0.02822 \text{X LW X DS} - 0.003167 \text{X DA X LW X DS}
\end{aligned}$$

The lower and upper bounds are

$$10 \leq \text{Die semi angle} \leq 15 \quad 5 \leq \text{Land width} \leq 10 \quad 4 \leq \text{Drawing speed} \leq 8$$

Importing these values as an input to PSO algorithm code written in Matlab, the optimum conditions of input parameters are obtained by analyzing the output of the algorithm. Optimization using Matlab coding for PSO is done. The minimum value of the springback obtained with particle swarm optimization is 0.0492 for the parameter setting of 15° die semi angle, 10 mm land width and 8 m/min drawing speed.

4.3.2   Optimization Using Simulated Annealing

A simulated annealing is a metropolis optimization algorithm. This algorithm starts with an initial solution which may be the best solution so far. The temperature set at the initial is denoted as high temperature Temp(i) which now becomes the current solution and the parent or active solution [7].

Optimization using Matlab coding is done for simulated annealing. The minimum value of springback obtained with simulated annealing is 0.0491868 as shown in Fig. 3. With optimization terminated, the best function values for the three parameters are 15° die semi angle, 10 mm land width and 8 m/min drawing speed.

Fig. 3 Output of simulated annealing

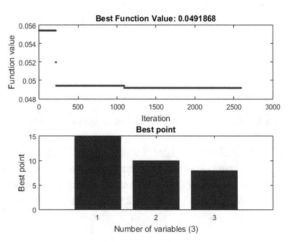

4.3.3 Optimization Using Genetic Algorithm

Genetic algorithms (GA) are an evolutionary optimization based on direct, parallel, stochastic method that imitates the evolution of the living beings. This algorithm has following steps:

1. Initial population generation
2. Maximization or minimization of the function.
3. Checking for termination of the algorithm by

   - Value of the function
   - Maximal number of iterations
   - Stall generation

4. Selection
5. Crossover
6. Mutation
7. New generation.

Optimization using Matlab for genetic algorithm as shown in Fig. 4 gives values of 15°, 10 mm and 8 m/min for die semi angle, land width and drawing speed, respectively, for least springback.

The comparative study of all these algorithms is tabulated as shown in Table 3.

Fig. 4 Output of genetic algorithm

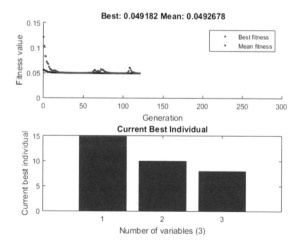

Table 3 Comparison of springback for initial and optimal parameter setting

Response	Initial parameter setting	Optimal parameter setting using PSO	Optimal parameter setting using SA	Optimal parameter setting using GA
Parameter values	Die semi angle 15	Die semi angle 15	Die semi angle 15	Die semi angle 15
	Land width 5	Land width 10	Land width 10	Land width 10
	Drawing speed 6	Drawing speed 8	Drawing speed 8	Drawing speed 8
Springback	0.055	0.0492	0.0491868	0.049182
Improvement		5.800E-3	5.813E-3	5.818E-3
% improvement		10.54%	10.57%	10.58%

5   Conclusions

The present work has disclosed an optimized parameter setting for cold drawing of seamless tubes. Based upon experimentation, modeling and optimization techniques following conclusions can be drawn.

- The results of ANOVA show that die semi angle is the major influencing parameter affecting the springback with a contribution of 42.18%.
- The optimal combination predicted for cold drawing of seamless tubes of EN 8 D(C-45) material is 15° die angle, 5 mm land width and 6 m/min drawing speed.
- Confirmation results show better improvement.
- Empirical equations are developed for springback using regression analysis.
- This parameter setting can be applied so as to minimize springback in cold drawing of seamless tubes.
- The optimization algorithm viz. PSO, SA and GA shows that 15° die angle, 10 mm land width and 8 m/min drawing speed gives least springback.
- The optimization algorithms have effectively proved for the optimization of springback and have shown 10.54% improvement by PSO, 10.57% improvement by SA and 10.58% improvement by GA.

Acknowledgments The authors are very much thankful to Yashashree Tubes Private Limited, F-48, M.I.D. C., Ahmednagar, M.S., India for financial assistance and permitting experimental work.

References

1. Karanjule, D., Bhamare, S., Rao, T.: Effect of Young's modulus on springback for low, medium and high carbon steels during cold drawing of seamlesss tubes. Mater. Sci. Eng. **346**, 1–13 (2018). https://doi.org/10.1088/1757-899X/346/1/012043/meta
2. Tosun, N., Pihtili, H.: The effect of cutting parameters on wire crater sizes in wire EDM. Int. J. Adv. Manuf. Technol. **21**(10), 857–865 (2003). https://doi.org/10.1007/s00170-002-1404-1

3. Das, R., Mandal, N., Doloi, B., Mondal, B.: Optimization of flank wear using Zirconia Toughened Alumina (ZTA) cutting tool: Taguchi method and regression analysis. Measurement **44**(10), 2149–2155 (2011). https://doi.org/10.1016/j.measurement.2011.07.022
4. Choudhury, A., Dingal, S., Pradhan, T., Sundar, J., Roy, S.: The application of Taguchi's method in the experimental investigation of the laser sintering process. Int. J. Adv. Manuf. Technol. **38**(9–10), 904–914 (2008). https://doi.org/10.1007/s00170-007-1154-1
5. Ficici, F., Koksal, S., Kayikci, R., Savas, O.: Experimental optimization of dry sliding wear behavior of in situ AlB2/Al composite based on Taguchi's method. Mater. Des. **42**, 124–130 (2012). https://doi.org/10.1016/j.matdes.2012.05.048
6. Tarng, Y., Hsiao, Y., Huang, W.: Optimization of plasma arc welding parameters by using the Taguchi method with the grey relational analysis. Mater. Manuf. Process. **23**(1), 51–58 (2007). https://doi.org/10.1080/10426910701524527
7. Kumar, A., Kumar, V., Datta, S., Mahapatra, S.: Parametric appraisal and optimization in machining of CFRP composites by using TLBO. J. Intell. Manuf. **28**(8), 1769–1785 (2017). https://doi.org/10.1007/s10845-015-1050-8
8. Patel, V., Rao, R.: Design optimization of shell and tube heat exchanger using particle swarm optimization technique. Appl. Therm. Eng. **30**(11–12), 1417–1425 (2010). https://doi.org/10.1016/j.applthermaleng.2010.03.001
9. Savsani, V., Rao, R., Vakharia, D.: Optimal weight design of a gear train using particle swarm optimization and simulated annealing algorithms. Mech. Mach. Theory **45**(3), 531–541 (2010). https://doi.org/10.1016/j.mechmachtheory.2009.10.010
10. Rao, R., Kalyankar, V.: Parameter optimization of modern machining processes using TLBO algorithm. Eng. Appl. Artif. Intell. **26**(1), 524–531 (2013). https://doi.org/10.1016/j.engappai.2012.06.007
11. Srinivas, N., Deb, K.: Multiobjective optimization using nondominated sorting in genetic algorithm. Evol. Comput. **2**(3), 221–248 (2007). https://doi.org/10.1162/evco.1994.2.3.221
12. Haddock, J., Mittenthal, J.: Simulation optimization using simulated annealing. Comput. Ind. Eng. **22**(4), 387–395 (1992). https://doi.org/10.1016/0360-8352(9290014-B)
13. Brooks, S., Morgan, B.: Optimization using simulated annealing. J. Royal Stat. Soc. **44**(2), 241–257 (1995). Doi: https://doi.org/Triyoga_Widiastomo/publication/305441015

Integrated Production Planning and Scheduling for Parallel Production Lines

K. C. Bhosale and P. J. Pawar

Abstract Production planning department prepares demand for the coming months considering the plant capacity, available time. Depending on this demand, inventory of the raw material is kept in the production unit. While, in the scheduling problem, the time horizon is selected as a shift, day or week. Scheduling model determines the start time, processing time, finish time and transition time. However, in most of the reported literature, production planning problem and scheduling problem are solved independently. But, to achieve the global optimum solution and minimise material flow and reduce the total cost, there is a need of integrating production planning and scheduling model. A case study based on the parallel line continuous process plant is selected and optimisation is obtained by real coded genetic algorithm (RCGA). Results show that RCGA outperforms the solutions obtained by previous researchers.

Keywords Integrated production planning and scheduling · Real coded genetic algorithm (RCGA)

1 Introduction

In most of the manufacturing industries, flexible manufacturing system (FMS) is becoming an important element. FMS is useful to achieve many outcomes like lead time reduction, increase in throughput, reduction of inventory and other benefits

K. C. Bhosale (✉) · P. J. Pawar
Department of Production Engineering, K. K. Wagh Institute of Engineering Education and Research, Nashik, Maharashtra, India
e-mail: bhosale_kailash@yahoo.co.in

P. J. Pawar
e-mail: pjpawar1@rediffmail.com

Savitribai Phule Pune University, Pune, Maharashtra, India

K. C. Bhosale
Department of Mechanical Engineering, Sanjivani College of Engineering, Kopargaon, Maharashtra, India

© Springer Nature Singapore Pte Ltd. 2020
R. Venkata Rao and J. Taler (eds.), *Advanced Engineering Optimization Through Intelligent Techniques*, Advances in Intelligent Systems and Computing 949,
https://doi.org/10.1007/978-981-13-8196-6_59

which will combine to assure that the system should be run at lower cost. However, the initial investment for FMS is high. The management expects that the payback period should be less. Hence, to achieve this, there is a need to minimise the total manufacturing cost. In manufacturing industries, production planning and scheduling are considered as two major decisions. Both these problems are solved independently with different objectives, constraints and time limit. In the production planning problem, the production plan is prepared for a long-time duration such as a week or month. Planning problem is solved on the basis of the aggregate information collected in the form of rough values of the real-world problem. In scheduling problem, an optimum schedule and sequence of operations are obtained. Even though the production planning and scheduling problems are solved in a sequence, the obtained solution is a local optimal because the constraints related to real-time world problem are only considered in scheduling stage [1]. So, to achieve a global optimal solution, production planning and scheduling problem is solved in an integrated way.

In the manufacturing industries, multiple products are produced on parallel lines. The market demand of these products is small and fluctuating. So, it is not possible to purchase a special processing unit or manufacturing machines for each and every product. It is required to complete the given demand in the planning period by scheduling the multiple products on the available processing units. The objective considered is minimisation of the total cost. In the manufacturing industries, multiple products are manufactured on the available processing units. So, to solve the problem, a mathematical model is developed in the form of mixed integer programming for the objective of minimising total cost.

Kopanos et al. [2, 3] have considered planning and scheduling problems simultaneously. They have grouped the products into families. This approach is useful when the products have fixed sequence of product changeover. Bhosale and Pawar [4] have considered an operation allocation problem under scheduling stage. A mathematical model is modified by adding the waiting time. The proposed model has considered production planning and scheduling in an integrated way. They used real coded genetic algorithm (RCGA) for optimisation of the problem. Bhosale and Pawar [5] have proposed RCGA to optimise material flow of flexible manufacturing system.

In this paper, a case study of [3] based on production planning and scheduling of a continuous process plant is considered. The objective function considered is to minimise total cost. Real coded genetic algorithm is used for optimisation of the objective function. The results obtained by RCGA are verified with that obtained by [3]. There is a 15% reduction is observed by using RCGA in the objective function value. In the next section, a brief information of RCGA is presented.

2   Real Coded Genetic Algorithm (RCGA)

Genetic algorithm (GA) is a population-based stochastic search algorithm. Binary coding or real coding types are used to represent the parameters of the solution for GA. In case of binary GA, due to the use of binary bits, the length of the chromosome is more. In the process of encoding and decoding, a discretization error occurs in binary GA. In the case of real coded genetic algorithm (RCGA), real coding parameters are used to represent the chromosome. So, RCGA is very easy to handle because there is no need for encoding and decoding of parameters. The chromosome length is equal to the length of the solution [6, 7]. Deb and Kumar [8] have used real coded genetic algorithm and expressed that it performs well or even better in the comparison of binary coded Genetic Algorithm. To obtain optimal solution selection, crossover and mutation are used and explained in the next section. A penalty function is used for degrading the objective function value. In this paper, for the voilation of constraints, a static penalty is used [9]. To obtain a feasible optimal solution, Kumar and Shankar [10] have used penalty function.

In the next section, a problem statement of case study of [3] is presented.

3 Problem Statement

Here, a case study of [3] is considered in which 15 products are grouped into five families, which will be processed on three different processing units. All the products can be processed on any available unit. The total available production horizon is of four days, and it is divided into four 24 h periods. Maintenance activity is scheduled on units 1, 2 and 3, in period n_2, n_3 and n_4, respectively. Processing data include the following: setup time $\delta_{ij} = 0.5$; setup cost $\theta_{ij} = 50$; production rate $\rho_{ij}^{\max} = 10$; operating cost $\lambda_{ij} = 0.1$; inventory cost $\xi_{in} = 0.5$; minimum processing time $\tau_{ij} = 0.2$; back log cost $\psi_{in} = 3$ for all products.

Changeover time $\gamma_{ff'j}$ and costs $\varphi_{ff'j}$ between product families are shown in Table 1. The demand of all products is given in Table 2. There is a predetermined sequence of processing of all the products.

4 Mathematical Formulation

The production planning and scheduling problem is presented as mixed linear programming. The mathematical model is used from [3].

$$P_{in} = \sum_{j \in J_i} Q_{ijn} \forall i, n \tag{1}$$

$$S_{in} - B_{in} = Sin - 1 - Bin - 1 + P_{in} - D_{in} \forall i, n \tag{2}$$

$$S_{in} \leq \text{product storage capacity } \forall i, n \tag{3}$$

$$Y_{fn}^{F} \geq Y_{ijn} \forall f, i, j, n \tag{4}$$

$$Y_{ffn}^{F} \leq \sum_{i \in I_f} Y_{ijn} \forall f, j, n \qquad (5)$$

$$\sum_{f' \neq f, f' \in F_j} X_{f'fjn} + WF_{fjn} = Y_{fjn}^{F} \forall f, j, n \qquad (6)$$

$$\sum_{f' \neq f, f' \in F_j} X_{ff'jn} + WL_{fjn} = Y_{fjn}^{F} \forall f, j, n \qquad (7)$$

$$\sum_{f \in F_j} \sum_{f' \neq f, f' \in F_j} X_{ff'jn} + 1 = \sum_{f \in F_j} Y_{fjn}^{F} \forall j, n \qquad (8)$$

$$C_{f'jn} \geq C_{fjn} + T_{f'jn}^{F} + \gamma_{ff'j} X_{ff'jn} - \omega_{jn}(1 - X_{ff'n}) \forall f, j, n \qquad (9)$$

Table 1 Initial population

	Day 1						Day 2					
Unit	j1	j2	j3	j1	j2	j3	j1	j2	j3	j1	j2	j3
Sol 1	1	5	4	3	2	4	0	3	2	0	5	1
Sol 2	5	2	4	3	1	4	0	1	2	0	3	5
Sol 3	4	2	3	5	1	4	0	2	3	0	1	5
Sol 4	5	2	1	4	3	4	0	5	2	0	1	3
Sol 5	1	3	4	5	2	1	0	4	3	0	2	5
	Day 3						Day 4					
Unit	j1	j2	j3	j1	j2	j3	j1	j2	j3	j1	j2	j3
Sol 1	1	0	5	3	0	2	4	3	0	2	4	1
Sol 2	1	0	4	3	0	5	2	4	0	2	3	1
Sol 3	3	0	4	5	0	1	2	4	0	5	3	1
Sol 4	4	0	5	1	0	3	2	4	0	2	3	1
Sol 5	2	0	1	5	0	4	3	2	0	3	1	5

Table 2 Fitness function and Roulette Wheel selection method

Sol. No.	Combined objective function (z)	Reciprocal of z	Fitness function	Cumulative	Random number	Solution replaced by
1	2393	0.000418	0.226858	0.226858	0.2573	2
2	2763	0.000362	0.196479	0.423337	0.6981	4
3	2809	0.000356	0.193262	0.616599	0.1595	1
4	2813	0.000355	0.192987	0.809586	0.9959	5
5	2851	0.000351	0.190414	1	0.1984	1
Total		0.001842				

$$WF_{fjn} = \sum_{f' \in F_j} \bar{X}_{f'fjn} \forall f_i j, n > 1 \tag{10}$$

$$WL_{fjn-1} = \sum_{f' \in F_j} \bar{X}_{ff'jn} \forall f, j, n > 1 \tag{11}$$

$$\bar{U}_{jn} + U_{jn-1} = \sum_{f \in F_j} \sum_{f' \in F_j, f \neq f'} \gamma_{ff'j} \bar{X}_{ff'jn} \forall j, n > 1 \tag{12}$$

$$\bar{U}_{jn} + U_{jn-1} + \sum_{f \in F_j} T^F_{fjn} + \sum_{f \in F_j} \sum_{f' \in F_j, f \neq f'} \gamma_{ff'j} \bar{X}_{ff'jn} \leq \omega_{jn} \forall j, n \tag{13}$$

$$\rho^{min}_{ij} T_{ijn} \leq Q_{ijn} \leq \rho^{max}_{ij} T_{ijn} \forall i, j, n \tag{14}$$

$$\tau^{min}_{ijn} Y_{ijn} \leq T_{ijn} \leq \tau^{max}_{ijn} Y_{ijn} \forall i, j, n \tag{15}$$

$$T^F_{fjn} = \sum_{i \in I_f} \left(T_{ijn} + \partial_{ij} Y_{ijn} \right) \forall f, j, n \tag{16}$$

The detailed notations and mathematical model is given in [3].

Objective Function

The goal is to minimise total of inventory, backlog, changeover (inside and across periods), setup and operating costs:

$$\text{Min} \sum_i \sum_n (\xi_{in} S_{in} + \psi_{in} B_{in}) + \sum_i \sum_{j \in J_i} \sum_n (\theta_{ij} Y_{ijn} + \lambda_{ij} Q_{ijn})$$
$$+ \sum_f \sum_{f' \neq f} \sum_j \sum_n \phi_{ff'j} (X_{ff'jn} + \bar{X}_{ff'jn}) \tag{17}$$

In the next section, the application of RCGA to the case study of [3] is presented.

5 Application of RCGA to the Case Study Problem

As discussed in Sect. 3, the working of RCGA in the presented approach is explored as follows:

Step 1 The parameters taken as follows:
The population size is decided by conducting various experimentations. Different sizes from 10 to 50 are considered. It is observed that after selecting population size 10, there is no significant change in result. Various experiments were conducted to decide the stopping criteria, and it is observed that after 100 generations no significant change was observed in the objective function value. Experiments were conducted to decide crossover and mutation probability for various combinations from 0 to 1. So, crossover probability is taken as 0.8 and mutation probability is taken as 0.2.

Step 2   There are five family of products manufactured from three processing units. Now, a random number is generated within 1–5 numbers. These random numbers is the sequence of products on different processing units. This one set of sequence of numbers is called *Chromosome*. In this way, ten *Chromosomes* are generated randomly. This is called as the *Initial Population*. In Table 1, only five initial solutions are shown due to space limitations.

In Table 1, solution 1 shows that family 1 to be processed on unit 1, family 5 on unit 2 and family 4 on unit 3. After processing these families, family 3 will be processed on unit 1, family 2 will be processed on unit 2. As this is a continuous process plant, even though family 4 is planned on day 2, it will be processed after finishing family 4 on unit 3. Number "0" indicates no allocation as there is a maintenance activity on unit 1 on day 2, on unit 2 on day 3 and on unit 3 on day 4.

Step 3   According to objective function which is shown in Eq. (17), the *Fitness Function* is calculated. There is constraint of finishing the processing of all the products within four days. So, if this constraint is violated then a static penalty is added in Eq. (17) and thus combined objective function is calculated. The calculations are shown in Table 2. Here, total of inventory cost, backlog cost, changeover cost, setup cost, operating costs are considered to calculate objective function value. There is maintenance on j1, j2 and j3 on day 2, day 3 and day 4, respectively. If the completion time of j1 on day 1 is more than 24 h, then a static penalty is added in the objective function value. Thus, a combined objective function value is obtained. Those solutions have large combined objective function value are having less chance to get selected in the next iteration.

Step 4   For selection operation, *Roulette Wheel* method is used. The Roulette wheel selection example is shown in Fig. 1. Suppose that five *Chromosomes* are generated randomly. Their *Fitness function* values are shown in Table 2. Roulette wheel shows the contribution of each chromosome in the population. When the Roulette wheel is spun, the probability of selection of *Chromosome* "1" is more as compared to the remaining *Chromosomes*.

Step 5   Here, the crossover operation is carried out on 80% of selected chromosomes. There are three units, four days and five products. So, the chromosome

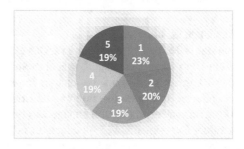

Fig. 1 Roulette wheel selection

(a) Parent 1 | 3 | 5 | 1 | 2 | 4 | 4 | 0 | 2 | 1 | 0 | 3 | 5
Parent 2 | 1 | 3 | 4 | 5 | 2 | 1 | 0 | 4 | 3 | 0 | 2 | 5

(b) Child 1 | 3 | 5 | 1 | 2 | 4 | 1 | 0 | 4 | 3 | 0 | 2 | 5
Child 2 | 1 | 3 | 4 | 5 | 2 | 4 | 0 | 2 | 1 | 0 | 3 | 5

Fig. 2 **a** Before crossover and **b** after crossover

Table 3 Optimum result

S. No.	Methodology	Objective function ($)
1	Kopanos et al. [4]	2630
2	Proposed methodology (RCGA)	2567

length is $3 \times 2 \times 4 \times 5 = 120$ numbers. In which 1–5 numbers are selected randomly. Here, one chromosome string is prepared for five families. If maintenance activity is there, it is shown by 0. In this case, four chromosome strings for four different days are prepared. For performing the crossover operation, two parent strings are selected randomly, and a random number is generated in between 1 and 20. Suppose that the number is 5. Then, the chromosomes are swapped to get the two offspring as shown in Fig. 2. However, in Fig. 2 only half portion of the chromosome string is shown due to space limitation.

Step 6   After crossover operation, it may be possible that the obtained solution will not be feasible and solution may trap in local optima. So, to avoid the solution to be trapped in local optima, mutation operation is performed. For demonstration purpose, only one set of chromosomes is presented. Here, the two numbers are generated randomly in between 1 and 5, suppose they are 2 and 3. Then the *Genes* at that second and third place from the *Chromosomes* are exchanged. It is shown below.

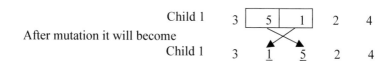

Child 1   3 | 5 | 1 | 2   4
After mutation it will become
Child 1   3   1   5   2   4

Step 7   When the crossover and mutation operations are completed, one iteration is completed and a new set of chromosome is generated. This set of chromosome is ready to take participation in the next iteration.

Step 8   In this way, the number of *Generations* is performed and at every *Generation* the best *Chromosome* values are recorded. After 100 trials, the best sequence of products on three finishing lines is obtained by the present RCGA approach and shown in Table 3.

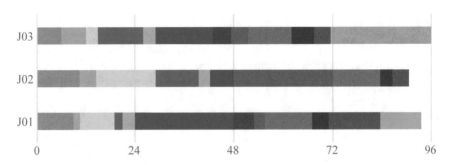

Fig. 3 Gantt's chart of the optimum solution

Proposed real coded genetic algorithm (RCGA) is applied on the case study of [3] and an optimal solution is obtained as shown in Table 3. The total cost obtained by [3] is $2630 whereas by proposed methodology the total cost obtained is $2567. The Gantt's chart for the optimal solution obtained is shown in Fig. 3.

6   Conclusion

In this paper, a case study on production planning and scheduling problem on single stage parallel lines of [3] is considered. In this paper, an objective is considered to minimise total of inventory cost, backlog cost, set up cost, processing cost and changeover cost. The total cost obtained by [3] is $2630. Real coded genetic algorithm is proposed to optimise the objective function value. The proposed methodology has obtained $2567 which is 15% less as compared to [3]. In [3], the total demand was not fully satisfied. So, a backlog cost was considered. But, in this paper, it is also observed that, the demand of all the family of products is fully satisfied. So, no backlog cost is given. RCGA is also useful in solving problems of the production planning and scheduling problem on single stage parallel lines.

References

1. Chu, Y., et al.: Integrated planning and scheduling under production uncertainties: Bi-level model formulation and hybrid solution method. Comput. Chem. Eng. **72**, 255–272 (2015). https://doi.org/10.1016/j.compchemeng.2014.02.023
2. Kopanos, G.M., Puigjaner, L., Georgiadis, M.C.: Optimal production scheduling and lot-sizing in dairy plants: the yogurt production line. Ind. Eng. Chem. Res. **49**(2), 701–718 (2010). https://doi.org/10.1021/ie901013k
3. Kopanos, G.M., Puigjaner, L., Maravelias, C.T.: Production planning and scheduling of parallel continuous processes with product families. Ind. Eng. Chem. Res **50**, 1369–1378 (2011). https://doi.org/10.1021/ie100790t

4. Bhosale, K.C., Pawar, P.J.: Flex Serv Manuf J (2018). https://doi.org/10.1007/s10696-018-9310-5

5. Bhosale, K.C., Pawar, P.J.: Optimisation of operation allocation problem in flexible manufacturing system by real coded genetic algorithm. Mater. Today: Proc. **5**, 7160–7167 (2018). https://doi.org/10.1016/j.matpr.2017.11.381

6. Herrera, F., Lozano, M., Verdegay, J.L.: Tackling real-coded genetic algorithms: operators and tools for behavioural analysis. Artif. Intell. Rev. **12**(4), 265–319 (1998). https://doi.org/10.1023/A:1006504901164

7. Chuang, Y.-C., Chen, C.-T., Hwang, C.: A simple and efficient real-coded genetic algorithm for constrained optimization. Appl. Soft Comput. **38**, 87–105 (2016). https://doi.org/10.1016/j.asoc.2015.09.036

8. Deb, K., Kumar, A.: Real-coded genetic algorithms with simulated binary crossover: studies on multimodal and multiobjective problems. Complex Syst. **9**, 431–454 (1995)

9. Homaifar, A., Qi, C.X., Lai, S.H.: Constrained optimization via genetic algorithms. Simulation **62**(4), 242–253 (1994)

10. Kumar, N., Shanker, K.: A genetic algorithm for FMS part type selection and machine loading. Int. J. Prod. Res. **38**(16), 3861–3887 (2000). https://doi.org/10.1080/00207540050176058

Optimization of Surface Modification Phenomenon for P20+Ni Die Steel Using EDM with P/M Composite Electrode

Jayantigar Lakhmangar Ramdatti and Ashishkumar Vitthaldas Gohil

Abstract In this experimental work, attempts have been made to study the improvement in microhardness of P20+Ni die steel, EDMed using powder metallurgy electrode. The powder metallurgy (P/M) electrode was manufactured from a combination of metallic powders such as tungsten (W), copper (Cu), and silicon (Si) with a particle size less than 325 mesh. The work has been performed using straight polarity (electrode negative and work as positive) and EDM oil as a dielectric fluid. Experiments were performed as per Taguchi's L9 orthogonal array. The effect of input process parameters such as compaction pressure (C_p), peak current (I_p), and pulse on time (T_{on}) has been investigated on microhardness. The result of experiments was indicated that the microhardness of the machined surface obtained by powder metallurgy electrode was improved by 282.6% as compared to the base material and by 168.7% as compared to the surface machined using conventional copper electrode. SEM analysis of EDMed samples indicated the formation of defects free surface. İmprovements in the percentage of Cu, Si, and W of machined surface confirmed migration of material from P/M electrode during EDS analysis.

Keywords Electric discharge machining (EDM) · Powder metallurgy electrode (P/M) · Microhardness · Taguchi methods · Design of experiments

1 Introduction

In a country like India, the application of components manufactured from plastic has been continuously increasing and data shows that the total consumption of plastic is higher than the steel since last few years. Different components of automobiles,

J. L. Ramdatti (✉)
Gujarat Technological University, Ahmedabad, Gujarat, India
e-mail: jlramdatti@gmail.com

A. V. Gohil
S. S. Engineering College, Bhavnagar, Gujarat, India
e-mail: Avgohil1@gmail.com

© Springer Nature Singapore Pte Ltd. 2020
R. Venkata Rao and J. Taler (eds.), *Advanced Engineering Optimization Through Intelligent Techniques*, Advances in Intelligent Systems and Computing 949,
https://doi.org/10.1007/978-981-13-8196-6_60

aerospace, and household appliances, previously manufactured from a different grade of steels have been replaced with plastic components. Besides, higher production of plastic components leads to increasing the consumption of steel used for manufacturing the die and tool. Replacements of the metallic component by plastic component is only possible when plastic exhibits mechanical and physical properties which are equivalent to those of steel. The properties of plastic have been improved by adding various hardest constituents like silica, mica, and glass fiber into plastic. Pressing of plastic in the die with such hard constituents subjected to high wear, scratch, fatigue, and corrosion. Hence, to improve the life of die and tool, various surface coating processes such as chemical vapor deposition (CVD), physical vapor deposition (PVD), and plasma arc spraying have been performed. However, these conventional surface treatments incurred high cost and processing time due to the use of expensive and special tooling equipment [1–6].

Recently, the use of EDM electrode manufactured through powder metallurgy process has been found a viable alternative to eliminate conventional surface modification treatments such as CVD and PVD [7, 8]. Researchers reported the feasibility of surface alloying using P/M electrode in EDM; however, the process is still at the experimental stage and many issues related to it yet to be resolved before the process is widely accepted by the die and mold making industries [9, 10]. After exploring various literature related to surface alloying, the following gaps have been identified and efforts have been made to address these aspects in this work: (1) No research work available related to surface modification of P20+Ni steel, which is widely used and most popular plastic mold and die steel. (2) Limited work has been reported with green powder metallurgy electrode manufactured using Cu (75% wt), W (23% wt), and Si (2% wt) powders.

Researchers have been exploring the feasibility of surface alloying or surface modification using EDM. Roethel and Garbajs have been concluded that the variations in the properties of the EDMed surface depend on phases formed in the alloyed surface layer due to the diffusion of material from tool electrode and breakdown of the dielectric. Gill and Kumar performed experiments to study surface modification of die steel using Cu–W powder metallurgy sintered electrode. They have reported that there was the formation of tungsten carbide as well as a significant increase in the percentage of tungsten and carbon in a modified surface. Mohri et al. studied the surface modification of aluminum and carbon steel using P/M electrode prepared from a fine metallic powder of tungsten carbide, copper, aluminum, and titanium. The results show that the surface produced has higher corrosion and wear resistance than the base metal. Simao et al. carried out comparative performance study of EDMed surface produce using TiC/WC/Co type green P/M tool and WC/Co type fully sintered electrode. They observed the presence of W, Co, and C in the alloyed layer formed and increased the surface hardness by almost 100% [11, 12].

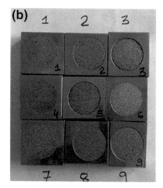

Fig. 1 **a** P/M tool electrode and **b** machined work material

2 Experimentation

2.1 P/M Tool Electrode Preparation

In this work, the experiments were performed using two different types of EDM electrodes such as powder metallurgy (P/M) electrode and a conventional electrolytic copper electrode (purity 99.9%). The comparative analysis of microhardness of EDMed surface produced using P/M electrode and the conventional copper electrode was carried out. P/M electrode was manufactured from a homogeneous mixture of metallic powders, i.e., copper, tungsten, and silicon (particle size 325 mesh) at 75, 23, and 2% of weight, respectively. Electrode assembly consists of two parts, P/M compacted pellet which actually works as an electrode and especially designs tool holding fixture. Further, tool clamping fixture has been comprised of two parts, one of them is electrode clamping fixture and second is an electrolytic copper rod. A homogeneous mixture of copper, tungsten, and silicon powders was obtained using ceramic mortar and pestle. A mixture of powders was pressed at a compaction pressure of 75, 125, and 175 kg/cm^2 using hand operated the hydraulic press. Special purpose compaction die assembly (ϕ19 mm) was manufactured to produce cylindrical shape powder metallurgy electrode. P/M tool electrode and machined samples are shown in Fig. 1.

2.2 Experimental Planning and Procedure

Experiments were planned as per design of experiments (DOE) based on Taguchi technique. Experiments were performed as per standard Taguchi L9 orthogonal array. The analysis of variance (ANOVA) was carried out for finding the contribution of individual parameters for obtaining maximum microhardness. ANOVA was per-

Table 1 Selected input parameters and their levels

Symbol	Parameters	Unit	Levels		
			L1	L2	L3
A	Compaction pressure (C_p)	(kg/cm^2)	75	125	175
B	Peak current (I_p)	(A)	4	8	12
C	Pulse on time (T_{on})	(μs)	30	65	100

formed based on the signal-to-noise (*S/N*) ratio using statistical software Minitab 16. Levels of input process parameters were decided based on literature and performing numbers of trial runs. During experimentation, levels of few parameters were maintained constant like duty factor = 50%, polarity = Normal, gap voltage = 55 V, and flushing pressure = 0.75 kg/cm^2. Different levels of variable parameters are shown in Table 1. For obtaining maximum microhardness, "larger–the–better" criteria were considered for calculating the signal-to-noise (*S/N*) ratio. The planned experiments were performed on die sinker EDM (Model: M22, Make: Maruti Machine Tools, Vadodara, India.).

The work material was cut in size of 25 mm × 25 mm × 10 mm for performing the individual experiment. P/M electrodes were manufactured in a cylindrical shape having 19 mm diameter using powder metallurgy. The microhardness of EDMed surface was measured using digital microhardness tester (Model No. 7005–B, Vaisesika, India). The obtained microhardness was a result of the mean of three trials under a 300 gf load applied for 15 s dwell time.

3   Results and Discussions

3.1   Analysis of Microhardness

The data shown in the Table 2 represents the mean value of three measurements of microhardness on each EDMed surface with it's *S/N ratio.*

In order to evaluate the significant effect and percentage contribution of the individual parameter to achieve higher microhardness, ANOVA was performed using *S/N* ratio. The main objective of the signal-to-noise ratio for high microhardness is "higher the better." The following equation was used to formulate *S/N* ratio for microhardness.

$$\frac{S}{N} = -10 \log \log_{10} \left[\frac{1}{n} \Sigma 1/y_{ij}^2 \right] \tag{1}$$

Table 2 Design of experiment matrix and machining characteristics

Exp. No.	Input process parameters						Responses	
	C_p		I_p		T_{on}		MH (VHN)	S/N ratio (dB)
	Actual	Coded	Actual	Coded	Actual	Coded		
1	75	1	4	1	30	1	546	54.7439
2	75	1	8	2	65	2	748	57.4780
3	75	1	12	3	100	3	910	59.1808
4	125	2	4	1	65	2	554	54.8702
5	125	2	8	2	100	3	790	57.9525
6	125	2	12	3	30	1	738	57.3611
7	175	3	4	1	100	3	589	55.4023
8	175	3	8	2	30	1	678	56.6246
9	175	3	12	3	65	2	771	57.7411

Table 3 ANOVA using S/N ratio data for microhardness

Factor	DOF	SS	V	F	P	% of contribution
Compaction pressure	2	0.4812	0.2406	15.27	0.061	2.59
Peak current	2	15.5978	7.7989	494.96	0.002	84.98
Pulse on time	2	2.4800	1.2400	78.70	0.013	13.34
Error	2	0.0315	0.0158	–	–	–
Total	8	18.5905	–	–	–	–

$S = 1.80958$ R-Sq = 99.03% R-Sq (adj) = 96.11%

where n is a number of results in a row, here, in this case, $n = 1$, and $y =$ value of ith experiment in a row. Outcome of ANOVA for microhardness is given in Table 3.

The result of ANOVA reflects that the peak current and pulse on time are two most significant parameters for improving microhardness. The mean value of microhardness of work material before and after the EDMed was measured as 322 and 910 VHN, respectively. The result shows that the three times improvement in microhardness as compared to microhardness of base material was observed using P/M electrode. Moreover, about 168% improvement in microhardness has been observed in EDMed surface as compared to base material when machined with the conventional copper electrode. The increase in microhardness was attributed to the effect of surface alloying. The high value of microhardness was observed at the higher level of peak current and pulse on time due to the effect of high heat energy. The contributions of EDM parameters for increasing microhardness are given below:

Effect of peak current It has been observed that the increase in microhardness from 546 to 910 HVN while increases peak current from 4 to 12 A. Generation of high discharge energy was observed at a high value of peak current. There will be high heating and quenching during a high value of peak current.

Effect of pulse on time It has been observed that an increase in pulse on time will lead to increase microhardness. The main reason behind increase in microhardness is due to longer discharge time. The same quantum of heat energy available for a longer duration shall results in thick and hard alloyed layer.

Effect of compaction pressure Density of P/M electrode mainly depends on the amount of compaction pressure applied. However, the porous structure of the electrode is a desire for a surface alloying process performed using EDM. P/M electrode is manufactured with high density than the transfer of electrode constituents on the work surface is not possible. Similarly, less density electrode is subjected to high wear leading to unstable EDM operation.

3.2 Surface Characterization and Microscopic Observation

EDMed samples were analyzed by SEM and EDS to confirm material migration from tool electrode and dielectric. The results of SEM and EDS confirmed the presence of significant amount of copper, tungsten, and silicon from powder metallurgy electrode. Tungsten (W) was observed on the EDMed surface in the form of the phases such as W, WC, and W_2C. Copper (Cu) was observed with the highest proportions, while silicon (Si) was observed in the form of the phases such as Si and SiC during XRD analysis. Increase in the carbon percentage was observed due to the breakdown of hydrocarbon from dielectric (EDM Oil). Dissociated carbon element from dielectric combined as an intermetallic bond with tungsten and silicon and surface was alloyed in the form of carbide. Proper dispersion of hard elements was observed in the microstructures. Hard particles dispersion and compound in the form of carbides (cementite formation) significantly increase the microhardness of EDMed surface.

Increase in microhardness of the EDMed surfaces of work material was attributed to the formation of carbides and migration of electrode materials. It is also important to quantify the dispersion of materials from powder metallurgy electrode and dielectric (EDM oil), EDS was also carried out. EDS and SEM analysis were performed on SEM/EDS model JSM 6010 LA manufactured by JEOL, Japan at TCR Advanced engineering testing laboratory, Vadodara.

Two machined samples (trial 3 and trial 8) were considered for SEM and EDS analyses. Figure 2a, b represent SEM images of trial 3 taken at two different magnification scale for work material machined at 12 A peak current, 100 μs pulse on time and powder metallurgy electrode with compaction pressure 75 kg/cm². The SEM analysis confirmed the formation of thick recast layers and smooth, homogeneous and defect-free machined surface were observed. No any major surface defects such as microcrack and microvoids were observed. Similarly, Fig. 2c, d represent SEM image of trial 8 taken at two different magnification scales for work material machined at 8 A peak current, 30 μs pulse on time and powder metallurgy electrode with compaction pressure 175 kg/cm². The reduction in thickness of the recast layer and less amount of waviness was observed due to a reduction in peak current as

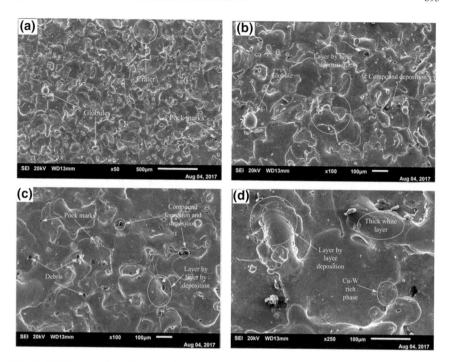

Fig. 2 SEM image of machined surface

compared to trial 3. In both the cases, a significant amount of the presence of copper deposition was observed on work material from powder metallurgy electrode [13].

The EDS analysis of the EDMed surfaces shows the variations in compositions of work material in the area of EDMed surface after machining. EDS was carried out for samples of trial 3 and trial 8 in the areas of EDMed surface. The EDS spectrum of trial 3 is shown in Fig. 3a. The result of EDS contains 0.66% Si, 0.62% Cr, 0.66% Mn, 0.42% C, 20.08% Cu, and 6.92% W. The EDS spectrum of trial 8 is shown in Fig. 3b. The results of EDS for trial 8 were found to be 0.44% Si, 1.06% Cr, 0.92% Cr, 0.53% C, 18.02% C, and 5.46% W. The results of EDS confirmed the transfer of material from powder metallurgy electrode to work material. Increase in carbon percentage of EDMed surface as compared to base material is result of migration of dissociated carbon element from hydrocarbon dielectric [14]. The presence of tungsten (W) on the EDMed surface (previously not part of compositions of work material) strongly confirmed migration of material from powder metallurgy electrode. EDS spectrum as shown in Fig. 3 represents elements identified during the EDS analysis [15].

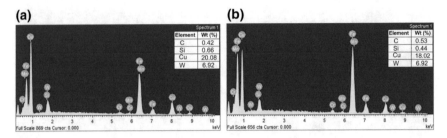

Fig. 3 EDS spectrum of trial 3 and trial 8

4 Conclusion

Microhardness of work material P20+Ni machined using powder metallurgy electrode on the EDM was investigated using Taguchi's design of experiments methodology. Based on the result of experiments, the following conclusions can be drawn:

1. The maximum microhardness was observed to be 910 HVN of the EDMed surface using powder metallurgy electrode, while 541 HVN using conventional electrolytic copper electrode. Hard particles dispersion and compound in the form of carbides (cementite formation) significantly increase the microhardness of EDMed surface.
2. No microvoids and microcracks were observed on the EDMed surface during SEM analysis, while machining was carried out using powder metallurgy electrode. Hence, it shows that there were not any adverse effects of powder metallurgy electrode.
3. Peak current and pulse on time were observed subsequently most significant factors for dominating high microhardness.
4. SEM and EDS analyses confirm the significant amount of material transfer from the powder metallurgy electrode and dissociated carbon from dielectric on the EDMed surface, resulting in improvements in microhardness and defect-free multilayer alloyed surface.

References

1. Gill, A.S., Kumar, S.: Surface alloying of H11 die steel by tungsten using EDM process. Int. J. Adv. Manuf. Technol. **78**, 1585–1593 (2015)
2. Jain VK (2004) Advanced machining processes. Allied Publishers, New Delhi. ISBN 1-7764-294-4
3. Singh, A.K., Kumar, S., Singh, V.P.: Optimization of parameters using conductive powder in dielectric for EDM of super Co 605 with multiple quality characteristics. Mater. Manuf. Process. **29**, 267–273 (2014)
4. Kumar, S., Singh, R., Singh, T.P., Sethi, B.L.: Surface modification by electrical discharge machining: a review. J. Mater. Process. Technol. **209**, 3675–3687 (2009)

5. Mohri, N., Saito, N., Tsunekawa, Y.: Metal surface modification by electrical discharge machining with composite electrode. CIRP Ann. Manuf. Technol. **42**, 219–222 (1993)

6. Simao, J., Aspinwall, D., Menshawy, F.E., et al.: Surface alloying using PM composite electrode materials when electrical discharge texturing hardened AISI D2. J. Mater. Process. Technol. **127**, 211–216 (2002)

7. Patowari, P.K., Saha, P., Mishra, P.K.: Artificial neural network model in surface modification by EDM using tungsten–copper powder metallurgy sintered electrodes. Int. J. Adv. Manuf. Technol. **51**, 627–638 (2010)

8. Ghosh, A., Mallik, A.K.: Manufacturing science. Affiliated East West Press Private Limited, New Delhi, India (2006)

9. Kansal, H.K., Singh, S., Kumar, P.: Parametric optimization of powder mixed electrical discharge machining by response surface methodology. J. Mater. Process. Technol. **169**, 427–436 (2005)

10. Pecas, P., Henriques, E.: Effect of the powder concentration and dielectric flow in the surface morphology in electrical discharge machining with powder-mixed dielectric (PMD-EDM). Int. J. Adv. Manuf. Technol. **37**, 1120–1132 (2008)

11. Samuel, M.P., Philip, P.K.: Power metallurgy tool electrodes for electrical discharge machining. Int. J. Mach. Tools Manuf. **37**(11), 1625–1633 (1997)

12. Furutani, K., Saneto, A., Takezawa, H., Mohri, N., Miyake, H.: Accretion of titanium carbide by electrical discharge machining with powder suspended in working fluid. Precision Eng. **25**, 138–144 (2001)

13. Soni, J.S., Chakraverti, G.: Experimental investigation on migration of material during EDM of T 215 Cr12 dies steel. J. Mater. Process. Technol. **56**, 439–451 (1996)

14. Shunmugam, M.S., Philip, P.K., Gangadhar, A.: Improvement of wear resistance by EDM with tungsten carbide P/M electrode. Wear **171**, 1–5 (1994)

15. Wang, Z.L., Fang, Y., Zhao, W.S., Cheng, K.: Surface modification process by electro discharge machining with Ti powder green compact electrode. J. Mater. Process. Technol. **129**, 139–142 (2002)

Optimization of Tool Wear and Surface Roughness of Hybrid Ceramic Tools

G. Amith Kumar and T. Jagadeesha

Abstract In this work, machinability tests were conducted on AISI 1030 steel with ceramic tools. TiN-coated and TiN-uncoated carbide tools were used. Actual tool wear is measured and actual roughness is measured. Using the response surface methodology (RSM), flank tool wear and surface finish prediction models are developed. The simulated values obtained from the model are compared with the experimental results. Theoretical values predicted by the response surface methodology matches with experimental values of tool wear along with surface finish. The study focuses on the influence of interaction between machinability factors such as feed given to work, surface roughness and hardness of ceramic insert that affects tool wear. Feed rate was found to be the influencing parameter that affects the roughness characteristics of the machined surface and flank tool wear followed by cutting speed. The process parameters are optimized using genetic algorithm for optimum surface finish and flank wear of the tool.

Keywords Ceramic tools · RSM · Tool wear · Surface roughness

1 Introduction

Newer class of materials demand newer cutting tools. Newer cutting tool material should have high hot hardness, good bending strength, high fracture toughness and high wear resistance [1–5]. Ceramics are commonly used as cutting tool and alumina combined with titanium nitride, titanium carbide, silicon carbide whiskers, titanium boride and zirconium oxide gives improved cutting performance [2, 6].

In machining high-strength materials, dry cutting, i.e. without any coolant or lubricant is most desirable. Dry cutting helps in preventing pollution of atmosphere

G. Amith Kumar
BTLIM Institute of Technology, Bengaluru, India

T. Jagadeesha (✉)
NIT, Calicut, Kerala, India
e-mail: jagdishsg@nitc.ac.in

© Springer Nature Singapore Pte Ltd. 2020
R. Venkata Rao and J. Taler (eds.), *Advanced Engineering Optimization Through Intelligent Techniques*, Advances in Intelligent Systems and Computing 949,
https://doi.org/10.1007/978-981-13-8196-6_61

and water resources and save machining cost as less or no lubricant is required [7, 8]. Some coolant poses health hazard to operator, which could be prevented using dry cutting. Coolants and lubricants are expensive and most of the coolants used today are not biodegradable [2, 9, 10]. Literature survey reveals the possibility of $Al_2O_3/SiC/CaF_2$ sintered ceramic tools that can be used effectively and efficiently in hard machining application [11, 12]. This work deals with machining studies and evaluation of responses such as flank wear of the cutting tool and surface roughness of AISI 1035 steel. An RSM predictive modelling was done to compare and validate the results with experimental values.

2   Experimental Details

AISI 1030 steel was selected as work material with a nominal composition of 0.3% carbon, 0.75% manganese, 0.05% sulphur and 0.10% silicon. The diameter of work-piece is 75 mm and length of the workpiece is 800 mm. Conventional engine lathe, Kirloskar make, was used for machining the workpieces. The maximum speed is 1200 RPM and feed rate is 0.114 rev/mm. This engine lathe uses three HP, three phase induction motor for drive. Inserts are purchased from Kennametals, Bangalore and designation of the insert is ISO DNGN150608. It is Al_2O_3/TiC insert coated with titanium nitrate (TiN). Uncoated tungsten carbide tool insert was also used for comparison purpose. Tool holder is purchased from WIDIA and designation is SVJBR2525M16D64.

2.1   Measurement of Tool Wear and Surface Roughness

Roughness measurement was done using Mitutoyo make stylus type roughness measuring instrument. Initially, calibration of surface measuring instrument was carried out by using specimen of known surface roughness. The calibrated roughness values must be within 5% of the tolerance of specimen, whose roughness is known. Surface roughness measurement was done by changing the sampling length as per the manufacturer's catalogue. Roughness was measured over five different locations of the workpiece. The tool wear, in particular, flank wear was observed with the help of a metallurgical microscope. Figure 1 shows the flank wear at a feed rate of 0.06 mm/rev, cutting speed of 12 m/min and using a cutting insert having a hardness of 2500 HV.

Fig. 1 Flank wear at feed $= 0.06$ mm/rev, cutting speed $= 12$ m/min and hardness $= 2500$ HV

Table 1 Three factors and two-level experiment

Factor used	Parameter	Units used	Level (low)	Level (high)
A	Hardness	HV	1800	2500
B	Feed	mm/min	0.06	0.1
C	Speed	m/min	12	110

3 Response Surface Methodology

Machinability of steel depends on several factors. Among them, flank tool wear and surface roughness are predominant. Larger flank wears give lower tool life. Higher surface roughness is not desirable in gas-tight applications. Flank wear and surface roughness strongly depend on process parameters of machining, geometry of tool, machine tool rigidity, rigidity of tool, hardness of insert, etc. The flank wear affects the cutting tool geometry and results in a substantial increase of thrust force, which further leads to deformation and wear of tool. The tool geometry refers to angles, edge length, thickness and radius of the inserts used. The factors are taken in account and their levels are shown in Table 1.

Response surface methodology being a popular technique for developing the predictive models and the results obtained can be used for further analysis. A 2^3 factorial design developed with two replicates was performed for each set of the experiments. The lower and upper limits of insert hardness value, feed and cutting speed for the tool workpiece combinations are given in Table 1.

By varying process parameters as presented in Table 1, a predictive model was developed at 95% confidence level to fit the surface response design. The flank wear and surface roughness values were obtained experimentally. These experimental values are used to obtain the predictive model for flank wear and surface roughness. The regression equations developed for both flank wear and surface roughness (Ra) are shown in Eqs. 1 and 2, respectively. Table 2 gives the quantitative comparison

Table 2 Comparsion of experimental values with predicted values

No of Expt.	Hardness (HV)	Cutting speed (m/min)	Feed (mm/rev)	Actual flank wear (mm)	Actual surface roughness (μm)	Flank wear predicted (mm)	Surface roughness predicted (μm)
1	1800	110	0.06	0.22	10.83	0.2100	10.84625
2	1800	110	0.10	0.20	11.01	0.2100	10.84625
3	2500	12	0.10	0.12	7.51	0.1265	7.65375
4	1800	12	0.06	0.133	7.65	0.1265	7.65375
5	2500	12	0.06	0.239	4.79	0.2465	4.95375
6	2500	110	0.10	0.254	4.89	0.2465	4.95375
7	1800	110	0.10	0.201	3.23	0.2110	3.17625
8	2500	12	0.06	0.221	3.35	0.2110	3.17625
9	2500	110	0.06	0.24	12.25	0.2355	12.36375
10	1800	12	0.06	0.231	12.33	0.2355	12.36375
11	1800	12	0.10	0.15	9.21	0.1460	9.17125
12	2500	12	0.10	0.142	9.28	0.1460	9.17125
13	2500	110	0.06	0.392	7.35	0.3965	7.28125
14	1800	110	0.06	0.410	7.44	0.3965	7.28125
15	2500	110	0.10	0.259	5.58	0.2570	5.50375
16	1800	12	0.10	0.255	5.2	0.2570	5.50475

of experimental values obtained with the predicted values using response surface methodology.

The flank wear model in terms of coded factors was obtained and is given as

$$\text{Flank wear} = -0.545 + 0.000366 \text{ Hardness} + 6.26 \text{ Feed} + 0.000243 \text{ Cutting Speed}$$
$$- 0.003464 \text{ Hardness} * \text{Feed} + 0.000001 \text{ Hardness} * \text{Cutting Speed}$$
$$- 0.00714 \text{ Feed} * \text{Cutting Speed} \qquad (1)$$

The surface roughness model in terms of coded factors was given as

$$Ra = 3.6 - 0.00070 \text{ Hardness} + 120 \text{ Feed} + 0.1585 \text{ Cutting Speed} - 0.0264 \text{ Hardness} * \text{Feed}$$
$$- 0.000037 \text{ Hardness} * \text{Cutting Speed} - 0.837 \text{ Feed} * \text{Cutting Speed} \qquad (2)$$

The equations show that feed rate is the most crucial parameter when compared to cutting speed and hardness of the cutting insert for both the responses considered. This is due to the fact that there will be an increase in the area of contact between the tool work interface which leads to the increment in cutting force, hence Ra value.

Figures 2 and 3 show the surface plots for Ra and flank wear, respectively. It is observed that for a constant hardness value of the cutting insert as the feed increases, there is corresponding increase in surface roughness along with flank wear of the

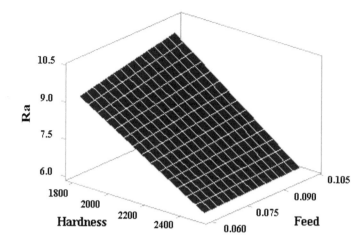

Fig. 2 Surface plot of surface roughness

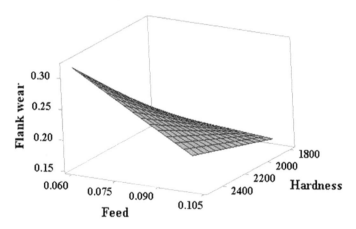

Fig. 3 Surface plot for flank wear

tool. As stated earlier, it is due to increase in the cutting force with increase in the feed rate.

4   Genetic Algorithm

Genetic algorithms are the most desired techniques used for the optimization of a system consisting of a large number of variables. In this study, multi-objective genetic algorithm was utilized for solving the problem as there were two responses: Minimization of flank wear of the tool and Ra value. The process parameters such as hardness (2500–1800 HV), feed (0.06–0.1 mm/rev) and cutting speed (12–110 m/min) were optimized using MATLAB R2017a optimization tool. The optimum process parameters obtained from genetic algorithm for surface finish were: Hardness 2173 HV, feed 0.063 mm/rev and cutting speed 12 m/min and for that of flank wear were: Hardness 1800 HV, feed 0.1 mm/rev and cutting speed 12 m/min.

5 Conclusions

In this work, machinability tests were conducted on AISI 1030 steel with ceramic tools, i.e. TiN-coated and TiN-uncoated carbide tools, for measuring the flank wear and Ra value. RSM techniques were employed to develop a predictive model for flank wear and Ra value. To enhance understanding of the study, the simulated values from the model were compared with the experimental results. The predictions made by the RSM model matched the experimental values of flank wear and Ra. For better understanding, the interaction of machining parameters such as cutting speed of the work, feed and hardness of ceramic insert on the effect of flank wear and Ra were studied under a constrained environment. It was observed that feed rate was the dominating parameter and high feed rate values result in higher flank wear and Ra value. This is because at higher feed rate, there is an increase in the contact area between workpiece and tool which increases the cutting force thus causing higher tool wear and Ra value. From the surface plot, it was observed that at certain hardness and cutting speed, increase in the feed rate would give a significant increase in the Ra value and tool wear. Use of genetic algorithm optimized the values of the process parameters in order to reduce Ra and flank wear.

References

1. Ozel, T., Karpat, Y.: Predictive modeling of surface roughness and tool wear in hard turning using regression and neural networks. Int. J. Mach. Tools Manuf. **45**, 467–479 (2005)
2. Jianxin, D., Tongkun, C., Xuefeng, Y., Jianhua, L.: Self lubrication of sintered ceramic tools with CaF_2 additions in dry cutting. Int. J. Mach. Tools Manuf. **46**, 957–963 (2006)
3. Senthil Kumar, A., Raja Duraia, A., Sornakumar, T.: Wear behaviour of alumina based ceramic cutting tools on machining steels. Tribol. Int. **39**, 1393–1401 (2006)
4. Jianxin, D., Lili, L., Jianhua, L., Jinlong, Z., Xuefeng, Y.: Failure mechanism of TiB_2 particles and SiC whisker reinforced Al_2O_3 ceramic cutting tools when machining nickel based alloys. Int. J. Mach. Tools Manuf. **45**, 1393–1401 (2005)
5. Huang, C.Z., Wang, J., Ai, X.: Development of new ceramic cutting tools with alumina coated carbide powders. Int. J. Mach. Tools Manuf. **40**, 823–832 (2000)
6. Errico, G.E., Bugliosi, S., Calzavarini, R., Cuppini, D.: Wear of advanced ceramic tool materials. Wear **225**, 267–272 (1999)
7. Jianxin, D., Xing, A.: Friction and wear behaviour of Al_2O_3/TiB_2 composite against cemented carbide in various temperature at elevated temperature. Wear **195**, 128–132 (1996)
8. Sethi, G., Upadhaya, A., Agarwal, D.: Microwave and conventional sintering of premixed and prealloyed Cu–12Sn bronze. Sci. Sinter. **35**, 49–65 (2003)

9. Parhami, F., McMeeking, R.M.: A network model for initial stage sintering. Mech. Mater. **27**, 111–124 (1998)
10. German, R.M.: Sintering theory and practice. Wiley, New York (1996)
11. Thummler, F., Oberacker, R.: Introduction to powder metallurgy. The Insitute of Materials, London (1993)
12. Canakci, M., Erdil, A., Arcakliog, E.: Performance and exhaust emission of a biodiesel engine using regression analysis and ANN. Appl. Energy **83**, 594–605 (2006)

Assessment of Viscosity of Coconut-Oil-Based CeO$_2$/CuO Hybrid Nano-lubricant Using Artificial Neural Network

Ayamannil Sajeeb and Perikinalil Krishnan Rajendrakumar

Abstract In the present work, coconut-oil-based hybrid nano-lubricants are prepared by dispersing CeO$_2$ and CuO nanoparticles in three different proportions 75/25, 50/50, and 25/75. Experimental studies on the viscosity of hybrid nano-lubricants have been carried out by varying the concentration of combined nanoparticles in weight % from 0 to 1% and temperature ranging from 30 to 60 °C for each proportion of CeO$_2$/CuO nanoparticles. A new empirical correlation and an optimal artificial neural network (ANN) for each proportion of CeO$_2$/CuO nanoparticles in terms of temperature and concentration are devised to assess the viscosity ratio of hybrid nano-lubricant, using 48 experimental data. The results showed that the output of correlation and optimal ANN have a margin of deviation of 2 and 1%, respectively, and hence, the optimal artificial neural network is better in predicting the viscosity of hybrid nano-lubricant in comparison with empirical correlation.

Keywords ANN · Hybrid nano-lubricant · Viscosity · Coconut oil · Correlation

1 Introduction

The lubricating oils are used in many applications for reducing friction and wear between mating parts. Vegetable oil like coconut oil is biodegradable, nontoxic and maintain better lubricity compared with mineral oils. Many researchers reported that nano-lubricants which are made by adding nanoparticles in oil can have more thermal conductivity, film coefficient, and heat flux compared to base oils [1–4]. Researchers also pointed out that the addition of nanoparticles will also enhance the physical properties of the oil, especially viscosity. The temperature and concentration

A. Sajeeb (✉)
Government Engineering College, Kozhikode, India
e-mail: sajeebamas@gmail.com

P. K. Rajendrakumar
National Institute of Technology, Calicut, India
e-mail: pkrkumar@nitc.ac.in

© Springer Nature Singapore Pte Ltd. 2020
R. Venkata Rao and J. Taler (eds.), *Advanced Engineering Optimization Through Intelligent Techniques*, Advances in Intelligent Systems and Computing 949,
https://doi.org/10.1007/978-981-13-8196-6_62

of nanoparticles are the two important parameters which influence viscosity significantly [5–9]. Moreover, the pumping power is greatly controlled by the viscosity of the oil.

Most of the literature on coconut-oil-based nano-lubricants emphasizes the thermal, rheological, and tribological properties of nano-lubricants with the dispersion of single nanoparticles. In recent years, hybrid nanofluids have got more attention which is prepared by mixing two or more nanoparticles combined in the base oil. They demonstrate the synergistic effect on properties that individual nanoparticles do not occupy [10]. The viscosity of several hybrid nanofluids, viz. Al_2O_3/Cu, Cu/TiO$_2$, Ag/MgO, SiO$_2$/MWCNT, MWCNT/Fe$_2$O$_3$ has been measured and reported by many researchers [11–15]. It is very costly and difficult to measure the viscosity of nanofluids experimentally for various temperatures and concentrations of nanoparticles. The numerical methods are the cheap, best, and reliable choices in foreseeing the viscosity of nanofluids. Esfe et al. [16] designed an artificial neural network (ANN) to envisage the dynamic viscosity of nanofluid based on the experimental results with temperature, concentration, and diameter of nanoparticles as input variables and showed the maximum error of 2.5% in predicting the dynamic viscosity. Afrand et al. [17] observed that ANN provides more accurate results than correlation in predicting the relative viscosity of MWCNTs-SiO$_2$/AE40 hybrid nano-lubricant.

There is no work appeared in the literature survey in predicting the viscosity of coconut-oil-based CeO$_2$/CuO hybrid nano-lubricant for various temperatures, concentrations of nanoparticles, and proportions of CeO$_2$/CuO nanoparticles using ANN. Hence, in the present study, a comparative study has been done between newly developed empirical correlation and optimal ANN in assessing the viscosity ratio of coconut-oil-based CeO$_2$/CuO hybrid nano-lubricant using 48 experimental data.

2 Methodology

2.1 Experimental

The hybrid nano-lubricant is prepared by dispersing CeO$_2$ and CuO nanoparticles combined in coconut oil [18]. The physical properties of the nanoparticles given by the supplier are shown in Table 1.

The dynamic viscosity of the nano-lubricant is measured by using modular compact rheometer (make: Anton paar). Forty eight experimental data are taken for developing models for the dynamic viscosity (μ) of a hybrid nano-lubricant for three different proportions of CeO$_2$/CuO nanoparticles, viz. 75/25, 50/50, and 25/75, by varying concentration of total nanoparticles (ϕ_w) from 0 to 1% and temperature (T) ranging from 30 to 60 °C. Figures 1, 2, and 3 indicate viscosity ratio versus temperature at different concentrations for each proportions of CeO$_2$/CuO nanoparticles.

Table 1 Physical properties of CeO$_2$ and CuO nanoparticles [18]

Property	Value	
	CeO$_2$	CuO
Purity	>99.5%	>99.5%
Average particle size	30–50 nm	50–70 nm
Color	Light yellow	Brownish black
Morphology	Spherical	Spherical
Specific surface area	40–45 m^2/g	>10 m^2/g
True density	7.65 g/cm^3	6.4 g/cm^3

Fig. 1 Variation of viscosity ratio with temperature for CeO$_2$/CuO:75/25

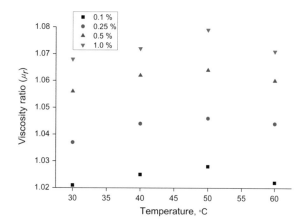

Fig. 2 Variation of viscosity ratio with temperature for CeO$_2$/CuO:50/50

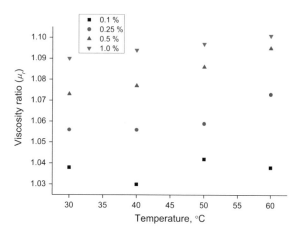

Fig. 3 Variation of viscosity
ratio with temperature for
CeO$_2$/CuO:25/75

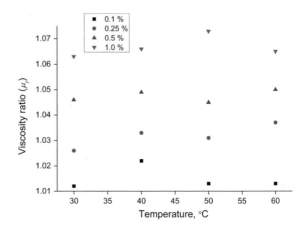

2.2 Curve Fitting by New Correlation

A new empirical correlation has been suggested to the experimentally measured
viscosity of the hybrid nano-lubricant for different temperatures, concentrations of
nanoparticles and proportions of CeO$_2$/CuO. Marquardt–Levenberg algorithm [19]
is employed to fit the curve and get the correct correlation in terms of the temperature
and concentration for various proportions of CeO$_2$/CuO. This algorithm ensures the
minimum deviation from the experimental data. The viscosity ratio of the hybrid
nano-lubricant is estimated using the correlation (Eqs. 1–3).

For 75/25 proportion of CeO$_2$/CuO,

$$\mu_r = \frac{\mu_{nf}}{\mu_{bf}} = \exp\left(0.4535 * T^{0.0096}\varphi_w^{0.0296}\right) - 0.527 \tag{1}$$

For 50/50 proportion of CeO$_2$/CuO,

$$\mu_r = \frac{\mu_{nf}}{\mu_{bf}} = \exp\left(0.3359 * T^{0.0367}\varphi_w^{0.0484}\right) - 0.373 \tag{2}$$

For 25/75 proportion of CeO$_2$/CuO,

$$\mu_r = \frac{\mu_{nf}}{\mu_{bf}} = \exp\left(0.08425 * T^{0.0615}\varphi_w^{0.2583}\right) - 0.0454 \tag{3}$$

where μ_{nf}, μ_{bf} and μ_r are the viscosity of nanofluid, base fluid, and viscosity ratio,
respectively, and T is the nanofluid temperature in °C and φ_w is the concentration of
nanoparticles in weight %.

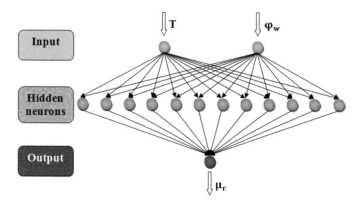

Fig. 4 Optimal ANN architecture

2.3 Fitting by Artificial Neural Network

Artificial neural network (ANN) is employed to map between numerical input data and output data in fitting problems. This neural network application is helpful in selecting data, constructing and training a network and assessing its attainment using mean square error (MSE) and margin of deviation (MOD) analysis with high precision, low cost and time [20, 21]. A two-layer feed-forward network with Sigmoid hidden neuron and linear output neurons is tried in mapping two input variables and one output for each proportions of CeO_2/CuO and is shown in Fig. 4.

The network is trained with a Levenberg–Marquardt backpropagation algorithm [22], to model the viscosity of CeO_2/CuO-coconut oil nano-lubricant. Temperature (°C) and concentration of nanoparticles (weight %) for each proportions of CeO_2/CuO nanoparticles are the two input variables and viscosity ratio (μ_r) is the output in ANN. Out of the 48 experimental data, 7 nos. (15%) set aside for validation and 7 nos. (15%) for testing. The remaining 34 data are subjected to training and network is tuned in accordance with the error. The network generalization is checked by using validation data and training will be stopped if generalization fails in improving. Though the testing data have no influence on training, it can be used for checking the overall performance of the network.

The performance of a network is measured by calculating mean square error (MSE).

$$\text{MSE} = \frac{1}{N} \sum_{i=1}^{N} \left(Y_{\text{pred}} - Y_{\text{exp}}\right)^2 \tag{4}$$

where N is the number of experiments, Y_{pred} and Y_{exp} are predicted output by using ANN/correlation and experimental output, respectively. Performance studies have done on the network by changing the number of hidden neurons from 6 to 20 and calculated MSE for each set of neurons (shown in Table 2). The lowest MSE is

Table 2 Performance evaluation of ANN through number of hidden neurons

No. of hidden neurons	MSE				R
	Training	Validation	Test	Overall	
6	0.000004761	0.00003423	0.0001695	0.0000338922	0.971
7	0.000008044	0.00003303	0.00004467	0.0000172858	0.985
8	0.000001266	0.00004487	0.0002907	0.0000512217	0.957
9	0.00001726	0.00002907	0.00002646	0.0000204115	0.982
10	0.000002699	0.00004587	0.00004128	0.0000149618	0.988
11	0.00003018	0.0001207	0.0001921	0.0000680460	0.945
12	**0.00000492**	**0.00001913**	**0.00003793**	**0.0000120062**	**0.989**
13	0.0001513	0.0001882	0.0002193	0.0001670350	0.85
14	0.00001724	0.0001215	0.0002479	0.0000674780	0.942
15	0.00000076	0.00005383	0.000184	0.0000362064	0.968
16	0.00003992	0.00002413	0.0001341	0.0000516785	0.953
17	3.321E−16	0.0001098	0.0001093	0.0000328650	0.972
18	0.00002153	0.0004581	0.0001278	0.0001029560	0.928
19	0.000000187	0.0002486	0.0001106	0.0000540109	0.956
20	0.00002544	0.00003079	0.00004697	0.0000294720	0.974

observed on 12 hidden neuron layer networks; hence, the same is selected as an optimal neural network.

Margin of deviation (MOD) is calculated for evaluating the accuracy of optimal neural network and correlation output by using Eq. (5).

$$\text{MOD} = \left[\frac{(\mu_\text{r})_\text{exp} - (\mu_\text{r})_\text{pred}}{(\mu_\text{r})_\text{exp}} \right] \times 100 \qquad (5)$$

3 Results and Discussion

The viscosity ratio obtained by experimental, correlation, and ANN methods are detailed in Fig. 5. The results show that the ANN predictions are in good agreement with that of experimental values. The viscosity ratio obtained from empirical correlation deviates a little from that of experimental data. Figure 6 compares the predicted output by correlation (using Eqs. 1–3) with the experimental data and shows a margin of deviation of about 2%. Figure 7 compares the predicted output by ANN with the experimental data and its margin of deviation is 1%. Hence, the optimal ANN model is superior to empirical correlation in terms of agreement with the experimental output.

Fig. 5 Output of experimental, correlation, and ANN

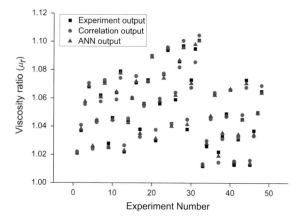

Fig. 6 Comparison of correlation output with experiment

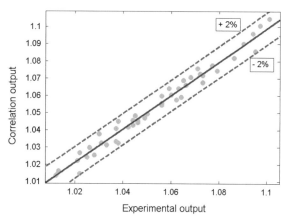

Fig. 7 Comparison of ANN output with experiment

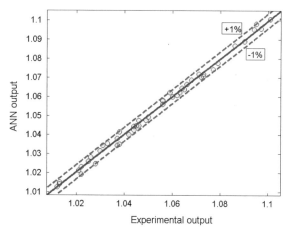

4   Conclusion

In this work, the influence of temperature, the concentration of nanoparticles, and the proportion of CeO$_2$/CuO nanoparticles on the viscosity ratio of coconut-oil-based CeO$_2$/CuO hybrid nano-lubricant were studied. Then, a new empirical correlation was developed based on experimental data to anticipate the viscosity ratio of the hybrid nano-lubricant as a function of temperature and concentration for each proportion of CeO$_2$/CuO nanoparticles. An optimal ANN was later proposed to foresee the viscosity ratio of the hybrid nano-lubricant. The number of neurons in the hidden layer was varied to obtain optimal ANN. The output of empirical correlation and ANN were compared with the experimental results. The comparison results showed that the optimal ANN method is better than an empirical correlation in predicting the viscosity ratio of the hybrid nano-lubricant.

References

1. Esfe, M.H., Saedodin, S., Yan, W.M., Afrand, M., Sina, N.: Study on thermal conductivity of water-based nanofluids with hybrid suspensions of CNTs/Al$_2$O$_3$ nanoparticles. J. Therm. Anal. Calorim. **124**(1), 455–460 (2016)
2. Soltanimehr, M., Afrand, M.: Thermal conductivity enhancement of COOH-functionalized MWCNTs/ethylene glycol–water nanofluid for application in heating and cooling systems. Appl. Therm. Eng. **105**, 716–723 (2016). https://doi.org/10.1016/j.applthermaleng. 2Q4622016.03.089
3. Toghraie, D., Chaharsoghi, V.A., Afrand, M.: Measurement of thermal conductivity of ZnO–TiO$_2$/EG hybrid nanofluid. J. Therm. Anal. Calorimetry **125**(1), 527–535 (2016). https://doi.org/10.1007/s10973-016-5436-4
4. Esfe, M.H., Refahi, A.H., Teimouri, H., Noroozi, M., Afrand, M., Karimiopour, A.: Mixed convection fluid flow and heat transfer of the Al$_2$O$_3$—water nanofluid with variable properties in a cavity with an inside quadrilateral obstacle. Heat Transf. Res. **46**(5) (2015)
5. Yiamsawas, T., Mahian, O., Dalkilic, A.S., Kaewnai, S., Wongwises, S.: Experimental studies on the viscosity of TiO$_2$ and Al$_2$O$_3$ nanoparticles suspended in a mixture of ethylene glycol and water for high temperature applications. Appl. Energy **111**, 40–45 (2013)
6. Esfe, M.H., Saedodin, S.: An experimental investigation and new correlation of viscosity of ZnO–EG nanofluid at various temperatures and different solid volume fractions. Exp. Therm. Fluid Sci. **55**, 1–5 (2014)
7. Esfe, M.H., Saedodin, S., Mahian, O., Wongwises, S.: Thermophysical properties, heat transfer and pressure drop of COOH-functionalized multi walled carbon nanotubes/water nanofluids. Int. Commun. Heat Mass Transf. **58**, 176–183 (2014)
8. Baratpour, M., Karimipour, A., Afrand, M., Wongwises, S.: Effects of temperature and concentration on the viscosity of nanofluids made of single-wall carbon nanotubes in ethylene glycol. Int. Commun. Heat Mass Transf. **74**, 108–113 (2016)
9. Eshgarf, H., Afrand, M.: An experimental study on rheological behavior of non-Newtonian hybrid nano-coolant for application in cooling and heating systems. Exp. Therm. Fluid Sci. **76**, 221–227 (2016)
10. Sarkar, J., Ghosh, P., Adil, A.: A review on hybrid nanofluids: recent research, development and applications. Renew. Sustain. Energy Rev. **43**, 164–177 (2015)
11. Suresh, S., Venkitaraj, K.P., Selvakumar, P., Chandrasekar, M.: Effect of Al$_2$O$_3$–Cu/water hybrid nanofluid in heat transfer. Exp. Therm. Fluid Sci. **38**, 54–60 (2012)

12. Madhesh, D., Parameshwaran, R., Kalaiselvam, S.: Experimental investigation on convective heat transfer and rheological characteristics of Cu–TiO$_2$ hybrid nanofluids. Exp. Thermal Fluid Sci. **52**, 104–115 (2014)

13. Esfe, M.H., Arani, A.A.A., Rezaie, M., Yan, W.M., Karimipour, A.: Experimental determination of thermal conductivity and dynamic viscosity of Ag–MgO/water hybrid nanofluid. Int. Commun. Heat Mass Transf. **66**, 189–195 (2015)

14. Munkhbayar, B., Tanshen, M.R., Jeoun, J., Chung, H., Jeong, H.: Surfactant-free dispersion of silver nanoparticles into MWCNT-aqueous nanofluids prepared by one-step technique and their thermal characteristics. Ceram. Int. **39**(6), 6415–6425 (2013)

15. Chen, L., Cheng, M., Yang, D., Yang, L.: Enhanced thermal conductivity of nanofluid by synergistic effect of multi-walled carbon nanotubes and Fe$_2$O$_3$ nanoparticles. Appl. Mech. Mater. **548–549**, 118–123 (2014)

16. Esfe, M.H., Saedodin, S., Sina, N., Afrand, M., Rostami, S.: Designing an artificial neural network to predict thermal conductivity and dynamic viscosity of ferromagnetic nanofluid. Int. Commun. Heat Mass Transf. **68**, 50–57 (2015)

17. Afrand, M., Najafabadi, K.N., Sina, N., Safaei, M.R., Kherbeet, A.S., Wongwises, S., Dahari, M.: Prediction of dynamic viscosity of a hybrid nano-lubricant by an optimal artificial neural network. Int. Commun. Heat Mass Transf. **76**, 209–214 (2016). https://doi.org/10.1016/j.icheatmasstransfer.2016.05.023

18. Sajeeb, A., Rajendrakumar, P.K.: Investigation on the rheological behavior of coconut oil based hybrid CeO$_2$/CuO nanolubricants. Proc. Inst. Mech. Eng. Part J J. Eng. Tribol. 1–8, 1350650118772149 (2018). https://doi.org/10.1177/1350650118772149

19. Marquardt, D.W.: An algorithm for least-squares estimation of nonlinear parameters. J. Soc. Ind. Appl. Math. **11**(2), 431–441 (1963)

20. Shakeri, S., Ghassemi, A., Hassani, M., Hajian, A.: Investigation of material removal rate and surface roughness in wire electrical discharge machining process for cementation alloy steel using artificial neural network. Int. J. Adv. Manuf. Technol. **82**(1–4), 549–557 (2016). https://doi.org/10.1007/s00170-015-7349-y

21. Shirani, M., Akbari, A., Hassani, M.: Adsorption of cadmium (ii) and copper (ii) from soil and water samples onto a magnetic organozeolite modified with 2-(3, 4-dihydroxyphenyl)-1, 3-dithiane using an artificial neural network and analysed by flame atomic absorption spectrometry. Anal. Methods **7**(14), 6012–6020 (2015)

22. Vaferi, B., Samimi, F., Pakgohar, E., Mowla, D.: Artificial neural network approach for prediction of thermal behavior of nanofluids flowing through circular tubes. Powder Technol. **267**, 1–10 (2014)

Elimination of Nick Defect by Process Optimization for Input Shaft Reverse Gear Section

Sagar U. Sapkal◉ **and Tejas A. Bhilawade**◉

Abstract This paper focuses on the input shaft of a gearbox assembly from a well-known automobile company. The company was facing a problem of noise and vibrations in the reverse gear section of the input shaft. The problem was the formation of nick at the gear root of the reverse gear. A nick is plus material anywhere on the part. The reverse gear section undergoes two machining operations, viz. shaping and shaving, and therefore, these two processes are focused in this work. The experiments are designed as per Taguchi's L9 orthogonal array with two replications of each experiment. Total 27 experiments were carried out for the process optimization by varying three control factors, viz. cutter span width, shaping over ball diameter, and shaving over ball diameter, and by considering nick value as the response. After the analysis, we found that the response is best when cutter span width is 69.450 mm, shaping over ball diameter is 33.636 mm, and shaving over ball diameter is 33.450 mm. The optimum condition for the best results is calculated and validated.

Keywords ANOVA · Nicks · Optimization · Reverse gear · Shaping · Shaving

1 Introduction

Gears are manufactured and used for various applications in enormous amounts all over the world. Various industries such as aerospace, construction machinery, agricultural machinery, and industrial gearing require huge amounts of gears. If it is possible to reduce the production cost of every gear being produced just by little percentage, then the total cost savings could reach millions of dollars. This shows that the processes of gear design and gear production need utmost care. Hence, there is a scope for gear design engineers and production companies to focus on gear

S. U. Sapkal (✉) · T. A. Bhilawade
Walchand College of Engineering, Sangli, India
e-mail: sagar.sapkal@walchandsangli.ac.in

T. A. Bhilawade
e-mail: bhilawadetb@gmail.com

© Springer Nature Singapore Pte Ltd. 2020
R. Venkata Rao and J. Taler (eds.), *Advanced Engineering Optimization Through Intelligent Techniques*, Advances in Intelligent Systems and Computing 949,
https://doi.org/10.1007/978-981-13-8196-6_63

Fig. 1 Input shafts before
soft machining

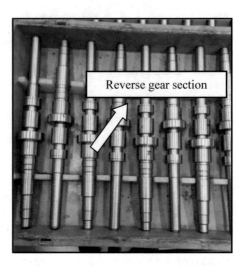

manufacturing processes in order to improve the quality of gears and reduce the cost
of production. Manufacturing of quality gears generally includes soft turning of the
raw material, followed by soft machining of the gear teeth, deburring, heat treatment,
hard machining, and finishing of the hardened gear teeth [1]. The major consumer of
gears is the automotive industry, and gears are critical components in this industry.
Gear manufacturing for automobiles is a very important process which has a rapidly
growing market for increasing fuel efficiency as well. Different machines are used
for machining of automobile gear teeth like hobbing, shaping, milling, broaching,
punching, and shear cutting. The automobile industries generally use spur and helical
gears in the transmission of the gearbox. Noise and vibration are one of the main
parameters to decide the quality of the gearbox for an automobile [2].

This paper focuses on reverse gear section of the input shaft of the gearbox assembly from an automobile company. The input shaft goes through various operations
in a sequential manner, and there are three basic sections during machining of input
shaft from the raw component into a finished product, viz. soft machining, heat treatment, and hard machining. Figures 1 and 2 show the input shafts kept in bins before
and after the soft machining, respectively.

The problem of noise and vibrations in the reverse gear section of the input shaft
was identified during inspection. This problem was due to the formation of nick at
the gear root of reverse gear. This nick is plus material anywhere on the part and can
form ridges, uneven surfaces, or burrs, and thus, it is a surface irregularity. Due to
the formation of nicks on the surfaces of the gear teeth of reverse gears, there were
chances of rejection of the input shafts. And hence, there was a concern about quality-
related issues of input shaft having various gears. Therefore, in this work, process
optimization for input shaft reverse gear section is carried out for the elimination
of nick defects. The manufacturing processes used for the production of reverse

Fig. 2 Input shafts after soft
machining

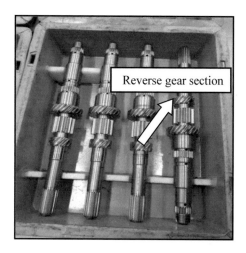

Reverse gear section

gear, identification and optimization of parameters, and related experimentation are
reported in subsequent sections.

2   Gear Manufacturing Process

The reverse gear section of the input shaft undergoes gear shaping operation followed
by gear shaving operation during soft machining, and it does not undergo any operation during hard machining. Therefore, gear shaping and gear shaving operations are
considered for further study and review. Gear shaping is a gear-cutting method for
generating teeth on a gear blank using a cutting tool which is shaped like a pinion.
During this operation, the cutter cuts while traversing across the face width and also
rolls with the gear blank at the same time as shown in Fig. 3. Gear shaving is a finishing operation which removes small amounts of metal from the working surfaces of
gear teeth using a shaving cutter as shown in Fig. 4. The cutter with closely spaced
grooves extending from the tip to the root of each tooth rotates with gear in close
mesh in both directions during the shaving operation. The centre distance between
the gear and the cutter is reduced in small controlled steps to remove metal from the
gear tooth surfaces till the final required size is achieved.

The complete geometry of a two-dimensional shaper cutter for the production of
spur gear is presented by Tsay et al. [3]. The prescribed tooth parameters of spur gears
are achieved by determining adequate shaper cutter parameters by using a numerical
method for optimization. Erkorkmaz et al. [4] presented a model for predicting the
kinematics, geometry, and cutting mechanics of gear shaping. Different aspects of the
model have been thoroughly verified with experiments. Kim and Kim [5] developed
application software used for shaving cutter design for the finishing of precision
gears. The simulation of the axial shaving of the gear shaving process is performed

Fig. 3 Shaping process

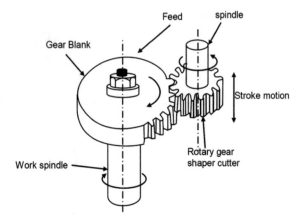

Gear Blank

Feed   spindle

Stroke motion

Work spindle

Rotary gear
shaper cutter

Fig. 4 Shaving process

by Hung et al. [6]. They also constructed the gear tooth surface of the shaved gear and reported the effects of machine set-up parameters and cutter assembly errors on the work gear surface.

Gear shaving removes the cutter marks, waviness, and surface irregularities of the gears generated by gear shaping operation. At the final stage, the inspection of gears is carried out to measure the various quality terms, tolerances, errors, and nick value. METREL electronic system rolling master which works on the principle of Parkinson's gear tester is used for the measurement of nick value. It measures the nick value by measuring the deviation in between the work gear axis and master gear axis which are in mesh with each other [7]. In order to find significant parameters of gear shaping and shaving operations for reverse gear of input shaft, Bhilawade and Sapkal [8] performed extensive experimentation (50 trials) by varying the parameters such as shaping root diameter, shaping tip diameter, shaping over ball diameter, shaving over ball diameter, and cutter span width.

3 Experimental Work

From the results of trail experiments carried out by Bhilawade and Sapkal [8], the most influencing parameters on nick value of reverse gears are identified as cutter span width, shaping over ball diameter (OBD), and shaving OBD. In order to systematically analyse the nick problem and optimize the process, the design of experiments is performed using Taguchi method considering these three control factors: cutter span width, shaping OBD, and shaving OBD. Based on lower and upper specification levels given in the process sheet, their values for three levels have been decided which are as mentioned in Table 1. The middle level is taken for the appropriate analysis of the pattern and results.

The aim of this experimentation is to check the effect of these parameters on the nick formation at the reverse gear root. Hence, the response variable is the nick value. This value should be zero but considering the tolerance limit, it is acceptable upto 20 µm. The shaping and shaving OBDs were provided on the MCU of the CNC shaping and shaving machines. The cutter span widths were measured over ten teeth by means of a special span micrometre. The nick values were recorded for each of these experiments by using METREL electronic system.

Nine experiments with two replications of each experiment were performed to minimize the effects of noise factors like cutting oil temperature and cutting oil

Table 1 Control factors and their levels

Control factor	Level 1 (mm)	Level 2 (mm)	Level 3 (mm)
Cutter span width (A)	69.450	70.025	70.600
Shaping OBD (B)	33.599	33.636	33.673
Shaving OBD (C)	33.450	33.487	33.524

Table 2 Control factors and their response values

S. No.	Cutter span width (mm)	Shaping OBD (mm)	Shaving OBD (mm)	Nicks for trial 1 (mm)	Nicks for trial 2 (mm)	Nicks for trial 3 (mm)	Average nick value
1	A1	B1	C1	0.011	0.012	0.011	0.011
2	A1	B2	C2	0.017	0.018	0.015	0.017
3	A1	B3	C3	0.015	0.013	0.013	0.014
4	A2	B1	C2	0.019	0.018	0.017	0.018
5	A2	B2	C3	0.012	0.012	0.012	0.012
6	A2	B3	C1	0.016	0.015	0.013	0.015
7	A3	B1	C3	0.018	0.017	0.017	0.017
8	A3	B2	C1	0.015	0.016	0.013	0.015
9	A3	B3	C2	0.017	0.018	0.019	0.018

viscosity. Thus, 27 experiments are carried out and the measured nick values for three trials of each experiment are reported in Table 2. The average value for each experiment is considered for further analysis. The nick value can never be negative and ideally should be zero. Therefore, smaller the better characteristic for the loss function is considered.

4 Results and Discussions

After performing the Taguchi's analysis, the main effects plot for SN ratios is generated. The main effect plot for SN ratios is plotted for three factors: cutter span width, shaping OBD, and shaving OBD, against their three respective levels as shown in Fig. 5.

From Fig. 5, it is clear that optimum condition occurs when the cutter span width is 69.450 mm, shaping OBD is 33.636 mm, and shaving OBD is 33.450 mm. The SN ratio for shaping OBD goes on increasing from 33.599 to 33.636 mm, becomes maximum at 33.636 mm, and then starts to decrease up to 33.673 mm. An increase in shaping OBD results in a decrease in nick value till 33.636, and then onwards, an increase in shaping OBD increases the nick value. The SN ratio for shaving OBD goes on decreasing from 33.450 to 33.487, becomes minimum at 33.487, and then starts to increase up to 33.524. An increase in shaving OBD will result in increase in nick value up to 33.487 mm. Then onwards, an increase in shaving OBD will result in a decrease in nick value. The SN ratio for cutter span width goes on decreasing from

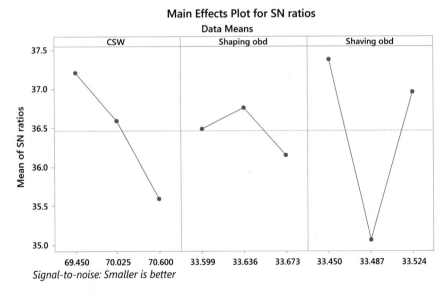

Fig. 5 Main effects plot for SN ratios

69.450 to 70.60 mm. We can say that as the cutter span width goes on increasing from 69.450 to 70.60 mm, there is an increase in the nick value. From the basic principle of shaving operation, we know that the shaving cutter meshes with the gear blank to remove a small amount of material. If the cutter span width goes on increasing, the serration from the cutter teeth will not be able to reach the gear root and would result in a poor quality of finishing at the gear root. It would result in surface irregularities like plus material which could cause the formation of nicks at the gear root. The reverse gear section undergoes the reverse gear shaping operation first and then reverse gear shaving process. The shaving operation removes about 80–100 μm of material after the shaping operation and acts as a finishing operation. The errors occurred during shaping operation can be corrected during the shaving operation. Hence, shaving operation is more significant for close accuracies and tolerances of a good-quality gear.

ANOVA is performed to find out the most significant factor and contribution of each parameter on the desired response. It is performed on MINITAB software using 95% confidence level. ANOVA performed for nick value is shown in Table 3.

The percentage contributions for each of the parameters are shown. It is clear that shaving OBD is contributing more than others. It is contributing about 53.45% which is highest amongst all. Hence, shaving OBD is the most influencing parameter for the response variable which is nick value. The next influencing parameter is the cutter span width which contributes 21.12% for the response variable. The least contributing parameter is the shaping OBD which contributes about 3.02%. The F value column shows that the F value is highest for the shaving OBD which is 2.38. Larger F value signifies that small change in process parameter results in a large effect on the response variable. The F value is least for shaping OBD, i.e. 0.13. It implies that even a large change in shaping OBD would not result in a large effect on the nick value.

Table 3 ANOVA for nick value

Source	DF	Seq SS	Contribution	Adj SS	Adj MS	F value	P value
CSW	2	0.000011	21.12%	0.000011	0.000005	0.94	0.047
Shaping OBD	2	0.000002	3.02%	0.000002	0.000001	0.13	0.206
Shaving OBD	2	0.000028	53.45%	0.000028	0.000014	2.38	0.035
Error	2	0.000012	22.41%	0.000012	0.00006		
Total	8	0.000052	100%				

$S = 0.0024037$, R-sq $= 81.46\%$, R-sq (adj) $= 85.83\%$

Table 4 Best result during confirmation

Cutter span width (mm)	Shaping OBD (mm)	Shaving OBD (mm)	Nick value (mm)
69.450	33.636	33.450	0.009

Table 5 Average value for the response variable

Trial experiments (mm)	Orthogonal array (mm)	Confirmation runs (mm)	Best result (mm)
0.016	0.015	0.012	0.009

5   Experimental Validations

The optimum condition obtained from the above experimentation is implemented for the confirmation of expected results. For this purpose, the implementation trials in two stages are carried out. During the first stage, hundred parts were inspected and no rejection was observed. And during the second stage, another two hundred parts were inspected and no rejection was observed. During both the stages, all parts showed the nick value below 20 μm. During confirmation runs, the best result of 0.009 mm nick value is observed and is as shown in Table 4. This is the lowest value compared to all trial experiments and experiments from the orthogonal array.

The average nick values measured during trial experiments, during orthogonal array experiments, during confirmation runs and the best value obtained are reported in Table 5. This indicates that the best results are obtained due to the process optimization.

6   Conclusions

From this experimental investigation, it can be concluded that nick problem under consideration can be eliminated by using optimized condition. The rejection of the input shaft due to nick problem in reverse gear section is brought down to zero due to the implementation of optimum condition. The process optimization is attempted for the gear shaping and gear shaving operations, and the following results are obtained.

- Gear shaving OBD is the most significant factor and has a large contribution (53.45%) for the formation of nicks at the reverse gear.
- By performing DOE and analysis, the optimum condition, i.e. cutter span width at 69.450 mm, shaping OBD at 33.636 mm, and shaving OBD at 33.450 mm, for nick elimination was calculated and confirmed through validation experiments in which we got 0.009 mm nick value as the best result.
- As all the nick values obtained during experimental validation were below limiting value of 20 μm, there was no rejection of components having nick defect. Hence,

the overall rejection of components is reduced which leads to the improvement in the productivity of input shafts manufacturing.

Acknowledgements Authors are thankful to authorities and staff of Fiat India Automobiles Limited, Pune, for providing the facilities and necessary support for experimentation.

References

1. Radzevich, S.: Practical Gear Design and Manufacture, 2nd edn. CRC Press, Taylor & Francis Group, New York, pp. 1–16 (2012)
2. Gupta, K., Laubscher, R., Jain, N.: Recent developments in sustainable manufacturing of gears: a review. J. Clean. Prod. **112**, 3320–3330 (2016)
3. Tsay, C.B., Liu, W.Y., Chen, Y.C.: Spur gear generation by shaper cutters. J. Mater. Process. Technol. **104**, 271–279 (2000)
4. Erkorkmaz, K., Katz, A., Hosseinkhani, Y., Plakhotnik, D., Stautner, M., Ismail, F.: Chip geometry and cutting forces in gear shaping. CIRP Ann. Manuf. Technol. **65**, 133–136 (2016)
5. Kim, J., Kim, D.: The development of software for shaving cutter design. J. Mater. Process. Technol. **59**, 359–366 (1996)
6. Hung, C., Liu, J., Chang, S., Lin, H.: Simulation of gear shaving with considerations of cutter assembly errors and machine setting parameters. Int. J. Adv. Manuf. Technol. **35**, 400–407 (2007)
7. Gosh, G.: Gear metrology. Ann. CIRP **52**, 659–695 (2003)
8. Bhilawade, T., Sapkal, S.: Experimental analysis on input shaft reverse gear section. In: Proceedings of 3rd National Conference on Recent Trends in Mechanical Engineering, pp. 240–245 (2018)

Experimental Investigation of Cutting Parameters on Surface Roughness in Hard Turning of AISI 4340 Alloy Steel

Vaishal J. Banker, Jitendra M. Mistry and Mihir H. Patel

Abstract The material AISI 4340 alloy steel is a heat-treatable steel containing Ni, Cr, and Mo, and has a variety of applications due to its unique properties. The input parameters considered are cutting speed, feed rate, and DOC while the output parameter is targeted as surface roughness (Ra). The machining experiments for the present study were conducted in wet conditions based on sequential approach by means of face-centered central composite design (FCC) and response surface methodology (RSM). The variation of surface roughness is best described by nonlinear quadratic model with the foremost influence of feed rate. The percentage contribution of feed rate (mm/rev) is 86.34%, cutting speed (m/min) is 0.62%, and depth of cut (mm) is 0.34% on surface roughness in unheat-treated and with coolant cutting condition.

Keywords AISI 4340 · Hard turning · Response surface methodology · Regression analysis · ANOVA

1 Introduction

Turning is a metal removal process used for the manufacturing of cylindrical surfaces. In lathe machine, the specimen is mounted in the chuck and then it is rotated on a spindle after then the tool is guided toward the workpiece to give required surface finish and dimensions. Hard turning is improved finish turning of hardened materials which have hardness above 45 HRC. It is typically used for components,

V. J. Banker (✉)
ADIT, Anand, Gujarat, India
e-mail: bvaishal@gmail.com

J. M. Mistry
SVIT, Vasad, Gujarat, India
e-mail: jitumistry16@gmail.com

M. H. Patel
CEAT Tyres, Vadodara, Gujarat, India
e-mail: mihirpatel0501@gmail.com

© Springer Nature Singapore Pte Ltd. 2020
R. Venkata Rao and J. Taler (eds.), *Advanced Engineering Optimization Through Intelligent Techniques*, Advances in Intelligent Systems and Computing 949,
https://doi.org/10.1007/978-981-13-8196-6_64

where requirements for accuracy and surface finish are not too demanding. Increasing awareness of the hard turning has led to great interest in improving the process capability so that surfaces with optical quality can be manufactured by hard turning. In the surface roughness evaluation, when these deviations are more, then the surface is rough and if these deviations are less, then the surface is said to be smooth [1–3]. AISI 4340 has the ability to develop high strength when it is heat treated and it is also known for its toughness. It was investigated that the values of surface roughness elevate with increase in feed rate [4].

Surface roughness grows gradually using multilayer coated carbide insert in hard turning of AISI 4340 steel when conditions are dry [5]. It was observed by studying surface roughness by multilayer hard coatings on cemented carbide substrate using CVD process for turning off the same steel that combination of small values of feed rate and large values of cutting speed is essential for reducing the surface roughness. Taguchi's technique and ANOVA study are utilized to observe the effects of input parameters on process aspects of turning process [6]. An effort was made to explore the outcomes of cutting speed, machining time, DOC, and feed rate on machinability characteristics like cutting force and surface roughness using RSM. The experiments were done using factorial design and sequential approach design method [7]. For various values of cutting speeds, the tool wear progresses the surface roughness elevates and at larger cutting speeds, the surface roughness values are found in a fine band [8].

It was observed that the feed rate was more effective by means of various regression and RSM models [3]. Quantile regression models and random forest models help to predict surface roughness more effectively [9]. Multilayer CVD tools tend to have a lesser value of MRR than PVD-coated tools [10]. MQL when used with MWCNT gives a very low value of surface roughness as compared to traditional machining [11].

In present work, an effort is done to inspect the outcome of cutting parameters like cutting speed (m/min), feed rate (mm/rev), and DOC (mm) on surface roughness during hard turning using TNMG insert under wet lubricating conditions.

2   Materials and Methods

2.1   Experimental Setup

Workpiece   The material which was used to conduct the surface roughness was AISI (American Iron and Steel Institute) 4340 alloy steel whose chemical composition is given in Table 1.

Experimental Setup   The experimentation was done on α-Alpha CNC lathe machine having maximum spindle 3500 rpm and cutting speed 250 m per minute as shown in Fig. 1. For the holding of workpiece, 3-jaw chuck was used. The maximum cutting power was 3 KW. The CNC high-speed precision lathe adopts GSK980TDB

Table 1 Chemical composition of AISI 4340 alloy

Element %w	C	S	Ni	Cr	Mn	Mo	P	Si
Composition	0.4	0.023	1.4	1.09	0.6	0.25	0.019	0.23

Fig. 1 α-Alpha CNC turning machine

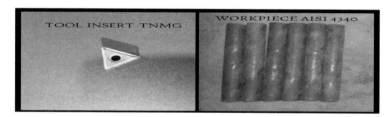

Fig. 2 Tool insert and workpieces

system to control different cutting actions of the machine. The longitudinal and cross feeds used servo-motors to drive.

As shown in Fig. 2, the cutting tool used in the experiment was tungsten carbide grade TNMG insert and the workpiece of AISI 4340 alloy steel before experimentation work had six rods which were used for the turning process. The material of the cutting tool holder was carbon steel.

After the turning process was concluded, surface roughness (Ra) of the workpiece was recorded by using a calibrated surf test SJ-201 P surface roughness tester as shown in Fig. 3. The measurement was taken at four positions (90° apart) about the periphery of the workpieces and then their mean values were considered as the final output. Height gauge was used with the setup of surface roughness tester for

(a)   **(b)**

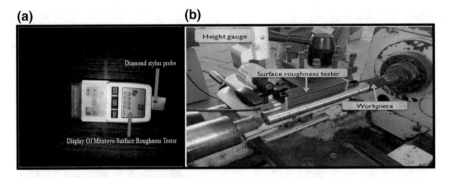

Fig. 3 **a** Surface roughness tester. **b** Setup created for measuring surface roughness

Table 2 Input parameters and their levels

Factor	Input parameters	Notation	Units	Lower level −1	Center points 0	Upper level 1
A	Cutting speed	V	(m/min)	150	200	250
B	Feed rate	f	(mm/rev)	0.15	0.25	0.35
C	Depth of cut	d	(mm)	0.1	0.2	0.3

quantifying the same for the cylindrical surface. The workpiece was fixed between the centers of the lathe machine to get more accurate results as shown in Fig. 3b.

Design of Experiments The two-level full factorial design methods are used for the design of experiments (DOE) or experimental design. There are two levels and three factors used in the DOE method which are shown in Table 2. The four center points were used with the replications of factorial points in the two-level full factorial points created using Design Expert 7 software. RSM is a tool used for solving problems in which an output parameter of significance is affected by many variables and the aim is to optimize this parameter [12].

i.   In the design matrix, use the factorial points (2^3), and 2–4 center points for checking the curvature effect (With 2 replications).
ii. Axial points and 2–4 center points in the design matrix (With 2 replications).

The runs in step 2 are to be done only if the curvature is detected after doing experimental analysis in step 1. Table 5 shows that general whole design matrix with responses with two-level full factorial (2^3) with replication, four center points and axial points with replication and plus 2 center points after getting curvature in design matrix that makes it 28 experiments [13].

Table 3 Design matrix with responses (surface roughness with different conditions)

Std	Run	Cutting speed (m/min)	Feed rate (mm/rev)	Depth of cut (mm)	Surface roughness (Ra)
1	3	150	0.15	0.1	2.3
2	13	150	0.15	0.1	2.23
3	16	250	0.15	0.1	3.5
4	17	250	0.15	0.1	3.58
5	18	150	0.35	0.1	5.13
6	20	150	0.35	0.1	5.15
7	19	250	0.35	0.1	5.39
8	12	250	0.35	0.1	5.19
9	8	150	0.15	0.3	2.25
10	1	150	0.15	0.3	2.49
11	15	250	0.15	0.3	1.81
12	2	250	0.15	0.3	2.29
13	9	150	0.35	0.3	6.29
14	6	150	0.35	0.3	6.35
15	11	250	0.35	0.3	3.17
16	5	250	0.35	0.3	6.45
17	7	200	0.25	0.2	3.55
18	14	200	0.25	0.2	3.75
19	10	200	0.25	0.2	3.6
20	4	200	0.25	0.2	3.39

3 Results and Discussions

3.1 Experimental Results

Results CCD Design is focused in the current work where investigations of surface roughness as a function of cutting parameters in hard turning are considered. The full factorial design was used to identify whether the curvature was present or not using center points. If the curvature was present in the model, then the model is quadratic in nature. ANOVA is used to check out the most dominant parameter influencing the response. The F-value of ANOVA table decides most dominating parameter. The higher the F-value, more it is significant. A general quadratic model was used to develop the model by using Design Expert-7 software with a confidence level was set at 95%. Table 3 shows that design matrix with responses with two-level full factorial/cube points (2^3), plus 4 center points to check the curvature in the surface roughness for UHT and C (Unheat-treated and coolant) conditions. If the curvature is found to be significant, then extra eight runs are needed to be carried for further analysis of curvature effects.

Table 4 ANOVA for surface roughness (Ra for first 20 experiments) (unheat-treated with coolant-UHT and C)

Source	Sum of squares	d.f	Mean square	F-value	Prob > f	%Contribution	Remark
Model	46.31404	7	6.616291	262.81199	< 0.0001		Significant
A-cutting speed	0.299756	1	0.299756	11.906901	0.0054	0.6284485	
B-feed rate	41.1843	1	41.1843	1635.9208	< 0.0001	86.34421	
C-depth of cut	0.166056	1	0.16056	6.5960774	0.0261	0.3481422	
AB	0.166056	1	0.16056	6.5960774	0.0261	0.3481422	
AC	0.770006	1	0.770006	30.586146	0.0002	1.6143426	
BC	3.213056	1	3.213056	127.62884	< 0.0001	6.7362748	
ABC	0.514806	1	0.514806	20.449106	0.0009	1.0793077	
Curvature	1.1068512	1	1.1068512	43.966286	< 0.0001	2.3205489	Significant
Pure error	0.276925	11	0.025175				
Cor. total	47.69782	19					
Std. dev.	0.1586663		R^2	0.9940562			
Mean	4.043		R^2 Adj	0.9902738			
C.V. %	3.9244698		R^2 Pred	0.9798551			

It was concluded from the ANOVA Table 4 that feed rate was the most affecting parameter and had the contribution of 86.344%, cutting speed and depth of cut had the contributions of 0.628 and 0.348% that were found to be less affecting parameters for surface roughness in unheat-treated condition of workpiece with coolant conditions (UHT and C). The interaction effect AC, BC, and ABC showed significant effects on surface roughness in UHT and C conditions. It was found that the curvature is significant in the model in UHT and C. Table 5 shows the values of surface roughness after conducting 20 experiments as listed in Tables 3 and 8 experiments with six axial points and two curvature points in the design matrix for eliminating the curvature effect (Tables 6 and 7).

Figure 4 shows that in normal probability plot, the residual are normally distributed and the model is significant in UHT and C conditions. Figure 5 shows that the values of predicted surface roughness values from the model are closer to actual values and there is less error in the model for UHT & C conditions.

3.2 Regression Analysis

Regression analysis is mostly used for prediction and forecasting. For estimating the relationships among input and output variables, regression analysis is used as a statistical process. The quadratic regression model equation is shown below:

Table 5 Matrix with responses using axial/star points with replication and plus 2 center points after getting curvature in design matrix

Std	Run	Cutting speed (m/min)	Feed rate (mm/rev)	Depth of cut (mm)	Surface roughness (Ra)
1	3	150	0.15	0.1	2.3
2	13	150	0.15	0.1	2.23
3	16	250	0.15	0.1	3.5
4	17	250	0.15	0.1	3.58
5	18	150	0.35	0.1	5.13
6	20	150	0.35	0.1	5.15
7	19	250	0.35	0.1	5.39
8	12	250	0.35	0.1	5.19
9	8	150	0.15	0.3	2.25
10	1	150	0.15	0.3	2.49
11	15	250	0.15	0.3	1.81
12	2	250	0.15	0.3	2.29
13	9	150	0.35	0.3	6.29
14	6	150	0.35	0.3	6.35
15	11	250	0.35	0.3	3.17
16	5	250	0.35	0.3	6.45
17	7	200	0.25	0.2	3.55
18	14	200	0.25	0.2	3.75
19	10	200	0.25	0.2	3.6
20	4	200	0.25	0.2	3.39
21	25	150	0.25	0.2	2.74
22	24	250	0.25	0.2	2.6
23	23	200	15	0.2	1.62
24	21	200	0.35	0.2	5.05
25	27	200	0.25	0.1	2.01
26	26	200	0.25	0.3	2.04
27	22	200	0.25	0.2	3.55
28	28	200	0.25	0.2	3.43

Table 6 Summary of model of surface roughness (unheat-treated with coolant)

Model	S.D	R^2	R^2 Adj	R^2 Pred	Remarks
Linear	0.621172065	0.842	0.821827445	0.761417822	
2FI	0.486083333	0.916	0.890896562	0.86128151	
Quadratic	0.372810427	0.958	0.935820981	0.898104046	Suggested
Cubic	0.365505504	0.969	0.938311414	−2.95344133	Aliased

Table 7 ANOVA for surface roughness (Ra) with center points and axial points (unheat-treated with coolant)

Source	Sum of squares	d.f	Mean square	F-value	Prob $>f$	Remark
Block	7.7289657	1	7.7289657			
Model	53.94343	9	5.9937145	43.12409	<0.0001	Significant
B: feed rate	47.045	1	47.045	338.4834	<0.0001	
AC	0.7700062	1	0.7700062	5.540107	0.0309	
BC	3.2130562	1	3.2130562	23.11757	0.0002	
B^2	1.7663232	1	1.7663232	12.70849	0.0024	
C^2	0.6977881	1	0.6977881	5.020506	0.0387	
Residual	2.3627894	17	0.1389876			
Lack of fit	2.0786644	5	0.4157325	17.55845	<0.0001	
Pure error	0.284125	12	0.023677			
Cor. total	64.035185	27				
SD.	0.3728104		R^2	0.958037		
Mean	3.7107142		R^2 Adj	0.935821		
C.V. (%)	10.046864		R^2 Pred	0.898104		

$$Ra = (+8.06038) - (0.036489E-003 * A) - (38.41943 * B) + (11.13522 * C)$$
$$- (0.020375 * A * B) - (0.043875 * A * C) + (44.81250 * B * C)$$
$$+ (1.31589E-004 * A^2) + (99.39720 * B^2) - (31.60280 * C^2)$$

The above regression equations show that the model is quadratic in nature and the curvature is significant. The nonlinear equations are used to get the R^2 (coefficient of variation) values near to unity to fit the model for adequacy.

3.3 Validation Tests

In order to validate the regression model developed, five experiments were conducted, having combination of process parameters not belonging to the actual experimental list which are shown in Table 8.

Table 8 shows the results obtained in which a comparison was created between the values from predicted model developed in the present work equations and values obtained by experimentation under UHT and C conditions. From the analysis of Table 8, it can be deduced that the calculated errors for surface roughness are (Ra) 4.73% (Max. Value) and 2.18% (Min. Value) for all conditions. Figure 6 shows that the predicted values are in good agreement to actual experimental values and it shows

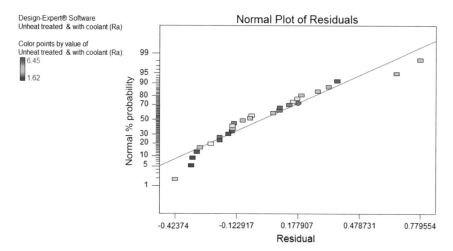

Fig. 4 Surface roughness (Ra) data of normal plot of residuals

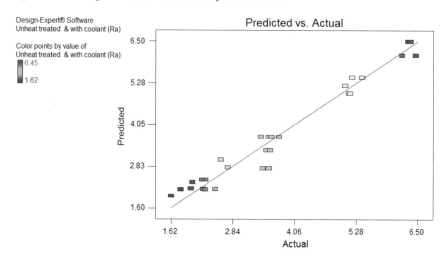

Fig. 5 Predicted values versus actual values of surface roughness (Ra)

good adequacy of the model in UHT and C conditions. The confirmation tests show that the error is less than 5%, so it shows good adequacy for the model.

4 Conclusions

From primary investigations of surface roughness after turning operation, it was found that the curvature is significant. So, further eight experiments were carried out

Table 8 Comparison of results obtained from experiment with predicted values for unheat-treated and with coolant

S. No.	Cutting speed (m/min)	Feed rate (mm/rev)	Depth of cut (mm)	Surface roughness (Ra) predicted	Surface roughness (Ra) experimental	Error (%)
1	150	0.25	0.3	3.13	3.01	3.98
2	175	0.2	0.15	2.53	2.45	3.26
3	200	0.35	0.3	5.85	5.62	4.09
4	225	0.3	0.25	4.42	4.22	4.73
5	250	0.15	0.2	2.8	2.74	2.18

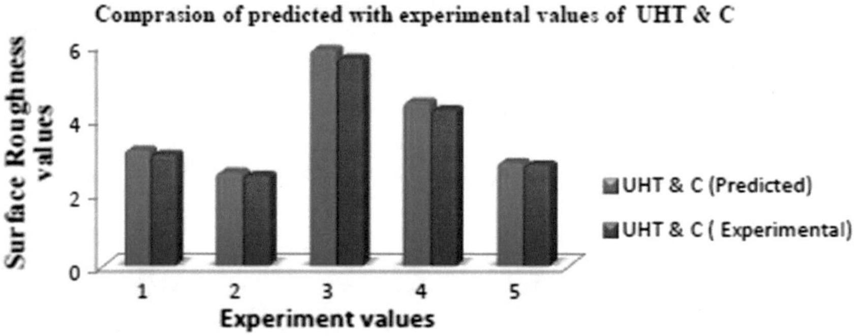

Fig. 6 Comparison chart between predicted with experimental values of UHT and coolant conditions

using sequential approach and RSM method. After performing eight experiments, the curvature is found to be insignificant in surface roughness. By performing ANOVA analysis, it is found that feed rate (mm/rev) is the primary dominant parameter as compared with DOC (mm) followed by cutting speed (m/min) on the surface roughness for all cutting conditions. The percentage contribution of feed rate (mm/rev) is 86.34%, cutting speed (m/min) is 0.62%, and depth of cut (mm) is 0.34% on surface roughness in unheat-treated and with coolant cutting condition. The variation of surface roughness is best described by nonlinear quadratic model with primary contribution of feed rate and secondary contributions of 2FI effect between DOC and feed rate, the quadratic effect in second order of feed rate, and interaction effect between DOC and speed. The regression model developed was in line with the validation results as error in actual and predicted values for the given conditions were less than 5%. Optimal condition of surface roughness can be found for values of cutting speed having a value of 250 m/min and depth of cut having a value of 0.3 mm at high level and feed rate having a value of 0.15 mm/rev.

References

1. Lima, J.G., Avila, R.F., Abrao, A.M., Faustino, M., Davim, J.P.: Hard turning: AISI 4340 high strength low alloy steel and AISI D2 cold work tool steel. J. Mater. Process. Technol. **169**, 388–395 (2005)
2. Sahoo, A.K., Sahoo, B.: Experimental investigations on machinability aspects in finish hard turning of AISI 4340 steel using uncoated and multilayer coated carbide inserts. Measurement **45**, 2153–2165 (2012)
3. Suresh, R., Basavarajappa, S., Samuel, G.L.: Some studies on hard turning of AISI 4340 steel using multilayer coated carbide tool. Measurement **45**, 1872–1884 (2012)
4. Suresh, R., Basavarajappa, S., Gaitonde, V.N., Samuel, G.L.: Machinability investigations on hardened AISI 4340 steel using coated carbide insert. Int. J. Refract Metal Hard. Mater. **33**, 75–86 (2012)
5. Lalbondre, R., Krishna, P., Mohankumar, G.C.: Machinability studies of low alloy steels by face turning method: an experimental investigation. Procedia Eng. **64**, 632–641 (2013)
6. Rashid, W.B., Goel, S., Luo, X., Ritchie, J.M.: The development of a surface defect machining method for hard turning processes. Wear **302**, 1124–1135 (2013)
7. Das, S.R., Kumar, A., Dhupal, D.: Effect of machining parameters on surface roughness in machining of hardened AISI 4340 steel using coated carbide inserts. Int. J. Innov. Appl. Stud. **2**, 445–453 (2013)
8. Adinarayana, M., Prasanthi, G., Krishnaiah, G.: Parametric analysis and multi objective optimization of cutting parameters in turning operation of AISI 4340 alloy steel with CVD cutting tool. Int. J. Res. Eng. Technol. **3**, 449–456 (2014)
9. Agarwal, A., Goel, S., Rashid, W.B., Pierce, M.: Prediction of surface roughness during hard turning of AISI 4340 steel (69 HRC). Appl. Soft Comput. **30**, 279–286 (2015)
10. Ginting A., Skein, R., Cuaca, D., Herdianto, P., Masyithah, Z.: The characteristics of CVD- and PVD-coated carbide tools in hard turning of AISI 4340. Measurement (2018)
11. Patole, P.B., Kulkarni, V.V.: Optimization of process parameters based on surface roughness and cutting force in mql turning of AISI 4340 using nano fluid. Mater. Today Proc. **5**, 104–112 (2018)
12. Montgomery, D.C.: Design and Analysis of Experiments 8th edn. Wiley (2001)
13. Lalwani, D.I., Mehta, N.K., Jain, P.K.: Experimental investigations of cutting parameters influence on cutting forces and surface roughness in finish hard turning of MDN250 steel. J. Mater. Process. Technol. **206**, 167–179 (2008)

Experimental Study of Process Parameters on Finishing of AISI D3 Steel Using Magnetorheological Fluid

V. S. Kanthale and D. W. Pande

Abstract In this work, a new finishing process has been proposed for polishing a work piece made of AISI D3 steel material with the use of magnetorheological (MR) fluid as working medium. Magnetorheological fluid finishing process is considered as one of the important non-conventional methods for polishing the complicated shapes as well as hard material components. The experimentation is carried out on predominating factors like the ratio of CIPs to Al_2O_3 abrasive particles, magnetizing current, the rotating speed of the tool, feed rate of work piece, working gap and hardness of work pieces to get the response as a percentage change in surface roughness. To design the experiments, a full factorial design is employed and the results are analyzed with the help of regression and ANOVA method. The results of experimentation reveal that the major contribution of the parameters to improve the surface finish is of working gap, current applied to the electromagnet and the hardness of the workpiece followed by the rotating speed of the tool, the ratio of CIP to abrasive and feed rate of the workpiece. Further, results showed that the average percentage deviation between the observed and the predicted value is less than 4%.

Keywords MR fluid · D3 steel · Process parameters · ANOVA · Percentage change in R_a

1 Introduction

In the past days, the manufacturers produce the machine components by conventional and unconventional methods of surface finishing, but this method of surface finish could not fulfill the functional requirement. Now, the manufacturing engineers are

V. S. Kanthale (✉)
Mechanical Engineering Department, MIT College of Engineering, Pune, India
e-mail: kanthalevilas@gmail.com

D. W. Pande
Mechanical Engineering Department, Government College of Engineering, Pune, India
e-mail: dwp.mech@coep.ac.in

© Springer Nature Singapore Pte Ltd. 2020
R. Venkata Rao and J. Taler (eds.), *Advanced Engineering Optimization Through Intelligent Techniques*, Advances in Intelligent Systems and Computing 949, https://doi.org/10.1007/978-981-13-8196-6_65

in a position to find an alternate method to make high-ultra surface finish on hard material components. Ultra precision is an essential requirement in die steel industry for most of the components like die, punch, stamping tools and injection mold [1]. Characterization of surface has a major contribution to decide the fatigue life of the finished parts [2]. To maintain the subsurface and avoid other defects during finishing, it is necessary to perform the operation on the parts under gentle conditions, means applying very low forces [3, 4]. The magnitude of the forces can be possible to change during finishing operation using MR fluid. This can be achieved by varying the strength of the magnetic field [5].

In magnetorheological finishing (MRF) process, the CeO_2 used as an abrasive to polish the workpieces [6]. The rheological behavior of the MR fluid changes under the activation of magnetic field. The liquid form of the MR fluid converts into the viscoplastic fluid [7, 8]. As the viscoplastic fluid comes in contact with the surface of workpiece, abrasive particles subjected to the different amount of forces during the finishing process, as a result of which material starts to remove from the workpiece [9].

To further improve the surface finish on hard material, a magnetorheological abrasive flow finishing (MRAFF) process was introduced and showed its capability to finish stainless steel material surfaces with intricate shape that was difficult to be finished using conventional finishing methods [10]. The effect of sizes of particles of carbonyl iron and SiC abrasive on surface roughness was studied using an MRAFF process [11]. In further study, the finite element method was employed to examine the effects of forces such as normal, tangential and axial. The surface finish was mainly affected due to the variation in the magnitude of normal force [12]. Variation in the initial roughness and the hardness of workpiece greatly affect the final R_a value [13]. In another study, it is reported that same size of CIPs and abrasive particles could obtain a better surface finish [14].

To enhance the effectiveness of MRAFF process, a new rotational technique was introduced and investigated the surfaces of different hardness material such as stainless steel, brass and En-8 [15]. Three different types of abrasives with various sizes were experimentally tested and found that the ultra-smooth surface roughness can be better achieved using abrasives CeO_2 and Al_2O_3 as compared to SiC [16]. The behavior of alumni abrasive particle was also studied for MEMS applications [17]. An MR finishing tool with ball end was used to evaluate the 3D surfaces for the responses of R_a and percentage change in R_a. After experimental results, working gap, finishing time and strength of the magnetic field around tool were identified as a most effective parameter to achieve better surface finish [18].

It is revealed from the literature study that the MR fluid can be applied for finishing of complex surfaces. However, the relationship between various parameters is very complex in nature; thereby, process performance is not yet becomes fully characterized. Also, the stagnated bath of the MR fluid and linear movement of the workpiece with rotating tool are not yet studied. Hence, there is a need to develop a suitable mechanism to make the polishing process effective and automate based on magnetorheological fluid.

In this work, customized set up is designed and developed by considering the specification of CNC machines with vertical machining center. Further, preliminary experiments conducted on the workpiece made of AISI D3 steel to identify the dominating input process parameters like ratio of CIPs to Al_2O_3 abrasive particles, magnetizing current, the rotating speed of the tool, feed rate of work piece, working gap and hardness of work pieces. However, the results are analyzed with the help of regression and ANOVA methods by designing the experiments using full factorial methods to get the response as a percentage change in surface roughness.

2   Experimental Details

2.1   Experimental Material

AISI D3 steel was selected as the work material because of its use in die making industries. The dimensions of the specimens were selected 70 mm × 25 mm × 5 mm as per the size of the flask in which specimen fixed to polish the surface. To study the characterization of the surface roughness and analyzing the effects of the input parameters of MR fluid-based process, the experiments were performed with hardness of 30 HRC and 65 HRC workpieces.

2.2   Design of Electromagnet

In order to get a stable and strong magnetic field with a uniform chain of the carbonyl iron particles, it was necessary to select the appropriate shape of the electromagnet. For this reason, various shapes of electromagnet were designed and analyzed the magnetic flux losses across the working zone. It is found that the H-shape core exhibits better results than the hollow circular, laminated plates and C-shape core due to the continuous shape of the electromagnet shoe. The details of the electromagnet are; DC voltage = 0–230 V, Maximum current supply = 2 A, Resistance = 115 Ω, Diameter of wire = 25 mm, Resistance of wire = 85.1 Ω/km, Weight of wire = 0.00181 kg/m, Length of wire for 115 ohms = 1351.35 m, No of turns = 7912.74 turns and Magnetic field strength $(B) = 0.45T$.

2.3   Selection of Input Process Parameters and Its Levels

Process parameters were selected on the performance of preliminary experiments. For an experiment, each input process parameter was varied at two different levels

Table 1 Identified parameters and their coded levels

S. No.	Processing parameters	Symbol	Unit	Actual levels	
				−1	+1
1.	Concentration of abrasives	C	% by wt.	20	40
2.	Magnetizing current	I	Amp	0.8	1.6
3.	Rotating speed of the tool	S	RPM	500	2000
4.	Feed rate of worktable	F	mm/min	1	3
5.	Working gap	G	mm	0.1	0.3
6.	Workpiece hardness	H	HRC	30	65

that are coded in low (−1) and high (+1). Table 1 shows the levels of selected input parameters and their values for conducting the experiments.

3 Experimental Procedure

Experiments were conducted on a CNC vertical milling center manufactured by Premier Machine Tools Ltd. Figures 1a, b show the customized experimental setup mounted on the bed of CNC machine. While clamping the setup on the bed of the machine, the utmost care was taken especially leveling of the surface of the work fixture so that workpiece surface and tool surface remain parallel to each other during operation and get a uniform polishing. The rotating tool firmly clamped in the tool arm in Z-direction of CNC machine for rotating the MR fluid over the polishing surface of the specimen. The longitudinal motion in Y-direction is given in the table on which specimen was mounted and rotating the MR fluid inside the flask using the rotary tool. The dry run was taken to locate the vertical distance between the tool and workpiece. As per the plan of experiments, MR fluid was prepared by mixing the proportion of CIPs and Al_2O_3 abrasives with heavy paraffin as a carrier fluid to perform the finishing operation. This prepared MR fluid poured into the flask in which already specimen fixed at the center of the flask. The strength of the magnetic field varied for experimentation using direct current supply. The variation of the magnetic field enables the CIPs to form the chain in which abrasive embedded. The gripping strength of abrasive particles in the chain of CIPs depends upon the intensity of the magnitude of the magnetic field across the finishing area. During the finishing operation, the chains of CIPs along with abrasive particles rotate with the rotation of the tool. The rotating tool acts the normal force on the abrasive particles, while the linear motion of the workpiece exhibits tangential force. Consequently, abrasive particles interlock with the peaks of the surface as a result of which the embedded particles remove the peaks from the surface of the work material and the surface gets flattened after the number of passes of abrasive particles due to the rotation of the tool.After completing the finishing process, the surface roughness is measured with

(a) (b)

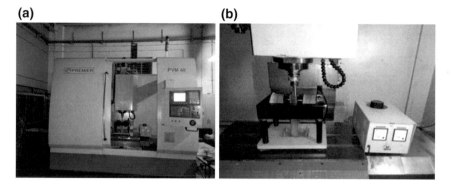

Fig. 1 **a** Set up with CNC machine, **b** Enlarged view of the setup

a surface analyzer (Mitutoyo SJ-210). The surface roughness is measured at three different places before and after the finishing operation using MR fluid, and then averaged the roughness value to quantify the finishing characteristics.

4 Results and Discussion

Full factorial design with two levels (2^k) technique was employed to conduct the experimental runs on the D3 steel workpieces to estimate the effects of input parameters on the response of percentage change in R_a ($\%\Delta R_a$). According to the two level full factorial design method for six input parameters (2^6), there were 64 independent runs. All 64 runs were carried out as per the plan of experiments and its responses noted. Once the finishing operation is completed then surface roughness of each specimen measured immediately after each run by the surface roughness tester and same data applied for statical analysis and inferences. The percentage change in R_a value is calculated as per the following Eq. (1).

$$\% \text{ change in } R_a = [R_{ai} - R_{af}/R_{ai}] \times 100 \tag{1}$$

where

R_{ai} Initial average surface roughness,
R_{af} Final average surface roughness.

4.1 Analysis of Variance (ANOVA)

In the present investigation, the ANOVA was employed to understand which input process parameter significantly affects quality characteristics such as percentage

Table 2 ANOVA for % change in R_a

Source	DF	Seq SS	Contribution (%)	Adj SS	Adj MS	F-value	P-value	Significance or not
Model	21	3328.99	70.68	3328.99	158.523	4.82	0.000	Y
Linear	6	1118.56	23.75	1118.56	186.426	5.67	0.000	Y
C	1	84.95	1.80	84.95	84.947	2.58	0.115	N
I	1	268.57	5.70	268.57	268.573	8.17	0.007	Y
S	1	141.30	3.00	141.30	141.299	4.30	0.044	Y
F	1	28.16	0.60	28.16	28.161	0.86	0.360	N
G	1	384.50	8.16	384.50	384.504	11.69	0.001	Y
H	1	211.07	4.48	211.07	211.074	6.42	0.015	N
2-FI	15	2210.43	46.93	2210.43	147.362	4.48	0.000	Y
C * I	1	258.82	5.50	258.82	258.824	7.87	0.008	Y
C * S	1	51.31	1.09	51.31	51.308	1.56	0.219	N
C * F	1	85.74	1.82	85.74	85.744	2.61	0.114	N
C * G	1	806.86	17.13	806.86	806.856	24.54	0.000	Y
C * H	1	65.81	1.40	65.81	65.807	2.00	0.165	N
I * S	1	8.97	0.19	8.97	8.974	0.27	0.604	N
I * F	1	280.26	5.95	280.26	280.256	8.52	0.006	Y
I * G	1	109.81	2.33	109.81	109.811	3.34	0.075	N
I * H	1	270.55	5.74	270.55	270.555	8.23	0.006	Y
S * F	1	8.26	0.18	8.26	8.257	0.25	0.619	N
S * G	1	25.84	0.55	25.84	25.842	0.79	0.380	N
S * H	1	0.61	0.01	0.61	0.608	0.02	0.893	N
F * G	1	43.49	0.92	43.49	43.491	1.32	0.257	N
F * H	1	2.95	0.06	2.95	2.954	0.09	0.766	N
G * H	1	191.14	4.06	191.14	191.141	5.81	0.020	Y
Error	42	1381.02	29.32	1381.02	32.882			
Total	63	4710.01	100					

change in R_a. Mathematical model generated based on regression technique. Table 2 presents the ANOVA for percentage change in R_a with the main effects of the input parameters and two variable interactions without eliminating the insignificant factors.

The 'F' value of the model is 4.82 indicates that the model is significant since the P-value is 0.000. ANOVA results show that the R-square value is 70.68%, and this indicates that the input parameters have a good correlation with the response of percentage change in R_a. R-square adjusted 56.02%, which means 56.02% variation is explained by the model. This value indicates that the variation in output is explained by input factors. It was observed from the results that the R-square and R-square adjusted are fairly nearby means model fitting was well. In this analysis, the significance level considered at $\alpha = 0.05$ means confidence level 0.95. The terms involved in the analysis considered as significant whose P-values were less than 0.05.

ANOVA results showed that the effect of interaction between the input parameter has more contribution in the performance of the model than the effect of individual parameters. It means that the MR fluid-based finishing process not only depends upon the individual effect of parameter but also depends upon the effect of interaction between the two parameters.

4.2 Mathematical Model for % Change in R_a

A mathematical model was developed on the basis of experimental data using regression method for predicting the % change in R_a from the ANOVA analysis as represented in Eq. (2). This model considers the effect of six individual parameters as well as the interaction between two input process parameters. This model was verified qualitatively and quantitatively on the basis of experimental results. It was observed that the average percentage deviation for 64 samples was 3.56%, which is less than 4%. Hence, it can be revealed that the process model was capable of polishing the hard D3 steel material at nano finishing level.

Mathematical model for percentage change in surface roughness ($\%\Delta R_a$)

$$
\begin{aligned}
\%\Delta R_a = {} & 8.64 - 8.17\,C - 2.77\,I + 0.00314\,S + 7.86\,F + 12.3\,G - 8.40\,H \\
& + 5.03\,C * I - 0.001194\,C * S - 1.157\,C * F + 35.51\,C * G + 1.014\,C * H \\
& + 0.00125\,I * S - 5.23\,I * F + 32.7\,I * G + 5.14\,I * H - 0.000479\,S * F \\
& - 0.00847\,S * G + 0.000130\,S * H - 8.24\,F * G + 0.215\,F * H + 17.28\,G * H
\end{aligned}
\tag{2}
$$

4.3 Main Effects Plot

The main effects plot was applied to evaluate the trend of an input parameter. Figure 2 represents the main effects plot for the response of percentage change in mean R_a.

The main effect plot shows strong relations for the percentage change in R_a. It can be seen from the main plots that more percentage of abrasive particle (40% by wt.) exhibits high percentage change in R_a than the lower percentage (20% by wt.) of abrasive particles. This is because maximum abrasives come in contact with the workpiece surface during finishing operation as a result of which a large proportion of abrasives helps to remove material as high as compared to less proportion of abrasives.

ANOVA results show that at low current (0.8 Amp), the response was better, while at high current (1.6 Amp), it was poor because the stiffness of the fluid goes on increasing at high current due to which the flow of material from the surface of the material was less; thereby, mean of percentage change in R_a was lower in

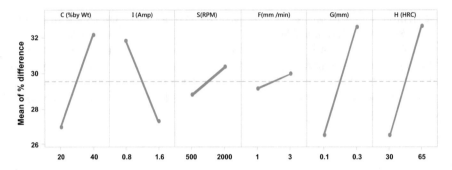

Fig. 2 Main effects plot for percentage change in surface roughness (R_a)

high current. The percentage change in R_a increases with increasing the speed of a rotating tool. It means that as the tool speed goes on increasing, the frequency of abrasive particles comes in contact with workpiece surface also increases while at low speed of the rotating tool abrasive comes in contact with lesser amounts. Main effects plot results reveal that the increasing or decreasing the feed rate (1 mm/min or 3 mm/min) does not affect the response appreciably. The response changed with small proportions with increasing the feed rate of the workpiece. This is because more abrasive particles can participate in the polishing process due to low feed rate of workpiece and as a result surface finish decreases. It was also observed from the result that the improvement in the percentage change in mean R_a is more (steeper graph) when the working gap (0.3 mm) was higher than the low working gap (0.1 mm). The high magnetic field was undesirable as tool attracts the specimen, as a result, low improvement observed in the percentage change in R_a in lower working gap. Variation in the range of working gap has a major change in the response of percentage change in R_a because magnetic flux and gap are inversely proportional. The maximum contribution to the improvement of percentage change occurred due to the working gap.

The percentage change in R_a was observed high on the surface of AISI D3 material with a hardness of 65 HRC compared to 30 HRC. When the hardness of specimen increases, abrasive particles remove the high peaks with smaller amount because resistance to the abrasive particle was high due to the hardness of material, and after a long period of contacting the abrasive particles with a hard surface, it can break an abrasive particle into smaller pieces or even can change the geometry of sharp particles into smoother ones.

5 Conclusion

A new approach has been developed for characterizing the performance of the finishing process on the surface of AISI D3 steel using magnetorheological fluid. In this work, the linear movement is fed to the workpiece, and MR fluid is rotated over the surface of workpiece by rotating the flat shaped tool tip. With this approach, changing the orientation of the abrasives and new cutting edges are getting continuously in the polishing operation. Further, the heating problem of MR fluid is reduced without providing cooling arrangement because the highest percentage change in R_a noticed with low ampere current.

The performance of the process is demonstrated by conducting the experiments using full factorial method. ANOVA is employed to analyze the effect of input process parameters on the workpiece made of AISI D3 steel in terms of percentage change in R_a. The results showed that the major contribution of the parameters to improve the surface finish is of working gap, current applied to the electromagnet and the hardness of the workpiece followed by the rotating speed of the tool, the ratio of CIP to abrasive and feed rate of the workpiece. Further, results showed that the percentage deviation between observed and predicted surface roughness R_a is 3.56 within an error band of 4%. Therefore, it confirms that the developed process is highly capable to polish the hard surfaces at nano level. The developed polishing setup is simple and compact in nature; it can be attached on any automated vertical machining center for polishing the surfaces of component using MR fluid.

References

1. Belgassim, O., Abusaada, A.: Investigation of the influence of EDM parameters on the overcut for AISI D3 tool steel. Proc. Inst. Mech. Eng. Part B J. Eng. Manuf. **226**, 365–370 (2011)
2. Sidpara, A., Jain, V.K.: Experimental investigations into surface roughness and yield stress in magnetorheological fluid based nano-finishing process. Int. J. Precis. Eng. Manuf. **13**, 855–860 (2012)
3. Mulik, R.S., Pandey, P.M.: Experimental investigations and modeling of finishing force and torque in ultrasonic assisted magnetic abrasive finishing. J. Manuf. Sci. Eng. **134**(5), 1008–1019 (2012)
4. Gorana, V.K., Jain, V.K., La, G.K.: Experimental investigation into cutting forces and active grain density during abrasive flow machining. Int. J. Mach. Tools Manuf. **44**, 201–211 (2004)
5. Sidpara, A., Jain, V.K.: Theoretical analysis of forces in magnetorheological fluid based finishing process. Int. J. Mech. Sci. **56**, 50–59 (2012)
6. Kordonski, W.I., Golini, D.: Fundamental of magnetorheological fluid utilization in high precision finishing. J. Intell. Mater. Syst. Struct. **10**, 683–689 (1999)
7. Cheng, H., Yeung, Y., Tong, H.: Viscosity behavior of magnetic suspensions in fluid assisted finishing. Prog. Nat. Sci. **18**, 91–96 (2008)
8. Kciuk, M., Kciuk, S., Turczyn, R.: Magnetorheological characterization of carbonyl iron based suspension. J. Achiev. Mater. Manuf. Eng. **33**(2), 135–141 (2009)
9. Sidpara, A., Jain, V.K.: Analysis of forces on the freeform surface in magnetorheological fluid based finishing process. Int. J. Mach. Tools Manuf. **69**, 1–10 (2013)
10. Jha, S., Jain, V.K.: Design and development of magnetorheological abrasive flow finishing (MRAFF) process. Int. J. Mach. Tool Manuf. **44**(10), 1019–1029 (2004)
11. Jha, S., Jain, V.K.: Modeling and simulation of surface roughness in magnetorheological abrasive flow finishing (MRAFF) process. Wear **261**, 856–866 (2006)
12. Jha, S., Jain, V.K.: Effect of extrusion pressure and number of finishing cycles on surface roughness in magnetorheological abrasive flow finishing (MRAFF) process. Int. J. Adv. Manuf. Technol. **33**, 725–729 (2007)

13. Jha, S., Jain, V.K.: Parametric analysis of magnetorheological abrasive flow finishing process. Int. Manuf. Technol. Manage. **13**, 308–323 (2008)
14. Jha, S., Jain, V.K.: Rheological characterization of magnetorheological polishing fluid for MRAFF. Int. J. Adv. Manuf. Technol. **42**, 656–668 (2009)
15. Das, M., Jain, V.K., Ghoshdastidar, P.S.: Parameteric study of process parameters and characterization of surface texture using rotational magnetorheological abrasive flow finishing (R-MRAFF). Proc. ASME **2**, 251–260 (2009)
16. Yan, Q., Yan, J., Lu, J., Gao, W.: Ultra smooth planarization polishing technique based on the cluster magnetorheological effect. Adv. Mater. Res. **135**, 18–23 (2010)
17. Kim, D.W., Cho, M.W., Seo, T.I., Shin, Y.J.: Experimental study on the effect of alumni abrasive particle behavior in MR polishing for MEMS applications. Sensors **8**, 222–235 (2008)
18. Singh, A.K., Jha, S., Pandey, P.M.: Design and development of nanofinishing processes for 3D surfaces using the ball end MR finishing tool. Int. J. Mach. Tools Manuf. **51**, 142–151 (2011)

Evaluation of Aeration Efficiency of Triangular Weirs by Using Gaussian Process and M5P Approaches

Akash Jaiswal and Arun Goel

Abstract Oxygen concentration being vital to sustain the aquatic life, different efforts have been made to maintain its required concentration in water. Triangular weir as a hydraulic structure acts as a natural aerator used to enhance dissolved oxygen in water body, e.g. streams, rivers, ponds. This paper investigates the modelling performance of experimentally observed aeration efficiency (E_{20}) of triangular weir with different angles and no. of keys. The aeration efficiency as an output is computed using Gaussian process (GP) (normalised poly kernel, poly kernel, PUK and RBF) and M5P (pruned and unpruned). To compare the aeration performance of weirs, standard statistical measures such as the coefficient of correlation (CC) and root mean square error (RMSE) have been applied. The outcome of test results shows that values by M5P (pruned) are nearly matching with observed values. It was observed that there exists good agreement between predicted and measured values of aeration efficiency meaning thereby the M5P (pruned) can be employed successfully to predict aeration parameters of triangular weirs.

Keywords Aeration efficiency · Triangular weirs · GP and M5P techniques

1 Introduction

Aeration is a process by which oxygen is absorbed from atmosphere to replenish the oxygen deficit in water. Efforts have been made to replenish the required oxygen using various methods, one of them is by using hydraulic structures. Use of weir to accelerate the aeration process was introduced by Gamenson [7]. Since then this work has been evolved and a number of researchers have worked to investigate factors affecting aeration, characteristics of weir, shapes of weir, mainly aiming to have best

A. Jaiswal (✉) · A. Goel
NIT Kurukshetra, Kurukshetra, India
e-mail: akaskjaiswal@gmail.com

A. Goel
e-mail: drarun_goel@yahoo.com

© Springer Nature Singapore Pte Ltd. 2020
R. Venkata Rao and J. Taler (eds.), *Advanced Engineering Optimization Through Intelligent Techniques*, Advances in Intelligent Systems and Computing 949,
https://doi.org/10.1007/978-981-13-8196-6_66

possible aeration efficiency. Investigations to evaluate the basic concept of aeration capacity and effect of drop height, configuration of weir, flow rate and temperature were considered by Van der Kroon and Schram [14], Apted and Novak [1], Avery and Novak [2], Gulliver and Rindels [9], Gulliver et al. [10], Watson et al. [15], Baylar and Bagatur [5, 3], Baylar et al. [4], Goel [8], Jaiswal and Goel [12], etc. Along with this, Baylar et al. [6] used ANFIS and Goel [8] used SVM-based modelling tools to evaluate the aeration performance of weirs. Both the studies observed that there exist a good agreement between the predicted value using these modelling tools and the measured value from experimentation.

In order to extend the investigation on comparison of measured value of aeration performance with that of predicted values utilising modelling tools, the present article aims to evaluate the aeration performance of triangular weir having different vertex angle and no. of keys by making use of laboratory experimental data. The modelling techniques such as Gaussian process (normalised poly kernel, poly kernel, PUK and RBF) and M5P (pruned and unpruned) are applied on the experimental data set by using WEKA.

1.1 Methodology and Experimental Set-up

The experiments were performed (in Hydraulics laboratory of the Department of Civil Engineering, NIT, Kurukshetra) on a rectangular flume (25 cm × 30 cm × 4 m). A layout of flume is shown in Fig. 1. The models of triangular weir were placed at the downstream end of the flume, and to keep fixed in the position, a weir stand is attached there. Tests were performed under discharge range of 0.5–6 l/s and drop height range of 60, 75 and 90 cm. DO concentration were measured at upstream

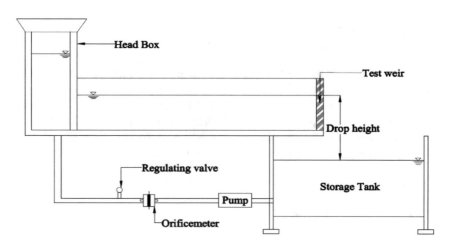

Fig. 1 Layout of experimental set-up

and downstream of the weir using Winkler's method. The results so obtained were utilised to evaluate the aeration performance of triangular weir using mentioned modelling tools.

2 Basic Equations of Weir Aeration

In air–water phase system, the rate of transfer of oxygen is entirely governed by the water phase. The rate of change in concentration of oxygen over time in any air–water phase system as water travels through a hydraulic structure can be expressed by Eq. (1) as purposed by Gulliver et al. [11].

$$dC/dt = K_L A/V \, (C_s - C) \tag{1}$$

where dC/dt = rate of change in concentration of oxygen over time, $K_L A/V$ = oxygen transfer coefficient, C_s = oxygen concentration in water at saturation, A = air and water contact area, V = volume of water and C = actual oxygen concentration in water at t time.

On integrating Eq. (1), we get aeration efficiency 'E' as,

$$E = \frac{C_t - C_o}{C_s - C_o} \tag{2}$$

where C_t = DO concentration at any time t, C_o = initial concentration of DO and C_s = saturation value of DO.

Aeration efficiency E_{20} is found to be quite sensitive to water temperature. Hence investigators have used a correction factor in the equation for applying temperature effect. For any hydraulic structure like weirs, Gulliver [11] proposed a following equation for applying temperature correction factor as given:

$$1 - E_{20} = (1 - E)^{1/f} \tag{3}$$

Meaning of these parameters are E = aeration efficiency at the water temperature, E_{20} = aeration efficiency at 20 °C and f = temperature correction factor as given by the following equation.

$$f = 1 + 0.02103 \, (T - 20) + 8.261 * 10^{-5} \, (T - 20)^2 \tag{4}$$

where T is designated for water temperature in °C.

3   Modelling Techniques

Each of the modelling tools utilises some user-defined parameters to process input data. The values of these user-defined parameters were chosen by hit and trial method.

3.1   Gaussian Processes (GP)

GP is the optimisation technique of an unknown function. In order to improve the modelling performance of a system and to overcome other associated problems, GP a non-parametrical probabilistic models proposed by Murray-Smith [13]. In GP models, evaluation of variance in predicted output was computed by the view of neural network. Any GP model can be easily characterised in the form of normalised poly kernel, poly kernel, RBF and PUK. These output is represented as mean and covariance function.

Normalised Poly Kernel: There are two user-defined parameters of GP used in normalised poly kernel, i.e. exponent and noise as mentioned in Table 1.
Poly Kernel: There are two user-defined parameters of GP used in poly kernel, i.e. E (exponent) and noise, shown in Table 2. The value of noise is taken as the maximum of the noise that comes in the normalised poly kernel.
PUK (Person VII Universal Kernel): The maximum noise of normalised poly kernel is taken as the noise in the PUK. Two user-defined parameters are in PUK kernel, i.e. σ (sigma) and ω (omega), shown in Table 3. These parameters maintain the width (called Person width) and the actual shape (tailing nature) of the Person VII function and SVM regulation.

Table 1 Optimal values of user-defined parameters used in normalised poly kernel function

S. No.	Parameters	Values
1	Noise	0.1
2	Exponent	2

Table 2 Optimal values of user-defined parameters used in poly kernel function

S. No.	Parameters	Values
1	Noise	0.1
2	Exponent	1

Table 3 Optimal values of user-defined parameters used in of PU kernel function

S. No.	Parameters	Values
1	σ (sigma)	2
2	ω (omega)	2

Table 4 Optimal values of user-defined parameters used in RBF kernel function

S. No.	Parameters	Values
1	γ (gamma)	0.1

Table 5 Optimal values of user-defined parameters used in M5P function

S. No.	Parameters	Values for	
		Pruned	Unpruned
1	Minimum number of Instances	4	4
2	Number of Iteration	10	10

RBF: There is one user-defined parameter in RBF Kernel, i.e. gamma, shown in Table 4. In this, also the maximum value of noise from normalised poly kernel is taken as the noise.

3.2 M5P

A model tree is basically used for the purpose of prediction of output for some input value after analysing any data series provided. A linear regression model is stored at each leaf of this model tree that predicts the missing value of the input data that reach to this leaf. It is performed in two stages; in first stage, a decision making tree is created and in second stage pruning of unnecessary branches and substitution of these subtrees with linear regression functions is done. The main objective is to construct a model that relates the output values of the training data to the input values. The optimal values of the user-defined parameters used are given in Table 5.

4 Analysis of Results

Experimental values of aeration efficiency calculated in laboratory are evaluated by using the above stated modelling techniques. The lab result is first divided into training and testing data set, but of total data, 70% data are considered for training and 30% data are considered for testing. Training data is utilised as input to these modelling tools to process, and the output of these tools are cross-checked by utilising testing data set. The predicted values of aeration efficiency of these modelling techniques are compared with the actual experimental values of the aeration efficiency. For making comparison of the predicted and actual aeration efficiency values, statistical measures such as CC and RMSE values of each modelling techniques are considered as specified in Table 6.

Table 6 shows the CC and RMSE value of the modelling techniques used. From correlation coefficient value and root mean square error value (generally testing data

Table 6 Detailing of performance evaluation parameters of aeration efficiency using modelling techniques for testing and training data set

S. No.	Techniques used		Training		Testing	
			CC	RMSE	CC	RMSE
1	GP	npoly	0.9897	0.0184	0.9659	0.0357
		Poly	0.9252	0.0496	0.924	0.0531
		PUK	0.9998	0.0025	0.9983	0.0088
		RBF	0.9998	0.0031	0.9997	0.0036
2	M5P	Pruned	0.9999	0.0019	0.9998	0.0026
		Unpruned	0.9998	0.0023	0.9997	0.0034

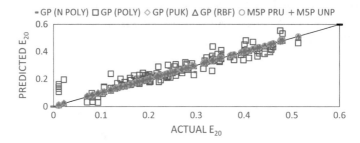

Fig. 2 Variation of predicted and actual values of aeration efficiency for training data set

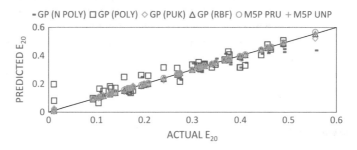

Fig. 3 Variation of predicted and actual values of aeration efficiency for testing data set

set value is taken for evaluation), it was observed that all the modelling techniques used are providing nearly same outcome. So it can be said that WEKA can be effectively utilised to predict aeration performance of triangular weirs. But out of various tools used, M5P (pruned) predicted the measured values are nearly same accurately with CC 0.9998 and RMSE 0.0026 for aeration efficiency.

Using the data set of actual values of aeration efficiency of triangular weirs calibrated in laboratory and the predicted values from different modelling techniques, the following Figs. 2, 3 and 4 are obtained.

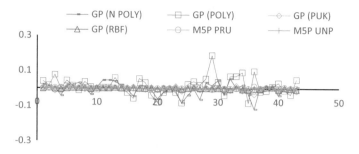

Fig. 4 Error lines of predicted E_{20} value from actual E_{20} values for testing data set

Figure 2 shows the variation of predicted and actual value of aeration efficiency in training data set. From this figure, it is clear that CC and RMSE value for training data set are 0.9897 and 0.0184 for GP (npoly), 0.9252 and 0.0496 for GP (poly), 0.9998 and 0.0025 for GP (PUK), 0.9998 and 0.0031 for GP (RBF), 0.9999 and 0.0019 for M5P (pruned) and 0.9998 and 0.0023 for M5P (unpruned) modelling techniques. Since maximum CC and minimum RMSE values are taken into account, M5P pruned is found to predict the measured values are nearly same accurately with maximum CC value and minimum RMSE value for training data set.

Figure 3 shows the variation of predicted and actual value of aeration efficiency in testing data set. From this figure, it can be observed that CC and RMSE value for training data set are 0.9659 and 0.0357 for GP (npoly), 0.924 and 0.0531 for GP (poly), 0.9983 and 0.0088 for GP (PUK), 0.9997 and 0.0036 for GP (RBF), 0.9998 and 0.0026 for M5P pruned and 0.9997 and 0.0034 for M5P unpruned modelling techniques. Since maximum CC and minimum RMSE values are taken into account, M5P pruned is found to predict the measured values are nearly same accurately with maximum CC value and minimum RMSE value for testing data set.

Figure 4 shows error lines of predicted E_{20} values as given by the modelling techniques used from actual E_{20} values for testing data set. From this figure, it is found that predicted E_{20} values of M5P pruned have minimum deviation from actual value of E_{20} with RMSE value as 0.0026.

5 Conclusion

In this paper, GP- (normalised poly kernel, poly kernel, PUK and RBF) and M5P- (pruned and unpruned) based modelling techniques are applied to determine the aeration efficiency of the triangular weirs (having different weir angles and no. of keys) and were compared. As the outcome of modelling tools used in this article is nearly same, it can be concluded that WEKA could be effectively and successfully used in computation of aeration efficiency of triangular weirs. The outcomes of test results show that out of all the different modelling tools used, M5P (pruned) predicted measured values of aeration efficiency are nearly same as that of observed value. A very good agreement between the predicted and measured values confirms that the M5P (pruned) can be used successfully used to predict aeration efficiency of triangular weirs.

References

1. Apted, R.W., Novak, P.: Oxygen uptake at weirs. In: Proceedings 15th IAHR Congress, Vol. 1, Paper No. B23, Istanbul, Turkey, pp. 177–186 (1973)
2. Avery, S.T., Novak, P.: Oxygen transfer at hydraulic structures. J. Hydr. Div. ASCE **104**(11), 1521–1540 (1978)
3. Baylar, A., Bagatur, T.: Experimental studies on air entrainment and oxygen content downstream of sharp crested weirs. Water Environ. J. **20**(4), 210–216 (2006)
4. Baylar, A., Bagatur, T., Tuna, A.: Aeration performance of triangular-notch weirs. Water Environ. J. **15**(3), 203–206 (2001)
5. Baylar, A., Bağatur, T.: Study of aeration efficiency at weirs. Turk. J. Eng. Environ. Sci. **24**(4), 255–264 (2000)
6. Baylar, A., Hanbay, D., Ozpolat, E.: An expert system for predicting aeration performance of weirs by using ANFIS. Expert Syst. Appl. **35**(3), 1214–1222 (2008)
7. Gameson, A.L.H.: Weirs and aeration of rivers. J. Inst. Water Eng. **11**(5), 477–490 (1957)
8. Goel, A.: Modeling aeration of sharp crested weirs by using support vector machines. WASET Int. J. Mech. Aerosp. Ind. Mech. Manuf. Eng. **7**(12), 2620–2625 (2015)
9. Gulliver, J.S., Rindels, A.J.: Measurement of air-water oxygen transfer at hydraulic structures. J Hydraul. Eng. **119**(3), 327–349 (1993)
10. Gulliver, J.S., Wilhelms, S.C., Parkhill, K.L.: Predictive capabilities in oxygen transfer at hydraulic structures. J. Hydraul. Eng. **124**(7), 664–671 (1998)
11. Gulliver, J.S., Thene, J.R., Rindels, A.J.: Indexing gas transfer in self-aerated flows. J. Environ. Eng. ASCE **116**(3), 503–523. Discussion, **117**, 866–869 (1990)
12. Jaiswal, A., Goel, A.: Aeration through weirs—a critical review. In: International Conference on Environmental Geo-Technology, Recycled Waste Materials and Sustainable Engineering, NIT Jalandhar (2018)
13. Murray-Smith, R., Johansen, T.A., Shorten, R.: On transient dynamics, off-equilibrium behaviour and identification in blended multiple model structures. In: Control Conference (ECC), 1999 European, pp. 3569–3574. IEEE (1999)
14. Van der karoon, G.T., Schram, A.H.: Weir aeration—part I: single free fall. Water **2**(22), 528–537 (1969)
15. Watson, C.C., Walters, R.W., Hogan, S.A.: Aeration performance of low drop weirs. J. Hydraul. Eng. **124**(1), 65–71 (1998)

Performance Evaluation of Kaplan Turbine with Different Runner Solidity Using CFD

Brijkishore, **Ruchi Khare** and **Vishnu Prasad**

Abstract In axial flow turbine, the extent of kinetic energy coming out of runner depends mainly on the design of runner, i.e. the blade profile and solidity. The hydropower plant efficiency is directly related to the efficiency of the turbine and efficiency of the turbine is mainly depends on runner design. The flow characteristics and losses are greatly affected due to change in the solidity of the runner. In the present work, the numerical flow simulations have been carried out in the whole flow passage of Kaplan turbine using CFD Analysis for four different solidities of runner to optimise the number of runner blade, i.e. solidity. The flow parameters and efficiency are computed to identify the optimum runner solidity.

Keywords Solidity · Runner · Kaplan turbine · Computational fluid dynamics (CFD) · Draft tube · Efficiency

Nomenclature

H_n	Net head (m)
N	Rotational speed (rpm)
Q	Discharge (m³/s)
g	Gravitational acceleration (m/s²)
TP_{CASI}	Total pressure at turbine inlet (Pa)
TP_{DTO}	Total pressure at turbine outlet (Pa)

Brijkishore (✉) · R. Khare · V. Prasad
Maulana Azad National Institute of Technology, Bhopal, India
e-mail: Brijkishore844@gmail.com

R. Khare
e-mail: ruchif1@yahoo.com

V. Prasad
e-mail: vpp7@yahoo.com

© Springer Nature Singapore Pte Ltd. 2020
R. Venkata Rao and J. Taler (eds.), *Advanced Engineering Optimization Through Intelligent Techniques*, Advances in Intelligent Systems and Computing 949,
https://doi.org/10.1007/978-981-13-8196-6_67

T Torque (N-m)
V Velocity (m/s)
ρ Mass density of water (kg/m^3)

1   Introduction

Hydropower has a very important role in the development of the country as it provides power at the cheapest rate being source of the natural energy. The basic idea behind the working of a hydropower plant is to convert energy stored within the water (mainly in the form of potential energy) into mechanical energy by a hydraulic turbine. This mechanical energy is further converted into electrical energy by a generator coupled with the turbine. The turbine generator set together forms a hydro unit. For generation of hydropower, water is moved from a higher elevation to lower elevation and during its movement, the available potential energy is converted either fully into kinetic energy or partially into kinetic and pressure energy at the inlet of the turbine depending upon the type of turbine. The rotating component of turbine called runner extracts hydraulic energy from water and converts into mechanical energy [1–3].

The flow through runner is very complex. It is three-dimensional, non-uniform and unsteady and difficult to be simulated precisely. Hence, the three-dimensional analysis techniques can be used to modify, optimise, or verify the capability of an existing or the proposed design [4]. The considerable effort has been made in recent years for the development of computing systems with varying degree of approximation for the problem of three-dimensional flows in rotating and stationary blade passage of turbomachinery [5]. The accurate treatment of three-dimensional flow in hydraulic machinery is monotonous and requires significant effort in creating the geometric, kinematic and dynamic inputs for CFD analysis.

The amount of energy extracted from the water depends mainly on the design of runner, i.e. the blade profile and solidity. The ratio of circumferential spacing (pitch, t) and stream-wise length (chord, l) of runner blade is known as pitch-chord ratio (t/l) and its inverse (l/t) is known as solidity. It affects the lift and drag on blade. The solidity varies from hub to tip. The range of solidity at tip is 0.6–1.8 and at hub, it is (1.1–1.2) times of solidity at tip [6, 7].

In the present paper, the numerical flow simulations of an existing axial flow Kaplan turbine have been done for four different solidities of runner using commercial CFD code at constant guide vanes opening and rotational speed. The flow parameters are calculated from the results of CFD simulation for four runner solidities of Kaplan turbine and the effect of solidity on pressure contours and streamline patterns are studied to assess the optimum solidity to get the best performance of the turbine.

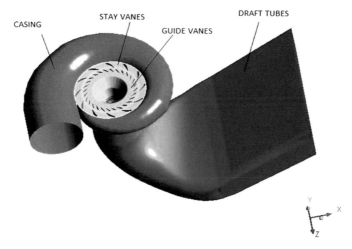

Fig. 1 Complete 3D assembly of axial flow Kaplan turbine

2 Geometric Modelling

The geometric modelling of complete Kaplan turbine in a single-flow domain is very complex. Hence, it is divided into five different domains, i.e. spiral casing, stay vanes, guide vanes, runner and draft tube. Therefore, 3D geometry of each component of turbine passage created separately using ANSYS Workbench and ANSYS ICEM CFD software and then assembled through proper interfaces for numerical simulation of complete turbine passage. The numerical flow simulation is done of an experimentally tested Kaplan turbine model. The specification of geometry of whole flow domain is an important input for the CFD analysis. The complete assembly of Kaplan turbine is shown in Fig. 1. The existing Kaplan turbine model has a spiral casing, a stay ring with 12 stay vanes, a distributor with 28 guide vanes, a runner with four blades and an elbow draft tube [5, 8, 9]. The variation in solidity has been done by varying the number of runner blades from 3 to 6, as shown in Fig. 2.

3 Meshing

In numerical computations, the quality of grid has paramount importance. An unstructured mesh with tetrahedral elements in all components has been used. The mesh quality is refined by using different mesh quality parameters. The complete assembly is discretized in 708,031 nodes and 3,793,611 elements.

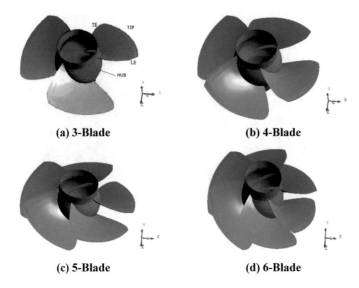

(a) 3-Blade   (b) 4-Blade

(c) 5-Blade   (d) 6-Blade

Fig. 2 Runner with different solidity

4 Boundary Conditions

The mass flow rate of 620 kg/s is given at casing inlet as inlet boundary condition and zero atmospheric pressure is specified at draft tube outlet as outlet boundary condition [8]. Due to curvature in flow passage and rotating flow, κ-ω based Shear Stress Transport (SST) turbulence model has been used. The walls of complete flow domain are taken as smooth impervious boundaries with no-slip conditions. The casing, stay vane, guide vane and draft tube domains are set stationary. The runner rotational speed is specified as 1155 rpm.

5 Computation of Parameters

The pressure and velocity at nodes in entire flow domain can be predicted through CFD analysis. The values of pressure and velocity obtained from CFD simulations are used to calculate the mass flow average values of local and global dimensionless parameters [9–12]. The following formulae are used for performance assessment of turbine:
Net head

$$H_n = \frac{TP_{CASI} - TP_{DTO}}{\rho g} \tag{1}$$

Input power

$$P_{\text{in}} = \rho g Q H_{\text{n}} \tag{2}$$

Output power

$$P_{\text{out}} = \frac{2\pi n T}{60} \tag{3}$$

Hydraulic efficiency

$$\eta_h = \frac{P_{\text{out}}}{P_{\text{in}}} * 100 \tag{4}$$

Velocity coefficient

$$C_{\text{v}} = \frac{V}{\sqrt{2g H_{\text{n}}}} \tag{5}$$

6 Results and Discussions

The CFD analysis result for constant guide vane opening of 40° and runner speed of 1155 rpm gives the pressure and velocity distribution within the flow domain at different solidity of the runner. The results are analysed to study the effect of runner solidity on different flow parameters.

6.1 Effect of Solidity on Pressure Distribution

The pressure variations on runner blade for different solidity are shown in Fig. 3. It is observed from pressure contours that the pressure is decreasing on the suction side with the increase in solidity due to decrease in flow area and thus increase in velocity.

The blade loading shown in Fig. 4 depicts that pressure peak at suction surface is reducing as the solidity increases. The variation in pressure is gradual from leading edge to trailing edge. There is smooth loading of blade in all cases, but there is sharp pressure drop towards trailing edge in case of higher solidity.

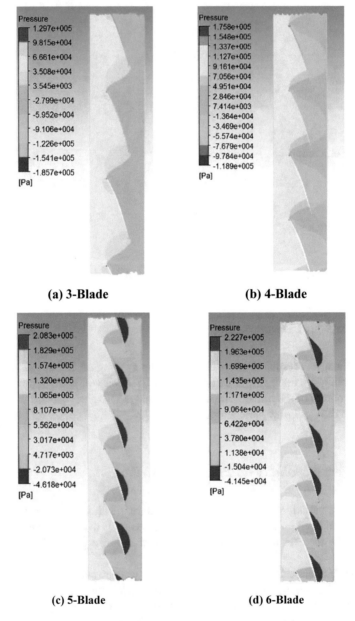

(a) 3-Blade   (b) 4-Blade

(c) 5-Blade   (d) 6-Blade

Fig. 3 Variation of pressure with solidity

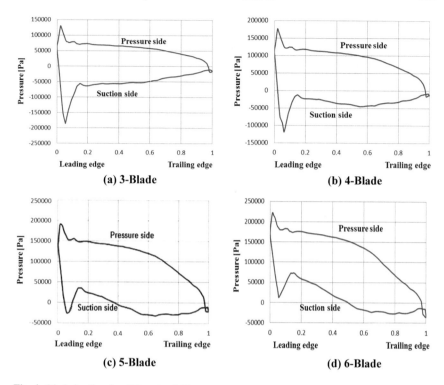

Fig. 4 Blade loading for different solidity

6.2 *Effect of Solidity on Velocity*

The streamline pattern on runner cascade is shown in Fig. 5. It is observed that the velocity is increasing at outlet with the increase in the solidity due to reduction in flow area.

The streamline from inlet to outlet of the runner with 3, 4, 5 and 6 blades is shown in Fig. 6. It is seen that the variation of velocity is more at higher solidity. It is increasing from hub to tip because of the twisted blade of runner.

The streamline pattern from inlet to outlet of draft tube in Fig. 7 for all solidities depicts that velocity at inlet increases with increase in solidity. This leads to an increase in negative pressure at the inlet of draft tube, and hence, turbine is more susceptible to cavitation at higher solidity. It is seen that the whirl velocity coming out of runner increases with increase in the solidity of the runner. The streamline pattern in draft tube is more uniform at designed solidity, i.e. runner with four blades in comparison with the other three solidities. It is observed that more whirls are emerging in draft tube with the increase in the solidity of the runner.

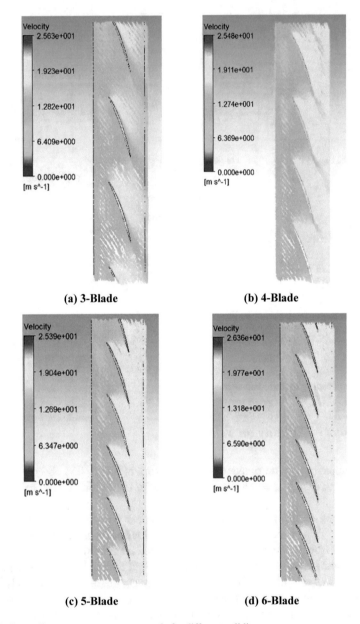

(a) 3-Blade   (b) 4-Blade

(c) 5-Blade   (d) 6-Blade

Fig. 5 Streamline pattern on runner cascade for different solidity

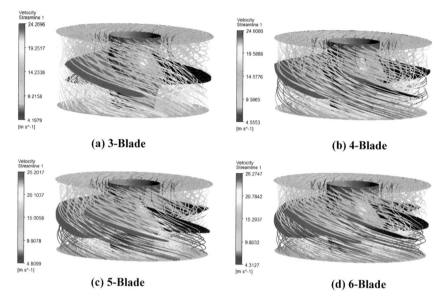

Fig. 6 Streamline pattern in runner with different solidity

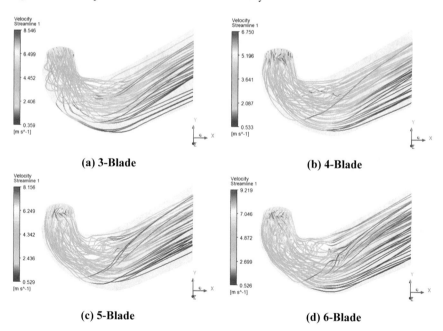

Fig. 7 Streamline pattern in draft tube with different solidity

Table 1 Average value of flow parameters

No of blade	Absolute velocity coefficient		Relative velocity coefficient		Meridional velocity coefficient		Whirl velocity coefficient		Flow angle in degree	
	Inlet	Outlet	Inlet	Outlet	Inlet	Outlet	Inlet	Outlet	Inlet	Outlet
3	0.624	0.442	0.922	1.149	0.404	0.414	0.491	0.038	33.025	22.206
4	0.513	0.349	0.755	1.098	0.328	0.331	0.404	0.052	32.118	17.704
5	0.457	0.340	0.671	1.075	0.290	0.294	0.352	0.144	33.135	15.550
6	0.424	0.348	0.628	1.063	0.269	0.279	0.324	0.192	33.135	14.472

6.3   Flow Parameters

The velocity coefficients are computed from mass-averaged flow velocities by normalising with the spouting velocity, i.e. $\sqrt{2gH_n}$. These coefficient and the flow angles at inlet and outlet of runner are mentioned in Table 1. The values of different velocities in Table 1 indicate that absolute and whirl velocities decrease from inlet to outlet while relative velocity increases from inlet to outlet. The flow angles at outlet are found to be less than inlet. The meridional velocity remains nearly constant from inlet to outlet. This variation of flow parameters from inlet to outlet confirms the characteristics of an axial flow turbine.

It is also observed that there is decrease in normalised velocities with increase in solidity both at inlet and outlet except whirl velocity at outlet. The flow angles at inlet are nearly independent of solidity while at outlet, flow angles also decrease with increase in solidity.

6.4   Efficiency with Different Runner Solidity

The variation of hydraulic efficiency of the turbine with number of blades, i.e. different solidity is shown in Fig. 8. The efficiency of the turbine is found to be maximum for runner with four blades in comparison with the other three solidities because more whirls are emerging at outlet of runner with the increase in solidity under the same operating conditions.

7   Conclusions

It is found that the velocity components both at inlet and outlet are affected due to solidity. The flow angle at outlet is only affected due to solidity. The efficiency of turbine is found to be maximum at runner solidity for four blades. It is observed that more whirls are emerging from runner with the increase in the solidity of the

Fig. 8 Variation of
efficiency with solidity

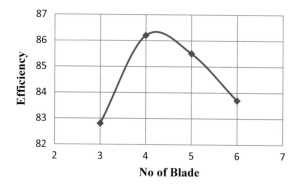

runner. The leading edge is more susceptible to cavitation at lower solidity while
at trailing edge, cavitation may occur at higher solidity. The flow into the runner is
more smooth at designed solidity i.e. runner with 4-blade in comparison with the
other three solidities.

References

1. Lal, J.: Hydraulic Machine, 6th ed., pp. 195–226. Metropolitan Book Co. Private Ltd., New Delhi (2011)
2. Pati, S.: A Textbook on Fluid Mechanics and Hydraulic Machines. Tata McGraw Hill, New Delhi (2012)
3. Mulu, B., Jonsson, P., Cervantes, M.: Experimental investigation of a Kaplan draft tube—part I: best efficiency point. Appl. Energy **93**, 695–706 (2012)
4. Jonsson, P., Mulu, B., Cervantes, M.: Experimental investigation of a Kaplan draft tube—part II: off-design conditions. Appl. Energy **94**, 71–83 (2012)
5. Khare, R., Prasad, V., Brijkishore: Effect of solidity on flow pattern in Kaplan turbine runner. Int. J. Sci. Eng. Res. IJSER **6**(2), 602–606 (2015)
6. Barlit, V.V., Krishnamachar, P., Deshmukh, M.M., Swaroop, A., Gahlot, V.K.: Hydraulic Turbine. Volume-II, pp. 1–29 (2005)
7. Consul, C.A., Willden, R.H.J., Ferrer, E., McCulloch, M.D.: Influence of solidity on the performance of a cross-flow turbine. In: Proceedings of the 8th European Wave and Tidal Energy Conference, Uppsala, Sweden (2009)
8. Prasad, V.: Numerical simulation for flow characteristics of axial flow hydraulic turbine runner. Energy Procedia **14**, 2060–2065 (2012)
9. Mourya, L., Mishra, R., Jain, S.: A review of literature on elbow draft tube geometry of kaplan turbine. Int. J. Res. Appl. Sci. Eng. Technol. (IJRASET) **5**(VI), 5 (2017)
10. Liu, S., Shao, J., Wu, S., Wu, Y.: Numerical simulation of pressure fluctuation in Kaplan turbine. Sci. China Ser. E: Technol. Sci. **51**(8), 1137–1148 (2008)
11. Khare, R., Prasad, V., Mittal, S.: Effect of runner solidity on performance of elbow draft tube. Energy Procedia **14**, 2054–2059 (2012)
12. Prasad, V., Khare, R.: CFD: An effective tool for flow simulation in hydraulic reaction turbines. Int. J. Eng. Res. Appl. (IJERA) **2**(4), 1029–1035 (2012)

Prediction of Occupational Risk at Workplace Using Artificial Neural Network—A Case Study

Dharmendra V. Jariwala and R. A. Christian

Abstract An optimum solution to complex environmental problems has been effectively found with the utilization of neural network (NN). The fundamental objective of the study is to build up a NN model to anticipate an occupational risk due to indoor environment at workplace. Feedforward multilayer perceptron NN with backpropagation was developed in this study. Four indoor environmental parameters such as wet bulb globe temperature index, noise, illumination, and air change per hour are considered as inputs. Two dyeing and printing units were selected, and the data had been collected for the input parameters. The data of one unit had been taken to train the network, and the data of another unit was used to predict the indoor workplace environmental condition. Root-mean-square and R value had been selected as performance indicator to assess the developed NN. The best expectation had been seen in a model with system structure 4-7-18-1 with R estimation of 0.966 and root-mean-square error (RMSE) of 0.0771.

Keywords Artificial neural network · Indoor environment · Occupational risk · Prediction

1 Introduction

Rapid urbanization and industrialization lead to increase in pollution level, particularly in developing nations, with negligible focus on adequate pollution control measures that resulted in degrading and distorting natural environment such as air, water, land [1]. Environment is a complex phenomenon; it includes so many components like air, water, land, noise. Each of them involves so many parameters in their

D. V. Jariwala (✉)
Dr. S & S.S. Gandhay College of Engineering and Technology, Surat, Gujarat, India
e-mail: dv_jariwala@yahoo.com

R. A. Christian
Sardar Vallabhbhai National Institute of Technology, Surat, Gujarat, India
e-mail: rac@ced.svnit.ac.in

© Springer Nature Singapore Pte Ltd. 2020
R. Venkata Rao and J. Taler (eds.), *Advanced Engineering Optimization Through Intelligent Techniques*, Advances in Intelligent Systems and Computing 949,
https://doi.org/10.1007/978-981-13-8196-6_68

constituents. Change in the concentration of single parameter can lead to change in the concentration of others due to their interrelationship. Change in the concentration of one parameter may lead to change in the concentration of only one other parameter than it is easy to predict the concentration and interrelationship of it. But due to involvment of so many parameters majority phenomena are multivariate, dynamic, and nonlinear in nature, so to predict concentration of constituents are very difficult where human having limitation. To solve any complex problem requires huge resources, large database, and time. The prediction with the help of simple basic mathematical formula is not adequate for the complex problem [2]. It is quite difficult to set nonlinear relationship among different variables involved due to the complexity of the data [3, 4]. So, to solve a complex problem, soft computing techniques have become the best tools which provide the best optimal solution [5].

In developing countries, workers in most of the industries are continuously working in the sluggish environment, where indoor temperature, humidity, noise level, illumination, ventilation, etc., are not maintained within the standards. Indoor environmental parameter at workplace beyond its permissible limit ultimately affects the health of the workers. Due to unhealthy environment, workers are frequently observed with irritation of eye, nose, and throat; skin infections; respiratory problems; headache; fatigue; and dizziness [6]. Many researches found a highly positive correlation between the indoor workplace environment and occupant symptoms. In this study, attempt has been made to predict the risk at indoor workplace environment associated with workers working in dyeing and printing mills.

2   Artificial Neural Network (ANN)

ANN is an artificial intelligence soft computing technique using biological nervous system as a base to solve complex problems [7]. ANN is a data-driven technique which learns from precedents. Properly trained ANN with sufficient data can help in correct prediction even if the relationship between various parameters is not properly known [6]. Hui Xie et al. made an attempt with multilayer feedforward network with backpropagation. Indoor air quality was determined in office building in terms of occupant symptom. In the study, the best expectation was observed in the network with two hidden layers. They also concluded that ANN gives a good output than multiple linear regression analysis. Benjamin et al. [8] also developed feedforward network in the prediction of air quality. Ghazali and Ismail [9] also developed ANN to predict air quality in terms of SO_2, NO_2, CO, and NO with seven input parameters which include four pollutants and three comfort variables. They also observed similar results as Hui Xie et al. [6] Singh et al. [10] also modeled ANN with the help of feedforward backpropagation algorithm for the prediction of DO and BOD levels in the River Gomti. They also concluded that neural network can be successfully used for river water quality prediction. Azid et al. [11] in their studies found that feedforward multilayer perceptron network with two hidden layers was effective on the prediction of air pollutant index. Saad et al. [12] developed feedforward ANN model with nine inputs which classify sources influencing indoor air quality. They also concluded that ANN was satisfactorily used in the prediction. Kim et al. [13] in their studies concluded that recurrent neural network gives a better performance to judge an indoor air quality.

3 Materials and Methods

3.1 Study Area

Studies had been carried out in several dyeing and printing houses located in the industrial estate in the Surat City. Parameters which had been selected to determine the workplace environmental conditions include ambient temperature, natural wet bulb temperature (NWBT), dry bulb temperature (DBT), globe temperature (GT), relative humidity (RH), noise level, indoor illumination, and air velocity. Observation locations were fixed on the basis of manufacturing process. A total of six locations were set up in the processing and non-processing areas such as printing area, jet dyeing area, stenter area, looping or color fixation area, washing area, and office area.

3.2 Data Collection

The data was collected at each location for two seasons, one during the winter season (December 16–January 17) and another during the summer season (April 17–May 17). Observations were recorded for the day (8.00 AM–8.00 PM) and night (8.00 PM–8.00 AM) for both shifts at an interval of an hour. QuestTemp 36 heat stress area monitor was used to observe NWB, DBT, RH, and GT in the particular area. Measurement was taken at 1.1 height in the center of the selected area. Sound level meter, anemometer, and lux meter were used to observe noise level, air velocity, and illumination at place, respectively. Noise and horizontal illumination were measured at nine locations in each selected area, and the average of these was recorded. Illuminance was taken at the working heights on the machines. Noise level was recorded at a height of 1.5 m. Air velocity at the entrance of window was taken.

WBGT index and ACH were calculated using Eqs. (1) and (2). A total of 864 observations on each input variable were gathered and used in the development of NN. Total available data was partitioned into three data sets, namely training, validation, and testing. 70% data was used to train the network, and remaining data was used for the validation and testing in equal proportion. Statistics of all four input variables is presented in Table 1.

Table 1 Statistic of inputs

Statistic	Min	Max	Mean	Median	Standard deviation
WBGT index (°C)	21.0	37.6	29.6	29.8	3.31
Noise (dB)	73.7	102.9	85.7	85.7	3.72
ACH (Nos)	0.5	464.5	71.4	14.4	105.99
Illumination (lux)	15	3450	156.5	106	300.13

$$\text{WBGT} = 0.7\, t_{nw} + 0.3\, t_g \tag{1}$$

where t_{nw} is the natural wet bulb temperature and t_g is the globe thermometer temperature.

$$\text{ACH} = \frac{\text{Total hourly air supply to room}}{\text{Cubic capacity of room}} \tag{2}$$

3.3 Development of ANN Model

Feedforward multilayer perceptron network having one or two hidden layers with backpropagation algorithm had been used in the study. Schematic diagram of network architecture is shown in Fig. 1. The first is the input layer, which contains four input variables that include indoor wet bulb globe temperature (WBGT) index, noise, illumination, and air changes per hour (ACH). The next layer is known as hidden layer, where actual process is carried out. The number of hidden layers and neurons in each hidden layer had been an important part in the development of model, which was chosen in the model after a number of trials. Therefore, one or two hidden layers and different values of neurons were tested to optimize the ANN performance. The last is the output layer, which comprises expectation as far as risk related at working environment. Risk level at workplace was first determined with the use of fuzzy rule-based system, and the same had been taken into ANN model.

NN toolbox in MATLAB R2014 had been used to develop ANN. The fundamental steps followed in the development of NN are shown in Fig. 2.

The performance of ANN was measured in terms of R values and RMSE. For the development of the most efficient ANN model, three transfer functions such as pureln, logsig, and tansig had been tried with a single-layer feedforward network. From the available results, hyperbolic tangent function gives maximum R value and minimum RMSE, which were then taken for further development of multilayer feedforward network. The developed ANN of one industry was used to stimulate the output of other industry.

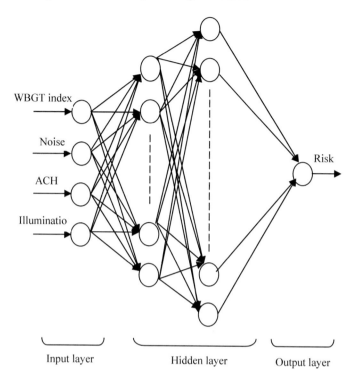

Fig. 1 ANN architecture for risk prediction

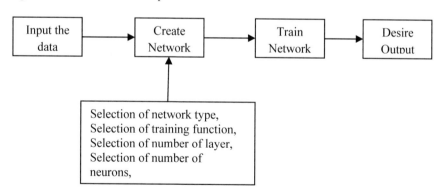

Fig. 2 Steps followed in the development of NN

Table 2 R value and RMSE of different multilayer feedforward neural networks

Network structure	Industry 1				Industry 2 (predicted)
	R value			RMSE	RMSE
	Training	Validation	Testing		
4-7-6-1	0.8663	0.8896	0.9336	0.0624	0.1121
4-7-7-1	0.8752	0.9470	0.9154	0.0589	0.0671
4-7-8-1	0.8639	0.8943	0.8735	0.0646	0.0753
4-7-18-1	0.9037	0.9710	0.9663	0.0492	0.0771
4-7-19-1	0.8082	0.9533	0.9146	0.0650	0.0791
4-7-20-1	0.8191	0.9010	0.9056	0.0691	0.0476

4 Results and Discussion

Optimum ANN was obtained by a number of trials using different nodes in hidden layer and transfer functions. Gradient descent momentum backpropagation had been considered in training phase to adjust weights and biases. Table 2 presents the different ANNs with their performance. ANN model shows R value more than 0.75. From Table 2, it was seen that R values are ranging from 0.8735 to 0.9663 for different ANNs. Network structure 4-7-18-1 gives higher R value of 0.9663 and lower RMSE of 0.0492 than other ANNs. Figure 3 shows the regression analysis plots of the network architecture 4-7-18-1. The output of industry 2 was predicted, and RMSE ranges from 0.0476 to 0.1121 for various ANN structures. It was confirmed that RMSE is very close for different network architectures.

5 Conclusion

In the study, feedforward backpropagation ANN with hyperbolic tangent function as a transfer function used to predict the risk associated with indoor workplace environment was developed considering four variables as inputs. The consequences of this study show maximum correlation coefficient (R) between the measured output and predicted output reaching up to 0.97. Therefore, the model created in this work has a satisfactory result and accuracy. The significance of the study is that one can easily predict the risk associated due to such environmental parameters. It is presumed that ANN shows an effective analyzing tool to understand and simulate the performance and uses a valuable performance assessment tool for the assessment of risk associated with workplace environmental condition in the same such industries.

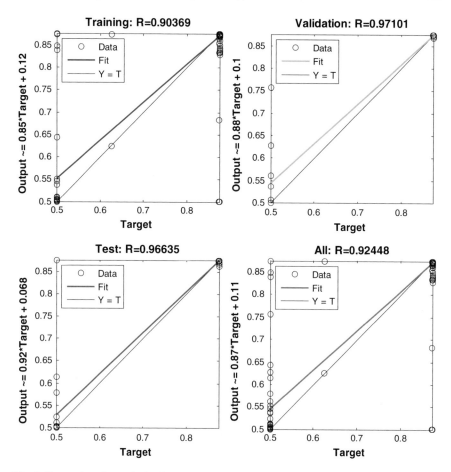

Fig. 3 Regression plot analysis of 4-7-18-1 ANN

References

1. Yadav, J., Kharat, V., Despande, A.: Fuzzy description of air quality using fuzzy inference system with degree of match via computing with words: a case study. Air Qual. Atmos. Health **7**, 325–334 (2014). https://doi.org/10.1007/s11869-014-0239-x
2. Azid, A., Juahir, H., Toriman, M.E., Kamarudin, M.K., Saudi, A.S., Hasman, C.N., Aziz, N.A., Azaman, F., Latif, M.T., Zainaddin, S.F., Osman, M.R., Yamin, M.: Prediction of the level of air pollution using principal component analysis and artificial neural network techniques: a case study in Malaysia. Water Air Soil Pollut. **225**(2063), 1–14 (2014). https://doi.org/10.1007/s11270-014-2063-1
3. Mahboubeh, A., Afsaneh, A., Gholamreza, Z.: The potential of artificial neural network technique in daily and monthly ambient air temperature prediction. Int. J. Environ. Sci. Dev. **3**(1), 33–38 (2012)

4. Mutalib, S.N.S.A., Juahir, H., Azid, A., Sharif, S.M., Latif, M.T., Aris, A.Z., Zain, S.M., Dominick, D.: Spatial and temporal air quality pattern recognition using environmetric techniques: a case study in Malaysia. Environ. Sci. Processes Impacts 1–12 (2013). https://doi.org/10.1039/c3em00161j
5. Das, S.K., Kumar, A., Das, B., Burnwal, A.P.: On soft computing techniques in various areas. Comput. Sci. Inf. Technol. 59–68 (2013). https://doi.org/10.5121/csit.2013.3206
6. Xie, H., Bai, Q.: Prediction of Indoor Air Quality Using Artificial Neural Networks. In: Fifth International Conference on Natural Computation, pp. 414–418 (2009). https://doi.org/10.1109/icnc.2009.502
7. Moustris, K.P., Ziomas, I.C., Paliatsos, A.G.: 3-day-ahead forecasting of regional pollution index for the pollutants NO_2, CO, SO_2, and O_3 using artificial neural networks in Athens, Greece. Water Air Soil Pollut. **209**, 29–43 (2010). https://doi.org/10.1007/s11270-009-0179-5
8. Benjamin, N.L., Sharma, S., Pandharkar, U., Shrivastava, J.K.: Air quality prediction using artificial neural network. Int. J. Chem. Stud. **2**(4), 7–9 (2014)
9. Ghazali, S., Ismail, L.H.: Air Quality Prediction Using Artificial Neural Network. In: Proceedings of the International Conference on Civil Environmental Engineering Sustainability. Johor Bahru, Malaysia. 3–5, pp. 1–5 (2012)
10. Singh, K.P., Basant, A., Malik, A., Jain, G.: Artificial Neural network of the river quality—a case study. Ecol. Model. **220**, 888–895 (2009). https://doi.org/10.1016/j.ecolmodel.2009.01.004
11. Azid, A., Juahir, H., Latif, M.T., Zain, S.M., Osman, M.R.: Feed-forward artificial neural network model for air pollutant index prediction in the southern region of Peninsular Malaysia. J. Environ. Prot. **4**, 1–10 (2013). https://doi.org/10.4236/jep.2013.412A001
12. Saad, S.M., Andrew, A.M., Md Shakaf, A.Y., Saad, A.R., Kamarudin, A.M., Zakaria, A.: Classifying sources influencing Indoor Air Quality (IAQ) using Artificial Neural Network (ANN). Sensors **15**, 11665–11684 (2015). https://doi.org/10.3390/s150511665
13. Kim, M., Kim, Y., Sung, S., Yoo, C.: Data-Driven Prediction Model of Indoor Air Quality by the Preprocessed Recurrent Neural Networks. In: ICROS-SICE International Joint Conference, pp. 1688–1692 (2009)

Quantification of Influence of Casting Variables for Minimizing Hot Spot Area in T Junction by ANOVA and Taguchi Method

Prachi Kanherkar, Chandra Pal Singh and Sanjay Katarey

Abstract Shrinkage porosity is common defect found at casting junctions. Slowly, cooling of molten metal causes hot spot area generation at the junction that is the cause of defect such as shrinkage porosity. Location of riser above the junction may be the one design change for minimization of defects. The hot spot area size and its extent depend on geometrical complexity like thickness, fillet radii, and the angle between walls. Alternatively, junction modification is one of the optimum solutions to minimize such hot areas before solidification. Pro Engineering was used for modeling and through STL, files format data exchange to AutoCAST. Taguchi and ANOVA methods were employed to evaluate the influence of parameters on hot spot area in casting junctions. Area of hot spot was selected as the response variable. It is concluded that upper thickness makes the maximum contribution, and lower thickness, radius, and angle followed upper thickness.

Keywords Green sand casting · Casting junction · Hot spot area · Taguchi method · ANOVA

1 Introduction

Metal casting is a versatile manufacturing process to make parts with intricate shapes and sizes ranging from few mm to meters. However, defects in castings have been a perpetual concern for foundry engineers. The casting junctions are always prone to shrinkage defect. Junctions are at high thermal concentration with a low surface area that causes this defect. General guidelines to reduce porosity defects in L, V, T, Y,

P. Kanherkar (✉) · C. P. Singh · S. Katarey
Samrat Ashok Technological Institute Vidisha, Vidisha 464001, India
e-mail: prachikanherkar0@gmail.com

C. P. Singh
e-mail: chandraid@gmail.com

S. Katarey
e-mail: sanjaykatarey@hotmail.com

© Springer Nature Singapore Pte Ltd. 2020
R. Venkata Rao and J. Taler (eds.), *Advanced Engineering Optimization Through Intelligent Techniques*, Advances in Intelligent Systems and Computing 949,
https://doi.org/10.1007/978-981-13-8196-6_69

and X junctions through minor changes in the design are to add fillets for reducing stress concentrations, add a central cored hole, and reduce thicknesses. Solidification simulations are modern tools to predict shrinkage extent of porosity defects and their locations [1]. Taguchi method is focused on the design of minimum required experiments to ascertain the effects of different parameters [1]. Density of the cast part depends on metal temperature, hydraulic pressure, piston velocity, filling time, and its hydraulic pressure that has maximum effect on density [2]. Casting of thin-walled parts made of magnesium alloy used in communication industries was studied to assess dependency on process parameters [3]. Taguchi method is a versatile optimization method and has been useful to reduce time in sense of reduced required experiments and associated money to optimize processes. Parameters with a certain range were optimized for green sand casting process and deep drawing process parameters [4, 5]. Evaporative pattern casting parameters such as fineness number, temperature at pouring influence tensile strength of Al alloy and optimum values of these parameters can improve tensile strength [6]. Complex castings junction design was improved by the vector element method [7]. Taguchi method was carried for optimization of parameters for micro-holes formation [8]. Casting defects of axle casing can be minimized by changing temperature and liquid flowing rate. Finite difference method and Z-cast software were used to analyze defect like shrinkage porosity, hot cracks, cold shut, etc. of nodular cast iron [9]. Investigation of defects was done by simulation technique along with optimization of parameters by Taguchi and ANOVA methods [10]. Location of feeder influences defects formation and efforts that have been made to change and improves feeder design so to reduce porosity and shrinkage cavity. Modification in feeder design was made to upgrade the quality and strength of cover plate [11]. Sand properties such as moisture content, hardness number, and compression strength with permeability have significant influence on defects formation. These properties were studied to improve quality [12]. Crack generation during solidification is common in casting. Different techniques both experiments and computer simulation have been effectively applied to investigate hot tearing caused by cracks and stress concentration was the reason for hot cracks [13]. In investment casting, hot cracking investigation was performed using Pro CAST software. Conclusions were made that the occurrence of large tensile stress at an area causes cavity appearance in that area and it is the cause of cracking. Findings showed that casting stress depends on pouring temperature, angle of fork, and preheating temperature [14].

2   Methodology

The role of CAD/CAM is increasing and become easier approach when comparing to old days. The study available easily on this area and many software are formed which can make some task "Fast accurate and in one time always." Low-yield casting results in less saleable castings. The extra material due to low yield has to be re-melted. Thus, low-yield casting decreases the production rate and increase the energy consumption.

Fig. 1 Geometry parameters considered in "T" junction showing t_1 as upper section thickness, t_2 as vertical section thickness, r as radius at junction, and θ as angle at junction

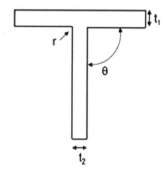

Focusing to reduce these wastes by analysis of junction by Taguchi method and ANOVA. These defects lead to 6–10% in which 80% of defects are shrinkage that defects most probable area in junctions. Figure 1 shows T junctions with geometric details.

2.1 Modeling of T Junction in Pro E and Solidification Simulation Using AutoCAST

In this research, we used Pro E software for different geometric parameters of green sand T casting junction for modeling purpose. Figure 2i–iii shows modeling of T junction in Pro Engineering software.

İn present study the various step follow for modeling and simulation first modeling of T junction is done in Pro E software than STL format of T junction model is import in AutoCAST software for simulation purpose. Simulation is perform in various process solidification feeding than locate than cooling after that hop spot is identified. Figure 3 shows the steps involved in modeling and simulation. Figure 4i–iii shows simulation T junction done in AutoCAST.

In Fig. 4, hot spot area has been shown in different model of T casting junctions. It has been observed that the size of hotspot area is changing due to change in geometric parameters such as thicknesses, radius, and angle of T casting junctions.

(i)   **(ii)**   **(iii)**

Fig. 2 i–iii Modeling of T junction in Pro E

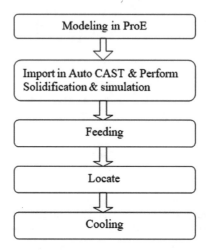

Fig. 3 Flowchart for simulation in AutoCAST

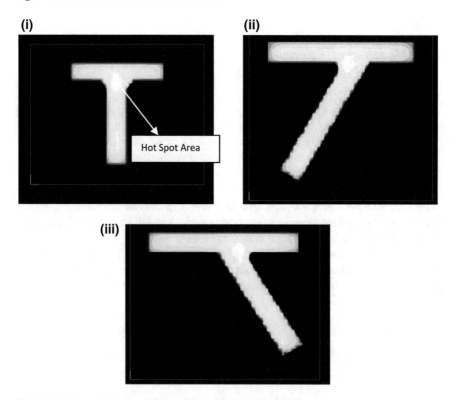

Fig. 4 i–iii Some of the simulations of T junctions in AutoCAST

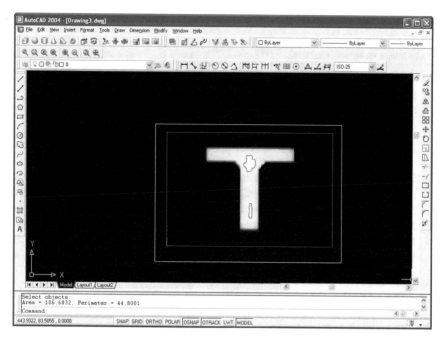

Fig. 5 Hot spot area measurement in auto CAD

2.2   *Measurement of Hot Spot Area*

Simulations of casting were performed in AutoCAST with various values of different parameters. The results which depict hot spot area at T junction can be clearly seen and these area are captured in the form of .jpg images as shown in Fig. 5. These images of hot spot obtained from AutoCAST were then imported to AutoCAD to measure area of hot spot. The process of measurement of area of cast part starts with insert command to format the image and then raster command is used to insert the image. To find actual dimensions of cast object, first image scaling is done and then spline command is used to identify the hot spot area. Now, simple area command gives hot spot area at junction. Further, as parameters are changed, there is change in size of hot spot area. It is evident from literature that raster scan can measure areas of image very near to actual size [15].

3   Analysis of Influence of Parameters

Taguchi and ANOVA methods were employed to assess parameters' influence. In Taguchi's approach, experiments are designed in such a manner that the individual factors' effect and interaction between the factors (combination) can be studied.

Table 1 Coded and actual levels of the independent variables (parameters) for design of experiments for T junction

Independent variables	Symbols	Level 1	Level 2	Level 3	Level 4
Thickness	t_1	30	36	42	48
Thickness	t_2	32	38	44	50
Radius	R	0	10	20	30
Angle	Θ	30	60	90	120

Design of experiments scheme is one of the important steps in Taguchi's approach. Taguchi used orthogonal arrays to design experiments. In orthogonal array, all factors are considered with all levels and are equally distributed. Signal-to-noise ration, i.e. logarithmic transformation of mean square deviation was used to quantify influence of parameters. Three signal-to-noise ratio approaches are commonly used.

Following relations have been used.

(i)   MS is hot spot area in T junction
(ii) The S/N ratio is expressed as $(S/N)_i = -10 \times \log(\text{MS})$
(iii) The overall mean S/N ratio is expressed as $\overline{S/N} = \frac{1}{16} \sum_{i=1}^{i=16} (S/N)_i$
(iv) The sum of squares due to variation about mean is expressed as

$$SS = \sum_{i=1}^{i=16} \left((S/N)_i - \left(\overline{S/N} \right) \right)^2$$

(v)   For ith process parameter, the sum of squares due to variation about mean is given as $(SS)_i = \sum_{j=1}^{j=4} \left((S/N)_{ij} - \left(\overline{S/N} \right) \right)^2$.

Table 1 shows process parameters with selected levels. Table 2 is showing orthogonal array for experimental design by Taguchi method. Anova data are shown in Tables 3, 4, 5, and 6 for upper thickness, lower thickness, radius at junction, and angle respectively. Table 7 shows percentage contribution of each variable. It is evident that thicknesses play vital role in casting process.

4 Conclusion

Analysis of geometrical parameters' influence on hot spot area has been studied by Taguchi and ANOVA methods for T junction. In this research, influence of casting variables for minimizing hot spot area in T junction has carried out four process variables such as two thicknesses, radius, and angle were considered. AutoCAST software was used to simulate casting process and hot spot area was measured for each simulation. Modeling was done in Pro E software than STL file format imported

Table 2 Orthogonal array L16

Experiments	Thickness (t_1)	Thickness (t_2)	Radius (r)	Angle (θ)
1	1	1	1	1
2	1	2	2	2
3	1	3	3	3
4	1	4	4	4
5	2	1	2	3
6	2	2	1	4
7	2	3	4	1
8	2	4	3	2
9	3	1	3	4
10	3	2	4	3
11	3	3	1	2
12	3	4	2	1
13	4	1	4	2
14	4	2	3	1
15	4	3	2	4
16	4	4	1	3

Table 3 ANOVA data for thickness (t_1)

No of runs	Hot spot area	S/N R	S/N$_{ij}$
1	158.74	44.01372752	35.215
2	38.52	31.71372557	
3	46.45	33.33971437	
4	38.91	31.80122462	
5	165.12	44.3559936	41.407
6	38.81	31.77887285	
7	148.44	43.43101891	
8	201.45	46.08334543	
9	165.24	44.36230373	41.71
10	46.75	33.3956323	
11	207.34	46.33366188	
12	138.44	42.82523181	
13	82.65	38.34485716	42.905
14	195.68	45.83092879	
15	156.63	43.89749896	
16	150.74	43.56457022	

Table 4 ANOVA data for thickness (t_2)

No of runs	Hot spot area	S/N R	S/N$_{ij}$
1	158.74	44.013	42.765
5	165.12	44.35	
9	165.24	44.36	
13	82.65	38.34	
2	38.52	31.71	35.66
6	38.81	31.77	
10	46.45	33.33	
14	95.68	45.83	
3	46.45	33.34	41.747
7	148.44	43.43	
11	207.34	46.33	
15	156.63	43.89	
4	38.91	31.8	41.065
8	201.45	46.08	
12	138.44	42.82	
16	150.74	43.56	

Table 5 ANOVA data for radius (r)

No of runs	Hot spot area	S/N R	S/N$_{ij}$
1	158.74	44.013	41.418
6	38.81	31.77	
11	207.34	46.33	
16	150.74	43.56	
2	38.52	31.71	40.402
5	165.12	44.35	
12	138.44	42.82	
15	156.63	43.89	
3	46.45	33.34	42.402
8	201.45	46.08	
9	165.24	44.36	
14	95.68	45.83	
4	38.91	31.8	36.725
7	148.44	43.43	
10	46.45	33.33	
13	82.65	38.34	

Table 6 ANOVA data for angle (θ)

No of runs	Hot spot area	S/N R	S/N$_{ij}$
1	158.74	44.013	44.023
7	148.44	43.43	
12	138.44	42.82	
14	95.68	45.83	
2	38.52	31.71	40.615
8	201.45	46.08	
11	207.34	46.33	
13	82.65	38.34	
3	46.45	33.34	38.645
5	165.12	44.35	
10	46.45	33.33	
16	150.74	43.56	
4	38.91	31.8	37.955
6	38.81	31.77	
9	165.24	44.36	
15	156.63	43.89	

Table 7 Contribution of independent variables

Process parameters	Sum of squares $(SS)_i$	Percentage contribution ($\frac{SS_i}{SS} \times 100$)
Thickness (t_1)	35.851	33.525
Thickness (t_2)	30.282	28.314
Radius (r)	18.607	17.4
Angle (θ)	22.195	20.755

in AutoCAST software simulation casting process of cast steel was done by defining all the parameters. After simulation the results are plotted in the form of images of cast T junction. In the images hot spot area was identified now these images are imported to Auto CAD software by Raster scan hot spot area was measured for each trail. Taguchi orthogonal array method was used to design no of simulations and ANOVA method was used to assess percentage of influence of each parameter. It has been concluded that thickness of T junction have maximum effect on the minimization of hot spot area.

Cumulative effect of thickness has been measured as 61.83% and other two factors have effectiveness as 17.4 and 20.76%; so in the present research, it is concluded that thickness has maximum influence on defectless casting.

References

1. Taguchi, G., Konishi, S.: Taguchi Method, Orthogonal Arrays and Linear Graphs. Tools for Quality Engineering, pp. 35–38. American Supplier Institute (1987)
2. Syrcos, G.P.: Die casting process optimization using Taguchi method. J. Mater. Process. Technol. **135**, 68–74 (2003)
3. Wu, D., Chang, M.: Use of Taguchi method to develop a robust design for the magnesium alloy die casting process. Mater. Sci. Eng. **379**, 366–371 (2004)
4. Guharaja, S., Haq, A., Karuppannan, K.: Optimization of green sand casting process parameters by using Taguchi's method. Int. J. Adv. Manuf. Technol. **30**, 1040–1048 (2006)
5. Padmanabhana, R., Oliveira, M.C., Alves, J.L., Menezes, L.F.: Influence of process parameters on the deep drawing of stainless steel. Finite Elem. Anal. Des. **43**, 1062–1067 (2007)
6. Kumar, S., Kumar, P., Shan, H.S.: Optimization of tensile properties of evaporative pattern casting process through Taguchi's method. J. Mater. Process. Technol. **204**, 59–69 (2008)
7. Joshi, D., Ravi, B.: Classification and simulation based design of 3D junction in casting. Trans. Am. Foundry Soc. **117**, 7–22 (2009)
8. Jung, J.H., Kwon, W.: Optimization of EDM process for multiple performance characteristics using Taguchi method and Grey relational analysis Springer. J. Mech. Sci. Technol. **24**(5), 1083–1090 (2010)
9. Sun, Y., Luo, J., Mi, G.F., Lin, X.: Numerical simulation and defect elimination in the casting of truck rear axle using a nodular cast iron. Mater. Des. **32**, 1623–1629 (2011)
10. Dabade, U.A., Bhedasgaonkar Rahul, C.: Casting Defect Analysis using Design of Experiments (DoE) and Computer Aided Casting Simulation Technique. In: Procedia CIRP, pp. 7616–7621 (2013)
11. Choudharia, C.M., Narkhede, B.E., Mahajan, S.K.: Casting design and simulation of cover plate using auto CAST-X software for defect minimization with experimental validation. Procedia Mater. Sci. **6**, 786–797 (2014)
12. Kumar, P., Mohan Das, N.: Reduce Casting defects in foundry by Taguchi method. Int. J. Adv. Eng. Technol. 833–835 (2016)
13. Gawronska, G.: Different techniques of determination of the cracking criterion for solidification in casting. Procedia Eng. **177**, 86–91 (2017)
14. Mingguang, W., Yong, P.: Numerical simulation and analysis of hot cracking in casting of Fork. Rare Metal Mater. Eng. **46**(4), 0946–0950 (2017)
15. Yeo, W.S., Berger, J.: Application of raster scanning method to image sonification, sound visualization, sound analysis and synthesis. In: Proceedings of the 9th International Conference on Digital Audio Effects, pp. 18–20 (2006)

Taguchi-Based Optimization of Machining Parameter in Drilling Spheroidal Graphite Using Combined TOPSIS and AHP Method

Pruthviraj Chavan and Ajinkya Patil

Abstract This paper envisages on enhancing the cutting environments for exterior irregularity (Ra), tool wear (Tw), and material removal rate (MRR) obtained in machining of Spheroidal graphite SG 500/7. Machining trials are accomplished at the VMC using Titanium Nitride (TiN), Titanium Aluminum Nitride (TiAlN), Titanium Carbo-Nitride (TiCN), and Zirconium Nitride (ZrN) covered carbide-cutting tools on SG 500/7 material. Cutting speed, tool material, and feed rate are preferred as the cutting factors. Taguchi L16 orthogonal array is used to design of a tryout. The weight of each criterion is required for calculating the weighted normalized matrix is determined by the analytic hierarchy process (AHP) method. The best cutting environments and their preferences depend on relative nearness value determined using technique for order preference by similarity to ideal solution (TOPSIS). This combined practice is a multi-objective optimization method which has been implemented to simultaneously minimize tool wear, exterior irregularity, and maximize material removal rate (MRR). The statistical outcome shows that significant improvement in the MRR and lowers the exterior irregularity and tool wear using the optimal combination of drilling process parameter obtained by this collective technique.

Keywords AHP · Spheroidal graphite SG 500/7 · Optimization · TOPSIS · Drilling

1 Introduction

Every industry starts with objective stays to produce the goods with lesser price and great worth in tiny duration. The apparent superiority is an essential constraint to estimate the throughput of machine tools as well as finished works. The appar-

P. Chavan (✉) · A. Patil
RIT, Islampur, India
e-mail: Pruthviraj9496@gmail.com

A. Patil
e-mail: aji.patil08@gmail.com

© Springer Nature Singapore Pte Ltd. 2020
R. Venkata Rao and J. Taler (eds.), *Advanced Engineering Optimization Through Intelligent Techniques*, Advances in Intelligent Systems and Computing 949, https://doi.org/10.1007/978-981-13-8196-6_70

787

ent irregularity and material removal rate are exaggerated by influences like tool dimensions, cutting speed, and feed rate, the microstructure of the workpiece and the inflexibility of the mechanism tool. Spheroidal graphite (SG007) is the most widely used ferrous materials in engineering. Ramesh et al. [1] employed Taguchi technique and ANOVA to obtain the finest hole-making factors of EN31 under dry conditions and found that the feed rate has the peak important factor intended for refining the metal removal rate with involvement percentage of 85.1 and 85.7%, respectively. Pradeep and Packiaraj [2] implemented an investigational effort with a goal to inaugurate a correspondence among speed, feed, and tool diameter with the OHNS material using high-speed steel spiral drill. Balasubramaniyan and Thangiah [3] studied hybrid use of TOPSIS and AHP method and investigated that peak blend of turning practice constraint for concurrent subtraction of microhardness, exterior irregularity, and betterment of MRR are feed rate 0.26 mm/rev, cutting speed 179 m/min, and depth of cut 1.8 mm with CVD-coated carbide tool. Sreenivasulu and Srinivasa Rao [4] reported that rendering to outcome generated in MADM proposes the appropriate substitute of Al alloys in a linkwise in both AHP and TOPSIS methods. Bhanu Prakash and Krishnaiah [5] employed MCDM procedures for enhancing cutting factors for AISI 1040 steel turning expending coated tools and revealed that the optimization results of AHP and TOPSIS are in a decent settlement with those achieved using AHP and VIKOR. Anand Babu et al. [6] employed new-fangled process, i.e., analytic hierarchy process—DENG's connection-grounded methodology for improving the wire EDM practice controls to acquire a practical or best response on Al 6061/2% SiCp/3 μm composite and superlative choice procedure for multi-criteria decision-making method in WEDM of AMM. Venkata Rao [7] studied machinability can be estimated in the turning process of titanium expending shared TOPSIS and AHP process. The revision of works shows that nearby works are available on the consequence of procedure constraints on exterior irregularity and hole measurement. However, inadequate works are accessible on the revision of the consequence of hole-making procedure factors on spheroidal graphite SG 500/7. Spheroidal graphite having a wide range of applications, an effort has been ready in this work to enhance drilling factors such as speed, feed, and tool material for least value of exterior irregularity and better MRR. This is able by expending shared TOPSIS and AHP method [1, 5, 7]. In this paper, the Taguchi way is presented first. The investigational particulars of using the shared TOPSIS and AHP method to decide and evaluate the best cutting considerations are defined next. Finally, the paper accomplishes with an instantaneous of this work and forthcoming effort.

2 Experimental Setup

The spheroidal graphite (SG 500/7) used as a workpiece and its composition as shown in Table 1. The cutting tools were such as TiN- , TiCN- , TiAlN- , ZN-coated carbide [1].

Table 1 Chemical composition (wt%) of SG 500/7

C%	Si	Mn	P	S	Mg	Cu
3.21–3.61	2.25–2.85	0.41–0.59	0.031–0.061	0.021–0.041	0.031–0.056	<0.41

Table 2 Machining factor levels

Parameters	Levels			
	L 1	L 2	L 3	L 4
Speed (rpm)	989	1194	1250	1432
Feed (mm/min)	151	177	192	238
Forms of carbide-coated tool	TiN-covered tool	TiCN-covered tool	TiAlN-covered tool	ZrN-covered tool

Table 3 Investigational outline and their outcomes

Sr. no.	Speed (rpm)	Feed rate (mm/min)	Carbide-coated tools	Surface roughness (μm)	Tool wear (gm × 10^{-3})	MRR (cm^3/min × 10^{-3})
1	989	151	TiN	4.71	2.92	30.0262
2	989	177	TiCN	3.09	3.25	35.1963
3	989	192	TiAlN	3.28	3.59	38.1790
4	989	238	ZrN	3.05	3.91	47.3261
5	1194	151	TiCN	3.01	2.83	36.2500
6	1194	177	TiN	3.15	3.24	42.4918
7	1194	192	ZrN	3.31	3.12	46.0928
8	1194	238	TiAlN	3.41	3.98	57.1359
9	1250	151	TiAlN	3.45	2.10	37.9503
10	1250	177	ZrN	2.97	2.28	44.4847
11	1250	192	TiN	2.69	2.32	48.2546
12	1250	238	TiCN	2.21	2.92	59.8156
13	1432	151	ZrN	2.89	1.61	43.4758
14	1432	177	TiAlN	2.71	1.81	50.9617
15	1432	192	TiCN	2.08	1.90	55.2805
16	1432	238	TiN	2.31	1.95	68.5248

In the proposed work, the recognized machining factors and their choice of standards are stated in Table 2. The demonstration features are determined by performing 16 trials as shown in Table 3. The drilling processes are accompanied under the dry condition on VMC as shown in Fig. 1.

Unevenness is the portion of quality of an apparent. It is computed by straight up nonconformities of the actual exterior from its model system. The measurement of exterior irregularity, Mitutoyo 178-561-02A irregularity tester is utilized to discover

Fig. 1 Experimental setup

the Ra value for each drill. The efficiency of some machining processes principally hinges on its percentage subtraction of material. The percentage subtraction of material for a hole-making process is specified by the combination of speed, feed rate, and tool material. Material removal rate in hole-making process is clear as the bulk of the material that is detached per unit time in cm^3/min [2].

$$MRR = \frac{\pi}{4}d^2 * fd * N \tag{1}$$

where material subtraction rate in cm^3/min, N is the speed (rpm), f is the feed rate (mm/min), and diameter of a drill bit (d) is in mm.

3 Methodology

The multi-criteria decision making (MCDM) denotes in building excellent of the superlative substitute between a restricted bunch of conclusion and another possibility in relationships of numerous, usually inconsistent benchmarks. The technique for order of preference by similarity to ideal solution (TOPSIS) was at the beginning developed by Hwang and Yoon. The traditional method depends on data on a characteristic from pronouncement fabricator, mathematical information; the answer is meant at estimating, listing, and choosing, and the only particular inputs are the load. TOPSIS permits trade-offs among benchmarks, where an unfortunate outcome in a single condition can be invalid by a decent outcome in additional condition. In multi-criteria decision-making method also analytic hierarchy process (AHP) is one was originally developed by Prof. Thomas L. Saaty. In miniature, it is a process to descend proportion balances from corresponding contrasts and used to conclude the weight conditions of numerous qualities. Hence, the aids of together are occupied and offered as a shared AHP and TOPSIS system. The succeeding stages are employed to decide the best substitutes in shared TOPSIS and AHP method [3, 4].

Stage 1: The strategy and the significant assessment characteristics are measured. For this specific case, MRR is maximized because measured as advantageous characteristic, while exterior irregularity and tool wear are minimized because measured as non-beneficial characteristics.

Stage 2: Decision matrix used to represent all the information.

$$D_{16 \times 3} = \begin{bmatrix} 4.71 & 2.92 & 30.0262 \\ 3.09 & 3.25 & 35.1963 \\ 3.28 & 3.59 & 38.1790 \\ 3.05 & 3.91 & 47.3261 \\ 3.01 & 2.83 & 36.2500 \\ 3.15 & 3.24 & 42.4918 \\ 3.31 & 3.12 & 46.0928 \\ 3.41 & 3.98 & 57.1359 \\ 3.45 & 2.10 & 37.9503 \\ 2.97 & 2.28 & 44.4847 \\ 2.69 & 2.32 & 48.2546 \\ 2.21 & 2.92 & 59.8156 \\ 2.89 & 1.61 & 43.4758 \\ 2.71 & 1.81 & 50.9617 \\ 2.08 & 1.90 & 55.2805 \\ 2.31 & 1.95 & 68.5248 \end{bmatrix} \tag{2}$$

The decision matrix $D_{16 \times 3}$ displayed by Eq. (2)
The regularized matrix M_{ij} is obtained by using the next equation,

$$M_{ij} = \frac{V_{ij}}{\left[\sum_i^j d_{ij}^2 \right]^{0.5}} \tag{3}$$

Equation (3) is used to calculate regularized decision matrix $M_{16 \times 3}$, given as

$$M_{16 \times 3} = \begin{bmatrix} 0.3826 & 0.2580 & 0.1584 \\ 0.2510 & 0.2871 & 0.1857 \\ 0.2664 & 0.3171 & 0.2015 \\ 0.2477 & 0.3454 & 0.2497 \\ 0.2445 & 0.2500 & 0.1913 \\ 0.2559 & 0.2862 & 0.2242 \\ 0.2689 & 0.2756 & 0.2432 \\ 0.2770 & 0.3516 & 0.3015 \\ 0.2802 & 0.1855 & 0.2003 \\ 0.2412 & 0.2014 & 0.2347 \\ 0.2185 & 0.2049 & 0.2546 \\ 0.1795 & 0.2580 & 0.3156 \\ 0.2347 & 0.1422 & 0.2294 \\ 0.2201 & 0.1599 & 0.2689 \\ 0.1690 & 0.1678 & 0.2917 \\ 0.1876 & 0.1723 & 0.3616 \end{bmatrix} \qquad (4)$$

Regularized decision matrix $M_{16 \times 3}$ shown by the Eq. (4).

Stage 3: The multiplication of associated weights to the normalized decision matrix provides weighted regularized decision matrix.

$$V_{ij} = M_{ij} * W_j \qquad (5)$$

where M_{ij} is the regularized matrix and W_j is the weight conditions.

The AHP process is used to calculate weight (W_j) of each condition that detailed procedure is given below,

(A) A pairwise contrast matrix is built to define the comparative significance of dissimilar characteristics with respect to the strategy.

A typical pairwise contrast matrix (Q) can be stated as,

$$D_{16 \times 3} = \begin{bmatrix} q11 & q12 & \ldots & q1n \\ q21 & q22 & \ldots & q2n \\ . & . & \ldots & . \\ . & . & \ldots & . \\ qn1 & qn2 & \ldots & qmn \end{bmatrix}$$

where q_{ij} (for $i, j = 1, 2, 3. \ldots n$) is the preferences strength of the q_i over q_j corresponding to the criterion ($a_{ij} = a_i/a_j$), also $q_{ji} = 1/q_{ij}$ and $q_{ii} = 1$ for all values of i and j. The values of q_i and q_j were taken from the ratio scale for pairwise contrast.

Pairwise contrast matrix

$$Q = \begin{bmatrix} 1 & 5 & 1 \\ 1/5 & 1 & 1/3 \\ 1 & 3 & 1 \end{bmatrix}$$ (6)

(B) The relative priorities for each decision alternative are estimated by adding the numbers in each pillar of the pairwise contrast matrix, secondly division individual component in the pairwise matrix by its pillar sum. The subsequent matrix is mentioned as the regularized pairwise contrast matrix. Finally, calculate the normal of the components in each row of the regularized matrix. These mean to provide an approximation of the comparative significances of the components being equated [5].

(C) Calculate the value of λ_{max}, is the average of values computed by dividing the components of the direction of subjective calculations gained by multiplying the individual value in the initial pillar of the pairwise contrast matrix by the comparative importance of the initial item measured. Identical trials for further substances. Total the values crosswise the rows to achieve a direction of standards characterized "weighted totality," by the conforming importance worth.

(D) Calculate the index of consistency (CI)

$$CI = \frac{\lambda_{max} - n}{n - 1}$$

where the quantity of substances being equated is n.

(E) Calculate the consistency ratio (CR):

$$CR = \frac{CI}{RI}$$

where RI is the arbitrary index, which is the consistency index of arbitrarily generated pairwise contrast matrix.

The regularized weights of each characteristic are; $W_{Ra} = 0, 48$; $W_{Tw} = 0.12$ and $W_{MRR} = 0.40$. The assessment of λ_{max} is 3.03 and ratio of consistency (CR) = 0.0344, which is abundant fewer than the permissible CR worth of 0.1. Thus, there is a decent uniformity in the decision of comparative significance matrix.

Equation (5) used to calculate weighted normalized value $W_{18 \times 3}$ is given as,

$$D_{16\times 3} = \begin{bmatrix} 0.1836 & 0.0310 & 0.0634 \\ 0.1205 & 0.0345 & 0.0743 \\ 0.1279 & 0.0381 & 0.0806 \\ 0.1189 & 0.0414 & 0.0999 \\ 0.1174 & 0.0300 & 0.0765 \\ 0.1228 & 0.0343 & 0.0897 \\ 0.1291 & 0.0331 & 0.0973 \\ 0.1330 & 0.0422 & 0.1206 \\ 0.1345 & 0.0223 & 0.0801 \\ 0.1158 & 0.0242 & 0.0939 \\ 0.1049 & 0.0246 & 0.1019 \\ 0.0862 & 0.0310 & 0.1263 \\ 0.1127 & 0.0171 & 0.0918 \\ 0.1057 & 0.0192 & 0.1076 \\ 0.0811 & 0.0201 & 0.1167 \\ 0.0901 & 0.0207 & 0.1446 \end{bmatrix} \tag{7}$$

Weighted regularized matrix $R_{16\times 3}$ shown by Eq. (7)

Stage 4: The ideal and adverse ideal answers are calculated by Eqs. (8) and (9)

- Ideal solution.

$$D^* = \{x_1^* \ldots \ldots X_n^*\} \tag{8}$$

where, $X_j^* = \{\max(X_{ij})$ if $j \in J$; $\min(X)$ if $j \in J'\}$

- Adverse ideal solution

$$D' = \{x_1' \ldots \ldots X_n'\} \tag{9}$$

where, $V' = \{\min(X_{ij})$ if $j \in J$; $\max(X_{ij})$ if $j \in J'\}$
$J = 1, 2, 3\ldots, n$, where J is linked with the profit conditions.
$J' = 1, 2, 3\ldots, n$, where J' is linked with the charge conditions.
$D_{Ra}^* = 0.0811$, $D_{Tw}^* = 0.0171$, $D_{MRR}^* = 0.1446$
$D_{Ra}' = 0.1836$, $D_{Tw}' = 0.0422$, $D_{MRR}' = 0.0634$

Stage 5: Parting trials for each substitute is calculated.
The parting from the ideal substitute is;

$$P^* = \left[\sum (x_j^* - X_{ij})^2\right]^{1/2}, \quad i = 1\ldots m.$$

Likewise, the parting from the adverse ideal substitute is;

Table 4 Parting measure of positive, negative ideal solutions, and comparative nearness value

Sr. no.	Separation quantity of positive ideal solution (P_i^*)	A Parting quantity of adverse ideal solution (P_i')	Comparative nearness (C_i^*)	Position
(i)	0.1315	0.1655	0.5572	15
(ii)	0.0824	0.1030	0.5556	16
(iii)	0.0820	0.1109	0.5749	13
(iv)	0.0634	0.1070	0.6279	7
(v)	0.0782	0.1006	0.5626	14
(vi)	0.0711	0.1080	0.6030	11
(vii)	0.0692	0.1161	0.6266	9
(viii)	0.0624	0.1281	0.6724	5
(ix)	0.0839	0.1190	0.5865	12
(x)	0.0618	0.1037	0.6266	8
(xi)	0.0495	0.0963	0.6605	6
(xii)	0.0235	0.0931	0.7985	2
(xiii)	0.0616	0.1016	0.6225	10
(xiv)	0.0445	0.1005	0.6931	4
(xv)	0.0281	0.0852	0.7520	3
(xvi)	0.0097	0.1105	0.9193	1

$$P^* = \left[\sum (x_j' - X_{ij})^2\right]^{1/2}, \quad i = 1 \ldots m.$$

Stage 6: Calculate the comparative nearness to the ideal solution;

$$CR_i^* = \frac{P_i'}{(P_i^* + P_i')}$$

For the better performance of the alternatives larger the CR_i^* value is preferred [6]. The parting quantity of positive, adverse ideal solution and comparative nearness value are mentioned in Table 4.

Stage 7: Position the comparative nearness value.

4 Results and Discussion

The current effort aims to improve the drilling factors by means of cohesive TOPSIS and AHP process for the respective strategy. The comparative nearness is maximum imperative quantity popular the cohesive TOPSIS and AHP process for investigating untried data. According to the TOPSIS method, the comparative nearness would have extreme worth to acquire best cutting surroundings is considered in this work. In this study, minimization and maximization performance characteristics are selected. To obtain the best machining result, the lesser distinctive for surface roughness, tool wear, and larger features for MRR have been occupied. Table 3 shows outcomes attained from the investigational runs and Table 4 displays statistics of comparative nearness for exterior irregularity, tool wear, and MRR.

The best blend of machining factors for reducing exterior roughness, tool wear, and exploiting MRR determined by comparative nearness. The maximum standards of MRR and the minimum values of exterior irregularity, tool wear is shown by extreme comparative nearness value and their values are the speed of 1432 rpm, feed rate of 238 mm/min and tool material of TiN coated carbide tool. As associated with other methods like genetic algorithm, artificial neural network, and gray relational analysis, this process is modest with fewer computational stages.

5 Conclusions

The work aims with minimization of surface roughness, tool wear and of MRR while drilling spheroidal graphite SG 500/7 using collective TOPSIS and AHP method to find the best blend of machining factors such as speed, feed rate, and tool material.

- Taguchi L16 orthogonal array is used to the strategy of these trials.
- Weight features elaborate for all replies calculated by analytic hierarchy process.
- The greatest blend of machining factors in the drilling process is selected by collective TOPSIS and AHP method. The possibility is organized in a positionwise set with their comparative nearness as 15-16-13-7-14-11-9-5-12-8-6-2-10-4-3-1.
- Choices are given based on relative closeness value.
- Optimum drilling process parameter such as speed, feed rate, and tool material obtained by collective TOPSIS and AHP gives minimization of surface roughness, tool wear, and significant improvement in MRR.
- Collective TOPSIS and AHP method consider weight benchmarks for enhanced and precise estimation of the choices.

References

1. Ramesh, V., Mohammed, H., Gopinath, T.: Taguchi method based optimization of drilling parameters of EN31 steel with PVD coated and uncoated drills. J. Manuf. Eng. **10**, 112–116 (2015)
2. Pradeep Kumar, J., Packiaraj, P.: Effect of drilling parameters on surface roughness, tool wear, material removal rate and hole diameter error in the drilling of OHNS. Int. J. Adv. Eng. Res. Stud. 150–154 (2012). E-ISSN2249–8974
3. Balasubramaniyan, S., Thangiah, S.: Optimization of machining parameters in turning operation using combined TOPSIS and AHP method. Tech. Gaz. **22**, 1475–1480 (2015). https://doi.org/10.17559/TV-20140530140610

4. Sreenivasulu, R., Srinivasa Rao, C.: Optimization of surface roughness, circularity deviation and selection of different aluminium alloys during drilling for automotive and aerospace industry. Independent J. Manag. Prod. **7**, 413–430 (2015). https://doi.org/10.14807/ijmp.v7i2.414
5. Bhanu Prakash, D., Krishnaiah, G.: Optimization of process parameters using AHP and VIKOR when turning AISI 1040 steel with coated tools. Int. J. Mech. Eng. Technol. **8**, 241–248 (2017). ISSN Online: 0976-6359
6. Anand Babu, K., Venkataramaiah, P., Dileep, P.: AHP-DENG'S similarity based optimization of WEDM process parameters of Al/SiCp composite. Am. J. Mater. Sci. Technol. **6**, 1–14 (2017). https://doi.org/10.7726/ajmst.2017.1001
7. Venkata Rao, R.: Decision Making in the Manufacturing Environment: Using Graph Theory and Fuzzy Multiple Attribute Decision Making Methods. Springer London Ltd, England, UK (2007)

Optimal Configuration of Viscous Dampers Connected to Adjacent Similar Buildings Using Particle Swarm Optimization

Uppari Ramakrishna, S. C. Mohan and K. Sai Pavan Kumar

Abstract Controlling the seismic response of adjacent buildings coupled with straight damper configuration is efficient when the buildings have different dynamic properties. If the dynamically similar buildings are connected with straight dampers, this method will become inefficient. So, the configurations other than the straight connection have to be used in case of adjacent similar buildings. It is not necessary to connect dampers on all floors, but there exists an optimal location. When the dampers are provided at optimal locations, then the response reduction is almost comparable to that of building with dampers provided at all the floors. This will save the cost of dampers required to reduce the seismic response of the buildings. In this study, one of the nature-inspired optimization approaches, particle swarm optimization (PSO), is used for the optimal configuration of viscous dampers to connect ten-storied adjacent similar RC buildings. The present study shows that the provided optimal dampers reduce the dynamic response of both the buildings simultaneously, instead of providing dampers for each building separately.

Keywords Dynamically similar building · Viscous damper · Particle swarm optimization · Optimal damper configuration

1 Introduction

We cannot stop the earthquake but we can reduce the vulnerability of buildings during an earthquake. All the buildings are not collapsed by the earthquake, but some buildings can survive provided they are designed to withstand. Also, it is very

U. Ramakrishna (✉) · S. C. Mohan · K. S. Pavan Kumar
Department of Civil Engineering, BITS Pilani Hyderabad Campus, Hyderabad, India
e-mail: p20170409@hyderabad.bits-pilani.ac.in

S. C. Mohan
e-mail: mohansc@hyderabad.bits-pilani.ac.in

K. S. Pavan Kumar
e-mail: h20171430033@hyderabad.bits-pilani.ac.in

© Springer Nature Singapore Pte Ltd. 2020
R. Venkata Rao and J. Taler (eds.), *Advanced Engineering Optimization Through Intelligent Techniques*, Advances in Intelligent Systems and Computing 949,
https://doi.org/10.1007/978-981-13-8196-6_71

important that the buildings have to be designed and constructed properly in order to mitigate the earthquake damage. Therefore, the performance of the building during an earthquake has to be studied well in order to reduce the damage in an economical way through the optimization technique. A group of buildings which serves the same facility, like educational buildings, institutional buildings, residential buildings, etc. are built adjacent to each other and have the same structural property. If seismic protection is to be provided individually in each of the building, then it will be a costly issue [1]. Instead, adopting the connected vibration control method that controls the response of both the buildings simultaneously prove to be cost-effective by providing more number of dampers. Many research works have been carried out under the coupled technique for dissimilar buildings [2–4] classified based on the factors: Number of degrees of freedom, type of the control technique adopted; type of the dampers used; type of excitation; type of damper configuration; number of buildings coupled, etc. In order to adopt the coupling technique in an economical way, optimization of the damper is necessary [5, 6]. The optimal configuration and mechanical properties of dampers connected to individual buildings using different optimization algorithms has been done studied [7, 8]. Two multi-degree-of-freedom (MDOF) structures of similar buildings are connected with optimal viscous dampers [9].

The present study involves connecting optimal damper to the adjacent similar building by using one of the nature-inspired algorithms, PSO technique. It was implemented by Russell Eberhart and Kennedy in the year 1995 [10]. Firstly, it was run by developing computer software simulation of birds gathering around food sources and then after realized how well this technique worked on the optimization problem. PSO is really a very simple algorithm to program and apply on engineering problems. Over a number of iterations, a group of variables has their values adjusted neared to the particle whose value is closest to the target at any given moment. The same concept of gathering birds in an optimal way toward their destination is implemented to get the optimal dampers configuration connected in single building [11]. In this study, PSO technique is used as connecting the optimal damper to the adjacent similar buildings to reduce the dynamic response.

2 Methodology

The numerical study involves damper connected control technique for ten-storied dynamically similar RC Buildings as shown in Fig. 1. The objective is to couple them with optimal viscous damper configuration by optimization techniques with the help of the MATLAB R2015a programming. The design and analysis of structure are done as per the IS 1893-2016 part-1 [12] with the help of SAP2000. Lumped mass model has been considered for the time history analysis of El Centro earthquake ground acceleration, which is carried out through the numerical method—Newmark's beta method. PSO technique is used to find the optimal configuration and optimal damper

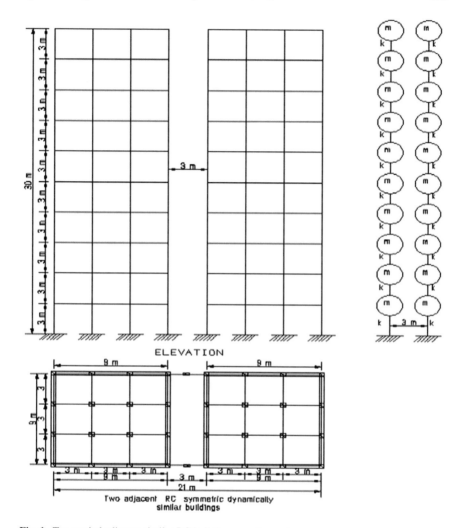

ELEVATION

Two adjacent RC symmetric dynamically similar buildings

Fig. 1 Ten-storied adjacent similar RC buildings as a lumped mass

coefficient to connect adjacent building. The step-by-step methodology involved in getting optimal damper configuration is given below.

2.1 Equation of Motion for the Coupled Buildings

The equation of motion of the coupled system is given by

$$M\ddot{X} + (C + C_{\mathrm{D}})\dot{X} + KX = -MI\ddot{x}_{\mathrm{g}} \tag{1}$$

Fig. 2 Lumped mass model of adjacent similar buildings with straight damper connectors

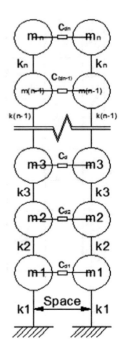

where $[M] = \begin{bmatrix} [m] & [0] \\ 0 & [m] \end{bmatrix}$ is the mass matrix, $[K] = \begin{bmatrix} [k] & [0] \\ [0] & [k] \end{bmatrix}$ is the stiffness matrix,

and $[C] = \begin{bmatrix} [c] & [0] \\ [0] & [c] \end{bmatrix}$ is the damping matrix of the coupled building. Here, C_D is the damping matrix for the dampers provided between the similar buildings. The C_D matrix formulation depends entirely on the configuration of the dampers which are connected in between the two buildings. Depending on the nodes to which ends of the dampers are connected (Fig. 2), the external damping matrix is formulated. Further, I = unit vector, $\ddot{x}_g$ = earthquake ground acceleration .

2.2 Partial Swarm Optimization

The PSO is a simple algorithm which can be implemented easily. This algorithm preserves three global variables, namely target value, global best (gBest) value showing which particle from all iterations is closer to the target value, and stopping value indicating the algorithm to stop when target reached. Here, each particle consists of data representing possible solutions. In the present study, the numbers of dampers are limited from all possible locations and damping value is restricted within a range. The moving velocity of birds, which can be obtained through relative displacement, represents how much the each particle's data can be improved. The personal best (pBest) value indicates the closest particle's data have ever come to the target in a particular iteration. The each particles' data representing each bird in the swarm could be located anywhere within the search space. In the gathering bird's example above the particles, data would be the X, Y, Z coordinates of each bird [11].

The same concept of flying birds or fish schooling can be adopted for finding the best possible location of dampers to connect the dynamically similar buildings. The flow diagram for the application of PSO on building structure is shown in Fig. 3. The minimization of roof displacement is taken as the objective function. A limited number of dampers and damping coefficient range are taken as constraints. This PSO technique is used on a single objective function with constraints of non-linear optimization.

Particle swarm optimization Pseudo code [10]

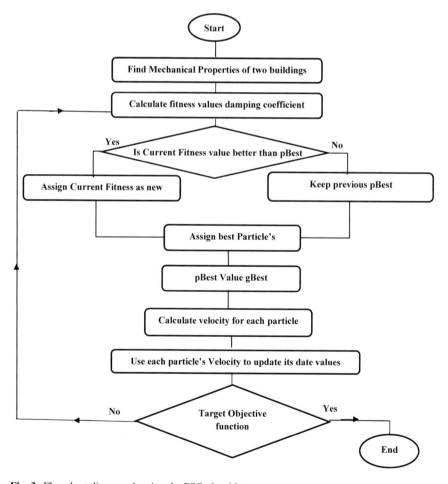

Fig. 3 Flowchart diagram showing the PSO algorithm

```
clc;
clear all;
m=28;                                        %No of configuration
objective function=@(Q) funct(Q);            % minimize the top floor displacement
nVar =m+1;
VarSize= [1 nVar];                           % variable size
Climit=[1e5;1e14];                           % Damping Coefficient
MaxIt=30;                                     % Maximum iteration
nPop=300;                                     % Population size
w=0.7; wdamp=0.99; c1=2.1; c2=1.9;           % Parameters
empty_particle. Position =[ ];               % Initialize locations (x)
empty particle. Velocity =[ ];               % Initialize velocity (v)
empty_particle. objt=[ ];
empty_particle.Best.Position=[ ];
empty_particle. Best. Objt=[ ];
particle=repmat(empty_particle,nPop,1);
GlobalBest.objt=inf;
While (criterion)
        For loop over all n particles and all the dimensions
        Generate new velocity
        Using equation from the relative displacement
        Calculate new locations
        Evaluate objective functions at new locations
        Find the current best for each particle
        End for
        Find the current global best
        Update (pseudo time or iteration counter)
End while
Output the final results
location=particle(f). Best.Position(1,1:28);
disp (['Damper locations= ', num2str(location)]);
loct=damper_location_function_file(location);
disp(loct);
opti_damp=particle(f).Best.Position(29);
disp(['Cd value=',num2str(opti_damp)]);
```

The PSO technique is found to reduce the computational time in finding the optimal parameters of the dampers along with its optimal location compared to point by point search method (by considering all the combinations). Hence, the PSO technique is employed for obtaining the optimal location of the dampers by limiting a number of dampers to be used for coupling. Also, the optimal damping coefficient of each damper for the given location of the damper is obtained through PSO. At first, the optimal location for the given number of dampers is obtained by considering all possible damper configurations, while keeping the same damping coefficient for all dampers. Then, the optimal damping coefficients to be provided are obtained at damper locations. PSO technique is adopted in which the search space is taken as a set of all the possible location at all the floors of the building. Then, by limiting a number of dampers and by considering all the configurations of the dampers, optimal locations are obtained. Finally, the different optimal damping coefficients for each damper provided at their locations are obtained using PSO.

3   Result and Discussion

In order to study the effect of optimal configuration of dampers connected to similar buildings for reducing the seismic response, at first, the dampers were provided throughout the building with various configurations as shown in Fig. 4. Modal analysis is done for the first four cases using SAP2000 and optimization valve obtained from the MATLAB R2015a code. The optimal damping coefficient is obtained by varying the damping coefficient value from zero to 1×10^{10} N-s/m. It is observed that after a particular damping coefficient range of 1×10^8 N-s/m, the seismic response of building remains unchanged. For this reason, the damping coefficient value of 1×10^6 N-s/m is chosen for the first four cases to see their effect on the variation of response. All these four cases are provided with dampers at all the location. But, when the number of dampers limited to three, the optimal damping coefficient obtained is 3.78×10^8 N-s/m. Figure 5 shows the top floor displacement response of building with each damper configuration case. From Fig. 5, it is understood that there exists an optimal damping coefficient value with a limited number of dampers at which the response is minimum.

When the damping coefficient value is 1×10^6 N-s/m, 71% reduction in the roof displacement is achieved in case-3 and case-4. There is no significant change observed in case-2. Later, the particle swarm optimization is employed for obtaining the optimal location of the damper when the numbers of dampers are to be fixed. Also, the optimal damping coefficient of each damper for the given location is obtained.

After optimization using PSO, the optimal locations are as shown in case-5 of Fig. 5. It is observed that the response was reduced by 82.45% compared to case-1. The El Centro earthquake response of uncoupled building and buildings coupled with optimal dampers connection (case-5) is shown in Fig. 6. Hence, providing the optimal dampers and their locations can be economical.

4   Conclusions

This study addresses the effect of various damper configurations on seismic response reduction of dynamically similar buildings. It is observed that straight damper connection for similar buildings is not effective. PSO technique is implemented in this study to identify the optimal damper location. Providing dampers to all the floors is uneconomical. Even with a limited number of dampers, the response can be reduced

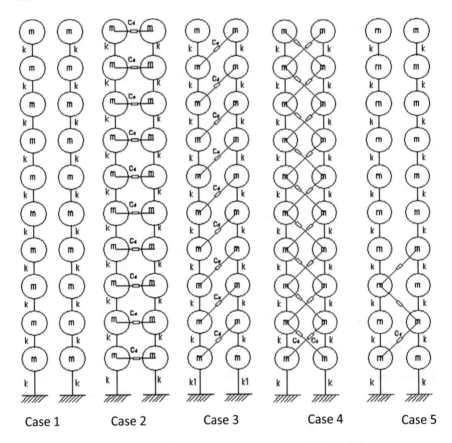

Fig. 4 A different configuration of damper connected to adjacent similar buildings

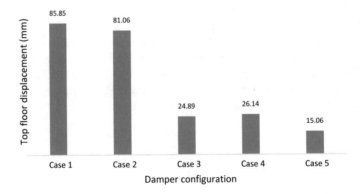

Fig. 5 Response curve of different configuration of damper connected to adjacent similar buildings

Fig. 6 Displacement response of uncoupled building and buildings coupled with optimal dampers connection (case-5)

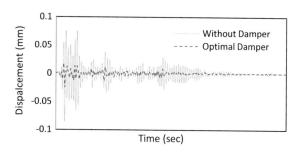

efficiently provided the optimal locations are chosen. The optimized locations of a limited number of dampers are obtained using PSO. There is a considerable reduction in the dynamic response of both the similar buildings with these optimal damper locations. Hence, it is concluded that it is not necessary to connect "n" degree of freedom adjacent similar buildings at all the floors, but few dampers at optimal locations are sufficient.

Acknowledgements The effort presented in this work is funded by the "Department of science and technology (DST)-Science and Engineering Board (SERB), India."

References

1. Daisuke, A., John, E.B., Stefano, B.: Design optimization of passive devices in multi-degree of freedom structures. In: 13th World Conference on Earthquake Engineering, Canada, Paper No. 1600, 1–6 August 2004
2. Athanassiadou, C.J., Kappos, A.J., Penelis, G.G.G.: Seismic response of adjacent buildings with similar or different dynamic characteristics. Earthq. Spectra **10**(2), 293–317 (1994)
3. Cimellaro, G., De Angelis, M: Theory ed experimentation on passive control of adjacent structures. In: Proceedings of the 13th World Conference on Earthquake Engineering, no. 1837 (2004)
4. Matsagar, V.A., Jangid, R.S.: Viscoelastic damper connected to adjacent structures involving seismic isolation. J. Civ. Eng. Manag. **11**(4), 309–322 (2005)
5. Tande, S.N., Krishnaswamy, K.T., Shinde, D.N.: Optimal seismic response of adjacent coupled buildings with dampers. J. Inst. Eng. Civ. Eng. Div. **90**(Nov), 19–24 (2009)
6. Makita, K., Christenson, R.E., Asce, M., Seto, K., Watanabe, T.: Optimal design strategy of connected control method. **133**(12), 1247–1257 (2008)
7. Bigdeli, K., Hare, W., Nutini, J., Tesfamariam, S.: Optimizing damper connectors for adjacent buildings. 1–25 (2015)
8. Hadi, M.N.S., Uz, M.E.: Improving the dynamic behaviour of adjacent buildings by connecting them with fluid viscous dampers. Building, no. June, 22–24 (2009)
9. Taylor, P., Patel, C.C., Jangid, R.S.: Seismic response of dynamically similar adjacent structures connected with viscous dampers. IES J. Part A : Civ. Struct. Eng., no. December 2014, 37–41 (2010)
10. Yang, X.: Nature-Inspired Optimization Algorithms (2014)
11. Zahrai, S.M., Akhlaghi, M.M., Rabipour, M.: Application of Particle Swarm Optimization for improving seismic response of structures with MR dampers. In: Proc. Int. Conf. Noise Vib. Eng. ISMA 2012, no. November 2014, pp. 441–448 (2012)
12. Indian Standard code: Earthquake resistant design of structures. Part-1, 1893–2016

Combined Economic Emission Dispatch Using Spider Monkey Optimization

Anand Singh Tomar, Hari Mohan Dubey and Manjaree Pandit

Abstract In this paper, combined economic emission dispatch (CEED) solution is presented using a newly proposed swarm intelligence-based spider monkey optimization (SMO). The model of SMO is based on collective behavior, decentralization and self organizing. The spider monkeys work in groups and are segregated in small groups with adult female as subgroup leader during search for their food. In general, spider monkey follows fission–fusion process during foraging. Here, mathematical mode of SMO is implemented for the solution of optimal dispatch problem having cubic cost and emission function to reduce the green house effect. To validate efficacy of SMO performance, comparison of simulation results are also made with other reported techniques available in literature.

Keywords Spider monkey optimization · Combined economic emission dispatch · Cubic cost function

Nomenclature

$Fc(P_g)$	Total fuel cost
a_i, b_i, c_i, d_i	Fuel cost coefficients
$\alpha_i, \beta_i, \gamma_i, \delta_i$	Emission coefficient
$\mathcal{P}_d$	Power demand
$Ec(P_g)$	Total emission released
TC	Total Cost

A. S. Tomar · H. M. Dubey (✉) · M. Pandit
Madhav Institute of Technology & Science, Gwalior, MP, India
e-mail: harimohandubeymits@gmail.com

A. S. Tomar
e-mail: sandyanand99@gmail.com

M. Pandit
e-mail: drmanjareep@gmail.com

© Springer Nature Singapore Pte Ltd. 2020
R. Venkata Rao and J. Taler (eds.), *Advanced Engineering Optimization Through Intelligent Techniques*, Advances in Intelligent Systems and Computing 949,
https://doi.org/10.1007/978-981-13-8196-6_72

809

gll	Global leader limit
lll	Local leader limit

1 Introduction

Electric power generation significantly depends on fossil fuel resources. Considering green house effect and global warming, nowadays researchers are paying more attention to areas where pollution reduction is an objective, hence the trend of economic emission dispatch. The key issue related with power generation using fossil fuel is to reduce total operational cost as well as hazardous emission content too. Both objectives cost and emission are conflicting in nature and difficult to solve due to additional operational constraints of generators. These problems can be solved by two methods, conventional method and stochastic biological inspired algorithm. Conventional method includes linear programming method [1], quadratic programming [2], Newton's method [3], etc. However, these methods are found to be not much effective for solution of complex constrained practical problems related to power system operation. Second optimization methods biologically inspired, generally they mimic the natural phenomenon in their analytical model inspired by evolution, swarm intelligence, ecological process, human intelligence or by physical science [4]. Among these category GA [5], bacterial foraging algorithm (BFA) [6], artificial bee colony (ABC) [7], invasive weed optimization (IWO) [8], (PSO) [9], backtracking search algorithm (BSA) [10], flower pollination algorithm (FPA) [11], stochastic fractal search algorithm (SFSA) [12], exchange market algorithm (EMA) [13], mine blast algorithm(MBA) [14], etc. are successfully applied to solve different types of environmental economic scheduling problems. A survey of various optimization techniques for environmental economic dispatch can be found in Ref. [15].

Here, a newly proposed biologically inspired swarm intelligence-based spider monkey optimization (SMO) is implemented to solve CEED problem with cubic cost function. Spider monkeys work in group and divide themselves in small group led by adult female spider monkey during food searching. Unique fission–fusion and unique technique of information sharing are the basic backbone of SMO. Here, formulation of problem is presented in Sect. 2, and idea behind SMO is presented in Sect. 3. Implementation process and outcome of simulation results are presented in Sect. 4, and finally conclusion is presented in Sect. 5.

2 Formulation of Problem

Minimization of total operating cost associated with committed power generating units and fulfill the required power demand under the specified operational constraints are the main objective of economic dispatch.

For a cubic cost function, total fossil fuel cost used for minimization is presented as:

$$f_1 = \text{Fc}(P_g) = \sum_{i=1}^{N} (a_i P_i^3 + b_i P_i^2 + c_i P_i + d_i) \tag{1}$$

P_i is the power output of ith unit and N is total number of generators.
It is subjected to the following operational constraints.

$$P_i^{\max} \leq P_i \leq P_i^{\min} \tag{2}$$

$$\sum_{i=1}^{N} P_i = P_d \tag{3}$$

The amount of emission released by burning of fossil fuel of thermal power generating units is directly related to power generation.

The economic emission dispatch (EED) problem described as minimization of emission contents released as in (4)

$$f_2 = \text{Ec}(P_g) = \sum_{i=1}^{N} (\alpha_i P_i^3 + \beta_i P_i^2 + \gamma_i P_i + \delta_i) \tag{4}$$

ED and EED are two dissimilar problems and also conflicting in nature. These problems combined together in a single objective using price penalty factor (h) as in (5)

$$\text{TC} = \text{Fc}(P_g) + h \times \text{Ec}(P_g) \tag{5}$$

The above equation must fulfill associated operational constraints (2) and (3).

The trade-off between conflicting objectives can be achieved using weighted sum approach technique as in (6):

Minimize

$$f_3 = \text{TC} = w \times \text{Fc}(P_g) + (1 - W) \times h \times \text{Ec}(P_g) \tag{6}$$

The weight factor (w) considered as 1 for ED problem and 0 for EED problem.
The price penalty factor (h) can be computed as in (7):

$$h_i = \frac{\sum_{i=1}^{N} (a_i P_i^3 + b_i P_i^2 + c_i P_i + d_i)}{\sum_{i=1}^{N} (\alpha_i P_i^3 + \beta_i P_i^2 + \gamma_i P_i + \delta_i)} \tag{7}$$

3 Spider Monkey Optimization

Spider Monkey optimization (SMO) is a population-based optimization approach proposed by J. C. Bansal in 2014 [16]. Spider monkey generally lives in large groups called parent group. In order to minimize competition for searching their food (foraging), they divide themselves in small groups. The subgroup members start searching food in different direction, share information regarding place and its quality of food source within group and outside the group. Analytical model of SMO has six main phases which are explained below.

Population is initialized within lower and upper bonds as (8):

$$x_{ij} = x_j^{\min} + \Re(0, 1) \times (x_j^{\max} - x_j^{\min}) \tag{8}$$

A. Local leader phase (*llp*):

It depends on the remarkable observation made by female local leader as well as remaining members of the group. Here, each member (x) updates their position as per the value of fitness. If the current fitness is found to be better, spider monkey (x) updates their position otherwise they retain their previous position as per (9).

$$x_{ij}^{\text{new}} = x_{ij} + \Re(0, 1) \times (ll_{kj} - x_{ij}) + \Re(-1, 1) \times (x_{rj} - x_{ij}) \tag{9}$$

where x_{ij} represents the ith spider monkey in jth dimension, ll_{kj} is the influence of kth local leader in jth dimension and x_{rj} represents the randomly selected rth spider monkey from kth group in such a way that for jth dimension r not equal to i.

B. Global leader phase (*glp*):

Similar to above *llp* in *glp* also depending on observation of global leader members of local group, spider monkey (x) updates their location as in (10):

$$x_{ij}^{\text{new}} = x_{ij} + \Re(0, 1) \times (gl_j - x_{ij}) + \Re(-1, 1) \times (x_{rj} - x_{ij}) \tag{10}$$

Here gl_j is the global leader position in jth dimension, $j = 1, 2, 3\text{--}D$ is the randomly selected index. Spider monkey (x_i) updates their position as per probability (ρ), which can be computed and calculated using fitness function as in (11).

$$\rho_i = 0.9 \times \frac{\text{fit}_i}{\text{fit}_{\max}} + 1 \tag{11}$$

Now the fitness value of newly generated position are calculated and compared with old one. If it is found to be better, then it is preserved.

C. **Global leader learning phase (*gllp*):**

Here, greedy selection procedure is adopted for updation of global leader position. The current position global leader is also verified here; if the current position is not updated, *glc* is increased by 1.

D. **Local leader learning phase (*lll*):**

Here also greedy selection is applied to find out the best position of local leader similar to above. If the current position of local leader is not found to be better than previous one, then the *llc* is increased by 1.

E. **Local leader decision phase (*lldp*):**

If the position of local leader is found to be un-updated within predefined number of iteration called local leader limit (*lll*), then all spider monkey (solution) updates their position randomly combined information of local and global leader as in (12).

$$x_{ij}^{\text{new}} = x_{ij} + \Re(0, 1) \times (gl_j - x_{ij}) + \Re(0, 1) \times (x_{ij} - ll_{kj}) \tag{12}$$

F. **Global leader decision phase (*gldp*):**

If position of global leader is not updated within predefined iteration, i.e., *glc*, then the leader divides the population in small subgroup till the achievement of group of maximum number (GM).

The optimization procedure using SMO is depicted using flowchart in Fig. 1.

4 Simulation Results and Discussion

The SMO approach is applied to solve CEED problem in MATLAB environment and executed on CPU having configuration as core 2 duo processor, 2.10 GHz and 2 GB RAM. It is tested on a standard test case having six thermal power generating unit systems with cubic cost and emission function. The cost coefficient, emission coefficient and upper and lower power generation limit are adopted as per [17]. The load demands of the system are considered as 150 and 250 MW. Transmission loss is not considered here.

The control parameters of SMO are considered as maximum group (MG) = NP/10, gll = NP/2, lll = NPXD and probability (ρ) = {0.1, 0.9} where NP is the population size. The effect of population is investigated with different size over thirty repeated runs and its effect is plotted in Fig. 2. Here, it is observed that with NP = 100 and 150, SMO can attain lowest average cost. However, NP = 100 has low CPU time too; therefore, it is considered for simulation analysis.

The best cost, best emission and best compromise solution (CEED) along with corresponding optimum generation schedule were tabulated in Tables 1 and 2 for power demand of 150 and 250 MW, respectively. While comparing the result of

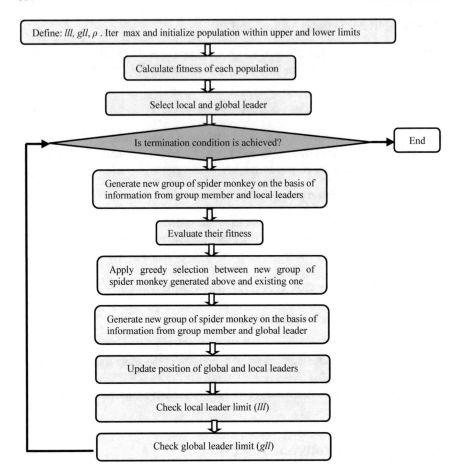

Fig. 1 Flow chart of SMO for constrained optimization

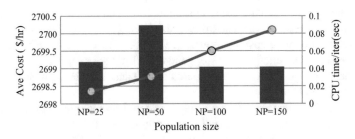

Fig. 2 Effect of population on cost and CPU time

Table 1 Economic dispatch (ED), economic emission dispatch (EED) and best compromise solution (BCS) for power demand of 150 MW

Objective	Method	P_1	P_2	P_3	P_4	P_5	P_6	Cost ($/h)	Emission (Kg/h)
ED	SMO	50.00	20.00	15.00	19.60	19.49	25.91	2699.04	2628.09
EED	SMO	50.00	28.58	19.14	30.28	10.00	12.00	2838.39	2415.12
CEED	SMO	50.00	20.00	15.00	32.53	16.41	16.06	2730.67	2497.31
CEED	QPSO [17]	50.00	20.00	15.0	22.9	20.4	22.3	2701.47	2583.64

Table 2 Economic dispatch (ED), economic emission dispatch (EED) and best compromise solution (BCS) for power demand of 250 MW

Objective	Method	P_1	P_2	P_3	P_4	P_5	P_6	Cost ($/h)	Emission (Kg/h)
ED	SMO	50.00	48.22	21.99	50.00	39.79	40.00	4962.56	4919.26
EED	SMO	50.00	55.06	44.83	50.00	26.58	23.53	5211.40	4686.95
CEED	SMO	50.00	50.83	33.04	50.00	30.96	35.17	5030.22	4757.49

CEED solution with QPSO [17], it is observed that SMO has slightly higher power generation cost 2730.67 $/h as compared to QPSO [17] 2701.47 $/h, however, emission content significantly reduced to 2497.31 kg/h in comparison with reported result of QPSO [17] 2583.64 kg/h. The smooth cost convergence curve obtained by SMO is plotted in Figs. 3 and 5.

The optimal pareto front obtained using (6) by assigning different weight is plotted in Figs. 4 and 6, respectively.

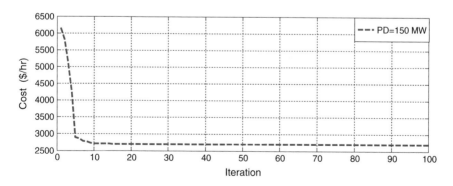

Fig. 3 Cost convergence characteristics of SMO with $P_d = 150$ MW

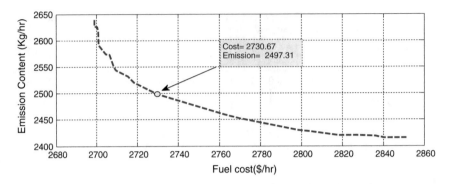

Fig. 4 Optimal pareto for six unit system ($P_d = 150$ MW)

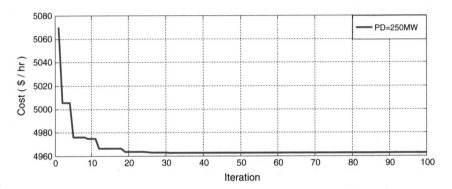

Fig. 5 Cost convergence characteristics of SMO with $P_d = 250$ MW

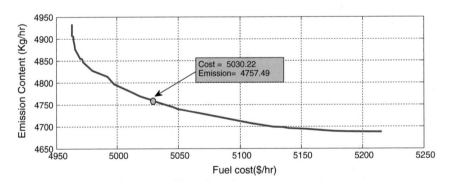

Fig. 6 Optimal pareto for six unit system ($P_d = 250$ MW)

5　Conclusion

In this paper, SMO, a newly developed swarm-based algorithm is implemented to solve CEED problem with cubic fuel cost and emission function. As per the results obtained for different objectives, it is clear that SMO has global search capability using its unique exploration and exploitation phases, and also it handles the constraints in an efficient manner. Further, it can be used for the solution of complex constrained real world problems.

Acknowledgements　The authors sincerely acknowledge the financial support provided by AICTE-RPS project File No. 8-36/RIFD/RPS/POLICY-1/2016-17 dated 2nd Sep 2017 and TEQIP III. The authors also thank the Director and management of M.I.T.S. Gwalior, India and IIT Delhi, India, for providing facilities for carrying out this work.

References

1. Farag, A., Baiyat, S.A., Cheng, T.C.: Economic load dispatch multi-objective optimization procedures using linear programming techniques. IEEE Trans. Power Syst. **10**, 731–738 (1995)
2. Ji-Yuan, F., Lan, Z.: Real-time economic dispatch with line flow and emission constraints using quadratic programming. IEEE Trans. Power Syst. **13**, 320–325 (1998)
3. Chen, S.D., Chen, J.F.: A direct Newton–Raphson economic emission dispatch. Electr. Power Energy Syst. **25**, 411–417 (2003)
4. Dubey, H.M., Pandit, M., Panigrahi, B.K.: An overview and comparative analysis of recent bio-inspired optimization techniques for wind integrated multi-objective power dispatch. Swarm Evol. Comput. **38**, 12–34 (2018)
5. Abido, M.A.: A Niched Pareto genetic algorithm for multi-objective environmental/economic dispatch. Electr. Power Energy Syst. **25**, 97–105 (2003)
6. Pandit, N., Tripathi, A., Tapaswi, S., Pandit, M.: An improved bacterial foraging algorithm for combined static/dynamic environmental economic dispatch. Appl. Soft Comput. **12**, 3500–3513 (2012)
7. Aydin, D., Özyön, S., Yasar, C., Liao, T.: Artificial bee colony algorithm with dynamic population size to combined economic and emission dispatch problem. Electr. Power Energy Syst. **54**, 144–153 (2014)
8. Panigrahi, B.K., Pandit, M., Dubey, H.M., Agarwal, A., Hong, W.C.: Invasive weed optimization for combined economic and emission problems. Int. J. Appl. Evol. Comput. **5**, 1–18 (2014)
9. Mandal, K.K., Mandal, S., Bhattacharya, B., Chakraborty, N.: Non-convex emission constrained economic dispatch using a new self-adaptive particle swarm optimization technique. Appl. Soft Comput. **28**, 188–195 (2015)
10. Bhattacharjee, K., Bhattacharya, A., Dey, S.H.N.: Backtracking search optimization based economic environmental power dispatch problems. Electr. Power Energy Syst. **73**, 830–842 (2015)
11. Dubey, H.M., Pandit, M., Panigrahi, B.K.: Hybrid flower pollination algorithm with time-varying fuzzy selection mechanism for wind integrated multi-objective dynamic economic dispatch. Renew. Energy **83**, 188–202 (2015)
12. Tyagi, T., Dubey, H.M., Pandit, M.: Multi-objective optimal dispatch solution of solar-wind-thermal system using improved stochastic fractal search algorithm. Int. J. Inf. Technol. Comput. Sci. **8**, 61–73 (2016)

13. Ghorbani, N., Babaei, E., Sadikoglu, F.: Exchange market algorithm for multi-objective economic emission dispatch and reliability. Proc. Comput. Sci. **120**, 633–640 (2017)
14. Ali, E.S., Abd Elazim, S.M.: Mine blast algorithm for environmental economic load dispatch with valve loading effect. Neural Comput. Appl. **30** (2018). https://doi.org/10.1007/s00521-016-2650-8
15. Qu, B.Y., Zhu, Y.S., Jiao, Y.C., Wu, M.Y., Suganthan, P.N., Liang, J.J.: A survey on multi-objective evolutionary algorithms for the solution of the environmental/economic dispatch problems. Swarm Evol. Comput. **38**, 1–11 (2018)
16. Bansal, J.C., Sharma, H., Jadon, S.S., Clerc, M.: Spider Monkey Optimization algorithm for numerical optimization. Memetic Comput. https://doi.org/10.1007/s12293-013-0128-0
17. Mahdi, F.P., Vasant, P., Al-Wadud, M.A., Rahman, M.M., Watada, J., Kallimani, V.: Quantum particle swarm optimization for multi objective combined economic emission dispatch problem using cubic criterion function. In: IEEE International Conference on Imaging, Vision & Pattern Recognition (icIVPR-2017). https://doi.org/10.1109/icivpr.2017.7890879

Solution of Non-convex Economic Dispatch Problems by Water Cycle Optimization

Vishal Chaudhary, Hari Mohan Dubey and Manjaree Pandit

Abstract This paper presents water cycle algorithm (WCA) for solving optimization of economic dispatch problems of modern power system. It is a newly developed algorithm which mimics the concept behind natural hydrological process and flow of water through rivers and streams to the ocean. Performance of WCA is tested on small- and medium-scale standard test cases having non-convex fuel cost characteristics of six and 13 thermal generating units. Comparative analysis is carried out to validate applicability and superiority over other reported methods. The new approach is found to provide optimal results for tested complex constrained optimization problems.

Keywords Water cycle optimization · Water stream · River · Ocean · Non-convex · Transmission loss

Nomenclature

a_i, b_i, c_i, e_i, f_i	Cost coefficients
$\mathcal{P}_D$	Power demand
$\mathcal{B}_{ij}, \mathcal{B}_{i0}, \mathcal{B}_{00}$	Loss coefficients
NP	Total Population size (rain drop)
D	Dimension of problem
N	Total no. of river + 1 (one sea)
N_{R_total}	Total no. of river
N_{S_total}	Total no. of water stream

V. Chaudhary · H. M. Dubey (✉) · M. Pandit
Madhav Institute of Technology & Science, Gwalior, MP, India
e-mail: harimohandubeymits@gmail.com

V. Chaudhary
e-mail: Vishal.chaudhary30@gmail.com

M. Pandit
e-mail: drmanjareep@gmail.com

© Springer Nature Singapore Pte Ltd. 2020
R. Venkata Rao and J. Taler (eds.), *Advanced Engineering Optimization Through Intelligent Techniques*, Advances in Intelligent Systems and Computing 949,
https://doi.org/10.1007/978-981-13-8196-6_73

N_{Rr}	Rest stream
$\mathcal{N}sn$	Number of stream flow to specific river
$\mathbb{C}$	Constant between 1 and 2
δ	Current distance between stream and river
C_i	Outcome of objective function
ub	Upper limit
lb	Lower limit
v	Variance

1 Introduction

During the last decade, various bioinspired (BI) computing techniques have come into existence for the solution of complex, constrained and real-world problems. It may be due to their ability to solve multimodal, non-convex as well as discontinuous functions in an efficient manner. Almost all BI techniques incorporate stochastic random operators and use exploration and exploitation processes to generate different solutions in reasonable time frame which help in providing better search results and avoidance of local minima. BI techniques are population-based search mechanism and their analytical model mimics one of the natural phenomena during optimization process. Broadly, BI computations can be classified into five groups, which are based on biological evolution, swarm intelligence, ecology, human intelligence and physical science [1]. For solution of different types of practical problems with complex operational constraints, researchers have successfully applied various BI optimization techniques. Brief review can be found for different stochastic optimizers such as particle swarm optimization(PSO) [2], differential evolution(DE) [3], artificial bee colony (ABC) [4], TLBO [5], cuckoo search algorithm (CSA) [6], etc.

Economic dispatch(ED) is one of the highly complex and constrained optimization problems of power system, where the objective is to meet load demand at lowest possible cost by the committed power generating units subject to the different operational constraints. Due to a variety of complexities involved in ED problem, the objective function is highly complex, discontinuous and multimodal, which is therefore quite difficult to solve. A comprehensive review of various BI optimization techniques used to solve ED problems is presented in Ref. [7].

Infact, any BI computing technique does not guarantee to solve all types of real-world problems in an efficient manner due to high level of randomness involved in it. Therefore, there is always a chance of improvement. Keeping this fact in mind, this paper presents a new water cycle optimization approach [8] to solve ED problems in order to find the most feasible solution in a reasonable time frame. Fundamentally, water cycle optimization mimics the observation of hydrological process that exists in nature and how movement of water through rivers and under ground streams takes it to the sea.

2 Problem Formulation

The objective function of ED problem, F_{total} is described as (1):

$$F_{\text{total}} = \min \sum_{i=1}^{n} f_i(\mathcal{P}_i) \tag{1}$$

Subjected to

$$\sum_{i=1}^{n} (\mathcal{P}_i) = \mathcal{P}_{\text{D}} + \mathcal{P}_{\text{Loss}} \tag{2}$$

where $\mathcal{P}_{\text{Loss}}$ is the transmission loss, by Krons formula it is presented as below (3):

$$\mathcal{P}_{Loss} = \sum_{i=1}^{n} \sum_{j=1}^{n} (\mathcal{P}_i \mathcal{B}_{ij} \mathcal{P}_j) + \sum_{i=1}^{n} \mathcal{B}_{i0} \mathcal{P}_i + \mathcal{B}_{00} \tag{3}$$

and

$$\mathcal{P}_i^{\min} \le \mathcal{P}_i \le \mathcal{P}_i^{\max} \tag{4}$$

Considering valve point loading (VPL), the generator cost function can be expressed as [7]:

$$f_i(\mathcal{P}_i) = a_i \mathcal{P}_i^2 + b_i \mathcal{P}_i + c_i + \left| e_i \times \sin\left(f_i \times \left(\mathcal{P}_i^{\min} - \mathcal{P}_i \right) \right) \right| \tag{5}$$

Practically, the thermal generators have steam turbine with multiple valve which produces large variation in fuel cost function. This valve point loading effect (VPL) introduces ripples in the heat rate curve [9]. VPL effect makes the objective function discontinuous non-convex with multiple minima.

3 Water Cycle Optimization

Water cycle algorithm (WCA) is a novel optimization method proposed by Eskandar et al. in 2012 [8]. It is basically a swarm intelligence-based search mechanism which replicates the phenomenon of water stream and river flow over downhill to the ocean. From water stream, the river water is evaporated, and this evaporated water goes to the atmosphere and forms clouds, which again returns back to earth in the form of raindrops and hence forms new water stream and then river. In this algorithm, raindrop represents the population and the sea represents the final solution. The hydrological

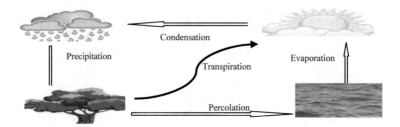

Fig. 1 Hydrological concept of nature

concept behind WCA is shown in Fig. 1. It has mainly four steps in its optimization process which are described as below.

Step 1. Initialization

Similar to other BI techniques initial population, i.e. water stream is generated within its lower and upper limits as below.

$$\mathcal{X}_{ij} = \text{lb} + r \times (\text{ub} - \text{lb}) \quad \text{where } r \in (0, 1) \tag{6}$$

where

$$\text{stream} = [\mathcal{X}_i]_{1 \times D}, \quad \forall i \in D \tag{7}$$

and

$$\text{NP} = [\mathcal{X}_{ij}]_{\text{np} \times D}, \quad i \in \text{np}, \ j \in D \tag{8}$$

Sum of all rivers and ocean (N) can be written as [8]:

$$N = N_{\text{R\_total}} + 1 \text{ (one ocean)} \tag{9}$$

$$N_{\text{S\_total}} = \text{NP} - N \tag{10}$$

$$\mathcal{N}sn = \left\{ \left| \frac{C_n}{\sum_{i=1}^{N} C_i} \right| \times N_{\text{S\_total}} \right\}, \quad \forall n \in N \tag{11}$$

Step 2. Flow of Stream to River or Ocean

The distance $\mathbb{X}$ between water stream and river given as:

$$\mathbb{X} \in (0, \mathbb{C} \times \delta), 1 < \mathbb{C} < 2, \tag{12}$$

with $\mathbb{C}$ more than 1 helps stream in disparate direction towards rivers.

New position of water streams and rivers can be represented by δ during exploration. It can be updated using (13)–(15) as below [8].

$$\mathcal{X}_S^{i+1} = \mathcal{X}_S^i + r \times \mathbb{C} \times \left(\mathcal{X}_R^i - \mathcal{X}_S^i\right) \tag{13}$$

$$\mathcal{X}_R^{i+1} = \mathcal{X}_R^i + r \times \mathbb{C} \times \left(\mathcal{X}_{sea}^i - \mathcal{X}_R^i\right) \tag{14}$$

Step 3. Evaporation Phase

If the distance between river and ocean is found to be very small, indicates river has merged at ocean/sea. Now evaporation process gets started and followed by precipitation (raining).

This distance represented by d_{max} is user-defined control parameter which adoptively decreases with respect to the iteration and finally helps to find out optimal solution as [8].

$$d_{max}^{i+1} = d_{max}^i + \frac{d_{max}^i}{\text{Iter\_max}} \tag{15}$$

Step 4. Raining Phase

After evaporation, newly generated rain drops form streams at different location as:

$$\mathcal{X}_{ij}^{new} = \text{lb} + r \times (\text{ub} - \text{lb}) \tag{16}$$

It is assumed that best raindrop converted into streams forms the river again which flows to the ocean.

To boost the rate of convergence, only those streams are considered which are merged into the ocean directly. It is achieved by the concept of variance (v) and presented as in (17).

$$\mathcal{X}_S^{new} = \mathcal{X}_{ocean} + \sqrt{v} \times r \times (1, D) \tag{17}$$

The implementation process of WCA for solution of practical ED problem is depicted using flowchart in Fig. 2.

4 Simulation Results of Test Cases

The WCA is coded and implemented to solve ED problems using MATLAB 2015b software with CPU having processor i-7-3770, 2 GB RAM and 3.40 GHz speed. In order to validate performance, two complex constrained problems with different dimensions are examined here, and to confirm the superiority of simulation results comparison are also made with other optimizer reported in the recent literatures.

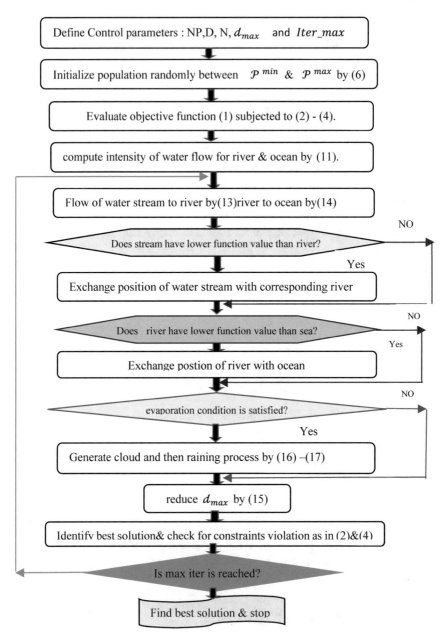

Fig. 2 Implementation of WCA for solution of ED problem

Table 1 Comparison of results for test case 1

Unit	GA [11]	PSO [9]	FPSOGSA [10]	WCA
P_{g1}	150.724	197.8648	199.5997	199.5997
P_{g2}	60.8707	50.3374	20.0000	20
P_{g3}	30.8965	15.0000	23.9896	23.7587
P_{g4}	14.2138	10.0000	18.8493	18.9226
P_{g5}	19.4888	10.0000	18.2153	18.0251
P_{g6}	15.9154	12.0000	13.8506	14.0892
$\sum P_g$	292.1096	295.2022	294.5045	294.3953
Total cost ($/h)	996.0369	925.7581	925.4137	924.8805
P_{Loss} (MW)	8.7060	11.8022	11.1044	10.9952
Time (s)	0.5780	0.3529	1.4108	1.138

Statistical comparison of results are also presented to confirm the robustness of algorithm. In our simulation analysis, population size considered as 100, Nsr *as* 4 and dmax *as* 1e−3.

4.1 Test Case 1: Six Generating Unit System

The six unit systems from IEEE 30 bus sytem having non-convex fuel cost characteristic are selected for testing and validating WCA. Transmission loss is also considered here with power demand of 283.4 MW. The fuel cost and B-loss coefficient data are adopted from Ref. [10].The best power schedule for committed generator as obtained by WCA has been presented in Table 1. With the comparison of results, it is clear that WCA is able to attain cost $/h 924.8805, which is found to be minimum as compared to genetic algorithm(GA) [11], particle swarm optimization (PSO) [9] and hybrid gravitational search-based PSO (GSAPSO) [10]. Also, generation schedule of this test case fully satisfies the operational constraints. The cost convergence curve obtained by WCA is presented in Fig. 3.

4.2 Test Case 2: Thirteen Generating Unit System

In this simulation study, a complex system with 13 power generating units is considered to analyse the effectiveness of WCA. The power generation limits, fuel cost coefficients and B-loss coefficient data are taken from [10]. In this case, power demand of the system is assumed to be 2520 MW. The best operating cost $/h 24513.8450 and corresponding generation schedule obtained are presented in Table 2. Comparison of results are also made with a hybrid PSOGSA-based fuzzy logic (FPSOGSA)

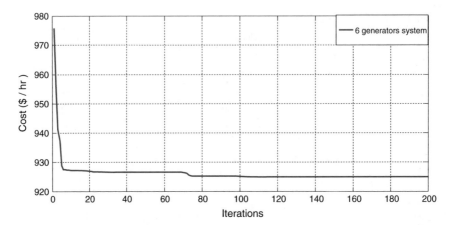

Fig. 3 Convergence curve for test case 1 obtained by WCA

Table 2 Comparison of results for test case 2

Unit	FPSOGSA [10]	GWO [12]	OIWO [13]	SDE [14]	WCA
P_{g1}	628.3185	628.1678	628.3185	628.32	628.3185
P_{g2}	299.1993	298.9229	299.1989	299.20	299.1992
P_{g3}	299.1993	298.2269	299.1991	299.20	299.1992
P_{g4}	159.7331	159.7232	159.7331	159.73	159.7331
P_{g5}	159.7331	159.7210	159.7331	159.73	159.7331
P_{g6}	159.7331	159.7270	159.7331	159.73	159.7330
P_{g7}	159.7331	159.7173	159.7330	159.73	159.7331
P_{g8}	159.7331	159.6793	159.7331	159.73	159.7331
P_{g9}	159.7331	159.6673	159.7330	159.73	159.7328
P_{g10}	76.9368	77.3971	77.3953	77.40	77.3988
P_{g11}	114.2795	114.6051	113.1079	113.12	112.0694
P_{g12}	92.2438	92.3886	92.3594	92.40	92.3985
P_{g13}	92.2007	92.3550	92.3911	92.40	92.3977
$\sum P_g$	2560.7765	2560.2985	2560.3686	2560.43	2559.3795
Total cost	24515.35543	24514.4774	24514.83	24514.88	24513.8450
P_{Loss} (MW)	40.7765	40.2983	40.3686	40.43	39.3794

[10], grey wolf optimization (GWO) [12], oppositional invasive weed optimization (OIWO) [13] and shuffled differential evolution (SDE) [14]. Here, it is observed that optimum generation cost obtained by WCA is found to be minimum and corresponding generation schedule fully satisfies the operational constraints. The cost convergence obtained by WCA is presented in Fig. 4.

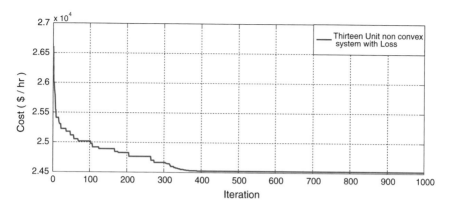

Fig. 4 Convergence curve for test case 2 obtained by WCA

4.3 *Effect of Control Paramters*

As WCA is heuristic optimization approach, with mainly three control parameters, which are NP, Nsr, dmax along with Iter_max as stopping criteria, are selected on the basis of trial and error approach. Selection of population size depends upon complexity of problem being solved; dmax is selected close to zero which helps to provide convergence in an efficient manner. As generator under analysis is rated in MW, value of dmax *is* kept constant at $1e-3$. With each combination of control parameter, tests were carried out over thirty repeated trials and statistical comparison of results are tabulated in Table 3. Here, it is clearly observed that NP = 100 and Nsr = 4 will give better results as compared to other combinations, hence considered for analysis. As we further increased the population size, there is no further improvement in results, however elapsed time gets increased, as depicted in Fig. 5.

Table 3 Comparison of results for test case 2

NP	Nsr	Min cost ($/h)	Ave cost ($/h)	Max cost ($/h)	SD
50	2	925.0435	962.716	984.9433	20.6876
	4	925.00175	960.716	981.8604	20.0016
	6	925.0202	961.512	983.2012	20.0001
100	2	924.8956	948.0758	967.4946	20.6883
	4	*924.8805*	*937.5611*	*966.4667*	*18.8854*
	6	924.9497	938.0519	966.9142	19.3017
150	2	924.9026	945.0332	965.6414	19.9126
	4	924.9162	943.5885	966.9984	19.0004
	6	925.1014	951.0524	969.2625	19.0026

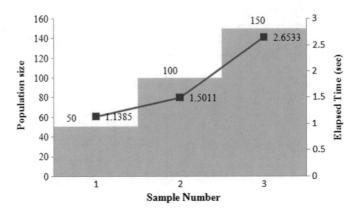

Fig. 5 Effect on elapsed time (s) with increase in population

5 Conclusion

In this paper, an efficient water cycle optimization is implemented for solution of complex constrained optimization problem related to power system. The performance of WCA was analysed on two thermal generating unit systems with non-convex fuel cost characteristics. The outcome of simulation results are also validated with other metaheuristic techniques reported in literature. Comparison of results clearly indicates that WCA has the ability to get robust optimal solution with smooth cost convergence in reasonable time frame. This novel metaheuristic is expected to work well for other problems also.

Acknowledgements The authors acknowledge financial support provided by AICTE-RPS project File No. 8-36/RIFD/RPS/POLICY-1/2016-17 dated 2.9.2017 and TEQIP III. The authors also thank the Director and management of M.I.T.S. Gwalior, India, for providing facilities for carrying out this work.

References

1. Dubey, H.M., Pandit, M., Panigrahi, B.K.: An overview and comparative analysis of recent bio-inspired optimization techniques for wind integrated multi-objective power dispatch. Swarm Evol. Comput. **38**, 12–34 (2018)
2. Mahor, A., Prasad, V., Rangnekar, S.: Economic dispatch using particle swarm optimization: a review. Renew. Sustain. Energy Rev. **13**, 2134–2141 (2009)
3. Das, S., Suganthan, P.N.: Differential evolution: a survey of the state-of-the-art. IEEE Trans. Evol. Comput. **15**(1), 4–31 (2011)
4. Karaboga, D., Gorkemli, B., Ozturk, C., Karaboga, N.: A comprehensive survey: artificial bee colony (ABC) algorithm and applications. Artif. Intell. Rev. **42**, 21–57 (2014)
5. Rao, R.V.: Review of applications of TLBO algorithm and a tutorial for beginners to solve the unconstrained and constrained optimization problems. Decis. Sci. Lett. **5**, 1–30 (2016)

6. Yang, X.S., Deb, S.: Cuckoo search: recent advances and applications. Neural Comput. Appl. **24**(1), 169–174 (2014)
7. Dubey, H.M., Panigrahi, B.K., Pandit, M.: Bio-inspired optimization for economic load dispatch: a review. Int. J. Bio-inspired Comput. **6**, 7–21 (2014)
8. Eskandar, H., Sadollah, A., Bahreininejad, A., Hamdi, M.: Water cycle algorithm—a novel metaheuristic optimization method for solving constrained engineering optimization problems. Comput. Struct. **110–111**, 151–166 (2012)
9. Yasar, C., Ozyon, S.: A new hybrid approach for nonconvex economic dispatch problem with valve-point effect. Energy **35**, 5838–5845 (2011)
10. Duman, S., Yorukeren, N., Altas, I.H.: A novel modified hybrid PSOGSA based on fuzzy logic for non-convex economic dispatch problem with valve-point effect. Electr. Power Energy Syst. **64**, 121–135 (2015)
11. Malik, T.N., ul Asar, A., Wyne, M.F., Akhtar, S.: A new hybrid approach for the solution of nonconvex economic dispatch problem with valve-point effects. Electr. Power Syst. Res. **80**, 1128–1136 (2010)
12. Pradhan, M., Roy, P.K., Pal, T.: Oppositional based grey wolf optimization algorithm for economic dispatch problem of power system. Ain Shams Eng. J. (2017). https://doi.org/10.1016/j.asej.2016.08.023
13. Barisal, A.K., Prusty, R.C.: Large scale economic dispatch of power systems using oppositional invasive weed optimization. Appl. Soft Comput. **29**, 122–137 (2015)
14. Reddya, A.S., Vaisakh, K.: Shuffled differential evolution for large scale economic dispatch. Electr. Power Syst. Res. **96**, 237–245 (2013)

Multi-objective Design Optimization of Shell-and-Tube Heat Exchanger Using Multi-objective SAMP-Jaya Algorithm

R. Venkata Rao, Ankit Saroj, Jan Taler and Pawel Oclon

Abstract This paper presents the application of a posteriori multi-objective version of self-adaptive multi-population (MO-SAMP) Jaya algorithm for the multi-objective optimization of a widely used heat exchanger known as shell-and-tube heat exchanger (STHE). The SAMP-Jaya algorithm is an improved form of Jaya algorithm. The objective functions considered in this work are the maximization of effectiveness, and minimization of pressure drop, total cost and number of entropy generation units of STHE. The design of STHE is subjected to six design variables which are number of tubes, tube diameter, length of tube, baffle cut ratio, baffle spacing ratio tube pitch ratio. MO-SAMP Jaya algorithm has generated sets of non-dominated solutions. Deviation index is used as a quantity measure index to find the finest set of solutions. The designs suggested by MO-SAMP Jaya algorithm are found to be having less deviation index value as compared to the results of the previous researchers.

Keywords MO-SAMP Jaya algorithm · Shell and tube · Deviation index · Design optimization

1 Introduction

Design optimization of heat exchangers (HEs) is one of the utmost issues in the field of design of an effective thermal system. Due to the involvement of large number of component and complexity in design process, design optimization of individual components or devices is carried out [1]. Shell-and-tube HE (STHE) is one of the most important parts of the thermal system which is used for the transfer of heat from one fluid to another. In STHE, one fluid is carried by tubes and another is carried by shell. Transfer of heat takes place between the fluid which is being either heated or

R. Venkata Rao (✉) · A. Saroj
Department of Mechanical Engineering, SV National Institute of Technology, Surat 395007, India
e-mail: ravipudirao@gmail.com

J. Taler · P. Oclon
Institute of Thermal Power Engineering, Cracow University of Technology, Kraków, Poland

© Springer Nature Singapore Pte Ltd. 2020
R. Venkata Rao and J. Taler (eds.), *Advanced Engineering Optimization Through Intelligent Techniques*, Advances in Intelligent Systems and Computing 949,
https://doi.org/10.1007/978-981-13-8196-6_74

831

cooled. Furthermore, baffles are one of the most important parts of the STHE which is used for the proper distribution of fluid coming into the shell over tubes. The sizes of tubes, shell, and baffles and placing of tubes and baffles are associated with the performance of STHE [2].

The literature review reveals that a lot of work is carried out in the field of design optimization of STHE design aiming to minimize the area of heat transfer, total cost, and maximization of effectiveness [2]. Most of the work found with single-objective optimization [2] and a very few with the multi-objective optimization. Differential evolution (DE) [3], genetic algorithm (GA) [4], TLBO [5], Jaya algorithm [2], and heat transfer search (HTS) algorithm [6] are used for the design optimization of STHE associated with multiple objectives.

To see whether there is any improvement in the design of STHE, in comparison to the previous design, MO-SAMP Jaya is used. Objective functions and design variables considered are the same which were used by the previous researchers [6]. The next section describes the working of MO-SAMP Jaya algorithm [7] which is applied for the multi-objective design optimization of STHE.

2   Working of MO-SAMP Jaya Algorithm

Following Eq. (1) is adopted in Jaya algorithm for updating the solutions in every iteration:

$$A'_{q,r,i} = A_{q,r,i} + r_1 * \left(A_{q,\text{best},i} - \left| A_{q,r,i} \right| \right) - r_2 * \left(A_{q,\text{worst},i} - \left| A_{q,r,i} \right| \right) \qquad (1)$$

Here, the symbols and notation are having their usual meaning. Readers may refer to book authored by Rao [9]. Figure 1 expresses the working of the algorithm used in this work. For the working of MO-SAMP Jaya algorithms, one may refer to[7–10]. The next section describes the details of thermal modeling of STHE design used in this paper.

3   Formulation of Objectives

In this work, design optimization is presented among the four conflicting objectives. There are two types of objectives considered in this design optimization of STHE design [6]. The details of the objectives are as follows:

(a) **Thermodynamic objectives**

In this case study, three objectives, namely maximization of effectiveness, minimization of pressure loss, and number of entropy generation units, are used. The effectiveness of the considered E-type TEMA STHE is given by the following equation [6]:

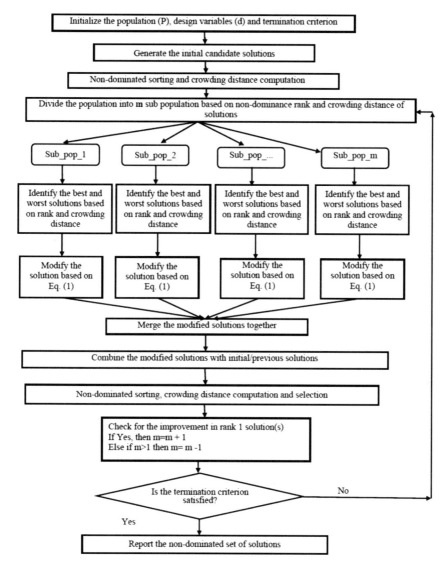

Fig. 1 Flowchart of SAMP-Jaya algorithm

$$\varepsilon = \frac{2}{(1 + R) + (1 + R^2)^{0.5} \coth\left(\frac{NTU}{2}(1 + R^2)^{0.5}\right)} \quad (2)$$

Here, R is defined as the ratio of heat capacity and NTU is known as number of transfer units. Similarly, total pressure drop is calculated as [6]:

$$\Delta P_{\text{total}} = \Delta P_t + \Delta P_s \tag{3}$$

Here, ΔP_t and ΔP_s are the drops in pressure of tube-side and shell-side fluid, respectively.

And the calculation of entropy generation units is given as (N_s) [6]:

$$N_s = \frac{1}{C_{\max}}\left[C_h \ln\left(\frac{T_{h,o}}{T_{h,i}}\right) + C_c \ln\left(\frac{T_{c,o}}{T_{c,i}}\right)\right] \tag{4}$$

Here, $C_{\max}$ is maximum heat capacity among cold fluid and hot fluid. $T_{h,o}$ and $T_{h,i}$ are temperatures of hot and cold fluid at inlet and outlet. $T_{c,o}$ and $T_{c,i}$ are the temperatures of cold fluid at inlet and outlet. C_c and C_h are the heat capacity of cold and hot fluids, respectively.

(b) **Economic objective**

Minimization of total cost of the STHE is considered as another objective. The objective function of the STHE cost is defined as follows [6].

Total cost (C_{tot}) of the STHE is defined as the sum of initial setup cost (C_{in}) and operational cost (C_{op}) and given as:

$$C_{\text{tot}} = C_{\text{in}} + C_{\text{op}} \tag{5}$$

Initial cost of the STHE is directly dependent upon the heat transfer area and calculated by an empirical relation which is expressed as:

$$C_{\text{in}} = 8500 + 409 * S_t \tag{6}$$

Furthermore, operational cost is directly dependent upon pressure drop of shell-side and tube-side fluid, respectively. The detailed calculation of operational cost can be retrieved from the literature [10].

(c) **Limitations on design**

In this design optimization problem of STHE, six design variables have been considered. These design variables are: diameter of tube (D), number of tubes (N), length of tubes (L), baffle cut ratio (bcr), tube pitch ratio (pt), and baffle spacing ratio (bsr). The ranges of these design variables are as follows [6]:

- D (mm): 11.2–15.3
- N: 100–600
- N (m): 3.0–12.0
- L: 0.19–0.32
- pt: 1.25–2
- bsr: 0.2–1.4.

In addition to the restriction of these design variables, design is subjected to some of the structural and operational constraints and these are defined as follows.

$$3 \leq \frac{L}{D_s} \leq 12 \tag{7}$$

$$\Delta P_s \leq \Delta P_{s,\max} \tag{8}$$

$$\Delta P_t \leq \Delta P_{t,\max} \tag{9}$$

4 Results and Discussion

The coding of the used algorithm is prepared in MATLAB R2014a in a laptop with 4 GB RAM and AMD processor. Sets of 50 non-dominated solutions are generated by using the MO-SAMP Jaya algorithms. The evaluation of the designs obtained by MO-SAMP Jaya algorithm and heat transfer search (HTS) algorithm [6] for individual objectives is shown in Table 1. This table reveals that the design recommended by MO-SAMP Jaya algorithm is better or equal in comparison to the design obtained by HTS algorithm [6]. The effectiveness of STHE given by MO-SAMP Jaya algorithm is improved by 6.5%. Similarly, total cost is reduced by 4.44% and the value of pressure loss is reduced by 70%. In case of minimization of entropy generation units, the value of optimal entropy generation unit is same as compared to the result obtained by HTS algorithm [6].

Furthermore, to retrieve the finest sets of non-dominated solution and its fair comparison with the non-dominated solution reported by the previous researchers, deviation index is calculated for all the solutions. The deviation index is defined as the divergence of solution from ideal solution and non-ideal solution. The equivalent distance from the ideal solution (d_+) is calculated as [7]:

$$d_+ = \sqrt{\left(\varepsilon_n - \varepsilon_{n,\text{ideal}}\right)^2 + (\Delta P_{\text{total},n} - \Delta P_{\text{total,ideal}})^2 + \left(N_{s,n} - N_{s,\text{ideal}}\right)^2 + \left(C_{\text{tot},n} - C_{\text{tot,ideal}}\right)^2} \tag{10}$$

The equivalent distance from the non-ideal solution (d_-) is calculated as:

$$d_- = \sqrt{\left(\varepsilon_n - \varepsilon_{n,\text{non-ideal}}\right)^2 + \left(\Delta P_{\text{total},n} - \Delta P_{\text{total,non-ideal}}\right)^2 + \left(N_{s,n} - N_{s,\text{non-ideal}}\right)^2 + \left(C_{\text{tot},n} - C_{\text{tot,non-ideal}}\right)^2} \tag{11}$$

Now, the deviation index from the ideal solution is calculated as follows:

$$d = \frac{d_+}{d_+ + d_-} \tag{12}$$

Table 2 shows multi-objective optimization results given by MO-SAMP Jaya algorithm and HTS algorithm [6]. The previous researchers had used LINMAP, TOPSIS, and fuzzy logic for identifying the best set of non-dominated solution which were obtained by using HTS algorithm. This table shows that the value of d for MO-SAMP Jaya algorithm is lowest as compared to the value of LINMAP, TOPSIS, and fuzzy logic. Fuzzy logic, LINMAP, and TOPSIS have obtained second, third,

Table 1 Comparison of individual objective results

| Output | Objective | | | | | | | |
| | Effectiveness | | Total cost ($) | | Pressure drop (Pa) | | N_s | |
	HTS [6]	MO-SAMP Jaya	HTS [6]	MOSAMP-Jaya	HTS [6]	MOSAMP-Jaya	HTS [6]	MOSAMP-Jaya
Effectiveness	0.7863	**0.8380**	0.0688	0.1235	0.2076	0.2835	0.0674	0.1114
Total cost ($)	56297.9	106809.79	13806.9	**13193.730**	30250.2	30421.6500	13869.5	13540.4840
Pressure drop (Pa)	5247.2	473541.49	5820.7	5125.2803	314.2	**81.0377**	5410.9	1895.3845
N_s	0.0028	0.0020	0.0008	0.0006	0.0013	0.0011	**0.0005**	**0.0005**

Table 2 Comparison of multi-objective optimization results

	FUZZY logic [6]	TOPSIS [6]	LINMAP [6]	MO-SAMP Jaya
D (m)	15.3	12	13.1	11.2000
N (m)	400	466	449	250.0000
L (m)	3	3	3	3.0000
Pt	1.25	1.25	1.25	1.8302
bcr	0.19	0.32	0.19	0.2260
bsr	0.2	0.2	0.2	0.7426
ε	0.5025	0.5097	0.5408	0.2976
Total cost ($)	24100.3	23618.8	24058.6	17981.5390
Pressure drop (Pa)	7808.6	12914.9	11681.7	10290.9420
Ns	0.0025	0.0026	0.0026	0.0012
Deviation index from ideal solution		0.817240519	0.818040407	**0.3528**

and fourth ranks, respectively. Total cost of STHE given by the present algorithm is reduced by 25.26, 23.86, and 25.39% in comparison to the results suggested by LINMAP, TOPSIS, and fuzzy logic. The value of tube diameter is reduced by 14.50, 6.67, and 26.79% as compared to the results of LINMAP, TOPSIS, and fuzzy logic. Furthermore, the number of tubes suggested by MO-SAMP Jaya algorithm is reduced by 44.32, 46.43, and 37.50% as compared to the results of LINMAP, TOPSIS, and fuzzy logic.

5   Conclusions

In this paper, the design optimization of STHE is carried out for multiple objectives by using a posteriori MO-SAMP Jaya algorithm. The objectives used are the maximization of effectiveness, and minimization of pressure drop, total cost, and entropy generation units of STHE. Deviation index from the ideal solution is used for selecting the best set of non-dominated solution and its comparison with the previous results. The non-dominated solution suggested by MO-SAMP Jaya algorithm is found best as compared to the solutions given by LINMAP, TOPSIS, and fuzzy logic. The value of total cost of STHE design obtained by MO-SAMP Jaya algorithm is reduced by 25.26, 23.86, and 25.39% in comparison to the results suggested by LINMAP, TOPSIS, and fuzzy logic. Furthermore, MO-SAMP Jaya algorithm can be easily integrated to other multi-objective optimization problems of thermal design such as chillers, evaporators, thermal cycles.

References

1. Rao, R.V., More, K.C.: Design optimization and analysis of selected thermal devices using self-adaptive Jaya algorithm. Energy Convers. Manage. **140**, 24–35 (2017)
2. Rao, R.V., Saroj, A.: Multi-objective design optimization of heat exchangers using Elitist-Jaya algorithm. Energy Syst. **9**(2), 305–341 (2018)
3. Ayala, H.V.H., Keller, P., de Fátima Morais, M., Mariani, V.C., dos Santos Coelho, L., Rao, R.V.: Design of heat exchangers using a novel multiobjective free search differential evolution paradigm. Appl. Therm. Eng. **94**, 170–177 (2016)
4. Sanaye, S., Hajabdollahi, H.: Multi-objective optimization of shell and tube heat exchangers. Appl. Therm. Eng. **30**(14–15), 1937–1945 (2010)
5. Rao, R.V., Patel, V.: Multi-objective optimization of heat exchangers using a modified teaching-learning-based optimization algorithm. Appl. Math. Model. **37**(3), 1147–1162 (2013)
6. Raja, B.D., Jhala, R.L., Patel, V.: Many-objective optimization of shell and tube heat exchanger. Thermal Sci. Eng. Prog **2**, 87–101 (2017)
7. Rao, R.V., Saroj, A., Oclon, P., Taler, J., Lakshmi, J.: A Posteriori multiobjective self-adaptive multipopulation Jaya algorithm for optimization of thermal devices and cycles. IEEE Access [In press]
8. Rao, R.: Jaya: a simple and new optimization algorithm for solving constrained and unconstrained optimization problems. Int. J. Ind. Eng. Comput. **7**(1), 19–34 (2016)
9. Rao, R.V.: Jaya: An Advanced Optimization Algorithm and its Engineering Applications. Springer International Publishing
10. Rao, R.V., Saroj, A.: A self-adaptive multi-population based Jaya algorithm for engineering optimization. Swarm Evol. Comput. **37**, 1–26 (2017)

An Improved Jaya Algorithm and Its Scope for Optimization in Communications Engineering

Ravipudi Jaya Lakshmi

Abstract This paper proposes an improved version of Jaya algorithm for optimization. The effectiveness of the proposed version is demonstrated by solving standard benchmark functions taken from the literature. The scope of applications of the proposed algorithm in communications engineering is also presented.

Keywords Optimization · Jaya algorithm · Information and communications engineering

1 Introduction

Nowadays, the field of optimization algorithms is congested with a number of research papers appearing every month proposing 'new' or 'novel' algorithms mimicking some physical, biological, or any phenomenon or event that occurs in this universe. There is no limit to the imagination of the 'researchers' who are proposing these 'new' or 'novel' optimization algorithms. Every researcher claims that his or her algorithm is the best one and it outperforms all other existing algorithms. These 'researchers' carefully 'manage' their 'new' or 'novel' algorithms to show better results than the other algorithms. The concerns of the real optimization problems of the industrial world are addressed only by a very few researchers. The characteristics of the optimization algorithms which the users generally look for include simplicity, less complexity, ability to solve the single as well as multi-objective problems, robustness, etc. However, many 'new' or 'novel' optimization algorithms claim that they possess these characteristics, in actual practice, these claims are not found valid.

Jaya algorithm is having the characteristics which a common user looks for. It is simple yet powerful, robust, least complex, and can be used for single as well as multi-objective optimization problems. The other salient feature of Jaya algorithm is that

R. J. Lakshmi (✉)
Department of Information Engineering and Computer Science, University of Trento, I-381243 Trento, Italy
e-mail: jayalakshmi.ravipudi@studenti.unitn.it

© Springer Nature Singapore Pte Ltd. 2020
R. Venkata Rao and J. Taler (eds.), *Advanced Engineering Optimization Through Intelligent Techniques*, Advances in Intelligent Systems and Computing 949,
https://doi.org/10.1007/978-981-13-8196-6_75

the user has no burden of tuning the algorithm-specific parameters as this algorithm does not have such parameters. The convergence rate of Jaya algorithm is also better. The optimization research community is fast-adapting Jaya algorithm, and various applications can be found in different disciplines of science and engineering [1, 2].

2 Jaya Algorithm

The first step in the implementation of Jaya algorithm is to generate initial solutions. Therefore, n initial solutions are generated in random fashion by keeping in consideration the design variables' range values, i.e., upper and lower bounds. Then, every solution's variable values are updated using Eq. (1). Let the objective function be denoted with f. For demonstration, it is considered that the function is to be minimized (or maximized, according to the design problem). Say, the no. of design variables is 'nd'. best_f is used to denote the best solution and worst_f to denote the worst solution.

$$A(x+1, y, z) = A(x, y, z) + r(x, y, 1)(A(x, y, b) - |A(x, y, z)|) - r(x, y, 2)(A(x, y, w) - |A(x, y, z)|) \quad (1)$$

Among the population, the best and worst solutions are identified by their indices b and w. The indices to denote iteration, variable, and the candidate solution x, y, z. $A(x, y, z)$ mean the yth variable of zth candidate solution in xth iteration. $r(x, y, 1)$ and $r(x, y, 2)$ are the randomly generated numbers in the range of [0, 1] and act as scaling factors and to ensure good diversification. Figure 1 shows the flowchart of Jaya algorithm.

In Jaya algorithm, the value of the objective function of every candidate solution of the population is improved. Values of variables of each solution are updated by the algorithm in order to move the corresponding function value closer to the best solution. After updating variable values, comparison between the new function value and corresponding old function value. Then, only the solutions having better objective function value (lower values for minimization and higher values for maximization problems) are included in the next generation.

3 Proposed Version of Jaya Algorithm

During the last three years, few researchers had proposed certain variants of Jaya algorithm. In this paper, another variant of Jaya algorithm is proposed. This variant is expected to help in improving the speed of convergence and also in improving the search space exploration without getting entrapped in the local optima. In the original Jaya algorithm, the solutions are updated using Eq. (1). In the proposed version, Eq. (1) is modified and is given as Eq. (2).

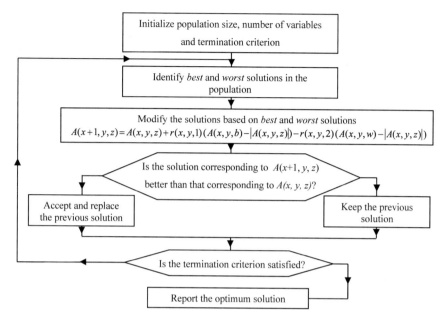

Fig. 1 Flowchart of Jaya algorithm

$$A(x + 1, y, z) = A(x, y, z)$$
$$+ \left(\frac{1}{n}\right) r(x, y, 1) \sum (A(x, y, b) - |A(x, y, z)|)$$
$$- \left(\frac{1}{n}\right) r(x, y, 2) \sum (A(x, y, w) - |A(x, y, z)|) \quad (2)$$

The only difference in the implementation of Jaya algorithm and the proposed version is that the proposed algorithm considers all the values $A(x, y, z)$ of a variable and subtracts those from the best and worst values corresponding to that variable. Then the average is considered. In Eq. (2), n is the population size.

4 Demonstration of the Proposed Version of Jaya Algorithm

The performance of the proposed version of Jaya algorithm is tested on ten unconstrained benchmark functions well documented in the optimization litera-ture. These unconstrained functions have different characteristics like unimodal-ity/multimodality, separability/non-separability, regularity/non-regularity, etc. The number of design variables and their ranges are different for each problem. A com-mon experimental platform is provided by setting the maximum number of function evaluations as 500,000 for each benchmark function. The proposed version of Jaya algorithm is executed 30 times for each benchmark function.

Table 1 shows the optimization results corresponding to ten unconstrained benchmark functions. The results are shown for different population sizes n. The best values, worst values, mean values, standard deviation (SD), and the mean function evaluations (MFE) are shown in Table 1.

It may be observed from Table 1 that the proposed version of Jaya algorithm is giving good results and reaching the known optimum solutions of the corresponding benchmark functions. A comparison of these results with the results obtained by Jaya algorithm and other optimization algorithms like GA, DE, PSO, ABC given in Rao [1] reveals that the proposed version is competitive and effective.

5   Scope of Applications in Communications Engineering

The proposed version of Jaya algorithm (for that matter, original Jaya algorithm as well) has lot of scope to solve the optimization related problems of the information and communications engineering field. Some of the potential areas (not exhaustive) include the following:

- Varieties of problems in telecommunications, computer communications, and network design and routing, signal processing
- Energy consumption optimization for green device-to-device multimedia communications
- Energy efficiency optimization of device-to-device communications in 5G networks
- Outage protection for cellular-mode users in device-to-device communications
- Throughput maximization for UAV-enabled full-duplex relay system in 5G communications
- Optimization of intensities and locations of diffuse spots in indoor optical wireless communications
- Network control and rate optimization for multi-user MIMO communications
- Joint radar communications design
- Sparse transmit array design for dual-function radar communications
- System reliability optimization
- Optimization of load-balanced routing for AMI with wireless mesh networks
- Optimization of IoT cloud energy consumption
- Data mining in IoT
- Compressive sensing in IoT and monitoring applications
- Combinatorial optimization in VLSI
- Range-free localization for three-dimensional wireless sensor networks
- Task scheduling problem in the phased array radar
- Design of different types of antenna arrays
- Frequency assignment problems

Table 1 Results of application of proposed improved Jaya algorithm on unconstrained benchmark functions

S. No.	Function	Known optimum	n	Best	Worst	Mean	SD	MFE
1	Rosenbrock	0	10	0	22.191719	1.479448	5.63022	474,318.33
			20	0	0	0	0	406,028
			25	0	0	0	0	497,746
			40	0	0.000121	0.000008	0.00003	497,634
			100	0.039206	90.240443	23.304398	33.008805	491,670
2	Sphere	0	10	0	0	0	0	499,987
			20	0	0	0	0	499,953
			25	0	0	0	0	499,889
			40	0	0	0	0	499,748
			100	0	0	0	0	498,770
3	SumSquares	0	10	0	0	0	0	499,989
			20	0	0	0	0	499,946
			25	0	0	0	0	499,920
			40	0	0	0	0	499,761
			100	0	0	0	0	498,953
4	Beale	0	10	0	0	0	0	3501.33
			20	0	0	0	0	9754
			25	0	0	0	0	13,879.167
			40	0	0	0	0	25,162.67
			100	0	0	0	0	79,063
5	Easom	−1	10	−1	0	−0.966667	0.182574	2092
			20	−1	−1	−1	0	6204
			25	−1	−1	−1	0	8080
			40	−1	−1	−1	0	14,673.33
			100	−1	−1	−1	0	53,603.33

(continued)

Table 1 (continued)

S. No.	Function	Known optimum	n	Best	Worst	Mean	SD	MFE
6	Matyas	0	10	0	0	0	0	33,903.67
			20	0	0	0	0	83,976.67
			25	0	0	0	0	111,365
			40	0	0	0	0	199,464
			100	0	0	0	0	499,056.67
7	Colville	0	10	0	0	0	0	100,925
			20	0	0	0	0	358,473
			25	0	0	0	0	489,805
			40	0	0	0	0	487,614
			100	0.000002	0.000193	0.000032	0.000039	459,143
8	Trid 6	−50	10	−50	−50	−50	0	67,435
			20	−50	−50	−50	0	77,929.33
			25	−50	−50	−50	0	80,560
			40	−50	−50	−50	0	133,041.33
			100	−50	−50	−50	0	241,546.67
9	Trid 10	−210	10	−210	−210	−210	0	87,636.33
			20	−210	−210	−210	0	113,577.33
			25	−210	−210	−210	0	164,991.67
			40	−210	−210	−210	0	198,116
			100	−210	−210	−210	0	355,733.33
10	Zakharov	0	10	0	0	0	0	257,952.67
			20	0	0	0	0	499,846.67
			25	0	0	0	0	499,747.5
			40	0	0	0	0	499,568
			100	0	0	0	0	498,676.67

- Route optimization based on clustering
- Automatic quadratic time-frequency distribution
- Defense against emulation attacks related to security in cognitive radio networks
- Security and quality of service optimization in a real-time DBMS
- Task scheduling in heterogeneous computing systems
- Problems related to big data optimization
- Coalition formation in large-scale UAV networks
- Image encryption
- Topological optimization of interconnection networks
- Web service combination problems
- FPGA implementation to detect optimal user by cooperative spectrum sensing
- Optimization of programmable data-flow crypto processors
- Large-scale planning of dense and robust industrial wireless networks
- Congestion control in wireless sensor networks
- Cluster head selection to prolong lifetime of wireless sensor networks
- Design of band-notched UWB antennas
- Multi-user Detection for UWB communications
- Reliability and topology-based network design
- MIMO broadcast scheduling
- Training of artificial neural networks
- Optimization problems in computer communications.

The above topics are indicative, and there is a lot of scope for application of Jaya algorithm and its improved version to the optimization problems related to information and communication engineering. An application of chaotic Jaya algorithm for antenna design can be found in [3].

6   Conclusions

In this paper, the basic Jaya algorithm is modified and the performance is demonstrated on ten unconstrained benchmark functions. It is observed that the proposed version performs comparatively better and is proved to have good search ability. The implementation of the proposed version is relatively simple and easier. A selective list of potential topics in the field of communications engineering is prepared, and these topics can be addressed by the proposed version. The future scope of work includes these topics.

Acknowledgements The author acknowledges the help provided by Prof. R. V. Rao and Mr. R. B. Pawar of S. V. National Institute of Technology, Surat, India, in understanding the Jaya algorithm and the computational procedures.

References

1. Rao, R.V.: Jaya: a simple and new optimization algorithm for solving constrained and unconstrained optimization problems. Int. J. Ind. Eng. Comput. **7**, 19–34 (2016)
2. Rao, R.V.: Jaya: An Advanced Optimization Algorithm and Its Engineering Applications. Springer Nature, Switzerland (2018)
3. Jaya Lakshmi, R., Mary Neebha, T.: Synthesis of linear antenna arrays using Jaya, self-adaptive Jaya and chaotic Jaya algorithms. AEU-Int. J. Electron. Commun. **92**, 54–63 (2018)

Performance Evaluation of TiN-Coated CBN Tools During Turning of Variable Hardened AISI 4340 Steel

Sanjeev Kumar, Dilbag Singh and Nirmal S. Kalsi

Abstract In precision hard turning, the machining performance is highly influenced by the frictional heat generated during the process. The high frictional heat generation causes the earlier tool wear, which tends to decrease the machining performance. The coating of hard materials on cutting tools is applied to enhance machining performance. In this research work, the performance of TiN-coated CBN tools was evaluated during hard turning. The workpiece hardness, tool nose radius, cutting speed and feed rate were selected as process parameters. The cutting force and chip–tool interface temperature were chosen as performance parameters. Center composite design was chosen for experimental design, and analysis of variance (ANOVA) was performed to analyze the significance of the process parameters. The outcomes are compared with uncoated CBN tools. The results showed that the machining performance of TiN-coated CBN tools was superior in all cutting conditions.

Keywords Interface temperature · Cutting force · Workpiece hardness · Nose radius · CBN · ANOVA

1 Introduction

Applications of hard turning increase day by day in the manufacturing industries. With the advent of new superhard cutting tool materials like CBN and PCBN, the hard turning has adequately replaced grinding and other finishing operations [1]. CBN tools are extensively used for machining of difficult-to-cut materials such as case hardened steels and superalloys due to their high thermal stability along with

S. Kumar · D. Singh · N. S. Kalsi (✉)
Beant College of Engineering and Technology, Gurdaspur 143521, Punjab, India
e-mail: ns_kalsi@yahoo.com

S. Kumar
e-mail: sk_74@rediffmail.com

D. Singh
e-mail: singh_dilbag@yahoo.com

© Springer Nature Singapore Pte Ltd. 2020
R. Venkata Rao and J. Taler (eds.), *Advanced Engineering Optimization Through Intelligent Techniques*, Advances in Intelligent Systems and Computing 949,
https://doi.org/10.1007/978-981-13-8196-6_76

847

high wear resistance [2]. CBN tools are mostly preferred during hard turning, because extreme thermomechanical loading conditions occur at the tool edge during machining. Some authors have evaluated the machining performance of CBN tools. Ozel et al. reported that minimum surface roughness with low cutting forces was observed during machining of AISI H13 steel with CBN inserts [3]. Sahin reported that the tool life of CBN tools was better than ceramic tools in all cutting conditions during hard turning [4]. Dogra et al. reported that the machining performance of cryogenically treated coated carbide was equivalent to CBN tools [5]. Aouici et al. reported that the feed rate was the significant parameter, which affects the surface roughness and cutting forces during hard turning of X38CrMoV5-1 steel (50 HRC) with CBN inserts [6]. Sobiyi et al. reported that minimum surface roughness with high-dimensional precision was accomplished with CBN tools [7]. Zhao et al. investigated that the tool edge radius has maximum effects on surface roughness and tool wear rate during machining of hardened AISI 52100 steel with CBN inserts [8].

The previous studies showed that the machining performance of CBN tools was superior to carbide and ceramic tools in all cutting conditions. However, Huang et al. reported that the rapid degradation and catastrophic failure of the CBN tool occurred with an increase in workpiece hardness [9]. The machining performance can only be enhanced by providing the coating of hard materials on cutting tools. Hence, in the present work, the machining performance of TiN-coated CBN tools is evaluated during dry hard turning of AISI 4340 steel. A few research work is available having usage of TiN coated CBN tools during hard turning of AIS 4340 steel.

2   Materials and Methods

2.1   Materials

AISI 4340 steel was selected as workpiece material for experimentations. AISI 4340 steel has high toughness and ductility due to the presence of nickel and chrome alloying elements. AISI 4340 steel has a broad range of applications in aerospace, automobile and general engineering industries. AISI 4340 steel has a chemical composition of 0.42% carbon, 1.47% nickel, 1.06% chromium, 0.22% molybdenum, 0.58% manganese and 0.27% silicon. The workpiece materials were thoroughly heat-treated to hardness 40, 45, 50, 55 and 60 within ±2 HRC; 65-mm diameter and 350-mm length of the workpiece were selected during the experimentation to keep the L/D ratio not more than ten as per ISO 3685 standard (1993) [10]. CBN insert with grade K5625, an ISO designation SNGA 431S0425MT, was selected as cutting tool material. The CBN inserts were coated with TiN material to 5-μm thickness. The inserts were clamped on a right-hand tool holder, an ISO designation MSSNR2525M12 N.

Table 1 Process variable with ranges

Factors	Symbol	Range				
Cutting speed	v (m/min)	75	100	125	150	175
Feed rate	f (mm/rev)	0.1	0.125	0.15	0.175	0.2
Workpiece hardness	h (HRC)	40	45	50	55	60
Tool nose radius	r (mm)	0.2	0.4	0.8	1.2	1.6

2.2 Process Parameters

Based on the previous research work and machining data handbook, the cutting parameters (cutting speed and feed rate), the geometrical parameter (tool nose radius) and workpiece hardness were selected for the experimentations. The ranges of each process parameter have been chosen based on the tool manufacturer's recommendation and industrial practices. The ranges of process parameters are shown in Table 1. A central composite design (CCD) based on RSM was selected to design the experimental layout. In RSM, a large number of variables influence the response of surface [11]. As per CCD, the total 30 experiments were performed. The cutting force and chip–tool interface temperature were taken as the response variables. The depth of cut was kept constant at 0.2 mm.

2.3 Cutting Force and Interface Temperature Measurements

The high rigidity of machine tool is the prime requirement for hard turning process. Therefore, highly rigid HMT-made lathe is selected for experimentations. The main component of cutting force generated on tool point in the hard turning was recorded online using a high-precision lathe tool dynamometer (Make DKM 2010 of TeLC, Germany, software XKM 2000) as shown in Fig. 1. The measurement system is built with strain gage sensors with minimal deflection, range 2000 N, resolution 1 N, data acquisition rate adjustable 5–100 SPS. The temperature in the chip was measured by a distance IR pyrometer, model Raynger 3i Plus [distance to spot 250:1, accuracy $\pm$ (0.5% of reading $+1$ °C)].

Fig. 1 Experimental setup

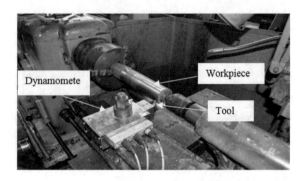

3   Results and Discussion

3.1   Analysis of Variance (ANOVA)

The ANOVA was performed to analyze the significance of the process parameters. The first-order models of cutting force and interface temperature were not significant. Therefore, the second-order models of cutting force and interface temperature were formulated. Some nonsignificant terms in the models were removed by using a backward elimination process. The results of the ANOVA for cutting force and interface temperature are shown in Tables 2 and 3, respectively.

The ANOVA tables indicate the percentage contributions of each parameter and their interactions. The calculated F of models in Tables 2 and 3 are 543.15 (p-value < 0.0001) and 511.28 (p-value < 0.0001), respectively. It represented that both these models are significant. The workpiece hardness is the main significant parameter, which has maximum effects on cutting force and interface temperature. Residual error represents the inconsistencies between calculated and predicted values.

3.2   Mathematical Models

The mathematical models of cutting force and interface temperature were formulated by using response surface methodology (RSM). The second order mathematical model for cutting force using TiN coated CBN cutting tool is:

$$
\begin{aligned}
F_2 = {} & 343.68 - 0.47 \times v + 1179.48 \times f - 5.54 \times h - 51.63 \\
& \times r - 1.60 \times v \times f + 0.03 \times v \times h - 24.0 \times f \times h + 137.50 \times f \\
& \times r - 0.006 \times v^2 + 1346.14 \times f^2 + 0.05 \times h^2 + 28.32 \times r^2
\end{aligned} \tag{1}
$$

Equation (2) represents the second-order mathematical model for cutting temperature using TiN-coated CBN cutting tool:

Table 2 ANOVA for cutting force

Source	Sum of squares	df	Mean square	F value	p-value Prob > F	% Contribution	Remarks
Model	6517.86	12	543.15	543.15	< 0.0001		Significant
v	1176	1	1176	978.33	< 0.0001	19.06	
f	1290.66	1	1290.66	1073.73	< 0.0001	20.92	
h	1536	1	1536	1277.82	< 0.0001	24.90	
r	699.55	1	699.55	581.97	< 0.0001	11.34	
$v*f$	16	1	16	13.31	< 0.0020	0.25	
$v*h$	380.25	1	380.25	316.33	< 0.0001	6.16	
$f*h$	144	1	144	119.79	< 0.0001	2.33	
$f*r$	30.25	1	30.25	25.16	< 0.0001	0.49	
v^2	397.28	1	397.28	330.50	< 0.0001	6.44	
f^2	19.64	1	19.64	16.34	< 0.0008	0.31	
h^2	59.66	1	59.66	49.63	< 0.0001	0.96	
r^2	397.34	1	397.34	330.55	< 0.0001	6.44	
Residual error	20.43	17	1.20			0.33	
Total	6167.06	29					

R-square $= 0.9969$

$$T_2 = 4254.36 - 5.1 \times v - 11311.76 \times f - 135.62 \times h + 472.63 \times r$$
$$- 14.30 \times v \times f + 1.56 \times v \times r - 793.75 \times f \times r - 17.03 \times h \times r$$
$$+ 0.03 \times v^2 + 53264.20 \times f^2 + 1.63 \times h^2 + 216.54 \times r^2 \qquad (2)$$

3.3 Functional Relationship

The mathematical models as represented by Eqs. (1–2) are used to predict the functional relationship between input and response variables.

Effect of cutting speed. Figure 2a indicates that the cutting force is decreased with an increase in cutting speed. This is because of the decrease in time for which the chip comes in contact with the tool. It causes the rise in the temperature of the shear zone as the heat is not rapidly carried away by the chips. It resulted in the decrease in cutting force. It is observed from Fig. 2a that after 150 m/min of cutting speed, the cutting force is increased. It is due to an increase in tool wear associated with an increase in cutting speed. Figure 2b shows that interface temperature is increased with an increase in cutting speed. This is due to an increase in temperature of shear zone temperature. Figure 2, TiN-coated CBN tools, shows the superior machining

Table 3 ANOVA for cutting temperature

Source	Sum of squares	df	Mean square	F value	p-value Prob > F	% Contribution	Remarks
Model	347,296.67	12	28,941.38	511.28	< 0.0001		Significant
v	34,277.04	1	34,277.04	605.54	< 0.0001	9.30	
f	75,600.37	1	75,600.37	1335.56	< 0.0001	20.50	
h	116,065.04	1	116,065.04	2050.41	< 0.0001	31.49	
r	6792.62	1	6792.62	119.99	< 0.0001	1.84	
v*f	1278.06	1	1278.06	22.57	< 0.0002	0.34	
v*r	3937.56	1	3937.56	69.56	< 0.0001	1.06	
f*r	1008.06	1	1008.06	17.80	< 0.0006	0.27	
h*r	18,564.06	1	18,564.06	327.95	< 0.0001	5.03	
v^2	9928.68	1	9928.68	175.40	< 0.0001	2.69	
f^2	30,754.17	1	30,754.17	543.30	< 0.0001	8.34	
h^2	46,172.48	1	46,172.48	815.68	< 0.0001	12.52	
r^2	23,227.24	1	23,227.24	410.33	< 0.0001	6.30	
Residual error	962.29	17	56.60			0.26	
Total	368,567.7	29					

R-square = 0.9972

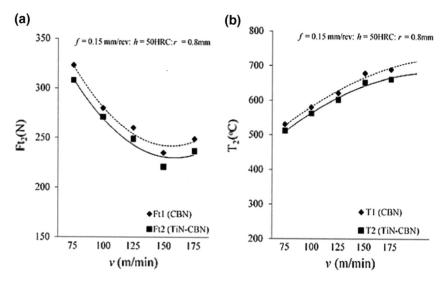

Fig. 2 Variation of **a** cutting force and **b** interface temperature with cutting speed

performance than the uncoated CBN tools. It is because TiN coating decomposes at high temperature and provides lubrication, which resulted in the improvement in machining performance.

Effect of feed rate. Figure 3a indicates that cutting force is increased with an increase in feed rate. Since with low feed rate, the small area of contact is available at tool–workpiece interface.

The small area of contact provides a very low resistance to the cutting tool in the direction of feed rate. It resulted in the decrease in cutting force [12]. Whereas with a higher feed rate, the workpiece and tool contact area increases, which causes an increase in frictional resistance per revolution and resulted in an increase in cutting force. Figure 3b shows that the interface temperature is increased with an increase in feed rate. This is because of high frictional heat generation at work material and tool flank, which is increased with an increase in feed rate.

Effect of workpiece hardness. Figure 4a illustrates the variation of cutting force with workpiece hardness. It indicates that the cutting force decreased up to 55 HRC of workpiece hardness. This is due to the thermal softening of the cutting area because of an increase in frictional heat generation during the process, which resulted in the decrease in cutting force. Figure 4a shows the cutting force increased as the workpiece hardness is more than 55 HRC.

It is because of the increase in flank wear which is due to the presence of hard martensite particle at the upper layer of the workpiece material [13]. Figure 4b shows that the interface temperature increased with increase in workpiece hardness. It is due to an increase in frictional heat generation with an increase in workpiece hardness.

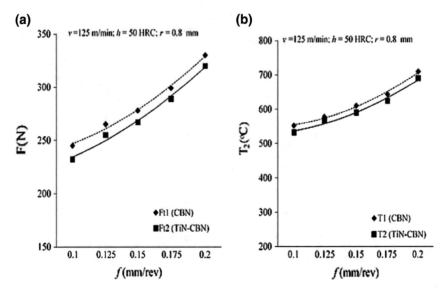

Fig. 3 Variation of **a** cutting force and **b** interface temperature with a feed rate

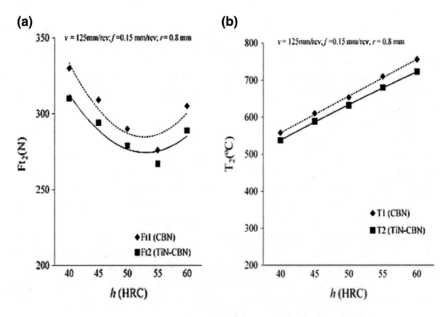

Fig. 4 Variation of **a** cutting force and **b** interface temperature with workpiece hardness

Table 4 Optimization results

S. No.	v	f	h	r	F (N)	T (°C)	Desirability	Remarks
1	150	0.1	40	1.2	202.25	552.62	0.822	
2	149.6	0.1	40	1.2	202.48	551.03	0.822	
3	150	0.1	40	1.2	202.24	551.85	0.821	
4	150	0.1	40	1.2	202.23	551.1	0.819	
5	150	0.1	40.18	1.2	202.53	551	0.819	
6	1483.56	0.1	40	1.2	203.07	546.95	0.819	
7	150	0.1	40.22	1.2	202.6	550.68	0.819	
8	148.17	0.1	40	1.2	203.3	545.41	0.818	
9	150	0.1	40	1.2	202.23	550.28	0.818	
10	147.38	0.1	40	1.2	203.73	542.38	0.817	

4 Prediction of Optimum Results

The optimal values of the process, which maximize the machining performance, are estimated from the desirability function of the RSM. The conditions, i.e., either maximize or minimize or to assign to a target value, of response variables are used to assess the desirability function. To estimate the desirability function, the cutting force and interface temperature are set to the minimum. The calculated response is converted into scale called desirability ranging from 0 to 1 and entirely dependent on proximity to the lower and upper limits of the process variables. If the desirability value shows 1, it indicates the ideal case, and if the desirability value is 0, it indicates that one or more response is being outside their acceptable limits. The optimum results calculated using Design-Expert software are given in Table 4.

It indicates that desirability is decreased with any change in the process parameter after solution no. 2. Therefore, the optimal value of parameters which provide minimum cutting force and interface temperature is 150 m/min of cutting speed, 0.1 mm/rev of feed rate, 1.2 mm of nose radius and 40 HRC workpiece hardness.

5 Chip Surface Morphology

The chip surface characterized the deformation and shearing of the cutting edge during machining [14, 15]. With the increase in cutting speed, high frictional heat is generated at the cutting area. The huge part of the generated heat is carried away by the intermediate chips as they are under high frictional force and plastic deformation. Figure 6 shows the chip surface micrograph as collected at optimal values of process parameters. Figure 6a indicates saw tooth-type chip is formed when machining CBN tools which indicate easy removal of the chip due to intensive plastic deformation.

(a)   **(b)**

Chip Micrographs with Uncoated CBN   Chip Micrographs with TiN-Coated CBN

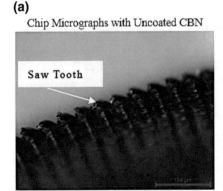
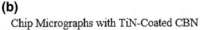
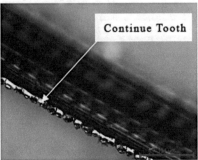

Fig. 6 Micrograph of chip surface (taken at 500X)

Figure 6b shows that the continuous chips are formed with TiN-coated CBN tools. It is because during high frictional heat generation the TiN acts as a lubricant, it tends to reduce the frictional heat and the chip found is in continuous formation.

6 Conclusions

In this research work, the performance of TiN-coated CBN tools was evaluated during hard turning of AISI 4340 steel. The following conclusion is drawn:

- The results of ANOVA indicate that workpiece hardness is the significant parameter, which affects the cutting force and cutting temperature generation during hard turning.
- The results indicate better machining performance obtained at 150 m/min of cutting speed with 0.1 mm/rev of feed rate, 40 HRC of workpiece hardness and 1.2 mm of nose radius in this experimentation. The reason is that with small feed rate, and low workpiece hardness, the low frictional heat is generated at the tool–workpiece interface. It maintains the stability of cutting tool and keeps the edge of cutting tool sharp, which results in the minimum cutting force and interface temperature during the machining. The higher cutting force and maximum interface temperature are observed at high workpiece hardness and large feed rate. This is due to increase in frictional heat generation and wear of flank surface.
- The minimum values of cutting force and interface temperature were observed at 1.2 mm of tool nose radius. This is because with the small nose the area of contact of tool and workpiece is small, and it causes higher cutting force and earlier wear of tooltip. For higher nose radius, the stability of cutting tool with the workpiece is increased. The 10–15% reduction in cutting force was seen with TiN-coated CBN tool compared to uncoated CBN.

Therefore, this research work offers significant benefits of using TiN-coated CBN tools during machining of variable hardened AISI 4340 steel. This work also highlights that the appropriate selection of the workpiece hardness, tool geometry and cutting parameters is essential for attaining of better performance in the hard turning process.

References

1. Tönshoff, H.K., Arendt, C., Ben Amor, R.: Cutting of hardened steel. CIRP Ann.-Manufact. Technol **49**(2), 547–566 (2000)
2. Davim, J.P. (ed.): Machining of Hard Materials. Springer Science & Business Media (2011)
3. Özel, Tugrul, Hsu, Tsu-Kong, Zeren, Erol: Effects of cutting edge geometry, workpiece hardness, feed rate and cutting speed on surface roughness and forces in finish turning of hardened AISI H13 steel. Int. J. Adv. Manufact. Technol. **25**(3–4), 262–269 (2005)
4. Sahin, Y.: Comparison of tool life between ceramic and cubic boron nitride (CBN) cutting tools when machining hardened steels. J. Mater. Process. Technol. **209**(7), 3478–3489 (2009)
5. Dogra, M., Sharma, V.S., Sachdeva, A., Suri, N.M., Dureja J.S.: Performance evaluation of CBN, coated carbide, cryogenically treated uncoated/coated carbide inserts in finish-turning of hardened steel. Int. J. Adv. Manufact. Technol. **57**(5–8), 541–553 (2011)
6. Aouici, H., Yallese, M.A., Belbah, A., Ameur, M.F., Elbah, M.: Experimental investigation of cutting parameters influence on surface roughness and cutting forces in hard turning of X38CrMoV5-1 with CBN tool. Sadhana **38**(3), 429–445 (2013)
7. Sobiyi, Kehinde, Sigalas, Iakovos, Akdogan, Guven, Turan, Yunus: Performance of mixed ceramics and CBN tools during hard turning of martensitic stainless steel. Int. J. Adv. Manufact. Technol. **77**(5–8), 861–871 (2015)
8. Zhao, T., Zhou, J.M., Bushlya, V., Ståhl, J.E.: Effect of cutting edge radius on surface roughness and tool wear in hard turning of AISI 52100 steel. Int. J. Adv. Manufact. Technol. **91**(9–12), 3611–3618 (2017)
9. Huang, Y., Chou, Y.K., Liang, S.Y.: CBN tool wear in hard turning: a survey on research progresses. Int. J. Adv. Manufact. Technol. **35**(5–6), 443–453 (2007)
10. ISO: 3685: Tool-Life Testing with Single-Point Turning Tools. International Organization for Standardization (ISO), Geneva, Switzerland (1993)
11. Montgomery, D.C.: Design and Analysis of Experiments. Wiley (2017)
12. Suresh, R., Basavarajappa, S., Gaitonde, V.N., Samuel, G.L.: Machinability investigations on hardened AISI 4340 steel using coated carbide insert. Int. J. Refract Metal Hard Mater. **33**, 75–86 (2012)
13. Shaffer, W.: Getting a better edge. Cutting Tool Eng. **52**(3), 44–48 (2000)
14. Benlahmidi, S., Aouici, H., Boutaghane, F., Khellaf, A., Fnides, B., Yallese, M.A.: Design optimization of cutting parameters when turning hardened AISI H11 steel (50 HRC) with CBN7020 tools. Int. J. Adv. Manufact. Technol. **89**(1–4), 803–820 (2017)
15. Shalaby, M.A., El Hakim, M.A., Veldhuis, S.C., Dosbaeva, G.K.: An investigation into the behavior of the cutting forces in precision turning. Int. J. Adv. Manufact. Technol. **90**(5–8), 1605–1615 (2017)

Author Index

© Springer Nature Singapore Pte Ltd. 2020
R. Venkata Rao and J. Taler (eds.), *Advanced Engineering Optimization Through Intelligent Techniques*, Advances in Intelligent Systems and Computing 949,
https://doi.org/10.1007/978-981-13-8196-6

Printed in the United States
By Bookmasters